T0202167

INFORMATION, PHYSICS, AND COMPUTATION

Information, Physics, and Computation

Marc Mézard

Laboratoire de Physique Théorique et Modèles Statistiques,
CNRS, and Université Paris Sud

Andrea Montanari

Department of Electrical Engineering and Department of Statistics,
Stanford University,
and Laboratoire de Physique Théorique de l'ENS, Paris

OXFORD
UNIVERSITY PRESS

OXFORD
UNIVERSITY PRESS

Great Clarendon Street, Oxford OX2 6DP

Oxford University Press is a department of the University of Oxford.
It furthers the University's objective of excellence in research, scholarship,
and education by publishing worldwide in

Oxford New York

Auckland Cape Town Dar es Salaam Hong Kong Karachi
Kuala Lumpur Madrid Melbourne Mexico City Nairobi
New Delhi Shanghai Taipei Toronto

With offices in

Argentina Austria Brazil Chile Czech Republic France Greece
Guatemala Hungary Italy Japan Poland Portugal Singapore
South Korea Switzerland Thailand Turkey Ukraine Vietnam

Oxford is a registered trade mark of Oxford University Press
in the UK and in certain other countries

Published in the United States
by Oxford University Press Inc., New York

© Oxford University Press 2009

The moral rights of the authors have been asserted
Database right Oxford University Press (maker)

First Published 2009

All rights reserved. No part of this publication may be reproduced,
stored in a retrieval system, or transmitted, in any form or by any means,
without the prior permission in writing of Oxford University Press,
or as expressly permitted by law, or under terms agreed with the appropriate
reprographics rights organization. Enquiries concerning reproduction
outside the scope of the above should be sent to the Rights Department,
Oxford University Press, at the address above

You must not circulate this book in any other binding or cover
and you must impose the same condition on any acquirer

British Library Cataloguing in Publication Data

Data available

Library of Congress Cataloging in Publication Data

Data available

Printed in Great Britain
on acid-free paper by
CPI Group (UK) Ltd, Croydon, CR0 4YY

ISBN 978–0–19–857083–7 (Hbk.)

Preface

Over the last few years, several research areas have witnessed important progress through the unexpected collaboration of statistical physicists, computer scientists, and information theorists. This dialogue between scientific disciplines has not been without difficulties, as each field has its own objectives and rules of behaviour. Nonetheless, there is increasing consensus that a common ground exists and that it can be fruitful. This book aims at making this common ground more widely accessible, through a unified approach to a selection of important research problems that have benefited from this convergence.

Historically, information theory and statistical physics have been deeply linked since Shannon, sixty years ago, used entropy to quantify the information content of a message. A few decades before, entropy had been the cornerstone of Boltzmann's statistical mechanics. However, the two topics separated and developed in different directions. Nowadays most statistical physicists know very little of information theory, and most information theosists know very little of statistical physics. This is particularly unfortunate, as recent progress on core problems in both fields has been bringing these two roads closer over the last two decades. In parallel, there has been growing interest in applying probabilistic concepts in computer science, both in devising and in analysing new algorithms. Statistical physicists have started to apply to this field the non-rigorous techniques they had developed to study disordered systems. Conversely, they have become progressively aware of the powerful computational techniques invented by computer scientists and applied them in large scale simulations.

In statistical physics, the last quarter of the twentieth century has seen the emergence of a new topic. The main focus until then had been on 'ordered' materials: crystals in which atoms vibrate around equilibrium positions arranged in a periodic lattice, or liquids and gases in which the density of particles is uniform. In the 1970s, the interest in strongly disordered systems started to grow, through studies of spin glasses, structural glasses, polymer networks, etc. The reasons for this development were the incredible richness of behaviour in these systems and their many applications in materials science, and also the variety of conceptual problems which are involved in the understanding of these behaviours. Statistical physics deals with the collective behavior of many interacting components. With disordered systems, it started to study collective behaviour of systems in which all of the components are heterogeneous. This opened the way to the study of a wealth of problems outside of physics, where heterogeneity is common currency.

Some of the most spectacular recent progress in information theory concerns error-correcting codes. More than fifty years after Shannon's theorems, efficient codes have now been found which approach Shannon's theoretical limit. Turbo codes and low-density parity-check (LDPC) codes have allowed large improvements in error correction. One of the main ingredients of these schemes is message-passing decoding strategies, such as the celebrated 'belief propagation' algorithm. These approaches are intimately related to the mean-field theories of disordered systems developed in statistical physics.

Probability plays an important role in theoretical computer science, from randomized algorithms to probabilistic combinatorics. Random ensembles of computational problems are studied as a way to model real-world situations, to test existing algorithms, or to develop new ones. In such studies one generally defines a family of instances and endows it with a probability measure, in the same way as one defines a family of samples in the case of spin glasses or LDPC codes. The discovery that the hardest-to-solve instances, with all existing algorithms, lie close to a phase transition boundary spurred a lot of interest. Phase transitions, or threshold phenomena, are actually found in all of these three fields, and play a central role in each of them. Predicting and understanding them analytically is a major challenge. It can also impact the design of efficient algorithms. Statistical physics suggests that the reason for the hardness of random constraint satisfaction problems close to phase transitions is a structural one: it hinges on the existence of a glass transition, a structural change in the geometry of the set of solutions. This understanding has opened up new algorithmic perspectives.

In order to emphasize the real convergence of interest and methods in all of these fields, we have adopted a unified approach. This book is structured in five large parts, focusing on topics of increasing complexity. Each part typically contains three chapters that present some core topics in each of the disciplines of information theory, statistical physics, and combinatorial optimization. The topics in each part have a common mathematical structure, which is developed in additional chapters serving as bridges.

- Part I (Chapters 1–4) contains introductory chapters to each of the three disciplines and some common probabilistic tools.
- Part II (Chapters 5–8) deals with problems in which independence plays an important role: the random energy model, the random code ensemble, and number partitioning. Thanks to the independence of random variables, classical techniques can be applied successfully to these problems. The part ends with a description of the replica method.
- Part III (Chapters 9–13) describes ensembles of problems on graphs: satisfiability, low-density parity-check codes, and spin glasses. Factor graphs and statistical inference provide a common language.
- Part IV (chapters 14–17) explains belief propagation and the related 'replica-symmetric' cavity method. These can be thought of as approaches to studying systems of correlated random variables on large graphs, when the correlations decay fast enough with distance. The part shows the success of this approach with three problems: decoding, assignment, and ferromagnets.
- Part V (Chapters 18–22) is dedicated to an important consequence of long-range correlations, namely the proliferation of pure states and 'replica symmetry breaking'. It starts with the simpler problem of random linear equations with Boolean variables, and then develops the general approach and applies it to satisfiability and coding. The final chapter reviews some open problems.

At the end of each chapter, a section of notes provides pointers to the literature. The notation and symbols are summarized in Appendix A. The definitions of new concepts are signalled by boldfaced fonts, both in the text and in the index. The book contains many examples and exercises of various difficulty, which are signalled by a light grey background. They are an important part of the book.

As the book develops, we venture into progressively less well-understood topics. In particular, the number of mathematically proved statements decreases and we rely on heuristic or intuitive explanations in some places. We have put special effort into distinguishing what has been proved from what has not, and into presenting the latter as clearly and as sharply as we could. We hope that this will stimulate the interest and contributions of mathematically minded readers, rather than alienate them.

This is a graduate-level book, intended to be useful to any student or researcher who wants to study and understand the main concepts and methods in this common research domain. The introductory chapters help to set up the common language, and the book should thus be understandable by any graduate student in science with some standard background in mathematics (probability, linear algebra, and calculus).

Our choice of presenting a selection of problems in some detail has left aside a number of other interesting topics and applications. Some of them are of direct common interest in information, physics, and computation, for instance source coding, multiple-input multiple-output communication, and learning and inference in neural networks. But the concepts and techniques studied in this book also have applications in a broader range of 'complex systems' studies, ranging from neurobiology, or systems biology, to economics and the social sciences. A few introductory pointers to the literature are provided in the Notes of Chapter 22.

The critical reading and the many comments of Heiko Bauke, Alfredo Braunstein, John Eduardo Realpe Gomez, Florent Krzakala, Frauke Liers, Stephan Mertens, Elchanan Mossel, Sewoong Oh, Lenka Zdeborová, have been very useful. We are grateful to them for their feedback. We have also been stimulated by the kind encouragements of Amir Dembo, Persi Diaconis, James Martin, Balaji Prabhakar, Federico Ricci-Tersenghi, Bart Selman and Rüdiger Urbanke, and the discreet attention and steady support of Sonke Adlung, from Oxford University Press.

This book is dedicated to Fanny, Mathias, Isabelle, Claudia, and Ivana.

Marc Mézard and Andrea Montanari, December 2008

Contents

Part I

Background

1
Introduction to information theory

This chapter introduces some of the basic concepts of information theory, as well as the definitions and notation of probability theory that will be used throughout the book. The notion of entropy, which is fundamental to the whole topic of this book, is introduced here. We also present two major questions of information theory, those of data compression and error correction, and state Shannon's theorems.

Sec. 1.1 introduces the basic notations in probability. The notion of entropy, and the entropy rate of a sequence are discussed in Sections 1.2 and 1.3. A very important concept in information theory is the mutual information of two random variables, which is introduced in Section 1.4. Then we move to the two main aspects of the theory, the compression of data, in Sec. 1.5, and the transmission of data in, Sec. 1.6.

1.1 Random variables

The main object of this book will be the behaviour of large sets of **discrete random variables**. A discrete random variable X is completely defined[1] by the set of values it can take, $\mathcal{X}$, which we assume to be a finite set, and its **probability distribution** $\{p_X(x)\}_{x \in \mathcal{X}}$. The value $p_X(x)$ is the probability that the random variable X takes the value x. The probability distribution $p_X : \mathcal{X} \to [0,1]$ is a non-negative function that satisfies the normalization condition

$$\sum_{x \in \mathcal{X}} p_X(x) = 1 . \qquad (1.1)$$

We shall denote by $\mathbb{P}(A)$ the probability of an **event** $A \subseteq \mathcal{X}$, so that $p_X(x) = \mathbb{P}(X = x)$. To lighten the notation, when there is no ambiguity, we shall use $p(x)$ to denote $p_X(x)$.

If $f(X)$ is a real-valued function of the random variable X, the **expectation value** of $f(X)$, which we shall also call the **average** of f, is denoted by

$$\mathbb{E} f = \sum_{x \in \mathcal{X}} p_X(x) f(x) . \qquad (1.2)$$

While our main focus will be on random variables taking values in finite spaces, we shall sometimes make use of **continuous random variables** taking values in $\mathbb{R}^d$ or in some smooth finite-dimensional manifold. The probability measure for an

---

[1]In probabilistic jargon (which we shall avoid hereafter), we take the probability space $(\mathcal{X}, \mathsf{P}(\mathcal{X}), p_X)$, where $\mathsf{P}(\mathcal{X})$ is the σ-field of the parts of $\mathcal{X}$ and $p_X = \sum_{x \in X} p_X(x) \, \delta_x$.

'infinitesimal element' dx will be denoted by $dp_X(x)$. Each time when p_X admits a density (with respect to the Lebesgue measure), we shall use the notation $p_X(x)$ for the value of this density at the point x. The total probability $\mathbb{P}(X \in \mathcal{A})$ that the variable X takes a value in some (measurable) set $\mathcal{A} \subseteq \mathcal{X}$ is given by the integral

$$\mathbb{P}(X \in \mathcal{A}) = \int_{x \in \mathcal{A}} dp_X(x) = \int \mathbb{I}(x \in \mathcal{A}) \, dp_X(x) \,, \qquad (1.3)$$

where the second form uses the **indicator function** $\mathbb{I}(s)$ of a logical statement s, which is defined to be equal to 1 if the statement s is true, and equal to 0 if the statement is false.

The expectation value $\mathbb{E} f(X)$ and the variance $\mathrm{Var}\, f(X)$ of a real-valued function $f(x)$ are given by

$$\mathbb{E} f(X) = \int f(x) \, dp_X(x) \quad , \quad \mathrm{Var}\, f(X) = \mathbb{E}\{f(X)^2\} - \{\mathbb{E} f(X)\}^2 \,. \qquad (1.4)$$

Sometimes, we may write $\mathbb{E}_X f(X)$ to specify the variable to be integrated over. We shall often use the shorthand **pdf** for the **probability density function** $p_X(x)$.

Example 1.1 A fair die with M faces has $\mathcal{X} = \{1, 2, \ldots, M\}$ and $p(i) = 1/M$ for all $i \in \{1, \ldots, M\}$. The average of x is $\mathbb{E} X = (1 + \cdots + M)/M = (M + 1)/2$.

Example 1.2 *Gaussian variable.* A continuous variable $X \in \mathbb{R}$ has a Gaussian distribution of mean m and variance σ^2 if its probability density is

$$p(x) = \frac{1}{\sqrt{2\pi}\sigma} \exp\left(-\frac{[x-m]^2}{2\sigma^2}\right) \,. \qquad (1.5)$$

We have $\mathbb{E}X = m$ and $\mathbb{E}(X - m)^2 = \sigma^2$.

Appendix A contains some definitions and notation for the random variables that we shall encounter most frequently

The notation of this chapter refers mainly to discrete variables. Most of the expressions can be transposed to the case of continuous variables by replacing sums $\sum_x$ by integrals and interpreting $p(x)$ as a probability density.

Exercise 1.1 *Jensen's inequality.* Let X be a random variable taking values in a set $\mathcal{X} \subseteq \mathbb{R}$ and let f be a convex function (i.e. a function such that $\forall x, y$ and $\forall \alpha \in [0, 1]$: $f(\alpha x + (1 - \alpha)y) \le \alpha f(x) + (1 - \alpha)f(y)$). Then

$$\mathbb{E}f(X) \ge f(\mathbb{E}X) \,. \qquad (1.6)$$

Supposing for simplicity that $\mathcal{X}$ is a finite set with $|\mathcal{X}| = n$, prove this equality by recursion on n.

1.2 Entropy

The **entropy** H_X of a discrete random variable X with probability distribution $p(x)$ is defined as

$$H_X \equiv -\sum_{x \in \mathcal{X}} p(x) \log_2 p(x) = \mathbb{E} \log_2 \left[\frac{1}{p(X)} \right] \,, \tag{1.7}$$

where we define, by continuity, $0 \log_2 0 = 0$. We shall also use the notation $H(p)$ whenever we want to stress the dependence of the entropy upon the probability distribution of X.

In this chapter, we use the logarithm to base 2, which is well adapted to digital communication, and the entropy is then expressed in **bits**. In other contexts, and in particular in statistical physics, one uses the natural logarithm (with base $e \approx 2.7182818$) instead. It is sometimes said that, in this case, entropy is measured in **nats**. In fact, the two definitions differ by a global multiplicative constant, which amounts to a change of units. When there is no ambiguity, we shall use H instead of H_X.

Intuitively, the entropy H_X is a measure of the uncertainty of the random variable X. One can think of it as the missing information: the larger the entropy, the less a priori information one has on the value of the random variable. It roughly coincides with the logarithm of the number of typical values that the variable can take, as the following examples show.

Example 1.3 A fair coin has two values with equal probability. Its entropy is 1 bit.

Example 1.4 Imagine throwing M fair coins: the number of all possible outcomes is 2^M. The entropy equals M bits.

Example 1.5 A fair die with M faces has entropy $\log_2 M$.

Example 1.6 *Bernoulli process.* A Bernoulli random variable X can take values $0, 1$ with probabilities $p(0) = q$, $p(1) = 1 - q$. Its entropy is

$$H_X = -q \log_2 q - (1 - q) \log_2 (1 - q) \,, \tag{1.8}$$

which is plotted as a function of q in Fig. 1.1. This entropy vanishes when $q = 0$ or $q = 1$ because the outcome is certain; it is maximal at $q = 1/2$, when the uncertainty of the outcome is maximal.

Since Bernoulli variables are ubiquitous, it is convenient to introduce the function $\mathcal{H}(q) \equiv -q \log q - (1 - q) \log (1 - q)$ for their entropy.

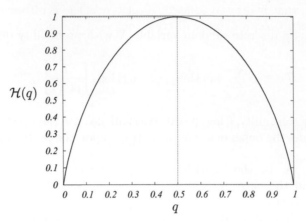

Fig. 1.1 The entropy $\mathcal{H}(q)$ of a binary variable with $p(X = 0) = q$ and $p(X = 1) = 1 - q$, plotted versus q.

Exercise 1.2 An unfair die with four faces and $p(1) = 1/2$, $p(2) = 1/4$, $p(3) = p(4) = 1/8$ has entropy $H = 7/4$, smaller than that of the corresponding fair die.

Exercise 1.3 DNA is built from a sequence of bases which are of four types, A, T, G, C. In the natural DNA of primates, the four bases have nearly the same frequency, and the entropy per base, if one makes the simplifying assumption of independence of the various bases, is $H = -\log_2(1/4) = 2$. In some genuses of bacteria, one can have big differences in concentrations: for example, $p(G) = p(C) = 0.38$, $p(A) = p(T) = 0.12$, giving a smaller entropy $H \approx 1.79$.

Exercise 1.4 In some intuitive way, the entropy of a random variable is related to the 'risk' or 'surprise' which is associated with it. Let us see how these notions can be made more precise.

Consider a gambler who bets on a sequence of Bernoulli random variables $X_t \in \{0, 1\}$, $t \in \{0, 1, 2, \ldots\}$, with mean $\mathbb{E}X_t = p$. Imagine he knows the distribution of the X_t's and, at time t, he bets a fraction $w(1) = p$ of his money on 1 and a fraction $w(0) = (1 - p)$ on 0. He loses whatever is put on the wrong number, while he doubles whatever has been put on the right one. Define the average doubling rate of his wealth at time t as

$$W_t = \frac{1}{t} \, \mathbb{E} \log_2 \left\{ \prod_{t'=1}^{t} 2w(X_{t'}) \right\}. \tag{1.9}$$

It is easy to prove that the expected doubling rate $\mathbb{E}W_t$ is related to the entropy of X_t: $\mathbb{E}W_t = 1 - \mathcal{H}(p)$. In other words, it is easier to make money out of predictable events.

Another notion that is directly related to entropy is the **Kullback–Leibler** (KL) **divergence** between two probability distributions $p(x)$ and $q(x)$ over the same finite space $\mathcal{X}$. This is defined as

$$D(q\|p) \equiv \sum_{x\in\mathcal{X}} q(x) \log \frac{q(x)}{p(x)}, \qquad (1.10)$$

where we adopt the conventions $0\log 0 = 0$ and $0\log(0/0) = 0$. It is easy to show that (*i*) $D(q\|p)$ is convex in $q(x)$; (*ii*) $D(q\|p) \geq 0$; and (*iii*) $D(q\|p) > 0$ unless $q(x) \equiv p(x)$. The last two properties derive from the concavity of the logarithm (i.e. the fact that the function $-\log x$ is convex) and Jensen's inequality (eqn (1.6)): if $\mathbb{E}$ denotes the expectation with respect to the distribution $q(x)$, then $-D(q\|p) = \mathbb{E}\log[p(x)/q(x)] \leq \log\mathbb{E}[p(x)/q(x)] = 0$. The KL divergence $D(q\|p)$ thus looks like a distance between the probability distributions q and p, although it is not symmetric.

The importance of the entropy, and its use as a measure of information, derives from the following properties:

1. $H_X \geq 0$.

2. $H_X = 0$ if and only if the random variable X is certain, which means that X takes one value with probability one.

3. Among all probability distributions on a set $\mathcal{X}$ with M elements, H is maximum when all events x are equiprobable, with $p(x) = 1/M$. The entropy is then $H_X = \log_2 M$. To prove this statement, note that if $\mathcal{X}$ has M elements, then the KL divergence $D(p\|\overline{p})$ between $p(x)$ and the uniform distribution $\overline{p}(x) = 1/M$ is $D(p\|\overline{p}) = \log_2 M - H(p)$. The statement is a direct consequence of the properties of the KL divergence.

4. If X and Y are two **independent** random variables, meaning that $p_{X,Y}(x,y) = p_X(x)p_Y(y)$, the total entropy of the pair X,Y is equal to $H_X + H_Y$:

$$H_{X,Y} = -\sum_{x,y} p_{X,Y}(x,y) \log_2 p_{X,Y}(x,y)$$

$$= -\sum_{x,y} p_{X,Y}(x,y) \left(\log_2 p_X(x) + \log_2 p_Y(y)\right) = H_X + H_Y. \quad (1.11)$$

5. For any pair of random variables, one has in general $H_{X,Y} \leq H_X + H_Y$, and this result is immediately generalizable to n variables. (The proof can be obtained by using the positivity of the KL divergence $D(p_1\|p_2)$, where $p_1 = p_{X,Y}$ and $p_2 = p_X p_Y$.)

6. *Additivity for composite events.* Take a finite set of events $\mathcal{X}$, and decompose it into $\mathcal{X} = \mathcal{X}_1 \cup \mathcal{X}_2$, where $\mathcal{X}_1 \cap \mathcal{X}_2 = \emptyset$. Denote by $q_1 = \sum_{x\in\mathcal{X}_1} p(x)$ the probability of $\mathcal{X}_1$, and denote by q_2 the probability of $\mathcal{X}_2$. For each $x \in \mathcal{X}_1$, define as usual the conditional probability of x, given that $x \in \mathcal{X}_1$, by $r_1(x) = p(x)/q_1$ and define $r_2(x)$ similarly as the conditional probability of x, given that $x \in \mathcal{X}_2$. The total entropy can then be written as the sum of two contributions $H_X = -\sum_{x\in\mathcal{X}} p(x) \log_2 p(x) = H(q) + \widetilde{H}(q,r)$, where

$$H(q) = -q_1 \log_2 q_1 - q_2 \log_2 q_2 \qquad (1.12)$$

$$\widetilde{H}(q,r) = -q_1 \sum_{x \in \mathcal{X}_1} r_1(x) \log_2 r_1(x) - q_2 \sum_{x \in \mathcal{X}_1} r_2(x) \log_2 r_2(x). \qquad (1.13)$$

The proof is straightforward and is done by substituting the laws r_1 and r_2 by their definitions. This property can be interpreted as the fact that the average information associated with the choice of an event x is additive, being the sum of the information $H(q)$ associated to a choice of subset, and the information $\widetilde{H}(q,r)$ associated with the choice of the event inside the subset (weighted by the probability of the subset). This is the main property of the entropy, which justifies its use as a measure of information. In fact, this is a simple example of the chain rule for conditional entropy, which will be illustrated further in Sec. 1.4.

Conversely, these properties, together with appropriate hypotheses of continuity and monotonicity, can be used to define the entropy axiomatically.

1.3 Sequences of random variables and their entropy rate

In many situations of interest, one deals with a random process which generates **sequences of random variables** $\{X_t\}_{t \in \mathbb{N}}$, each of them taking values in the same finite space $\mathcal{X}$. We denote by $P_N(x_1, \ldots, x_N)$ the joint probability distribution of the first N variables. If $A \subset \{1, \ldots, N\}$ is a subset of indices, we denote by $\overline{A}$ its complement $\overline{A} = \{1, \ldots, N\} \setminus A$ and use the notation $\underline{x}_A = \{x_i, i \in A\}$ and $\underline{x}_{\overline{A}} = \{x_i, i \in \overline{A}\}$ (the set subscript will be dropped whenever it is clear from the context). The **marginal distribution** of the variables in A is obtained by summing P_N over the variables in $\overline{A}$:

$$P_A(\underline{x}_A) = \sum_{\underline{x}_{\overline{A}}} P_N(x_1, \ldots, x_N) . \qquad (1.14)$$

Example 1.7 The simplest case is when the X_t's are independent. This means that $P_N(x_1, \ldots, x_N) = p_1(x_1)p_2(x_2) \ldots p_N(x_N)$. If all the distributions p_i are identical, equal to p, the variables are **independent identically distributed**, and abbreviated as **i.i.d.** The joint distribution is

$$P_N(x_1, \ldots, x_N) = \prod_{t=1}^{N} p(x_i) . \qquad (1.15)$$

Example 1.8 The sequence $\{X_t\}_{t\in\mathbb{N}}$ is said to be a **Markov chain** if

$$P_N(x_1,\ldots,x_N) = p_1(x_1) \prod_{t=1}^{N-1} w(x_t \to x_{t+1}).$$ (1.16)

Here $\{p_1(x)\}_{x\in\mathcal{X}}$ is called the **initial state**, and the $\{w(x \to y)\}_{x,y\in\mathcal{X}}$ are the **transition probabilities** of the chain. The transition probabilities must be non-negative and normalized:

$$\sum_{y\in\mathcal{X}} w(x \to y) = 1, \quad \text{for any } x \in \mathcal{X}.$$ (1.17)

When we have a sequence of random variables generated by a process, it is intuitively clear that the entropy grows with the number N of variables. This intuition suggests that we should define the **entropy rate** of a sequence $\underline{x}_N \equiv \{X_t\}_{t\in\mathbb{N}}$ as

$$h_X = \lim_{N\to\infty} H_{\underline{X}_N}/N,$$ (1.18)

if the limit exists. The following examples should convince the reader that the above definition is meaningful.

Example 1.9 If the X_t's are i.i.d. random variables with distribution $\{p(x)\}_{x\in\mathcal{X}}$, the additivity of entropy implies

$$h_X = H(p) = -\sum_{x\in\mathcal{X}} p(x) \log p(x).$$ (1.19)

Example 1.10 Let $\{X_t\}_{t\in\mathbb{N}}$ be a Markov chain with initial state $\{p_1(x)\}_{x\in\mathcal{X}}$ and transition probabilities $\{w(x \to y)\}_{x,y\in\mathcal{X}}$. Call $\{p_t(x)\}_{x\in\mathcal{X}}$ the marginal distribution of X_t and assume the following limit to exist independently of the initial condition:

$$p^*(x) = \lim_{t\to\infty} p_t(x).$$ (1.20)

As we shall see in Chapter 4, this indeed turns out to be true under quite mild hypotheses on the transition probabilities $\{w(x \to y)\}_{x,y\in\mathcal{X}}$. It is then easy to show that

$$h_X = -\sum_{x,y\in\mathcal{X}} p^*(x) \, w(x \to y) \log w(x \to y).$$ (1.21)

If you imagine, for instance, that a text in English is generated by picking letters randomly from the alphabet $\mathcal{X}$, with empirically determined transition probabilities $w(x \to y)$, then eqn (1.21) gives a rough estimate of the entropy of English.

A more realistic model can be obtained using a Markov chain *with memory*. This means that each new letter x_{t+1} depends on the past through the values of the k previous letters $x_t, x_{t-1}, \ldots, x_{t-k+1}$. Its conditional distribution is given by the transition probabilities $w(x_t, x_{t-1}, \ldots, x_{t-k+1} \to x_{t+1})$. Computing the corresponding entropy rate is easy. For $k = 4$, one obtains an entropy of 2.8 bits per letter, much smaller than the trivial upper bound $\log_2 27$ (there are 26 letters, plus the space symbol), but many words so generated are still not correct English words. Better estimates of the entropy of English, obtained through guessing experiments, give a number around 1.3.

1.4   Correlated variables and mutual information

Given two random variables X and Y taking values in $\mathcal{X}$ and $\mathcal{Y}$, we denote their joint probability distribution as $p_{X,Y}(x, y)$, which is abbreviated as $p(x, y)$, and we denote the conditional probability distribution for the variable y, given x, as $p_{Y|X}(y|x)$, abbreviated as $p(y|x)$. The reader should be familiar with the classical Bayes' theorem

$$p(y|x) = p(x, y)/p(x) \ . \tag{1.22}$$

When the random variables X and Y are independent, $p(y|x)$ is independent of x. When the variables are dependent, it is interesting to have a measure of their degree of dependence: how much information does one obtain about the value of y if one knows x? The notions of conditional entropy and mutual information will answer this question.

We define the **conditional entropy** $H_{Y|X}$ as the entropy of the law $p(y|x)$, averaged over x:

$$H_{Y|X} \equiv - \sum_{x \in \mathcal{X}} p(x) \sum_{y \in \mathcal{Y}} p(y|x) \log_2 p(y|x) \ . \tag{1.23}$$

The joint entropy $H_{X,Y} \equiv - \sum_{x \in \mathcal{X}, y \in \mathcal{Y}} p(x, y) \log_2 p(x, y)$ of the pair of variables x, y can be written as the entropy of x plus the conditional entropy of y given x, an identity known as the **chain rule**:

$$H_{X,Y} = H_X + H_{Y|X} \ . \tag{1.24}$$

In the simple case, where the two variables are independent, $H_{Y|X} = H_Y$, and $H_{X,Y} = H_X + H_Y$. One way to measure the correlation of the two variables is to use the **mutual information** $I_{X,Y}$, which is defined as

$$I_{X,Y} \equiv \sum_{x \in \mathcal{X}, y \in \mathcal{Y}} p(x, y) \log_2 \frac{p(x, y)}{p(x)p(y)} \ . \tag{1.25}$$

It is related to the conditional entropies by

$$I_{X,Y} = H_Y - H_{Y|X} = H_X - H_{X|Y} \ . \tag{1.26}$$

This shows that the mutual information $I_{X,Y}$ measures the reduction in the uncertainty of x due to the knowledge of y, and is symmetric in x, y.

Proposition 1.11 $I_{X,Y} \geq 0$. *Moreover, $I_{X,Y} = 0$ if and only if X and Y are independent variables.*

Proof Write $I_{X,Y} = \mathbb{E}_{x,y} - \log_2\{p(x)p(y)/p(x,y)\}$. Consider the random variable $u = (x,y)$ with probability distribution $p(x,y)$. As the function $-\log(\cdot)$ is convex, one can apply Jensen's inequality (eqn (1.6)). This gives the result $I_{X,Y} \geq 0$ $\square$

Exercise 1.5 A large group of friends plays the following game ('telephone without cables'). The person number zero chooses a number $X_0 \in \{0,1\}$ with equal probability and communicates it to the person number one without letting the others hear, and so on. The first person communicates the number to the second person, without letting anyone else hear. Call X_n the number communicated from the n-th to the $(n+1)$-th person. Assume that, at each step a person may become confused and communicate the wrong number with probability p. How much information does the n-th person have about the choice of the first person?

We can quantify this information through $I_{X_0,X_n} \equiv I_n$. Show that $I_n = 1 - \mathcal{H}(p_n)$ with p_n given by $1 - 2p_n = (1-2p)^n$. In particular, as $n \to \infty$,

$$I_n = \frac{(1-2p)^{2n}}{2\log 2} \left[1 + O((1-2p)^{2n})\right] . \tag{1.27}$$

The 'knowledge' about the original choice decreases exponentially along the chain.

Mutual information is degraded when data is transmitted or processed. This is quantified as follows

Proposition 1.12 *Data-processing inequality. Consider a Markov chain $X \to Y \to Z$ (so that the joint probability of the three variables can be written as $p_1(x)w_2(x \to y)w_3(y \to z)$). Then $I_{X,Z} \leq I_{X,Y}$. In particular, if we apply this result to the case where Z is a function of Y, $Z = f(Y)$, we find that applying f degrades the information: $I_{X,f(Y)} \leq I_{X,Y}$.*

Proof We introduce the mutual information of two variables conditioned on a third one: $I_{X,Y|Z} = H_{X|Z} - H_{X|(YZ)}$. The mutual information between a variable X and a pair of variables (YZ) can be decomposed using the following chain rule: $I_{X,(YZ)} = I_{X,Z} + I_{X,Y|Z} = I_{X,Y} + I_{X,Z|Y}$. If we have a Markov chain $X \to Y \to Z$, X and Z are independent when we condition on the value of Y, and therefore $I_{X,Z|Y} = 0$. The result follows from the fact that $I_{X,Y|Z} \geq 0$. $\square$

The conditional entropy also provides a lower bound on the probability of guessing a random variable. Suppose you want to guess the value of the random variable X, but you observe only the random variable Y (which can be thought of as a noisy version of X). From Y, you compute a function $\widehat{X} = g(Y)$, which is your estimate for X. What is the probability P_e that you guessed incorrectly? Intuitively, if X and Y are strongly correlated, one can expect that P_e is small, whereas it increases for less well-correlated variables. This is quantified as follows.

Proposition 1.13 *Fano's inequality. Consider a random variable X taking values in the alphabet $\mathcal{X}$, and the Markov chain $X \to Y \to \widehat{X}$, where $\widehat{X} = g(Y)$ is an estimate*

for the value of X. Define the probability of making an error as $P_e = \mathbb{P}(\widehat{X} \neq X)$. This is bounded from below as follows:

$$\mathcal{H}(P_e) + P_e \log_2(|\mathcal{X}| - 1) \geq H(X|Y). \tag{1.28}$$

Proof Define a random variable $E = \mathbb{I}(\widehat{X} \neq X)$, equal to 0 if $\widehat{X} = X$ and to 1 otherwise, and decompose the conditional entropy $H_{X,E|Y}$ using the chain rule in two ways: $H_{X,E|Y} = H_{X|Y} + H_{E|X,Y} = H_{E|Y} + H_{X|E,Y}$. Then notice that (i) $H_{E|X,Y} = 0$ (because E is a function of X and Y); (ii) $H_{E|Y} \leq H_E = \mathcal{H}(P_e)$ and (iii) $H_{X|E,Y} = (1 - P_e)H_{X|E=0,Y} + P_e H_{X|E=1,Y} = P_e H_{X|E=1,Y} \leq P_e \log_2(|\mathcal{X}| - 1)$. $\square$

Exercise 1.6 Suppose that X can take k values, and that its distribution is $p(1) = 1 - p$, $p(x) = p/(k-1)$ for $x \geq 2$. If X and Y are independent, what is the value of the right-hand side of Fano's inequality? Assuming that $1 - p > \frac{p}{k-1}$, what is the best guess one can make about the value of X? What is the probability of error? Show that Fano's inequality holds as an equality in this case.

1.5   Data compression

Imagine an information source which generates a sequence of symbols $\underline{X} = \{X_1, \ldots, X_N\}$ taking values in a finite alphabet $\mathcal{X}$. We assume a probabilistic model for the source, meaning that the X_i are random variables. We want to store the information contained in a given realization $\underline{x} = \{x_1 \ldots x_N\}$ of the source in the most compact way.

This is the basic problem of **source coding**. Apart from being an issue of the utmost practical interest, it is a very instructive subject. It in fact allows us to formalize in a concrete fashion the intuitions of 'information' and 'uncertainty' which are associated with the definition of entropy. Since entropy will play a crucial role throughout the book, we present here a little detour into source coding.

1.5.1   Codewords

We first need to formalize what is meant by 'storing the information'. We define a **source code** for the random variable $\underline{X}$ to be a mapping w which associates with any possible information sequence in $\mathcal{X}^N$ a string in a reference alphabet, which we shall assume to be $\{0, 1\}$:

$$w : \mathcal{X}^N \to \{0, 1\}^*$$
$$\underline{x} \mapsto w(\underline{x}). \tag{1.29}$$

Here we have used the convention of denoting by $\{0, 1\}^*$ the set of binary strings of arbitrary length. Any binary string which is in the image of w is called a **codeword**.

Often, the sequence of symbols $X_1 \ldots X_N$ is a part of a longer stream. The compression of this stream is realized in three steps. First, the stream is broken into blocks of length N. Then, each block is encoded separately using w. Finally, the codewords are glued together to form a new (hopefully more compact) stream. If the original stream consists of the blocks $\underline{x}^{(1)}, \underline{x}^{(2)}, \ldots, \underline{x}^{(r)}$, the output of the encoding process will be the concatenation of $w(\underline{x}^{(1)}), \ldots, w(\underline{x}^{(r)})$. In general, there is more than one way of parsing this concatenation into codewords, which may cause troubles when one wants to recover the compressed data. We shall therefore require the code w to be such that any concatenation of codewords can be parsed unambiguously. The mappings w satisfying this property are called **uniquely decodable codes**.

Unique decodability is certainly satisfied if, for any $\underline{x}, \underline{x}' \in \mathcal{X}^N$, $w(\underline{x})$ is not a prefix of $w(\underline{x}')$ (see Fig. 1.2). In such a case the code is said to be **instantaneous**. Hereafter, we shall focus on instantaneous codes, since they are both practical and slightly simpler to analyse.

Now that we have specified how to store information, namely using a source code, it is useful to introduce a figure of merit for source codes. If $l_w(x)$ is the length of the string $w(x)$, the average length of the code is

$$L(w) = \sum_{\underline{x} \in \mathcal{X}^N} p(\underline{x})\, l_w(\underline{x}) . \tag{1.30}$$

Example 1.14 Take $N = 1$, and consider a random variable X which takes values in $\mathcal{X} = \{1, 2, \ldots, 8\}$ with probabilities $p(1) = 1/2$, $p(2) = 1/4$, $p(3) = 1/8$, $p(4) = 1/16$, $p(5) = 1/32$, $p(6) = 1/64$, $p(7) = 1/128$, and $p(8) = 1/128$. Consider the two codes w_1 and w_2 defined by the table below:

x	$p(x)$	$w_1(x)$	$w_2(x)$
1	1/2	000	0
2	1/4	001	10
3	1/8	010	110
4	1/16	011	1110
5	1/32	100	11110
6	1/64	101	111110
7	1/128	110	1111110
8	1/128	111	11111110

(1.31)

These two codes are instantaneous. For instance, looking at the code w_2, the encoded string 10001101110010 can be parsed in only one way, since each symbol 0 ends a codeword. It thus corresponds to the sequence $x_1 = 2, x_2 = 1, x_3 = 1, x_4 = 3, x_5 = 4, x_6 = 1, x_7 = 2$. The average length of code w_1 is $L(w_1) = 3$, and the average length of code w_2 is $L(w_2) = 247/128$. Notice that w_2 achieves a shorter average length because it assigns the shortest codeword (namely 0) to the most probable symbol (i.e. $x = 1$).

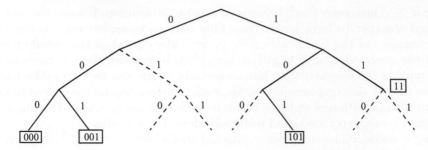

Fig. 1.2 An instantaneous source code: each codeword is assigned to a node in a binary tree in such a way that none of them is the ancestor of another one. Here, the four codewords are framed.

Example 1.15 A useful graphical representation of a source code can be obtained by drawing a binary tree and associating each codeword with the corresponding node in the tree. In Fig. 1.2, we represent a source code with $|\mathcal{X}^N| = 4$ in this way. It is quite easy to recognize that the code is indeed instantaneous. The codewords, which are framed, are such that no codeword is the ancestor of any other codeword in the tree. Given a sequence of codewords, parsing is immediate. For instance, the sequence 00111000101001 can be parsed only into $001, 11, 000, 101, 001$.

1.5.2 Optimal compression and entropy

Suppose that you have a 'complete probabilistic characterization' of the source you want to compress. What is the 'best code' w for this source?

This problem was solved (to a large extent) by Shannon in his celebrated 1948 paper, by connecting the best achievable average length to the entropy of the source. Following Shannon, we assume that we know the probability distribution of the source $p(\underline{x})$. Moreover, we interpret 'best code' as 'code with the shortest average length'.

Theorem 1.16 *Let L_N^* be the shortest average length achievable by an instantaneous code for the variable $\underline{X} = \{X_1, \ldots, X_N\}$, which has entropy $H_{\underline{X}}$. Then:*

1. *For any $N \geq 1$*

$$H_{\underline{X}} \leq L_N^* \leq H_{\underline{X}} + 1. \tag{1.32}$$

2. *If the source has a finite entropy rate $h = \lim_{N \to \infty} H_{\underline{X}}/N$, then*

$$\lim_{N \to \infty} \frac{1}{N} L_N^* = h. \tag{1.33}$$

Proof The basic idea of the proof of eqn (1.32) is that if the codewords were too short, the code would not be instantaneous. **Kraft's inequality** makes this simple remark more precise. For any instantaneous code w, the lengths $l_w(\underline{x})$ satisfy

$$\sum_{\underline{x} \in \mathcal{X}^N} 2^{-l_w(\underline{x})} \leq 1. \tag{1.34}$$

This fact is easily proved by representing the set of codewords as a set of leaves on a binary tree (see Fig. 1.2). Let L_M be the length of the longest codeword . Consider the set of all the 2^{L_M} possible vertices in the binary tree at the generation L_M; let us call them the 'descendants'. If the information $\underline{x}$ is associated with a codeword at generation l (i.e. $l_w(\underline{x}) = l$), there can be no other codewords in the branch of the tree rooted at this codeword, because the code is instantaneous. We 'erase' the corresponding 2^{L_M-l} descendants, which cannot be codewords. The subsets of erased descendants associated with each codeword are not overlapping. Therefore the total number of erased descendants, $\sum_{\underline{x}} 2^{L_M - l_w(\underline{x})}$, must be less than or equal to the total number of descendants, 2^{L_M}. This establishes Kraft's inequality.

Conversely, for any set of lengths $\{l(\underline{x})\}_{\underline{x} \in \mathcal{X}^N}$ which satisfy Kraft's inequality (1.34), there exists at least one code whose codewords have lengths $\{l(\underline{x})\}_{\underline{x} \in \mathcal{X}^N}$. A possible construction is obtained as follows. Consider the smallest length $l(\underline{x})$ and take the first allowed binary sequence of length $l(\underline{x})$ to be the codeword for $\underline{x}$. Repeat this operation with the next shortest length and so on, until all the codewords have been exhausted. It is easy to show that this procedure is successful if eqn (1.34) is satisfied.

The problem is therefore reduced to finding the set of codeword lengths $l(\underline{x}) = l^*(\underline{x})$ which minimize the average length $L = \sum_{\underline{x}} p(\underline{x}) l(\underline{x})$ subject to Kraft's inequality (1.34). Supposing first that $l(\underline{x})$ can take arbitrary non-negative real values, this is easily done with Lagrange multipliers, and leads to $l(\underline{x}) = -\log_2 p(\underline{x})$. This set of optimal lengths, which in general cannot be realized because some of the $l(\underline{x})$ are not integers, gives an average length equal to the entropy H_X. It implies the lower bound in eqn (1.32). In order to build a real code with integer lengths, we use

$$l^*(\underline{x}) = \lceil -\log_2 p(\underline{x}) \rceil . \tag{1.35}$$

Such a code satisfies Kraft's inequality, and its average length is less than or equal than $H_X + 1$, proving the upper bound in eqn (1.32).

The second part of the theorem is a straightforward consequence of the first part. $\square$

The code that we have constructed in the proof is often called a **Shannon code**. For long strings ($N \gg 1$), it is close to optimal. However, it has no reason to be optimal in general. For instance, if only one $p(x)$ is very small, it will assign x to a very long codeword, while shorter codewords are available. It is interesting to know that, for a given source $\{X_1, \ldots, X_N\}$, there exists an explicit construction of the optimal code, called Huffman's code.

At first sight, it may appear that Theorem 1.16, together with the construction of Shannon codes, completely solves the source coding problem. Unhappily, this is far from true, as the following arguments show.

From a computational point of view, the encoding procedure described above is unpractical when N is large. One can build the code once for all, and store it somewhere, but this requires $\Theta(|\mathcal{X}|^N)$ memory. On the other hand, one could reconstruct the code every time a string required to be encoded, but this takes $\Theta(|\mathcal{X}|^N)$ operations. One can use the same code and be a little smarter in the encoding procedure, but this does not yield a big improvement. (The symbol Θ means 'of the order of'; the precise definition is given in Appendix A.)

From a practical point of view, the construction of a Shannon code requires an accurate knowledge of the probabilistic law of the source. Suppose now that you want to compress the complete works of Shakespeare. It is exceedingly difficult to construct a good model for the source 'Shakespeare'. Even worse, when you will finally have such a model, it will be of little use for compressing Dante or Racine.

Happily, source coding has made tremendous progresses in both directions in the last half-century. However, in this book, we shall focus on another crucial aspect of information theory, the transmission of information.

1.6 Data transmission

We have just seen how to encode information in a string of symbols (we used bits, but any finite alphabet is equally good). Suppose now that we want to communicate this string. When the string is transmitted, it may be corrupted by noise, which depends on the physical device used for the transmission. One can reduce this problem by adding redundancy to the string. This redundancy is to be used to correct some of the transmission errors, in the same way as redundancy in the English language could be used to correct some of the typos in this book. This is the domain of **channel coding**. A central result in information theory, again due to Shannon's pioneering work of 1948, relates the level of redundancy to the maximal level of noise that can be tolerated for error-free transmission. As in source coding, entropy again plays a key role in this result. This is not surprising, in view of the duality between the two problems. In data compression, one wants to reduce the redundancy of the data, and the entropy gives a measure of the ultimate possible reduction. In data transmission, one wants to add some well-tailored redundancy to the data.

1.6.1 Communication channels

A typical flowchart of a communication system is shown in Fig. 1.3. It applies to situations as diverse as communication between the earth and a satellite, cellular phones, and storage within the hard disk of a computer. Alice wants to send a message m to Bob. Let us assume that m is an M-bit sequence. This message is first encoded into a longer one, an N-bit message denoted by $\underline{x}$, with $N > M$, where the added bits will provide the redundancy used to correct transmission errors. The encoder is a map from $\{0,1\}^M$ to $\{0,1\}^N$. The encoded message is sent through a communication channel. The output of the channel is a message $\underline{y}$. In the case of a noiseless channel, one would simply have $\underline{y} = \underline{x}$. In the case of a realistic channel, $\underline{y}$ is in general a string of symbols different from $\underline{x}$. Note that $\underline{y}$ is not necessarily a string of bits. The **channel** is described by a transition probability $Q(\underline{y}|\underline{x})$. This is the probability that the received signal is $\underline{y}$, conditional on the transmitted signal being $\underline{x}$. Different physical channels are described by different functions $Q(\underline{y}|\underline{x})$. The decoder takes the message $\underline{y}$ and deduces from it an estimate m' of the sent message.

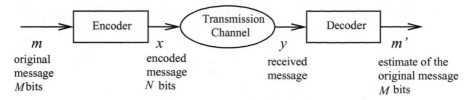

Fig. 1.3 Typical flowchart of a communication device.

Exercise 1.7 Consider the following example of a channel with **insertions**. When a bit x is fed into the channel, either x or x0 is received with equal probability 1/2. Suppose that you send the string 111110. The string 1111100 will be received with probability $2 \cdot 1/64$ (the same output can be produced by an error in either the fifth or the sixth digit). Notice that the output of this channel is a bit string which is always longer than or equal in length to the transmitted string.

A simple code for this channel is easily constructed: use the string 100 for each 0 in the original message and 1100 for each 1. Then, for instance, we have the encoding

$$01101 \mapsto 100110011001001100 . \qquad (1.36)$$

The reader is invited to define a decoding algorithm and verify its effectiveness.

Hereafter, we shall consider **memoryless** channels. In this case, for any input $\underline{x} = (x_1, ..., x_N)$, the output message is a string of N letters $\underline{y} = (y_1, ..., y_N)$ from an alphabet $\mathcal{Y} \ni y_i$ (not necessarily binary). In a memoryless channel, the noise acts independently on each bit of the input. This means that the conditional probability $Q(\underline{y}|\underline{x})$ factorizes, i.e.

$$Q(\underline{y}|\underline{x}) = \prod_{i=1}^{N} Q(y_i|x_i) , \qquad (1.37)$$

and the transition probability $Q(y_i|x_i)$ is independent og i.

Example 1.17 Binary symmetric channel (BSC). The input x_i and the output y_i are both in $\{0, 1\}$. The channel is characterized by one number, the probability p that the channel output is different from the input, called the **crossover** (or **flip**) probability. It is customary to represent this type of channel by the diagram on the left of Fig. 1.4.

Example 1.18 Binary erasure channel (BEC). In this case some of the input bits are erased instead of being corrupted: x_i is still in $\{0, 1\}$, but y_i now belongs to $\{0, 1, *\}$, where $*$ means that the symbol has been erased. In the symmetric case, this channel is described by a single number, the probability ϵ that a bit is erased, see Fig. 1.4, middle.

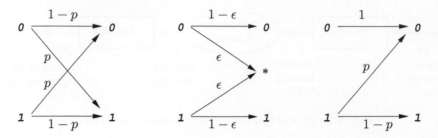

Fig. 1.4 Three communication channels. *Left*: the binary symmetric channel. An error in the transmission, in which the output bit is the opposite of the input one, occurs with probability *p*. *Middle*: the binary erasure channel. An error in the transmission, signaled by the output ∗, occurs with probability ϵ. *Right*: the Z channel. An error occurs with probability *p* whenever a 1 is transmitted.

Example 1.19 Z channel. In this case the output alphabet is again {0, 1}. Now, however, a 0 is always transmitted correctly, whereas a 1 becomes a 0 with probability *p*. The name of this channel comes from its graphical representation: see Fig. 1.4, right.

A very important characteristic of a channel is the **channel capacity** C. This is defined in terms of the mutual information $I_{X,Y}$ of the variables X (the bit which was sent) and Y (the signal which was received), through

$$C = \max_{p(x)} I_{X,Y} = \max_{p(x)} \sum_{x \in \mathcal{X}, y \in \mathcal{Y}} p(x, y) \log_2 \frac{p(x, y)}{p(x)p(y)} \ . \tag{1.38}$$

We recall that in our case $p(x, y) = p(x)Q(y|x)$, and $I_{X,Y}$ measures the reduction in the uncertainty of x due to the knowledge of y. The capacity C gives a measure of how faithful a channel can be. If the output of the channel is pure noise, x and y are uncorrelated and $C = 0$. At the other extreme if $y = f(x)$ is known for sure, given x, then $C = \max_{\{p(x)\}} H(p) = 1$ bit (for binary inputs). The reason for our interest in the capacity will become clear in Section 1.6.3 with Shannon's coding theorem, which shows that C characterizes the amount of information which can be transmitted faithfully through a channel.

Example 1.20 Consider a binary symmetric channel with flip probability p. Let us call the probability that the source sends $x = 0$ q, and the probability of $x = 1$ $1 - q$. It is easy to show that the mutual information in eqn (1.38) is maximized when zeros and ones are transmitted with equal probability (i.e. when $q = 1/2$).

Using eqn (1.38), we get $C = 1 - \mathcal{H}(p)$ bits, where $\mathcal{H}(p)$ is the entropy of a Bernoulli process with parameter p (plotted in Fig. 1.1).

Example 1.21 Consider now a binary erasure channel with error probability ϵ. The same argument as above applies. It is therefore easy to obtain $C = 1 - \epsilon$.

Exercise 1.8 Compute the capacity of a Z channel.

1.6.2 Error-correcting codes

We need one last ingredient in order to have a complete definition of the channel coding problem: the behaviour of the information source. We shall assume that the source produces a sequence of uncorrelated, unbiased bits. This may seem at first a very crude model for any real information source. Surprisingly, Shannon's source–channel separation theorem ensures that there is indeed no loss of generality in treating this case.

The sequence of bits produced by the source is divided into blocks $m_1, m_2, m_3, \ldots$ of length M. The **encoding** is a mapping from $\{0,1\}^M \ni m$ to $\{0,1\}^N$, with $N \geq M$. Each possible M-bit message m is mapped to a **codeword** $\underline{x}(m)$, which can be seen as a point in the N-dimensional unit hypercube. The codeword length N is also called the **block length**. There are 2^M codewords, and the set of all possible codewords is called the **codebook**. When a message is transmitted, the corresponding codeword $\underline{x}$ is corrupted to $\underline{y} \in \mathcal{Y}^N$ with probability $Q(\underline{y}|\underline{x}) = \prod_{i=1}^N Q(y_i|x_i)$. The output alphabet $\mathcal{Y}$ depends on the channel. The **decoder** is a mapping from $\mathcal{Y}^N$ to $\{0,1\}^M$ which takes the received message $\underline{y} \in \mathcal{Y}^N$ and maps it to one of the possible original messages $m' = d(\underline{y}) \in \{0,1\}^M$.

An **error-correcting code** is defined by a pair of functions, the encoding $\underline{x}(m)$ and the decoding $d(\underline{y})$. The ratio

$$R = \frac{M}{N} \tag{1.39}$$

of the original number of bits to the transmitted number of bits is called the **rate** of the code. The rate is a measure of the redundancy of the code. The smaller the rate, the more redundancy is added to the code, and the more errors one should be able to correct.

The **block error probability** of a code on an input message m, denoted by $P_B(m)$, is the probability that the decoded message differs from the message which was sent:

$$P_B(m) = \sum_{\underline{y}} Q(\underline{y}|\underline{x}(m)) \; \mathbb{I}(d(\underline{y}) \neq m) \,. \tag{1.40}$$

Knowing the error probability for each possible transmitted message amounts to an exceedingly detailed characterization of the performance of the code. One can therefore introduce a **maximal block error probability** as

$$P_B^{\max} \equiv \max_{m \in \{0,1\}^M} P_B(m) \,. \tag{1.41}$$

This corresponds to characterizing the code by its 'worst case' performances. A more optimistic point of view corresponds to averaging over the input messages. Since we have assumed all of them to be equiprobable, we introduce the **average block error probability**, defined as

$$P_B^{av} \equiv \frac{1}{2^M} \sum_{m \in \{0,1\}^M} P_B(m). \tag{1.42}$$

Since this is a very common figure of merit for error-correcting codes, we shall call it simply the block error probability, and use the symbol P_B without further specification hereafter.

Example 1.22 Repetition code. Consider a BSC which transmits a wrong bit with probability p. A simple code consists in repeating each bit k times, with k odd. Formally, we have $M = 1$, $N = k$, and

$$\underline{x}(0) = \underbrace{000\ldots00}_{k}, \tag{1.43}$$

$$\underline{x}(1) = \underbrace{111\ldots11}_{k}. \tag{1.44}$$

This code has rate $R = M/N = 1/k$. For instance, with $k = 3$, the original stream 0110001 is encoded as 000111111000000000111. A possible decoder consists in parsing the received sequence into groups of k bits, and finding the message m' using a majority rule among the k bits. In our example with $k = 3$, if the received group of three bits is 111 or 110 or any permutation, the corresponding input bit is assigned to 1, otherwise it is assigned to 0. For instance, if the channel output is 000101111011000010111, this decoder returns 0111001.

Exercise 1.9 The k-repetition code corrects up to $\lfloor k/2 \rfloor$ errors per group of k bits. Show that the block error probability for general k is

$$P_B = \sum_{r=\lceil k/2 \rceil}^{k} \binom{k}{r} (1-p)^{k-r} p^r. \tag{1.45}$$

Note that, for any finite k and $p > 0$, P_B is strictly positive. In order to have $P_B \to 0$, we must consider $k \to \infty$. Since the rate is $1/k$, the price to pay for a vanishing block error probability is a vanishing communication rate!

Happily, however, we shall see that much better codes exist.

1.6.3   The channel coding theorem

Consider a communication channel whose capacity (eqn (1.38)) is C. In his seminal 1948 paper, Shannon proved the following theorem.

Theorem 1.23 *For every rate $R < C$, there exists a sequence of codes $\{\mathcal{C}_N\}$, of block length N, rate R_N, and block error probability $P_{B,N}$, such that $R_N \to R$ and*

$P_{B,N} \to 0$ *as* $N \to \infty$. *Conversely, if, for a sequence of codes* $\{\mathcal{C}_N\}$*, one has* $R_N \to R$ *and* $P_{B,N} \to 0$ *as* $N \to \infty$*, then* $R < C$.

In practice, for long messages (i.e. large N), reliable communication is possible if and only if the communication rate remains below the channel capacity. The direct part of the proof will be given in Sec. 6.4 using the random code ensemble. We shall not give a full proof of the converse part in general, but only in the case of a BSC, in Sec. 6.5.2. Here we confine ourselves to some qualitative comments and provide the intuitive idea underlying this theorem.

First of all, the result is rather surprising when one meets it for the first time. As we saw in the example of repetition codes above, simple-minded codes typically have a positive error probability for any non-vanishing noise level. Shannon's theorem establishes that it is possible to achieve a vanishing error probability while keeping the communication rate bounded away from zero.

One can get an intuitive understanding of the role of the capacity through a qualitative argument, which uses the fact that a random variable with entropy H 'typically' takes 2^H values. For a given codeword $\underline{x}(m) \in \{0,1\}^N$, the channel output $\underline{y}$ is a random variable with an entropy $H_{\underline{y}|\underline{x}} = N H_{y|x}$. There exist about $2^{N H_{y|x}}$ such outputs. For perfect decoding, one needs a decoding function $d(\underline{y})$ that maps each of them to the original message m. Globally, the typical number of possible outputs is $2^{N H_y}$, and therefore one can distinguish at most $2^{N(H_y - H_{y|x})}$ codewords. In order to have a vanishing maximal error probability, one needs to be able to send all of the $2^M = 2^{NR}$ codewords. This is possible only if $R < H_y - H_{y|x} \leq C$.

Notes

There are many textbooks that provide introductions to probability and to information theory. A classic probability textbook is Feller (1968). For a more recent reference see Durrett (1995). The original Shannon paper (Shannon, 1948) is universally recognized as the foundation of information theory. A very nice modern introduction to the subject is the book by Cover and Thomas (1991). The reader may find in there a description of Huffman codes, which we did not treat in the present Chapter, as well as more advanced topics in source coding.

We did not show that the six properties listed in Section 1.2 in fact provide an alternative (axiomatic) definition of entropy. The interested reader is referred to Csiszár abd Körner (1981). An advanced book on information theory with much space devoted to coding theory is Gallager (1968). The recent and very rich book by MacKay (2002) discusses the relations with statistical inference and machine learning.

The information-theoretic definition of entropy has been used in many contexts. It can be taken as a founding concept in statistical mechanics. This approach, pioneered by Jaynes (1957), is discussed by Balian (1992).

2
Statistical physics and probability theory

One of the greatest achievements of science has been to realize that matter is made out of a small number of simple elementary components. This result seems to be in striking contrast to our experience. Both at a simply perceptual level and with more refined scientific experience, we come into touch with an ever-growing variety of states of matter with disparate properties. The ambitious purpose of statistical physics (and, more generally, of a large branch of condensed matter physics) is to understand this variety. It aims at explaining how complex behaviours can emerge when large numbers of identical elementary components are allowed to interact.

We have, for instance, experience of water in three different states (solid, liquid and gaseous). Water molecules and their interactions do not change when passing from one state to the other. Understanding how the same interactions can result in qualitatively different macroscopic states, and what governs the change of state, is a central topic of statistical physics.

The foundations of statistical physics rely on two important steps. The first one consists in passing from the deterministic laws of physics, such as Newton's laws, to a probabilistic description. The idea is that a precise knowledge of the motion of each molecule in a macroscopic system is inessential to an understanding of the system as a whole: instead, one can postulate that the microscopic dynamics, because of its chaoticity, allows a purely probabilistic description. The detailed justification of this basic step has been achieved in only a small number of concrete cases. Here we shall bypass any attempt at such a justification: we directly adopt a purely probabilistic point of view, as a basic postulate of statistical physics.

The second step starts from the probabilistic description and recovers determinism at a macroscopic level by some sort of law of large numbers. We all know that water boils at 100° C (at atmospheric pressure) and that its density (at 25° C and atmospheric pressure) is $1\,\mathrm{g/cm}^3$. The regularity of these phenomena is not related to the deterministic laws which rule the motions of water molecules. It is instead a consequence of the fact that, because of the large number of particles involved in any macroscopic system, fluctuations are 'averaged out'. We shall discuss this kind of phenomenon in Section 2.4 and, more mathematically, in Chapter 4.

The purpose of this chapter is to introduce the most basic concepts of this discipline for an audience of non-physicists with a mathematical background. We adopt a somewhat restrictive point of view, which keeps to classical (as opposed to quantum) statistical physics, and basically describes it as a branch of probability theory (Sections 2.1–2.3). In Section 2.4 we focus on large systems, and stress that the statistical-physics approach becomes particularly meaningful in this regime. Theoretical statistical physics often deals with highly idealized mathematical models of real materials. The most interesting (and challenging) task is in fact to understand the *qualitative* behaviour of such systems. With this aim, one can discard any 'irrelevant' microscopic detail from the mathematical description of the model. In Section 2.5, the study of ferromagnetism through the introduction of the Ising model gives an example of this modelling procedure. Compared with the case of Ising ferromagnets, the theoretical understanding of spin glasses is much less developed. Section 2.6 presents a rapid preview of this fascinating subject.

2.1 The Boltzmann distribution

The basic ingredients for a probabilistic description of a physical system are:

- A **space of configurations** $\mathcal{X}$. One should think of $x \in \mathcal{X}$ as giving a complete microscopic determination of the state of the system under consideration. We are not interested in defining the most general mathematical structure for $\mathcal{X}$ such that a statistical-physics formalism can be constructed. Throughout this book we shall in fact consider only two very simple types of configuration spaces: (*i*) finite sets, and (*ii*) smooth, compact, finite-dimensional manifolds. If the system contains N 'particles', the configuration space is a product space:

$$\mathcal{X}_N = \underbrace{\mathcal{X} \times \cdots \times \mathcal{X}}_{N} . \tag{2.1}$$

  The configuration of the system has the form $\underline{x} = (x_1, \ldots, x_N)$. Each coordinate $x_i \in \mathcal{X}$ is meant to represent the state (position, orientation, etc.) of one of the particles. Except for a few examples, we shall focus on configuration spaces of type (*i*). We shall therefore adopt a discrete-space notation for $\mathcal{X}$. The generalization to continuous configuration spaces is in most cases intuitively clear (although it may present some technical difficulties).

- A set of **observables**, which are real-valued functions on the configuration space $\mathcal{O} : x \mapsto \mathcal{O}(x)$. If $\mathcal{X}$ is a manifold, we shall limit ourselves to observables which are smooth functions of the configuration x. Observables are physical quantities which can be measured through an experiment (at least in principle).

- Out of all the observables, a special role is played by the **energy function** $E(x)$. When the system is an N-particle system, the energy function generally takes the form of sums of terms involving few particles. An energy function of the form

$$E(\underline{x}) = \sum_{i=1}^{N} E_i(x_i) \tag{2.2}$$

corresponds to a **non-interacting** system. An energy of the form

$$E(\underline{x}) = \sum_{i_1, \ldots, i_k} E_{i_1, \ldots, i_k}(x_{i_1}, \ldots, x_{i_k}) \tag{2.3}$$

is called a k-**body** interaction. In general, the energy will contain some pieces involving k-body interactions, with $k \in \{1, 2, \ldots, K\}$. An important feature of real physical systems is that K is never a large number (usually $K = 2$ or 3), even when the number of particles N is very large. The same property holds for all measurable observables. However, for the general mathematical formulation which we shall use here, the energy can be any real-valued function on $\mathcal{X}$.

Once the configuration space $\mathcal{X}$ and the energy function are fixed, the probability $\mu_\beta(x)$ for the system to be found in the configuration x is given by the **Boltzmann distribution**:

$$\mu_\beta(x) = \frac{1}{Z(\beta)}\, e^{-\beta E(x)}\,, \qquad Z(\beta) = \sum_{x \in \mathcal{X}} e^{-\beta E(x)}\,. \qquad (2.4)$$

The real parameter $T = 1/\beta$ is the **temperature** (and β is referred to as the inverse temperature). Note that the temperature is usually defined as $T = 1/(k_B\beta)$, where the value of k_B, Boltzmann's constant, depends on the unit of measure for the temperature. Here we adopt the simple choice $k_B = 1$. The normalization constant $Z(\beta)$ is called the **partition function**. Notice that eqn (2.4) indeed defines the density of the Boltzmann distribution with respect to some reference measure. The reference measure is usually the counting measure if $\mathcal{X}$ is discrete or the Lebesgue measure if $\mathcal{X}$ is continuous. It is customary to denote the expectation value with respect to the Boltzmann measure by angle brackets: the expectation value $\langle \mathcal{O}(x) \rangle$ of an observable $\mathcal{O}(x)$, also called its **Boltzmann average**, is given by

$$\langle \mathcal{O} \rangle = \sum_{x \in \mathcal{X}} \mu_\beta(x)\mathcal{O}(x) = \frac{1}{Z(\beta)} \sum_{x \in \mathcal{X}} e^{-\beta E(x)}\mathcal{O}(x)\,. \qquad (2.5)$$

Example 2.1 One intrinsic property of elementary particles is their spin. For 'spin-1/2' particles, the spin σ takes only two values: $\sigma = \pm 1$. A localized spin-1/2 particle, whose only degree of freedom is the spin, is described by $\mathcal{X} = \{+1, -1\}$, and is called an **Ising spin**. The energy of the spin in a state $\sigma \in \mathcal{X}$ in a magnetic field B is

$$E(\sigma) = -B\,\sigma\,. \qquad (2.6)$$

The Boltzmann probability of finding the spin in the state σ is

$$\mu_\beta(\sigma) = \frac{1}{Z(\beta)}\, e^{-\beta E(\sigma)} \qquad Z(\beta) = e^{-\beta B} + e^{\beta B} = 2\cosh(\beta B)\,. \qquad (2.7)$$

The average value of the spin, called the **magnetization**, is

$$\langle \sigma \rangle = \sum_{\sigma \in \{1, -1\}} \mu_\beta(\sigma)\, \sigma = \tanh(\beta B)\,. \qquad (2.8)$$

At high temperatures, $T \gg |B|$, the magnetization is small. At low temperatures, the magnetization its close to its maximal value: $\langle \sigma \rangle = 1$ if $B > 0$. Section 2.5 will discuss the behaviour of many Ising spins, with some more complicated energy functions.

Example 2.2 Some spin variables can have a larger space of possible values. For instance, a **Potts spin** with q states takes values in $\mathcal{X} = \{1, 2, \ldots, q\}$. In the presence of a magnetic field of intensity h pointing in the direction $r \in \{1, \ldots, q\}$, the energy of the Potts spin is

$$E(\sigma) = -B\,\mathbb{I}(\sigma = r)\,. \tag{2.9}$$

In this case, the average value of the spin in the direction of the field is

$$\langle \mathbb{I}(\sigma = r) \rangle = \frac{\exp(\beta B)}{\exp(\beta B) + (q - 1)}\,. \tag{2.10}$$

Example 2.3 Let us consider a single water molecule inside a closed container: for instance, inside a bottle. A water molecule H_2O is already a complicated object, but in a first approximation, we can neglect its structure and model the molecule as a point inside the bottle. The space of configurations then reduces to

$$\mathcal{X} = \texttt{BOTTLE} \subset \mathbb{R}^3\,, \tag{2.11}$$

where we have denoted by $\texttt{BOTTLE}$ the region of $\mathbb{R}^3$ delimited by the container. Note that this description is not very accurate at a microscopic level.

The description of the precise form of the bottle can be quite complex. On the other hand, it is a good approximation to assume that all positions of the molecule are equiprobable: the energy is independent of the particle's position $x \in \texttt{BOTTLE}$. One then has:

$$\mu(x) = \frac{1}{Z}\,, \qquad Z = |\mathcal{X}|\,, \tag{2.12}$$

and the Boltzmann average of the particle's position, $\langle x \rangle$, is the barycentre of the bottle.

Example 2.4 In assuming that all the configurations in the previous example are equiprobable, we neglected the effect of gravity on the water molecule. In the presence of gravity our water molecule at position x has an energy

$$E(x) = w\,\mathrm{h}(x)\,, \tag{2.13}$$

where h(x) is the height corresponding to the position x and w is a positive constant, determined by terrestrial attraction, which is proportional to the mass of the molecule. Given two positions x and y in the bottle, the ratio of the probabilities to find the particle at these positions is

$$\frac{\mu_\beta(x)}{\mu_\beta(y)} = \exp\{-\beta w[\mathrm{h}(x) - \mathrm{h}(y)]\}. \tag{2.14}$$

For a water molecule at a room temperature of 20° C ($T = 293$ K), one has $\beta w \approx 7 \times 10^{-5}\,\mathrm{m}^{-1}$. Given a point x at the bottom of the bottle and y at a height of 20 cm, the probability to find a water molecule 'near' x is approximately 1.000014 times larger than the probability to find it 'near' y. For a tobacco-mosaic virus, which is about 2×10^6 times heavier than a water molecule, the ratio is $\mu_\beta(x)/\mu_\beta(y) \approx 1.4 \times 10^{12}$, which is very large. For a grain of sand, the ratio is so large that one never observes the grain floating at around y. Note that, while these ratios of probability densities are easy to compute, the partition function and therefore the absolute values of the probability densities can be much more complicated to estimate, and depend on the shape of the bottle.

Example 2.5 In many important cases, we are given the space of configurations $\mathcal{X}$ and a stochastic dynamics defined on it. The most interesting probability distribution for such a system is the stationary state $\mu_{\mathrm{st}}(x)$ (we assume that it is unique). For the sake of simplicity, we can consider a finite space $\mathcal{X}$ and a discrete-time Markov chain with transition probabilities $\{w(x \to y)\}$ (in Chapter 4 we shall recall some basic definitions concerning Markov chains). It happens sometimes that the transition rates satisfy, for any pair of configurations $x, y \in \mathcal{X}$, the relation

$$f(x)w(x \to y) = f(y)w(y \to x), \tag{2.15}$$

for some positive function $f(x)$. As we shall see in Chapter 4, when this condition, called **detailed balance**, is satisfied (together with a few other technical conditions), the stationary state has the Boltzmann form (2.4) with $e^{-\beta E(x)} = f(x)$.

Exercise 2.1 As a particular realization of the above example, consider an 8×8 chessboard and a special piece sitting on it. At any time step, the piece will stay still (with probability $1/2$) or move randomly to one of the neighbouring positions (with probability $1/2$). Does this process satisfy the condition (2.15)? Which positions on the chessboard have lower and higher 'energy'? Compute the partition function.

From a purely probabilistic point of view, one can wonder why one bothers to decompose the distribution $\mu_\beta(x)$ into the two factors $e^{-\beta E(x)}$ and $1/Z(\beta)$. Of course the motivations for writing the Boltzmann factor $e^{-\beta E(x)}$ in exponential form come essentially from physics, where one knows (either exactly or to within some level of approximation) the form of the energy. This also justifies the use of the inverse temperature β (after all, one could always redefine the energy function in such a way as to set $\beta = 1$).

However, even if we adopt a mathematical viewpoint, and if we are interested in only a particular distribution $\mu(x)$ which corresponds to a particular value of the temperature, it is often illuminating to embed it into a one-parameter family as is done in the Boltzmann expression (2.4). Indeed, eqn (2.4) interpolates smoothly between several interesting situations. As $\beta \to 0$ (**high-temperature limit**), one recovers the uniform probability distribution

$$\lim_{\beta \to 0} \mu_\beta(x) = \frac{1}{|\mathcal{X}|}\,. \tag{2.16}$$

Both the probabilities $\mu_\beta(x)$ and the expectation values $\langle \mathcal{O}(x) \rangle$ of the observables can be expressed as convergent Taylor expansions around $\beta = 0$. For small β the Boltzmann distribution can be seen as a 'softening' of the original distribution.

In the limit $\beta \to \infty$ (**low-temperature limit**), the Boltzmann distribution concentrates on the global maxima of the original distribution. More precisely, a configuration $x_0 \in \mathcal{X}$ such that $E(x) \geq E(x_0)$ for any $x \in \mathcal{X}$ is called a **ground state**. The minimum value of the energy $E_0 = E(x_0)$ is called the **ground state energy**. We shall denote the set of ground states by $\mathcal{X}_0$. It is elementary to show that, for a discrete configuration space,

$$\lim_{\beta \to \infty} \mu_\beta(x) = \frac{1}{|\mathcal{X}_0|}\, \mathbb{I}(x \in \mathcal{X}_0)\,, \tag{2.17}$$

where $\mathbb{I}(x \in \mathcal{X}_0) = 1$ if $x \in \mathcal{X}_0$ and $\mathbb{I}(x \in \mathcal{X}_0) = 0$ otherwise. The above behaviour is summarized in physicists' jargon by saying that, at low temperature, 'low energy configurations dominate' the behaviour of the system.

2.2   Thermodynamic potentials

Several properties of the Boltzmann distribution (eqn (2.4)) are conveniently summarized through the thermodynamic potentials. These are functions of the temperature $1/\beta$ and of the various parameters defining the energy $E(x)$. The most important thermodynamic potential is the **free energy**

$$F(\beta) = -\frac{1}{\beta} \log Z(\beta)\,, \tag{2.18}$$

where $Z(\beta)$ is the partition function already defined in eqn (2.4). The factor $-1/\beta$ in eqn (2.18) is due essentially to historical reasons. In calculations, it is often more convenient to use the **free entropy**[1] $\Phi(\beta) = -\beta F(\beta) = \log Z(\beta)$.

---

[1] Unlike the other potentials, there is no universally accepted name for $\Phi(\beta)$; however, because this potential is very useful, we have adopted the name 'free entropy' for it.

Two more thermodynamic potentials are derived from the free energy: the **internal energy** $U(\beta)$ and the **canonical entropy** $S(\beta)$:

$$U(\beta) = \frac{\partial}{\partial\beta}(\beta F(\beta)), \qquad S(\beta) = \beta^2 \frac{\partial F(\beta)}{\partial\beta}. \tag{2.19}$$

By direct computation, one obtains the following identities concerning the potentials defined so far:

$$F(\beta) = U(\beta) - \frac{1}{\beta} S(\beta) = -\frac{1}{\beta} \Phi(\beta), \tag{2.20}$$

$$U(\beta) = \langle E(x) \rangle, \tag{2.21}$$

$$S(\beta) = -\sum_x \mu_\beta(x) \log \mu_\beta(x), \tag{2.22}$$

$$-\frac{\partial^2}{\partial\beta^2}(\beta F(\beta)) = \langle E(x)^2 \rangle - \langle E(x) \rangle^2. \tag{2.23}$$

For discrete $\mathcal{X}$, eqn (2.22) can be rephrased by saying that the canonical entropy is the Shannon entropy of the Boltzmann distribution, as we defined it in Chapter 1. This implies that $S(\beta) \geq 0$. Equation (2.23) implies that the free entropy is a convex function of the temperature. Finally, eqn (2.21) justifies the name 'internal energy' for $U(\beta)$.

In order to have some intuition of the content of these definitions, let us reconsider the high- and low-temperature limits already treated in the previous section. In the high-temperature limit, $\beta \to 0$, one finds

$$F(\beta) = -\frac{1}{\beta} \log |\mathcal{X}| + \langle E(x) \rangle_0 + \Theta(\beta), \tag{2.24}$$

$$U(\beta) = \langle E(x) \rangle_0 + \Theta(\beta), \tag{2.25}$$

$$S(\beta) = \log |\mathcal{X}| + \Theta(\beta). \tag{2.26}$$

(Recall that Θ stands for 'of the order of-'; see Appendix A.) The interpretation of these formulae is straightforward. At high temperature, the system can be found in any possible configuration with similar probabilities (the probabilities being exactly equal when $\beta = 0$). The entropy counts the number of possible configurations. The internal energy is just the average value of the energy over the configurations with uniform probability.

While the high-temperature expansions (2.24)–(2.26) have the same form for both a discrete and a continuous configuration space $\mathcal{X}$, in the low-temperature case we must be more careful. If $\mathcal{X}$ is finite, we can meaningfully define the **energy gap** $\Delta E > 0$ as follows (recall that we have denoted by E_0 the ground-state energy):

$$\Delta E = \min\{E(y) - E_0 : y \in \mathcal{X} \backslash \mathcal{X}_0\}. \tag{2.27}$$

With this definition, we get

$$F(\beta) = E_0 - \frac{1}{\beta} \log |\mathcal{X}_0| + \Theta(e^{-\beta\Delta E}), \tag{2.28}$$

$$E(\beta) = E_0 + \Theta(e^{-\beta\Delta E}), \tag{2.29}$$

$$S(\beta) = \log |\mathcal{X}_0| + \Theta(e^{-\beta\Delta E}). \tag{2.30}$$

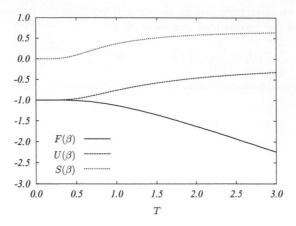

Fig. 2.1 Thermodynamic potentials for a two-level system with $\epsilon_1 = -1$ and $\epsilon_2 = +1$ as a function of the temperature $T = 1/\beta$.

The interpretation is that, at low temperature, the system is found with equal probability in any of the ground states, and nowhere else. Once again, the entropy counts the number of available configurations, and the internal energy is the average of their energies (which coincide with that of the ground state).

Exercise 2.2 *A two-level system.* This is the simplest non-trivial example: $\mathcal{X} = \{1, 2\}$, $E(1) = \epsilon_1$, $E(2) = \epsilon_2$. Without loss of generality, we assume $\epsilon_1 < \epsilon_2$. This example can be used as a mathematical model for many physical systems, such as the spin-1/2 particle discussed above.

Derive the following results for the thermodynamic potentials (where $\Delta = \epsilon_2 - \epsilon_1$ is the energy gap):

$$F(\beta) = \epsilon_1 - \frac{1}{\beta} \log(1 + e^{-\beta \Delta}) \,, \tag{2.31}$$

$$U(\beta) = \epsilon_1 + \frac{e^{-\beta \Delta}}{1 + e^{-\beta \Delta}} \Delta \,, \tag{2.32}$$

$$S(\beta) = \frac{e^{-\beta \Delta}}{1 + e^{-\beta \Delta}} \beta \Delta + \log(1 + e^{-\beta \Delta}) \,. \tag{2.33}$$

The behaviour of these functions is presented in Fig. 2.1. The reader can work out the asymptotics, and check the general high- and low-temperature behaviour given above.

Exercise 2.3 We return to the example of the previous section: one water molecule, modelled as a point, in a bottle. We consider the case of a cylindrical bottle of base $B \subset \mathbb{R}^2$ (surface area $|B|$) and height d.

Using the energy function in eqn (2.13), derive the following explicit expressions for the thermodynamic potentials:

$$F(\beta) = -\frac{1}{\beta} \log |B| - \frac{1}{\beta} \log \frac{1 - e^{-\beta wd}}{\beta w}, \tag{2.34}$$

$$U(\beta) = \frac{1}{\beta} - \frac{wd}{e^{\beta wd} - 1}, \tag{2.35}$$

$$S(\beta) = \log |Bd| + 1 - \frac{\beta wd}{e^{\beta wd} - 1} - \log \left(\frac{\beta wd}{1 - e^{-\beta wd}} \right). \tag{2.36}$$

Note that the internal-energy formula can be used to compute the average height of the molecule $\langle h(x) \rangle = U(\beta)/w$. This is a consequence of the definition of the energy (see eqn (2.13)) and of eqn (2.21). If we plug in the correct constant w, we can find that the average height falls below 49.99% of the height of the bottle $d = 20$ cm only when the temperature is below 3.2 K.

Exercise 2.4 Using eqns (2.34)–(2.36), derive the low-temperature expansions

$$F(\beta) = -\frac{1}{\beta} \log \left(\frac{|B|}{\beta w} \right) + \Theta(e^{-\beta wd}), \tag{2.37}$$

$$U(\beta) = \frac{1}{\beta} + \Theta(e^{-\beta wd}), \tag{2.38}$$

$$S(\beta) = \log \left(\frac{|B| e}{\beta w} \right) + \Theta(e^{-\beta wd}). \tag{2.39}$$

In this case $\mathcal{X}$ is continuous, and the energy has no gap. Nevertheless, these results can be understood as follows: at low temperature, the molecule is confined to a layer of height of order $1/(\beta w)$ above the bottom of the bottle. It therefore occupies a volume of size $|B|/(\beta w)$. Its entropy is approximately given by the logarithm of such a volume.

Exercise 2.5 Let us reconsider the above example and assume the bottle to have a different shape, for instance a sphere of radius R. In this case it is difficult to compute explicit expressions for the thermodynamic potentials, but one can easily compute the low-temperature expansions. For the entropy, one gets at large β

$$S(\beta) = \log \left(\frac{2\pi e^2 R}{\beta^2 w^2} \right) + \Theta(1/\beta). \tag{2.40}$$

The reader should try to understand the difference between this result and eqn (2.39) and provide an intuitive explanation, as in the previous example. Physicists say that the low-temperature thermodynamic potentials reveal the 'low-energy structure' of the system.

2.3 The fluctuation–dissipation relations

It often happens that the energy function depends smoothly upon some real parameters. These can be related to the experimental conditions under which a physical system is studied, or to some fundamental physical quantity. For instance, the energy of a water molecule in a gravitational field (see eqn (2.13)) depends upon the weight w of the molecule itself. Although this is a constant number in the physical world, it is useful, in the theoretical treatment, to consider it as an adjustable parameter.

It is therefore interesting to consider an energy function $E_\lambda(x)$ which depends smoothly upon some parameter λ and admits the following Taylor expansion in the neighbourhood of $\lambda = \lambda_0$:

$$E_\lambda(x) = E_{\lambda_0}(x) + (\lambda - \lambda_0) \left. \frac{\partial E}{\partial \lambda} \right|_{\lambda_0}(x) + O((\lambda - \lambda_0)^2). \tag{2.41}$$

The dependence of the free energy and of other thermodynamic potentials upon λ in the neighbourhood of λ_0 is easily related to the explicit dependence of the energy function itself. Let us consider the partition function, and expand it to first order in $\lambda - \lambda_0$:

$$\begin{aligned} Z(\lambda) &= \sum_x \exp\left(-\beta \left[E_{\lambda_0}(x) + (\lambda - \lambda_0) \left. \frac{\partial E}{\partial \lambda} \right|_{\lambda_0}(x) + O((\lambda - \lambda_0)^2) \right] \right) \\ &= Z(\lambda_0) \left[1 - \beta(\lambda - \lambda_0) \left\langle \left. \frac{\partial E}{\partial \lambda} \right|_{\lambda_0} \right\rangle_0 + O((\lambda - \lambda_0)^2) \right], \end{aligned} \tag{2.42}$$

where we have denoted by $\langle \cdot \rangle_0$ the expectation with respect to the Boltzmann distribution at $\lambda = \lambda_0$.

This shows that the free entropy behaves as

$$\left. \frac{\partial \Phi}{\partial \lambda} \right|_{\lambda_0} = -\beta \left\langle \left. \frac{\partial E}{\partial \lambda} \right|_{\lambda_0} \right\rangle_0, \tag{2.43}$$

One can also consider the λ dependence of the expectation value of a generic observable $A(x)$. Using again the Taylor expansion, one finds that

$$\left. \frac{\partial \langle A \rangle_\lambda}{\partial \lambda} \right|_{\lambda_0} = -\beta \left\langle A \, ; \left. \frac{\partial E}{\partial \lambda} \right|_{\lambda_0} \right\rangle_0. \tag{2.44}$$

where we have denoted by $\langle A; B \rangle$ the **connected correlation function**: $\langle A; B \rangle = \langle AB \rangle - \langle A \rangle \langle B \rangle$. A particular example of this relation was given in eqn (2.23).

The result (2.44) has important practical consequences and many generalizations. Imagine you have an experimental apparatus that allows you to tune some parameter λ (for instance the pressure of a gas, or the magnetic or electric field acting on some material) and to monitor the value of an observable $A(x)$ (the volume of the gas, or the polarization or magnetization of the material). The quantity on the left-hand side of eqn (2.44) is the response of the system to an infinitesimal variation of the tunable parameter. On the right-hand side, we find some correlation function within the 'unperturbed' system. One possible application is to measure correlations within a system by monitoring its response to an external perturbation. The relation (2.44) between a correlation and a response is called the **fluctuation–dissipation theorem**.

2.4 The thermodynamic limit

The main purpose of statistical physics is to understand the macroscopic behaviour of a large number, $N \gg 1$, of simple components (atoms, molecules, etc.) when they are brought together.

To be concrete, let us consider a few drops of water in a bottle. A configuration of the system is given by the positions and orientations of all the H_2O molecules inside the bottle. In this case $\mathcal{X}$ is the set of positions and orientations of a single molecule, and N is typically of order 10^{23} (more precisely, $18\,\mathrm{g}$ of water contains approximately 6×10^{23} molecules). The sheer magnitude of such a number leads physicists to focus on the $N \to \infty$ limit, also called the **thermodynamic limit**.

As shown by the examples below, for large N, the thermodynamic potentials are often proportional to N. One is thus led to introduce the **intensive thermodynamic potentials** as follows. Let us denote by $F_N(\beta)$, $U_N(\beta)$, and $S_N(\beta)$ the free energy, internal energy, and canonical entropy, respectively. for a system with N 'particles'. The **free energy density** is defined by

$$f(\beta) = \lim_{N \to \infty} F_N(\beta)/N\,, \tag{2.45}$$

if the limit exists, which is usually the case (at least if the forces between particles decrease fast enough at large distance). One defines analogously the **energy density** $u(\beta)$ and the **entropy density** $s(\beta)$.

The free energy $F_N(\beta)$, is, quite generally, an analytic function of β in a neighbourhood of the real β axis. This is a consequence of the fact that $Z(\beta)$ is analytic throughout the entire β plane, and strictly positive for real β's. A question of great interest is whether analyticity is preserved in the thermodynamic limit (2.45), under the assumption that the limit exists. Whenever the free energy density $f(\beta)$ is non-analytic, one says that a **phase transition** occurs. Since the free entropy density $\phi(\beta) = -\beta f(\beta)$ is convex, the free energy density is necessarily continuous whenever it exists.

In the simplest cases, the non-analyticities occur at isolated points. Let β_c be such a point. Two particular types of singularities occur frequently:

- The free energy density is continuous, but its derivative with respect to β is discontinuous at β_c. This singularity is called a **first-order phase transition**.
- The free energy and its first derivative are continuous, but the second derivative is discontinuous at β_c. This is called a **second-order phase transition**.

Higher-order phase transitions can be defined as well, along the same lines.

Apart from being interesting mathematical phenomena, phase transitions correspond to *qualitative* changes in the underlying physical system. For instance, the transition from water to vapor at 100 °C at normal atmospheric pressure is modelled mathematically as a first-order phase transition in the above sense. A great part of this book will be devoted to the study of phase transitions in many different systems, where the interacting 'particles' can be very diverse objects such as bits of information or occupation numbers on the vertices of a graph.

When N grows, the volume of the configuration space increases exponentially: $|\mathcal{X}_N| = |\mathcal{X}|^N$. Of course, not all of the configurations are equally important under the Boltzmann distribution: the lowest-energy configurations have greater probability. What is important is therefore the number of configurations at a given energy. This information is encoded in the **energy spectrum** of the system,

$$\mathcal{N}_\Delta(E) = |\Omega_\Delta(E)|, \qquad \Omega_\Delta(E) \equiv \{x \in \mathcal{X}_N : E \le E(x) < E + \Delta\}. \qquad (2.46)$$

In many systems of interest, the energy spectrum diverges exponentially as $N \to \infty$, if the energy is scaled linearly with N. More precisely, there exists a function $s(e)$ such that, given two numbers e and $\delta > 0$,

$$\lim_{N \to \infty} \frac{1}{N} \log \mathcal{N}_{N\delta}(Ne) = \sup_{e' \in [e, e+\delta]} s(e'). \qquad (2.47)$$

The function $s(e)$ is called the **microcanonical entropy density**. The statement (2.47) is often rewritten in the more compact form

$$\mathcal{N}_\Delta(E) \doteq_N \exp\left[Ns\left(\frac{E}{N}\right)\right]. \qquad (2.48)$$

The notation $A_N \doteq_N B_N$ will be used throughout the book to denote that two quantities A_N and B_N (which behave exponentially in N) are equal **to leading exponential order**, meaning $\lim_{N \to \infty}(1/N) \log(A_N/B_N) = 0$. We shall often use $\doteq$ without an index when there is no ambiguity about the large variable N.

The microcanonical entropy density $s(e)$ conveys a great amount of information about the system. Furthermore, it is directly related to the intensive thermodynamic potentials through a fundamental relation

Proposition 2.6 *If the microcanonical entropy density (2.47) exists for any e and if the limit in eqn (2.47) is uniform in e, then the free entropy density (2.45) exists and is given by*

$$\phi(\beta) = \max_e [s(e) - \beta e]. \qquad (2.49)$$

If the maximum of $s(e) - \beta e$ is unique, then the internal-energy density equals $\arg\max[s(e) - \beta e]$.

Proof The basic idea is to write the partition function as

$$Z_N(\beta) \doteq \sum_{k=-\infty}^{\infty} \mathcal{N}_\Delta(k\Delta)\, e^{-\beta \Delta} \doteq \int \exp\{Ns(e) - N\beta e\}\, de, \qquad (2.50)$$

and to evaluate the last integral by the saddle point method. The reader will find references in the Notes section at the end of this chapter. □

Example 2.7 Let us consider N identical two-level systems, i.e. $\mathcal{X}_N = \mathcal{X} \times \cdots \times \mathcal{X}$, with $\mathcal{X} = \{1, 2\}$. We take the energy to be the sum of the single-system energies: $E(x) = E_{\text{single}}(x_1) + \cdots + E_{\text{single}}(x_N)$, with $x_i \in \mathcal{X}$. As in the previous section, we set $E_{\text{single}}(1) = \epsilon_1$, $E_{\text{single}}(2) = \epsilon_2 > \epsilon_1$, and $\Delta = \epsilon_2 - \epsilon_1$.

The energy spectrum of this model is quite simple. For any energy $E = N\epsilon_1 + n\Delta$, there are $\binom{N}{n}$ configurations x with $E(x) = E$. Therefore, using the definition (2.47), we get

$$s(e) = \mathcal{H}\left(\frac{e - \epsilon_1}{\Delta}\right). \tag{2.51}$$

Equation (2.49) can now be used to get

$$f(\beta) = \epsilon_1 - \frac{1}{\beta}\log(1 + e^{-\beta\Delta}), \tag{2.52}$$

which agrees with the result obtained directly from the definition (2.18).

The great attention paid by physicists to the thermodynamic limit is extremely well justified by the huge number of degrees of freedom involved in a macroscopic piece of matter. Let us stress that the interest of the thermodynamic limit is more general than these huge numbers might suggest. First of all, it often happens that fairly small systems are well approximated by the thermodynamic limit. This is extremely important for numerical simulations of physical systems: one cannot, of course, simulate 10^{23} molecules on a computer! Even the cases in which the thermodynamic limit *is not* a good approximation are often fruitfully analysed as *violations* of this limit. Finally, the insight gained from analysing the $N \to \infty$ limit is always crucial in understanding moderate-size systems.

2.5   Ferromagnets and Ising models

Magnetic materials contain molecules with a magnetic moment, a three-dimensional vector which tends to align with the magnetic field felt by the molecule. Moreover, the magnetic moments of two different molecules interact with each other. Quantum mechanics plays an important role in magnetism. Because of quantum effects, the space of possible configurations of a magnetic moment becomes discrete. Quantum effects are also the origin of the 'exchange interaction' between magnetic moments. In many materials, the effect of the exchange interaction is such that the energy is lower when two moments align. While the behaviour of a single magnetic moment in an external field is qualitatively simple, when we consider a bunch of interacting moments, the problem is much richer, and exhibits remarkable collective phenomena.

A simple mathematical model for such materials is the Ising model. This describes the magnetic moments by Ising spins localized at the vertices of a certain region of a d-dimensional cubic lattice. To keep things simple, let us consider a region $\mathbb{L}$ which is a cube of side L: $\mathbb{L} = \{1, \ldots, L\}^d$. On each site $i \in \mathbb{L}$, there is an Ising spin $\sigma_i \in \{+1, -1\}$ (see Fig. 2.2).

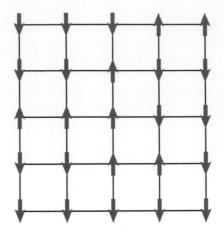

Fig. 2.2 A configuration of a two-dimensional Ising model with $L = 5$. There is an Ising spin σ_i on each vertex i, shown by an arrow pointing up if $\sigma_i = +1$ and pointing down if $\sigma_i = -1$. The energy (2.53) is given by the sum of two types of contributions: (i) a term $-\sigma_i \sigma_j$ for each edge (ij) of the graph, such that the energy is minimized when the two neighbouring spins σ_i and σ_j point in the same direction; and (ii) a term $-B\sigma_i$ for each site i, due to the coupling to an external magnetic field. The configuration depicted here has an energy $-8 + 9B$.

A configuration $\underline{\sigma} = (\sigma_1 \dots \sigma_N)$ of the system is given by assigning the values of all the spins in the system. Therefore, the space of configurations $\mathcal{X}_N = \{+1, -1\}^{\mathbb{L}}$ has the form (2.1), with $\mathcal{X} = \{+1, -1\}$ and $N = L^d$.

The definition of a ferromagnetic Ising model is completed by the definition of the energy function. A configuration $\underline{\sigma}$ has an energy

$$E(\underline{\sigma}) = -\sum_{(ij)} \sigma_i \sigma_j - B \sum_{i \in \mathbb{L}} \sigma_i \,, \tag{2.53}$$

where the sum over (ij) runs over all the (unordered) pairs of sites $i, j \in \mathbb{L}$ which are nearest neighbours. The real number B measures the applied external magnetic field.

Determining the free energy density $f(\beta)$ in the thermodynamic limit for this model is a non-trivial task. The model was invented by Wilhem Lenz in the early 1920s, who assigned the task of analysing it to his student Ernst Ising. In his thesis of 1924, Ising solved the $d = 1$ case and showed the absence of phase transitions. In 1948, Lars Onsager brilliantly solved the $d = 2$ case, exhibiting the first soluble 'finite-dimensional' model with a second-order phase transition. In higher dimensions, the problem is unsolved, although many important features of the solution are well understood.

Before embarking on any calculations, let us discuss some qualitative properties of this model. Two limiting cases are easily understood. At infinite temperature, $\beta = 0$, the energy (2.53) no longer matters and the Boltzmann distribution weights all the configurations with the same factor 2^{-N}. We have therefore an assembly of completely independent spins. At zero temperature, $\beta \to \infty$, the Boltzmann distribution concentrates onto the ground state(s). If there is no magnetic field, i.e. $B = 0$, there are two degenerate ground states: the configuration $\underline{\sigma}^{(+)}$ with all the spins pointing up, $\sigma_i = +1$, and the configuration $\underline{\sigma}^{(-)}$ with all the spins pointing down, $\sigma_i = -1$. If the magnetic field is set to some non-zero value, one of the two configuration dominates: $\underline{\sigma}^{(+)}$ if $B > 0$ and $\underline{\sigma}^{(-)}$ if $B < 0$.

Notice that the reaction of the system to the external magnetic field B is quite different in the two cases. To see this fact, define a 'rescaled' magnetic field $x = \beta B$ and take the limits $\beta \to 0$ and $\beta \to \infty$ keeping x fixed. The expected value of any spin in $\mathbb{L}$, in the two limits, is

$$\langle \sigma_i \rangle = \begin{cases} \tanh(x) & \text{for } \beta \to , \\ \tanh(Nx) & \text{for } \beta \to \infty . \end{cases} \tag{2.54}$$

Each spin reacts independently for $\beta \to 0$. In contrast, they react as a whole as $\beta \to \infty$: one says that the response is cooperative.

A useful quantity for describing the response of the system to the external field is the **average magnetization**,

$$M_N(\beta, B) = \frac{1}{N} \sum_{i \in \mathbb{L}} \langle \sigma_i \rangle . \tag{2.55}$$

Because of the symmetry between the up and down directions, $M_N(\beta, B)$ is an odd function of B. In particular, $M_N(\beta, 0) = 0$. A cooperative response can be emphasized by considering the **spontaneous magnetization**

$$M_+(\beta) = \lim_{B \to 0+} \lim_{N \to \infty} M_N(\beta, B) . \tag{2.56}$$

It is important to understand that a non-zero spontaneous magnetization can appear only in an infinite system: the order of the limits in eqn (2.56) is crucial. Our analysis so far has shown that a spontaneous magnetization exists at $\beta = \infty$: $M_+(\infty) = 1$. On the other hand, $M_+(0) = 0$. It can be shown that the spontaneous magnetization $M_+(\beta)$ is always zero in a high temperature phase defined by $\beta < \beta_c(d)$ (such a phase is called **paramagnetic**). In one dimension ($d = 1$), we shall show below that $\beta_c(1) = \infty$. The spontaneous magnetization is always zero, except at zero temperature ($\beta = \infty$): one speaks of a zero-temperature phase transition. In dimensions $d \geq 2$, $\beta_c(d)$ is finite, and $M_+(\beta)$ becomes non-zero in the **ferromagnetic phase** , i.e. for $\beta > \beta_c$: a phase transition takes place at $\beta = \beta_c$. The temperature $T_c = 1/\beta_c$ is called the **critical temperature**. In the following, we shall discuss the $d = 1$ case, and a variant of the model, called the Curie–Weiss model, where each spin interacts with all the other ones: this is the simplest model which exhibits a finite-temperature phase transition.

2.5.1 The one-dimensional case

The $d = 1$ case has the advantage of being simple to solve. We want to compute the partition function (2.4) for a system of N spins with energy $E(\underline{\sigma}) = -\sum_{i=1}^{N-1} \sigma_i \sigma_{i+1} - B \sum_{i=1}^{N} \sigma_i$. We shall use the **transfer matrix method**, which belongs to the general dynamic programming strategy familiar to computer scientists.

We introduce a partial partition function, where the configurations of all spins $\sigma_1, \ldots, \sigma_p$ have been summed over, at fixed σ_{p+1}:

$$z_p(\beta, B, \sigma_{p+1}) \equiv \sum_{\sigma_1, \ldots, \sigma_p} \exp\left[\beta \sum_{i=1}^{p} \sigma_i \sigma_{i+1} + \beta B \sum_{i=1}^{p} \sigma_i\right]. \qquad (2.57)$$

The partition function (2.4) is given by $Z_N(\beta, B) = \sum_{\sigma_N} z_{N-1}(\beta, B, \sigma_N) \exp(\beta B \sigma_N)$. Obviously, z_p satisfies the recursion relation

$$z_p(\beta, B, \sigma_{p+1}) = \sum_{\sigma_p = \pm 1} T(\sigma_{p+1}, \sigma_p) z_{p-1}(\beta, B, \sigma_p) \qquad (2.58)$$

where we have defined the **transfer matrix** $T(\sigma, \sigma') = \exp\left[\beta \sigma \sigma' + \beta B \sigma'\right]$. This is the 2×2 matrix

$$T = \begin{pmatrix} e^{\beta + \beta B} & e^{-\beta - \beta B} \\ e^{-\beta + \beta B} & e^{\beta - \beta B} \end{pmatrix} \qquad (2.59)$$

Introducing the two-component vectors $\psi_{\mathrm{L}} = \begin{pmatrix} \exp(\beta B) \\ \exp(-\beta B) \end{pmatrix}$ and $\psi_{\mathrm{R}} = \begin{pmatrix} 1 \\ 1 \end{pmatrix}$, and the standard scalar product between vectors $(a, b) = a_1 b_1 + a_2 b_2$, the partition function can be written in matrix form:

$$Z_N(\beta, B) = (\psi_{\mathrm{L}}, T^{N-1} \psi_{\mathrm{R}}). \qquad (2.60)$$

Let us call the eigenvalues of T λ_1, λ_2, and the corresponding eigenvectors ψ_1, ψ_2. It is easy to realize that ψ_1, ψ_2 can be chosen to be linearly independent, and hence ψ_{R} can be decomposed as $\psi_{\mathrm{R}} = u_1 \psi_1 + u_2 \psi_2$. The partition function is then expressed as

$$Z_N(\beta, B) = u_1 (\psi_{\mathrm{L}}, \psi_1) \lambda_1^{N-1} + u_2 (\psi_{\mathrm{L}}, \psi_2) \lambda_2^{N-1}. \qquad (2.61)$$

The diagonalization of the matrix T gives

$$\lambda_{1,2} = e^{\beta} \cosh(\beta B) \pm \sqrt{e^{2\beta} \sinh^2 \beta B + e^{-2\beta}}. \qquad (2.62)$$

For β finite, in the large-N limit, the partition function is dominated by the largest eigenvalue λ_1, and the free-entropy density is given by $\phi = \log \lambda_1$:

$$\phi(\beta, B) = \log\left[e^{\beta} \cosh(\beta B) + \sqrt{e^{2\beta} \sinh^2 \beta B + e^{-2\beta}}\right]. \qquad (2.63)$$

Using the same transfer matrix technique, we can compute expectation values of observables. For instance, the expected value of a given spin is

$$\langle \sigma_i \rangle = \frac{1}{Z_N(\beta, B)} (\psi_{\mathrm{L}}, T^{i-1} \hat{\sigma} T^{N-i} \psi_{\mathrm{R}}), \qquad (2.64)$$

where $\hat{\sigma}$ is the following matrix:

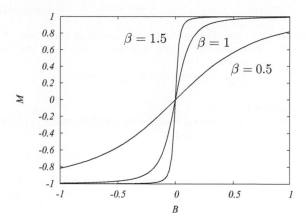

Fig. 2.3 The average magnetization of the one-dimensional Ising model, as a function of the magnetic field B, at inverse temperatures $\beta = 0.5, 1, 1.5$.

$$\hat{\sigma} = \begin{pmatrix} 1 & 0 \\ 0 & -1 \end{pmatrix}. \tag{2.65}$$

Averaging over the position i, one can compute the average magnetization $M_N(\beta, B)$. In the thermodynamic limit, we get

$$\lim_{N \to \infty} M_N(\beta, B) = \frac{\sinh \beta B}{\sqrt{\sinh^2 \beta h + e^{-4\beta}}} = \frac{1}{\beta} \frac{\partial \phi}{\partial B}(\beta, B). \tag{2.66}$$

Both the free energy and the average magnetization turn out to be analytic functions of β and B for $\beta < \infty$. In particular the spontaneous magnetization vanishes at any non-zero temperature:

$$M_+(\beta) = 0, \quad \forall \beta < \infty. \tag{2.67}$$

In Fig. 2.3, we plot the average magnetization $M(\beta, B) \equiv \lim_{N \to \infty} M_N(\beta, B)$ as a function of the applied magnetic field B for various values of the temperature β. The curves become steeper and steeper as β increases. This statement can be made more quantitative by computing the **susceptibility** associated with the average magnetization,

$$\chi_M(\beta) = \frac{\partial M}{\partial h}(\beta, 0) = \beta \, e^{2\beta}. \tag{2.68}$$

This result can be interpreted as follows. A single spin in a field has a susceptibility $\chi(\beta) = \beta$. If we consider N spins constrained to take the same value, the corresponding susceptibility will be $N\beta$, as in eqn (2.54). In the present case, the system behaves as if the spins were grouped into blocks of $\chi_M(\beta)/\beta$ spins each. The spins in each group are constrained to take the same value, while spins belonging to different blocks are independent.

This qualitative interpretation can be given further support by computing a **correlation function**.

Exercise 2.6 Consider the one-dimensional Ising model in zero field, $B = 0$. Show that when $\delta N < i < j < (1 - \delta)N$, the correlation function $\langle \sigma_i \sigma_j \rangle$ is, in the large-N limit,

$$\langle \sigma_i \sigma_j \rangle = e^{-|i-j|/\xi(\beta)} + \Theta(e^{-\alpha N}) , \tag{2.69}$$

where $\xi(\beta) = -1/\log\tanh\beta$.

[Hint: You can either use the general transfer matrix formalism or, more simply, use the identity $e^{\beta\sigma_i\sigma_{i+1}} = \cosh\beta(1 + \sigma_i\sigma_{i+1}\tanh\beta)$.]

Note that, in eqn (2.69), $\xi(\beta)$ gives the typical distance below which two spins in the system are well correlated. For this reason, it is usually called the **correlation length** of the model. This correlation length increases as the temperature decreases: spins become correlated at larger and larger distances. The result (2.69) is clearly consistent with our interpretation of the susceptibility. In particular, as $\beta \to \infty$, $\xi(\beta) \approx e^{2\beta}/2$ and $\chi_M(\beta) \approx 2\beta\xi(\beta)$.

The connection between correlation length and susceptibility is very general and can be understood as a consequence of the fluctuation–dissipation theorem (2.44):

$$\chi_M(\beta) = \beta N \left\langle \left(\frac{1}{N} \sum_{i=1}^{N} \sigma_i \right) ; \left(\frac{1}{N} \sum_{i=1}^{N} \sigma_i \right) \right\rangle$$

$$= \frac{\beta}{N} \sum_{i,j=1}^{N} \langle \sigma_i ; \sigma_j \rangle = \frac{\beta}{N} \sum_{i,j=1}^{N} \langle \sigma_i \sigma_j \rangle , \tag{2.70}$$

where the last equality comes from the fact that $\langle \sigma_i \rangle = 0$ when $B = 0$. Using eqn (2.69), we get

$$\chi_M(\beta) = \beta \sum_{i=-\infty}^{+\infty} e^{-|i|/\xi(\beta)} + \Theta(e^{-\alpha N}) . \tag{2.71}$$

It is therefore evident that a large susceptibility must correspond to a large correlation length.

2.5.2   The Curie–Weiss model

The exact solution of the one-dimensional model led Ising to think that there could not be a phase transition for any dimension. Some thirty years earlier, a qualitative theory of ferromagnetism had been put forward by Pierre Curie. Such a theory assumed the existence of a phase transition at a non-zero temperature T_c (the 'Curie point') and a non-vanishing spontaneous magnetization for $T < T_c$. The dilemma was eventually solved by Onsager's solution of the two-dimensional model.

Curie's theory is realized exactly within a rather abstract model: the **Curie–Weiss model**. We shall present it here as one of the simplest solvable models with a finite-temperature phase transition. Once again, we have N Ising spins $\sigma_i \in \{\pm 1\}$, and a configuration is given by $\underline{\sigma} = (\sigma_1, \ldots, \sigma_N)$. However, the spins no longer sit on a d-dimensional lattice: they all interact in pairs. The energy function, in the presence of a magnetic field B, is given by

$$E(\underline{\sigma}) = -\frac{1}{N} \sum_{(ij)} \sigma_i \sigma_j - B \sum_{i=1}^{N} \sigma_i \,, \tag{2.72}$$

where the sum over (ij) runs over all of the $N(N-1)/2$ couples of spins. Notice the peculiar $1/N$ scaling in front of the exchange term. The exact solution presented below shows that this is the only choice which yields a non-trivial free energy density in the thermodynamic limit. This can be easily understood intuitively as follows. The sum over (ij) involves $\Theta(N^2)$ terms of order $\Theta(1)$. In order to get an energy function that scales as N, we need to put a coefficient $1/N$ in front.

In adopting the energy function (2.72), we gave up an attempt to describe any finite-dimensional geometrical structure. This is a severe simplification, but has the advantage of making the model exactly soluble. The Curie–Weiss model is the first example of a large family: the **mean-field models**. We shall explore many instances of this family throughout the book.

A possible approach to the computation of the partition function consists in observing that the energy function can be written in terms of a simple observable, the **instantaneous** (or **empirical**) **magnetization**,

$$m(\underline{\sigma}) = \frac{1}{N} \sum_{i=1}^{N} \sigma_i \,. \tag{2.73}$$

Notice that this is a function of the configuration $\underline{\sigma}$, and should not be confused with its expected value, the average magnetization (see eqn (2.55)). It is a 'simple' observable because it is equal to a sum of observables depending upon a single spin.

We can write the energy of a configuration in terms of its instantaneous magnetization:

$$E(\underline{\sigma}) = \frac{1}{2} N - \frac{1}{2} N m(\underline{\sigma})^2 - N B \, m(\underline{\sigma}) \,. \tag{2.74}$$

This implies the following formula for the partition function:

$$Z_N(\beta, B) = e^{-N\beta/2} \sum_m \mathcal{N}_N(m) \exp \left\{ \frac{N\beta}{2} m^2 + N\beta B m \right\} \,, \tag{2.75}$$

where the sum over m runs over all of the possible instantaneous magnetizations of N Ising spins: $m = -1 + 2k/N$ with $0 \le k \le N$, where k is an integer number, and $\mathcal{N}_N(m)$ is the number of configurations that have a given instantaneous magnetization m. This is a binomial coefficient whose large-N behaviour can be expressed in terms of the entropy function of Bernoulli variables:

$$\mathcal{N}_N(m) = \binom{N}{N(1+m)/2} \doteq \exp \left[N \mathcal{H} \left(\frac{1+m}{2} \right) \right] \,. \tag{2.76}$$

To leading exponential order in N, the partition function can thus be written as

$$Z_N(\beta, B) \doteq \int_{-1}^{+1} e^{N\phi_{\mathrm{mf}}(m;\beta,B)} \, dm \,, \qquad (2.77)$$

where we have defined

$$\phi_{\mathrm{mf}}(m; \beta, B) = -\frac{\beta}{2}(1 - m^2) + \beta B m + \mathcal{H}\left(\frac{1+m}{2}\right). \qquad (2.78)$$

The integral in eqn (2.77) is easily evaluated by the Laplace method, to get the final result for the free energy density

$$\phi(\beta, B) = \max_{m \in [-1,+1]} \phi_{\mathrm{mf}}(m; \beta, B). \qquad (2.79)$$

One can see that the maximum is obtained away from the boundary points, so that the corresponding m must be a stationary point of $\phi_{\mathrm{mf}}(m; \beta, B)$, which satisfies the **saddle point equation** $\partial\phi_{\mathrm{mf}}(m; \beta, B)/\partial m = 0$:

$$m_* = \tanh(\beta m_* + \beta B). \qquad (2.80)$$

In the above derivation, we were slightly sloppy in two steps: substituting the binomial coefficient by its asymptotic form, and changing the sum over m into an integral. The mathematically minded reader is invited to show that these passages are indeed correct.

With a little more work, the above method can be extended to expectation values of observables. Let us consider, for instance, the average magnetization $M(\beta, B)$. It can be easily shown that, whenever the maximum of $\phi_{\mathrm{mf}}(m; \beta, B)$ over m is non-degenerate,

$$M(\beta, B) \equiv \lim_{N \to \infty} \langle m(\underline{\sigma}) \rangle = m_*(\beta, B) \equiv \arg\max_m \phi_{\mathrm{mf}}(m; \beta, B). \qquad (2.81)$$

We can now examine the implications that can be drawn from Eqs. (2.79) and (2.80). Let us consider first the $B = 0$ case (see Fig. 2.4). The function $\phi_{\mathrm{mf}}(m; \beta, 0)$ is symmetric in m. For $0 \leq \beta \leq 1 \equiv \beta_{\mathrm{c}}$, it is also concave and achieves its unique maximum at $m_*(\beta) = 0$. For $\beta > 1$, $m = 0$ remains a stationary point but becomes a local minimum, and the function develops two degenerate global maxima at $m_\pm(\beta)$, with $m_+(\beta) = -m_-(\beta) > 0$. These two maxima bifurcate continuously from $m = 0$ at $\beta = \beta_{\mathrm{c}}$.

A phase transition takes place at β_{c}. Its meaning can be understood by computing the expectation value of the spins. Notice that the energy function (2.72) is symmetric under a spin-flip transformation which maps $\sigma_i \to -\sigma_i$ for all i. Therefore $\langle \sigma_i \rangle = \langle (-\sigma_i) \rangle = 0$, and the average magnetization vanishes, i.e. $M(\beta, 0) = 0$. On the other hand, the spontaneous magnetization, defined in eqn (2.56), is zero in the paramagnetic phase, i.e. for $\beta < \beta_{\mathrm{c}}$, and equal to $m_+(\beta)$ in the ferromagnetic phase for $\beta > \beta_{\mathrm{c}}$. The physical interpretation of this phase is the following: for any finite N, the pdf of the instantaneous magnetization $m(\underline{\sigma})$ has two symmetric peaks, at $m_\pm(\beta)$,

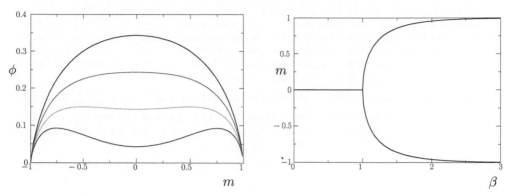

Fig. 2.4 *Left*: the function $\phi_{\mathrm{mf}}(m; \beta, B = 0)$ plotted versus m, for $\beta = 0.7, 0.9, 1.1, 1.3$ (from *top* to *bottom*). For $\beta < \beta_{\mathrm{c}} = 1$, there is a unique maximum at $m = 0$; for $\beta < \beta_{\mathrm{c}} = 1$, there are two degenerate maxima at two symmetric values $\pm m_{+}(\beta)$. *Right*: the values of m which maximize $\phi_{\mathrm{mf}}(m; \beta, B = 0)$, plotted versus β. The phase transition at $\beta_{\mathrm{c}} = 1$ is signalled by a bifurcation.

which become sharper and sharper as N increases. Any external perturbation which breaks the symmetry between the peaks, for instance a small positive magnetic field B, favours one peak with respect to the other one, and therefore the system develops a spontaneous magnetization. Let us stress that the occurrence of a phase transition is a property of systems in the thermodynamic limit $N \to \infty$.

In physical magnets, symmetry breaking can arise, for instance, from impurities, subtle effects of dipolar interactions together with the shape of the magnet, or an external magnetic field. The result is that at low enough temperatures some systems, the ferromagnets, develop a spontaneous magnetization. If a magnet made of iron is heated, its magnetization disappears at a critical temperature $T_{\mathrm{c}} = 1/\beta_{\mathrm{c}} \approx 770°$ C. The Curie Weiss model is a simple solvable case exhibiting this phase transition.

Exercise 2.7 Compute the expansions of $m_{+}(\beta)$ and of $\phi(\beta, B = 0)$ near $\beta = \beta_{\mathrm{c}}$, and show that the transition is of second order. Compute the low-temperature behaviour of the spontaneous magnetization.

Exercise 2.8 *Inhomogeneous Ising chain.* The one-dimensional Ising problem does not have a finite-temperature phase transition, as long as the interactions are short-range and translationally invariant. On the other hand, if the couplings in the Ising chain grow fast enough at large distance, one can have a phase transition. This is not a very realistic model from the point of view of physics, but it is useful as a solvable example of a phase transition.

Consider a chain of Ising spins $\sigma_0, \sigma_1, \ldots, \sigma_N$ with energy $E(\underline{\sigma}) = -\sum_{n=0}^{N-1} J_n \sigma_n \sigma_{n+1}$. Suppose that the coupling constants J_n form a positive, monotonically increasing sequence, growing logarithmically. More precisely, we assume that $\lim_{n\to\infty} J_n/\log n = 1$. Denote by $\langle \cdot \rangle_+$ and $\langle \cdot \rangle_-$ the expectation value with respect to the Boltzmann probability distribution when the spin σ_N is fixed at $\sigma_N = +1$ and -1, respectively.

(a) Show that, for any $n \in \{0, \ldots, N-1\}$, the magnetization is $\langle \sigma_n \rangle_\pm = \prod_{p=n}^{N-1} \tanh(\beta J_p)$

(b) Show that the critical inverse temperature $\beta_c = 1/2$ separates two regimes, such that for $\beta < \beta_c$, one has $\lim_{N\to\infty} \langle \sigma_n \rangle_+ = \lim_{N\to\infty} \langle \sigma_n \rangle_- = 0$, and for $\beta > \beta_c$, one has $\lim_{N\to\infty} \langle \sigma_n \rangle_\pm = \pm M(\beta)$, with $M(\beta) > 0$.

Notice that in this case, the role of the symmetry-breaking field is played by the choice of boundary condition.

2.6   The Ising spin glass

In real magnetic materials, localized magnetic moments are subject to several sources of interaction. Apart from the exchange interaction mentioned in the previous section, they may interact through intermediate conduction electrons, for instance. As a result, depending on the material considered, their interaction can be either ferromagnetic (their energy is minimized when they are parallel) or **antiferromagnetic** (their energy is minimized when they point *opposite* to each other). **Spin glasses** are a family of materials whose magnetic properties are particularly complex. They can be produced by diluting a small fraction of a magnetic transition metal such as manganese into a noble metal such as copper in a ratio, say, of $1:100$. In such an alloy, magnetic moments are localized at manganese atoms, which are placed at random positions in a copper background. Depending on the distance between two manganese atoms, the net interaction between their magnetic moments can be either ferromagnetic or antiferromagnetic.

The **Edwards–Anderson model** is a widely accepted mathematical abstraction of these physical systems. Once again, the basic degrees of freedom are Ising spins $\sigma_i \in \{-1, +1\}$ sitting on the vertices of a d-dimensional cubic lattice $\mathbb{L} = \{1, \ldots, L\}^d$, $i \in \mathbb{L}$. The configuration space is therefore $\{-1, +1\}^{\mathbb{L}}$. As in the ferromagnetic Ising model, the energy function reads

$$E(\underline{\sigma}) = -\sum_{(ij)} J_{ij} \sigma_i \sigma_j - B \sum_{i \in \mathbb{L}} \sigma_i, \qquad (2.82)$$

where $\sum_{(ij)}$ runs over each edge of the lattice. Unlike the case of the Ising ferromagnet, however, a different coupling constant J_{ij} is now associated with each edge (ij), and its sign can be positive or negative. The interaction between spins σ_i and σ_j is ferromagnetic if $J_{ij} > 0$ and antiferromagnetic if $J_{ij} < 0$.

A pictorial representation of this energy function is given in Fig. 2.5. The Boltzmann distribution is given by

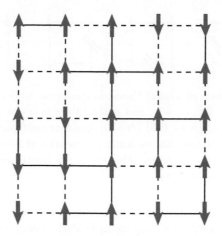

Fig. 2.5 A configuration of a two-dimensional Edwards–Anderson model with $L = 5$. Spins are coupled by two types of interactions: ferromagnetic ($J_{ij} = +1$), indicated by a continuous line, and antiferromagnetic ($J_{ij} = -1$), indicated by a dashed line. The energy of the configuration shown here is $-14 - 7h$.

$$\mu_\beta(\underline{\sigma}) = \frac{1}{Z(\beta)} \exp\left\{\beta \sum_{(ij)} J_{ij}\sigma_i\sigma_j + \beta B \sum_{i\in\mathbb{L}} \sigma_i\right\}, \tag{2.83}$$

$$Z(\beta) = \sum_{\underline{\sigma}} \exp\left\{\beta \sum_{(ij)} J_{ij}\sigma_i\sigma_j + \beta B \sum_{i\in\mathbb{L}} \sigma_i\right\}. \tag{2.84}$$

It is important to notice that the couplings $\{J_{ij}\}$ play a completely different role from the spins $\{\sigma_i\}$. The couplings are just parameters involved in the definition of the energy function, like the magnetic field B, and they are not summed over when computing the partition function. In principle, for any particular sample of a magnetic material, one should estimate experimentally the values of the J_{ij}'s and then compute the partition function. We could have made explicit the dependence of the partition function and of Boltzmann distribution on the couplings by using notation such as $Z(\beta, B; \{J_{ij}\})$, $\mu_{\beta,B;\{J_{ij}\}}(\underline{\sigma})$. However, when these explicit mentions are not necessary, we prefer to keep to lighter notation.

The present understanding of the Edwards–Anderson model is much poorer than for the ferromagnetic models introduced in the previous section. The basic reason for this difference is **frustration**, which is illustrated in Fig. 2.6 for an $L = 2$, $d = 2$ model (a model consisting of just four spins).

A spin glass is frustrated whenever there exist local constraints that are in conflict, meaning that it is not possible to satisfy all of them simultaneously. In the Edwards–Anderson model, a plaquette is the name given to a group of four neighbouring spins forming a square (i.e. a cycle of length four). A plaquette is frustrated if and only if the product of the J_{ij}'s along all four edges of the plaquette is negative. As shown in Fig. 2.6, it is then impossible to minimize simultaneously all of the four local energy

Fig. 2.6 Four configurations of a small Edwards–Anderson model: continuous lines indicate ferromagnetic interactions ($J_{ij} = +1$), and dashed lines indicate antiferromagnetic interactions ($J_{ij} = -1$). In zero magnetic field ($B = 0$), the four configurations are degenerate and have energy $E = -2$. The double bar indicates an unsatisfied interaction. Notice that there is no configuration with a lower energy. This system is frustrated since it is impossible to satisfy simultaneously all constraints.

terms associated with each edge. In a spin glass, the presence of a finite density of frustrated plaquettes generates a very complicated energy landscape. The resulting effect of all the interactions is not obtained by 'summing' the effects of each of them separately, but is the outcome of a complex interplay. The ground state spin configuration (the one satisfying the largest possible number of interactions) is difficult to find: it cannot be guessed on symmetry grounds. It is also frequent to find in a spin glass a configuration which is very different from the ground state but has an energy very close to the ground state energy. We shall explore these and related issues throughout the book.

Notes

There are many good introductory textbooks on statistical physics and thermodynamics, for instance Reif (1965) and Huang (1987). Going towards more advanced texts, we can suggest the books by Ma (1985) and Parisi (1988). A more mathematically minded presentation can be found in the books by Gallavotti (1999) and Ruelle (1999). The reader will find there a proof of Proposition 2.6.

The two-dimensional Ising model in a vanishing external field can also be solved by a transfer matrix technique: see for instance Baxter (1982). The transfer matrix, which links one column of the lattice to the next, is a $2^L \times 2^L$ matrix, and its dimension diverges exponentially with the lattice size L. Finding its largest eigenvalue is therefore a complicated task. No one has found the solution so far for $B \neq 0$.

Spin glasses will be a recurring theme in this book, and more will be said about them in the following chapters. An introduction to this subject from a physicist's point of view is provided by the book by Fischer and Hertz (1993) and the review by Binder and Young (1986). The concept of frustration was introduced by Toulouse (1977).

3
Introduction to combinatorial optimization

This chapter provides an elementary introduction to some basic concepts in theoretical computer science. Which computational tasks can/cannot be accomplished efficiently by a computer? How much resources (time, memory, etc.) are needed for solving a specific problem? What is the performance of a specific solution method (an algorithm), and, whenever more than one method is available, which one is preferable? Are some problems intrinsically harder than others? These are some of the questions one would like to answer.

One large family of computational problems is formed by combinatorial optimization problems. These consist in finding an element of a finite set which maximizes (or minimizes) an easy-to-evaluate objective function. Several features make such problems particularly interesting. First of all, most of the time they are equivalent to decision problems (questions which require a yes/no answer), which is the most fundamental class of problems within computational-complexity theory. Second, optimization problems are ubiquitous both in applications and in the pure sciences. In particular, there exist some evident connections both with statistical mechanics and with coding theory. Finally, they form a very large and well-studied family, and therefore an ideal context for understanding some advanced issues. One should, however, keep in mind that computation is more than just combinatorial optimization. A larger family, which we shall also discuss later on, contains the counting problems: one wants to count how many elements of a finite set have some easy-to-check property. There are also other important families of computational problems that we shall not address at all, such as continuous optimization problems.

The study of combinatorial optimization is introduced in Section 3.1 through the simple example of the minimum spanning tree. This section also contains the basic definitions of graph theory that we shall use throughout the book. General definitions and terminology are given in Section 3.2. These definitions are illustrated further in Section 3.3 with several additional examples. Section 3.4 provides an informal introduction to some basic concepts in computational complexity: we define the classes P and NP, and the notion of NP-completeness. As mentioned above, combinatorial optimization problems often appear bothe in the pure sciences and in applications. The examples of statistical physics and coding are briefly discussed in Sections 3.5 and 3.6.

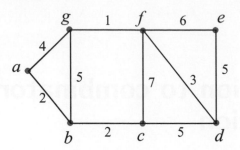

Fig. 3.1 This graph has seven vertices (labelled a to g) and 10 edges. The 'cost' of each edge is indicated next to it. In the minimum-spanning-tree problem, one seeks a loop-free subgraph of minimum cost connecting all vertices.

3.1   A first example: The minimum spanning tree

The minimum-spanning-tree problem is easily stated and may appear in many practical applications. Suppose, for instance, you have a collection of computers in a building. You may want to connect them pairwise in such a way that the resulting network is connected and the amount of cable used is a minimum.

3.1.1   Definition and basics of graph theory

A mathematical abstraction of the above practical problem requires a few basic definitions from graph theory. A **graph** is a set $\mathcal{V}$ of vertices, labelled by $\{1, 2, \ldots, |\mathcal{V}|\}$, together with a set $\mathcal{E}$ of edges connecting them: $\mathcal{G} = (\mathcal{V}, \mathcal{E})$. The vertex set can be any finite set but one often takes the set of the first $|\mathcal{V}|$ integers $\mathcal{V} = \{1, 2, \ldots, |\mathcal{V}|\}$. The edges are simply unordered pairs of distinct vertices $\mathcal{E} \subseteq \mathcal{V} \times \mathcal{V}$. For instance, an edge joining vertices i and j is identified by $e = (i, j)$. A **weighted graph** is a graph where a cost (a real number) is associated with every edge. The **degree** of a vertex is the number of edges connected to it. A **path** between two vertices i and j is a set of edges $\{(j, i_2); (i_2, i_3); (i_3, i_4); \ldots; (i_{r-1}, i_r); (i_r, j)\} \subseteq \mathcal{E}$. A graph is **connected** if, for every pair of vertices, there is a path which connects them. A **completely connected** graph, or **complete** graph, also called a **clique**, is a graph where all of the $|\mathcal{V}|(|\mathcal{V}| - 1)/2$ edges are present. A **cycle** is a path starting and ending on the same vertex. A **tree** is a connected graph without cycles.

Consider the graph in Fig. 3.1. You are asked to find a tree (a subset of the edges forming a cycle-free subgraph) such that any two vertices are connected by exactly one path (in this case the tree is said to be 'spanning'). To find such a subgraph is an easy task. The edges $\{(a, b); (b, c); (c, d); (b, g); (d, e)\}$, for instance, do the job. However in our problem a cost is associated with each edge. The cost of a subgraph is assumed to be equal to the sum of the costs of its edges, and you want to minimize it. This is a non-trivial problem.

In general, an instance of the **minimum-spanning-tree** (MST) problem is given by a connected weighted graph (where each edge e has a cost $w(e) \in \mathbb{R}$). The optimization problem consists in finding a spanning tree with minimum cost. What one seeks is an algorithm which, given an instance of the MST problem, outputs the spanning tree with lowest cost.

3.1.2 An efficient algorithm

The simple-minded approach would consist in enumerating all the spanning trees for the given graph, and comparing their weights. However, the number of spanning trees grows very rapidly with the size of the graph. Consider, as an example, the complete graph on N vertices. The number of spanning trees of such a graph is, according to the Cayley formula, N^{N-2}. Even if the cost of any such tree were evaluated in 10^{-3} s, it would take two years to find the MST of an $N = 12$ graph, and half a century for $N = 13$. At the other extreme, if the graph is very simple, it may contain only a small number of spanning trees, a single one in the extreme case where the graph is itself a tree. Nevertheless, in most of the interesting examples, the situation is nearly as dramatic as in the complete-graph case.

A much better algorithm can be obtained from the following theorem

Theorem 3.1 *Let $\mathcal{U} \subset \mathcal{V}$ be a proper subset of the vertex set $\mathcal{V}$ (such that neither $\mathcal{U}$ nor $\mathcal{V} \backslash \mathcal{U}$ is empty). Consider the subset $\mathcal{F}$ of edges which connect a vertex in $\mathcal{U}$ to a vertex in $\mathcal{V} \backslash \mathcal{U}$, and let $e \in \mathcal{F}$ be an edge of lowest cost in this subset: $w(e) \leq w(e')$ for any $e' \in \mathcal{F}$. If there are several such edges, e can be any one of them. There exists then a minimum spanning tree which contains e.*

Proof Consider an MST $\mathcal{T}$, and suppose that it does not contain the edge e mentioned in the statement of the theorem. This edge is such that $e = (i, j)$, with $i \in \mathcal{U}$ and $j \in \mathcal{V} \backslash \mathcal{U}$. The spanning tree $\mathcal{T}$ must contain a path between i and j. This path contains at least one edge f connecting a vertex in $\mathcal{U}$ to a vertex in $\mathcal{V} \backslash \mathcal{U}$, and f is distinct from e. Now consider the subgraph $\mathcal{T}'$ built from $\mathcal{T}$ by removing the edge f and adding the edge e. We leave to the reader the exercise of showing that $\mathcal{T}'$ is a spanning tree. If we denote by $E(\mathcal{T})$ the cost of tree $\mathcal{T}$, $E(\mathcal{T}') = E(\mathcal{T}) + w(e) - w(f)$. Since $\mathcal{T}$ is an MST, $E(\mathcal{T}') \geq E(\mathcal{T})$. On the other hand, e has the minimum cost within $\mathcal{F}$, and hence $w(e) \leq w(f)$. Therefore $w(e) = w(f)$ and $\mathcal{T}'$ is an MST containing e. $\square$

This result allows one to construct a minimum spanning tree of $\mathcal{G}$ incrementally. One starts from a single vertex. At each step a new edge is added to the tree, where the cost of this edge is minimum among all the edges connecting the already existing tree with the remaining vertices. After $N - 1$ iterations, the tree will be spanning. The pseudo-code for this algorithm is as follows.

MST ALGORITHM (graph $\mathcal{G} = (\mathcal{V}, \mathcal{E})$, weight function $w : \mathcal{E} \to \mathbb{R}_+$)

1: Set $\mathcal{U} := \{1\}$, $\mathcal{T} := \emptyset$ and $E = 0$;
2: **while** $\mathcal{V} \backslash \mathcal{U}$ is not empty:
3:      Let $\mathcal{F} := \{e = (i, j) \in \mathcal{E} \text{ such that } i \in \mathcal{U}, j \in \mathcal{V} \backslash \mathcal{U}\}$;
4:      Find $e_* = (i_*, j_*) := \arg \min_{e \in \mathcal{F}} \{w(e)\}$;
5:      Set $\mathcal{U} := \mathcal{U} \cup j_*$, $\mathcal{T} := \mathcal{T} \cup e_*$, and $E := E + w(e_*)$;
6: **end**
7: **return** the spanning tree $\mathcal{T}$ and its cost E.

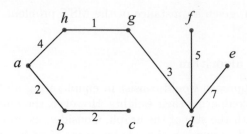

Fig. 3.2 A minimum spanning tree for the graph defined in Fig. 3.1. The cost of this tree is $E = 17$.

Exercise 3.1 Write a code for this algorithm, and find an MST for the problem described in Fig. 3.1. A solution is given in Fig. 3.2.

Exercise 3.2 Show explicitly that the algorithm MST always outputs a minimum spanning tree.

Theorem 3.1 establishes that, for any $\mathcal{U} \subset \mathcal{V}$ and any lowest-cost edge e among the ones connecting $\mathcal{U}$ to $\mathcal{V}\backslash\mathcal{U}$, there exists an MST containing e. This does not guarantee, however, that when two different sets $\mathcal{U}_1$ and $\mathcal{U}_2$ and the corresponding lowest-cost edges e_1 and e_2 are considered, there exists an MST containing *both* e_1 and e_2. The above algorithm works by constructing a sequence of such $\mathcal{U}$'s and adding to the tree the corresponding lowest-weight edges. It is therefore not obvious a priori that it will output an MST (unless this is unique).

Let us analyse the number of elementary operations required by the algorithm to construct a spanning tree on a graph with N vertices. By 'elementary operation' we mean comparisons, sums, multiplications, etc., all of them counting as one operation. Of course, the number of such operations depends on the graph, but we can find a simple upper bound by considering the completely connected graph. Most of the operations in the above algorithm are comparisons of edge weights for finding e_* in step 4. In order to identify e_*, one has to scan at most $|\mathcal{U}| \times |\mathcal{V}\backslash\mathcal{U}| = |\mathcal{U}| \times (N - |\mathcal{U}|)$ edges connecting $\mathcal{U}$ to $\mathcal{V}\backslash\mathcal{U}$. Since $|\mathcal{U}| = 1$ at the beginning and is augmented by one element at each iteration of the cycle 2–6, the number of comparisons is bounded from above by $\sum_{U=0}^{N} U(N - U) \leq N^3/6$.[1] This is an example of a polynomial algorithm, whose computing time grows like a power of the number of vertices. The insight gained from the above theorem provides an algorithm which is much better than the naive one, at least when N is large.

---

[1] The algorithm can easily be improved by keeping an ordered list of the edges already encountered

Exercise 3.3 Suppose you are given a weighted graph $(\mathcal{V}, \mathcal{E})$ in which the weights are all different, and the edges are ordered in such a way that their weights form an increasing sequence $w(e_1) < w(e_2) < w(e_3) < \cdots$. Another graph with the same $(\mathcal{V}, \mathcal{E})$ has different weights $w'(e)$, but they are also increasing in the same sequence, i.e. $w'(e_1) < w'(e_2) < w'(e_3) < \cdots$. Show that the MST is the same in these two graphs.

3.2 General definitions

The MST problem is an example of a **combinatorial optimization problem**. This is defined by a set of possible instances. An instance of the MST problem is defined by a connected weighted graph. In general, an **instance** of a combinatorial optimization problem is described by a finite set $\mathcal{X}$ of allowed **configurations**, and a **cost function** E defined on this set and taking values in $\mathbb{R}$. The optimization problem consists in finding the **optimal** configuration $C \in \mathcal{X}$, namely the configuration with the smallest cost $E(C)$. Any set of such instances defines a combinatorial optimization problem. For a particular instance of the MST problem, the space of configurations $\mathcal{X}$ is simply the set of spanning trees on the given graph, and the cost function associated with each spanning tree is the sum of the costs of its edges.

We shall say that an algorithm solves an optimization problem if, for every instance of the optimization problem, it gives the optimal configuration or if it computes its cost. In all the problems which we shall discuss, there is a 'natural' measure of the size of the problem N (typically the number of variables used to define a configuration, such as the number of edges of the graph in the MST problem), and the number of configurations scales, at large N like c^N or, in some cases, even faster, for example like $N!$ or N^N. Notice that, quite generally, evaluating the cost function of a particular configuration is an easy task. The difficulty of solving the combinatorial optimization problem therefore comes essentially from the size of the configuration space.

It is generally accepted practice to estimate the **complexity** of an algorithm by the number of 'elementary operations' required to solve the problem. Usually one focuses on the asymptotic behaviour of this quantity as $N \to \infty$. It is obviously of great practical interest to construct algorithms whose complexity is as small as possible.

One can solve a combinatorial optimization problem at several levels of refinement. Usually, one distinguishes three types of problems:

- The **optimization** problem: Find an optimal configuration C^*.
- The **evaluation** problem: Determine the cost $E(C^*)$ of an optimal configuration.
- The **decision** problem: Answer the question 'Is there a configuration of cost less than a given value E_0?'

3.3 More examples

The general setting described in the previous section includes a large variety of problems of both practical and theoretical interest. In the following, we shall provide a few selected examples.

3.3.1 Eulerian circuit

One of the oldest documented examples goes back to the eighteenth century. The old city of Königsberg had seven bridges (see Fig. 3.3), and its inhabitants were wondering whether it was possible to cross each of these bridges once and get back home. This can be generalized and translated into graph-theoretic language as the following decision problem. We define a **multigraph** exactly like a graph except for the fact that two given vertices can be connected by several edges. The problem consists in finding whether there is there a circuit which goes through all edges of the graph only once, and returns to its starting point. Such a circuit is now called an **Eulerian circuit**, because this problem was solved by Euler in 1736, when he proved the following nice theorem. As for ordinary graphs, we define the **degree** of a vertex as the number of edges which have the vertex as an end-point.

Theorem 3.2 *Given a connected multigraph, there exists an Eulerian circuit if and only if every vertex has even degree.*

This theorem directly provides an algorithm for the decision problem whose complexity grows linearly with the number of vertices of the graph: just go through all the vertices of the graph and check their degree.

Exercise 3.4 Show that if an Eulerian circuit exists, the degrees are necessarily even.

Proving the inverse implication is more difficult. A possible approach consists in showing the following slightly stronger result. If all the vertices of a connected graph $\mathcal{G}$ have even degree except for i and j, then there exists a path from i to j that visits each edge in $\mathcal{G}$ once. This can be proved by induction on the number of vertices.

[Hint: Start from i and take a step along the edge (i, i'). Show that it is possible to choose i' in such a way that the residual graph $\mathcal{G} \backslash (i, i')$ is connected.]

3.3.2 Hamiltonian cycle

More than a century after Euler's theorem, the great scientist Sir William Hamilton introduced in 1859 a game called the icosian game. In its generalized form, it basically asks whether there exists, in a graph, a **Hamiltonian cycle**, that is a path that goes once through every vertex of the graph, and gets back to its starting point. This is another decision problem, and, at first sight, it seems very similar to the Eulerian circuit. However, it turns out to be much more complicated. The best existing algorithms for determining the existence of a Hamiltonian cycle in a given graph run in a time which grows exponentially with the number of vertices N. Moreover, the theory of computational complexity, which we shall describe in Section 3.4, strongly suggests that this problem is in fact intrinsically difficult.

3.3.3 Travelling salesman problem

Given a complete graph with N points, and the distances d_{ij} between all pairs of points $1 \leq i < j \leq N$, the famous **travelling salesman problem** (TSP) is an optimization problem: find a Hamiltonian cycle of minimum total length. One can consider the case

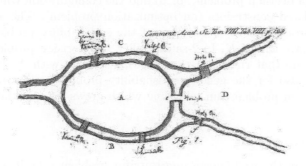 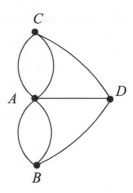

Fig. 3.3 *Left*: a map of the old city of Königsberg, with its seven bridges, as drawn in Euler's paper of 1736. The problem is whether one can walk through the city, crossing each bridge exactly once, and get back home. *Right*: a graph summarizing the problem. The vertices A, B, C, D are the various areas of land separated by a river; an edge exists between two vertices whenever there is a bridge. The problem is to make a closed circuit on this graph, going exactly once through every edge.

where the points are in a portion of the plane, and the distances are Euclidean distances (we then speak of a Euclidean TSP), but of course the problem can be stated more generally, with d_{ij} representing general costs, which are not necessarily distances. As for the Hamiltonian cycle problem, the best algorithms known so far for the TSP have a running time which grows exponentially with N at large N. Nevertheless, Euclidean problems with thousands of points can be solved.

3.3.4 Assignment

Given N persons and N jobs, and a matrix C_{ij} giving the affinity of person i for job j, the **assignment** problem consists in finding the assignment of the jobs to the persons (an exact one-to-one correspondence between jobs and persons) which maximizes the total affinity. A configuration is characterized by a permutation of the N indices (there are thus $N!$ configurations), and the cost of a permutation π is $\sum_i C_{i\pi(i)}$. This is an example of a polynomial problem: there exist algorithms that solve it in a time growing like N^3.

3.3.5 Satisfiability

In the **satisfiability** problem, one has to find the values of N Boolean variables $x_i \in \{T, F\}$ which satisfy a set of logical constraints. Since each variable can be either true or false, the space of configurations has a size $|\mathcal{X}| = 2^N$. Each logical constraint, called in this context a **clause**, takes a special form: it is a logical OR (for which we use the symbol $\vee$) of some variables or their negations. For instance $x_1 \vee \overline{x}_2$ is a 2-clause (a '2-clause' is a clause of length 2, i.e. one which involves exactly two variables), which is satisfied if either $x_1 = T$ or $x_2 = F$, or both. Analogously, $\overline{x}_1 \vee \overline{x}_2 \vee x_3$ is a 3-clause, which is satisfied by all configurations of the three variables except $x_1 = x_2 = T, x_3 = F$. The problem is to determine whether there exists a configuration which satisfies all constraints (a decision problem), or to find the configuration which minimizes the number of violated constraints (an optimization problem). The K-**satisfiability** (or 'K-SAT') problem is the restriction of the satisfiability problem to the case where all clauses have length K. In 2-satisfiability, the decision problem is easy: there exists an algorithm that runs in a time growing linearly with N. For K-satisfiability, and therefore also for the general satisfiability problem, all known algorithms that solve the decision problem run in a time which grows exponentially with N.

3.3.6 Colouring and vertex covering

Given a graph and an integer q, the well-known q-**colouring** problem asks if it is possible to colour the vertices of the graph using q colours, in such a way that two vertices connected by an edge have different colours. In the same spirit, the **vertex-covering** problem asks if it is possible to cover the vertices with 'pebbles', using the smallest possible number of pebbles, in such a way that every edge of the graph has at least one of its two end-points covered by a pebble.

3.3.7 Number partitioning

Number partitioning is an example which does not come from graph theory. An instance of the problem consists in a set $\mathcal{S}$ of N integers $\mathcal{S} = \{x_1, \dots, x_N\}$. A configuration is a partition of these numbers into two groups $\mathcal{A}$ and $\mathcal{S} \setminus \mathcal{A}$. Is there a partition such that $\sum_{i \in \mathcal{A}} x_i = \sum_{i \in \mathcal{S} \setminus \mathcal{A}} x_i$?

3.4 Elements of the theory of computational complexity

One of the main branches of theoretical computer science aims at constructing an intrinsic theory of computational complexity. One would like, for instance, to establish which problems are harder than others. By 'harder problem', we mean a problem that takes a longer running time to be solved. In order to discuss rigorously the computational complexity of a problem, however, we would need to define a precise *model of computation* (introducing, for instance, Turing machines). This would take us too far. We shall instead evaluate the running time of an algorithm in terms of 'elementary operations': comparisons, sums, multiplications, etc. This informal approach is essentially correct as long as the size of the operands remains uniformly bounded.

3.4.1 The worst-case scenario

As we already mentioned in Section 3.2, a combinatorial optimization problem is defined by the set of its possible instances. Given an algorithm that solves the problem, its running time will vary from instance to instance, even if the 'size' of the instance is fixed. How should we quantify the overall hardness of the problem? A crucial choice in computational-complexity theory consists in considering the 'worst' instance (i.e. the one which takes the longest time to be solved) out of all instances of the same size.

This choice has two advantages: (i) it allows one to construct a 'universal' theory; and (ii) once the worst-case running time of a given algorithm has been estimated, this provides a performance guarantee for any instance of the problem.

3.4.2 Polynomial or not?

A second crucial choice consists in classifying algorithms into two classes. (*i*) **Polynomial**, if the running time is bounded from above by a fixed polynomial in the size of the instance. In mathematical terms, let T_N be the number of operations required for solving an instance of size N in the worst case. The algorithm is polynomial when there exists a constant k such that $T_N = O(N^k)$. (*ii*) **Super-polynomial**, if no such upper bound exists. This is, for instance, the case if the time grows exponentially with the size of the instance (we shall call algorithms of this type **exponential**), i.e. $T_N = \Theta(k^N)$ for some constant $k > 1$.

Example 3.3 In Section 3.1.2, we were able to show that the running time of the MST algorithm is bounded from above by N^3, where N is the number of vertices in the graph. This implies that such an algorithm is polynomial.

Notice that we have not given a precise definition of the 'size' of a problem. One may wonder whether, by changing the definition, a particular problem could be classified both as polynomial and as super-polynomial. Consider, for instance, the assignment problem with $2N$ points. One can define the size as being N, or $2N$, or even N^2 which is the number of possible person–job pairs. The last definition would be relevant if one were to count, for instance, the number of entries in the person–job cost matrix. However, all of these 'natural' definitions of size are polynomial functions of the others. Therefore the choice of one or another of these definitions does not affect the classification of an algorithm as polynomial or super-polynomial. We shall discard other definitions (such as e^N or $N!$) as 'unnatural', without further ado. The reader can convince him/herself of this for each of the examples of the previous Section.

3.4.3 Optimization, evaluation, and decision

In order to get a feeling for their relative levels of difficulty, let us come back for a while to the three types of optimization problem defined in Section 3.2, and study which one is the hardest.

Clearly, if the cost of any configuration can be computed in polynomial time, the evaluation problem is not harder than the optimization problem: if one can find the optimal configuration in polynomial time, one can also compute its cost in polynomial time. The decision problem (deciding whether there exists a configuration with a cost smaller than a given E_0) is not harder than the evaluation problem. So the order of increasing difficulty is decision, evaluation, optimization.

However, in many cases where the costs take discrete values, the evaluation problem is not harder than the decision problem, in the following sense. Suppose that we have a polynomial algorithm that solves the decision problem, and that the costs of all configurations can be scaled so as to be integers in an interval $[0, E_{\max}]$ of length $E_{\max} = \exp\{O(N^k)\}$ for some $k > 0$. An algorithm that solves the decision problem can be used to solve the evaluation problem by dichotomy. One first takes $E_0 = E_{\max}/2$. If there exists a configuration of energy smaller than E_0, one iterates taking E_0 to be the centre of the interval $[0, E_{\max}/2]$. In the opposite case, one iterates taking E_0 to be the centre of the interval $[E_{\max}/2, E_{\max}]$. Clearly, this procedure finds the cost of the optimal configuration(s) in a time which is also polynomial.

3.4.4 Polynomial reduction

One would like to compare the levels of difficulty of various *decision problems*. The notion of polynomial reduction formalizes the sentence 'not harder than' which we have used so far, and helps us to obtain a classification of decision problems.

Roughly speaking, we say that a problem $\mathcal{B}$ is not harder than $\mathcal{A}$ if any efficient algorithm for $\mathcal{A}$ (if such an algorithm exists) could be used as a subroutine of an algorithm that solves $\mathcal{B}$ efficiently. More precisely, given two decision problems $\mathcal{A}$ and $\mathcal{B}$, one says that $\mathcal{B}$ is **polynomially reducible** to $\mathcal{A}$ if the following conditions hold:

1. There exists a mapping R which transforms any instance I of problem $\mathcal{B}$ into an instance $R(\text{I})$ of problem $\mathcal{A}$, such that the solution (yes/no) of the instance $R(\text{I})$ of $\mathcal{A}$ gives the solution (yes/no) of the instance I of $\mathcal{B}$.

2. The mapping $\text{I} \mapsto R(\text{I})$ can be computed in a time which is polynomial in the size of I.

3. The size of $R(\text{I})$ is polynomial in the size of I. This is in fact a consequence of the previous assumptions, but there is no harm in stating it explicitly.

A mapping R that satisfies the above requirements is called a polynomial reduction. Constructing a polynomial reduction between two problems is an important achievement, since it effectively reduces their study to the study of just one of them. Suppose for instance that we have a polynomial algorithm $\text{Alg}_{\mathcal{A}}$ for solving $\mathcal{A}$. A polynomial reduction of $\mathcal{B}$ to $\mathcal{A}$ can then be used for to construct a polynomial algorithm for solving $\mathcal{B}$. Given an instance I of $\mathcal{B}$, the algorithm just computes $R(\text{I})$, feeds it into $\text{Alg}_{\mathcal{A}}$, and outputs the output of $\text{Alg}_{\mathcal{A}}$. Since the size of $R(\text{I})$ is polynomial in the size of I, the resulting algorithm for $\mathcal{B}$ is still polynomial.

Let us work out an explicit example. We shall show that the problem of the existence of a Hamiltonian cycle in a graph is polynomially reducible to the satisfiability problem.

Example 3.4 An instance of the Hamiltonian cycle problem is a graph with N vertices, labelled by $i \in \{1, \ldots, N\}$. If there exists a Hamiltonian cycle in the graph, it can be characterized by N^2 Boolean variables $x_{ri} \in \{0, 1\}$, where $x_{ri} = 1$ if vertex number i is the r-th vertex in the cycle, and $x_{ri} = 0$ otherwise (one can take, for instance, $x_{11} = 1$). We shall now write down a number of constraints that the variables x_{ri} must satisfy in order for a Hamiltonian cycle to exist, and ensure that these constraints take the form of the clauses used in the satisfiability problem (identifying $x = 1$ as true and $x = 0$ as false):

- Each vertex $i \in \{1, \ldots, N\}$ must belong to the cycle: this can be written as the clause $x_{1i} \vee x_{2i} \vee \cdots \vee x_{Ni}$, which is satisfied only if at least one of the numbers $x_{1i}, x_{2i}, \ldots, x_{Ni}$ equals one.
- For every $r \in \{1, \ldots, N\}$, one vertex must be the r-th visited vertex in the cycle: $x_{r1} \vee x_{r2} \vee \cdots \vee x_{rN}$.
- Each vertex $i \in \{1, \ldots, N\}$ must be visited only once. This can be implemented through the $N(N-1)/2$ clauses $\bar{x}_{rj} \vee \bar{x}_{sj}$, for $1 \le r < s \le N$.
- For every $r \in \{1, \ldots, N\}$, there must be only one r-th visited vertex in the cycle. This can be implemented through the $N(N-1)/2$ clauses $\bar{x}_{ri} \vee \bar{x}_{rj}$, for $1 \le i < j \le N$.
- If two vertices $i < j$ are not connected by an edge of the graph, these vertices should not appear consecutively in the list of vertices belonging to the cycle. Therefore we add, for every such pair and for every $r \in \{1, \ldots, N\}$, the clauses $\bar{x}_{ri} \vee \bar{x}_{(r+1)j}$ and $\bar{x}_{rj} \vee \bar{x}_{(r+1)i}$ (with the 'cyclic' convention $N + 1 = 1$).

It is straightforward to show that the size of the satisfiability problem constructed in this way is polynomial in the size of the Hamiltonian cycle problem. We leave it as an exercise to show that the set of all the above clauses is a sufficient set: if the N^2 variables satisfy all the above constraints, they describe a Hamiltonian cycle.

3.4.5 Complexity classes

Let us continue to focus on decision problems. The classification of these problems with respect to polynomiality is as follows:

- **Class P.** These are the **polynomial** problems, which can be solved by an algorithm running in polynomial time. An example (see Section 3.1) is the decision version of the minimum-spanning-tree problem (which asks for a yes/no answer to the following question: given a graph with costs on the edges, and a number E_0, is there a spanning tree with total cost less than E_0?).
- **Class NP.** This is the class of **non-deterministic polynomial** problems, which can be solved in polynomial time by a 'non-deterministic' algorithm. Roughly speaking, such an algorithm can run in parallel on an arbitrarily large number of processors. We shall not explain this notion in detail here, but rather use an alternative, equivalent characterization. We say that a problem is in the class NP if there exists a 'short' certificate which allows one to check a 'yes' answer to the problem. A short certificate means a certificate that can be checked in polynomial time.

A polynomial problem such as the MST problem described above is automatically in NP, and so P $\subseteq$ NP. The decision version of the TSP is also in NP: if there is a TSP tour with cost smaller than E_0, the short certificate is simple: just give the tour, and its cost can be computed in linear time, allowing one to check that it is smaller than E_0. The satisfiability problem also belongs to NP: a certificate is obtained from the assignment of variables satisfying all clauses. Checking that all clauses are satisfied is linear in the number of clauses, taken here as the size of the system. In fact, there are many important problems in the class NP, with a broad spectrum of applications ranging from routing to scheduling, chip verification, and protein folding.

- **Class NP-complete.** These are the hardest problems in the class NP. A problem is **NP-complete** if (*i*) it is in NP, and (*ii*) any other problem in NP can be polynomially reduced to it, using the notion of polynomial reduction defined in Section 3.4.4. If $\mathcal{A}$ is NP-complete, then for any other problem $\mathcal{B}$ in NP, there is a polynomial reduction that maps $\mathcal{B}$ to $\mathcal{A}$. In other words, if we had a polynomial algorithm to solve $\mathcal{A}$, then all the problems in the broad class NP could be solved in polynomial time.

It is not a priori obvious whether there exist any NP-complete problems. A major achievement of the theory of computational complexity is the following theorem, obtained by Cook in 1971.

Theorem 3.5 *The satisfiability problem is NP-complete.*

We shall not give the proof of the theorem here. Let us just mention that the satisfiability problem has a very universal structure (an example of which was shown above, in the polynomial reduction of the Hamiltonian cycle problem to satisfiability). A clause is built as a logical OR (denoted by $\vee$) of some variables or their negations. A set of several clauses to be satisfied simultaneously is the logical AND (denoted by $\wedge$) of those clauses. Therefore a satisfiability problem is written in general in the form $(a_1 \vee a_2 \vee \dots) \wedge (b_1 \vee b_2 \vee \dots) \wedge \dots$, where the a_i, b_i are 'literals', i.e. any of the original variables or their negations. This form is called a **conjunctive normal form** (CNF), and it is easy to see that any logical statement between Boolean variables can be written as a CNF. This universal decomposition gives an idea of why the satisfiability problem plays a central role.

3.4.6   P = NP?

When an NP-complete problem $\mathcal{A}$ is known, one can find other NP-complete problems relatively easily: if there exists a polynomial reduction from $\mathcal{A}$ to another problem $\mathcal{B} \in$ NP, then $\mathcal{B}$ is also NP-complete. In fact, whenever $R_{\mathcal{A} \leftarrow \mathcal{P}}$ is a polynomial reduction from a problem $\mathcal{P}$ to $\mathcal{A}$ and $R_{\mathcal{B} \leftarrow \mathcal{A}}$ is a polynomial reduction from $\mathcal{A}$ to $\mathcal{B}$, then $R_{\mathcal{B} \leftarrow \mathcal{A}} \circ R_{\mathcal{A} \leftarrow \mathcal{P}}$ is a polynomial reduction from $\mathcal{P}$ to $\mathcal{B}$. Starting from the satisfiability problem, it has been possible to find, by this method, thousands of NP-complete problems. To quote a few of them, among the problems we have encountered so far, the Hamiltonian cycle problem, the TSP, and the 3-satisfiability problem are NP-complete. Actually, most NP problems can be classified either as being in P or as NP-complete (see Fig. 3.4). The precise status of some NP problems, however, such as graph isomorphism, is still unknown.

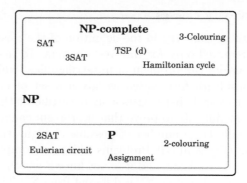

Fig. 3.4 Classification of some well known decision problems. If P $\neq$ NP, the classes P and NP-complete are disjoint. But if it happened that P=NP, all the problems in NP, and in particular all those mentioned here, would be solvable in polynomial time.

Exercise 3.5 Show that the 3-satisfiability problem is NP-complete, by constructing a polynomial reduction from the satisfiability problem. The idea is to transform all possible clauses into sets of 3-clauses, using the following steps:

- A 2-clause $x_1 \vee x_2$ can be written as two 3-clauses $(x_1 \vee x_2 \vee y) \wedge (x_1 \vee x_2 \vee \overline{y})$ with one extra variable y.
- Write a 1-clause with four 3-clauses and two extra variables.
- Show that a k-clause $x_1 \vee x_2 \vee \cdots \vee x_k$ with $k \geq 4$ can be written with $k - 3$ auxiliary variables as $(x_1 \vee x_2 \vee y_1) \wedge (x_3 \vee \overline{y}_1 \vee y_2) \wedge \cdots \wedge (x_{k-2} \vee \overline{y}_{k-4} \vee y_{k-3}) \wedge (x_{k-1} \vee x_k \vee \overline{y}_{k-3})$.

Finally, those problems which, not being in NP, are at least as hard as NP-complete problems, are usually called **NP-hard**. These include both decision problems for which a short certificate does not exist, and non-decision problems. For instance, the optimization and evaluation versions of the TSP are NP-hard. However, in such cases, we shall choose between the expressions 'the TSP is NP-complete' and 'the TSP is NP-hard' rather freely.

One major open problem in the theory of computational complexity is whether the classes P and NP are distinct or not. It might be that P = NP = NP-complete: this would be the case if someone found a polynomial algorithm for one NP-complete problem. This would imply that any problem in the broad class NP could be solved in polynomial time.

It is a widespread conjecture that there exists no polynomial algorithm for NP-complete problems. In this case the classes P and NP-complete would be disjoint. Moreover, it is known that if P $\neq$ NP, then there are NP problems which are neither in P nor in NP-complete.

3.4.7 Other complexity classes

Notice the fundamental asymmetry in the definition of the class NP: the existence of a short certificate is required only for the 'yes' answers. To understand the meaning of this asymmetry, consider the problem of unsatisfiability (which is the complement of the satisfiability problem), formulated as 'given a set of clauses, is the problem unsatisfiable?' It is not clear if there exists a short certificate that allows one to check a 'yes' answer: it is very difficult to prove that a problem cannot be satisfied without checking an exponentially large number of possible configurations. So it is not at all obvious that unsatisfiability is in NP. Problems which are complements of those in NP define the class of co-NP problems, and it is not known whether NP = co-NP or not, although it is widely believed that co-NP is different from NP. This consideration opens a Pandora's box containing many other classes of complexity, but we shall immediately close it, since leaving it open would carry us too far.

3.5 Optimization and statistical physics

3.5.1 General relation

There exists a natural mapping from optimization to statistical physics. Consider an optimization problem defined by a finite set $\mathcal{X}$ of allowed configurations, and a cost function E defined on this set with values in $\mathbb{R}$. Although optimization consists in finding the configuration $C \in \mathcal{X}$ with the smallest cost, one can introduce a probability measure of the Boltzmann type on the space of configurations: for any β, each C is assigned a probability

$$\mu_\beta(C) = \frac{1}{Z(\beta)}\, \mathrm{e}^{-\beta E(C)}\ , \qquad Z(\beta) = \sum_{C \in \mathcal{X}} \mathrm{e}^{-\beta E(C)}\ . \tag{3.1}$$

The positive parameter β plays the role of an inverse temperature. In the limit $\beta \to \infty$, the probability distribution μ_β concentrates on the configurations of minimum energy (ground states in statistical-physics jargon). This is the relevant limit for optimization problems. Note that there exist many alternatives to the straightforward generalization (3.1). In some problems, it may be useful to use more than one inverse temperature parameter β. Some of these parameters can be used to 'soften' constraints. For instance in the TSP, one might like to relax the constraint that a configuration is a tour, by summing over all length-N paths of the salesman, with an extra cost each time the path does not make a full tour, associated with an inverse temperature β_1. The length of the path would be another cost, associated with an inverse temperature β_2. The original problem is recovered when both β_1 and β_2 go to infinity.

In the statistical-physics approach, one generalizes the optimization problem to study properties of the distribution μ_β at finite β. In many cases, it is useful to follow μ_β as β increases (for instance by monitoring the thermodynamic properties, i.e. the internal energy, entropy, ...). This may be particularly useful, both for analytical and for algorithmic purposes, when the thermodynamic properties evolve smoothly. An example of a practical application is the simulated-annealing method, which samples the configuration space at larger and larger values of β until it finds a ground state.

This method will be described in Chapter 4. As we shall see, the occurrence of phase transitions poses major challenges to this kind of approach.

3.5.2   Spin glasses and maximum cuts

To give a concrete example, let us go back to the spin glass problem of Section 2.6. This involves N Ising spins $\sigma_1, \ldots, \sigma_N$ in $\{\pm 1\}$, located on the vertices of a graph, and the energy function is

$$E(\underline{\sigma}) = -\sum_{(ij)} J_{ij} \sigma_i \sigma_j, \qquad (3.2)$$

where the sum $\sum_{(ij)}$ runs over all edges of the graph, and the variables J_{ij} are exchange couplings, which can be either positive or negative. Given the graph and the exchange couplings, what is the ground state of the corresponding spin glass? This is a typical optimization problem. In fact, it very well known in computer science in a slightly different form.

Each spin configuration partitions the set of vertices into two complementary subsets $V_\pm = \{i \mid \sigma_i = \pm 1\}$. Let us call the set of edges with one end-point in V_+ and the other in V_- $\gamma(V_+)$. The energy of the configuration can be written as

$$E(\underline{\sigma}) = -C + 2 \sum_{(ij) \in \gamma(V_+)} J_{ij}, \qquad (3.3)$$

where $C = \sum_{(ij)} J_{ij}$. Finding the ground state of the spin glass is thus equivalent to finding a partition of the vertices $V = V_+ \cup V_-$ such that $\sum_{(ij) \in \gamma(V_+)} c_{ij}$ is maximum, where $c_{ij} \equiv -J_{ij}$. This problem is known as the **maximum cut** (max-cut) problem: the set of edges $\gamma(V_+)$ is a cut, each cut is assigned a weight $\sum_{(ij) \in \gamma(V_+)} c_{ij}$, and one seeks the cut with maximal weight.

Standard results on the max-cut problem immediately apply. In general, this is an NP-hard problem, but there are some categories of graphs for which it is polynomially solvable. In particular, the max-cut of a planar graph can be found in polynomial time, providing an efficient method to obtain the ground state of a spin glass on a square lattice in two dimensions. The three-dimensional spin glass problem falls into the general NP-hard class, but efficient 'branch and bound' methods, based on its max-cut formulation, have been developed for this problem in recent years.

Another well-known application of optimization to physics is the random-field Ising model, which is a system of Ising spins with ferromagnetic couplings (all J_{ij} are positive), but with a magnetic field h_i which varies from site to site, taking both positive and negative values. Its ground state can be found in polynomial time thanks to the equivalence with the problem of finding a maximal flow in a graph.

3.6   Optimization and coding

Computational-complexity issues are also crucial in all problems of information theory. We shall see this recurrently throughout the book, but let us just give here some small examples in order to fix our ideas.

Consider the error-correcting-code problem of Chapter 1. We have a code which maps an original message to a codeword $\underline{x}$, which is a point in the N-dimensional hypercube $\{0, 1\}^N$. There are 2^M codewords (with $M < N$), which we assume to be a priori equiprobable. When the message is transmitted through a noisy channel, the codeword $\underline{x}$ is corrupted to –say– a vector $\underline{y}$ with probability $Q(\underline{y}|\underline{x})$. The decoder maps the received message $\underline{y}$ to one of the possible input codewords $\underline{x}' = d(\underline{y})$.

As we saw a measure of performance is the average block error probability

$$\mathrm{P}_{\mathrm{B}}^{\mathrm{av}} \equiv \frac{1}{2^M} \sum_{\underline{x}} \sum_{\underline{y}} Q(\underline{y}|\underline{x}) \ \mathbb{I}(d(\underline{y}) \neq \underline{x}) . \tag{3.4}$$

A simple decoding algorithm would be the following: for each received message $\underline{y}$, consider all the 2^M codewords, and determine the most likely one: $d(\underline{y}) = \arg\max_{\underline{x} \in \mathrm{Code}} Q(\underline{y}|\underline{x})$. It is clear that this algorithm minimizes the average block error probability.

For a general code, there is no better way to maximize $Q(\underline{y}|\underline{x})$ than to go through all codewords and compute their likelihood one by one. This takes a time of order 2^M, which is definitely too large. Recall that in fact, to achieve reliable communication, M and N have to be large (in data transmission applications, one may use an N as large as 10^5). One might object that 'decoding a general code' is too general a problem. Just to specify a single instance, we would need to specify all the codewords, which takes $N\,2^M$ bits. Therefore, the complexity of decoding could be a trivial consequence of the fact that even reading the input takes a huge time. However, it can be proved that, for codes that admit a concise specification (polynomial in the block length), also, decoding is NP-hard. We shall see some examples, namely linear codes, in the following chapters.

Notes

We have left aside most of the algorithmic issues in this chapter. A general approach to design efficient algorithms consists in finding a good 'convex relaxation' of the problem. The idea is to enlarge the space of feasible solutions in such a way that the problem translates into minimizing a convex function, a task that can be performed efficiently. A general introduction to combinatorial optimization, including all these aspects, is provided by Papadimitriou and Steiglitz (1998). Convex optimization is the topic of many textbooks, for instance Boys and Vandenberghe (2004).

The MST algorithm described in Section 3.1 was found by Prim (1957).

A complete treatment of computational complexity theory can be found in Garey and Johnson (1979), or in the more recent book by Papadimitriou (1994). The seminal theorem by Cook (1971) was independently rediscovered by Levin in 1973. The reader can find a proof in the above books.

Euler discussed the problem of the seven bridges of Könisberg in Euler (1736).

The TSP, which is simple to state and difficult to solve, and lends itself to nice representations in figures, has attracted much attention. The interested reader can find many references and also pictures of optimal tours with thousands of vertices, including tours of the main cities in various countries, applets, etc. on the Web, starting for instance from Applegate *et al.* (2008).

The book by Hartmann and Rieger (2002) focuses on the use of optimization algorithms for solving some problems in statistical physics. In particular, it explains the determination of the ground state of a random-field Ising model with a maximum-flow algorithm. A recent volume edited by the same authors (Hartmann and Rieger, 2004) addresses several algorithmic issues connecting optimization and physics. Chapter 4 of that volume describes the branch-and-cut approach to the maximum cut problem used for spin glass studies. The book by Hartmann and Weigt (2005) contains an introduction to combinatorial optimization considered as a physics problem, with particular emphasis on the vertex cover problem.

Standard computational problems in coding theory are reviewed in Barg (1998). Some more recent issues are addressed by Spielman (1997). Finally, the first proof of NP-completeness for a decoding problem was obtained by Berlekamp *et al.* (1978).

4

A probabilistic toolbox

The three fields that form the subject of this book all deal with large sets of discrete variables. Often, a probability distribution can be defined naturally on such variables. Not surprisingly, the problems arising in these three fields possess common underlying structures, and some general probabilistic techniques can be applied to all three domains. This chapter describes some of them, concentrating on the mathematical structures, namely large deviations on the one hand, and Markov chains for Monte Carlo computations on the other hand. These tools will reappear several times in the following chapters.

Since this chapter is more technical than the previous ones, we devote the whole of Section 4.1 to a qualitative introduction to the subject. In Section 4.2, we consider the large-deviation properties of simple functions of many independent random variables. In this case, many explicit results can be easily obtained. We present a few general tools for correlated random variables in Section 4.3 and the idea of the Gibbs free energy in Section 4.4. Section 4.5 provides a simple introduction to the Monte Carlo Markov chain method for sampling configurations from a given probability distribution. Finally, in Section 4.6, we show how simulated annealing exploits Monte Carlo techniques to solve optimization problems.

4.1 Many random variables: A qualitative preview

Consider a set of N random variables $\underline{x} = (x_1, x_2, \ldots, x_N)$, with $x_i \in \mathcal{X}$ and a probability distribution

$$P_N(\underline{x}) = P_N(x_1, \ldots, x_N).\qquad(4.1)$$

This could be, for instance, the Boltzmann distribution for a physical system with N degrees of freedom. The entropy of this distribution is $H_N = -\mathbb{E}\log P_N(\underline{x})$. It often happens that this entropy grows linearly with N at large N. This means that the entropy per variable $h_N = H_N/N$ has a finite limit: $\lim_{N\to\infty} h_N = h$. It is then natural to characterize any particular realization of the random variables $(x_1, \ldots, x_N)$ by computing the quantity

$$r(\underline{x}) = \frac{1}{N} \log\left[\frac{1}{P_N(\underline{x})}\right],\qquad(4.2)$$

which measures how probable the event $(x_1, \ldots, x_N)$ is. The expectation of r is $\mathbb{E}r(\underline{x}) = h_N$. One may wonder whether $r(\underline{x})$ fluctuates a lot, or whether its distribution is strongly peaked around $r = h_N$. The latter hypothesis turns out to be the correct one in many cases. It often happens that the probability distribution of $r(\underline{x})$ behaves exponentially with N, i.e.

$$\mathbb{P}\{r(\underline{x}) \approx \rho\} \doteq e^{-NI(\rho)} . \tag{4.3}$$

where $I(\rho)$ has a non-degenerate minimum at $\rho = \lim_{N\to\infty} h_N = h$, and $I(h) = 0$. This means that, with large probability, a randomly chosen configuration $\underline{x}$ has an empirical entropy $r(\underline{x})$ 'close to' h, and, because of the definition (4.2), its probability is approximately $\exp(-Nh)$. Since the total probability of realizations $\underline{x}$ such that $f(\underline{x}) \approx h$ is close to one, their number must behave as $\mathcal{N} \doteq \exp(Nh)$. In other words, the whole probability is carried by a small fraction of all configurations (since their number, $\exp(Nh)$, is in general exponentially smaller than $|\mathcal{X}|^N$), and these configurations all have the same probability. This property, called asymptotic equipartition, holds in many of the problems we are interested in, and its consequences are very important.

Suppose, for instance, that one is interested in compressing the information contained in the variables $(x_1, \ldots, x_N)$, which are a sequence of symbols produced by an information source. Clearly, one should focus on 'typical' sequences $\underline{x}$ such that $f(\underline{x})$ is close to h, because all other sequences have a vanishing small probability. Since there are about $\exp(Nh)$ such typical sequences, one must be able to encode them in $Nh/\log 2$ bits by simply numbering them.

Another general problem consists in sampling from the probability distribution $P_N(\underline{x})$. With n samples $\underline{x}^1, \ldots, \underline{x}^n$ drawn independently with distribution $P_N(\underline{x})$, one can estimate expectation values $\mathbb{E}\, \mathcal{O}(\underline{x}) \equiv \sum_{\underline{x}} P_N(\underline{x}) \mathcal{O}(\underline{x})$ as $\frac{1}{n} \sum_{k=1}^{r} \mathcal{O}(\underline{x}^k)$. This avoids summing over $|\mathcal{X}|^N$ terms, and the precision usually improves like $1/\sqrt{n}$ at large n.

A naive sampling algorithm could be the following. First, 'propose' a configuration $\underline{x}$ from the uniform probability distribution $P_N^{\text{unif}}(\underline{x}) = 1/|\mathcal{X}|^N$. This is simple to do using a pseudorandom generator: $\lceil N \log_2 |\mathcal{X}| \rceil$ unbiased random bits are sufficient to sample from $P_N^{\text{unif}}(\underline{x})$. Then 'accept' the configuration with probability $P_N(\underline{x})$. Such an algorithm is very inefficient. It is clear that, to estimate expectation of 'well-behaved' observables, we need to generate typical configurations $\underline{x}$, i.e. configurations such that $r(\underline{x})$ is close to h. However, such configurations are exponentially rare, and the above algorithm will require a time of order $\exp[N(\log |\mathcal{X}| - h)]$ to generate one of them. The Monte Carlo Markov chain method provides a better alternative.

4.2   Large deviations for independent variables

A behaviour of the type (4.3) is an example of a large-deviation principle. We shall first study the simplest case where such a behaviour is found, namely the case of independent random variables. This case is instructive because all properties can be controlled in great detail.

4.2.1   How typical is a series of observations?

Suppose that you are given given the values $s_1, \ldots, s_N$ of N i.i.d. random variables drawn from a finite space $\mathcal{X}$ according to a known probability distribution $\{p(s)\}_{s \in \mathcal{X}}$. The s_i could be produced, for instance, by an information source, or by repeated measurements on a physical system. You would like to know if the sequence $\underline{s} = (s_1, \ldots, s_N)$ is a typical one, or if you have found a rare event. If N is large, one can expect that the number of appearances of a given $x \in \mathcal{X}$ in a typical sequence will be close to $Np(x)$. The method of types allows us to quantify this statement.

The **type** $q_{\underline{s}}(x)$ of a sequence $\underline{s}$ is the frequency of appearance of the symbol x in the sequence:

$$q_{\underline{s}}(x) = \frac{1}{N} \sum_{i=1}^{N} \mathbb{I}(x = s_i).$$

(4.4)

The type $q_{\underline{s}}(x)$, considered as a function of x, has the properties of a probability distribution over $\mathcal{X}$: $q_{\underline{s}}(x) \geq 0$ for any $x \in \mathcal{X}$, and $\sum_x q_{\underline{s}}(x) = 1$. In the following, we shall denote by $\mathfrak{M}(\mathcal{X})$ the space of probability distributions over $\mathcal{X}$: $\mathfrak{M}(\mathcal{X}) \equiv \{q \in \mathbb{R}^{\mathcal{X}}$ such that $q(x) \geq 0$, $\sum_x q(x) = 1\}$. In particular, $q_{\underline{s}} \in \mathfrak{M}(\mathcal{X})$.

The expectation of the type $q_{\underline{s}}(x)$ coincides with the original probability distribution:

$$\mathbb{E}\, q_{\underline{s}}(x) = p(x).$$

(4.5)

Sanov's theorem estimates the probability that the type of the sequence differs from $p(x)$.

Theorem 4.1. (Sanov) *Let $s_1, \ldots, s_N \in \mathcal{X}$ be N i.i.d. random variables drawn from the probability distribution $p(x)$, and let $K \subset \mathfrak{M}(\mathcal{X})$ be a compact set of probability distributions over $\mathcal{X}$. If q is the type of $(s_1, \ldots, s_N)$, then*

$$\mathbb{P}\,[q \in K] \doteq \exp[-ND(q^*||p)],$$

(4.6)

where $q_* = \arg\min_{q \in K} D(q||p)$, and $D(q||p)$ is the Kullback–Leibler divergence defined in eqn (1.10) .

Informally, this theorem means that the probability of finding a sequence with type q behaves at large N like $\exp[-ND(q||p)]$. For large N, typical sequences have a type $q(x) \approx p(x)$, and those with a different type are exponentially rare. The proof of the theorem is a straightforward application of Stirling's formula and is left as an exercise for the reader. In Section 4.7 we give a derivation using a 'field-theoretical' manipulation as used in physics. This may be an instructive, simple example for the reader who wants to get used to this kind of technique, frequently used by physicists.

Example 4.2 Let the s_i be the outcome of a biased coin: $\mathcal{X} = \{\texttt{head}, \texttt{tail}\}$, with $p(\texttt{head}) = 1 - p(\texttt{tail}) = 0.8$. What is the probability of getting 50 heads and 50 tails in 100 throws of the coin? Using eqns (4.6) and (1.10) with $N = 100$ and $q(\texttt{head}) = q(\texttt{tail}) = 0.5$, we get Prob[50 tails] $\approx 2.04 \times 10^{-10}$.

Example 4.3 Consider the reverse case: we take a fair coin ($p(\mathtt{head}) = p(\mathtt{tail}) = 0.5$) and ask what the probability of getting 80 heads and 20 tails is. Sanov's theorem provides the estimate $\mathrm{Prob}[80\ \mathtt{heads}] \approx 4.27 \times 10^{-9}$, which is much higher than the value computed in the previous example.

Example 4.4 A simple model of a column of the atmosphere is obtained by considering N particles in the earth's gravitational field. The state of particle $i \in \{1, \ldots, N\}$ is given by a single coordinate $z_i \geq 0$ which measures its height with respect to ground level. For the sake of simplicity, we assume the z_i to be integer numbers. We can, for instance, imagine that the heights are discretized in terms of some small unit length (e.g. millimetres). The N-particle energy function reads, in properly chosen units,

$$E = \sum_{i=1}^{N} z_i \,. \tag{4.7}$$

The type of a configuration $\{z_1, \ldots, z_N\}$ can be interpreted as the density profile $\rho(z)$ of the configuration:

$$\rho(z) = \frac{1}{N} \sum_{i=1}^{N} \mathbb{I}(z = z_i) \,. \tag{4.8}$$

Using the Boltzmann probability distribution (2.4), it is simple to compute the expected density profile, which is usually called the 'equilibrium' profile:

$$\rho_{\mathrm{eq}}(z) \equiv \langle \rho(z) \rangle = (1 - e^{-\beta}) e^{-\beta z} \,. \tag{4.9}$$

If we take a snapshot of the N particles at a given instant, their density will show some deviations with respect to $\rho_{\mathrm{eq}}(z)$. The probability of seeing a density profile $\rho(z)$ is given by eqn (4.6) with $p(z) = \rho_{\mathrm{eq}}(z)$ and $q(z) = \rho(z)$. For instance, we can compute the probability of observing an exponential density profile like that in eqn (4.9) but with a different parameter λ: $\rho_\lambda(z) = (1 - e^{-\lambda}) e^{-\lambda z}$. Using eqn (1.10), we get:

$$D(\rho_\lambda || \rho_{\mathrm{eq}}) = \log \left(\frac{1 - e^{-\lambda}}{1 - e^{-\beta}} \right) + \frac{\beta - \lambda}{e^\lambda - 1} \,. \tag{4.10}$$

The function $I_\beta(\lambda) \equiv D(\rho_\lambda || \rho_{\mathrm{eq}})$ is depicted in Fig. 4.1.

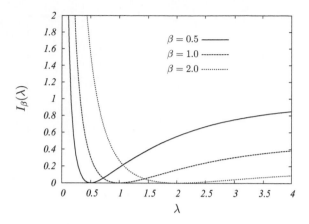

Fig. 4.1 In an atmosphere where the equilibrium density profile is $\rho_{eq}(z) \propto e^{-\beta z}$, the probability of observing an atypical profile $\rho(z) \propto e^{-\lambda z}$ is, for a large number of particles N, $\exp[-NI_\beta(\lambda)]$. The curves of $I_\beta(\lambda)$ plotted here show that small values of λ are very rare.

Exercise 4.1 The previous example is easily generalized to the density profile of N particles in an arbitrary potential $V(x)$. Show that the Kullback–Leibler divergence takes the form

$$D(\rho\|\rho_{eq}) = \beta \sum_x V(x)\rho(x) + \sum_x \rho(x) \log \rho(x) + \log z(\beta) \,. \tag{4.11}$$

4.2.2   How typical is an empirical average?

The result (4.6) contains detailed information concerning the large fluctuations of the random variables $\{s_i\}$. Often one is interested in measuring some property of the particles, that is described by a real number $f(s_i)$. The empirical average of the measurement is given by

$$\overline{f} \equiv \frac{1}{N} \sum_{i=1}^{N} f(s_i) \,. \tag{4.12}$$

Of course $\overline{f}$, will be 'close' to $\mathbb{E} f(x)$ with high probability. The following result quantifies the probability of rare fluctuations.

Corollary 4.5 *Let $s_1, \ldots, s_N$ be N i.i.d. random variables drawn from a probability distribution $p(.)$. Let $f : \mathcal{X} \to \mathbb{R}$ be a real-valued function and let $\overline{f}$ be its empirical average. If $A \subset \mathbb{R}$ is a closed interval of the real axis, then*

$$\mathbb{P}\left[\overline{f} \in A\right] \doteq \exp[-NI(A)] \,, \tag{4.13}$$

where

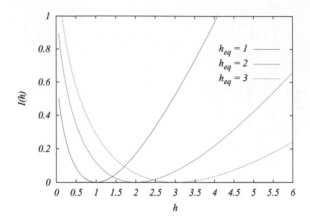

Fig. 4.2 Probability of an atypical average height for N particles with the energy function (4.7).

$$I(A) = \min_q \left[D(q\|p) \ \bigg|\ \sum_{x \in \mathcal{X}} q(x)f(x) \in A \right].$$ (4.14)

Proof: $\overline{f}$ is related to the type of $x_1, \dots, x_N$ through $\overline{f} = \sum_x q(x)f(x)$. Therefore we can apply Theorem 4.1, using for K the set of distributions such that the expectation value of f is in A:

$$K = \left\{ q \in \mathfrak{M}(\mathcal{X}) \ \bigg|\ \sum_{x \in \mathcal{X}} q(x)f(x) \in A \right\}.$$ (4.15)

This implies eqn (4.13) and (4.14) directly. □

Exercise 4.2 Let $s_1, \dots, s_N$ be N i.i.d. random variables drawn from a probability distribution $p(\cdot)$ with bounded support. Show that, to leading exponential order, $\mathbb{P}\{s_1 + \cdots + s_N \le 0\} \doteq \{\inf_{z \ge 0} \mathbb{E}\,[\mathrm{e}^{-z s_1}]\}^N$.

Example 4.6 We look again at N particles in a gravitational field, and consider the average height of the particles

$$\overline{z} = \frac{1}{N} \sum_{i=1}^N z_i.$$ (4.16)

The expected value of this quantity is $\mathbb{E}(\overline{z}) = z_{\mathrm{eq}} = (e^{\beta} - 1)^{-1}$. The probability of a fluctuation in $\overline{z}$ is easily computed using the above corollary. For $z > z_{\mathrm{eq}}$, we obtain $P[\overline{z} > z] \doteq \exp[-N\,I(z)]$, where

$$I(z) = (1+z)\log\left(\frac{1+z_{\mathrm{eq}}}{1+z}\right) + z\log\left(\frac{z}{z_{\mathrm{eq}}}\right). \tag{4.17}$$

Analogously, for $z < z_{\mathrm{eq}}$, $P[\overline{z} < z] \doteq \exp[-N\,I(z)]$, with the same rate function $I(z)$. The function $I(z)$ is depicted in Fig. 4.2.

Exercise 4.3 One can construct a thermometer using a system of N particles with the energy function (4.7). Whenever the temperature is required, one takes a snapshot of the N particles, computes $\overline{x}$, and estimates the inverse temperature β_{est} using the formula $(e^{\beta_{\mathrm{est}}} - 1)^{-1} = \overline{x}$. What (for $N \gg 1$) is the probability of getting a result $\beta_{\mathrm{est}} \neq \beta$?

4.2.3 Asymptotic equipartition

The above tools can also be used for *counting* the number of configurations $\underline{s} = (s_1, \ldots, s_N)$ with either a given type $q(x)$ or a given empirical average of some observable $\overline{f}$. One finds, for instance, the following.

Proposition 4.7 *The number $\mathcal{N}_{K,N}$ of sequences $\underline{s}$ which have a type belonging to the compact set $K \subset \mathfrak{M}(\mathcal{X})$ behaves as $\mathcal{N}_{K,N} \doteq \exp\{NH(q_*)\}$, where $q_* = \arg\max\{H(q)\,|\,q \in K\}$.*

This result can be stated informally by saying that 'there are approximately $e^{NH(q)}$ sequences with type q'.

Proof The idea is to apply Sanov's theorem, taking the 'reference' distribution $p(x)$ to be the uniform probability distribution $p_{\mathrm{unif}}(x) = 1/|\mathcal{X}|$. Using eqn (4.6), we get

$$\mathcal{N}_{K,N} = |\mathcal{X}|^N \mathbb{P}_{\mathrm{unif}}[q \in K] \doteq \exp\{N\log|\mathcal{X}| - ND(q_*\|p_{\mathrm{unif}})\} = \exp\{NH(q_*)\}. \tag{4.18}$$

$\square$

We now return to a generic sequence $\underline{s} = (s_1, \ldots, s_N)$ of N i.i.d. variables with a probability distribution $p(x)$. As a consequence of Sanov's theorem, we know that the most probable type is $p(x)$ itself, and that deviations are exponentially rare in N. We expect that almost all the probability will be concentrated into sequences that have a type close to $p(x)$ in some sense. On the other hand, because of the above proposition, the number of such sequences is exponentially smaller than the total number of possible sequences $|\mathcal{X}|^N$.

These remarks can be made more precise by defining what is meant by a sequence having a type 'close to $p(x)$'. Given a sequence $\underline{s}$, we introduce the quantity

$$r(\underline{s}) \equiv -\frac{1}{N}\log P_N(\underline{s}) = -\frac{1}{N}\sum_{i=1}^{N}\log p(x_i). \tag{4.19}$$

Clearly, $\mathbb{E}\, r(\underline{s}) = H(p)$. The sequence $\underline{s}$ is said to be ε-**typical** if and only if $|r(\underline{s}) - H(p)| \leq \varepsilon$.

Theorem 4.8 *Let $T_{N,\varepsilon}$ be the set of ε-typical sequences. Then we have the following properties.*

(i) $\lim_{N \to \infty} \mathbb{P}[\underline{s} \in T_{N,\varepsilon}] = 1$.

(ii) For N large enough, $\mathrm{e}^{N[H(p)-\varepsilon]} \leq |T_{N,\varepsilon}| \leq \mathrm{e}^{N[H(p)+\varepsilon]}$.

(iii) For any $\underline{s} \in T_{N,\varepsilon}$, $\mathrm{e}^{-N[H(p)+\varepsilon]} \leq P_N(\underline{s}) \leq \mathrm{e}^{-N[H(p)-\varepsilon]}$.

Proof Since $r(\underline{s})$ is an empirical average, we can apply Corollary 4.5. This allows us to estimate the probability of the sequence *not* being typical as $\mathbb{P}[\underline{s} \notin T_{N,\varepsilon}] \doteq \exp(-NI)$. The exponent is given by $I = \min_q D(q\|p)$, the minimum being taken over the set K of all probability distributions $q(x)$ such that $\left| \sum_{x \in \mathcal{X}} q(x) \log[1/q(x)] - H(p) \right| \geq \varepsilon$. But $D(q\|p) > 0$ unless $q = p$, and p does not belong to K. Therefore $I > 0$ and $\lim_{N \to \infty} \mathbb{P}[\underline{s} \notin T_{N,\varepsilon}] = 0$, which proves (*i*).

The condition for $q(x)$ to be the type of an ε-typical sequence can be rewritten as $|D(q\|p) + H(q) - H(p)| \leq \varepsilon$. Therefore, for any ε-typical sequence, $|H(q) - H(p)| \leq \varepsilon$, and Proposition 4.7 leads to (*ii*). Finally, (*iii*) is a direct consequence of the definition of ε-typical sequences. $\square$

The behaviour described in this theorem is usually referred to as the **asymptotic equipartition property**. Although we have proved it only for i.i.d. random variables, it holds in a much broader context. In fact, it will be found in many interesting systems throughout the book.

4.3 Correlated variables

In the case of independent random variables in finite spaces, the probability of a large fluctuation is easily computed by combinatorics. It would be nice to have some general result for large deviations of non-independent random variables. In this section, we shall describe the use of Legendre transforms and saddle point methods to study the general case. This method corresponds to a precise mathematical statement: the Gärtner–Ellis theorem. We shall first describe the approach informally and apply it to a few examples. Then we shall state the theorem and discuss it.

4.3.1 Legendre transformation

We consider a set of N random variables $\underline{x} = (x_1, \ldots, x_N)$, with $x_i \in \mathcal{X}$ and a probability distribution

$$P_N(\underline{x}) = P_N(x_1, \ldots, x_N). \tag{4.20}$$

Let $f : \mathcal{X} \to \mathbb{R}$ be a real-valued function. We are interested in estimating, for large N, the probability distribution of the empirical average

$$\overline{f}(\underline{x}) = \frac{1}{N} \sum_{i=1}^{N} f(x_i). \tag{4.21}$$

In the previous section, we studied the particular case in which the x_i are i.i.d. random variables. We proved that, quite generally, a finite fluctuation of $\overline{f}(\underline{x})$ is exponentially unlikely. It is natural to expect that the same statement will hold true if the x_i are 'weakly correlated'. Whenever $P_N(x)$ is the Boltzmann distribution for some physical system, this expectation is supported by physical intuition. We can think of the x_i as the microscopic degrees of freedom that make up the system, and of $\overline{f}(x)$ as a macroscopic observable (pressure, magnetization, etc.). It is a common observation that the relative fluctuations of macroscopic observables are very small.

Let us thus *assume* that the distribution of $\overline{f}$ follows a **large-deviation principle**, meaning that the asymptotic behaviour of the distribution at large N is

$$P_N(\overline{f}) \doteq \exp[-NI(\overline{f})] \,, \tag{4.22}$$

with a **rate function** $I(\overline{f}) \geq 0$.

In order to determine $I(\overline{f})$, a useful method is to 'tilt' the measure $P_N(\cdot)$ in such a way that the rare events responsible for $O(1)$ fluctuations of $\overline{f}$ become likely. In practice, we define the **(logarithmic) moment-generating function** of $\overline{f}$ as follows

$$\psi_N(t) = \frac{1}{N} \log \left(\mathbb{E} \, e^{Nt\overline{f}(x)} \right) \,, \qquad t \in \mathbb{R} \,. \tag{4.23}$$

When the large-deviation principle (4.22) holds, we can evaluate the large-N limit of $\psi_N(t)$ using the saddle point method:

$$\lim_{N\to\infty} \psi_N(t) = \lim_{N\to\infty} \frac{1}{N} \log \left\{ \int e^{Nt\overline{f}-NI(\overline{f})} \mathrm{d}\overline{f} \right\} = \psi(t) \,, \tag{4.24}$$

where

$$\psi(t) = \sup_{\overline{f}\in\mathbb{R}} \left[t\overline{f} - I(\overline{f}) \right] \,. \tag{4.25}$$

In other words, $\psi(t)$ is the Legendre transform of $I(\overline{f})$, and it is a convex function of t by construction (this can be proved by differentiating eqn (4.23) twice). It is therefore natural to invert the Legendre transform (4.25) as follows:

$$I_\psi(\overline{f}) = \sup_{t\in\mathbb{R}} \left[t\overline{f} - \psi(t) \right] \,, \tag{4.26}$$

and we expect $I_\psi(\overline{f})$ to coincide with the convex envelope of $I(\overline{f})$. This procedure is useful whenever computing $\psi(t)$ is easier than a direct estimate of the probability distribution $P_N(\overline{f})$. It is useful to gain some insight by considering a few examples.

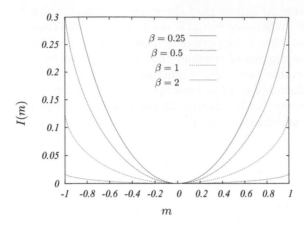

Fig. 4.3 Rate function for the magnetization of the one-dimensional Ising model. Notice that, as the temperature is lowered (β is increased), the probability of large fluctuations increases.

Exercise 4.4 Consider the one-dimensional Ising model without an external magnetic field (see Section 2.5.1). The variables are N Ising spins $\sigma_i \in \{+1, -1\}$, and their distribution is the Boltzmann distribution $P_N(\underline{\sigma}) = \exp[-\beta E(\underline{\sigma})]/Z$ with an energy function $E(\underline{\sigma}) = -\sum_{i=1}^{N-1} \sigma_i \sigma_{i+1}$. We want to compute the large-deviation properties of the magnetization,

$$m(\underline{\sigma}) = \frac{1}{N} \sum_{i=1}^{N} \sigma_i, \qquad (4.27)$$

using the moment-generating function.

(a) Show that $\psi(t) = \phi(\beta, t/\beta) - \phi(\beta, 0)$, where $\phi(\beta, B)$ is the free-energy density of the model in an external magnetic field B found in eqn (2.63). Explicitly, this gives

$$\psi(t) = \log \left(\frac{\cosh t + \sqrt{\sinh^2 t + e^{-4\beta}}}{1 + e^{-2\beta}} \right). \qquad (4.28)$$

(b) The function $\psi(t)$ is convex and analytic for any $\beta < \infty$. By applying eqn (4.26), obtain numerically the rate function $I_\psi(m)$. The result is shown in Fig. 4.3 for several inverse temperatures β.

Example 4.9 Consider a Markov chain $X_0, X_1, \ldots, X_i, \ldots$ taking values in a finite state space $\mathcal{X}$, and assume all the elements of the transition matrix $w(x \to y)$ to be strictly positive. Let us study the large-deviation properties of the empirical average $(1/N) \sum_i f(X_i)$.

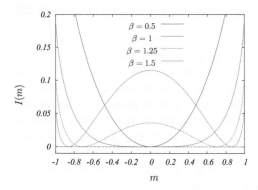 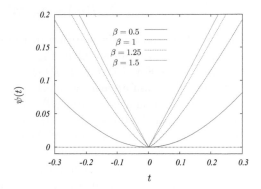

Fig. 4.4 The rate function for large fluctuations of the magnetization in the Curie–Weiss model (*left*), and the corresponding generating function (*right*).

One can show that the limit moment-generating function $\psi(t)$ (see eqn (4.23)) exists, and can be computed using the following recipe. Define the 'tilted' transition probabilities as $w_t(x \to y) = w(x \to y) \exp[t\, f(y)]$. Let $\lambda(t)$ be the largest solution of the eigenvalue problem

$$\sum_{x \in \mathcal{X}} \phi_t^l(x)\, w_t(x \to y) = \lambda(t)\, \phi_t^l(y) \qquad (4.29)$$

(which is unique and positive because of the Perron–Frobenius theorem). The moment-generating function is given simply by $\psi(t) = \log \lambda(t)$.

An interesting new phenomenon emerges if we consider now the Curie–Weiss model with a vanishing external field. We again study N Ising spins $\sigma_i \in \{+1, -1\}$ with a distribution $P_N(\underline{\sigma}) = \exp[-\beta E(\underline{\sigma})]/Z$, and the energy function is

$$E(\underline{\sigma}) = -\frac{1}{N} \sum_{(ij)} \sigma_i \sigma_j \,. \qquad (4.30)$$

We are interested in the large fluctuations of the global magnetization (4.27). In this case one can compute the large-deviation function directly as explained in the following exercise.

Exercise 4.5 Using the arguments of Section 2.5.2, show that, for any $m_1 < m_2$,

$$P_N\{m(\underline{\sigma}) \in [m_1, m_2]\} \doteq \frac{1}{Z_N(\beta)} \int_{m_1}^{m_2} e^{N\phi_{\mathrm{mf}}(m;\beta)} \, \mathrm{d}m \,, \qquad (4.31)$$

where $\phi_{\mathrm{mf}}(m; \beta) = \beta m^2/2 - \log[2\cosh(\beta m)]$.

This exercise shows that the large-deviation principle (4.22) holds, with

$$I(m) = \phi_{\mathrm{mf}}(m^*; \beta) - \phi_{\mathrm{mf}}(m; \beta), \tag{4.32}$$

and $m^*(\beta)$ is the largest solution of the Curie–Weiss equation $m = \tanh(\beta m)$. The function $I(m)$ is represented in Fig. 4.4, left frame, for several values of the inverse temperature β. For $\beta < \beta_c = 1$, $I(m)$ is convex and has its unique minimum at $m = 0$.

A new situation appears when $\beta > \beta_c$. The function $I(m)$ is non-convex, with two degenerate minima at $m = \pm m^*(\beta)$. In words, the system can be found in either of two well-distinguished 'states': the states of positive and negative magnetization. There is no longer a *unique* typical value of the magnetization such that large fluctuations away from this value are exponentially rare.

Let us now look at what happens if one uses the generating-function approach. It is easy to see that the limit (4.23) exists and is given by

$$\psi(t) = \sup_{m \in [-1,1]} [mt - I(m)]. \tag{4.33}$$

While at high temperature, $\beta < 1$, $\psi(t)$ is convex and analytic, for $\beta > 1$ $\psi(t)$ develops a singularity at $t = 0$. In particular, we have $\psi'(0+) = m^*(\beta) = -\psi'(0-)$. We now compute $I_\psi(m)$ using eqn (4.26). A little thought shows that, for any $m \in [-m^*(\beta), m^*(\beta)]$, the supremum is achieved for $t = 0$, which yields $I_\psi(m) = 0$. Outside this interval, the supremum is achieved at the unique solution of $\psi'(t) = m$, and $I_\psi(m)$. As anticipated, $I_\psi(m)$ is the convex envelope of $I(m)$. In the range $(-m^*(\beta), m^*(\beta))$, an estimate of the magnetization fluctuations through the function $\doteq \exp(-NI_\psi(m))$ would *overestimate* the fluctuations.

4.3.2   The Gärtner–Ellis theorem

The Gärtner–Ellis theorem has several formulations, which usually require some technical definitions beforehand. Here we shall state it in a simplified (and somewhat weakened) form. We need only the definition of an **exposed point**: $x \in \mathbb{R}$ is an exposed point of the function $F : \mathbb{R} \to \mathbb{R}$ if there exists $t \in \mathbb{R}$ such that $ty - F(y) > tx - F(x)$ for any $y \neq x$. If, for instance, F is convex, a sufficient condition for x to be an exposed point is that F is twice differentiable at x, with $F''(x) > 0$.

Theorem 4.10. (Gärtner–Ellis) *Consider a function $\overline{f}(\underline{x})$ (not necessarily of the form (4.21)) and assume that the moment-generating function $\psi_N(t)$ defined in eqn (4.23) exists and has a finite limit $\psi(t) = \lim_{N \to \infty} \psi_N(t)$ for any $t \in \mathbb{R}$. Define $I_\psi(\cdot)$ as the inverse Legendre transform of eqn (4.26), and let $\mathcal{E}$ be the set of exposed points of $I_\psi(\cdot)$.*

1. *For any closed set $F \in \mathbb{R}$,*

$$\limsup_{N \to \infty} \frac{1}{N} \log P_N(\overline{f} \in F) \leq - \inf_{f \in F} I_\psi(f). \tag{4.34}$$

2. *For any open set $G \in \mathbb{R}$,*

$$\limsup_{N \to \infty} \frac{1}{N} \log P_N(\overline{f} \in G) \geq - \inf_{f \in G \cap \mathcal{E}} I_\psi(f). \tag{4.35}$$

3. *If, moreover, $\psi(t)$ is differentiable for any $t \in \mathbb{R}$, then the last statement holds true if the inf is taken over the whole set G (rather than over $G \cap \mathcal{E}$).*

Informally, the inverse Legendre transform (4.26) generically yields an upper bound on the probability of a large fluctuation of the macroscopic observable. This upper bound is tight unless a 'first-order phase transition' occurs, corresponding to a discontinuity in the first derivative of $\psi(t)$, as we saw in the low-temperature phase of the Curie–Weiss model.

It is worth mentioning that $\psi(t)$ can be non-analytic at a point t_* even though its first derivative is continuous at t_*. This corresponds, in statistical-mechanics jargon, to a 'higher-order' phase transition. Such phenomena have interesting probabilistic interpretations too.

4.3.3 Typical sequences

Let us return to the concept of typical sequences introduced in Section 4.2. More precisely, we want to investigate the large deviations of the probability itself, measured by $r(\underline{x}) = -(1/N) \log P(\underline{x})$. For independent random variables, the study in Section 4.2.3 led to the concept of ε-typical sequences. What can one say about general sequences?

Let us compute the corresponding moment-generating function (4.23):

$$\psi_N(t) = \frac{1}{N} \log \left\{ \sum_{\underline{x}} P_N(\underline{x})^{1-t} \right\}. \tag{4.36}$$

Without loss of generality, we can assume $P_N(\underline{x})$ to have the Boltzmann form

$$P_N(\underline{x}) = \frac{1}{Z_N(\beta)} \exp\{-\beta E_N(\underline{x})\}, \tag{4.37}$$

with energy function $E_N(\underline{x})$. Inserting this into eqn (4.36), we get

$$\psi_N(t) = \beta f_N(\beta) - \beta f_N(\beta(1-t)), \tag{4.38}$$

where $f_N(\beta) = -(1/N) \log Z_N(\beta)$ is the free-energy density of the system with energy function $E_N(\underline{x})$ at inverse temperature β. Let us assume that the thermodynamic limit $f(\beta) = \lim_{N \to \infty} f_N(\beta)$ exists and is finite. It follows that the limiting generating function $\psi(t)$ exists and we can apply the Gärtner–Ellis theorem to compute the probability of a large fluctuation of $r(\underline{x})$. As long as $f(\beta)$ is analytic, large fluctuations are exponentially depressed and the asymptotic equipartition property of independent random variables is essentially recovered. On the other hand, if there is a phase transition at $\beta = \beta_c$, where the first derivative of $f(\beta)$ is discontinuous, then the likelihood $r(\underline{x})$ may take several distinct values with a non-vanishing probability. This is what happened in our Curie–Weiss example.

4.4 The Gibbs free energy

4.4.1 Variational principle

In the introduction to statistical physics in Chapter 2, we assumed that the probability distribution of the configurations of a physical system was the Boltzmann distribution.

It turns out that this distribution can be obtained from a variational principle. This is interesting, both as a matter of principle and for the purpose of finding approximation schemes.

Consider a system with a configuration space $\mathcal{X}$, and a real-valued energy function $E(x)$ defined on this space. The Boltzmann distribution is $\mu_\beta(x) = \exp[-\beta(E(x) - F(\beta))]$, where $F(\beta)$, the 'free energy', is a function of the inverse temperature β defined by the fact that $\sum_{x \in \mathcal{X}} \mu_\beta(x) = 1$. We define the **Gibbs free energy** $G[P]$ as the following real-valued functional over the space of probability distributions $P(x)$ on $\mathcal{X}$:

$$G[P] = \sum_{x \in \mathcal{X}} P(x)E(x) + \frac{1}{\beta} \sum_{x \in \mathcal{X}} P(x) \log P(x) \,. \tag{4.39}$$

(The Gibbs free energy should not be confused with $F(\beta)$.) It is easy to rewrite the Gibbs free energy in terms of the KL divergence between $P(x)$ and the Boltzmann distribution $\mu_\beta(x)$:

$$G[P] = \frac{1}{\beta} D(P\|\mu_\beta) + F(\beta) \,. \tag{4.40}$$

This representation implies straightforwardly the following proposition (the **Gibbs variational principle**).

Proposition 4.11 *The Gibbs free energy $G[P]$ is a convex functional of $P(x)$, and it achieves its unique minimum on the Boltzmann distribution $P(x) = \mu_\beta(x)$. Moreover, $G[\mu_\beta] = F(\beta)$, where $F(\beta)$ is the free energy.*

The relation between the Gibbs free energy and the Kullback–Leibler divergence in eqn (4.40) implies a simple probabilistic interpretation of the Gibbs variational principle. Imagine that a large number $\mathcal{N}$ of copies of the same physical system have been prepared. Each copy is described by the same energy function $E(\underline{x})$. Now consider the empirical distribution $P(\underline{x})$ of the $\mathcal{N}$ copies. Typically, $P(\underline{x})$ will be close to the Boltzmann distribution $\mu_\beta(\underline{x})$. Sanov's theorem implies that the probability of an 'atypical' distribution is exponentially small in $\mathcal{N}$: $\mathbb{P}[P] \doteq \exp[-\mathcal{N}(G[P] - F(\beta))]$.

When the partition function of a system cannot be computed exactly, the above result suggests a general line of approach for estimating the free energy: one can minimize the Gibbs free energy in some restricted subspace of 'trial probability distributions' $P(x)$. These trial distributions should be simple enough that $G[P]$ can be computed, but the restricted subspace should also contain distributions which are able to give a good approximation to the true behaviour of the physical system. For each new physical system one will thus need to find a good restricted subspace.

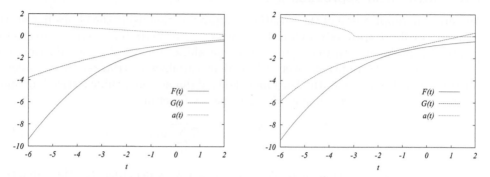

Fig. 4.5 Variational estimates of the free energy of the model (4.41), using the Gibbs free energy with the trial distribution (4.42) on the *left* and (4.44) on the *right*.

Exercise 4.6 Consider a system with configuration space $\mathcal{X} = \mathbb{R}$ and energy

$$E(x) = \frac{1}{2} t x^2 + \frac{1}{4} x^4, \qquad (4.41)$$

with $t \in \mathbb{R}$. Let us find an upper bound on its free energy at temperature $\beta = 1$, $F(t) = -\log\left(\int e^{-E(x)} dx\right)$, using the Gibbs variational principle. We first use the following family of Gaussian trial probability distributions:

$$Q_a(x) = \frac{1}{\sqrt{2\pi a}} e^{-x^2/2a}. \qquad (4.42)$$

Show that the corresponding Gibbs free-energy at $\beta = 1$ is:

$$G[Q_a] = \frac{1}{2} ta + \frac{3}{4} a^2 - \frac{1}{2} (1 + \log 2\pi a) \equiv G(a, t). \qquad (4.43)$$

The Gibbs principle implies that $F(t) \leq \min_a G(a, t)$. In Fig. 4.5, we have plotted the optimal value of a, $a_{\text{opt}}(t) = \text{argmin}_a G(a, t)$, and the corresponding estimate $G_{\text{opt}}(t) = G(a_{\text{opt}}(t), t)$.

Exercise 4.7 Consider the same problem with the family of trial distributions

$$Q_a(x) = \frac{1}{\sqrt{2\pi}} e^{-(x-a)^2/2}. \qquad (4.44)$$

Determine the optimal value of a_{opt}, and the corresponding upper bound on $F(t)$ (see Fig. 4.5). Notice the bifurcation at $t_{\text{cr}} = -3$. For $t > t_{\text{cr}}$, one finds that $a_{\text{opt}}(t) = 0$, while $G[Q_a]$ has two degenerate local minima $a = \pm a_{\text{opt}}(t)$ for $t \leq t_{\text{cr}}$.

4.4.2 Mean-field approximation

An important application of the variational approximation using the Gibbs free energy is the **mean-field approximation**. Usually, we do not know how to compute the free energy of a problem with many interacting variables. A natural approximation is to take the class of distributions over independent variables as trial family. In order to illustrate this idea, consider the Ising model on a d-dimensional cubic lattice $\mathbb{L}$ of linear size L (i.e. $\mathbb{L} = [L]^d$) (see Section 2.5). The energy function is

$$E(\underline{\sigma}) = -\frac{1}{d} \sum_{(ij)} \sigma_i \sigma_j - B \sum_{i \in \mathbb{L}} \sigma_i \,. \tag{4.45}$$

Notice that we have chosen a particular scale of energy by deciding that the strength of the interaction between the spins is $J = 1/d$. This choice will allow us to make direct contact with the Curie–Weiss model. For the sake of simplicity, we assume **periodic boundary conditions**. This means that two sites $i = (i_1, \ldots, i_d)$ and $j = (j_1, \ldots, j_d)$ are considered nearest neighbours if, for some $l \in \{1, \ldots, d\}$, $i_l - j_l = \pm 1 \pmod{L}$ and $i_{l'} = j_{l'}$ for any $l' \neq l$. The sum over (ij) in eqn (4.45) runs over all nearest-neighbour pairs in $\mathbb{L}$.

In order to obtain a variational estimate of the free energy $F(\beta)$ at inverse temperature β, we evaluate the Gibbs free energy on the following trial distribution:

$$Q_m(\underline{\sigma}) = \prod_{i \in \mathbb{L}} q_m(\sigma_i) \,, \tag{4.46}$$

where $q_m(+) = (1 + m)/2$, $q_m(-) = (1 - m)/2$, and $m \in [-1, +1]$. Note that, under $Q_m(\underline{\sigma})$, the σ_i are i.i.d. random variables with expectation m.

It is easy to evaluate the density of the Gibbs free energy, $g(m; \beta, B) \equiv G[Q_m]/L^d$, on this distribution:

$$g(m; \beta, B) = -\frac{1}{2} m^2 - B m - \frac{1}{\beta} \mathcal{H}\left(\frac{1+m}{2}\right) \,. \tag{4.47}$$

The Gibbs variational principle implies an upper bound on the free-energy density $f_d(\beta, B) \leq \inf_m g(m; \beta, B)$. Notice that, apart from an additive constant, this expression (4.47) has the same form as the solution of the Curie–Weiss model (see eqn (2.79)). This immediately implies that $f_d(\beta, B) \leq f_{\mathrm{CW}}(\beta, h) - \frac{1}{2}$. The minimization of $g(m; \beta, B)$ has already been discussed in Section 2.5.2; we shall not repeat it here. Let us just mention that, in zero magnetic field ($B = 0$), at low enough temperature, the mean-field approximation always predicts that the Ising model has a phase transition to a ferromagnetic phase. While this prediction is qualitatively correct when $d \geq 2$, it is wrong when $d = 1$, as we saw through the exact solution in Section 2.5.1. Indeed, the mean-field approximation becomes better the larger the dimension d, and it is asymptotically exact for $d \to \infty$.

4.5 The Monte Carlo method

The Monte Carlo method is an important generic tool which is common to probability theory, statistical physics, and combinatorial optimization. In all of these fields, we are often confronted with the problem of sampling a configuration $\underline{x} \in \mathcal{X}^N$ (here we assume $\mathcal{X}$ to be a finite space) from a given distribution $P(\underline{x})$. This can be quite difficult when N is large, because there are too many configurations, because the typical configurations are exponentially rare, and/or because the distribution $P(\underline{x})$ is specified by the Boltzmann formula with an unknown normalization (the partition function).

A general approach is to construct a Markov chain which is guaranteed to converge to the desired $P(\underline{x})$, and then simulate it on a computer. The computer is of course assumed to have access to some source of randomness: in practice, pseudo-random number generators are used. If the chain is simulated for a long enough time, the final configuration has a distribution 'close' to $P(\underline{x})$. In practice, the Markov chain is defined by a set of transition rates $w(\underline{x} \rightarrow \underline{y})$ with $\underline{x}, \underline{y} \in \mathcal{X}^N$ which satisfy the following conditions:

1. The chain is **irreducible**, i.e. for any pair of configurations $\underline{x}$ and $\underline{y}$ there exists a path $(\underline{x}_0, \underline{x}_1, \ldots \underline{x}_n)$ of length n, connecting $\underline{x}$ to $\underline{y}$ with non-zero probability. This means that $\underline{x}_0 = \underline{x}$, $\underline{x}_n = \underline{y}$ and $w(\underline{x}_i \rightarrow \underline{x}_{i+1}) > 0$ for $i = 0, \ldots, n - 1$.

2. The chain is **aperiodic**: for any pair $\underline{x}$ and $\underline{y}$, there exists a positive integer $n(\underline{x}, \underline{y})$ such that, for any $n \geq n(\underline{x}, \underline{y})$, there exists a path of length n connecting $\underline{x}$ to $\underline{y}$ with non-zero probability. Note that, for an irreducible chain, aperiodicity is easily enforced by allowing the configuration to remain unchanged with non-zero probability: $w(\underline{x} \rightarrow \underline{x}) > 0$.

3. The distribution $P(\underline{x})$ is **stationary** with respect to the probabilities $w(\underline{x} \rightarrow \underline{y})$:

$$\sum_{\underline{x}} P(\underline{x}) \, w(\underline{x} \rightarrow \underline{y}) = P(\underline{y}) \,. \tag{4.48}$$

Sometimes a stronger condition (implying stationarity) is satisfied by the transition probabilities. In this case, for each pair of configurations $\underline{x}, \underline{y}$ such that either $w(\underline{x} \rightarrow \underline{y}) > 0$ or $w(\underline{y} \rightarrow \underline{x}) > 0$, one has

$$P(\underline{x}) \, w(\underline{x} \rightarrow \underline{y}) = P(\underline{y}) \, w(\underline{y} \rightarrow \underline{x}) \,. \tag{4.49}$$

This condition is referred to as **reversibility** or **detailed balance**.

The strategy of designing and simulating such a process in order to sample from $P(\underline{x})$ goes under the name of the **dynamic Monte Carlo** method or the **Markov chain Monte Carlo** (MCMC) method (hereafter we shall refer to it simply as the Monte Carlo method). The theoretical basis for such an approach is provided by a classic theorem.

Theorem 4.12 *Assume the rates $w(\underline{x} \rightarrow \underline{y})$ to satisfy the conditions 1–3 above. Let $\underline{X}_0, \underline{X}_1, \ldots, \underline{X}_t, \ldots$ be random variables distributed according to the Markov chain with rates $w(\underline{x} \rightarrow \underline{y})$ and initial condition $\underline{X}_0 = \underline{x}_0$. Let $f : \mathcal{X}^N \rightarrow \mathbb{R}$ be any-real valued function. Then:*

1. *The probability distribution of X_t converges to the stationary distribution:*

$$\lim_{t \rightarrow \infty} \mathbb{P}[\underline{X}_t = \underline{x}] = P(\underline{x}) \,. \tag{4.50}$$

2. *Time averages converge to averages over the stationary distribution:*

$$\lim_{t \to \infty} \frac{1}{t} \sum_{s=1}^{t} f(\underline{X}_s) = \sum_{\underline{x}} P(\underline{x}) f(\underline{x}) \quad \textit{almost surely.} \tag{4.51}$$

The proof of this theorem can be found in any textbook on Markov processes. Here we shall illustrate it by considering some simple Monte Carlo algorithms which are frequently used in statistical mechanics (although they are by no means the most efficient ones).

Consider a system of N Ising spins $\underline{\sigma} = (\sigma_1 \dots \sigma_N)$ with energy function $E(\underline{\sigma})$ and inverse temperature β. We are interested in sampling the Boltzmann distribution μ_β. The **Metropolis algorithm** is defined as follows. We denote by $\underline{\sigma}^{(i)}$ the configuration which coincides with $\underline{\sigma}$ except for the site i ($\sigma_i^{(i)} = -\sigma_i$), and let $\Delta E_i(\underline{\sigma}) \equiv E(\underline{\sigma}^{(i)}) - E(\underline{\sigma})$. At each step, an integer $i \in [N]$ is chosen randomly with a uniform distribution, and the spin σ_i is flipped with probability

$$w_i(\underline{\sigma}) = \exp\{-\beta \max[\Delta E_i(\underline{\sigma}), 0]\}. \tag{4.52}$$

More explicitly, the transition probabilities are given by

$$w(\underline{\sigma} \to \underline{\tau}) = \frac{1}{N} \sum_{i=1}^{N} w_i(\underline{\sigma}) \, \mathbb{I}(\underline{\tau} = \underline{\sigma}^{(i)}) \; + \; \left[1 - \frac{1}{N} \sum_{i=1}^{N} w_i(\underline{\sigma})\right] \mathbb{I}(\underline{\tau} = \underline{\sigma}). \tag{4.53}$$

It is easy to check that this definition satisfies both the irreducibility and the stationarity conditions for any energy function $E(\underline{\sigma})$ and inverse temperature $\beta < 1$. Furthermore, the chain satisfies the detailed balance condition

$$\mu_\beta(\underline{\sigma}) \, w_i(\underline{\sigma}) = \mu_\beta(\underline{\sigma}^{(i)}) \, w_i(\underline{\sigma}^{(i)}). \tag{4.54}$$

Whether the condition of aperiodicity is fulfilled depends on the energy function. It is easy to construct systems for which it does not hold. Take, for instance, a single spin, $N = 1$, and let $E(\sigma) = 0$: the spin is flipped at each step, and there is no way to have a transition from $\sigma = +1$ to $\sigma = -1$ in an even number of steps. (Luckily, this kind of pathology is easily cured by modifying the algorithm as follows. At each step, with probability $1 - \varepsilon$, a site i is chosen and a spin flip is proposed as above. With probability ε, nothing is done, i.e. a null transition $\underline{\sigma} \to \underline{\sigma}$ is realized.)

Exercise 4.8 Variants of this chain can be obtained by changing the flipping probabilities given by eqn (4.52). A popular choice is the **heat bath** algorithm (also referred to as **Glauber dynamics**):

$$w_i(\underline{\sigma}) = \frac{1}{2}\left[1 - \tanh\left(\frac{\beta \Delta E_i(\underline{\sigma})}{2}\right)\right]. \tag{4.55}$$

Prove irreducibility, aperiodicity, and stationarity for these transition probabilities.

One reason that makes the heat bath algorithm particularly interesting is that it can be easily generalized for the purpose of sampling from a distribution $P(\underline{x})$ in any system whose configuration space has the form $\mathcal{X}^N$. In this algorithm, one chooses a variable index i, fixes all the other variables, and assigns a new value to the i-th one according to its conditional distribution. A more precise description is provided by the following pseudocode (Recall that, given a vector $\underline{x} \in \mathcal{X}^N$, we denote by $\underline{x}_{\sim i}$ the $N-1$-dimensional vector obtained by removing the i-th component of $\underline{x}$.)

HEAT BATH ALGORITHM $(P(\underline{x})$, number of iterations $r)$

1: Generate $\underline{x}^{(0)}$ uniformly at random in $\mathcal{X}^N$;
2: **for** $t = 1$ to $t = r$:
3:      Draw a uniformly random integer $i \in \{1, \dots, N\}$
4:      For each $z \in \mathcal{X}$, compute

$$P(X_i = z | \underline{X}_{\sim i} = \underline{x}_{\sim i}^{(t-1)}) = \frac{P(X_i = z, \underline{X}_{\sim i} = \underline{x}_{\sim i}^{(t-1)})}{\sum_{z' \in \mathcal{X}} P(X_i = z', \underline{X}_{\sim i} = \underline{x}_{\sim i}^{(t-1)})} \; ;$$

5:      Set $x_j^{(t)} = x_j^{(t-1)}$ for each $j \neq i$, and $x_i^{(t)} = z$,

            where z is drawn from the distribution $P(X_i = z | \underline{X}_{\sim i} = \underline{x}_{\sim i}^{(t-1)})$;
6: **end**
7: **return** the sequence $\underline{x}^{(t)}$, $t = 1, \cdots, r$.

Let us stress that this algorithm only requires one to be able to compute the probability $P(\underline{x})$ up to a multiplicative constant. If, for instance, $P(\underline{x})$ is given by Boltzmann's law (see Section 2.1) it is enough to be able to compute the energy $E(\underline{x})$ of a configuration, and it is not necessary to compute the partition function $Z(\beta)$.

This is a very general method for defining a Markov chain with the desired properties. The proof is left as exercise.

Exercise 4.9 Assuming for simplicity that $\forall \underline{x}$, $P(\underline{x}) > 0$, prove irreducibility, aperiodicity and stationarity for the heat bath algorithm.

Theorem 4.12 confirms that the Monte Carlo method is indeed a viable approach for sampling from a given probability distribution. However, it does not provide any information concerning its computational efficiency. In order to discuss this issue, it is convenient to assume that simulating a single step $\underline{X}_t \to \underline{X}_{t+1}$ of the Markov chain has a time cost equal to one. This assumption is a good one as long as sampling a new configuration requires a number of operations that depends only mildly (e.g. polynomially) on the system size. This is the case in the two examples provided above, and we shall stick here to this simple scenario.

Computational efficiency reduces therefore to the following question: how many steps of the Markov chain should be simulated? Of course, there is no unique answer to such a generic question. We shall limit ourselves to introducing two important figures of merit. The first concerns the following problem: how many steps should be simulated in order to produce a single configuration $\underline{x}$ which is distributed *approximately* according to $P(\underline{x})$? In order to specify what is meant by 'approximately' we have to introduce a notion of distance between two distributions $P_1(\cdot)$ and $P_2(\cdot)$ on $\mathcal{X}^N$. A widely used definition is given by the **variation distance**

$$||P_1 - P_2|| = \frac{1}{2} \sum_{\underline{x} \in \mathcal{X}^N} |P_1(\underline{x}) - P_2(\underline{x})|. \tag{4.56}$$

Consider now a Markov chain satisfying the conditions 1–3 above with respect to a stationary distribution $P(\underline{x})$, and denote by $P_t(\underline{x}|\underline{x}_0)$ the distribution of $\underline{X}_t$ conditional on the initial condition $\underline{X}_0 = \underline{x}_0$. Let $d_{\underline{x}_0}(t) = ||P_t(\cdot|\underline{x}_0) - P(\cdot)||$ be the distance from the stationary distribution. The **mixing time** (or **variation threshold time**) is defined as

$$\tau_{\mathrm{eq}}(\varepsilon) = \min\{t > 0 \,:\, \sup_{\underline{x}_0} d_{\underline{x}_0}(t) \leq \varepsilon\}. \tag{4.57}$$

In this book, we shall often refer informally to this quantity as the **equilibration time**. The number ε can be chosen arbitrarily, a change in ε usually implying a simple multiplicative change in $\tau_{\mathrm{eq}}(\varepsilon)$. For of this reason, the convention $\varepsilon = 1/\mathrm{e}$ is sometimes adopted.

Rather than producing a single configuration with the prescribed distribution, one is often interested in computing the expectation value of some observable $\mathcal{O}(\underline{x})$. In principle, this can be done by averaging over many steps of the Markov chain as suggested by eqn (4.51). It is therefore natural to ask the following question. Assume that the initial condition $\underline{X}_0$ is distributed according to the stationary distribution $P(\underline{x})$. This can be obtained (to a good approximation) by simulating $\tau_{\mathrm{eq}}(\varepsilon)$ steps of the chain in a preliminary (equilibration) phase. We shall denote by $\langle \cdot \rangle$ the expectation with respect to the Markov chain with this initial condition. How many steps should we average over in order to get expectation values within some prescribed accuracy? In other words, we estimate $\sum P(\underline{x})\mathcal{O}(\underline{x}) \equiv \mathbb{E}_P \mathcal{O}$ by

$$\overline{\mathcal{O}}_T \equiv \frac{1}{T} \sum_{t=0}^{T-1} \mathcal{O}(\underline{X}_t). \tag{4.58}$$

It is clear that $\langle \overline{\mathcal{O}}_T \rangle = \sum P(\underline{x})\mathcal{O}(\underline{x})$. Let us compute the variance of this estimator:

$$\mathrm{Var}(\overline{\mathcal{O}}_T) = \frac{1}{T^2} \sum_{s,t=0}^{T-1} \langle \mathcal{O}_s ; \mathcal{O}_t \rangle = \frac{1}{T^2} \sum_{t=0}^{T-1} (T-t)\langle \mathcal{O}_0 ; \mathcal{O}_t \rangle, \tag{4.59}$$

where we have used the notation $\mathcal{O}_t \equiv \mathcal{O}(\underline{X}_t)$. Let us introduce the **autocorrelation function** $C_{\mathcal{O}}(t-s) \equiv \langle \mathcal{O}_s ; \mathcal{O}_t \rangle / \langle \mathcal{O}_0 ; \mathcal{O}_0 \rangle$, so that $\mathrm{Var}(\overline{\mathcal{O}}_T) = (\langle \mathcal{O}_0 ; \mathcal{O}_0 \rangle / T^2) \sum_{t=0}^{T-1}(T-t)\,C_{\mathcal{O}}(t)$. General results for Markov chains on finite state spaces imply that $C_{\mathcal{O}}(t)$ decreases exponentially as $t \to \infty$. Therefore, for large T, we have

$$\mathrm{Var}(\overline{\mathcal{O}}_T) = \frac{\tau_{\mathrm{int}}^{\mathcal{O}}}{T} \left[\mathbb{E}_P \mathcal{O}^2 - (\mathbb{E}_P \mathcal{O})^2 \right] + O(T^{-2}). \tag{4.60}$$

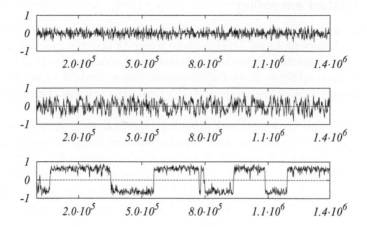

Fig. 4.6 Monte Carlo simulation of the Curie–Weiss model at three different temperatures: from *top* to *bottom* $\beta = 0.8$, 1.0, and 1.2. Here, we reproduce the global magnetization $m(\underline{\sigma})$ as a function of the number of iterations.

The **integrated autocorrelation time** $\tau_{\text{int}}^{\mathcal{O}}$ is given by

$$\tau_{\text{int}}^{\mathcal{O}} \equiv \sum_{t=0}^{\infty} C_{\mathcal{O}}(t) \,, \tag{4.61}$$

and provides a reference for estimating how long the Monte Carlo simulation should be run in order to achieve some prescribed accuracy. equation (4.60) can be interpreted by saying that one statistically independent estimate of $\mathbb{E}_P \mathcal{O}$ is obtained every $\tau_{\text{int}}^{\mathcal{O}}$ iterations.

Example 4.13 Consider the Curie–Weiss model (see Section 2.5.2), at inverse temperature β, and use the heat bath algorithm described above in order to sample from the Boltzmann distribution. In Fig. 4.6 we reproduce the evolution of the global magnetization $m(\underline{\sigma})$ during three different simulations at inverse temperatures $\beta = 0.8$, 1.0, and 1.2 for a model of $N = 150$ spin. In all cases, we initialized the Markov chain by drawing a random configuration with uniform probability.

A spectacular effect occurs at the lowest temperature, $\beta = 1.2$. Although the Boltzmann average of the global magnetization vanishes, i.e. $\langle m(\underline{\sigma}) \rangle = 0$, the sign of the magnetization remains unchanged over extremely long time scales. It is clear that the equilibration time is at least as large as these scales. An order-of-magnitude estimate would be $\tau_{\text{eq}} > 10^5$. Furthermore, this equilibration time diverges exponentially at large N. Sampling from the Boltzmann distribution using the present algorithm becomes exceedingly difficult at low temperature.

4.6   Simulated annealing

As we mentioned in Section 3.5, any optimization problem can be 'embedded' in a statistical-mechanics problem. The idea is to interpret the cost function $E(\underline{x})$, $\underline{x} \in \mathcal{X}^N$, as the energy of a statistical-mechanics system and consider the Boltzmann distribution $\mu_\beta(\underline{x}) = \exp[-\beta E(\underline{x})]/Z$. In the low-temperature limit $\beta \to \infty$, the distribution becomes concentrated on the minima of $E(\underline{x})$, and the original optimization setting is recovered.

Since the Monte Carlo method provides a general technique for sampling from the Boltzmann distribution, one may wonder whether it can be used, in the $\beta \to \infty$ limit, as an optimization technique. A simple-minded approach would be to take $\beta = \infty$ at the outset. Such a strategy is generally referred to as a **quench** in statistical physics and a **greedy search** in combinatorial optimization, and is often bound to fail. Consider the stationarity condition (4.48) and rewrite it using the Boltzmann formula:

$$\sum_{\underline{x}} \mathrm{e}^{-\beta\,[E(\underline{x}) - E(\underline{y})]}\, w(\underline{x} \to \underline{y}) = 1\,. \tag{4.62}$$

All the terms in the sum on the left-hand side must be less than or equal to one. This implies $0 \le w(\underline{x} \to \underline{y}) \le \exp\{-\beta\,[E(\underline{y}) - E(\underline{x})]\}$. Therefore, for any pair of configurations $\underline{x}$, $\underline{y}$ such that $E(\underline{y}) > E(\underline{x})$, we have $w(\underline{x} \to \underline{y}) \to 0$ in the $\beta \to \infty$ limit. In other words, the energy is always non-increasing along the trajectories of a zero-temperature Monte Carlo algorithm. As a consequence, the corresponding Markov chain is not irreducible (although it is irreducible at any $\beta < \infty$) and is not guaranteed to converge to the equilibrium distribution, i.e. to find a global minimum of $E(x)$. In fact, if the energy landscape $E(x)$ has local minima, this algorithm can get trapped in such a local minimum forever.

Another simple-minded approach would be to set β to some large but finite value. Although the Boltzmann distribution gives some weight to near-optimal configurations, the algorithm will also visit, from time to time, optimal configurations which are the most probable ones. How large should β be? How much time must we wait before an optimal configuration is visited? We can assume without loss of generality that the minimum of the cost function (the ground state energy) is zero: $E_0 = 0$. A meaningful quantity to look at is the probability for $E(\underline{x}) = 0$ under the Boltzmann distribution at an inverse temperature β. We can easily compute the logarithmic moment-generating function of the energy:

$$\psi_N(t) = \frac{1}{N} \log\left[\sum_{\underline{x}} \mu_\beta(\underline{x})\, \mathrm{e}^{tE(\underline{x})} \right] = \frac{1}{N} \log\left[\frac{\sum_x \mathrm{e}^{-(\beta - t)E(x)}}{\sum_x \mathrm{e}^{-\beta E(x)}} \right]. \tag{4.63}$$

This is given by $\psi_N(t) = \phi_N(\beta - t) - \phi_N(\beta)$, where $\phi_N(\beta)$ is the free-entropy density at inverse temperature β. Clearly $\mu_\beta[E(x) = 0] = \exp[N\psi_N(-\infty)] = \exp\{N[\phi_N(\infty) - \phi_N(\beta)]\}$, and the average time that we must wait before visiting the optimal configuration is $1/\mu_\beta[E(x) = 0] = \exp[-N\psi_N(-\infty)]$.

Exercise 4.10 Assume that the cost function takes integer values $E = 0, 1, 2 \ldots$ and denote $\mathcal{X}_E$ the set of configurations $\underline{x}$ such that $E(\underline{x}) = E$. You want the Monte Carlo trajectories to spend a fraction $(1 - \varepsilon)$ of the time on optimal solutions. Show that the temperature must be chosen such that

$$\beta = \log\left(\frac{|\mathcal{X}_1|}{\varepsilon|\mathcal{X}_0|}\right) + \Theta(\varepsilon). \tag{4.64}$$

In Section 2.4 we argued that, for many statistical-mechanics models, the free-entropy density has a finite thermodynamic limit $\phi(\beta) = \lim_{N\to\infty} \phi_N(\beta)$. In the following chapters, we shall show that this is also the case for several interesting optimization problems. This implies that $\mu_\beta[E(x) = 0]$ vanishes in the $N \to \infty$ limit. In order to have a non-negligibile probability of hitting a solution of the optimization problem, β must be scaled with N in such a way that $\beta \to \infty$ as $N \to \infty$. On the other hand, if we let $\beta \to \infty$, we are going to face the reducibility problem mentioned above. Although the Markov chain is formally irreducible, its equilibration time will diverge as $\beta \to \infty$.

The idea of **simulated annealing** consists in letting β vary with time. More precisely, one decides on an **annealing schedule** $\{(\beta_1, n_1); (\beta_2, n_2); \ldots (\beta_L, n_L)\}$, with inverse temperatures $\beta_i \in [0, \infty]$ and integers $n_i > 0$. The algorithm is initialized on a configuration $\underline{x}_0$ and executes n_1 Monte Carlo steps at temperature β_1, n_2 steps at temperature β_2, ..., and n_L steps at temperature β_L. The final configuration of each cycle i (with $i = 1, \ldots, L-1$) is used as the initial configuration of the next cycle. Mathematically, such a process is a **time-dependent Markov chain**. The common wisdom about the simulated-annealing algorithm is that varying the temperature with time should help one to avoid the two problems encountered above. Usually, one takes the β_i to be an increasing sequence. In the first stages, a small β should help in equilibrating across the space of configurations $\mathcal{X}^N$. As the temperature is lowered, the probability distribution becomes concentrated on the lowest-energy regions of this space. Finally, in the late stages, a large β forces the system to fix the few wrong details, and to find a solution. Of course, this image is somewhat simplistic, but we shall see later with some examples, in particular in Chapter 15, how the method works in practice, and what its limitations are.

4.7 Appendix: A physicist's approach to Sanov's theorem

Let us show how the formulae of Sanov's theorem can be obtained using the 'field-theoretic' type of approach used in statistical physics. The theorem is easy to prove by standard techniques. The aim of this section is not so much to give a proof, but rather to show, with a simple example, a type of calculation that is very common in physics, and which can be powerful. We shall not aim at a rigorous derivation.

The probability that the type of the sequence $x_1, \cdots, x_N$ is equal to $q(x)$ can be written as

$$\mathbb{P}[q(x)] = \mathbb{E}\left\{ \prod_{x \in \mathcal{X}} \mathbb{I}\left(q(x) = \frac{1}{N}\sum_{i=1}^{N}\delta_{x,x_i}\right)\right\}$$

$$= \sum_{x_1 \cdots x_N} p(x_1) \cdots p(x_N)\, \mathbb{I}\left(q(x) = \frac{1}{N}\sum_{i=1}^{N}\delta_{x,x_i}\right), \qquad (4.65)$$

where we have introduced the **Kronecker symbol** $\delta_{x,y}$, equal to 1 if $x = y$ and 0 otherwise. A typical approach in field theory is to introduce some auxiliary variables in order to enforce the constraint that $q(x) = (1/N)\sum_{i=1}^{N}\delta_{x,x_i}$. For each $x \in \mathcal{X}$, one introduces a variable $\lambda(x)$, and uses an 'integral representation' of the constraint in the form

$$\mathbb{I}\left(q(x) = \frac{1}{N}\sum_{i=1}^{N}\delta_{x,x_i}\right) = \int_0^{2\pi}\frac{\mathrm{d}\lambda(x)}{2\pi}\ \exp\left[i\lambda(x)\left(Nq(x) - \sum_{i=1}^{N}\delta_{x,x_i}\right)\right]. \qquad (4.66)$$

Dropping q-independent factors, we get

$$\mathbb{P}[q(x)] = C \int \prod_{x \in \mathcal{X}} \mathrm{d}\lambda(x)\ \exp\{NS[\lambda]\}, \qquad (4.67)$$

where C is a normalization constant, and the **action** S is given by

$$S[\lambda] = i\sum_x \lambda(x)q(x) + \log\left[\sum_x p(x)\,\mathrm{e}^{-i\lambda(x)}\right]. \qquad (4.68)$$

In the large-N limit, the integral in eqn (4.67) can be evaluated by the saddle point method. The saddle point $\lambda(x) = \lambda^*(x)$ is found by solving the stationarity equations $\partial S/\partial\lambda(x) = 0$ for any $x \in \mathcal{X}$. One obtains a family of solutions $-i\lambda(x) = C + \log(q(x)/p(x))$, with C arbitrary. The freedom in the choice of C comes from the fact that $\sum_x (\sum_i \delta_{x,x_i}) = N$ for any configuration $x_1 \ldots x_N$, and therefore one of the constraints is in fact redundant. This choice can be made arbitrarily: regardless of the choice, the action at the saddle point is

$$S[\lambda^*] = S_0 - \sum_x q(x)\log\frac{q(x)}{p(x)}, \qquad (4.69)$$

where S_0 is a q-independent constant. One thus gets $\mathbb{P}[q(x)] \doteq \exp[-ND(q\|p)]$.

The reader who has never encountered this type of reasoning before may wonder why use such an indirect approach. It turns out that it is a very common formalism in statistical physics, where similar methods are also applied, under the name 'field theory', to continuous spaces $\mathcal{X}$ (some implicit discretization is then usually assumed at intermediate steps, and the correct definition of a continuum limit is often not obvious). In particular, the reader interested in the statistical-physics approach to optimization problems or information theory will often find this type of calculation in research papers. One of the advantages of this approach is that it provides a formal solution to a large variety of problems. The quantity to be computed is expressed in an integral form as in eqn (4.67). In problems that have a 'mean-field' structure, the dimension of the space over which the integration is performed does not depend upon N. Therefore its leading exponential behaviour at large N can be obtained by saddle point methods. The reader who wants to get some practice with this approach is invited to 'derive' the various theorems and corollaries of this chapter in this way.

Notes

The theory of large deviations is set out in the book by Dembo and Zeitouni (1998), and its connections to statistical physics can be found in Ellis (1985).

A discussion of statistical physics from a computational perspective, with a nice introduction to Monte Carlo methods, is provided by Krauth (2006). Markov chains on discrete state spaces are treated by Norris (1997), and a more formal presentation of the Monte Carlo method is given by Sokal (1996).

Simulated annealing was introduced by Kirkpatrick *et al.* (1983). One of its main qualities is that it is very versatile. In some sense it is a completely 'universal' optimization algorithm: it can be defined without reference to any particular problem. Fo this reason, however, it often overlooks important structures that may help in solving a specific problem.

Part II

Independence

5
The random energy model

The random energy model (REM) is probably the simplest statistical-physics model of a disordered system which exhibits a phase transition. It is not supposed to give a realistic description of any physical system, but it provides an example with which various concepts and methods can be studied in full detail. Moreover, owing to its simplicity, the same mathematical structure appears in a large number of contexts. This is witnessed by the examples from information theory and combinatorial optimization presented in the next two chapters. The model is defined in Section 5.1 and its thermodynamic properties are studied in Section 5.2. The simple approach developed in these sections turns out to be useful in a large variety of problems. A more detailed, and more involved, study of the low-temperature phase is developed in Section 5.3. Section 5.4 provides an introduction to the 'annealed approximation,' which will be useful in more complicated models. Finally, in Section 5.5 we consider a variation of the REM that provides a 'cartoon' for the structure of the set of solutions of random constraint satisfaction problems.

5.1   Definition of the model

A statistical-mechanics model is defined by a set of configurations and an energy function defined on this space. In the REM, there are $M = 2^N$ configurations (as in a system of N Ising spins), denoted by indices $i, j, \ldots \in \{1, \ldots, 2^N\}$. The REM is a **disordered model**: the energy is not a deterministic function, but rather a stochastic process. A particular realization of such a process is usually called a **sample** (or **instance**). In the REM, one makes the simplest possible choice for this process: the energies $\{E_i\}$ are i.i.d. random variables (the energy of a configuration is also called an **energy level**). For definiteness, we shall keep here to the case where the energies have Gaussian distribution with zero mean and variance $N/2$, but other distributions could be studied as well. The scaling with N of the distribution should always be chosen in such a way that thermodynamic potentials are extensive. The pdf for the energy E_i of the state i is given by

$$P(E) = \frac{1}{\sqrt{\pi N}} \, e^{-E^2/N} \, . \tag{5.1}$$

Given an instance of the REM, defined by the 2^N real numbers $\{E_1, E_2, \ldots, E_{2^N}\}$, one assigns to each configuration j a Boltzmann probability $\mu_\beta(j)$ in the usual way:

$$\mu_\beta(j) = \frac{1}{Z} \exp\left(-\beta E_j\right) \, . \tag{5.2}$$

where $\beta = 1/T$ is the inverse of the temperature, and the normalization factor Z, the partition function, is given by

$$Z = \sum_{j=1}^{2^N} \exp\left(-\beta E_j\right). \tag{5.3}$$

Notice that Z depends upon the temperature β, the 'sample size' N, and the particular realization of the energy levels $E_1, \ldots, E_M$. We shall write $Z = Z_N(\beta)$ to emphasize the dependence of the partition function upon N and β.

It is important not to be confused by the existence of two levels of probability in the REM, as in all disordered systems. We are interested in the properties of a probability distribution, the Boltzmann distribution (5.2), which is itself a random object because the energy levels are random variables.

Physically, a particular realization of the energy function corresponds to a given sample of some substance whose microscopic features cannot be controlled experimentally. This is what happens, for instance, in a metallic alloy: only the proportions of the various components can be controlled. The precise positions of the atoms of each species can be described by random variables. The expectation value with respect to the sample realization will be denoted in the following by $\mathbb{E}(\cdot)$. For a given sample, Boltzmann's law (5.2) gives the probability of occupying the various possible configurations, according to their energies. The average with respect to the Boltzmann distribution will be denoted by $\langle \cdot \rangle$. In experiments one deals with a single (or a few) sample(s) of a given disordered material. One might therefore be interested in computing the various thermodynamic potentials (the free energy F_N, internal energy U_N, or entropy S_N) for *one given* sample. This is an extremely difficult task. However, in most cases, as $N \to \infty$, the probability distributions of intensive thermodynamic potentials concentrate around their expected values. Formally, for any tolerance $\theta > 0$,

$$\lim_{N \to \infty} \mathbb{P}\left[\left| \frac{X_N}{N} - \mathbb{E}\left(\frac{X_N}{N} \right) \right| \geq \theta \right] = 0, \tag{5.4}$$

where X is a thermodynamic potential ($X = F, S, U, \ldots$). In statistical physics, the quantity X is then said to be **self-averaging** (in probability theory, one says that it **concentrates**). This essential property can be summarized in plain language by saying that almost all large samples 'behave' in the same way. This is the reason why different samples of alloys with the same chemical composition have the same thermodynamic properties. Often the convergence is exponentially fast in N (this happens, for instance, in the REM): this means that the expected value $\mathbb{E}\, X_N$ provides a good description of the system even at moderate sizes.

5.2   Thermodynamics of the REM

In this section we compute the thermodynamic potentials of the REM in the thermodynamic limit $N \to \infty$. Our strategy consists in estimating the microcanonical entropy density, which was introduced in Section 2.4. This knowledge is then used for computing the partition function Z to leading exponential order for large N.

5.2.1 Direct evaluation of the entropy

Let us consider an interval of energy $\mathcal{I} = [N\varepsilon, N(\varepsilon + \delta)]$, and denote by $\mathcal{N}(\varepsilon, \varepsilon + \delta)$ the number of configurations i such that $E_i \in \mathcal{I}$. Each energy level E_i belongs to $\mathcal{I}$ independently, with probability

$$P_{\mathcal{I}} = \sqrt{\frac{N}{\pi}} \int_{\varepsilon}^{\varepsilon+\delta} e^{-Nx^2} \, dx \, . \tag{5.5}$$

Therefore $\mathcal{N}(\varepsilon, \varepsilon + \delta)$ is a binomial random variable, and its expectation and variance are given by

$$\mathbb{E}\,\mathcal{N}(\varepsilon, \varepsilon + \delta) = 2^N P_{\mathcal{I}}, \quad \mathrm{Var}\,\mathcal{N}(\varepsilon, \varepsilon + \delta) = 2^N P_{\mathcal{I}}[1 - P_{\mathcal{I}}] \, . \tag{5.6}$$

Because of the appropriate scaling of the interval $\mathcal{I}$ with N, the probability $P_{\mathcal{I}}$ depends exponentially upon N. To exponential accuracy, we thus have

$$\mathbb{E}\,\mathcal{N}(\varepsilon, \varepsilon + \delta) \doteq \exp\left\{ N \max_{x \in [\varepsilon, \varepsilon+\delta]} s_{\mathrm{a}}(x) \right\} \, , \tag{5.7}$$

$$\frac{\mathrm{Var}\mathcal{N}(\varepsilon, \varepsilon + \delta)}{[\mathbb{E}\,\mathcal{N}(\varepsilon, \varepsilon + \delta)]^2} \doteq \exp\left\{ -N \max_{x \in [\varepsilon, \varepsilon+\delta]} s_{\mathrm{a}}(x) \right\} \, , \tag{5.8}$$

where $s_{\mathrm{a}}(x) \equiv \log 2 - x^2$. Note that $s_{\mathrm{a}}(x) \geq 0$ if and only if $x \in [-\varepsilon_*, \varepsilon_*]$, where $\varepsilon_* = \sqrt{\log 2}$.

The intuitive content of these equalities is the following. When ε is outside the interval $[-\varepsilon_*, \varepsilon_*]$, the average density of energy levels is exponentially small in N: for a typical sample, there is no configuration at energy $E_i \approx N\varepsilon$. In contrast, when $\varepsilon \in]-\varepsilon_*, \varepsilon_*[$, there is an exponentially large density of levels, and the fluctuations of this density are very small. This result is illustrated by a small numerical experiment in Fig. 5.1. We now give a more formal version of this statement.

Proposition 5.1 *Define the entropy function*

$$s(\varepsilon) = \begin{cases} s_{\mathrm{a}}(\varepsilon) = \log 2 - \varepsilon^2 & \text{if } |\varepsilon| \leq \varepsilon_*, \\ -\infty & \text{if } |\varepsilon| > \varepsilon_*. \end{cases} \tag{5.9}$$

Then, for any pair ε and δ, with probability one,

$$\lim_{N \to \infty} \frac{1}{N} \log \mathcal{N}(\varepsilon, \varepsilon + \delta) = \sup_{x \in [\varepsilon, \varepsilon+\delta]} s(x) \, . \tag{5.10}$$

Proof The proof makes simple use of the two moments of the number of energy levels in $\mathcal{I}$, given in eqns (5.7), and (5.8).

Assume first that the interval $[\varepsilon, \varepsilon+\delta]$ is disjoint from $[-\varepsilon_*, \varepsilon_*]$. Then $\mathbb{E}\,\mathcal{N}(\varepsilon, \varepsilon+\delta) \doteq e^{-AN}$, where $A = -\sup_{x \in [\varepsilon, \varepsilon+\delta]} s_{\mathrm{a}}(x) > 0$. As $\mathcal{N}(\varepsilon, \varepsilon + \delta)$ is an integer, we have the simple inequality

$$\mathbb{P}[\mathcal{N}(\varepsilon, \varepsilon + \delta) > 0] \leq \mathbb{E}\,\mathcal{N}(\varepsilon, \varepsilon + \delta) \doteq e^{-AN} \, . \tag{5.11}$$

In words, the probability of having an energy level in any fixed interval outside $[-\varepsilon_*, \varepsilon_*]$ is exponentially small in N. An inequality of the form (5.11) goes under the name of

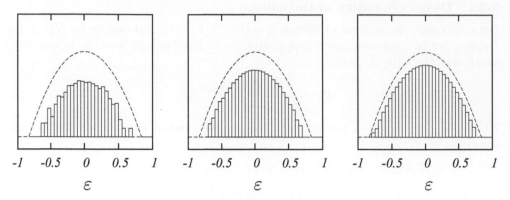

Fig. 5.1 Histogram of the energy levels for three samples of the random energy model with increasing size: from *left* to *right* $N = 10, 15$, and 20. Here we have plotted $N^{-1} \log \mathcal{N}(\varepsilon, \varepsilon + \delta)$ versus ε, with $\delta = 0.05$. The dashed curve gives the $N \to \infty$ analytical prediction (5.9).

Markov inequality, and the general strategy is sometimes called the **first-moment method**.

Assume now that the intersection between $[\varepsilon, \varepsilon + \delta]$ and $[-\varepsilon_*, \varepsilon_*]$ is an interval of non-zero length. In this case $\mathcal{N}(\varepsilon, \varepsilon + \delta)$ is tightly concentrated around its expectation $\mathbb{E}\,\mathcal{N}(\varepsilon, \varepsilon + \delta)$, as can be shown using the Chebyshev inequality. For any fixed $C > 0$, one has

$$\mathbb{P}\left\{\left|\frac{\mathcal{N}(\varepsilon, \varepsilon + \delta)}{\mathbb{E}\,\mathcal{N}(\varepsilon, \varepsilon + \delta)} - 1\right| > C\right\} \le \frac{\operatorname{Var}\mathcal{N}(\varepsilon, \varepsilon + \delta)^2}{C^2[\mathbb{E}\,\mathcal{N}(\varepsilon, \varepsilon + \delta)]^2} \doteq \mathrm{e}^{-BN}, \qquad (5.12)$$

where $B = \sup_{x \in [\varepsilon, \varepsilon + \delta]} s_{\mathrm{a}}(x) > 0$.

Finally, the statement (5.10) follows from the previous estimates by a straightfoward application of the Borel–Cantelli lemma. □

Exercise 5.1 (*Large deviations*) Let $\mathcal{N}_{\mathrm{out}}(\delta)$ be the total number of configurations j such that $|E_j| > N(\varepsilon_* + \delta)$, with $\delta > 0$. Use Markov inequality to show that the fraction of samples in which there exist such configurations is exponentially small.

Besides being an interesting mathematical statement, Proposition 5.1 provides a good quantitative estimate. As shown in Fig. 5.1, even at $N = 20$, the outcome of a numerical experiment is quite close to the asymptotic prediction. Note that, for energies in the interval $]-\varepsilon_*, \varepsilon_*[$, most of the discrepancy is due to the fact that we have dropped subexponential factors in $\mathbb{E}\,\mathcal{N}(\varepsilon, \varepsilon + \delta)$: this produces corrections of order $\Theta(\log N / N)$ to the asymptotic behaviour (5.10). The contribution due to fluctuations of $\mathcal{N}(\varepsilon, \varepsilon + \delta)$ around its average is exponentially small in N.

5.2.2 Thermodynamics and phase transition

From the previous result on the microcanonical entropy density, we now compute the partition function $Z_N(\beta) = \sum_{i=1}^{2^N} \exp(-\beta E_i)$. In particular, we are interested in intensive thermodynamic potentials such as the free-entropy density $\phi(\beta) = \lim_{N\to\infty} (1/N) \log Z_N(\beta)$. We start with a quick (and loose) argument, using the general approach outlined in Section 2.4. This amounts to discretizing the energy axis using some step size δ, and counting the energy levels in each interval using eqn (5.10). Taking at the end the limit $\delta \to 0$ (after the limit $N \to \infty$), one expects to get, to leading exponential order,

$$Z_N(\beta) \doteq \int_{-\varepsilon_*}^{\varepsilon_*} \exp\left[N\left(s_{\mathrm{a}}(\varepsilon) - \beta\varepsilon\right)\right] \mathrm{d}\epsilon . \tag{5.13}$$

A rigorous formulation of this result can be obtained by analogy[1] with the general equivalence relation stated in Proposition 2.6. We find the following free-entropy density:

$$\phi(\beta) = \max_{\varepsilon \in [-\varepsilon_*, \varepsilon_*]} [s_{\mathrm{a}}(\varepsilon) - \beta\varepsilon] . \tag{5.14}$$

Note that although every sample of the REM is a new statistical-physics system, with its own thermodynamic potentials, we have found that, with high probability, a random sample has a free-entropy (or free-energy) density arbitrarily close to (5.14). A little more work shows that the internal energy and entropy density concentrate as well. More precisely, for any fixed tolerance $\theta > 0$, we have $|(1/N) \log Z_N(\beta) - \phi(\beta)| < \theta$ with probability approaching one as $N \to \infty$.

Let us now discuss the physical content of the result (5.14). The optimization problem on the right-hand side can be solved through the geometrical construction illustrated in Fig. 5.2. One has to find a tangent to the curve $s_{\mathrm{a}}(\varepsilon) = \log 2 - \varepsilon^2$ with slope $\beta \geq 0$. We denote by $\varepsilon_{\mathrm{a}}(\beta) = -\beta/2$ the abscissa of the tangent point. If $\varepsilon_{\mathrm{a}}(\beta) \in [-\varepsilon_*, \varepsilon_*]$, then the 'max' in eqn (5.14) is realized in $\varepsilon_{\mathrm{a}}(\beta)$. In the other case $\varepsilon_{\mathrm{a}}(\beta) < -\varepsilon_*$ (because $\beta \geq 0$), and the 'max' is realized in $-\varepsilon_*$. Therefore we have the following result

Proposition 5.2 *The free-energy density of the REM, $f(\beta) = -\phi(\beta)/\beta$, is equal to*

$$f(\beta) = \begin{cases} -\frac{1}{4}\beta - \log 2/\beta & \text{if } \beta \leq \beta_{\mathrm{c}}, \\ -\sqrt{\log 2} & \text{if } \beta > \beta_{\mathrm{c}}, \end{cases} \quad \text{where} \quad \beta_{\mathrm{c}} = 2\sqrt{\log 2} . \tag{5.15}$$

This shows that a phase transition (i.e. a non-analyticity of the free-energy density) takes place at the inverse critical temperature $\beta_{\mathrm{c}} = 1/T_{\mathrm{c}} = 2\sqrt{\log 2}$. It is a second-order phase transition in the sense that the derivative of $f(\beta)$ is continuous, but because of the condensation phenomenon which we shall discuss in Section 5.3, it is often called a 'random first-order' transition. The other thermodynamic potentials are obtained through the usual formulae (see Section 2.2). They are plotted in Fig. 5.3.

---

[1]The task is, however, more difficult here, because the density of energy levels $\mathcal{N}(\varepsilon, \varepsilon + \delta)$ is a random function whose fluctuations must be controlled.

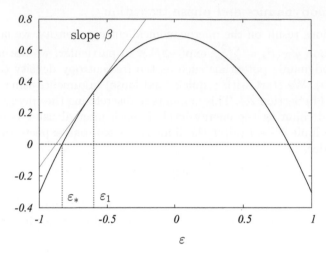

Fig. 5.2 The 'annealed' entropy density $s_a(\varepsilon) = \log 2 - \varepsilon^2$ of the REM as a function of the energy density ε. The canonical entropy density $s(\beta)$ is the ordinate of the point with slope $\mathrm{d}s_a/\mathrm{d}\varepsilon = \beta$ when this point lies within the interval $[-\varepsilon_*, \varepsilon_*]$ (this is for instance the case at $\varepsilon = \varepsilon_1$ in the plot), and $s(\beta) = 0$ otherwise. This gives rise to a phase transition at $\beta_c = 2\sqrt{\log 2}$.

The two temperature regimes or 'phases', $\beta < \beta_c$ and $\beta > \beta_c$ have distinct qualitative properties, which are most easily characterized through the thermodynamic potentials.

- In the high-temperature phase, i.e. for $\beta \leq \beta_c$, the energy and entropy densities are $u(\beta) = -\beta/2$ and $s(\beta) = \log 2 - \beta^2/4$, respectively. The Boltzmann measure is dominated by configurations with an energy $E_i \approx -N\beta/2$. There is an exponentially large number of configurations having such an energy density (the microcanonical entropy density $s(\varepsilon)$ is strictly positive at $\varepsilon = -\beta/2$), and the Boltzmann measure is roughly equidistributed among such configurations. In the infinite-temperature limit $\beta \to 0$ the Boltzmann measure becomes uniform, and one finds, as expected, $u(\beta) \to 0$ (because nearly all configurations have an energy E_i/N close to 0) and $s \to \log 2$.

- In the low-temperature phase, i.e. for $\beta > \beta_c$, the thermodynamic potentials are constant: $u(\beta) = -\varepsilon_*$ and $s(\beta) = 0$. The relevant configurations are the ones with the lowest energy density, namely those with $E_i/N \approx -\varepsilon_*$. The Boltzmann measure is dominated by a relatively small set of configurations, which is not exponentially large in N (the entropy density vanishes).

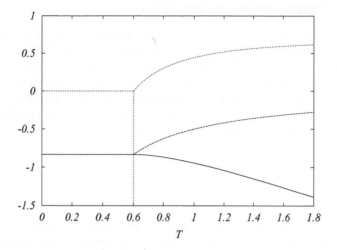

Fig. 5.3 Thermodynamics of the REM: the free-energy density (full line), the energy density (dashed line) and the entropy density (dotted line), are plotted versus the temperature $T = 1/\beta$. A phase transition takes place at $T_c = 1/(2\sqrt{\log 2}) \approx 0.6005612$.

Exercise 5.2 The REM was originally motivated by the attempt to provide a simple model of a spin glass. One can generalize it by introducing the effect of a magnetic field B. The 2^N configurations are divided into $N+1$ groups. Each group is labelled by its 'magnetization' $M \in \{-N, -N+2, \ldots, N-2, N\}$, and includes $\binom{N}{(N+M)/2}$ configurations. Their energies $\{E_j\}$ are independent Gaussian variables with variance $\sqrt{N/2}$ as in eqn (5.1) and mean $\mathbb{E}\, E_j = -MB$ which depends upon the group that j belongs to. Show that there exists a phase transition line $\beta_c(B)$ in the plane β, B such that

$$\frac{1}{N}\mathbb{E}\, M = \begin{cases} \tanh[\beta B] & \text{when} \quad \beta \le \beta_c(B), \\ \tanh[\beta_c(B)B] & \text{when} \quad \beta > \beta_c(B), \end{cases} \tag{5.16}$$

and plot the magnetic susceptibility $(\partial M/\partial B)_\beta = 0$ versus $T = 1/\beta$.

Exercise 5.3 Consider a generalization of the REM where the pdf of the energies, instead of being Gaussian, is given by $P(E) \propto \exp\left[-C|E|^\delta\right]$, where $\delta > 0$. Show that in order to have extensive thermodynamic potentials, one must scale C as $C = N^{1-\delta}\widehat{C}$ (i.e. the thermodynamic limit $N \to \infty$ should be taken at fixed $\widehat{C}$). Compute the critical temperature and the ground state energy density. What is the qualitative difference between the cases $\delta > 1$ and $\delta < 1$?

5.3   The condensation phenomenon

In the low-temperature phase, a smaller-than-exponential set of configurations dominates the Boltzmann measure: we say that a **condensation** of the measure onto these configurations takes place. This is a scenario which is typical of the appearance of a glass phase, and we shall encounter it in several other problems, including, for instance, the satisfiability and colouring problems. It usually leads to many difficulties in finding the relevant configurations. In order to characterize the condensation, one can compute a **participation ratio** $Y_N(\beta)$, defined in terms of the Boltzmann weights (5.2), by

$$Y_N(\beta) \equiv \sum_{j=1}^{2^N} \mu_\beta(j)^2 = \left[\sum_j e^{-2\beta E_j}\right] \left[\sum_j e^{-\beta E_j}\right]^{-2}. \tag{5.17}$$

One can think of $1/Y_N(\beta)$ as giving some estimate of the 'effective' number of configurations which contribute to the measure. If the measure were equidistributed on r levels, one would have $Y_N(\beta) = 1/r$.

The participation ratio can be expressed as $Y_N(\beta) = Z_N(2\beta)/Z_N(\beta)^2$, where $Z_N(\beta)$ is the partition function at inverse temperature β. The analysis in the previous section showed that $Z_N(\beta) \doteq \exp[N(\log 2 + \beta^2/4)]$ with very small fluctuations when $\beta < \beta_c$, while $Z_N(\beta) \doteq \exp[N\beta\sqrt{\log 2}]$ when $\beta > \beta_c$. This indicates that $Y_N(\beta)$ is exponentially small in N for almost all samples in the high-temperature phase, i.e. for $\beta < \beta_c$, in agreement with the fact that the measure is not condensed at high temperatures. In the low-temperature phase, in contrast, we shall see that $Y_N(\beta)$ is finite and fluctuates from sample to sample.

The computation of $\mathbb{E}Y$ (we drop the arguments N and β from now on) for the low-temperature phase is slightly involved. It requires us to control those energy levels E_i such that $E_i/N \approx -\varepsilon_*$. We give here only a sketch of the computation, and leave the details to the reader as an exercise. Using the integral representation $1/Z^2 = \int_0^\infty t \exp(-tZ)\, dt$, one gets (using $M = 2^N$)

$$\mathbb{E}Y = M\, \mathbb{E} \int_0^\infty t\, \exp\left[-2\beta E_1\right]\, \exp\left[-t \sum_{i=1}^M e^{-\beta E_i}\right]\, dt \tag{5.18}$$

$$= M \int_0^\infty t\, a(t)\, [1 - b(t)]^{M-1}\, dt\,, \tag{5.19}$$

where

$$a(t) \equiv \int \exp\left[-2\beta E - t e^{-\beta E}\right]\, dP(E)\,, \tag{5.20}$$

$$b(t) \equiv \int [1 - \exp(-t e^{-\beta E})]\, dP(E)\,, \tag{5.21}$$

and $P(E)$ is the Gaussian distribution (5.1). For large N, the leading contributions to $\mathbb{E}Y$ come from the regions where E is close to $-N\varepsilon_0$ and $\log t$ is close to $-N\beta\varepsilon_0$, where

$$\varepsilon_0 = \varepsilon_* - \frac{1}{2\varepsilon_*} \log \sqrt{\pi N} \tag{5.22}$$

is fixed by the condition $2^N P(-N\varepsilon_0) = 1$. This can be thought of as a refined estimate for the energy density of the lowest-energy configuration.

We thus change variables from E, t to u, θ using $E = -N\varepsilon_0 + u$ and $t = \theta \exp(-N\beta\varepsilon_0)$, and we study the regime where u and θ are finite as $N \to \infty$. In this regime, the function $P(E)$ can be replaced by $2^{-N} \mathrm{e}^{\beta_c u}$. One gets

$$a(t) \simeq \frac{1}{M} \mathrm{e}^{2N\beta\varepsilon_0} \int_{-\infty}^{+\infty} \mathrm{d}u \ \mathrm{e}^{\beta_c u - 2\beta u - z \mathrm{e}^{-\beta u}} = \frac{\mathrm{e}^{2N\beta\varepsilon_0}}{M\beta} \ z^{\beta_c/\beta - 2} \ \Gamma\left(2 - \frac{\beta_c}{\beta}\right), \tag{5.23}$$

$$b(t) \simeq \frac{1}{M} \int_{-\infty}^{+\infty} \mathrm{d}u \ \mathrm{e}^{\beta_c u} \left[1 - \exp(-z\mathrm{e}^{-\beta u})\right] = -\frac{1}{M\beta} \ z^{\beta_c/\beta} \ \Gamma\left(-\frac{\beta_c}{\beta}\right), \tag{5.24}$$

where $\Gamma(x)$ is Euler's gamma function. Note that the substitution $P(E) \simeq 2^{-N} \mathrm{e}^{\beta_c u}$ is harmless because the resulting integrals (5.23) and (5.24) converge at large u.

For large N, the expression $[1 - b(t)]^{M-1}$ in eqn (5.19) can be approximated by $\mathrm{e}^{-Mb(t)}$, and one finally obtains

$$\mathbb{E}\,Y = M \int_0^\infty \mathrm{d}t \ t \ a(t) \ \mathrm{e}^{-Mb(t)} \tag{5.25}$$

$$= \frac{1}{\beta}\Gamma\left(2 - \frac{\beta_c}{\beta}\right) \int_0^\infty \mathrm{d}z \ z^{\beta_c/\beta - 1} \ \exp\left[\frac{1}{\beta}\Gamma\left(-\frac{\beta_c}{\beta}\right) z^{\beta_c/\beta}\right] = 1 - \beta_c/\beta\,,$$

where we have used the approximate expressions (5.23) and (5.24), and the equalities are understood to hold up to corrections which vanish as $N \to \infty$.

We therefore obtain the following.

Proposition 5.3 *In the REM, when $N \to \infty$, the expectation value of the participation ratio is*

$$\mathbb{E}\,Y = \begin{cases} 0 & \text{when } T > T_c\,, \\ 1 - T/T_c & \text{when } T \le T_c\,. \end{cases} \tag{5.26}$$

This gives a quantitative measure of the degree of condensation of the Boltzmann measure: when T decreases, condensation starts at the phase transition temperature T_c. At lower temperatures, the participation ratio Y increases, meaning that the measure becomes concentrated onto fewer and fewer configurations, until at $T = 0$ only one configuration contributes and $Y = 1$.

With the participation ratio, we have a first qualitative and quantitative characterization of the low-temperature phase. Actually, the energies of the relevant configurations in this phase have many interesting probabilistic properties, to which we shall return in Chapter 8.

5.4 A comment on quenched and annealed averages

In the previous section, we found that the self-averaging property holds in the REM, and this result allowed us to discuss the thermodynamics of a generic sample.

Self-averaging of the thermodynamic potentials is a very frequent property, but in more complicated systems it is often difficult to compute their expectation. We discuss here an approximation which is frequently used in such cases, the 'annealed average'. When the free-energy density is self-averaging, the value of f_N is roughly the same for almost all samples and can be estimated by its expectation, called the **quenched average**, $f_{N,\mathrm{q}}$:

$$f_{N,\mathrm{q}} = \mathbb{E}\, f_N = -\frac{T}{N}\mathbb{E}\log Z_N \,. \tag{5.27}$$

This is the quantity that we computed in eqn (5.15). In general, it is hard to compute the expectation of the logarithm of the partition function. A much easier task is to compute the **annealed average**,

$$f_{N,\mathrm{a}} = -\frac{T}{N}\log(\mathbb{E}\, Z_N)\,. \tag{5.28}$$

Let us compute this quantity for the REM. Starting from the partition function (5.3), we find:

$$\mathbb{E}\, Z_N = \mathbb{E}\sum_{i=1}^{2^N} \mathrm{e}^{-\beta E_i} = 2^N \mathbb{E}\, \mathrm{e}^{-\beta E} = 2^N \mathrm{e}^{N\beta^2/4}\,, \tag{5.29}$$

yielding $f_{N,\mathrm{a}}(\beta) = -\beta/4 - \log 2/\beta$.

Let us compare this with the correct free-energy density found in eqn (5.15). Jensen's inequality (eqn (1.6)) shows that the annealed free-energy density $f_\mathrm{a}(\beta)$ is always smaller than or equal to the quenched value (remember that the logarithm is a concave function). In the REM, and a few other particularly simple problems, the annealed average gives the correct result in the high-temperature phase, i.e. for $T > T_\mathrm{c}$, but fails to identify the phase transition, and predicts wrongly a free-energy density in the low-temperature phase which is the analytic prolongation of the free-energy density at $T > T_\mathrm{c}$. In particular, it yields a *negative entropy density* $s_\mathrm{a}(\beta) = \log 2 - \beta^2/4$ for $T < T_\mathrm{c}$ (see Fig. 5.2).

A negative entropy is impossible in a system with a finite configuration space, as can be seen from the definition of entropy. It thus signals a failure, and the reason is easily understood. For a given sample with a free-energy density f, the partition function behaves as $Z_N = \exp(-\beta N f_N)$. Self-averaging means that f_N has small sample-to-sample fluctuations. However, these fluctuations, exist and are amplified in the partition function because of the factor N in the exponent. This implies that the annealed average of the partition function can be dominated by some very rare samples (those with an anomalously low value of f_N). Consider, for instance, the low-temperature limit. We already know that in almost all samples the configuration with the lowest energy density is found at $E_i \approx -N\varepsilon_*$. However, there exist exceptional samples where one configuration has a smaller energy $E_i = -N\varepsilon$ where $\varepsilon > \varepsilon_*$. These samples are exponentially rare (they occur with probability $\doteq 2^N \mathrm{e}^{-N\varepsilon^2}$), they are irrelevant as far as the quenched average is concerned, but they dominate the annealed average.

Let us add a short semantic note. The terms 'quenched' and 'annealed' originate from the thermal processing of materials, for instance in the context of the metallurgy of alloys: a quench corresponds to preparing a sample by bringing it suddenly from a high to low a temperature. During a quench, atoms do not have time to change position (apart from some small vibrations). A given sample is formed by atoms in some random positions. In contrast in an annealing process, one gradually cools down the alloy, and the various atoms will find favourable positions. In the REM, the energy levels E_i are quenched: for each given sample, they take certain fixed values (like the positions of atoms in a quenched alloy). In the annealed approximation, one treats the configurations i and the energies E_i on the same footing. One says that the variables E_i are thermalized (like the positions of atoms in an annealed alloy).

In general, the annealed average can be used to find a lower bound on the free energy for any system with a finite configuration space. Useful results can be obtained, for instance, using the following two simple relations, valid for all temperatures $T = 1/\beta$ and sizes N:

$$f_{N,\mathrm{q}}(T) \geq f_{N,\mathrm{a}}(T) \,, \qquad \frac{\mathrm{d}f_{N,\mathrm{q}}(T)}{\mathrm{d}T} \leq 0 \,. \tag{5.30}$$

The first relation one follows from Jensen's inequality as mentioned above, and the second can be obtained from the positivity of the canonical entropy (see eqn (2.22)), after averaging over the quenched disorder.

In particular, if one is interested in optimization problems (i.e. in the limit of vanishing temperature), the annealed average provides the following general bound.

Proposition 5.4 *The ground state energy density*

$$u_N(T = 0) \equiv \frac{1}{N}\mathbb{E}\left[\min_{\underline{x} \in \mathcal{X}^N} E(\underline{x})\right] \tag{5.31}$$

satisfies the bound $u_N(0) \geq \max_{T \in [0,\infty]} f_{N,\mathrm{a}}(T)$.

Proof Consider the annealed free-energy density $f_{N,\mathrm{a}}(T)$ as a function of the temperature $T = 1/\beta$. For any given sample, the free energy is a concave function of T because of the general relation (2.23). It is easy to show that the same property holds for the annealed average. Let T_* be the temperature at which $f_{N,\mathrm{a}}(T)$ achieves its maximum, and let $f_{N,\mathrm{a}}^*$ be its maximum value. If $T_* = 0$, then $u_N(0) = f_{N,\mathrm{q}}(0) \geq f_{N,\mathrm{a}}^*$. If $T_* > 0$, using the two inequalities (5.30), one gets

$$u_N(0) = f_{N,\mathrm{q}}(0) \geq f_{N,\mathrm{q}}(T_*) \geq f_{\mathrm{a}}(T_*) \,. \tag{5.32}$$

□

In the REM, this result immediately implies that $u(0) \geq \max_\beta[-\beta/4 - \log 2/\beta] = -\sqrt{\log 2}$, which is actually a tight bound.

5.5 The random subcube model

In the spirit of the REM, it is possible to construct a toy model for the set of solutions of a random constraint satisfaction problem. The **random subcube model** is defined by three parameters N, α, p. It has 2^N configurations: the vertices $\underline{x} = (x_1, \cdots, x_N)$ of the unit hypercube $\{0,1\}^N$. An instance of the model is defined by a subset $\mathcal{S}$ of the hypercube, the 'set of solutions'. Given an instance, the analogue of the Boltzmann measure is defined as the uniform distribution $\mu(\underline{x})$ over $\mathcal{S}$.

The solution space $\mathcal{S}$ is the union of $M = \lfloor 2^{(1-\alpha)N} \rfloor$ random subcubes which are i.i.d. subsets. Each subcube $\mathcal{C}_r$, $r \in \{1, \ldots, M\}$, is generated through the following procedure:

1. Generate the vector $t(r) = (t_1(r), t_2(r), \ldots, t_N(r))$, with independent entries

$$t_i(r) = \begin{cases} 0 \text{ with probability } (1-p)/2 \,, \\ 1 \text{ with probability } (1-p)/2 \,, \\ * \text{ with probability } p. \end{cases} \tag{5.33}$$

2. Given the values of $\{t_i(r)\}$, $\mathcal{C}_r$ is a subcube constructed as follows. For all i's such that $t_i(r)$ is 0 or 1, fix $x_i = t_i(r)$. Such variables are said to be 'frozen' for the subcube $\mathcal{C}_r$. For all other i's, x_i can be either 0 or 1. These variables are said to be 'free'.

A configuration $\underline{x}$ may belong to several subcubes. Whenever it belongs to at least one subcube, it is in $\mathcal{S}$.

To summarize, $\alpha < 1$ fixes the number of subcubes, and $p \in [0,1]$ fixes their size. The model can be studied using exactly the same methods as for the REM. Here we shall just describe the main results, omitting all proofs. It is a good exercise to work out the details and prove the various assertions.

Let us denote by σ_r the entropy density of the r-th cluster in bits: $\sigma_r = (1/N) \log_2 |\mathcal{C}_r|$. It is clear that σ_r coincides with the fraction of $*$'s in the vector $t(r)$. In the large-N limit, the number of clusters $\mathcal{N}(\sigma)$ with an entropy density σ obeys a large-deviation principle:

$$\mathcal{N}(\sigma) \doteq 2^{N\Sigma(\sigma)} \,. \tag{5.34}$$

The function $\Sigma(\sigma)$ is given as follows. Let $D(\sigma \| p)$ denote the Kullback–Leibler divergence between a Bernoulli distribution with mean σ and a Bernoulli distribution with mean p. As we saw in Section 1.2, this is given by

$$D(\sigma \| p) = \sigma \log_2 \frac{\sigma}{p} + (1-\sigma) \log_2 \frac{1-\sigma}{1-p} \,. \tag{5.35}$$

We define $[\sigma_1(p, \alpha), \sigma_2(p, \alpha)]$ as the interval in which $D(\sigma \| p) \leq 1 - \alpha$. Then

$$\Sigma(\sigma) = \begin{cases} 1 - \alpha - D(\sigma \| p) & \text{when } \sigma \in [\sigma_1(p, \alpha), \sigma_2(p, \alpha)] \,, \\ -\infty & \text{otherwise.} \end{cases} \tag{5.36}$$

We can now derive the phase diagram (see Fig. 5.4). We denote by s the total entropy density of the solution space, i.e. $s = (1/N) \log_2 |\mathcal{S}|$. Consider a configuration $\underline{x}$. The expected number of clusters to which it belongs is $2^{N(1-\alpha)}((1+p)/2)^N$. Therefore, if $\alpha < \alpha_{\rm d} \equiv \log_2(1+p)$, the solution space contains all but a vanishing fraction of the configurations, with high probability: $s = \log 2$. On the other hand, if $\alpha > \alpha_{\rm d}$, the probability that a configuration in $\mathcal{S}$ belongs to at least two distinct clusters is very small. In this regime, $s = \max_\sigma (\Sigma(\sigma) + \sigma)$. As in the REM, there are two cases. (*i*) The maximum of $\Sigma(\sigma) + \sigma$ is achieved for $\sigma = \sigma_*(p, \alpha) \in\,]s_1(p, \alpha), s_2(p, \alpha)[$. This happens

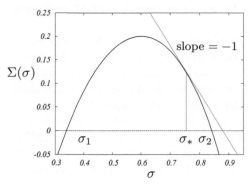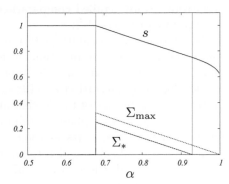

Fig. 5.4 *Left*: the function $\Sigma(\sigma)$ of the random subcube model, for $p = 0.6$ and $\alpha = 0.8 \in]\alpha_{\mathrm{d}}, \alpha_{\mathrm{c}}[$. The maximum of the curve gives the total number of clusters $\Sigma_{\max}$. A 'typical' random solution $\underline{x} \in \mathcal{S}$ belongs to one of the $e^{N\Sigma(\sigma_*)}$ clusters with entropy density σ_*, with $\Sigma'(\sigma_*) = -1$. As α increases above α_{c}, random solutions condense into a few clusters with entropy density s_2. *Right*: thermodynamic quantities plotted versus α for $p = 0.6$: the total entropy s, the total number of clusters $\Sigma_{\max}$, and the number of clusters where typical configurations are found, Σ_*.

when $\alpha < \alpha_{\mathrm{c}}(p) \equiv \log_2(1+p) + (1-p)/(1+p)$. In this case $s = (1-\alpha)\log 2 + \log(1+p)$.
(*ii*) The maximum of $\Sigma(\sigma) + \sigma$ is obtained for $\sigma = \sigma_2(p, \alpha)$. In this case $s = \sigma_2(p, \alpha)$.
Altogether, we have found three phases:

- For $\alpha < \alpha_{\mathrm{d}}$, subcubes overlap and one big cluster contains most of the configurations: $s_{\mathrm{tot}} = 1$.
- For $\alpha_{\mathrm{d}} < \alpha < \alpha_{\mathrm{c}}$, the solution space $\mathcal{S}$ is split into $2^{N(1-\alpha)}$ non-overlapping clusters of configurations (every subcube is a cluster of solutions, separated from the others). Most configurations of $\mathcal{S}$ are in the $e^{N\Sigma(s_*)}$ clusters which have an entropy density close to $s_*(p, \alpha)$. Note that the majority of clusters have entropy density $1 - p < s_*$. There is a tension between the number of clusters and their size (i.e. their internal entropy). The result is that the less numerous, but larger, clusters with entropy density s_* dominate the uniform measure.
- For $\alpha > \alpha_{\mathrm{c}}$, the solution space $\mathcal{S}$ is still partitioned into $2^{N(1-\alpha)}$ non-overlapping clusters of configurations. However, most solutions are in clusters with entropy density close to $s_2(p, \alpha)$. The number of such clusters is not exponentially large. In fact, the uniform measure over $\mathcal{S}$ shows a condensation phenomenon, which is completely analogous to that in the REM. One can define a participation ratio $Y = \sum_r \mu(r)^2$, where $\mu(r)$ is the probability that a configuration of $\mathcal{S}$ chosen uniformly at random belongs to cluster r; $\mu(r) = e^{N\sigma_r} / \sum_{r'} e^{N\sigma_{r'}}$. This participation ratio is finite, and equal to $1 - m$, where m is the slope $m = -(\mathrm{d}\Sigma/\mathrm{d}\sigma)$, evaluated at $s_2(p, \alpha)$.

Notes

The REM was invented by Derrida (1980), as an extreme case of family of spin glass models. Here we have followed his original analysis, which makes use of the microcanonical entropy. More detailed computations can be found in Derrida (1981), including the solution to Exercise 5.2.

The condensation formula (5.3) appeared first in Gross and Mézard (1984) as an application of replica computations which we shall discuss in Chapter 8. The direct estimate of the participation ratio presented here and the analysis of its fluctuations were developed by Mézard *et al.* (1985a) and Derrida and Toulouse (1985). We shall return to the properties of the fascinating condensed phase in Chapter 8.

Exercise 5.3 shows a phase transition which goes from second-order when $\delta > 1$, to first-order when $\delta < 1$. Its solution can be found in Bouchaud and Mézard (1997).

The random subcube model was introduced by Achlioptas (2007) and studied in detail by Mora and Zdeborová (2007). We refer to that paper for the derivations omitted from Section 5.5.

As a final remark, note that in most of the physics literature, authors do not explicitly write down all of the mathematical steps leading, for instance, to eqn (5.13), preferring a more synthetic presentation which focuses on the basic ideas. In more complicated problems, it may be very difficult to fill in the corresponding mathematical gaps. In many of the models studied in this book, this is still beyond the range of rigorous techniques. The recent book by Talagrand (2003) adopts a fully rigorous point of view, and it starts with a presentation of the REM which nicely complements the one given here and in Chapter 8.

6
The random code ensemble

As already explained in Section 1.6, one of the basic problems of information theory consists in communicating reliably over a noisy communication channel. Error-correcting codes achieve this task by systematically introducing some form of redundancy into the message to be transmitted. One of the major breakthroughs accomplished by Claude Shannon was to understand the importance of code *ensembles*. He realized that it is much easier to construct ensembles of codes which have good properties with high probability than to exhibit explicit examples that achieve the same performance. In a nutshell, 'stochastic' design is much easier than 'deterministic' design.

At the same time, he defined and analysed the simplest such ensembles, which has been named thereafter the random code ensemble (or, sometimes, the Shannon ensemble). Despite its great simplicity, the random code ensemble (RCE) has very interesting properties and, in particular, it achieves the optimal error-correcting performance. It provides therefore a proof of the 'direct' part of the channel coding theorem: it is possible to communicate with vanishing error probability as long as the communication rate is smaller than the channel capacity. Furthermore, it is the prototype of a code based on a random construction. In the following chapters we shall explore several examples of this approach, and the random code ensemble will serve as a reference.

We introduce the idea of code ensembles and define the RCE in Section 6.1. Some properties of this ensemble are described in Section 6.2, and its performance over the BSC is estimated in Section 6.3. We generalize these results to a general discrete memoryless channel in Section 6.4. Finally, in Section 6.5, we show that the RCE is optimal by a simple sphere-packing argument.

6.1 Code ensembles

An error-correcting code is defined as a pair of an encoding and a decoding map. The encoding map is applied to the information sequence to get an encoded message which is transmitted through the channel. The decoding map is applied to the (noisy) channel output. For the sake of simplicity, we shall assume throughout this chapter that the message to be encoded is given as a sequence of M bits and that encoding produces a redundant sequence of $N > M$ bits. The possible codewords (i.e. the 2^M points in the space $\{0, 1\}^N$ which are all possible outputs of the encoding map) form the codebook $\mathfrak{C}_N$. We denote by $\mathcal{Y}$ the output alphabet of the communication channel. We use the notation

$$\underline{x} : \{0,1\}^M \to \{0,1\}^N \quad \text{encoding map}, \tag{6.1}$$
$$\underline{x}^{\mathrm{d}} : \quad \mathcal{Y}^N \to \{0,1\}^N \quad \text{decoding map}. \tag{6.2}$$

Notice that the definition of the decoding map here is slightly different from the one given in Section 1.6. Here we consider only the difficult part of the decoding procedure, namely how to reconstruct from the received message the codeword which was sent. To complete the decoding as defined in Section 1.6, one needs to get back the original message by knowing the codeword, but this is assumed to be an easier task (encoding is assumed to be injective).

The customary recipe for designing a **code ensemble** is the following. (*i*) Define a subset of the space of encoding maps (6.1). (*ii*) Endow this set with a probability distribution. (*iii*) Finally, for each encoding map in the ensemble, define the associated decoding map. In practice, this last step is accomplished by declaring that one of a few general 'decoding strategies' is adopted. We shall introduce two such strategies below.

Our first example is the **random code ensemble**. Note that there exist 2^{N2^M} possible encoding maps of the type (6.1): one must specify N bits for each of the 2^M codewords. In the RCE, any of these encoding maps is picked with uniform probability. The code is therefore constructed as follows. For each of the possible information messages $m \in \{0,1\}^M$, we obtain the corresponding codeword $\underline{x}^{(m)} = (x_1^{(m)}, x_2^{(m)}, \ldots, x_N^{(m)})$ by tossing an unbiased coin N times: the i-th outcome is assigned to the i-th coordinate $x_i^{(m)}$.

Exercise 6.1 Note that, with this definition, the code is not necessarily injective: there could be two information messages $m_1 \neq m_2$ with the same codeword: $\underline{x}^{(m_1)} = \underline{x}^{(m_2)}$. This is an annoying property for an error-correcting code: any time that we send either of the messages m_1 or m_2, the receiver will not be able to distinguish between them, even in the absence of noise. Happily enough, though, these unfortunate coincidences occur rarely, i.e. their number is much smaller than the total number of codewords 2^M. What is the expected number of pairs m_1, m_2 such that $\underline{x}^{(m_1)} = \underline{x}^{(m_2)}$? What is the probability that all the codewords are distinct?

Let us now turn to the definition of the decoding map. We shall introduce two of the most important decoding schemes here: word MAP (MAP stands for 'maximum a posteriori probability' here) and symbol MAP decoding. Both schemes can be applied to any code. In order to define them, it is useful to introduce the probability distribution $P(\underline{x}|\underline{y})$ for $\underline{x}$ to be the channel input conditional on the received message $\underline{y}$. For a memoryless channel with a transition probability $Q(y|x)$, this probability has an explicit expression as a consequence of the Bayes rule:

$$\mathbb{P}(\underline{x}|\underline{y}) = \frac{1}{Z(\underline{y})} \prod_{i=1}^{N} Q(y_i|x_i) \, \mathbb{P}(\underline{x}). \tag{6.3}$$

Here $Z(\underline{y})$ is fixed by the normalization condition $\sum_{\underline{x}} \mathbb{P}(\underline{x}|\underline{y}) = 1$, and $\mathbb{P}(\underline{x})$ is the a priori probability for $\underline{x}$ to be the transmitted message. Throughout this book, we shall assume that the transmitter chooses the codeword to be transmitted with uniform probability. Therefore $\mathbb{P}(\underline{x}) = 1/2^M$ if $\underline{x} \in \mathfrak{C}_N$ and $\mathbb{P}(\underline{x}) = 0$ otherwise. This can also be written formally as

$$\mathbb{P}(\underline{x}) = \frac{1}{|\mathfrak{C}_N|} \, \mathbb{I}(\underline{x} \in \mathfrak{C}_N) \,. \tag{6.4}$$

Indeed, it is not hard to realize that $Z(\underline{y})$ is the probability of observing $\underline{y}$ when a random codeword is transmitted. Hereafter, we shall use $\mu_y(\cdot)$ to denote the a posteriori distribution (6.3) and $\mu_0(\cdot)$ for the a priori one (6.4). We can thus rewrite eqn (6.3) as

$$\mu_y(\underline{x}) = \frac{1}{Z(\underline{y})} \prod_{i=1}^N Q(y_i|x_i) \, \mu_0(\underline{x}) \,. \tag{6.5}$$

It is also useful to define the marginal distribution $\mu_y^{(i)}(x_i) = \mathbb{P}(x_i|\underline{y})$ of the i-th bit of the transmitted message conditional on the output message. This is obtained from the distribution (6.5) by marginalizing over all the bits x_j with $j \neq i$:

$$\mu_y^{(i)}(x_i) = \sum_{\underline{x}_{\backslash i}} \mu_y(\underline{x}) \,, \tag{6.6}$$

where we have introduced the shorthand $\underline{x}_{\backslash i} \equiv \{x_j : j \neq i\}$. **Word MAP** decoding outputs the most probable transmitted codeword, i.e. it maximizes the distribution (6.5):

$$\underline{x}^{\mathrm{w}}(\underline{y}) = \arg \max_{\underline{x}} \mu_y(\underline{x}) \,. \tag{6.7}$$

We do not specify what to do in the case of a tie (i.e. if the maximum is degenerate), since this is irrelevant for all the coding problems that we shall consider. Scrupulous readers can chose their own convention in such cases.

A strongly related decoding strategy is **maximum-likelihood** decoding. In this case, one maximizes $Q(\underline{y}|\underline{x})$ over $\underline{x} \in \mathfrak{C}_N$. This coincides with word MAP decoding whenever the a priori distribution over the transmitted codeword $\mathbb{P}(\underline{x}) = \mu_0(\underline{x})$ is taken to be uniform as in eqn (6.4). From now on, we shall therefore blur the distinction between these two strategies.

Symbol (or **bit**) **MAP** decoding outputs the sequence of the most probable transmitted bits, i.e. it maximizes the marginal distribution (6.6):

$$\underline{x}^{\mathrm{b}}(\underline{y}) = \left(\arg \max_{x_1} \mu_y^{(1)}(x_1), \ldots, \arg \max_{x_N} \mu_y^{(N)}(x_N) \right) \,. \tag{6.8}$$

Exercise 6.2 Consider a code of block length $N = 3$ and codebook size $|\mathfrak{C}| = 4$, with codewords $\underline{x}^{(1)} = 001$, $\underline{x}^{(1)} = 101$, $\underline{x}^{(1)} = 110$, $\underline{x}^{(1)} = 111$. What is the code rate? This code is used to communicate over a binary symmetric channel with flip probability $p < 0.5$. Suppose that the channel output is $\underline{y} = 000$. Show that the word MAP decoding outputs the codeword 001. Now apply symbol MAP decoding to decode the first bit x_1. Show that the result coincides with that of word MAP decoding only when p is small enough.

It is important to note that each of the above decoding schemes is optimal with respect a different criterion. Word MAP decoding minimizes the average block error probability P_B defined in Section 1.6.2. This is the probability, with respect to the channel distribution $Q(\underline{y}|\underline{x})$, that the decoded codeword $\underline{x}^d(\underline{y})$ is different from the transmitted one, averaged over the transmitted codeword:

$$P_B \equiv \frac{1}{|\mathcal{C}|} \sum_{\underline{x} \in \mathcal{C}} \mathbb{P}[\underline{x}^d(\underline{y}) \neq \underline{x}]. \qquad (6.9)$$

Bit MAP decoding minimizes the **bit error probability**, or **bit error rate** (BER), P_b. This is the fraction of incorrect bits, averaged over the transmitted code-word:

$$P_b \equiv \frac{1}{|\mathcal{C}|} \sum_{\underline{x} \in \mathcal{C}} \frac{1}{N} \sum_{i=1}^{N} \mathbb{P}[x_i^d(\underline{y}) \neq x_i]. \qquad (6.10)$$

Exercise 6.3 Show that word MAP and symbol MAP decoding are indeed optimal with respect to the above criteria.

6.2   The geometry of the random code ensemble

We begin our study of the RCE by first working out some of its geometrical properties. A code from this ensemble is defined by a codebook, i.e. a set $\mathcal{C}_N$ of 2^M points (the codewords) in the **Hamming space** $\{0,1\}^N$. Each of these points is drawn with uniform probability from the Hamming space. The simplest question one may ask about $\mathcal{C}_N$ is the following. Suppose you sit on one of the codewords and look around you. How many other codewords are there at a given distance? We shall use here the **Hamming distance**: the distance between two points $\underline{x}, \underline{y} \in \{0,1\}^N$ is the number of coordinates in which they differ.

This question is addressed through the **distance enumerator** $\mathcal{N}_{\underline{x}^{(0)}}(d)$ with respect to a codeword $\underline{x}^{(0)} \in \mathcal{C}_N$. This is defined as the number of codewords in $\underline{x} \in \mathcal{C}_N$ whose Hamming distance from $\underline{x}^{(0)}$ is equal to d: $d(\underline{x}, \underline{x}^{(0)}) = d$.

We shall now compute the typical properties of the distance enumerator for a random code. The simplest quantity to look at is the average distance enumerator $\mathbb{E}\mathcal{N}_{\underline{x}^{(0)}}(d)$, the average being taken over the code ensemble. In general one should also specify *which one* of the codewords is $\underline{x}^{(0)}$. Since, in the, RCE all codewords are drawn independently, each one with uniform probability over the Hamming space, such a specification is irrelevant and we can in fact fix $\underline{x}^{(0)}$ to be the **all-zeros codeword**, $\underline{x}^{(0)} = 000 \cdots 00$. Therefore we are asking the following question: take $2^M - 1$ points at random with uniform probability in the Hamming space $\{0,1\}^N$; what is the average number of points at distance d form the corner $00 \cdots 0$? This is simply the number of points $(2^M - 1)$, times the fraction of the Hamming space 'volume' at a distance d from $000 \cdots 0$ $(2^{-N}\binom{N}{d})$:

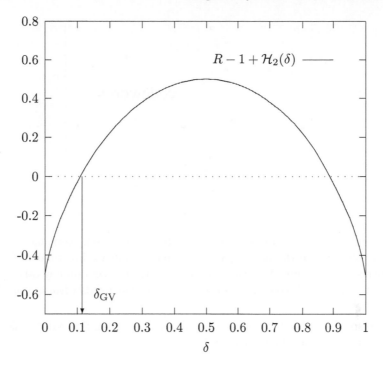

Fig. 6.1 Growth rate of the distance enumerator for the random code ensemble, with rate $R = 1/2$, as a function of the Hamming distance $d = N\delta$.

$$\mathbb{E}\,\mathcal{N}_{\underline{x}^{(0)}}(d) = (2^M - 1)\, 2^{-N} \binom{N}{d} \;\doteq\; 2^{N[R-1+\mathcal{H}_2(\delta)]} \,. \tag{6.11}$$

In the second expression above, we have introduced the fractional distance $\delta \equiv d/N$ and the rate $R \equiv M/N$, and considered the $N \to \infty$ asymptotics with these two quantities kept fixed. We plot the function $R - 1 + \mathcal{H}_2(\delta)$ (which is sometimes called the **growth rate** of the distance enumerator) in Fig. 6.1. For δ small enough, i.e. $\delta < \delta_{GV}$, the growth rate is negative: the average number of codewords at a small distance from $\underline{x}^{(0)}$ vanishes exponentially with N. By the Markov inequality, the probability of having any codeword at all at such a short distance vanishes as $N \to \infty$. The distance $\delta_{GV}(R)$, called the **Gilbert–Varshamov distance**, is the smallest root of $R - 1 + \mathcal{H}_2(\delta) = 0$. For instance, we have $\delta_{GV}(1/2) \approx 0.110278644$.

Above the Gilbert–Varshamov distance, i.e. $\delta > \delta_{GV}$, the average number of codewords is exponentially large, with the maximum occurring at $\delta = 1/2$: $\mathbb{E}\,\mathcal{N}_{\underline{x}^{(0)}}(N/2) \doteq 2^{NR} = 2^M$. It is easy to show that the distance enumerator $\mathcal{N}_{\underline{x}^{(0)}}(d)$ is sharply concentrated around its average in the whole regime $\delta_{GV} < \delta < 1 - \delta_{GV}$. This can be done using arguments similar to those developed in Section 5.2 for the random energy model (REM configurations become codewords in the present context, the role of the energy is played by the Hamming distance, and the Gaussian distribution of the energy levels is replaced here by the binomial distribution). A pictorial interpretation of the above result is shown in Fig. 6.2 (but note that it is often misleading to interpret phenomena

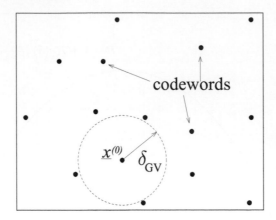

Fig. 6.2 A pictorial view of a typical code from the random code ensemble. The codewords are random points in the Hamming space. If we pick a codeword at random from the code and consider a ball of radius $N\delta$ around it, the ball will not contain any other codeword as long as $\delta < \delta_{\text{GV}}(R)$, it will contain exponentially many codewords when $\delta > \delta_{\text{GV}}(R)$.

occurring in spaces with a large number of dimensions using finite-dimensional images: such images must be handled with care!).

Exercise 6.4 The random code ensemble can easily be generalized to other (non-binary) alphabets. Consider, for instance, a q-ary alphabet, i.e. an alphabet with letters $\{0, 1, 2, \ldots, q-1\} \equiv \mathcal{A}$. A code $\mathfrak{C}_N$ is constructed by taking 2^M codewords with uniform probability in $\mathcal{A}^N$. We can define the distance between any two codewords $d_q(\underline{x}, \underline{y})$ as the number of positions in which the sequences $\underline{x}$, $\underline{y}$ differ. Show that the average distance enumerator is now

$$\mathbb{E}\,\mathcal{N}_{\underline{x}^{(0)}}(d) \doteq 2^{N[R - \log_2 q + \delta \log_2(q-1) + \mathcal{H}_2(\delta)]}, \tag{6.12}$$

where $\delta \equiv d/N$ and $R \equiv M/N$. The maximum of the above function is no longer at $\delta = 1/2$. How can we explain this phenomenon in simple terms?

6.3 Communicating over a binary symmetric channel

We shall now analyse the performance of the RCE when used for communicating over a binary symmetric channel (BSC) as defined in Fig. 1.4. We start by considering a word MAP (or, equivalently, maximum-likelihood) decoder, and then analyse the slightly more complicated symbol MAP decoder. Finally, we introduce another decoding strategy, inspired by the statistical-physics analogy, that generalizes word MAP and symbol MAP decoding.

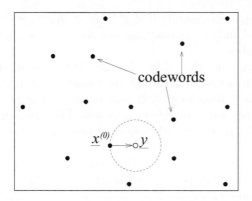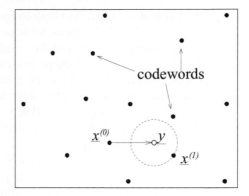

Fig. 6.3 A pictorial view of word MAP decoding for a BSC. A codeword $\underline{x}^{(0)}$ is chosen and transmitted through a noisy channel. The channel output is $\underline{y}$. If the distance between $\underline{x}^{(0)}$ and $\underline{y}$ is small enough (*left frame*), the transmitted message can be safely reconstructed by looking for the codeword closest to $\underline{y}$. In the opposite case (*right frame*), the closest codeword $\underline{x}_1$ does not coincide with the transmitted one.

6.3.1 Word MAP decoding

For a BSC, both the channel input $\underline{x}$ and the output $\underline{y}$ are sequences of bits of length N. The probability for the codeword $\underline{x}$ to be the channel input conditional on the output $\underline{y}$, defined in eqns (6.5) and (6.4), depends uniquely on the Hamming distance $d(\underline{x}, \underline{y})$ between these two vectors. Denoting by p the channel flip probability, we have

$$\mu_y(\underline{x}) = \frac{1}{Z'(\underline{y})} \, p^{d(\underline{x},\underline{y})} (1-p)^{N-d(\underline{x},\underline{y})} \, \mathbb{I}(\underline{x} \in \mathfrak{C}_N) \,, \tag{6.13}$$

where $Z'(\underline{y})$ is a normalization constant which depends uniquely upon $\underline{y}$ (up to a factor, this coincides with the normalization $Z(\underline{y})$ in eqn (6.5)). Without loss of generality, we can assume that $p < 1/2$. Therefore word MAP decoding, which prescribes that $\mu_y(\underline{x})$ is maximized with respect to $\underline{x}$, outputs the codeword which is the closest to the channel output.

We have obtained a purely geometrical formulation of the original communication problem. A random set of points $\mathfrak{C}_N$ is drawn in the Hamming space $\{0,1\}^N$, and one of them (let us call it $\underline{x}^{(0)}$) is chosen for communicating. Noise perturbs this vector, yielding a new point $\underline{y}$. Decoding consists in finding the closest point to $\underline{y}$ among all the points in $\mathfrak{C}_N$, and fails every time this is not $\underline{x}^{(0)}$. The block error probability is simply the probability for such an event to occur. This formulation is illustrated in Fig. 6.3.

This description should immediately make it clear that the block error probability vanishes, (in the $N \to \infty$ limit) as soon as p is below some finite threshold. In the previous Section we saw that, with high probability, the closest codeword $\underline{x}' \in \mathfrak{C}_N \backslash \underline{x}^{(0)}$ to $\underline{x}^{(0)}$ lies at a distance $d(\underline{x}', \underline{x}^{(0)}) \simeq N\delta_{\mathrm{GV}}(R)$. On the other hand, $\underline{y}$ is obtained from $\underline{x}^{(0)}$ by flipping each bit independently with probability p, and therefore $d(\underline{y}, \underline{x}^{(0)}) \simeq Np$ with high probability. By the triangle inequality, $\underline{x}^{(0)}$ is certainly the closest codeword to $\underline{y}$ (and therefore word MAP decoding is successful) if $d(\underline{x}^{(0)}, \underline{y}) < d(\underline{x}^{(0)}, \underline{x}')/2$. If $p < \delta_{\mathrm{GV}}(R)/2$, this happens with probability approaching one as $N \to \infty$, and therefore the block error probability vanishes.

However, the above argument overestimates the effect of noise. Although about $N\delta_{\mathrm{GV}}(R)/2$ incorrect bits may cause an unsuccessful decoding, they must occur in the appropriate positions for $\underline{y}$ to be closer to $\underline{x}'$ than to $\underline{x}^{(0)}$. If they occur at uniformly random positions (as happens in a BSC), they will probably be harmless. The difference between the two situations is most significant in large-dimensional spaces, as shown by the analysis provided below.

The distance between $\underline{x}^{(0)}$ and $\underline{y}$ is a sum of N i.i.d. Bernoulli variables with parameter p (each bit is flipped with probability p). By the central limit theorem, $N(p-\varepsilon) < d(\underline{x}^{(0)}, \underline{y}) < N(p+\varepsilon)$ with probability approaching one in the $N \to \infty$ limit, for any $\varepsilon > 0$. As for the remaining $2^M - 1$ codewords, they are completely uncorrelated with $\underline{x}^{(0)}$ and, therefore, with $\underline{y}$: $\{\underline{y}, \underline{x}^{(1)}, \cdots, \underline{x}^{(2^M-1)}\}$ are 2^M i.i.d. random points drawn from the uniform distribution over $\{0, 1\}^N$. The analysis of the previous section shows that, with probability approaching one as $N \to \infty$, none of the codewords $\{\underline{x}^{(1)}, \cdots, \underline{x}^{(2^M-1)}\}$ lies within a ball of radius $N\delta$ centred on $\underline{y}$ when $\delta < \delta_{\mathrm{GV}}(R)$. In the opposite case, if $\delta > \delta_{\mathrm{GV}}(R)$, there is an exponential (in N) number of these codewords within a ball of radius $N\delta$.

The performance of the RCE is easily deduced (see Fig. 6.4). If $p < \delta_{\mathrm{GV}}(R)$, the transmitted codeword $\underline{x}^{(0)}$ lies at a shorter distance than all the other ones from the received message $\underline{y}$: decoding is successful. At a larger noise level, if $p > \delta_{\mathrm{GV}}(R)$, there is an exponential number of codewords closer to $\underline{y}$ than the transmitted one is: decoding is unsuccessful. Note that the condition $p < \delta_{\mathrm{GV}}(R)$ can be rewritten as $R < C_{\mathrm{BSC}}(p)$, where $C_{\mathrm{BSC}}(p) = 1 - \mathcal{H}_2(p)$ is the capacity of a BSC with flip probability p.

6.3.2   Symbol MAP decoding

In symbol MAP decoding, the i-th bit is decoded by first computing the marginal $P^{(i)}(x_i | \underline{y})$ and then maximizing it with respect to x_i. Using eqn (6.13), we get

$$\mu_{\underline{y}}^{(i)}(x_i) = \sum_{\underline{x}\backslash i} \mu_{\underline{y}}(\underline{x}) = \frac{1}{Z} \sum_{\underline{x}\backslash i} \exp\{-2B\,d(\underline{x}, \underline{y})\}, \tag{6.14}$$

where we have introduced the parameter

$$B \equiv \frac{1}{2} \log \left(\frac{1-p}{p} \right) \tag{6.15}$$

and the normalization constant

$$Z \equiv \sum_{\underline{x} \in \mathcal{C}_N} \exp\{-2B\,d(\underline{x}, \underline{y})\}. \tag{6.16}$$

Equation (6.14) shows that the marginal distribution $\mu_{\underline{y}}^{(i)}(x_i)$ sums contributions from all the codewords, not only the one closest to $\underline{y}$. This makes the analysis of symbol MAP decoding slightly more involved than the word MAP decoding case.

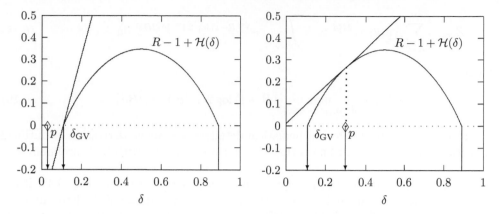

Fig. 6.4 Logarithm of the distance enumerator $\widehat{\mathcal{N}}_{\underline{y}}(d)$ (which counts the number of code-words at a distance $d = N\delta$ from the received message) divided by the block length N. Here, the rate is $R = 1/2$. We also show the distance of the transmitted codeword for two different noise levels: $p = 0.03 < \delta_{\mathrm{GV}}(1/2) \approx 0.110278644$ (*left*) and $p = 0.3 > \delta_{\mathrm{GV}}(R)$ (*right*). The tangent lines with slope $2B = \log[(1-p)/p]$ determine which codewords dominate the symbol MAP decoder.

Let us start by estimating the normalization constant Z. It is convenient to separate the contribution arising from the transmitted codeword $\underline{x}^{(0)}$ from that arising from the *incorrect* codewords $\underline{x}^{(1)}, \dots, \underline{x}^{(2^M - 1)}$:

$$Z = \mathrm{e}^{-2Bd(\underline{x}^{(0)}, \underline{y})} + \sum_{d=0}^{N} \widehat{\mathcal{N}}_{\underline{y}}(d)\, \mathrm{e}^{-2Bd} \equiv Z_{\mathrm{corr}} + Z_{\mathrm{err}}, \qquad (6.17)$$

where we have denoted by $\widehat{\mathcal{N}}_{\underline{y}}(d)$ the number of incorrect codewords at a distance d from the vector $\underline{y}$. The contribution from $\underline{x}^{(0)}$ in the above expression is easily estimated. By the law of large numbers, $d(\underline{x}^{(0)}, \underline{y}) \simeq Np$ and therefore Z_{corr} is close to e^{-2NBp} with high probability. More precisely, for any $\varepsilon > 0$, $\mathrm{e}^{-N(2Bp+\varepsilon)} \leq Z_{\mathrm{corr}} \leq \mathrm{e}^{-N(2Bp-\varepsilon)}$ with probability approaching one in the $N \to \infty$ limit.

For Z_{err}, we proceed in two steps: we first compute the distance enumerator $\widehat{\mathcal{N}}_{\underline{y}}(d)$, and then sum over d. The distance enumerator was computed in Section 6.2. As in the analysis of word MAP decoding, the fact that the distances are measured with respect to the channel output $\underline{y}$ and not with respect to a codeword does not change the result, because $\underline{y}$ is independent of the incorrect codewords $\underline{x}^{(1)}, \dots, \underline{x}^{(2^M - 1)}$. Therefore $\widehat{\mathcal{N}}_{\underline{y}}(d)$ is exponentially large in the interval $\delta_{\mathrm{GV}}(R) < \delta \equiv d/N < 1 - \delta_{\mathrm{GV}}(R)$, while it vanishes with high probability outside the same interval. Moreover, if $\delta_{\mathrm{GV}}(R) < \delta < 1 - \delta_{\mathrm{GV}}(R)$, $\widehat{\mathcal{N}}_{\underline{y}}(d)$ is tightly concentrated around its mean, given by eqn (6.11). The summation over d in eqn (6.17) can then be evaluated by the saddle point method. This calculation is very similar to the estimation of the free energy of the random energy model (see Section 5.2). Roughly speaking, we have

$$Z_{\text{err}} = \sum_{d=0}^{N} \widehat{\mathcal{N}}_{\underline{y}}(d) \, e^{-2Bd} \simeq N \int_{\delta_{\text{GV}}}^{1-\delta_{\text{GV}}} e^{N[(R-1)\log 2 + \mathcal{H}(\delta)2B\delta]} \, d\delta \doteq e^{N\phi_{\text{err}}}, \qquad (6.18)$$

where

$$\phi_{\text{err}} \equiv \max_{\delta \in [\delta_{\text{GV}}, 1-\delta_{\text{GV}}]} [(R-1)\log 2 + \mathcal{H}(\delta) - 2B\delta]. \qquad (6.19)$$

The reader can complete the mathematical details of the above derivation as outlined in Section 5.2. The bottom line is that Z_{err} is close to $e^{N\phi_{\text{err}}}$ with high probability as $N \to \infty$.

Let us examine the resulting expression given in eqn (6.19) (see Fig. 6.4). If the maximum is achieved in the interior of $[\delta_{\text{GV}}, 1-\delta_{\text{GV}}]$, its location δ_* is determined by the stationarity condition $\mathcal{H}'(\delta_*) = 2B$, which implies $\delta_* = p$. In the opposite case, the maximum must be realized at $\delta_* = \delta_{\text{GV}}$ (remember that $B > 0$). Evaluating the right hand side of eqn (6.19) in these two cases, we get

$$\phi_{\text{err}} = \begin{cases} -\delta_{\text{GV}}(R) \log\left((1-p)/p\right) & \text{if } p < \delta_{\text{GV}}, \\ (R-1)\log 2 - \log(1-p) & \text{otherwise.} \end{cases} \qquad (6.20)$$

We can now compare Z_{corr} and Z_{err}. At low noise levels (small p), the transmitted codeword $\underline{x}^{(0)}$ is close enough to the received message $\underline{y}$ to dominate the sum in eqn (6.17). At higher noise levels, the exponentially more numerous incorrect codewords overcome the term due to $\underline{x}^{(0)}$. More precisely, with high probability we have

$$Z = \begin{cases} Z_{\text{corr}}[1 + e^{-\Theta(N)}] & \text{if } p < \delta_{\text{GV}}, \\ Z_{\text{err}}[1 + e^{-\Theta(N)}] & \text{otherwise,} \end{cases} \qquad (6.21)$$

where the exponents $\Theta(N)$ are understood to be positive.

We now consider eqn (6.14), and once again separate out the contribution of the transmitted codeword,

$$P^{(i)}(x_i|\underline{y}) = \frac{1}{Z} [Z_{\text{corr}} \, \mathbb{I}(x_i = x_i^{(0)}) + Z_{\text{err},x_i}], \qquad (6.22)$$

where we have introduced the quantity

$$Z_{\text{err},x_i} = \sum_{\underline{z} \in \mathfrak{C}_N \setminus \underline{x}^{(0)}} e^{-2Bd(\underline{z},\underline{y})} \, \mathbb{I}(z_i = x_i) . \qquad (6.23)$$

Note that $Z_{\text{err},x_i} \leq Z_{\text{err}}$. Together with eqn (6.21), this implies, if $p < \delta_{\text{GV}}(R)$, that $\mu_{\underline{y}}^{(i)}(x_i = x_i^{(0)}) = 1 - e^{-\Theta(N)}$ and $\mu_{\underline{y}}^{(i)}(x_i \neq x_i^{(0)}) = e^{-\Theta(N)}$. Therefore, in this regime, the symbol MAP decoder correctly outputs the transmitted bit $x_i^{(0)}$. It is important to stress that this result holds with probability approaching one as $N \to \infty$. Concretely, there exist bad choices of the code $\mathfrak{C}_N$ and particularly unfavourable channel realizations $\underline{y}$ such that $\mu_{\underline{y}}^{(i)}(x_i = x_i^{(0)}) < 1/2$ and the decoder fails. However, the probability of such an event (i.e. the bit-error rate P_b) vanishes in the large-block length limit $N \to \infty$.

What happens for $p > \delta_{\text{GV}}(R)$? Arguing as for the normalization constant Z, it is easy to show that the contribution of incorrect codewords dominates the marginal distribution (6.22). Intuitively, this suggests that the decoder fails. A more detailed computation, sketched below, shows that the bit error rate in the $N \to \infty$ limit is

$$P_b = \begin{cases} 0 & \text{if } p < \delta_{\text{GV}}(R), \\ p & \text{if } \delta_{\text{GV}}(R) < p < 1/2. \end{cases} \tag{6.24}$$

Note that, above the threshold $\delta_{\text{GV}}(R)$, the bit error rate is the same as if the information message were transmitted without coding through the BSC: the code is useless.

A complete calculation of the bit error rate P_b in the regime $p > \delta_{\text{GV}}(R)$ is rather lengthy. We shall provide here a heuristic, albeit essentially correct, justification, and leave a more detailed derivation to the exercise below. As already stressed, the contribution Z_{corr} of the transmitted codeword can be safely neglected in eqn (6.22). Assume, without loss of generality, that $x_i^{(0)} = 0$. The decoder will be successful if $Z_{\text{err},0} > Z_{\text{err},1}$ and will fail in the opposite case. Two cases must be considered: either $y_i = 0$ (this happens with probability $1 - p$) or $y_i = 1$ (probability p). In the first case, we have

$$Z_{\text{err},0} = \sum_{\underline{z} \in \mathcal{C}_N \backslash \underline{x}^{(0)}} \mathbb{I}(z_i = 0) \, \mathrm{e}^{-2Bd_i(\underline{y},\underline{z})} \, ,$$

$$Z_{\text{err},1} = \mathrm{e}^{-2B} \sum_{\underline{z} \in \mathcal{C}_N \backslash \underline{x}^{(0)}} \mathbb{I}(z_i = 1) \, \mathrm{e}^{-2Bd_i(\underline{y},\underline{z})} \, , \tag{6.25}$$

where we have denoted by $d_i(\underline{x}, \underline{y})$ the number of of positions j, distinct from i, such that $x_j \neq y_j$. The sums in the above expressions are independent identically distributed random variables. Moreover, they are tightly concentrated around their mean. Since $B > 0$, this implies $Z_{\text{err},0} > Z_{\text{err},1}$ with high probability. Therefore the decoder is successful in the case $y_i = 0$. Analogously, the decoder fails with high probability if $y_i = 1$, and hence the bit error rate converges to $P_b = p$ for $p > \delta_{\text{GV}}(R)$.

Exercise 6.5 From a rigorous point of view, the weak point of the above argument is the lack of any estimate of the fluctuations of $Z_{\text{err},0/1}$. The reader may complete the derivation along the following lines:

(a) Define $X_0 \equiv Z_{\text{err},0}$ and $X_1 \equiv \mathrm{e}^{2B} Z_{\text{err},1}$. Prove that X_0 and X_1 are independent and identically distributed.

(b) Define the correct distance enumerators $\mathcal{N}_{0/1}(d)$ such that a representation of the form $X_{0/1} = \sum_d \mathcal{N}_{0/1}(d) \exp(-2Bd)$ holds.

(c) Show that a significant fluctuation of $\mathcal{N}_{0/1}(d)$ from its average is highly (more than exponentially) improbable (within an appropriate range of d).

(d) Deduce that a significant fluctuation of $X_{0/1}$ is highly improbable (the last two points can be treated along the lines already discussed for the random energy model in Chapter 5).

6.3.3 Finite-temperature decoding

The expression (6.14) for the marginal $\mu_y^{(i)}(x_i)$ is strongly reminiscent of a Boltzmann average. This analogy suggests a generalization which interpolates between the two 'classical' MAP decoding strategies discussed so far: **finite-temperature decoding**. We first define this new decoding strategy in the context of the BSC. Let β be a non-negative number playing the role of an inverse temperature, and let $\underline{y} \in \{0,1\}^N$ be the channel output. We define the probability distribution $\mu_{y,\beta}(\underline{x})$ to be given by

$$\mu_{y,\beta}(\underline{x}) = \frac{1}{Z(\beta)}\, e^{-2\beta B d(\underline{y},\underline{x})}\, \mathbb{I}(x \in \mathfrak{C}_N)\,, \qquad Z(\beta) \equiv \sum_{\underline{x} \in \mathfrak{C}_N} e^{-2\beta B d(\underline{x},\underline{y})}\,, \quad (6.26)$$

where B is always related to the noise level p through eqn (6.15). This distribution depends upon the channel output $\underline{y}$: for each received message $\underline{y}$, the finite-temperature decoder constructs the appropriate distribution $\mu_{y,\beta}(\underline{x})$. For the sake of simplicity, we shall not write this dependence explicitly. Let $\mu_{y,\beta}^{(i)}(x_i)$ be the marginal distribution of x_i when $\underline{x}$ is distributed according to $\mu_{y,\beta}(\underline{x})$. The new decoder outputs

$$\underline{x}^\beta = \left(\arg\max_{x_1} \mu_{y,\beta}^{(1)}(x_1)\,,\ldots, \arg\max_{x_N} \mu_{y,\beta}^{(N)}(x_N)\right)\,. \tag{6.27}$$

As in the previous sections, readers are free to choose their favourite convention in the case of ties (i.e. for those i's such that $\mu_{y,\beta}^{(i)}(0) = \mu_{y,\beta}^{(i)}(1)$).

Two values of β are particularly interesting: $\beta = 1$ and $\beta = \infty$. If $\beta = 1$, the distribution $\mu_{y,\beta}(\underline{x})$ coincides with the distribution $\mu_y(\underline{x})$ of the channel input conditional on the output (see eqn (6.13)). Therefore, for any $\underline{y}$, symbol MAP decoding coincides with finite-temperature decoding at $\beta = 1$: $\underline{x}_i^{\beta=1} = \underline{x}^{\mathrm{b}}$.

If $\beta = \infty$, the distribution (6.26) is concentrated on those codewords which are the closest to $\underline{y}$. In particular, if there is a unique closest codeword to $\underline{y}$, finite-temperature decoding at $\beta = \infty$ coincides with word MAP decoding: $\underline{x}^{\beta=\infty} = \underline{x}^{\mathrm{w}}$.

Exercise 6.6 Using the approach developed in the previous section, analyse the performances of finite-temperature decoding for the RCE at any β.

The results of the above exercise are summarized in Fig. 6.5, which gives the finite-temperature decoding phase diagram. There exist three regimes, which are three distinct phases with very different behaviours.

1. A 'completely ordered' phase at low noise ($p < \delta_{\mathrm{GV}}(R)$) and low temperature (large enough β). In this regime, the decoder works: the probability distribution $\mu_{y,\beta}(\underline{x})$ is dominated by the transmitted codeword $\underline{x}^{(0)}$. More precisely, $\mu_{y,\beta}(\underline{x}^{(0)}) = 1 - \exp\{-\Theta(N)\}$. The bit and block error rates vanish as $N \to \infty$.
2. A 'glassy' phase at higher noise ($p > \delta_{\mathrm{GV}}(R)$) and low temperature (large enough β). The transmitted codeword has a negligible weight $\mu_{y,\beta}(\underline{x}^{(0)}) = \exp\{-\Theta(N)\}$. The bit error rate is bounded away from 0, and the block error rate converges

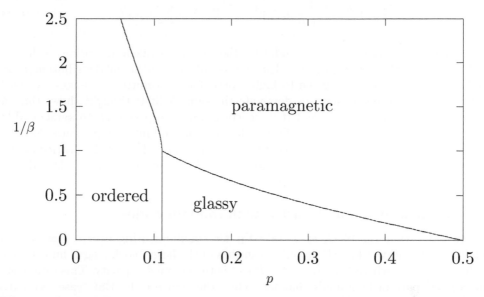

Fig. 6.5 Phase diagram for the rate-1/2 random code ensemble over a binary symmetric channel, using finite temperature decoding. Word MAP and bit MAP decoding correspond to $1/\beta = 0$ and $1/\beta = 1$, respectively. Note that the phase boundary of the error-free (ordered) phase is vertical in this interval of temperature.

to 1 as $N \to \infty$. The measure $\mu_{y,\beta}(\underline{x})$ is dominated by the codewords closest to the received message $\underline{y}$ (which are distinct from the correct codeword, since $p > \delta_{\mathrm{GV}}(R)$). Its Shannon entropy $H(\mu_{y,\beta})$ is sublinear in N. This situation is closely related to the 'measure condensation' phenomenon that occurs in the low-temperature phase of the random energy model.

3. An 'entropy-dominated' (paramagnetic) phase at high temperature (small enough β). The bit and block error rates behave as in the glassy phase, and $\mu_{y,\beta}(\underline{x}^{(0)}) = \exp\{-\Theta(N)\}$. However, the measure $\mu_{y,\beta}(\underline{x})$ is now dominated by codewords whose distance $d \simeq N\delta_*$ from the received message is larger than the minimal distance: $\delta_* = p^\beta/[p^\beta + (1-p)^\beta]$. In particular, $\delta_* = p$ if $\beta = 1$, and $\delta_* = 1/2$ if $\beta = 0$. In the first case we recover the result already obtained for symbol MAP decoding. In the second case, $\mu_{y,\beta=0}(\underline{x})$ is the uniform distribution over the codewords, and the distance from the received message under this distribution is, with high probability, close to $N/2$. In this regime, the Shannon entropy $H(\mu_\beta)$ is linear in N.

Finite-temperature decoding can be generalized to other channel models. Let $\mu_y(\underline{x})$ be the distribution of the transmitted message conditional on the channel output, given explicitly in eqn (6.5). For $\beta > 0$, we define the distribution[1]

---

[1] The partition function $Z(\beta)$ defined here differs by a multiplicative constant from the one defined in eqn (6.26) for a BSC.

$$\mu_{y,\beta}(\underline{x}) = \frac{1}{Z(\beta)} \, \mu_y(\underline{x})^\beta, \qquad Z(\beta) \equiv \sum_{\underline{x}} \mu_y(\underline{x})^\beta. \qquad (6.28)$$

Once more, the decision of the decoder for the i-th bit is taken according to the rule (6.27). The distribution $\mu_{y,\beta}(\underline{x})$ is a 'deformation' of the conditional distribution $\mu_y(\underline{x})$. At large β, more weight is given to highly probable transmitted messages. At small β, the most numerous codewords dominate the sum. A little thought shows that, as for a BSC, the cases $\beta = 1$ and $\beta = \infty$ correspond, respectively, to symbol MAP and word MAP decoding. The qualitative features of the finite-temperature-decoding phase diagram are easily generalized to any memoryless channel. In particular, the three phases described above can be found in such a general context. Decoding is successful in the low-noise, large-β phase.

6.4 Error-free communication with random codes

As we have seen, the block error rate P_B for communicating over a BSC with a random code and word MAP decoding vanishes in the large-block-length limit as long as $R < C_{\mathrm{BSC}}(p)$, where $C_{\mathrm{BSC}}(p) = 1 - \mathcal{H}_2(p)$ is the channel capacity. This establishes the 'direct' part of Shannon's channel coding theorem for the BSC case: error-free communication is possible at rates below the channel capacity. This result is in fact much more general. We describe here a proof for general memoryless channels, always based on random codes.

For the sake of simplicity we shall restrict ourselves to memoryless channels with binary input and discrete output. Such are defined by a transition probability $Q(y|x)$, where $x \in \{0,1\}$ and $y \in \mathcal{Y}$, with $\mathcal{Y}$ being a finite alphabet. In order to handle this case, we must generalize the RCE: each codeword $\underline{x}^{(m)} \in \{0,1\}^N$, $m = 0, \ldots, 2^M - 1$, is again constructed independently as a sequence of N i.i.d. bits $x_1^{(m)} \cdots x_N^{(m)}$. Unlike the case of symmetric channels, $x_i^{(m)}$ is now drawn from an arbitrary distribution $P(x)$, $x \in \{0,1\}$ instead of being uniformly distributed. It is important to distinguish $P(x)$, which is an arbitrary single-bit distribution defining the code ensemble and will be chosen at our convenience for optimizing it, from the a priori source distribution $\mu_0(\underline{x})$ of eqn (6.5), which is a distribution over the codewords and models the behaviour of the information source. As in the previous sections, we shall assume the source distribution μ_0 to be uniform over the codewords (see eqn (6.4)). On the other hand, the codewords themselves have been constructed using the single-bit distribution $P(x)$.

We shall first analyse the RCE for a generic distribution $P(x)$, under word MAP decoding. The main result is the following.

Theorem 6.1 *Consider communication over a binary-input discrete memoryless channel with transition probability $Q(y|x)$, using a code from the RCE with input bit distribution $P(x)$ and word MAP decoding. If the code rate is smaller than the mutual information $I_{X,Y}$ between two random variables X, Y with joint distribution $P(x)Q(y|x)$, then the block error rate vanishes in the large-block-length limit.*

Using this result, one can optimize the performance of the ensemble over the choice of the distribution $P(\cdot)$. More precisely, we maximixe the achievable rate for error-free communication, $I_{X,Y}$. The corresponding optimal distribution $P^*(\cdot)$ depends upon that channel: it is the best distribution adapted to the channel. Since the channel capacity is in fact defined as the maximum mutual information between the channel input and the channel output (see eqn (1.38)), the RCE with input bit distribution $P^*(\cdot)$ allows one to communicate without errors up to the channel capacity. The above theorem implies therefore the 'direct part' of Shannon's theorem (Theorem 1.23).

Proof Assume that the codeword $\underline{x}^{(0)}$ is transmitted through the channel and the message $\underline{y} \in \mathcal{Y}^N$ is received. The decoder constructs the probability for $\underline{x}$ to be the channel input, conditional on the output $\underline{y}$ (see eqn (6.5)). Word MAP decoding consists in minimizing the cost function

$$E(\underline{x}) = -\sum_{i=1}^{N} \log_2 Q(y_i|x_i) \qquad (6.29)$$

over the codewords $\underline{x} \in \mathcal{C}_N$ (note that we are using natural logarithms here). Decoding will be successful if and only if the minimum of $E(\underline{x})$ is realized over the transmitted codeword $\underline{x}^{(0)}$. The problem therefore consists in understanding the behaviour of the 2^M random variables $E(\underline{x}^{(0)}), \ldots, E(\underline{x}^{(2^M-1)})$.

Once more, it is necessary to single out $E(\underline{x}^{(0)})$. This is the sum of N i.i.d. random variables $-\log Q(y_i|x_i^{(0)})$, and it is therefore well approximated by its mean

$$\mathbb{E}\, E(\underline{x}^{(0)}) = -N \sum_{x,y} P(x)Q(y|x) \log_2 Q(y|x) = NH_{Y|X}\,. \qquad (6.30)$$

In particular $(1-\delta)NH_{Y|X} < E(\underline{x}^{(0)}) < (1+\delta)NH_{Y|X}$ with probability approaching one as $N \to \infty$.

As for the $2^M - 1$ incorrect codewords, the corresponding 'log-likelihoods' $E(\underline{x}^{(1)})$, $\ldots, E(\underline{x}^{(2^M-1)})$ are i.i.d. random variables. We can therefore estimate the smallest of them by following the approach developed for the REM and already applied to the RCE for a BSC. In Section 6.8, we prove the following large-deviation result for the distribution of these variables.

Lemma 6.2 *Let $\varepsilon_i = E(\underline{x}^{(i)})/N$. Then $\varepsilon_1, \ldots, \varepsilon_{2^M-1}$ are i.i.d. random variables and their distribution satisfies a large-deviation principle of the form $\mathbb{P}(\varepsilon) \doteq 2^{-N\psi(\varepsilon)}$. The rate function is given by*

$$\psi(\varepsilon) \equiv \min_{\{p_y(\cdot)\} \in \mathfrak{P}_\varepsilon} \left[\sum_y Q(y)D(p_y||P) \right]\,, \qquad (6.31)$$

where the minimum is taken over the set of probability distributions $\{p_y(\cdot),\ y \in \mathcal{Y}\}$ in the subspace $\mathfrak{P}_\varepsilon$ defined by the constraint

$$\varepsilon = -\sum_{xy} Q(y)p_y(x) \log_2 Q(y|x)\,, \qquad (6.32)$$

and where we have defined $Q(y) \equiv \sum_x Q(y|x)P(x)$.

The solution of the minimization problem formulated in this lemma is obtained through a standard Lagrange multiplier technique:

$$p_y(x) = \frac{1}{z(y)} P(x)Q(y|x)^\gamma, \qquad (6.33)$$

where the (ε-dependent) constants $z(y)$ and γ are chosen in order to satisfy the normalizations $\sum_x p_y(x) = 1$ for all $y \in \mathcal{Y}$, and the constraint (6.32).

The rate function $\psi(\varepsilon)$ is convex, with a global minimum (corresponding to $\gamma = 0$) at $\varepsilon_* = -\sum_{x,y} P(x)Q(y)\log_2 Q(y|x)$, where its value is $\psi(\varepsilon_*) = 0$. This implies that, with high probability, all incorrect codewords will have costs $E(\underline{x}^{(i)}) = N\varepsilon$ in the range $\varepsilon_{\min} - \delta \le \varepsilon \le \varepsilon_{\max} + \delta$ for all $\delta > 0$, where $\varepsilon_{\min}$ and $\varepsilon_{\max}$ are the two solutions of $\psi(\varepsilon) = R$. Moreover, for any ε inside thar interval, the number of codewords with $E(\underline{x}^{(i)}) \simeq N\varepsilon$ is exponentially large, and close to $2^{NR-N\psi(\varepsilon)}$. So, with high probability, the incorrect codeword with minimum cost has a cost close to $N\varepsilon_{\min}$, while the correct codeword has a cost close to $NH_{Y|X}$. Therefore MAP decoding will find the correct codeword if and only if $H_{Y|X} < \varepsilon_{\min}$.

Let us now show that the condition $H_{Y|X} < \varepsilon_{\min}$ is in fact equivalent to $R < I_{X,Y}$. It turns out that the value $\varepsilon = H_{Y|X}$ is obtained using $\gamma = 1$ in eqn (6.33), and therefore $p_y(x) = P(x)Q(y|x)/Q(y)$. The corresponding value of the rate function is $\psi(\varepsilon = H_{Y|X}) = H_Y - H_{Y|X} = I_{Y|X}$. The condition for error-free communication, $H_{Y|X} < \varepsilon_{\min}$, can thus be rewritten as $R < \psi(H_{Y|X})$, or $R < I_{X,Y}$. $\square$

Example 6.3 Reconsider a BSC with flip probability p. We have

$$E(\underline{x}) = -(N - d(\underline{x}, \underline{y}))\log(1-p) - d(\underline{x}, \underline{y})\log p. \qquad (6.34)$$

Up to a rescaling the cost coincides with the Hamming distance from the received message. If we take $P(0) = P(1) = 1/2$, the optimal types are

$$p_0(1) = 1 - p_0(0) = \frac{p^\gamma}{(1-p)^\gamma + p^\gamma} \qquad (6.35)$$

(see eqn (6.33)) and analogously for $p_1(x)$. The corresponding cost is

$$\varepsilon = -(1-\delta)\log(1-p) - \delta\log p, \qquad (6.36)$$

where have we defined $\delta = p^\gamma/[(1-p)^\gamma + p^\gamma]$. The large-deviation rate function is given, parametrically, by $\psi(\varepsilon) = \log 2 - \mathcal{H}(\delta)$. The reader will easily recognize the results already obtained in the previous section.

Exercise 6.7 Consider communication over a discrete memoryless channel with finite input and output alphabets $\mathcal{X}$ and $\mathcal{Y}$, and transition probability $Q(y|x)$, $x \in \mathcal{X}$, $y \in \mathcal{Y}$. Check that the above proof remains valid in this context.

6.5 Geometry again: Sphere packing

Coding has a lot to do with the optimal packing of spheres, which is a mathematical problem of considerable interest in various branches of science. Consider, for instance, communication over a BSC with flip probability p. A code of rate R and block length N consists of 2^{NR} points $\{\underline{x}^{(1)} \cdots \underline{x}^{(2^{NR})}\}$ in the hypercube $\{0,1\}^N$. With each possible channel output $\underline{y} \in \{0,1\}^N$, the decoder associates one of the codewords $\underline{x}^{(i)}$. Therefore we can think of the decoder as realizing a partition of the Hamming space into 2^{NR} decision regions $\mathfrak{D}^{(i)}$, $i \in \{1, \ldots, 2^{NR}\}$, each one associated with a distinct codeword. If we require each decision region $\{\mathfrak{D}^{(i)}\}$ to contain a sphere of radius ρ, the resulting code is *guaranteed* to correct *any* error pattern such that fewer than ρ bits are flipped. One often defines the **minimum distance** of a code as the smallest distance between any two codewords. If a code has a minimum distance d, the Hamming spheres of radius $\rho = \lfloor (d-1)/2 \rfloor$ do not overlap, and the code can correct ρ errors, whatever their positions.

6.5.1 The densest packing of Hamming spheres

We are thus led to consider the general problem of sphere packing on the hypercube $\{0,1\}^N$. A (Hamming) sphere of centre $\underline{x}^{(0)}$ and radius r is defined as the set of points $\underline{x} \in \{0,1\}^N$ such that $d(\underline{x}, \underline{x}^{(0)}) \leq r$. A packing of spheres of radius r and cardinality $\mathcal{N}_S$ is specified by a set of centres $\underline{x}_1, \ldots, \underline{x}_{\mathcal{N}_S} \in \{0,1\}^N$ such that the spheres of radius r centred on these points are disjoint. Let $\mathcal{N}_N^{\max}(\delta)$ be the maximum cardinality of a packing of spheres of radius $N\delta$ in $\{0,1\}^N$. We define the corresponding rate as $R_N^{\max}(\delta) \equiv N^{-1} \log_2 \mathcal{N}_N^{\max}(\delta)$ and would like to compute this quantity in the infinite-dimensional limit

$$R^{\max}(\delta) \equiv \lim_{N \to \infty} \sup R_N^{\max}(\delta). \tag{6.37}$$

The problem of determining the function $R^{\max}(\delta)$ is open: only upper and lower bounds are known. Here we shall derive the simplest of these bounds.

Proposition 6.4

$$1 - \mathcal{H}_2(2\delta) \leq R^{\max}(\delta) \leq 1 - \mathcal{H}_2(\delta) \tag{6.38}$$

The lower bound is often called the Gilbert–Varshamov bound, *the upper bound is called the* Hamming bound.

Proof Lower bounds can be proved by analysing good packing strategies. A simple such strategy is to take the centres of the spheres as 2^{NR} random points with uniform probability in the Hamming space. The minimum distance between any pair of points must be larger than $2N\delta$. This can be estimated by defining the distance enumerator $\mathcal{M}_2(d)$, which counts how many pairs of points have a distance d between them. It is straightforward to show that if $d = 2N\delta$ and δ is kept fixed as $N \to \infty$,

$$\mathbb{E}\, \mathcal{M}_2(d) = \binom{2^{NR}}{2} 2^{-N} \binom{N}{d} \doteq 2^{N[2R-1+\mathcal{H}_2(2\delta)]}. \tag{6.39}$$

As long as $R < [1 - \mathcal{H}_2(2\delta)]/2$, the exponent in the above expression is negative. Therefore, by the Markov inequality, the probability of having any pair of centres at a distance smaller than 2δ is exponentially small in the size. This implies that

$$R^{\max}(\delta) \geq \frac{1}{2}[1 - \mathcal{H}_2(2\delta)]. \tag{6.40}$$

A better lower bound can be obtained by a closer examination of the above (random) packing strategy. In Section 6.2, we derived the following result. If 2^{NR} points are chosen from the uniform distribution in the Hamming space $\{0,1\}^N$, and one of them is considered, its closest neighbour is with high probability at a Hamming distance close to $N\delta_{\mathrm{GV}}(R)$. In other words, if we draw around each point a sphere of radius δ, with $\delta < \delta_{\mathrm{GV}}(R)/2$, and one of these spheres is selected randomly, with high probability it will not intersect any other sphere. This remark suggests the following trick (sometimes called **expurgation** in coding theory). Go through all of the spheres one by one and check if the chosen sphere intersects any other one. If the answer is positive, simply eliminate the sphere. This reduces the cardinality of the packing, but only by a fraction approaching 0 as $N \to \infty$: the packing rate is thus unchanged. As $\delta_{\mathrm{GV}}(R)$ is defined by $R = 1 - \mathcal{H}_2(\delta_{\mathrm{GV}}(R))$, this proves the lower bound in eqn (6.38).

The upper bound can be obtained from the fact that the total volume occupied by the spheres is not larger than the volume of the hypercube. If we denote by $\Lambda_N(\delta)$ the volume of an N-dimensional Hamming sphere of radius $N\delta$, we get $\mathcal{N}_S \Lambda_N(\delta) \leq 2^N$. Since $\Lambda_N(\delta) \doteq 2^{N\mathcal{H}_2(\delta)}$, this implies the upper bound in eqn (6.38). $\square$

Better upper bounds can be derived using more sophisticated mathematical tools. An important result of this type is the *linear-programming bound*,

$$R^{\max}(\delta) \leq \mathcal{H}_2\left(\frac{1}{2} - \sqrt{2\delta(1 - 2\delta)}\right), \tag{6.41}$$

whose proof goes beyond our scope. On the other hand, no better lower bound than the Gilbert–Varshamov result is known. It is a widespread conjecture that this bound is indeed tight: in high dimensions, there is no better way to pack spheres than placing them randomly and expurgating the small fraction of them that are 'squeezed'. The various bounds are shown in Fig. 6.6.

Exercise 6.8 Derive two simple alternative proofs of the Gilbert–Varshamov bound using the following hints:

(a) Given a constant $\bar{\delta}$, we look at all of the 'dangerous' pairs of points whose distance is smaller than $2N\bar{\delta}$. For each dangerous couple, we can expurgate one of its two points. The number of points expurgated is less than or equal to the number of dangerous pairs, which can be bounded using $\mathbb{E}\,\mathcal{M}_2(d)$. What is the largest value of $\bar{\delta}$ such that this expurgation procedure does not reduce the rate?

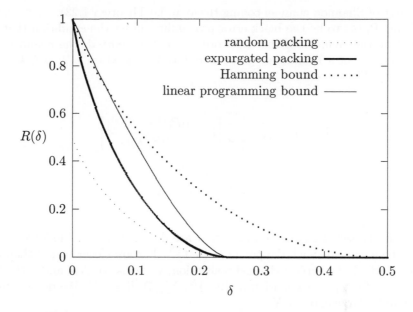

Fig. 6.6 Upper and lower bounds on the maximum packing rate $R^{\max}(\delta)$ of Hamming spheres of radius $N\delta$. Random packing and expurgated random packing provide lower bounds. The Hamming and linear-programming bounds are upper bounds.

(b)   Construct a packing $\underline{x}_1 \ldots \underline{x}_{\mathcal{N}}$ as follows. The first centre $\underline{x}_1$ can be placed anywhere in $\{0, 1\}^N$. The second centre can be placed anywhere outside a sphere of radius $2N\delta$ centred on $\underline{x}^{(0)}$. In general, the i-th centre $\underline{x}_i$ can be at any point outside the spheres centred on $\underline{x}_1 \ldots \underline{x}_{i-1}$. This procedure stops when the spheres of radius $2N\delta$ cover all the space $\{0, 1\}^N$, giving a packing of cardinality $\mathcal{N}$ equal to the number of steps, and with a radius $N\delta$.

6.5.2   Sphere packing and decoding over a BSC

Let us now see the consequences of Proposition 6.4 for coding over a BSC. If the transmitted codeword is $\underline{x}^{(i)}$, the channel output will be (with high probability) at a distance close to Np from $\underline{x}^{(i)}$. Clearly, $R \leq R^{\max}(p)$ is a necessary and sufficient condition for the existence of a code which corrects *any* error pattern such that fewer than Np bits are flipped. Note that this correction criterion is much stronger than requiring a vanishing (bit or block) error rate. The direct part of Shannon's theorem shows the existence of codes with a vanishing (as $N \to \infty$) block error probability for $R < 1 - \mathcal{H}_2(p) = C_{\mathrm{BSC}}(p)$. As shown by the linear-programming bound in Fig. 6.6, $C_{\mathrm{BSC}}(p)$ lies above $R^{\max}(p)$ for large enough p. Therefore, for such values of p, there is a non-vanishing interval of rates $R^{\max}(p) < R < C_{\mathrm{BSC}}(p)$ such that one can correct Np errors with high probability but one cannot correct *all* error patterns involving that many bits.

Let us show, for the BSC case, that the condition $R < 1 - \mathcal{H}_2(p)$ is actually a necessary one for achieving a vanishing block error probability (this is nothing but the converse part of Shannon channel coding theorem, i.e Theorem 1.23).

We define $P_B(k)$ to be the block error probability under the condition that k bits are flipped by the channel. If the codeword $\underline{x}^{(i)}$ is transmitted, the channel output lies on the border of a Hamming sphere of radius k centred on $\underline{x}^{(i)}$: $\partial B_i(k) \equiv \{\underline{z} : d(\underline{z}, \underline{x}^{(i)}) = k\}$. Therefore

$$P_B(k) = \frac{1}{2^{NR}} \sum_{i=1}^{2^{NR}} \left[1 - \frac{|\partial B_i(k) \cap \mathfrak{D}^{(i)}|}{|\partial B_i(k)|} \right] \tag{6.42}$$

$$\geq 1 - \frac{1}{2^{NR}} \sum_{i=1}^{2^{NR}} \frac{|\mathfrak{D}^{(i)}|}{|\partial B_i(k)|} , \tag{6.43}$$

where $\mathfrak{D}^{(i)}$ is the set of channel outputs that are decoded to $\underline{x}^{(i)}$ (under word MAP decoding, these coincide with the points in $\{0, 1\}^N$ that are closer to $\underline{x}^{(i)}$ than to any other codeword). For a typical channel realization, k is close to Np, and $|\partial B_i(Np)| \doteq 2^{N\mathcal{H}_2(p)}$. Since $\{\mathfrak{D}^{(i)}\}$ is a partition of $\{0, 1\}^N$, $\sum_i |\mathfrak{D}^{(i)}| = 2^N$. We deduce that for any $\varepsilon > 0$ and a large enough N,

$$P_B \geq 1 - 2^{N(1 - R - \mathcal{H}_2(p) + \varepsilon)} , \tag{6.44}$$

and thus reliable communication is possible only if $R \leq 1 - \mathcal{H}_2(p)$.

6.6 Other random codes

A major drawback of the random code ensemble is that specifying a particular code (an element of the ensemble) requires $N2^{NR}$ bits. This information has to be stored somewhere when the code is used in practice, and the memory requirement goes soon beyond the capabilities of hardware. A much more compact specification is possible for the **random linear code** (RLC) ensemble. In this case the encoder is required to be a linear map, and all such maps are equiprobable. Concretely, the code is fully specified by an $N \times M$ binary matrix $\mathbb{G} = \{G_{ij}\}$ (the **generator matrix**), and encoding is performed by left multiplication by $\mathbb{G}$:

$$\underline{x} : \{0, 1\}^M \to \{0, 1\}^N , \tag{6.45}$$

$$\underline{z} \mapsto \mathbb{G}\underline{z} , \tag{6.46}$$

where the multiplication has to be carried modulo 2. Endowing the set of linear codes with a uniform probability distribution is essentially equivalent to assuming the entries of $\mathbb{G}$ to be i.i.d. random variables, with $G_{ij} = 0$ or 1 with probability $1/2$. Note that only MN bits are required for specifying a code within this ensemble.

Exercise 6.9 Consider a linear code with $N = 4$ and $|\mathcal{C}| = 8$ defined by

$$\mathcal{C} = \{(z_1 \oplus z_2, \, z_2 \oplus z_3, \, z_1 \oplus z_3, \, z_1 \oplus z_2 \oplus z_3) \mid z_1, z_2, z_3 \in \{0, 1\}\}, \qquad (6.47)$$

where we have denoted by $\oplus$ the sum modulo 2. For instance, (0110) $\in \mathcal{C}$ because we can take $z_1 = 1$, $z_2 = 1$, and $z_3 = 0$, but (0010) $\notin \mathcal{C}$. Compute the distance enumerator for $\underline{x}^{(0)} = (0110)$.

It turns out that the RLC has extremely good performance. Like the original Shannon ensemble, it allows one to communicate without errors below capacity. Moreover, the rate at which the block error probability P_B vanishes is faster for the RLC than for the RCE. This justifies the considerable effort devoted so far to the design and analysis of specific ensembles of linear codes satisfying additional computational requirements. We shall discuss some of the best such codes in the following chapters.

6.7 A remark on coding theory and disordered systems

We would like to stress here the fundamental similarity between the analysis of random code ensembles and the statistical physics of disordered systems. As should already be clear, there are several sources of randomness in coding theory:

- First of all, the *code* used is chosen randomly from an ensemble. This was the original idea used by Shannon to prove the channel coding theorem.
- The *codeword* to be transmitted is chosen with uniform probability from the code. This hypothesis is supported by the source–channel separation theorem.
- The *channel output* is distributed, once the transmitted codeword is fixed, according to a probabilistic process which accounts for the channel noise.
- Once all the above elements have been given, one is left with the *decoding* problem. As we have seen in Section 6.3.3, both classical MAP decoding strategies and finite-temperature decoding can be defined in a unified framework. The decoder constructs a probability distribution $\mu_{y,\beta}(\underline{x})$ over the possible channel inputs, and estimates its single-bit marginals $\mu_{y,\beta}^{(i)}(x_i)$. The decision about the i-th bit depends upon the distribution $\mu_{y,\beta}^{(i)}(x_i)$.

The analysis of a particular coding system can therefore be regarded as the analysis of the properties of the distribution $\mu_{y,\beta}(\underline{x})$ when the code, the transmitted codeword and the noise realization are distributed as described above.

In other words, we are distinguishing between two levels of randomness: on the first level, we deal with the first three sources of randomness, and on the second level, we use the distribution $\mu_{y,\beta}(\underline{x})$. The deep analogy with the theory of disordered systems should be clear at this point. The code, channel input, and noise realization play the role of *quenched disorder* (the sample), while the distribution $\mu_{y,\beta}(\underline{x})$ is the analogue of the *Boltzmann distribution*. In both cases, the problem consists in studying the properties of a probability distribution which is itself a random object.

6.8   Appendix: Proof of Lemma 6.2

We estimate (to the leading exponential order in the large-N limit) the probability $\mathbb{P}_N(\varepsilon)$ for one of the incorrect codewords, $\underline{x}$, to have cost $E(\underline{x}) = N\varepsilon$. The channel output $\underline{y} = (y_1 \cdots y_N)$ is a sequence of N i.i.d. symbols distributed according to

$$Q(y) \equiv \sum_x Q(y|x)P(x), \tag{6.48}$$

and the cost can be rewritten as

$$E(\underline{x}) \equiv -\sum_{i=1}^{N} \log Q(y_i|x_i)$$

$$= -N \sum_{x,y} Q(y) \log Q(y|x) \frac{1}{NQ(y)} \sum_{i=1}^{N} \mathbb{I}(x_i = x, y_i = y). \tag{6.49}$$

There are approximatively $NQ(y)$ positions i such that $y_i = y$, for $y \in \mathcal{Y}$. We assume that there are *exactly* $NQ(y)$ such positions, and that $NQ(y)$ is an integer (of course, this hypothesis is false in general: it is, however, a routine exercise, left to the reader, to show that it can be avoided by a small technical detour). Furthermore, we introduce

$$p_y(x) \equiv \frac{1}{NQ(y)} \sum_{i=1}^{N} \mathbb{I}(x_i = x, \ y_i = y). \tag{6.50}$$

Under the above assumptions, the function $p_y(x)$ is a probability distribution over $x \in \{0, 1\}$ for each $y \in \mathcal{Y}$. Looking at the subsequence of positions i such that $y_i = y$, this function counts the fraction of the x_i's such that $x_i = x$. In other words $p_y(\cdot)$ is the type of the subsequence $\{x_i | y_i = y\}$. Because of eqn (6.49), the cost can be written in terms of these types as follows:

$$E(\underline{x}) = -N \sum_{xy} Q(y) p_y(x) \log Q(y|x). \tag{6.51}$$

Therefore $E(\underline{x})$ depends upon $\underline{x}$ uniquely through the types $\{p_y(\cdot) : y \in \mathcal{Y}\}$, and this dependence is linear in $p_y(x)$. Moreover, according to our definition of the RCE, $x_1, \ldots, x_N$ are i.i.d. random variables with distribution $P(x)$. The probability $\mathbb{P}_N(\varepsilon)$ that $E(\underline{x})/N = \varepsilon$ can therefore be deduced from Corollary 4.5. To the leading exponential order, we get

$$\mathbb{P}_N(\varepsilon) \doteq \exp\{-N\psi(\varepsilon)\log 2\}, \tag{6.52}$$

$$\psi(\varepsilon) \equiv \min_{p_y(\cdot)} \left[\sum_y Q(y) D(p_y \| P) \text{ such that } \varepsilon = -\sum_{xy} Q(y) p_y(x) \log_2 Q(y|x) \right] \tag{6.53}$$

Notes

The random code ensemble dates back to Shannon (1948), who used it (somewhat implicitly) in his proof of the channel coding thorem. A more explicit (and complete) proof was provided by Gallager (1965). The reader can find alternative proofs in standard textbooks such as Cover and Thomas (1991), Csiszár and Körner (1981), and Gallager (1968).

The distance enumerator is a code property that has been extensively investigated in coding theory. We refer for instance to Csiszár and Körner (1981), and Gallager (1968). A treatment of the random code ensemble by analogy with the random energy model was presented by Montanari (2001b). More detailed results in the same spirit can be found in Barg and Forney (2002) and Forney and Montanari (2001). The analogy between coding theory and the statistical physics of disordered systems was put forward by Sourlas (1989). Finite temperature decoding was introduced in by Rujan (1993).

A key ingredient of our analysis was the assumption, already mentioned in Section 1.6.2, that any codeword is a priori equiprobable. The fundamental motivation for such an assumption is the source–channel separation theorem. In simple terms, one does not lose anything in constructing an encoding system in two blocks. First, an ideal source code compresses the data produced by the information source and outputs a sequence of i.i.d. unbiased bits. Then, a channel code adds redundancy to this sequence in order to counteract the noise on the channel. The theory of error-correcting codes focuses on the design and analysis of this second block, leaving the first one to source coding. The interested reader may find proofs of the separation theorem in Cover and Thomas (1991), Csiszár and Körner (1981), and Gallager (1968).

Sphere packing is a classical problem in mathematics, with applications in various branches of science. The book by Conway and Sloane (1998) provides both a very good introduction to this problem and some far-reaching results related to it and its connections to other fields, in particular to coding theory. Finding the densest packing of spheres in $\mathbb{R}^n$ is an open problem when $n \geq 4$.

7
Number partitioning

Number partitioning is one of the most basic optimization problems. It is very easy to state: 'Given the values of N assets, is there a fair partition of them into two sets?' Nevertheless, it is very difficult to solve: it is NP-complete, and the known heuristics are often not very good. It is also a problem with practical applications, for instance in multiprocessor scheduling.

In this chapter, we shall pay special attention to the partitioning of a list of i.i.d. random numbers. It turns out that most heuristics perform poorly on this ensemble of instances. This motivates their use as a benchmark for new algorithms, as well as their analysis. On the other hand, it is relatively easy to characterize analytically the structure of random instances. The main result is that low-cost configurations (those with a small imbalance between the two sets) can be seen as independent energy levels. The model behaves very much like the random energy model of Chapter 5, although with a different energy distribution.

The problem is defined in Section 7.1. Section 7.2 discusses algorithmic aspects: it introduces a complete algorithm and a smart heuristic. These are used in Section 7.3 in order to study numerically the partitioning of i.i.d. random numbers. Then we discuss, in Section 7.4, a simple model in the REM family, the random cost model, which can be analysed in detail by elementary methods. Section 7.5 provides a short description of rigorous results on the partitioning of i.i.d. random numbers: these results show that the random cost model provides a very good approximation to the original problem.

7.1 A fair distribution into two groups?

An instance of the number-partitioning problem is defined by a set of N positive integers $\mathcal{S} = \{a_1, \ldots, a_N\}$ indexed by $i \in [N] \equiv \{1, \ldots, N\}$. One would like to **partition** the integers into two subsets $\{a_i : i \in \mathcal{A}\}$ and $\{a_i : i \in \mathcal{B} \equiv [N] \setminus \mathcal{A}\}$ in such a way as to minimize the discrepancy between the sums of the elements in the two subsets. In other words, a configuration is given by $\mathcal{A} \subseteq [N]$, and its cost is defined by

$$E_{\mathcal{A}} = \left| \left(\sum_{i \in \mathcal{A}} a_i \right) - \left(\sum_{i \in \mathcal{B}} a_i \right) \right| . \tag{7.1}$$

A **perfect partition** is such that the total numbers in each subset equilibrate, which means that $E_{\mathcal{A}} \leq 1$ (actually, $E_{\mathcal{A}} = 0$ if $\sum_i a_i$ is even, and $E_{\mathcal{A}} = 1$ if $\sum_i a_i$ is odd). As usual, one can define several versions of the problem, among which there are the following: *(i) The decision problem*: does there exist a perfect partition? *(ii) The optimization problem*: find a partition of lowest cost.

There are several variants of the problem. So far, we have left the size of $\mathcal{A}$ free. This is called the **unconstrained** version. One can also study a constrained version where the cardinality difference $|\mathcal{A}| - |\mathcal{B}|$ between the two subsets is fixed at some number D. If $D = 0$, the partition is said to be balanced. Here, for simplicity, we shall keep mainly to the unconstrained case.

Exercise 7.1 As a small warm-up, the reader can show that the following are true (maybe by writing a simple exhaustive search program):

(a) The set $\mathcal{S}_1 = \{10, 13, 23, 6, 20\}$ has a perfect partition.

(b) The set $\mathcal{S}_2 = \{6, 4, 9, 14, 12, 3, 15, 15\}$ has a perfect balanced partition.

(c) In the set $\mathcal{S}_3 = \{93, 58, 141, 209, 179, 48, 225, 228\}$, the lowest possible cost is 5.

(d) In the set $\mathcal{S}_4 = \{2474, 1129, 1388, 3752, 821, 2082, 201, 739\}$, the lowest possible cost is 48.

7.2   Algorithmic issues

7.2.1   An NP-complete problem

In order to understand the complexity of the problem, one must first measure its size. This is in turn given by the number of characters required for specifying a particular instance. In number partitioning, this depends crucially on how large the integers can be. Imagine that we restrict ourselves to the case

$$a_i \in \{1, \ldots, 2^M\} \quad \forall\, i \in \{1, \ldots, N\}, \tag{7.2}$$

so that each of the N integers can be encoded with M bits. The entire instance can then be encoded in NM bits. An exhaustive search obviously finds a solution in 2^N operations for unbounded numbers (any M). On the other hand, for bounded numbers (fixed M) and N going to infinity, the algorithm defined in the exercise below finds a solution in a time of order $N^2\, 2^M$. It turns out that no known algorithm solves the number-partitioning problem in a time bounded from above by a power of its size, NM. In fact, number partitioning is NP-complete and is considered to be among the fundamental problems in this class.

Exercise 7.2 Consider the following 'transfer matrix' (or 'dynamic-programming') approach to number partitioning with bounded numbers $\{a_i\}$. For any $k \in \{1, \ldots, N\}$ and any $q \in \{1, \ldots, N2^M\}$, define

$$Z_k(q) = \begin{cases} 1 & \text{if } q \text{ can be written as } \sum_{r=1}^{k} n_r a_r \text{ with } n_r \in \{0, 1\}, \\ 0 & \text{otherwise.} \end{cases} \tag{7.3}$$

Write a recursion relation expressing Z_{k+1} in terms of Z_k. Show that the set of numbers $\{Z_N(q)\}, q \in \{1, \ldots, N2^M\}$ can be computed in a time of order $N^2\, 2^M$. How could you then use Z_N to solve the number-partitioning problem?

7.2.2 A simple heuristic and a complete algorithm

There is no good algorithm for the number-partitioning problem. One of the best heuristics, due to Karmarkar and Karp (KK), uses the following idea. We start from a list $a_1, \ldots, a_N$ which coincides with the original set of integers, and reduce it by erasing two elements a_i and a_j from the list and replacing them by the difference $|a_i - a_j|$, if this difference is non-zero. This substitution means that a decision has been made to place a_i and a_j in two different subsets (but without fixing which subsets they are in). One then iterates this procedure as long as the list contains two or more elements. If in the end one finds either an empty list or the list $\{1\}$, then there exists a perfect partitioning. In the opposite case, the remaining integer is the cost of one particular partitioning, but the problem could have better solutions. Of course, there is a lot of flexibility and ingenuity involved in the best choice of the elements a_i and a_j selected at each step. In the KK algorithm one chooses the two largest numbers.

> **Example 7.1** Let us see how the KK algorithm works with the first list in Exercise 7.1: $\{10, 13, 23, 6, 20\}$. In the first iteration, we substitute 23 and 20 by 3, giving the list $\{10, 13, 6, 3\}$. The next step gives $\{3, 6, 3\}$, then $\{3, 3\}$, and then $\emptyset$, showing that there exists a perfect partition. Readers can find out for themselves how to systematically reconstruct the partition.

A modification due to Korf transforms the KK heuristic into a complete algorithm, which will return the best partitioning (possibly in exponential time). Each time one eliminates two elements a_i and a_j, two new lists are built: a 'left' list, which contains $|a_i - a_j|$ (this corresponds to placing a_i and a_j in different groups), and a 'right' list which contains $a_i + a_j$ (it corresponds to placing a_i and a_j in the same group). By iterating in this way, one constructs a tree with 2^{N-1} terminal nodes, each one containing the cost of a valid partition. Vice versa, the cost of each possible partition is reported at one of the terminal nodes (note that each of the 2^N possible partitions $\mathcal{A}$ is equivalent to its complement $[N] \setminus \mathcal{A}$). If one is interested only in the decision 'is there a perfect partition?', the tree can be pruned as follows. Each time one encounters a list whose largest element is larger than the sum of all other elements plus 1, this list cannot lead to a perfect partition. One can therefore avoid constructing the subtree whose root is such a list. Figure 7.1 shows a simple example of application of this algorithm.

7.3 Partition of a random list: Experiments

A natural way to generate random instances of the number-partitioning problem is to choose the N input numbers a_i as i.i.d. random variables. Here we will be interested in the case where they are uniformly distributed in the set $\{1, \ldots, 2^M\}$. As we discussed in Chapter 3, one can use these random instances in order to test the typical performance of algorithms, but we will also be interested in natural probabilistic issues, such as the distribution of the optimal cost, in the limits where N and M go to ∞.

It is useful to first get an intuitive feeling for the respective roles of N (the size of the set) and M (the number of digits of each a_i in base 2). Consider the instances

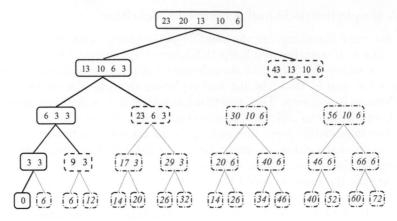

Fig. 7.1 A complete search algorithm. Starting from a list, one erases the two largest numbers a_i and a_j and generates two new lists: the left one contains $|a_i - a_j|$, and the right one contains $a_i + a_j$. At the bottom of the tree, every leaf contains the cost of a valid partition. In a search for a perfect partition, the tree can be pruned at the dashed leaves because the largest number is bigger than the sum of the others: the dash–dotted lists are not generated. The KK heuristic picks up only the left branch. In this example, it is successful and finds the unique perfect partition.

$\mathcal{S}_2, \mathcal{S}_3, \mathcal{S}_4$ of Exercise 7.1. Each of them contains $N = 8$ random numbers, but they are randomly generated with $M = 4$, $M = 8$, and $M = 16$, respectively. Clearly, the larger M is, the larger is the typical value of the a_i, and the more difficult it is to distribute them fairly. Consider the costs of all possible partitions: it is reasonable to expect that in about half of the partitions, the most significant bit of the cost is 0. Among these, about one-half should have the second significant bit equal to 0. The number of partitions is 2^{N-1}, and this qualitative argument can thus be iterated roughly N times. This leads one to expect that, in a random instance with large N, there will be a significant chance of having a perfect partition if $N > M$. In contrast, for $N < M$, the typical cost of the best partition should behave like 2^{M-N}.

This intuitive reasoning turns out to be essentially correct, as far as the leading exponential behaviour in N and M is concerned. Here, we first provide some numerical evidence, obtained with the complete algorithm of Section 7.2.2 for relatively small systems. In the next section, we shall validate our conclusions by a sharper analytical argument.

Figure 7.2 shows a numerical estimate of the probability $p_{\mathrm{perf}}(N, M)$ that a randomly generated instance has a perfect partition, plotted versus N. This was obtained by sampling n_{stat} instances of the problem for each pair N, M considered (here $n_{\mathrm{stat}} = 10^4$, 10^3, and 10^2 when $M = 8$, 16, and 24, respectively), and solving each instance by complete enumeration. The probability $p_{\mathrm{perf}}(N, M)$ was estimated as the fraction of the sampled instances for which a perfect partitioning was found. The standard deviation of such an estimate is $\sqrt{p_{\mathrm{perf}}(1 - p_{\mathrm{perf}})/n_{\mathrm{stat}}}$.

For a fixed value of M, $p_{\mathrm{perf}}(N, M)$ crosses over from a value close to 0 at small N to a value close to 1 at large N. The typical values of N where the crossover takes

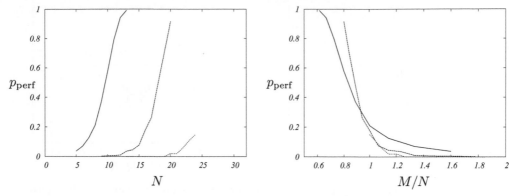

Fig. 7.2 A numerical study of randomly generated number-partitioning instances, where the a_i are uniformly distributed in $\{1, \dots 2^M\}$, with $\sum_i a_i$ even. The fraction of samples with a perfect balanced partition is plotted versus N (*left plot*: from left to right, $M = 8, 16, 24$), and versus $\kappa = M/N$ (*right plot*). In the limit $N \to \infty$ at fixed κ, it turns out that the probability becomes a step function, equal to 1 for $\kappa < 1$, and to 0 for $\kappa > 1$ (see also Fig. 7.4).

place seem to grow proportionally to M. It is useful to look at the same data from a slightly different perspective by defining the ratio

$$\kappa = \frac{M}{N}, \tag{7.4}$$

and considering p_{perf} as a function of N and κ. The plot of $p_{\mathrm{perf}}(\kappa, N)$ versus κ at fixed N shows a very interesting behaviour; see Fig. 7.2, right frame. A careful analysis of the numerical data indicates that $\lim_{N \to \infty} p_{\mathrm{perf}}(\kappa, N) = 1$ for $\kappa < 1$, and 0 for $\kappa > 1$. We stress that the limit $N \to \infty$ is taken with κ kept fixed (and therefore letting $M \to \infty$ in proportion to N). As we shall see in the following, we face here a typical example of a phase transition, in the sense introduced in Chapter 2. The behaviour of a generic large instance changes completely when the control parameter κ crosses a critical value $\kappa_{\mathrm{c}} \equiv 1$. For $\kappa < 1$, almost all instances of the problem have a perfect partition (in the large-N limit), for $\kappa > 1$ almost none of them can be partitioned perfectly. This phenomenon has important consequences for the computational difficulty of the problem. A good measure of the performance of Korf's complete algorithm is the number R of lists generated in the tree before the optimal partition is found. In Fig. 7.3, we plot the quantity $\log_2 R$ averaged over the same instances as those which we used for the estimation of p_{perf} in Fig. 7.2. The size of the search tree first grows exponentially with N and then reaches a maximum around $N \approx M$. We see a peak in $\log_2 R$ plotted as a function of κ, somewhere around $\kappa = \kappa_{\mathrm{c}} = 1$: problems close to the critical point are the hardest ones for the algorithm considered. A similar behaviour is found with other algorithms, and in fact we shall encounter it in many other decision problems, such as the satisfiability and colouring problems. When a class of random instances shows a phase transition as a function of one parameter, it is generally the case that the most difficult instances are found in the neighbourhood of the phase transition.

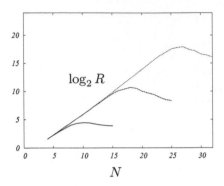 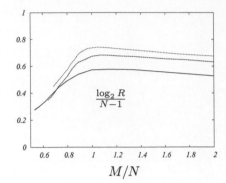

Fig. 7.3 *Left*: average of $\log_2 R$, where R is the size of the search tree for the KK algorithm. The three curves correspond to $M = 8$, 16, and 24 (from *left* to *right*). The size grows exponentially with N, and then reaches a maximum for $N \approx M$. *Right*: the average of $\log_2 R/(N-1)$ plotted versus $\kappa = M/N$.

7.4 The random cost model

7.4.1 Definition of the model

Consider, as before, the probability space of random instances constructed by taking the numbers a_j as i.i.d., uniformly distributed in $\{1, \ldots, 2^M\}$. For a given partition $\mathcal{A}$, the cost $E_\mathcal{A}$ is a random variable with a probability distribution $\mathcal{P}_\mathcal{A}$. Obviously, the costs of two partitions $\mathcal{A}$ and $\mathcal{A}'$ are correlated random variables. The random cost approximation consists in neglecting these correlations. Such an approximation can be applied to any kind of problem, but it is not always a good one. Remarkably, as discovered by Mertens, the random cost approximation turns out to be 'essentially exact' for the partitioning of i.i.d. random numbers.

In order to state precisely the above-mentioned approximation, one defines a random cost model (RCM), which is similar to the REM of Chapter 5. A sample is defined by the costs of all 2^{N-1} 'partitions' (here we identify the two complementary partitions $\mathcal{A}$ and $[N] \backslash \mathcal{A}$). The costs are supposed to be *i.i.d. random variables* drawn from the probability distribution $\mathcal{P}$. In order to mimic the random number-partitioning problem, $\mathcal{P}$ is taken to be the same as the distribution of the cost of a random partition $\mathcal{A}$ in the original problem:

$$\mathcal{P} \equiv \frac{1}{2^{N-1}} \sum_\mathcal{A} \mathcal{P}_\mathcal{A} . \tag{7.5}$$

Here $\mathcal{P}_\mathcal{A}$ is the distribution of the cost of a partition $\mathcal{A}$ in the original number-partitioning problem.

Let us analyse the behaviour of $\mathcal{P}$ for large N. The cost of a randomly chosen partition in the original problem is given by $|\sum_i \sigma_i a_i|$, where the σ_i are i.i.d. variables taking values ± 1 with probability $1/2$. For large N, the distribution of $\sum_i \sigma_i a_i$ is characterized by the central limit theorem, and $\mathcal{P}$ is obtained by restricting it to the positive domain. In particular, the cost of a partition will be, with high probability, of order $\sqrt{N \alpha_M^2}$, where

$$\alpha_M^2 \equiv \mathbb{E}\, a^2 = \frac{1}{3}\, 2^{2M} + \frac{1}{2}\, 2^M + \frac{1}{6}\,.$$ (7.6)

Moreover, for any $0 \le x_1 < x_2$,

$$\mathcal{P}\left(\frac{E}{\sqrt{N\alpha_M^2}} \in [x_1, x_2]\right) \simeq \sqrt{\frac{2}{\pi}} \int_{x_1}^{x_2} e^{-x^2/2} \, \mathrm{d}x\,.$$

Finally, the probability of a perfect partition $\mathcal{P}(E = 0)$ is just the probability of return to the origin for a random walk with steps $\sigma_i a_i \in \{-2^M, \dots, -1\} \cup \{1, \dots, 2^M\}$. Assuming for simplicity that $\sum_i a_i$ is even, we get

$$\mathcal{P}(0) \simeq 2\, \frac{1}{\sqrt{2\pi N\alpha_M^2}} \simeq \sqrt{\frac{6}{\pi N}} 2^{-M}\,,$$ (7.7)

where $1/\sqrt{2\pi N\alpha_M^2}$ is the density of a normal random variable of mean 0 and variance $N\alpha_M^2$ near the origin, and the extra factor of 2 comes from the fact that the random walk is on even integers only.

As we shall show in the next sections, the RCM is a good approximation to the original number-partitioning problem. An intuitive explanation of this property can be found in the exercise below.

Exercise 7.3 Consider two random, uniformly distributed, independent partitions $\mathcal{A}$ and $\mathcal{A}'$. Let $\mathcal{P}(E, E')$ denote the joint probability of their energies when the numbers $\{a_i\}$ are i.i.d. and uniformly distributed over $\{1, \dots, 2^M\}$. Show that $\mathcal{P}(E, E') = \mathcal{P}(E)\mathcal{P}(E')[1+o(1)]$ in the large-N, M limit, if $E, E' < C\, 2^M$ for some fixed C.

7.4.2 Phase transition

We can now proceed with the analysis of the RCM. We shall first determine the phase transition, then study the phase $\kappa > 1$, where, typically, no perfect partition can be found, and finally study the phase $\kappa < 1$, where an exponential number of perfect partitions exist.

Consider a random instance of the RCM. The probability that *no* perfect partition exist is just the probability that each partition has a strictly positive cost. Since, within the RCM, the 2^{N-1} partitions have i.i.d. costs with distribution $\mathcal{P}$, we have

$$1 - p_{\text{perf}}(\kappa, N) = [1 - \mathcal{P}(0)]^{2^{N-1}}\,.$$ (7.8)

In the large-N limit with fixed κ, the zero-cost probability is given by eqn (7.7). In particular, $\mathcal{P}(0) \ll 1$. Therefore,

$$p_{\text{perf}}(\kappa, N) = 1 - \exp[-2^{N-1}\mathcal{P}(0)] + o(1) = 1 - \exp\left[-\sqrt{\frac{3}{2\pi N}}\, 2^{N\,(1-\kappa)}\right] + o(1)\,.$$ (7.9)

This expression predicts a phase transition for the RCM at $\kappa_c = 1$. Notice, in fact, that $\lim_{N\to\infty} p_{\text{perf}}(\kappa, N) = 1$ if $\kappa < 1$, and 0 if $\kappa > 1$. Moreover, eqn (7.9) describes

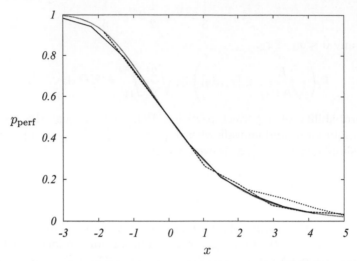

Fig. 7.4 Finite-size scaling plot for the data of Fig. 7.2. The (estimated) probability of perfect partition $p_{\mathrm{perf}}(N, M)$ is plotted versus the rescaled variable $x = N(\kappa - \kappa_c) + (1/2)\log_2 N$. The agreement with the theoretical prediction (7.10) is very good.

the precise behaviour of $p_{\mathrm{perf}}(\kappa, N)$ around the critical point κ_c for finite N. Let us define the variable $x = N(\kappa - \kappa_c) + (1/2)\log_2 N$. In the limit $N \to \infty$ and $\kappa \to \kappa_c$ at *fixed* x, one finds the following crossover behaviour:

$$\lim_{\substack{N \to \infty \\ \kappa \to \kappa_c}} p_{\mathrm{perf}}(\kappa, N) = 1 - \exp\left[-\sqrt{\frac{3}{2\pi}}\, 2^{-x}\right]. \tag{7.10}$$

This is an example of **finite-size scaling** behaviour.

In order to compare the above prediction with our numerical results for the original number-partitioning problem, we plot in Fig. 7.4 $p_{\mathrm{perf}}(\kappa, N)$ versus the scaling variable x. Here we have used the same data as that presented in Fig. 7.2, just changing the horizontal scale from N to x. The good collapse of the curves for various values of M provides evidence for the claim that the number-partitioning problem is indeed asymptotically equivalent to the RCM and undergoes a phase transition at $\kappa = 1$.

Exercise 7.4 The argument above assumes that $\sum_i a_i$ is even. This is a condition which was imposed in the simulation whose results are presented in Fig. 7.4. How should one modify the estimate of $\mathcal{P}(0)$ in eqn (7.7) when $\sum_i a_i$ is odd? Show that, in this case, if one keeps the definition $x = N(\kappa - \kappa_c) + (1/2)\log_2 N$, the scaling function becomes $1 - \exp\left[-\sqrt{6/\pi}\, 2^{-x}\right]$. Run a simulation to check this prediction.

7.4.3 Study of the two phases

Let us now study the minimum cost in the phase $\kappa > 1$. The probability that all configurations have a cost larger than E is

$$\mathbb{P}(\forall \mathcal{A} : E_{\mathcal{A}} > E) = \left(1 - \sum_{E'=0}^{E} \mathcal{P}(E') \right)^{2^{N-1}} . \tag{7.11}$$

This probability is non-trivial (i.e. it is bounded away from 0 and 1) if $\sum_{E'=0}^{E} \mathcal{P}(E') = O(2^{-N})$. This sum can be estimated by substituting $\mathcal{P}(E') \to \mathcal{P}(0)$, which gives the condition $E \simeq 1/(\mathcal{P}(0)2^{N-1}) \sim 2^{M-N}\sqrt{N}$ (note that, as this value of E is much smaller than the scale over which $\mathcal{P}(E)$ varies significantly (see eqn (7.7)), the substitution of $\mathcal{P}(0)$ for $\mathcal{P}(E')$ is indeed consistent). We therefore get, from eqn (7.11)

$$\lim_{N \to \infty} \mathbb{P}\left(\forall \mathcal{A} : E_{\mathcal{A}} > \frac{\varepsilon}{\mathcal{P}(0)2^{N-1}} \right) = e^{-\varepsilon}\, \mathbb{I}(\varepsilon \geq 0) . \tag{7.12}$$

In particular, the mean of the distribution on the right-hand side is equal to 1. This implies that the expectation of the lowest cost in the problem is $\mathbb{E}\, E_{\mathrm{gs}} = \sqrt{2\pi N/3}\, 2^{N(\kappa-1)}$. These predictions also fit the numerical results for number partitioning very well.

Exercise 7.5 Show that the probability density of the k-th lowest-cost configuration, in terms of the rescaled variable ε, is $(\varepsilon^{k-1}/(k-1)!)\exp(-\varepsilon)\,\mathbb{I}(\varepsilon > 0)$. This is a typical case of extreme value statistics for bounded i.i.d. variables.

For the phase $\kappa < 1$, we already know that, for all but a vanishing fraction of samples, there exists at least one configuration with zero cost. It is instructive to count the number of zero-cost configurations. Since each configuration has zero cost independently with probability $\mathcal{P}(0)$, the number Z of zero-cost configurations is a binomial random variable with distribution

$$P(Z) = \binom{2^{N-1}}{Z} \mathcal{P}(0)^Z \, [1 - \mathcal{P}(0)]^{2^{N-1}-Z} . \tag{7.13}$$

In particular, for large N, Z concentrates around its average value $Z_{\mathrm{av}} \doteq 2^{N(1-\kappa)}$. One can define an entropy density of the ground state as

$$s_{\mathrm{gs}} = \frac{1}{N} \log_2 Z . \tag{7.14}$$

The RCM result in eqn (7.13) predicts that for $\kappa < 1$, the entropy density is close to $(1 - \kappa)$ with high probability. Once again, numerical simulations of the original number-partitioning problem confirm this expectation.

Exercise 7.6 Using the integral representation of the logarithm

$$\log_2 x = \int_0^\infty \frac{1}{t} \left(\mathrm{e}^{-t \log 2} - \mathrm{e}^{-tx} \right) \, \mathrm{d}t \,, \tag{7.15}$$

compute $\mathbb{E} s_{\mathrm{gs}}$ directly. It will be useful to notice that the integral over t is dominated by very small values of t, of order $1/(2^{N-1} \mathcal{P}(0))$. One then easily finds $\mathbb{E} s_{\mathrm{gs}} \simeq (1/N) \log_2(2^{N-1} \mathcal{P}(0)) \simeq 1 - \kappa$.

7.5 Partition of a random list: Rigorous results

A detailed rigorous characterization of the phase diagram for the partitioning of random numbers confirms the predictions of the RCM. We shall first state some of the exact results known for the balanced partitioning of N numbers. For definiteness we keep as before to the case where the a_i are i.i.d. and uniformly distributed in $\{1, \ldots, 2^M\}$, and both N and $\sum_{i=1}^N a_i$ are even. The following results hold in the 'thermodynamic limit' $N, M \to \infty$ with fixed $\kappa = M/N$.

Theorem 7.2 *There is a phase transition at $\kappa = 1$. For $\kappa < 1$, with high probability, a randomly chosen instance has a perfect balanced partition. For $\kappa > 1$, with high probability, a randomly chosen instance does not have a perfect balanced partition.*

Theorem 7.3 *In the phase $\kappa < 1$, the entropy density (7.14) of the number of perfect balanced partitions converges in probability to $s = 1 - \kappa$.*

Theorem 7.4 *Define $\overline{E} = 2^{N(\kappa-1)} \sqrt{2\pi N/3}$, and let $E_1 \leq \cdots \leq E_k$ be the k lowest costs, with k fixed. The k-tuple $\left(\varepsilon_1 = E_1/\overline{E}, \ldots, \varepsilon_k = E_k/\overline{E} \right)$ then converges in distribution to $(W_1, W_1 + W_2, \ldots, W_1 + \ldots W_k)$, where the W_i are i.i.d. random variables with distribution $P(W_i) = \mathrm{e}^{-W_i} \, \mathbb{I}(W_i \geq 0)$. In particular, the (rescaled) optimal cost distribution converges to $P(\varepsilon_1) = \mathrm{e}^{-\varepsilon_1} \, \mathbb{I}(\varepsilon_1 \geq 0)$.*

Note that these results all agree with the RCM. In particular, Theorem 7.4 states that, for fixed k and $N \to \infty$, the lowest k costs are i.i.d. variables, as assumed in the RCM. This explains why the random cost approximation is so good.

The proofs of these theorems (and of more detailed results concerning the scaling in the neighbourhood of the phase transition point $\kappa = 1$) are all based on the analysis of an integral representation for the number of partitions with a given cost, which we shall derive below. We shall then outline the general strategy by proving the existence of a phase transition, as in Theorem 7.2, and we refer the reader to the original literature for the other proofs.

7.5.1 Integral representation

For simplicity, we keep to the case where $\sum_i a_i$ is even: similar results can be obtained in the case of an odd sum (but the lowest cost is then equal to 1).

Proposition 7.5 *Given a set* $\mathcal{S} = \{a_1, \ldots, a_N\}$ *with* $\sum_i a_i$ *even, the number* Z *of partitions with cost* $E = 0$ *can be written as*

$$Z = 2^{N-1} \int_{-\pi}^{\pi} \prod_{j=1}^{N} \cos(a_j x) \frac{\mathrm{d}x}{2\pi}. \qquad (7.16)$$

Proof We represent the partition $\mathcal{A}$ by writing $\sigma_i = 1$ if $i \in \mathcal{A}$, and $\sigma_i = -1$ if $i \in \mathcal{B} = [N] \setminus \mathcal{A}$. We can write $Z = \frac{1}{2} \sum_{\sigma_1, \ldots, \sigma_N} \mathbb{I}\left(\sum_{j=1}^{N} \sigma_j a_j = 0\right)$, where the factor $1/2$ comes from the symmetry between $\mathcal{A}$ and $\mathcal{B}$ (the same partition is represented by the sequence $\sigma_1, \ldots, \sigma_N$ and by $-\sigma_1, \ldots, -\sigma_N$). We use the integral representation valid for any integer number a,

$$\mathbb{I}(a = 0) = \int_{-\pi}^{\pi} \mathrm{e}^{ixa} \frac{\mathrm{d}x}{2\pi}, \qquad (7.17)$$

which gives

$$Z = \frac{1}{2} \sum_{\sigma_1, \ldots, \sigma_N} \int_{-\pi}^{\pi} \mathrm{e}^{ix(\sum_j \sigma_j a_j)} \frac{\mathrm{d}x}{2\pi}. \qquad (7.18)$$

The sum over the σ_i gives the integral representation (7.16) $\square$

Exercise 7.7 Show that a similar representation holds for the number of partitions with cost $E \geq 1$, with an extra factor $2\cos(Ex)$ in the integrand. For the case of balanced partitions, find a similar representation with a two-dimensional integral.

The integrand of eqn (7.16) is typically exponential in N and oscillates wildly. It is thus tempting to compute the integral by the method of steepest descent. This strategy yields correct results for the phase $\kappa \leq 1$, but it is not easy to control it rigorously. Hereafter, we shall use simple first- and second-moment estimates of the integral, which are powerful enough to derive the main features of the phase diagram. Finer control gives more accurate predictions which go beyond this presentation.

7.5.2 Moment estimates

We start by evaluating the first two moments of the number of perfect partitions Z.

Proposition 7.6 *In the thermodynamic limit, the first moment of* Z *behaves as*

$$\mathbb{E} Z = 2^{N(1-\kappa)} \sqrt{\frac{3}{2\pi N}} (1 + \Theta(1/N)). \qquad (7.19)$$

Proof The expectation value is taken over choices of a_i where $\sum_i a_i$ is even. Let us use a modified expectation, denoted by $\mathbb{E}_i$, over all choices of $a_1, \ldots, a_N$, without any parity constraint, so that the a_i are i.i.d. Clearly, $\mathbb{E}_i Z = (1/2)\mathbb{E} Z$, because a perfect partition can be obtained only in the case where $\sum_i a_i$ is even, and this happens with probability $1/2$.

Because of the independence of the a_i in the expectation $\mathbb{E}_i$, one gets from eqn (7.16)

$$\mathbb{E}\,Z = 2\mathbb{E}_i Z = 2^N \int_{-\pi}^{\pi} [\mathbb{E}_i \cos(a_1 x)]^N \, \frac{dx}{2\pi}\,. \tag{7.20}$$

The expectation of the cosine is

$$\mathbb{E}_i \cos(a_1 x) = 2^{-M} \cos\left(\frac{x}{2}(2^M + 1)\right) \frac{\sin(2^M x/2)}{\sin(x/2)} \equiv g(x)\,. \tag{7.21}$$

A little thought shows that the integral in eqn (7.20) is dominated in the thermodynamic limit by values of x very near to 0. More precisely, we can rescale the variable as $x = \hat{x}/(2^M \sqrt{N})$. We then have $g(x) = 1 - \hat{x}^2/(6N) + \Theta(1/N^2)$. The leading behaviour of the integral (7.20) at large N is thus given by

$$\mathbb{E}\,Z \simeq 2^{N-M} \frac{1}{\sqrt{N}} \int_{-\infty}^{\infty} \exp\left(-\frac{\hat{x}^2}{6}\right) \frac{d\hat{x}}{2\pi} \simeq 2^{N-M} \sqrt{\frac{3}{2\pi N}}\,, \tag{7.22}$$

up to corrections of order $1/N$. $\square$

Exercise 7.8 Show that, for E even, with $E \leq C2^M$ for a fixed C, the number of partitions with cost E is also given by eqn (7.19) in the thermodynamic limit.

Proposition 7.7 *When $\kappa < 1$, the second moment of Z behaves in the thermodynamic limit as*

$$\mathbb{E}\,Z^2 = [\mathbb{E}\,Z]^2 \, (1 + \Theta(1/N))\,. \tag{7.23}$$

Proof We again release the constraint of an even $\sum_i a_i$, so that

$$\mathbb{E}\,Z^2 = 2^{2N-1} \int_{-\pi}^{\pi} \frac{dx_1}{2\pi} \int_{-\pi}^{\pi} \frac{dx_2}{2\pi} \, [\mathbb{E} \cos(a_1 x_1) \cos(a_1 x_2)]^N \tag{7.24}$$

The expectation of the product of the two cosines is

$$\mathbb{E} \cos(a_1 x_1) \cos(a_1 x_2) = \frac{1}{2} \, [g(x_+) + g(x_-)]\,, \tag{7.25}$$

where $x_\pm = x_1 \pm x_2$. In order to find out which regions of the integration domain are important in the thermodynamic limit, one must be careful because the function $g(x)$ is periodic with a period of 2π. The double integral is performed in the square $[-\pi, +\pi]^2$. The region of this square where g can be very close to 1 are the 'centre' where $x_1, x_2 = \Theta(1/(2^M \sqrt{N}))$, and the four corners, close to $(\pm\pi, \pm\pi)$, obtained from the centre by a shift of $\pm\pi$ in x_+ or in x_-. Because of the periodicity of $g(x)$, the total contribution of the four corners equals that of the centre. Therefore one can first compute the integral near the centre, using the change of variables $x_{1(2)} = \hat{x}_{1(2)}/(2^M \sqrt{N})$. The correct value of $\mathbb{E} Z^2$ is equal to twice the result of this integral. The remaining part of the computation is straightforward, and indeed gives $\mathbb{E} Z^2 \simeq 3 \cdot 2^{2N(1-\kappa)}/(2\pi N)$.

In order for this argument to be correct, one must show that the contributions from outside the centre are negligible in the thermodynamic limit. The leading correction comes from regions where $x_+ = \Theta(1/(2^M \sqrt{N}))$, while x_- is arbitrary. One can explicitly evaluate the integral in such a region by using the saddle point approximation. The result is of order $\Theta(2^{N(1-\kappa)}/N)$. Therefore, for $\kappa < 1$ the relative contributions from outside the centre (and the corners) are exponentially small in N. A careful analysis of the above two-dimensional integral can be found in the literature. $\square$

Propositions 7.6 and 7.7 above have the following important implications. For $\kappa > 1$, $\mathbb{E} Z$ is exponentially small in N. Since Z is a non-negative integer, this implies (by the first-moment method) that, in most instances, Z is indeed 0. For $\kappa < 1$, $\mathbb{E} Z$ is exponentially large. Moreover, the normalized random variable $Z/\mathbb{E} Z$ has a small second moment, and therefore small fluctuations. An analysis similar to the one we did in Section 5.2.1 for the REM then shows that Z is positive with high probability. We have thus proved the existence of a phase transition at $\kappa_c = 1$, i.e. Theorem 7.2.

Exercise 7.9 Define as usual the partition function at inverse temperature β as $Z(\beta) = \sum_{\mathcal{A}} e^{-\beta E_{\mathcal{A}}}$. Using the integral representation

$$e^{-|U|} = \int_{-\infty}^{\infty} \frac{dx}{\pi} \frac{1}{1+x^2} e^{-ixU} \tag{7.26}$$

and the relation $\sum_{k \in \mathbb{Z}} 1/(1 + x^2 k^2) = \pi/(x \tanh(\pi/x))$, show that the 'annealed average' for i.i.d. numbers a_i is

$$\mathbb{E}_i(Z) = 2^{N(1-\kappa)} \sqrt{\frac{3}{2\pi N}} \frac{1}{\tanh(\beta/2)} (1 + \Theta(1/N)). \tag{7.27}$$

Notes

A nice elementary introduction to number partitioning is provided by Hayes (2002). The NP-complete nature of this problem is a classical result which can be found in textbooks such as Papadimitriou (1994) and Garey and Johnson (1979). The Karmarkar–Karp algorithm was introduced in a technical report (Karmarkar and Karp, 1982). Korf's complete algorithm is given in Korf (1998).

There has been a lot of work on the partitioning of random i.i.d. numbers. In particular, the case where a_i is uniform in $[0, 1]$ has been studied in detail. This can be regarded as equivalent to uniform costs in $\{1, \ldots, 2^M\}$, in the large-M limit. The scaling of the cost of the optimal solution in this case was studied as early as 1986 by Karmarkar *et al.* (1986). On the algorithmic side, this is a very challenging problem. As we have seen, the optimal partition has a cost $O(\sqrt{N}2^{-N})$; however, all known heuristics perform badly on this problem. For instance, the KK heuristic finds a solution with a cost $\Theta\left(\exp\left[-0.72(\log N)^2\right]\right)$, which is very far from the optimal scaling (Yakir, 1996).

The phase transition was identified numerically by Gent and Walsh (1998), and studied by statistical-physics methods by Ferreira and Fontanari (1998) and Mertens (1998). Mertens also introduced the random cost model (Mertens, 2000). His review paper (Mertens, 2001) provides a good summary of these publications, and will help to solve Exercises 7.3, 7.5, and 7.8. The parity questions discussed in Exercise 7.4 have been studied by Bauke (2002).

Elaborating on these statistical-mechanics treatments, detailed rigorous results have been obtained for the unconstrained problem (Borgs *et al.*, 2001) and more recently, for the constrained case (Borgs *et al.*, 2003). These results go far beyond the theorems which we have stated here, and the interested reader is encouraged to study the original papers. Readers will also find there all the technical details needed to fully control the integral representation used in Section 7.5, and the solutions to Exercises 7.6 and 7.7.

8

Introduction to replica theory

Over the past 30 years, the replica method has evolved into a rather sophisticated tool for attacking theoretical problems as diverse as spin glasses, protein folding, vortices in superconductors, and combinatorial optimization. In this book, we adopt a different (but equivalent and, in our view, more concrete) approach: the 'cavity method'. In fact, readers can skip this chapter without great harm to their understanding of the rest of this book.

It can be instructive, nevertheless, to have some knowledge of replicas: the replica method is an amazing construction which is incredibly powerful. It is not yet a rigorous method: it involves some formal manipulations, and a few prescriptions which may appear arbitrary. Nevertheless, these prescriptions are fully specified, and the method can be regarded as an 'essentially automatic' analytical tool. Moreover, several of its most important predictions have been confirmed rigorously through alternative approaches. Among its most interesting aspects is the role played by 'overlaps' among replicas. It turns out that the subtle probabilistic structure of the systems under study are often most easily phrased in terms of such variables.

Here we shall take advantage of the simplicity of the random energy model defined in Chapter 5 to introduce replicas. This is the topic of Section 8.1. A more complicated spin model is introduced and discussed in Section 8.2. In Section 8.3, we study the relationship between the simplest replica-symmetry-breaking scheme and extreme value statistics. Finally, in Section 8.4, we briefly explain how to perform a local stability analysis in replica space. This is one of the most common consistency checks in the replica method.

8.1 Replica solution of the random energy model

As we saw in Section 5.1, a sample (or instance) of the REM is given by the values of 2^N energy levels E_j, with $j \in \{1, \dots, 2^N\}$. The energy levels are i.i.d. Gaussian random variables with mean 0 and variance $N/2$. A configuration of the REM is specified by the index j of one energy level. The partition function for a sample with energy levels $\{E_1 \dots, E_{2^N}\}$ is

$$Z = \sum_{j=1}^{2^N} \exp\left(-\beta E_j\right) , \qquad (8.1)$$

and is itself a random variable (in physicists' language, 'Z fluctuates from sample to sample'). In Chapter 5 we showed that intensive thermodynamic potentials are self-averaging, meaning that their distribution is sharply concentrated around the mean value in the large-N limit. Among these quantities, a prominent role is played by the free-energy density $f = -(\beta N)^{-1} \log Z$. Other potentials can in fact be computed from derivatives of the free energy. Unlike thermodynamic potentials, the partition function has a broad distribution even for large sizes. In particular, its average is dominated, in the low temperature phase, by extremely rare samples. In order to describe typical samples, one has to compute the average of the log-partition function, $\mathbb{E} \log Z$, i.e., up to a constant, the average free-energy density.

It turns out that computing integer moments of the partition function $\mathbb{E} Z^n$, with $n \in \mathbb{N}$, is much easier than computing the average log-partition function $\mathbb{E} \log Z$. This happens because Z is the sum of a large number of 'simple' terms.

If, on the other hand, we were able to compute $\mathbb{E} Z^n$ for any *real* n or, at least, for n small enough, the average log-partition function could be determined using, for instance, the relation

$$\mathbb{E} \log Z = \lim_{n \to 0} \frac{1}{n} \log(\mathbb{E} Z^n) \ . \tag{8.2}$$

The idea is to carry out the calculation of $\mathbb{E} Z^n$ 'as if' n were an integer. At a certain point, after obtaining a manageable enough expression, we 'remember' that n has in fact to be a real number and take this into account. As we shall see this whole line of approach has some of the flavour of an analytic continuation, but in fact it has quite a few extra grains of salt...

The first step consists in noticing that Z^n can be written as an n-fold sum

$$Z^n = \sum_{i_1 \ldots i_n = 1}^{2^N} \exp\left(-\beta E_{i_1} - \cdots - \beta E_{i_n}\right) \ . \tag{8.3}$$

This expression can be interpreted as the partition function of a new system. A configuration of this system is given by the n-tuple $(i_1, \ldots, i_n)$, with $i_a \in \{1, \ldots, 2^N\}$, and its energy is $E_{i_1 \ldots i_n} = E_{i_1} + \cdots + E_{i_n}$. In other words, the new system is formed from n statistically independent copies of the original one. We shall refer to such copies as **replicas**.

In order to evaluate the average of eqn (8.3), it is useful to first rewrite it as

$$Z^n = \sum_{i_1 \ldots i_n = 1}^{2^N} \prod_{j=1}^{2^N} \exp\left[-\beta E_j \left(\sum_{a=1}^{n} \mathbb{I}(i_a = j)\right)\right] \ . \tag{8.4}$$

By exploiting the linearity of the expectation, the independence of the E_j's, and the fact that they are Gaussian, one gets

$$\mathbb{E} Z^n = \sum_{i_1 \ldots i_n = 1}^{2^N} \exp\left(\frac{\beta^2 N}{4} \sum_{a,b=1}^{n} \mathbb{I}(i_a = i_b)\right) \ . \tag{8.5}$$

$\mathbb{E} Z^n$ can also be interpreted as the partition function of a new 'replicated' system. As before, a configuration is given by the n-tuple $(i_1, \ldots, i_n)$, but now its energy is $E_{i_1 \ldots i_n} = -N\beta/4 \sum_{a,b=1}^{n} \mathbb{I}(i_a = i_b)$.

This replicated system has several interesting properties. First of all, it is no longer a disordered system: the energy is a deterministic function of the configuration. Second, replicas are no longer statistically independent: they do 'interact'. This is due to the fact that the energy function cannot be written as a sum of single replica terms. The interaction amounts to an attraction between different replicas. In particular, the lowest-energy configurations are obtained by setting $i_1 = \cdots = i_n$. Their energy is $E_{i_1 \ldots i_n} = -N\beta n^2/4$. Third, the energy itself depends upon the temperature, although in a very simple fashion. The effect of interaction is stronger at low temperature.

The origin of the interaction among replicas is easily understood. For one given sample of the original problem, the Boltzmann distribution is concentrated at low temperature ($\beta \gg 1$) on the lowest energy levels: all the replicas will tend to be in the same configuration with large probability. When averaging over sample realizations (i.e. over the energy levels $E_1, \ldots, E_{2^N}$), we do not see any longer which configuration $i \in \{1, \ldots, 2^N\}$ has the lowest energy, but we still see that the replicas prefer to stay in the same state. There is no mystery in these remarks. The elements of the n-tuple $(i_1 \ldots i_n)$ are independent *conditional* on the sample, that is on the realization of the energy levels E_j, $j \in \{1, \ldots, 2^N\}$. If we do not condition on the realization, $i_1, \ldots, i_n$ become dependent.

Given the configurations $(i_1 \ldots i_n)$ of the replicas, it is convenient to introduce an $n \times n$ matrix $Q_{ab} = \mathbb{I}(i_a = i_b)$, with elements in $\{0, 1\}$. We shall refer to this matrix as the **overlap matrix**. The summand in eqn (8.5) depends upon the configuration $(i_1 \ldots i_n)$ only through the overlap matrix. We can therefore rewrite the sum over configurations as

$$\mathbb{E}\, Z^n = \sum_Q \mathcal{N}_N(Q) \exp\left(\frac{N\beta^2}{4} \sum_{a,b=1}^{n} Q_{ab}\right). \qquad (8.6)$$

Here $\mathcal{N}_N(Q)$ denotes the number of configurations $(i_1 \ldots i_n)$ whose overlap matrix is $Q = \{Q_{ab}\}$, and the sum $\sum_Q$ runs over the symmetric $\{0, 1\}$ matrices with ones on the diagonal. The number of such matrices is $2^{n(n-1)/2}$, while the number of configurations of the replicated system is 2^{Nn}. It is therefore natural to guess that the number of configurations with a given overlap matrix satisfies a large-deviation principle of the form $\mathcal{N}_N(Q) \doteq \exp(Ns(Q))$.

Exercise 8.1 Show that the overlap matrix always has the following form: There exists a partition $\mathcal{G}_1, \mathcal{G}_2, \ldots, \mathcal{G}_{n_g}$ of the n replicas (this means that $\mathcal{G}_1 \cup \mathcal{G}_2 \cup \cdots \cup \mathcal{G}_{n_g} = \{1 \ldots n\}$ and $\mathcal{G}_i \cap \mathcal{G}_j = \emptyset$) into n_g groups such that $Q_{ab} = 1$ if a and b belong to the same group, and $Q_{ab} = 0$ otherwise. Prove that $\mathcal{N}_N(Q)$ satisfies the large-deviation principle described above, with $s(Q) = n_g \log 2$.

Using this form of $\mathcal{N}_N(Q)$, the replicated partition function can be written as

$$\mathbb{E}\, Z^n \doteq \sum_Q \exp\left(Ng(Q)\right), \qquad g(Q) \equiv \frac{\beta^2}{4} \sum_{a,b=1}^{n} Q_{ab} + s(Q). \qquad (8.7)$$

The strategy of the replica method is to estimate the above sum using the saddle point method.[1] The 'extrapolation' to non-integer values of n is discussed afterward. Note that this programme is completely analogous to the treatment of the Curie–Weiss model in Section 2.5.2 (see also Section 4.3 for related background), with the extra step of extrapolating to non-integer n.

8.1.1 The replica-symmetric saddle point

The function $g(Q)$ is symmetric under permutation of replicas. Let $\pi \in S_n$ be a permutation of n objects, and denote by Q^π the matrix with elements $Q^\pi_{ab} = Q_{\pi(a)\pi(b)}$. Then $g(Q^\pi) = g(Q)$. This is a simple consequence of the fact that the n replicas were equivalent from the beginning. This symmetry is called **replica symmetry**, and is a completely generic feature of the replica method.

When the dominant saddle point possesses this symmetry (i.e. when $Q^\pi = Q$ for any permutation π), one says that the system is **replica-symmetric** (RS). In the opposite case, replica symmetry is spontaneously broken in the large-N limit. This is analogous to the spontaneous breaking of $+/-$ symmetry in the Curie–Weiss model, which we discussed in Chapter 2 (see Section 2.5.2).

In view of this permutation symmetry, the simplest idea is to seek a replica-symmetric saddle point. If Q is invariant under permutation, then, necessarily, $Q_{aa} = 1$, and $Q_{ab} = q_0$ for any pair $a \neq b$. We are left with two possibilities:

- The matrix $Q_{\mathrm{RS},0}$ is defined by $q_0 = 0$. In this case $\mathcal{N}_N(Q_{\mathrm{RS},0}) = 2^N(2^N - 1)\ldots(2^N - n + 1)$, which yields $s(Q_{\mathrm{RS},0}) = n \log 2$ and $g(Q_{\mathrm{RS},0}) = n\left(\beta^2/4 + \log 2\right)$.
- The matrix $Q_{\mathrm{RS},1}$ is defined by $q_0 = 1$. This means that $i_1 = \cdots = i_n$. There are, of course, $\mathcal{N}_N(Q_{\mathrm{RS},1}) = 2^N$ choices of the n-tuple $(i_1 \ldots i_n)$ compatible with this constraint, which yields $s(Q_{\mathrm{RS},1}) = \log 2$ and $g(Q_{\mathrm{RS},1}) = n^2\beta^2/4 + \log 2$.

Keeping for the moment to these RS saddle points, we need to find which one dominates the sum. In Fig. 8.1 we plot, for $n = 3$ and $n = 0.5$, $g_0(n, \beta) \equiv g(Q_{\mathrm{RS},0})$ and $g_1(n, \beta) \equiv g(Q_{\mathrm{RS},1})$ as functions of $T = 1/\beta$. Note that the expressions we have obtained for $g_0(n, \beta)$ and $g_1(n, \beta)$ are polynomials in n, which we can plot for non-integer values of n.

When $n > 1$, the situation is always qualitatively the same as the one shown for the $n = 3$ case. If we let $\beta_c(n) = \sqrt{4\log 2/n}$, we have $g_1(\beta, n) > g_0(\beta, n)$ for $\beta > \beta_c(n)$, while $g_1(\beta, n) < g_0(\beta, n)$ for $\beta < \beta_c(n)$. Assuming for the moment that the sum in eqn (8.7) is dominated by replica-symmetric terms, we have $\mathbb{E}\, Z^n \doteq \exp\{N \max[g_0(\beta, n), g_1(\beta, n)]\}$. The point $\beta_c(n)$ can therefore be interpreted as a phase transition in the system of n replicas. At high temperatures ($\beta < \beta_c(n)$), the $q_0 = 0$ saddle point dominates the sum: replicas are essentially independent. At low temperature, the partition function is dominated by $q_0 = 1$: replicas are locked together. This fits nicely within our qualitative discussion of the replicated system in the previous section.

The problems appear when we consider the $n < 1$ situation. In this case we still have a phase transition at $\beta_c(n) = \sqrt{4\log 2/n}$, but the high- and low-temperature regimes

---

[1]Speaking of 'saddle points' is a bit sloppy in this case, since we are dealing with a *discrete* sum. By this, we mean that we aim at estimating the sum in eqn (8.7) through a single 'dominant' term.

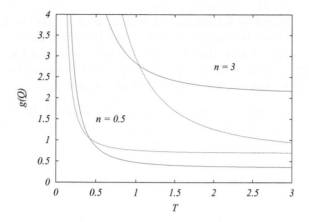

Fig. 8.1 Rate function $g(Q)$ for the REM (see eqn (8.7)) versus temperature; $g(Q)$ is evaluated here at the two replica-symmetric saddle points $Q_{\mathrm{RS},0}$ (continuous curves) and $Q_{\mathrm{RS},1}$ (dashed curves), in the cases $n = 3$ and $n = 0.5$.

exchange their roles. At low temperature ($\beta > \beta_c(n)$), we have $g_1(\beta, n) < g_0(\beta, n)$, and at high temperature ($\beta < \beta_c(n)$), we have $g_1(\beta, n) > g_0(\beta, n)$. If we were to apply the usual prescription and pick out the saddle point which maximizes $g(Q)$, we would obtain nonsense, physically (replicas become independent at low temperatures, and correlated at high temperatures, contrary to our general discussion) as well as mathematically (for $n \to 0$, the function $\mathbb{E}\, Z^n$ does not go to one, because $g_1(\beta, n)$ is not linear in n at small n). As a matter of fact, the replica method prescribes that, in this regime $n < 1$, one must estimate the sum (8.7) using the *minimum* of $g(Q)$! There is no mathematical justification for this prescription in the present context. In the next example and the following chapters we shall outline some of the arguments employed by physicists in order to rationalize this choice.

Example 8.1 In order to get some understanding of this prescription, consider the following toy problem. We want to apply the replica recipe to the quantity $Z_{\mathrm{toy}}(n) = (2\pi/N)^{n(n-1)/4}$ (for a generic real n). For n integer, we have the following integral representation:

$$Z_{\mathrm{toy}}(n) = \int e^{-(N/2)\sum_{(ab)} Q_{ab}^2} \prod_{(ab)} \mathrm{d}Q_{ab} \equiv \int e^{Ng(Q)} \prod_{(ab)} \mathrm{d}Q_{ab}, \qquad (8.8)$$

where (ab) runs over all of the unordered pairs of indices $a, b \in \{1 \ldots n\}$ with $a \neq b$, and the integrals over Q_{ab} run over the real line. Now we try to evaluate the above integral by the saddle point method, and begin with the assumption that it is dominated by a replica-symmetric point $Q_{ab}^* = q_0$ for any $a \neq b$, yielding $g(Q^*) = -n(n-1)q_0^2/2$. Next, we have to fix the value of $q_0 \in \mathbb{R}$. It is clear that the correct result is recovered by setting $q_0 = 0$, which yields $Z_{\text{toy}}(n) \doteq 1$. Moreover this is the unique choice such that $g(Q^*)$ is stationary. However, for $n < 1$, $q_0 = 0$ corresponds to a *minimum*, rather than to a maximum of $g(Q^*)$. A formal explanation of this odd behaviour is that the number of degrees of freedom, i.e. the number of matrix elements Q_{ab} with $a \neq b$, becomes *negative* for $n < 1$.

This is one of the strangest aspects of the replica method, but it is unavoidable. Another puzzle which we shall discuss later, concerns the exchange of the order of the $N \to \infty$ and $n \to 0$ limits.

Let us therefore select the saddle point $q_0 = 0$, and use the trick (8.2) to evaluate the free-energy density. Assuming that the $N \to \infty$ and $n \to 0$ limits commute, we get the RS free energy

$$-\beta f \equiv \lim_{N \to \infty} \frac{1}{N} \mathbb{E} \log Z = \lim_{N \to \infty} \lim_{n \to 0} \frac{1}{Nn} \log(\mathbb{E}\, Z^n)$$

$$= \lim_{n \to 0} \frac{1}{n} g_0(n, \beta) = \frac{\beta^2}{4} + \log 2 \,. \tag{8.9}$$

Comparing this result with the exact free-energy density (eqn (5.15)), we see that the RS result is correct only for the high-temperature phase $\beta < \beta_c = 2\sqrt{\log 2}$. It misses the phase transition. Within the RS framework, there is no way to get the correct solution for $\beta > \beta_c$.

8.1.2 One-step replica symmetry breaking

For $\beta > \beta_c$, the sum (8.7) is dominated by matrices Q which are not replica symmetric. The problem is to find these new saddle points, and they must make sense in the $n \to 0$ limit. In order to improve on the RS result, one can the subspace of matrices to be optimized over (i.e. weaken the requirement of replica symmetry). The **replica-symmetry-breaking** (RSB) scheme, initially proposed by Parisi in the more complicated case of spin glass mean-field theory, prescribes a recursive procedure for defining larger and larger spaces of matrices Q where one searches for saddle points.

The first step of this procedure is called **one-step replica symmetry breaking** (1RSB). In order to describe it, let us suppose that n is a multiple of x, divide the n replicas into n/x groups of x elements each, and set

$$
\begin{aligned}
Q_{aa} &= 1 \,, \\
Q_{ab} &= q_1 \quad \text{if } a \text{ and } b \text{ are in the same group,} \\
Q_{ab} &= q_0 \quad \text{if } a \text{ and } b \text{ are in different groups.}
\end{aligned}
\tag{8.10}
$$

Since, in the case of the REM, the matrix elements are in $\{0, 1\}$, this Ansatz is distinct from the RS Ansatz only if $q_1 = 1$ and $q_0 = 0$. This corresponds, after a relabeling of the replica indices, to $i_1 = \cdots = i_{\mathsf{x}}$, $i_{\mathsf{x}+1} = \cdots = i_{2\mathsf{x}}$, etc. The number of choices of $(i_1, \ldots i_n)$ which satisfy these constraints is $\mathcal{N}_N(Q) = 2^N(2^N - 1) \cdots (2^N - n/\mathsf{x} + 1)$, and therefore we get $s(Q) = (n/\mathsf{x}) \log 2$. The rate function in eqn (8.7) is given by $g(Q_{\text{RSB}}) = g_{\text{RSB}}(\beta, n, \mathsf{x})$:

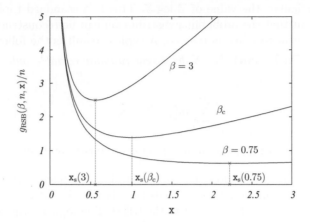

Fig. 8.2 The rate function $g(Q)$ (see eqn (8.7)), evaluated at the one-step replica-symmetry-breaking point, as a function of the replica-symmetry-breaking parameter $\mathbf{x}$.

$$g_{\mathrm{RSB}}(\beta, n, \mathbf{x}) = \frac{\beta^2}{4}n\mathbf{x} + \frac{n}{\mathbf{x}}\log 2 \ . \tag{8.11}$$

Following the discussion in Section 8.1.1, we should minimize $g_{\mathrm{RSB}}(\beta, n, \mathbf{x})$ with respect to $\mathbf{x}$, and then take the $n \to 0$ limit. Notice that eqn (8.11) can be interpreted as an analytic function both of n and of $\mathbf{x} \neq 0$. We shall therefore forget hereafter that n and $\mathbf{x}$ are integers with n a multiple of $\mathbf{x}$. The derivative $\partial g_{\mathrm{RSB}}(\beta, n, \mathbf{x})/\partial \mathbf{x}$ vanishes if $\mathbf{x} = \mathbf{x}_{\mathrm{s}}(\beta)$, where

$$\mathbf{x}_{\mathrm{s}}(\beta) \equiv \frac{2\sqrt{\log 2}}{\beta} = \frac{\beta_{\mathrm{c}}}{\beta} \ . \tag{8.12}$$

Substituting this into eqn (8.11), and assuming again that we can exchange the order of the limits $n \to 0$ and $N \to \infty$, we get

$$-\beta f = \lim_{n \to 0} \frac{1}{n} \min_{\mathbf{x}} g_{\mathrm{RSB}}(\beta, n, \mathbf{x}) = \beta\sqrt{\log 2}, \tag{8.13}$$

which is the correct result for $\beta > \beta_{\mathrm{c}}$: $f = -\sqrt{\log 2}$. In fact, we can recover the correct free energy of the REM over the whole temperature range if we accept that the inequality $1 \leq \mathbf{x} \leq n$, valid for $n, \mathbf{x}$ integers, becomes $n = 0 \leq \mathbf{x} \leq 1$ in the limit $n \to 0$ (we shall see later on some other arguments supporting this prescription). If the minimization is constrained to $\mathbf{x} \in [0, 1]$, we get a fully consistent answer: $\mathbf{x} = \beta_{\mathrm{c}}/\beta$ is the correct saddle point in the phase $\beta > \beta_{\mathrm{c}}$, while for $\beta < \beta_{\mathrm{c}}$ the parameter $\mathbf{x}$ remains at the value $\mathbf{x} = 1$. In Fig. 8.2, we sketch the function $g_{\mathrm{RSB}}(\beta, n, \mathbf{x})/n$ for a few values of the temperature β.

8.1.3 Comments on the replica solution

One might think that the replica method is just a fancy way of reconstructing a probability distribution from its integer moments. We know how to compute the integer moments of the partition function $\mathbb{E}\,Z^n$, and we would like to infer the full distribution of Z and, in particular, the value of $\mathbb{E} \log Z$. This is a standard topic in probability theory. It turns out that the probability distribution can be reconstructed if its integer moments do not grow too fast as $n \to \infty$. A typical result is the following.

Theorem 8.2. (Carleman) *Let X be a real random variable with moments $m_n = \mathbb{E} X^n$ such that*

$$\sum_{n=1}^{\infty} m_{2n}^{-1/2n} = \infty. \tag{8.14}$$

Then any variable with the same moments is distributed identically to X.

For instance, if the moments do not grow faster than exponentially, i.e. $\mathbb{E} X^n \sim e^{\alpha n}$, a knowledge of them completely determines the distribution of X.

Let us try to apply the above result to the REM. The replica-symmetric calculation of Section 8.1.1 is easily turned into a lower bound:

$$\mathbb{E} Z^n \geq e^{ng(Q_{\text{RS},0})} \geq e^{N\beta^2 n^2/4}. \tag{8.15}$$

Therefore the sum in eqn (8.14) converges, and the distribution of Z is not necessarily fixed by its integer moments.

Exercise 8.2 Assume that $Z = e^{-F}$, where F is a Gaussian random variable, with probability density

$$p(F) = \frac{1}{\sqrt{2\pi}} e^{-F^2/2}. \tag{8.16}$$

Compute the integer moments of Z. Do they satisfy the hypothesis of Carleman's theorem? Show that the moments are unchanged if $p(F)$ is replaced by the density $p_a(F) = p(F)[1 + a\sin(2\pi F)]$, with $|a| < 1$ (from Feller (1968)).

In our replica approach, there exist several possible analytic continuations to non-integer n's, and the whole issue is to find the correct one. Parisi's Ansatz (and its generalization to higher-order RSB that we shall discuss below) gives a well-defined class of analytic continuations, which turns out to be the correct one in many different problems.

The suspicious reader will notice that the moments of the REM partition function would not grow so rapidly if the energy levels had a distribution with bounded support. If for instance, we considered the E_i to be Gaussian random variables truncated to $E_i \in [-E_{\text{max}}, E_{\text{max}}]$, the partition function would be bounded from above by the constant $Z_{\text{max}} = 2^N e^{\beta E_{\text{max}}}$. Consequently, we would have $\mathbb{E} Z^n \leq Z_{\text{max}}^n$, and the whole distribution of Z could be recovered from its integer moments. In order to achieve such a goal, we would, however, need to know exactly all the moments $1 \leq n < \infty$ at fixed N (the system size). What we are instead able to compute, in general, is the large-N behaviour at any fixed n. In most cases, this information is insufficient to ensure a unique continuation to $n \to 0$.

In fact, one can think of the replica method as a procedure for computing the quantity

$$\psi(n) = \lim_{N\to\infty} \frac{1}{N} \log \mathbb{E}\, Z^n \,, \tag{8.17}$$

whenever the limit exists. In the frequent case where $f = -(1/\beta N)\log Z$ satisfies a large-deviation principle of the form $\mathbb{P}_N(f) \doteq \exp[-NI(f)]$, we have

$$\mathbb{E}\, Z^n \doteq \int \exp[-NI(f) - N\beta n f]\, \mathrm{d}f \doteq \exp\{-N\inf[I(f) + \beta n f]\}\,. \tag{8.18}$$

Therefore $\psi(n) = -\inf[I(f) + \beta n f]$. In turn, the large-deviation properties of f_N can be inferred from $\psi(n)$ through the Gärtner–Ellis theorem (Theorem 4.10). The typical value of the free-energy density is given by the location of the absolute minimum of $I(f)$. In order to compute it, one must in general use values of n which go to 0, and one cannot infer it from the integer values of n.

8.1.4 Condensation

As we discussed in Chapter 5, the appearance of a low-temperature 'glass' phase is associated with a condensation of the probability measure onto few configurations. We described this phenomenon quantitatively by the participation ratio Y. For the REM we obtained $\lim_{N\to\infty} \mathbb{E}\, Y = 1 - \beta_c/\beta$ for any $\beta > \beta_c$ (see Proposition 5.3). Let us see how this result can be recovered in just a few lines from a replica computation.

The participation ratio is defined by $Y = \sum_{j=1}^{2^N} \mu(j)^2$, where $\mu(j) = \mathrm{e}^{-\beta E_j}/Z$ is the Boltzmann probability of the j-th energy level. Therefore,

$$\mathbb{E}\, Y = \lim_{n\to 0} \mathbb{E}\left[Z^{n-2} \sum_{i=1}^{2^N} \mathrm{e}^{-2\beta E_i} \right] \qquad \text{[definition of } Y\text{]}$$

$$= \lim_{n\to 0} \mathbb{E}\left[\sum_{i_1\ldots i_{n-2}} \mathrm{e}^{-\beta(E_{i_1}+\cdots+E_{i_{n-2}})} \sum_{i=1}^{2^N} \mathrm{e}^{-2\beta E_i} \right] \qquad \text{[assume } n \in \mathbb{N}\text{]}$$

$$= \lim_{n\to 0} \mathbb{E}\left[\sum_{i_1\ldots i_n} \mathrm{e}^{-\beta(E_{i_1}+\cdots+E_{i_n})} \mathbb{I}(i_{n-1} = i_n) \right]$$

$$= \lim_{n\to 0} \frac{1}{n(n-1)} \sum_{a\neq b} \mathbb{E}\left[\sum_{i_1\ldots i_n} \mathrm{e}^{-\beta(E_{i_1}+\cdots+E_{i_n})} \, \mathbb{I}(i_a = i_b) \right] \qquad \text{[symmetrize]}$$

$$= \lim_{n\to 0} \frac{1}{n(n-1)} \sum_{a\neq b} \frac{\mathbb{E}\left[\sum_{i_1\ldots i_n} \mathrm{e}^{-\beta(E_{i_1}+\cdots+E_{i_n})} \, \mathbb{I}(i_a = i_b) \right]}{\mathbb{E}\left[\sum_{i_1\ldots i_n} \mathrm{e}^{-\beta(E_{i_1}+\cdots+E_{i_n})} \right]} \qquad \text{[denominator} \to 1\text{]}$$

$$= \lim_{n\to 0} \frac{1}{n(n-1)} \sum_{a\neq b} \langle Q_{ab}\rangle_n \,, \tag{8.19}$$

where the sums over the replica indices a, b run over $a, b \in \{1, \ldots, n\}$, while the configuration indices i_a are summed over $\{1, \ldots, 2^N\}$. In the last step, we have introduced the notation

$$\langle f(Q) \rangle_n \equiv \frac{\sum_Q f(Q) \, \mathcal{N}_N(Q) e^{(N\beta^2/4) \sum_{a,b} Q_{ab}}}{\sum_Q \mathcal{N}_N(Q) e^{(N\beta^2/4) \sum_{a,b} Q_{ab}}}, \qquad (8.20)$$

and noticed that the sum over $i_1, \ldots, i_n$ can be split into a sum over the overlap matrices Q and a sum over the n-tuples $i_1 \ldots i_n$ that have an overlap matrix Q. Note that $\langle \cdot \rangle_n$ can be interpreted as an expectation in the 'replicated system'.

In the large-N limit, $\mathcal{N}_N(Q) \doteq e^{Ns(Q)}$, and the expectation value in eqn (8.20) is given by a dominant[2] (saddle point) term: $\langle f(Q) \rangle_n \simeq f(Q^*)$. As argued in the previous sections, in the low-temperature phase $\beta > \beta_c$, the saddle point matrix is given by the 1RSB expression (8.10):

$$\mathbb{E}\, Y = \lim_{n \to 0} \frac{1}{n(n-1)} \sum_{a \neq b} Q_{ab}^{1\text{RSB}} \qquad \qquad \text{[saddle point]}$$

$$= \lim_{n \to 0} \frac{1}{n(n-1)} \, n[(n - \mathsf{x})q_0 + (\mathsf{x} - 1)q_1] \qquad \text{[eqn (8.10)]}$$

$$= 1 - \mathsf{x} = 1 - \frac{\beta_c}{\beta}. \qquad \qquad [q_0 = 0,\ q_1 = 1] \quad (8.21)$$

This is exactly the result that we found in Proposition 5.3, using a direct combinatorial approach. It also confirms that the 1RSB Ansatz (8.10) makes sense only if $0 \le \mathsf{x} \le 1$ (the participation ratio Y is positive by definition). Compared with the computation in Section 5.3, the simplicity of the replica derivation is striking.

At first sight, the manipulations in eqn (8.19) seem to require new assumptions with respect to the free-energy computation in the previous sections. Replicas are introduced in order to write the factor Z^{-2} in the participation ratio as the analytic continuation of a positive power Z^{n-2}. It turns out that this calculation is in fact equivalent to the one in eqn (8.2). This follows from the basic observation that expectation values can be obtained as derivatives of $\log Z$ with respect to some parameters.

Exercise 8.3 Using the replica method, show that, for $T < T_c$,

$$\mathbb{E} \left(\sum_{j=1}^{2^N} \mu(j)^r \right) = \frac{\Gamma(r - \mathsf{x})}{\Gamma(r)\Gamma(1 - \mathsf{x})} = \frac{(r - 1 - \mathsf{x})(r - 2 - \mathsf{x}) \ldots (1 - \mathsf{x})}{(r - 1)(r - 2) \ldots (1)}, \qquad (8.22)$$

where $\Gamma(x)$ denotes Euler's gamma function.

---

[2]If the dominant term corresponds to a non-replica-symmetric matrix Q^*, all the terms obtained by permuting the replica indices contribute with an equal weight. Because of this fact, it is a good idea to compute averages of symmetric functions $f(Q) = f(Q^\pi)$. This is what we have done in eqn (8.19).

Exercise 8.4 Using the replica method, show that, for $T < T_c$,

$$\mathbb{E}\left(Y^2\right) = \frac{3 - 5\mathbf{x} + 2\mathbf{x}^2}{3} . \tag{8.23}$$

8.2 The fully connected p-spin glass model

The replica method provides a compact and efficient way to compute –in a non-rigorous way– the free-energy density of the REM. The result proves to be exact, once replica symmetry breaking is used in the low-temperature phase. However, its power can be better appreciated on more complicated problems which cannot be solved by direct combinatorial approaches. In this section we shall apply the replica method to the 'p-spin glass' model. This model was invented in the course of the theoretical study of spin glasses. Its distinguishing feature is interactions which involve groups of p spins, with $p \geq 2$. It generalizes ordinary spin glass models (see Section 2.6) in which interactions involve pairs of spins (i.e. $p = 2$). This provides an additional degree of freedom, the value of p, and different physical scenarios appear depending on whether $p = 2$ or $p \geq 3$. Moreover, some pleasing simplifications show up for large p.

In the p-**spin model**, one considers the space of 2^N configurations of N Ising spins. The energy of a configuration $\sigma = \{\sigma_1, \ldots, \sigma_N\}$ is defined as

$$E(\sigma) = - \sum_{i_1 < i_2 < \cdots < i_p} J_{i_1 \ldots i_p} \sigma_{i_1} \cdots \sigma_{i_p} , \tag{8.24}$$

where $\sigma_i \in \{\pm 1\}$. This is a disordered system: a sample is characterized by the set of all couplings $J_{i_1 \ldots i_p}$, with $1 \leq i_1 < \cdots < i_p \leq N$. These are taken as i.i.d. Gaussian random variables with zero mean and variance $\mathbb{E} J_{i_1 \ldots i_p}^2 = p!/(2N^{p-1})$. Their probability density reads

$$P(J) = \sqrt{\frac{\pi p!}{N^{p-1}}} \exp\left(-\frac{N^{p-1}}{p!} J^2\right) . \tag{8.25}$$

The p-spin model is an **infinite-range interaction** (or **mean-field**) model: there is no notion of a Euclidean distance between the positions of the spins. It is also called a **fully connected** model, since each spin interacts directly with all the others. The last feature is the origin of the special scaling of the variance of the distribution of J in eqn (8.25). A simple criterion for arguing that the proposed scaling is the correct one consists in requiring that a flip of a single spin generates an energy change of order 1 (i.e. finite when $N \to \infty$). More precisely, we denote by $\sigma^{(i)}$ the configuration obtained from σ by reversing the spin i, and we define $\Delta_i \equiv [E(\sigma^{(i)}) - E(\sigma)]/2$. It is easy to see that $\Delta_i = \sum_{i_2 \ldots i_p} J_{i i_1 \ldots i_p} \sigma_i \sigma_{i_1} \cdots \sigma_{i_p}$. The sum is over $\Theta(N^{p-1})$ terms, and, if σ is a random configuration, the product $\sigma_i \sigma_{i_1} \cdots \sigma_{i_p}$ in each term is $+1$ or -1 with probability $1/2$. The scaling in eqn (8.25) ensures that Δ_i is finite as $N \to \infty$ (in contrast, the factor $p!$ is just a matter of convention).

Why is it important that the Δ_i's are of order 1? The intuition is that Δ_i estimates the interaction between a spin and the rest of the system. If Δ_i were much larger than 1, the spin σ_i would be completely frozen in the direction which makes Δ_i positive, and temperature would not have any role. On the other hand, if Δ_i were much smaller than one, the spin i would be effectively independent of the others.

Exercise 8.5 An alternative argument can be obtained as follows. Show that, at high temperature, i.e. $\beta \ll 1$, $Z = 2^N[1 + 2^{-1}\beta^2 \sum_{i_1 < \cdots < i_p} J_{i_1 \ldots i_p}^2 + O(\beta^3)]$. This implies $N^{-1}\mathbb{E}\log Z = \log 2 + C_N \beta^2/2 + O(\beta^3)$, with $C_N = 1$. What would happen with a different scaling of the variance? What scaling is required in order for C_N to have a finite $N \to \infty$ limit?

The special case of $p = 2$ is the closest to the original spin glass problem and is known as the **Sherrington–Kirkpatrick** (or SK) model.

8.2.1 The replica calculation

Let us start by writing Z^n as the partition function for n non-interacting replicas σ_i^a, with $i \in \{1, \ldots, N\}$, $a \in \{1, \ldots, n\}$:

$$Z^n = \sum_{\{\sigma_i^a\}} \prod_{i_1 < \cdots < i_p} \exp\left(\beta J_{i_1 \ldots i_p} \sum_{a=1}^n \sigma_{i_1}^a \cdots \sigma_{i_p}^a\right). \tag{8.26}$$

The average over the couplings $J_{i_1 \ldots i_p}$ is easily evaluated by using their independence and the well-known identity

$$\mathbb{E}\,e^{\lambda X} = e^{\Delta \lambda^2/2}, \tag{8.27}$$

which holds for a Gaussian random variable X with zero mean and variance $\mathbb{E}\,X^2 = \Delta$. One gets

$$\mathbb{E}\,Z^n = \sum_{\{\sigma_i^a\}} \exp\left(\frac{\beta^2}{4}\frac{p!}{N^{p-1}} \sum_{i_1 < \cdots < i_p} \sum_{a,b} \sigma_{i_1}^a \sigma_{i_1}^b\, \sigma_{i_2}^a \sigma_{i_2}^b \cdots \sigma_{i_p}^a \sigma_{i_p}^b\right)$$

$$\doteq \sum_{\{\sigma_i^a\}} \exp\left[\frac{\beta^2}{4}\frac{1}{N^{p-1}} \sum_{a,b}\left(\sum_i \sigma_i^a \sigma_i^b\right)^p\right] \tag{8.28}$$

where we have neglected corrections due to coincident indices $i_l = i_k$ in the first term, since they are irrelevant to the leading exponential order. We introduce the variables λ_{ab} and Q_{ab} for each $a < b$ by using the identity

$$1 = \int \delta\left(Q_{ab} - \frac{1}{N}\sum_{i=1}^N \sigma_i^a \sigma_i^b\right) dQ_{ab} = N \int\!\!\int e^{-i\lambda_{ab}\left(NQ_{ab} - \sum_i \sigma_i^a \sigma_i^b\right)} \frac{d\lambda_{ab}}{2\pi}\, dQ_{ab}, \tag{8.29}$$

with all of the integrals running over the real line. Using this identity in eqn (8.28), we get

$$\mathbb{E}\,Z^n \doteq \int \prod_{a<b} \mathrm{d}Q_{ab} \sum_{\{\sigma_i^a\}} \exp\left(\frac{N\beta^2}{4}\,n + \frac{N\beta^2}{2}\sum_{a<b} Q_{ab}^p\right) \delta\left(Q_{ab} - \frac{1}{N}\sum_{i=1}^{N} \sigma_i^a \sigma_i^b\right)$$

$$\doteq \mathrm{e}^{N\beta^2 n/4} \int \prod_{a<b} (\mathrm{d}Q_{ab}\,\mathrm{d}\lambda_{ab})\, \mathrm{e}^{\frac{N\beta^2}{2}\sum_{a<b}Q_{ab}^p - iN\sum_{a<b}\lambda_{ab}Q_{ab}} \sum_{\{\sigma_i^a\}} \mathrm{e}^{i\sum_{a<b}\lambda_{ab}\sum_i \sigma_i^a \sigma_i^b}$$

$$\doteq \int \prod_{a<b} (\mathrm{d}Q_{ab}\,\mathrm{d}\lambda_{ab})\, \mathrm{e}^{-NG(Q,\lambda)} \tag{8.30}$$

where we have introduced the function

$$G(Q,\lambda) = -n\frac{\beta^2}{4} - \frac{\beta^2}{2}\sum_{a<b} Q_{ab}^p + i\sum_{a<b}\lambda_{ab}Q_{ab} - \log\left[\sum_{\{\sigma_a\}} \mathrm{e}^{\sum_{a<b} i\lambda_{ab}\sigma_a\sigma_b}\right], \tag{8.31}$$

which depends upon the $n(n-1)/2 + n(n-1)/2$ variables $Q_{ab}, \lambda_{ab}, 1 \leq a < b \leq n$.

Exercise 8.6 An alternative route consists in noticing that the right-hand side of eqn (8.28) depends upon the spin configuration only through the overlap matrix $Q_{ab} = N^{-1}\sum_i \sigma_i^a \sigma_i^b$, with $a < b$. The sum can be therefore decomposed into a sum over the overlap matrices and a sum over configurations with a given overlap matrix:

$$\mathbb{E}\,Z^n \doteq \sum_Q \mathcal{N}_N(Q)\, \exp\left(\frac{N\beta^2}{4}\,n + \frac{N\beta^2}{2}\sum_{a<b} Q_{ab}^p\right). \tag{8.32}$$

Here $\mathcal{N}_N(Q)$ is the number of spin configurations with a given overlap matrix Q. In analogy to the REM case, it is natural to guess a large-deviation principle of the form $\mathcal{N}_N(Q) \doteq \exp[Ns(Q)]$. Use the Gärtner–Ellis theorem (Theorem 4.10) to obtain an expression for the 'entropic' factor $s(Q)$. Compare the resulting formula for $\mathbb{E}\,Z^n$ with eqn (8.28).

Following our general approach, we shall estimate the integral (8.30) at large N by the saddle point method. The stationarity conditions of G are most conveniently written in terms of the variables $\omega_{ab} = i\lambda_{ab}$. By differentiating eqn (8.31) with respect to its arguments, we get, $\forall\, a < b$,

$$\omega_{ab} = \frac{1}{2}p\beta^2\, Q_{ab}^{p-1}, \qquad Q_{ab} = \langle \sigma_a \sigma_b \rangle_n, \tag{8.33}$$

where we have introduced the average within the replicated system

$$\langle f(\sigma)\rangle_n \equiv \frac{1}{z(\omega)}\sum_{\{\sigma^a\}} f(\sigma)\, \exp\left(\sum_{a<b} \omega_{ab}\sigma_a\sigma_b\right), \qquad z(\omega) \equiv \sum_{\{\sigma^a\}} \exp\left(\sum_{a<b} \omega_{ab}\,\sigma_a\sigma_b\right), \tag{8.34}$$

for any function $f(\sigma) = f(\sigma^1,\ldots,\sigma^n)$.

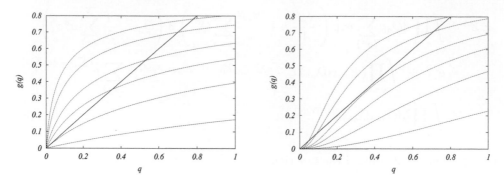

Fig. 8.3 Graphical solution of the RS equations for the p-spin model, with $p = 2$ (SK model, *left*) and $p = 3$ (*right*). The various curves correspond to inverse temperatures $\beta = 4, 3, 2,$ 1.5, 1, and 0.5 (from *top* to *bottom*).

We start by considering an RS saddle point $Q_{ab} = q$, $\omega_{ab} = \omega$ for any $a \neq b$. Using the Gaussian identity (8.27), we find that the saddle point equations (8.33) become

$$\omega = \frac{1}{2}p\beta^2 \, q^{p-1} \,, \qquad q = \mathsf{E}_z \tanh^2\left(z\sqrt{\omega}\right) , \qquad (8.35)$$

where E_z denotes the expectation with respect to a Gaussian random variable z of zero mean and unit variance. Eliminating ω, we obtain an equation for the overlap parameter $q = r(q)$, with $r(q) \equiv \mathsf{E}_z \tanh^2(z\sqrt{p\beta^2 \, q^{p-1}/2})$. In Fig. 8.3, we plot the function $r(q)$ for $p = 2$ and 3 and various temperatures. The equations (8.35) always admit the solution $q = \omega = 0$. Substituting into eqn (8.31), and using the trick (8.2), this solution would yield a free-energy density

$$f_{\mathrm{RS}} = \lim_{n \to 0} \frac{1}{\beta n} G(Q^{\mathrm{RS}}, \lambda^{\mathrm{RS}}) = -\beta/4 - (1/\beta) \log 2 \,. \qquad (8.36)$$

At low enough temperature, other RS solutions appear. For $p = 2$, a single such solution departs continuously from 0 at $\beta_{\mathrm{c}} = 1$, (see Fig. 8.3, left frame). For $p \geq 3$, a pair of non-vanishing solutions appear discontinuously for $\beta \geq \beta_*(p)$ and merge as $\beta \downarrow \beta_*(p)$, (see Fig. 8.3, right frame). However two arguments allow us to discard these saddle points:

- *Stability argument.* One can compute the Taylor expansion of $G(Q, \lambda)$ around such RS saddle points. The saddle point method can be applied only if the matrix of second derivatives has a defined sign. As discussed in Section 8.4, this condition does not hold for the non-vanishing RS saddle points.

- *Positivity of the entropy.* As explained in Chapter 2, because of the positivity of the entropy, the free energy of a physical system with discrete degrees of freedom must be a decreasing function of the temperature. Once again, one can show that this condition is not satisfied by the non-vanishing RS saddle points. On the other hand, the $q = 0$ saddle point also violates this condition at low enough temperature (as the reader can see from eqn (8.36)).

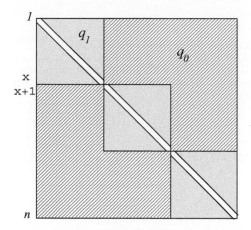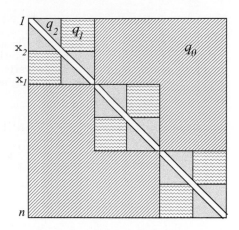

Fig. 8.4 Structure of the matrix Q_{ab} when replica symmetry is broken. *Left*: 1RSB Ansatz. The $n(n-1)/2$ values of Q_{ab} are the non-diagonal elements of a symmetric $n \times n$ matrix. The n replicas are divided into n/x blocks of size x. When a and b are in the same block, $Q_{ab} = q_1$; otherwise, $Q_{ab} = q_0$. *Right*: 2RSB Ansatz: an example with $n/\mathrm{x}_1 = 3$ and $\mathrm{x}_1/\mathrm{x}_2 = 2$.

The above arguments are very general. The second condition, in particular, is straightforward to check and must always be satisfied by the correct saddle point. The conclusion is that none of the RS saddle points is correct at low temperatures. This motivates us to look for 1RSB saddle points. We partition the set of n replicas into n/x groups of x replicas each and seek a saddle point of the following 1RSB form:

$$Q_{ab} = q_1, \quad \omega_{ab} = \omega_1, \quad \text{if } a \text{ and } b \text{ belong to the same group,}$$
$$Q_{ab} = q_0, \quad \omega_{ab} = \omega_0, \quad \text{if } a \text{ and } b \text{ belong to different groups.} \tag{8.37}$$

In practice, one can relabel the replicas in such a way that the groups are formed by successive indices $\{1 \ldots \mathrm{x}\}, \{\mathrm{x}+1 \ldots 2\mathrm{x}\}, \ldots, \{n - \mathrm{x} + 1 \ldots n\}$ (see Fig. 8.4).

The computation of $G(Q, \lambda)$ at this saddle point makes repeated use of the identity (8.27) and is left as an exercise. One gets

$$G(Q^{1\mathrm{RSB}}, \lambda^{1\mathrm{RSB}}) = -n\frac{\beta^2}{4} + n\frac{\beta^2}{4}\left[(1-\mathrm{x})q_1^p + \mathrm{x}q_0^p\right] - \frac{n}{2}\left[(1-\mathrm{x})q_1\omega_1 + \mathrm{x}q_0\omega_0\right]$$
$$+ \frac{n}{2}\omega_1 - \log\left\{\mathsf{E}_{z_0}\left[\mathsf{E}_{z_1}(2\cosh(\sqrt{\omega_0}\,z_0 + \sqrt{\omega_1 - \omega_0}\,z_1))^{\mathrm{x}}\right]^{n/\mathrm{x}}\right\}, \tag{8.38}$$

where E_{z_0} and E_{z_1} denote expectations with respect to the independent Gaussian random variables z_0 and z_1 with zero mean and unit variance.

Exercise 8.7 Show that the limit $G_{1\mathrm{RSB}}(q_0, q_1, \omega_0, \omega_1; \mathsf{x}) = \lim_{n \to 0} n^{-1} G(Q^{1\mathrm{RSB}}, \lambda^{1\mathrm{RSB}})$ exists, and compute it. Determine the stationarity condition for the parameters $q_0, q_1, \omega_0, \omega_1$, and x by computing the partial derivatives of $G_{1\mathrm{RSB}}(q, \omega; \mathsf{x})$ with respect to its arguments and setting them to 0. Show that these equations are always consistent with $q_0 = \omega_0 = 0$, and that

$$
G_{1\mathrm{RSB}}\big|_{q_0, \omega_0 = 0} = -\frac{1}{4}\beta^2[1 - (1 - \mathsf{x})q_1^p] + \frac{1}{2}\omega_1[1 - (1 - \mathsf{x})q_1]
$$
$$
- \frac{1}{\mathsf{x}} \log \mathsf{E}_z\left[(2\cosh(\sqrt{\omega_1}\, z))^{\mathsf{x}}\right].
\tag{8.39}
$$

If we choose the solution $q_0 = \omega_0 = 0$, the stationarity conditions for the remaining parameters q_1 and ω_1 are most easily obtained by differentiating eqn (8.39) with respect to q_1 and ω_1. These conditions read

$$
\omega_1 = \frac{1}{2}p\beta^2\, q_1^{p-1}, \qquad q_1 = \frac{\mathsf{E}_z\left[(2\cosh(\sqrt{\omega_1}\, z))^{\mathsf{x}}(\tanh(\sqrt{\omega_1}\, z))^2\right]}{\mathsf{E}_z\left[(2\cosh(\sqrt{\omega_1}\, z))^{\mathsf{x}}\right]}.
\tag{8.40}
$$

These equations always admit the solution $q_1 = \omega_1 = 0$: this choice reduces in fact to a replica-symmetric Ansatz, as can be seen from eqn (8.37). Let us now consider the case $p \geq 3$. At low enough temperature, two non-vanishing solutions appear. A local stability analysis shows that the larger one, which we shall call ω_1^{sp}, q_1^{sp}, must be chosen.

The next step consists in optimizing $G_{1\mathrm{RSB}}(Q_0 = 0, q_1^{\mathrm{sp}}, \omega_0 = 0, \omega_1^{\mathrm{sp}}; \mathsf{x})$ with respect to $\mathsf{x} \in [0, 1]$ (notice that $G_{1\mathrm{RSB}}$ depends on x both explicitly and through $q^{\mathrm{sp}}, \omega^{\mathrm{sp}}$). It turns out that a unique stationary point $\mathsf{x}_{\mathrm{s}}(\beta)$ exists, but $\mathsf{x}_{\mathrm{s}}(\beta)$ belongs to the interval $[0, 1]$ only at low enough temperature, i.e. for $\beta > \beta_{\mathrm{c}}(p)$. We refer to the literature for an explicit characterization of $\beta_{\mathrm{c}}(p)$. At the transition temperature $\beta_{\mathrm{c}}(p)$, the free energy of the 1RSB solution becomes equal to that of the RS solution. There is a phase transition from an RS phase for $\beta < \beta_{\mathrm{c}}(p)$ to a 1RSB phase for $\beta > \beta_{\mathrm{c}}(p)$.

These calculations are greatly simplified (and can be carried out analytically) in the large-p limit. The leading terms in a large p expansion are:

$$
\beta_{\mathrm{c}}(p) = 2\sqrt{\log 2} + \mathrm{e}^{-\Theta(p)}, \quad \mathsf{x}_{\mathrm{s}}(\beta) = \frac{\beta_{\mathrm{c}}(p)}{\beta} + \mathrm{e}^{-\Theta(p)}, \quad q_1 = 1 - \mathrm{e}^{-\Theta(p)}.
\tag{8.41}
$$

The corresponding free-energy density is constant in the whole low-temperature phase, and equal to $-\sqrt{\log 2}$. The reader will notice that several features of the REM are recovered in this large-p limit. One can get a hint that this should be the case from the following exercise.

Exercise 8.8 Consider a p-spin glass problem, and take an arbitrary configuration $\sigma = \{\sigma_1, \ldots, \sigma_N\}$. Let $P_\sigma(E)$ denote the probability that this configuration has energy E when a sample (i.e. a choice of couplings $J_{i_1 \ldots i_p}$) is chosen at random with the distribution (8.25). Show that $P_\sigma(E)$ is independent of σ, and is a Gaussian distribution with mean 0 and variance $N/2$. Now take two configurations σ and σ', and show that the joint probability distribution of their energies E and E', respectively, in a randomly chosen sample, is

$$P_{\sigma,\sigma'}(E, E') = C \exp\left[-\frac{(E+E')^2}{2N(1+q_{\sigma,\sigma'}^p)} - \frac{(E-E')^2}{2N(1-q_{\sigma,\sigma'}^p)} \right], \qquad (8.42)$$

where $q_{\sigma,\sigma'} = (1/N) \sum_i \sigma_i \sigma_i'$, and C is a normalization constant. When $|q_{\sigma,\sigma'}| < 1$, the energies of the two configurations become uncorrelated as $p \to \infty$ (i.e. $\lim_{p\to\infty} P_{\sigma,\sigma'}(E, E') = P_\sigma(E) P_{\sigma'}(E')$), suggesting an REM-like behaviour.

In order to know if the 1RSB solution which we have just found is the correct one, we should first check its stability by verifying that the eigenvalues of the Hessian (i.e. the matrix of second derivatives of $G(Q, \lambda)$ with respect to its arguments) have the correct sign. Although straightforward in principle, this computation becomes rather cumbersome and we shall just give the result, due to Elizabeth Gardner. The 1RSB solution is stable only in some intermediate phase defined by $\beta_c(p) < \beta < \beta_u(p)$. At the inverse temperature $\beta_u(p)$, there is a second transition to a new phase which involves a more complex replica-symmetry-breaking scheme.

The 1RSB solution was generalized by Parisi to higher orders of RSB. His construction is a hierarchical one. In order to define the structure of the matrix Q_{ab} with two steps of replica symmetry breaking (2RSB), one starts from the 1RSB matrix of Fig. 8.4 (left panel). The off-diagonal blocks with matrix elements q_0 are left unchanged. The diagonal blocks are changed as follows. Let us consider any diagonal block of size $x_1 \times x_1$ (we now call $x = x_1$). In the 1RSB case, all its matrix elements are equal to q_1. In the 2RSB case, the x_1 replicas are split into x_1/x_2 blocks of x_2 replicas each. The matrix elements in the off-diagonal blocks remain equal to q_1. The ones in the diagonal blocks become equal to a new number q_2 (see Fig. 8.4, right panel). The matrix is parameterized by five numbers: q_0, q_1, q_2, x_1, x_2. This construction can obviously be generalized by splitting the diagonal blocks again, grouping x_2 replicas into x_2/x_3 groups of x_3 replicas. The **full replica-symmetry-breaking** (FRSB) Ansatz Ansatz corresponds to iterating this procedure R times, and eventually taking R to infinity. Notice that, although the construction makes sense for n integer only when $n \geq x_1 \geq x_2 \geq \cdots \geq x_R \geq 1$, in the $n \to 0$ limit, this order is reversed to $0 \leq x_1 \leq x_2 \leq \cdots \leq x_R < 1$. Once one assumes an R-RSB Ansatz, computing the rate function G and solving the saddle point equations is a matter of calculus (special tricks have been developed for $R \to \infty$). It turns out that, in order to find a stable solution in the phase $\beta > \beta_u(p)$, an FRSB Ansatz is required. The same situation is also encountered in the case of the SK model ($p = 2$), in the whole of the phase $\beta > 1$, but its description would take us too far.

8.2.2   Overlap distribution

Replica symmetry breaking appeared in the previous subsections as a formal trick for computing some partition functions. One of the fascinating features of spin glass theory is that RSB has a very concrete physical, as well as a probabilistic, interpretation. One of the main characteristics of a system displaying RSB is the existence, in a typical sample, of some spin configurations which are very different from the lowest-energy (ground state) configuration, but are very close to it in energy. One can get a measure of this property through the distribution of overlaps between configurations. Given two spin configurations $\sigma = \{\sigma_1, \ldots, \sigma_N\}$ and $\sigma' = \{\sigma'_1, \ldots, \sigma'_N\}$, the **overlap** between σ and σ' is

$$q_{\sigma\sigma'} = \frac{1}{N} \sum_{i=1}^{N} \sigma_i \sigma'_i \,, \tag{8.43}$$

so that $N(1 - q_{\sigma\sigma'})/2$ is the Hamming distance between σ and σ'. For a given sample of the p-spin glass model, which we denote by J, the **overlap distribution** $P_J(q)$ is the probability density that two configurations, randomly chosen with the Boltzmann distribution, have overlap q:

$$\int_{-1}^{q} P_J(q') \, \mathrm{d}q' = \frac{1}{Z^2} \sum_{\sigma,\sigma'} \exp\left[-\beta E(\sigma) - \beta E(\sigma')\right] \mathbb{I}\left(q_{\sigma\sigma'} \leq q\right) \,. \tag{8.44}$$

Let us compute the expectation of $P_J(q)$ in the thermodynamic limit,

$$P(q) \equiv \lim_{N \to \infty} \mathbb{E} \, P_J(q), \tag{8.45}$$

using replicas. One finds:

$$\int_{-1}^{q} P(q') \, \mathrm{d}q' = \lim_{n \to 0} \sum_{\sigma^1 \ldots \sigma^n} \mathbb{E}\left[\exp\left(-\beta \sum_a E(\sigma^a)\right)\right] \mathbb{I}(q_{\sigma^1 \sigma^2} \leq q) \,. \tag{8.46}$$

The calculation is very similar to that of $\mathbb{E}(Z^n)$; the only difference is that now the overlap between replicas 1 and 2 is fixed to be $\leq q$. Following the same steps as before, we obtain an expression for $P(q)$ in terms of the saddle point matrix Q_{ab}^{sp}. The only delicate point is that there may be several RSB saddle points related by a permutation of the replica indices. If $Q = \{Q_{ab}\}$ is a saddle point, any matrix $(Q^\pi)_{ab} = Q_{\pi(a),\pi(b)}$ (where π is a permutation in S_n) is also a saddle point, with the same weight: $G(Q^\pi) = G(Q)$. When computing $P(q)$, we need to sum over all the equivalent distinct saddle points, which gives in the end

$$\int_{-1}^{q} P(q') \, \mathrm{d}q' = \lim_{n \to 0} \frac{1}{n(n-1)} \sum_{a \neq b} \mathbb{I}(Q_{ab}^{\mathrm{sp}} \leq q) \,. \tag{8.47}$$

In the case of an RS solution, one has,

$$\int_{-1}^{q} P(q') \, \mathrm{d}q' = \mathbb{I}(q^{\mathrm{RS}} \leq q) \,, \tag{8.48}$$

where q^{RS} is the solution of the saddle point equations (8.35). In words, if two configurations σ and σ' are drawn according to the Boltzmann distribution, their overlap will be q^{RS} with high probability. Since the overlap is a sum of many 'simple' terms, the fact that its distribution concentrates around a typical value is somehow to be expected.

In a 1RSB phase characterized by the numbers $q_0, q_1, \lambda_0, \lambda_1, \mathbf{x}$, one finds

$$\int_{-1}^{q} P(q') \, \mathrm{d}q' = (1 - \mathbf{x}) \, \mathbb{I}(q_1 \leq q) + \mathbf{x} \, \mathbb{I}(q_0 \leq q) \ . \tag{8.49}$$

The overlap can take, with positive probability, two values: q_0 or q_1. This has a very nice geometrical interpretation. When configurations are sampled randomly with the Boltzmann probability, at an inverse temperature $\beta > \beta_{\mathrm{c}}(p)$, the configurations will typically be grouped into clusters, such that any two configurations in the same cluster have an overlap close to q_1, while configurations in different clusters have an overlap $q_0 < q_1$, and thus a larger Hamming distance The probability that they fall into the same cluster is equal to $1 - \mathbf{x}$. The clustering property is a rather non-trivial one: it would have been difficult to anticipate it without a detailed calculation. We shall encounter later several other models where it also occurs. Although the replica derivation presented here is non-rigorous, the clustering phenomenon can be proved rigorously.

In a solution with higher-order RSB, the distribution $P(q)$ develops new peaks. The geometrical interpretation is that clusters contain subclusters, which themselves contain subclusters etc. This hierarchical structure leads to the property of **ultrametricity**. Consider a triangle formed by three independent configurations drawn from the Boltzmann distribution, and let the lengths of its sides be measured using the Hamming distance. With high probability, such a triangle will be either equilateral, or isosceles with the two equal sides larger than the third one. In the case of full RSB, $P(q)$ has a continuous part, showing that the clustering property is not as sharp, because clusters are no longer well separated; but ultrametricity still holds.

Exercise 8.9 For a given sample of a p-spin glass in its 1RSB phase, we define Y as the probability that two configurations fall into the same cluster. More precisely: $Y = \int_{q}^{1} P_J(q') \, \mathrm{d}q'$, where $q_0 < q < q_1$. The previous analysis shows that $\lim_{N \to \infty} \mathbb{E} \, Y = 1 - \mathbf{x}$. Show that, in the large-N limit, $\mathbb{E}\left(Y^2\right) = (3 - 5\mathbf{x} + 2\mathbf{x}^2)/3$, as in the REM. Show that all moments of Y are identical to those of the REM. This result depends only on the 1RSB structure of the saddle point, and not on any of its details.

8.3 Extreme value statistics and the REM

Exercise 8.9 suggests that there exist universal properties which hold in the glass phase, independently of the details of the model.

In systems with a 1RSB phase, this universality is related to the universality of extreme value statistics. In order to clarify this point, we shall consider in this section a slightly generalized version of the REM. Here, we assume the energy levels to be $M = 2^N$ i.i.d. random variables admitting a probability density function $P(E)$ with the following properties:

1. $P(E)$ is continuous.
2. $P(E)$ is strictly positive on a semi-infinite domain $-\infty < E \leq E_0$.
3. In the $E \to -\infty$ limit, $P(E)$ vanishes more rapidly than any power law. We shall keep here to the simple case in which

$$P(E) \simeq A \exp\left(-B|E|^\delta\right) \qquad \text{as} \quad E \to -\infty, \tag{8.50}$$

for some positive constants A, B, δ.

We allow for such a general probability distribution because we want to check which properties of the corresponding REM are universal.

As we have seen in Chapter 5, the low-temperature phase of the REM is controlled by a few low-energy levels. Let us therefore begin by computing the distribution of the lowest energy level among $E_1, \ldots, E_M$ (which we call E_{gs}). Clearly,

$$\mathbb{P}[E_{\mathrm{gs}} > E] = \left[\int_E^\infty P(x)\, \mathrm{d}x\right]^M. \tag{8.51}$$

Let $E^*(M)$ be the value of E such that $\mathbb{P}[E_i < E] = 1/M$ for one of the energy levels E_i. For $M \to \infty$, one gets

$$|E^*(M)|^\delta = \frac{\log M}{B} + O(\log \log M). \tag{8.52}$$

Let us focus on energies close to $E^*(M)$, such that $E = E^*(M) + \varepsilon/(B\delta|E^*(M)|^{\delta-1})$, and consider the limit $M \to \infty$ with ε fixed. Then,

$$\mathbb{P}[E_i > E] = 1 - \frac{A}{B\delta|E|^{\delta-1}} e^{-B|E|^\delta} \left[1 + o(1)\right]$$

$$= 1 - \frac{1}{M} e^\varepsilon \left[1 + o(1)\right]. \tag{8.53}$$

Therefore, if we define a rescaled ground state energy through $E_{\mathrm{gs}} = E^*(M) + \varepsilon_{\mathrm{gs}}/(B\delta|E^*(M)|^{\delta-1})$, we get

$$\lim_{N \to \infty} \mathbb{P}[\varepsilon_{\mathrm{gs}} > \varepsilon] = \exp\left(-e^\varepsilon\right). \tag{8.54}$$

In other words, the pdf of the rescaled ground state energy converges to $P_1(\varepsilon) = \exp(\varepsilon - e^\varepsilon)$. This limit distribution, known as the **Gumbel distribution**, is *universal*. The form of the energy level distribution $P(E)$ enters only into the values of the shift and the scale, and not into the form of $P_1(\varepsilon)$. The following exercises show that several other properties of the glass phase in the REM are also universal.

Exercise 8.10 Let $E_1 \leq E_2 \leq \cdots \leq E_k$ be the k lowest energies. Show that universality also applies to the joint distribution of these energies, in the limit $M \to \infty$ at fixed k. More precisely, define the rescaled energies $\varepsilon_1 \leq \cdots \leq \varepsilon_k$ through $E_i = E^*(M) + \varepsilon_i/(B\delta|E^*(M)|^{\delta-1})$. Prove that the joint distribution of $\varepsilon_1, \ldots, \varepsilon_k$ admits a density and converges (as $M \to \infty$) to

$$P_k(\varepsilon_1, \ldots, \varepsilon_k) = \exp\left(\varepsilon_1 + \cdots + \varepsilon_k - e^{\varepsilon_k}\right) \mathbb{I}\left(\varepsilon_1 \leq \cdots \leq \varepsilon_k\right). \tag{8.55}$$

As the previous exercise shows, any finite number of energy levels, properly rescaled, has a well-defined distributional limit. Remarkably, one can obtain a compact description of this limit for an *infinite* number of enegy levels. Recall that a **Poisson point process** with a continuous density ρ is defined as a (random) collection of points on the real line, such that the expected number of points in a small interval $[\varepsilon, \varepsilon + d\varepsilon]$ is $\rho(\varepsilon)d\varepsilon + O((d\varepsilon)^2)$, and the numbers of points in two disjoint intervals are independent random variables.

Now think of the energy levels as points on a line. The claim is that they are distributed as a Poisson point process with density e^ε. Since $\int e^\varepsilon d\varepsilon = \infty$, one can show that this process has an infinite number of points with high probability. This is indeed quite natural, since we want a model for the energy levels as $N \to \infty$. It is not difficult to check that the first (smallest) point of such a process has the distribution (8.54) and that the first k points have the distribution (8.55).

Out of these energy levels, one can construct a probability distribution, by letting $Z(\mathsf{x}) = \sum_j e^{-\varepsilon_j/\mathsf{x}}$ (this converges for $\mathsf{x} \in [0,1)$) and $\mu(j) = e^{-\varepsilon_j/\mathsf{x}}/Z(\mathsf{x})$. To make contact with previous calculations, one should choose $\mathsf{x} = \beta_c/\beta$. This defines a random probability distribution $\mu(\cdot)$ over the integers that is usually referred to as the Poisson–Dirichlet process.

Exercise 8.11 Consider an REM where the pdf of the energies satisfies the assumptions 1–3 above, and $M = 2^N$. Show that, in order for the ground state energy to be extensive (i.e. $E_1 \sim N$ in the large-N limit), one must have $B \sim N^{1-\delta}$. Show that the system has a phase transition at the critical temperature $T_c = \delta \, (\log 2)^{(\delta-1)/\delta}$.

We define the participation ratios $Y_r \equiv \sum_{j=1}^{2^N} \mu(j)^r$. Prove that, for $T < T_c$, these quantities signal a condensation phenomenon. More precisely,

$$\lim_{N \to \infty} \mathbb{E} Y_r = \frac{\Gamma(r - \mathsf{x})}{\Gamma(r)\Gamma(1 - \mathsf{x})}, \tag{8.56}$$

where $\mathsf{x} = (T/T_c) \min\{\delta, 1\}$, as in the standard REM (see Section 8.3).

[Hint: One can prove this equality by direct probabilistic means using the methods of Section 5.3. For $\delta > 1$, one can also use the replica approach of Section 8.1.4.]

To summarize, in the condensed phase only the configurations with low energy matter. Within the class of random energy models defined above, the distribution of their energies converges to a Poisson–Dirichlet process. All the details of the model are hidden in the temperature-dependent parameter x. This universality is expected to hold in a wide class of models for which the 1RSB approach is asymptotically exact (with the caveat that configurations should be replaced by 'configuration lumps' or 'pure states'). This explains the success of 1RSB in many systems with a glass phase.

Indeed, this universality is expected to hold also when higher-order RSB applies. Namely, energy levels, properly rescaled, have the same distribution as that described above. The difference is that in this case they are endowed with an ultrametric distance structure, that corresponds to the ultrametric matrix Q_{ab} that appears in the replica method.

8.4 Appendix: Stability of the RS saddle point

In order to establish if a replica saddle point is correct, one widely used criterion is its local stability. In order to explain the basic idea, let us take a step backward and express the replicated free energy as an integral over the overlap parameters uniquely:

$$\mathbb{E} Z^n \doteq \int e^{N\widehat{G}(Q)} \prod_{(a,b)} dQ_{ab}. \tag{8.57}$$

Such an expression can be obtained either from eqn (8.30) by integrating over $\{\lambda_{ab}\}$ or as described in Exercise 8.6. Following the latter approach, we get

$$\widehat{G}(Q) = -n\frac{\beta^2}{4} - \frac{\beta^2}{2} \sum_{a<b} Q_{ab}^p - s(Q), \tag{8.58}$$

where

$$s(Q) = -\sum_{a<b} \omega_{ab} Q_{ab} + \psi(\omega)\bigg|_{\omega=\omega^*(Q)}, \qquad \psi(\omega) = \log\left[\sum_{\{\sigma_a\}} e^{\sum_{a<b} \omega_{ab}\sigma_a\sigma_b}\right], \tag{8.59}$$

and $\omega^*(Q)$ solves the equation $Q_{ab} = (\partial\psi(\omega)/\partial\omega_{ab})$. In other words, $s(Q)$ is the Legendre transform of $\psi(\omega)$ (apart from an overall minus sign). An explicit expression for $s(Q)$ is not available, but we shall need only the following well-known property of Legendre transforms:

$$\frac{\partial^2 s(Q)}{\partial Q_{ab} \partial Q_{cd}} = -C^{-1}_{(ab)(cd)}, \qquad C_{(ab)(cd)} \equiv \frac{\partial^2 \psi(\omega)}{\partial\omega_{ab} \partial\omega_{cd}}\bigg|_{\omega=\omega^*(Q)}, \tag{8.60}$$

where C^{-1} is the inverse of C in the matrix sense. The right-hand side is, in turn, easily written down in terms of averages over the replicated system (see eqn (8.34)):

$$C_{(ab)(cd)} = \langle\sigma_a\sigma_b\sigma_c\sigma_d\rangle_n - \langle\sigma_a\sigma_b\rangle_n\langle\sigma_c\sigma_d\rangle_n. \tag{8.61}$$

Assume now that $(Q^{\text{sp}}, \lambda^{\text{sp}})$ is a stationary point of $G(Q, \lambda)$. This is equivalent to saying that Q^{sp} is a stationary point of $\widehat{G}(Q)$ (the corresponding value of ω coincides with $i\lambda^{\mathrm{sp}}$). We would like to estimate the sum (8.57) as $\mathbb{E}Z^n \doteq e^{N\widehat{G}(Q^{\mathrm{sp}})}$. A necessary condition for this to be correct is that the matrix of second derivatives of $\widehat{G}(Q)$ is positive semidefinite at $Q = Q^{\mathrm{sp}}$. This is referred to as the **local stability** condition. Using eqns (8.58) and (8.61), we get the explicit condition

$$M_{(ab)(cd)} \equiv \left[-\frac{1}{2}\beta^2 p(p-1)Q_{ab}^{p-2}\,\delta_{(ab),(cd)} + C_{(ab)(cd)}^{-1} \right] \succeq 0 \,, \tag{8.62}$$

where we write $A \succeq 0$ to denote that the matrix A is positive semidefinite.

In this technical appendix, we sketch this computation in two simple cases: the stability of the RS saddle point for the general p-spin glass in zero magnetic field, and the SK model in a field.

We consider first the RS saddle point $Q_{ab} = 0$, $\omega_{ab} = 0$ in the p-spin glass. In this case

$$\langle f(\sigma) \rangle_n = \frac{1}{2^n} \sum_{\{\sigma_a\}} f(\sigma) \,. \tag{8.63}$$

It is then easy to show that $M_{(ab)(cd)} = \delta_{(ab),(cd)}$ for $p \geq 3$ and $M_{(ab)(cd)} = (1 - \beta^2)\delta_{(ab),(cd)}$ for $p = 2$. The situations for $p \geq 3$ and for $p = 2$ are very different:

- If $p = 2$ (the SK model), the RS solution is stable for $\beta < 1$, and unstable for $\beta > 1$.
- If $p \geq 3$, the RS solution is always stable.

Let us now look at the SK model in a magnetic field. This is the $p = 2$ case but with an extra term $-B\sum_i \sigma_i$ added to the energy (8.24). It is straightforward to repeat all the replica computations with this extra term. The results are formally identical if the average within the replicated system (8.34) is changed to

$$\langle f(\sigma) \rangle_{n,B} \equiv \frac{1}{z(\omega)} \sum_{\{\sigma^a\}} f(\sigma)\, \exp\left(\sum_{a<b} \omega_{ab}\,\sigma_a\sigma_b + \beta B \sum_a \sigma_a \right) \,, \tag{8.64}$$

$$z(\omega) \equiv \sum_{\{\sigma^a\}} \exp\left(\sum_{a<b} \omega_{ab}\,\sigma_a\sigma_b + \beta B \sum_a \sigma_a \right) \,. \tag{8.65}$$

The RS saddle point equations (8.35) are changed to

$$\omega = \beta^2\, q \,, \qquad q = \mathsf{E}_z \tanh^2\left(z\sqrt{\omega} + \beta B \right) \,. \tag{8.66}$$

When $B \neq 0$, the values of q, ω are non-zero at any positive β. This complicates the stability analysis.

Since $p = 2$, we have $M_{(ab)(cd)} = -\beta^2 \delta_{(ab)(cd)} + C_{(ab)(cd)}^{-1}$. Let $\{\lambda_j\}$ be the eigenvalues of $C_{(ab)(cd)}$. Since $C \succeq 0$, the condition $M \succeq 0$ is in fact equivalent to $1 - \beta^2\lambda_j \geq 0$, for all the eigenvalues λ_j.

The matrix elements $C_{(ab)(cd)}$ take three different forms, depending on the number of common indices in the two pairs (ab), (cd):

$$C_{(ab)(ab)} = 1 - \left[\mathsf{E}_z \tanh^2 \left(z\sqrt{\omega} + \beta B\right)\right]^2 \equiv U\,,$$

$$C_{(ab)(ac)} = \mathsf{E}_z \tanh^2 \left(z\sqrt{\omega} + \beta B\right) - \left[\mathsf{E}_z \tanh^2 \left(z\sqrt{\omega} + \beta B\right)\right]^2 \equiv V\,,$$

$$C_{(ab)(cd)} = \mathsf{E}_z \tanh^4 \left(z\sqrt{\omega} + \beta B\right) - \left[\mathsf{E}_z \tanh^2 \left(z\sqrt{\omega} + \beta B\right)\right]^2 \equiv W\,,$$

where $b \neq c$ is assumed in the second line, and all indices are distinct in the last line. We want to solve the eigenvalue equation $\sum_{(cd)} C_{(ab)(cd)} x_{cd} = \lambda x_{(ab)}$.

The first eigenvector is the uniform vector $x_{(ab)} = x$. Its eigenvalue is $\lambda_1 = U + 2(n-2)V + (n-2)(n-3)/2W$. Next we consider eigenvectors which depend on one special value θ of the replica index in the form $x_{(ab)} = x$ if $a = \theta$ or $b = \theta$, and $x_{(ab)} = y$ in all other cases. Orthogonality to the uniform vector is enforced by choosing $x = (1 - n/2)y$, and we find the eigenvalue $\lambda_2 = U + (n-4)V + (3-n)W$. This eigenvalue has degeneracy $n - 1$. Finally, we consider eigenvectors which depend on two special values θ, ν of the replica index: $x_{(\theta,\nu)} = x$, $x_{(\theta,a)} = x_{(\nu,a)} = y$, $x_{(ab)} = z$, where a and b are distinct from θ, ν. Orthogonality to the previously found eigenvectors imposes $x = (2-n)y$ and $y = [(3-n)/2]z$. Plugging this into the eigenvalue equation, we obtain the eigenvalue $\lambda_3 = U - 2V + W$, with degeneracy $n(n-3)/2$.

In the limit $n \to 0$, the matrix C has two distinct eigenvalues: $\lambda_1 = \lambda_2 = U - 4V + 3W$ and $\lambda_3 = U - 2V + W$. Since $V \geq W$, the most dangerous eigenvalue is λ_3 (called the **replicon eigenvalue**). This implies that the RS solution of the SK model is locally stable if and only if

$$\mathsf{E}_z \left[1 - \tanh^2 \left(z\sqrt{\omega} + \beta B\right)\right]^2 \leq T^2 \tag{8.67}$$

This inequality is saturated on a line in the T, B plane, called the **AT line**. This behaves like $T = 1 - \left(\frac{3}{4}\right)^{2/3} B^{2/3} + o(B^{2/3})$ for $B \to 0$ and like $T \simeq 4\,\mathrm{e}^{-B^2/2}/(3\sqrt{2\pi})$ for $B \gg 1$.

Exercise 8.12 Readers who want to test their understanding of these replica computations can study the SK model in zero field ($B = 0$), but in the case where the couplings have a ferromagnetic bias: the J_{ij} are i.i.d. and Gaussian distributed, with mean J_0/N and variance $1/N$.

(a) Show that the RS equations (8.35) are modified to

$$\omega = \beta^2 q\,, \quad q = \mathsf{E}_z \tanh^2 \left(z\sqrt{\omega} + \beta J_0 m\right)\,, \quad m = \mathsf{E}_z \tanh \left(z\sqrt{\omega} + \beta J_0 m\right)\,. \tag{8.68}$$

(b) Solve these equations numerically. Notice that, depending on the values of T and J_0, three types of solutions can be found: (1) a paramagnetic solution $m = 0, q = 0$, (2) a ferromagnetic solution $m > 0, q > 0$, and (3) a spin glass solution $m = 0, q > 0$.

(c) Show that the AT stability condition becomes

$$\mathsf{E}_z \left[1 - \tanh^2 \left(z\sqrt{\omega} + \beta J_0 m \right) \right]^2 < T^2 \,, \tag{8.69}$$

and deduce that the RS solution found in (a) and (b) is stable only in the paramagnetic phase and in a part of the ferromagnetic phase.

Notes

The replica solution of the REM was derived in the original work of Derrida that introduced the model (Derrida, 1980; Derrida, 1981). His motivation for introducing the REM came actually from the large-p limit of p-spin glasses.

The problem of moments is discussed, for instance, by Shohat and Tamarkin (1943).

The first universally accepted model of spin glasses was that of Edwards and Anderson (1975). The mean-field theory was defined by Sherrington and Kirkpatrick (1975) and Kirkpatrick and Sherrington (1978), who considered the RS solution. The instability of this solution in the $p = 2$ case was found by de Almeida and Thouless (1978), who first computed the location of the AT line. The solution to Exercise 8.12 can be found in Kirkpatrick and Sherrington (1978) and de Almeida and Thouless (1978).

Parisi's Ansatz was introduced in a few very inspired works starting in 1979 (Parisi, 1979; Parisi, 1980a; Parisi 1980b). His original motivation came from his reflections on the meaning of the permutation group S_n when $n < 1$, particularly in the $n \to 0$ limit. Unfortunately, there have not been any mathematical developments along these lines. The replica method, in the presence of RSB, is still waiting for a proper mathematical framework. On the other hand, it is a very well-defined computational scheme, which applies to a wide variety of problems. The physical interpretation of RSB in terms of condensation was found by Parisi (1983), and developed by Mézard *et al.* (1985c), who discussed the distribution of weights in the glass phase and its ultrametric organization. The p-spin model was analysed at large p using replicas by Gross and Mézard (1984). The clustering phenomenon was discovered in that work. The finite-p case was later studied by Gardner (1985). A rigorous treatment of the clustering effect in the p-spin glass model was developed by Talagrand (2000) and can be found in his book (Talagrand, 2003).

Ruelle (1987) introduced the asymptotic model for the low-lying energy levels of the REM. The connection between 1RSB and Gumbel's statistics of extremes is discussed in Bouchaud and Mézard (1997). For a description of the mathematical structure of (free) energy levels in models with higher-order RSB, we suggest the review by Aizenman *et al.* (2005). Poisson–Dirichlet processes are discussed from a probabilistic point of view in Pitman and Yor (1997).

A more detailed presentation of the replica method, together with reprints of most of the above papers, can be found in Mézard *et al.* (1987).

Part III

Models on graphs

9
Factor graphs and graph ensembles

Systems involving a large number of simple variables with mutual dependencies (or constraints or interactions) appear recurrently in several fields of science. It is often the case that such dependencies can be 'factorized' in a non-trivial way, and distinct that variables interact only 'locally'. In statistical physics, the fundamental origin of such a property can be traced back to the locality of physical interactions. In computer vision it is due to the two-dimensional character of the retina and the locality of reconstruction rules. In coding theory, it is a useful property for designing a system with fast encoding/decoding algorithms. This important structural property plays a crucial role in many interesting problems.

There exist several possibilities for expressing graphically the structure of dependencies among random variables: graphical models, Bayesian networks, dependency graphs, normal realizations, etc. We adopt here the *factor graph* language, because of its simplicity and flexibility.

As argued in the previous chapters, we are particularly interested in *ensembles* of probability distributions. These may emerge either from ensembles of error-correcting codes, in the study of disordered materials, or in the study of random combinatorial optimization problems. Problems drawn from these ensembles are represented by factor graphs, which are themselves *random*. The most common examples are random hypergraphs, which are a simple generalization of the well-known random graphs.

Section 9.1 introduces factor graphs and provides a few examples of their utility. In Section 9.2, we define some standard ensembles of random graphs and hypergraphs. We summarize some of their important properties in Section 9.3. One of the most surprising phenomena in random graph ensembles is the sudden appearance of a 'giant' connected component as the number of edges crosses a threshold. This is the subject of Section 9.4. Finally, in Section 9.5, we describe the local structure of large random factor graphs.

9.1 Factor graphs

9.1.1 Definitions and general properties

Example 9.1 We begin with a toy example. A country elects its president from two candidates $\{A, B\}$ according to the following peculiar system. The country is divided into four regions $\{1, 2, 3, 4\}$, grouped into two states: North (regions 1 and 2) and South (3 and 4). Each of the regions chooses its favourite candidate according

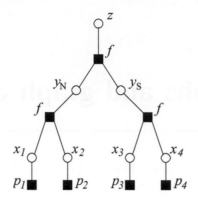

Fig. 9.1 Factor graph representation of the electoral process described in Example 1.

to the popular vote: we denote this candidate by $x_i \in \{A, B\}$, with $i \in \{1, 2, 3, 4\}$. Then, a North candidate y_N and a South candidate y_S are decided according to the following rule. If the preferences x_1 and x_2 in regions 1 and 2 agree, then y_N takes this value. If they do not agree, y_N is decided according to a fair-coin trial. The same procedure is adopted for the choice of y_S, given x_3, x_4. Finally, the president $z \in \{A, B\}$ is decided on the basis of the choices y_N and y_S in the two states using the same rule as inside each state.

A polling institute has obtained fairly good estimates of the probabilities $p_i(x_i)$ for the popular vote in each region i to favour the candidate x_i. They ask you to calculate the odds for each of the candidates to become the president.

It is clear that the electoral procedure described above has important 'factorization' properties. More precisely, the probability distribution for a given realization of the random variables $\{x_i\}$, $\{y_j\}$, z has the form

$$P(\{x_i\}, \{y_j\}, z) = f(z, y_N, y_S)\, f(y_N, x_1, x_2)\, f(y_S, x_3, x_4) \prod_{i=1}^{4} p_i(x_i). \qquad (9.1)$$

We leave it to the reader to write explicit forms for the function f. The election process and the above probability distribution can be represented graphically as in Fig. 9.1. Can this particular structure be exploited for computing the chances for each candidate to become president?

Abstracting from the above example, let us consider a set of N variables $x_1, \ldots, x_N$ taking values in a finite alphabet $\mathcal{X}$. We assume that their joint probability distribution takes the form

$$P(\underline{x}) = \frac{1}{Z} \prod_{a=1}^{M} \psi_a(\underline{x}_{\partial a}). \qquad (9.2)$$

Here we have used the shorthands $\underline{x} \equiv \{x_1, \ldots, x_N\}$ and $\underline{x}_{\partial a} \equiv \{x_i \,|\, i \in \partial a\}$, where $\partial a \subseteq [N]$. The set of indices ∂a, with $a \in [M]$, has a size $k_a \equiv |\partial a|$. When necessary,

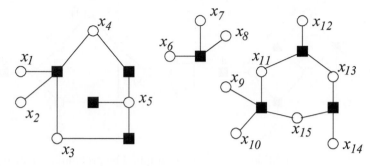

Fig. 9.2 A generic factor graph is formed by several connected components. Variables belonging to distinct components (for instance x_3 and x_{15} in the graph shown here) are statistically independent.

we shall use the notation $\{i_1^a, \ldots, i_{k_a}^a\} \equiv \partial a$ to denote the indices of the variables which correspond to the factor a, and $\underline{x}_{i_1^a, \ldots, i_{k_a}^a} \equiv \underline{x}_{\partial a}$ for the corresponding variables. The **compatibility functions** $\psi_a : \mathcal{X}^{k_a} \to \mathbb{R}$ are non-negative, and Z is a positive constant. In order to completely determine the form of eqn (9.2), we need to specify both the functions $\psi_a(\cdot)$ and an ordering of the indices in ∂a. In practice, this last specification will always be clear from the context.

Factor graphs provide a graphical representation of distributions of the form (9.2), which are also referred to as undirected **graphical models**. The factor graph for the distribution (9.2) contains two types of nodes: N **variable nodes**, each one associated with a variable x_i (represented by circles), and M **function nodes**, each one associated with a function ψ_a (represented by squares). An edge joins a variable node i and a function node a if the variable x_i is among the arguments of $\psi_a(\underline{x}_{\partial a})$ (in other words if $i \in \partial a$). The set of function nodes that are adjacent to (share an edge with) the variable node i is denoted by ∂i. The graph is bipartite: an edge always joins a variable node to a function node. The reader can easily check that the graph in Fig. 9.1 is indeed the factor graph corresponding to the factorized form (9.1). The degree of a variable node $|\partial i|$ or of a factor node $|\partial a|$ is defined as usual as the number of edges incident on it. In order to avoid trivial cases, we shall assume that $|\partial a| \geq 1$ for any factor node a. The basic property of the probability distribution (9.2), encoded in its factor graph, is that two 'well-separated' variables interact uniquely through those variables which are interposed between them. A precise formulation of this intuition is given by the following observation, named the **global Markov property**.

Proposition 9.2 *Let $A, B, S \subseteq [N]$ be three disjoint subsets of the variable nodes, and denote by $\underline{x}_A$, $\underline{x}_B$, and $\underline{x}_S$ the corresponding sets of variables. If S 'separates' A and B (i.e. if there is no path in the factor graph joining a node of A to a node of B without passing through S), then*

$$P(\underline{x}_A, \underline{x}_B | \underline{x}_S) = P(\underline{x}_A | \underline{x}_S) \, P(\underline{x}_B | \underline{x}_S) \,. \tag{9.3}$$

In such a case, the variables $\underline{x}_A, \underline{x}_B$ are said to be conditionally independent.

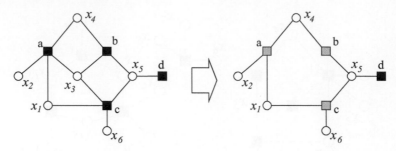

Fig. 9.3 The action of conditioning on a factor graph. The probability distribution on the *left* has the form $P(\underline{x}_{1\ldots 6}) \propto f_a(\underline{x}_{1\ldots 4})f_b(\underline{x}_{3,4,5})f_c(\underline{x}_{1,3,5,6})f_d(\underline{x}_5)$. After conditioning on x_3, we get $P(\underline{x}_{1\ldots 6}|x_3 = x_*) \propto f_a'(\underline{x}_{1,2,4})f_b'(\underline{x}_{4,5})f_c'(\underline{x}_{1,5,6})f_d(\underline{x}_5)$. Note that the functions $f_a'(\cdot)$, $f_b'(\cdot)$, $f_c'(\cdot)$ (grey nodes on the *right*) are distinct from $f_a(\cdot)$, $f_b(\cdot)$, $f_c(\cdot)$ and depend upon the value x_*.

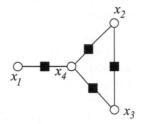

Fig. 9.4 A factor graph with four variables; $\{x_1\}$ and $\{x_2, x_3\}$ are independent, conditional on x_4. The set of variables $\{x_2, x_3, x_4\}$ and the three function nodes connecting two points in this set form a clique.

Proof It is easy to provide a 'graphical' proof of this statement. Notice that, if the factor graph is disconnected, then variables belonging to distinct components are independent; see Fig. 9.2. Conditioning on a variable x_i is equivalent to eliminating the corresponding variable node from the graph and modifying the adjacent function nodes accordingly; see Fig. 9.3. Finally, when one conditions on $\underline{x}_S$ as in eqn (9.3), the factor graph becomes split in such a way that A and B belong to distinct components. We leave to the reader the exercise of filling in the details. □

It is natural to wonder whether any probability distribution which is 'globally Markov' with respect to a given graph can be written in the form (9.2). In general, the answer is negative, as can be shown with a simple example. Consider the small factor graph in Fig. 9.4. The global Markov property has a non-trivial content only for the following choice of subsets: $A = \{1\}$, $B = \{2, 3\}$, $S = \{4\}$. The most general probability distribution such that x_1 is independent of $\{x_2, x_3\}$ conditional on x_4 is of the type $f_a(x_1, x_4)f_b(x_2, x_3, x_4)$. The probability distribution encoded by the factor graph is a special case where $f_b(x_2, x_3, x_4) = f_c(x_2, x_3)f_d(x_3, x_4)f_e(x_4, x_2)$.

The factor graph of our counterexample in Fig. 9.4 has a peculiar property: it contains a subgraph (the one with variables $\{x_2, x_3, x_4\}$) such that, for any pair of variable nodes, there is a function node adjacent to both of them. We call any factor subgraph possessing this property a **clique** (this definition generalizes the notion of a clique in the usual type of graphs with pairwise edges). It turns out that once one gets rid of cliques, the converse of Proposition 9.2 can be proved. We 'get rid' of cliques by completing the factor graph. Given a factor graph F, its **completion** $\overline{F}$ is obtained by adding one factor node for each clique in the graph and connecting it to each variable node in the clique and to no other node.

Theorem 9.3. (Hammersley–Clifford) *Let $P(\cdot)$ be a strictly positive probability distribution over the variables $\underline{x} = (x_1,\ldots,x_N) \in \mathcal{X}^N$, satisfying the global Markov property (9.3) with respect to a factor graph F. Then P can be written in the factorized form (9.2) with respect to the completed graph $\overline{F}$.*

Roughly speaking, the only assumption behind the factorized form (9.2) is the rather weak notion of locality encoded by the global Markov property. This may serve as a general justification for studying probability distributions that have a factorized form. Note that the positivity hypothesis $P(x_1,\ldots,x_N) > 0$ is not just a technical assumption: there exist counterexamples to the Hammersley–Clifford theorem if P is allowed to vanish.

9.1.2 Examples

Let us look at a few examples.

We start with Markov chains. The random variables $X_1,\ldots,X_N$ taking values in a finite state space $\mathcal{X}$ form a **Markov chain of order** r (with $r < N$) if

$$P(x_1\ldots x_N) = P_0(x_1\ldots x_r) \prod_{t=r}^{N-1} w(x_{t-r+1}\ldots x_t \to x_{t+1}), \qquad (9.4)$$

for some non-negative transition probabilities $\{w(x_{-r}\ldots x_{-1} \to x_0)\}$ and initial condition $P_0(x_1\ldots x_r)$ satisfying the normalization conditions

$$\sum_{x_1\ldots x_r} P_0(x_1\ldots x_r) = 1, \qquad \sum_{x_0} w(x_{-r}\ldots x_{-1} \to x_0) = 1. \qquad (9.5)$$

The parameter r is the 'memory range' of the chain. Ordinary Markov chains have $r = 1$. Higherorder Markov chains allow one to model more complex phenomena. For instance, in order to get a reasonable probabilistic model of the English language with the usual alphabet $\mathcal{X} = \{$a,b,$\ldots$ z, blank$\}$ as the state space, it is reasonable to choose r to be of the order of the average word length.

It is clear that eqn (9.4) is a particular case of the factorized form (9.2). The corresponding factor graph includes N variable nodes, one for each variable x_i; $N-r$ function nodes, one for each of the factors $w(\cdot)$; and one function node for the initial condition $P_0(\cdot)$. In Fig. 9.5, we present a small example with $N = 6$ and $r = 2$.

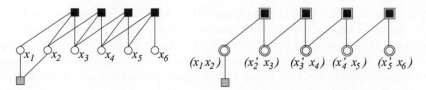

Fig. 9.5 *Left*: factor graph for a Markov chain of length $N = 6$ and memory range $r = 2$. *Right*: by adding auxiliary variables, the same probability distribution can be written as a Markov chain with memory range $r = 1$.

Exercise 9.1 Show that a Markov chain with memory range r and state space $\mathcal{X}$ can always be rewritten as a Markov chain with memory range 1 and state space $\mathcal{X}^r$.
   [Hint: The transition probabilities $\hat{w}$ of the new chain are given in terms of the original ones by

$$\hat{w}(\vec{x} \to \vec{y}) = \begin{cases} w(x_1, \dots, x_r \to y_r) & \text{if } x_2 = y_1,\, x_3 = y_2, \dots x_r = y_{r-1}, \\ 0 & \text{otherwise,} \end{cases} \qquad (9.6)$$

where we have used the shorthands $\vec{x} \equiv (x_1, \dots, x_r)$ and $\vec{y} = (y_1, \dots, y_r)$.]
   Figure 9.5 shows a reduction to an order-1 Markov chain in factor graph language.

What is the content of the global Markov property for Markov chains? Let us start from the case of order-1 chains. Without loss of generality, we can choose S as containing only one variable node (let us say the i-th one) while A and B are, respectively the nodes to the left and to the right of i: $A = \{1, \dots, i-1\}$ and $B = \{i+1, \dots, N\}$. The global Markov property reads

$$P(x_1 \dots x_N | x_i) = P(x_1 \dots x_{i-1} | x_i)\, P(x_{i+1} \dots x_N | x_i), \qquad (9.7)$$

which is just a rephrasing of the usual Markov condition: $X_{i+1} \dots X_N$ depend upon $X_1 \dots X_i$ uniquely through X_i. We invite the reader to discuss the global Markov property for order-r Markov chains.

Our second example is borrowed from coding theory. Consider a 'Hamming' code $\mathfrak{C}$ of block length $N = 7$ defined by the codebook

$$\mathfrak{C} = \{(x_1, x_2, x_3, x_4) \in \{0, 1\}^4 \mid x_1 \oplus x_3 \oplus x_5 \oplus x_7 = 0, \qquad (9.8)$$

$$x_2 \oplus x_3 \oplus x_6 \oplus x_7 = 0,\ x_4 \oplus x_5 \oplus x_6 \oplus x_7 = 0\}.$$

Let $\mu_0(\underline{x})$ be the uniform probability distribution over the codewords. Then,

$$\mu_0(\underline{x}) = \frac{1}{Z_0}\, \mathbb{I}(x_1 \oplus x_3 \oplus x_5 \oplus x_7 = 0)\, \mathbb{I}(x_2 \oplus x_3 \oplus x_6 \oplus x_7 = 0) \qquad (9.9)$$

$$\times\, \mathbb{I}(x_4 \oplus x_5 \oplus x_6 \oplus x_7 = 0),$$

where $Z_0 = 16$ is a normalization constant. This distribution has the form (9.2), and the corresponding factor graph is reproduced in Fig. 9.6.

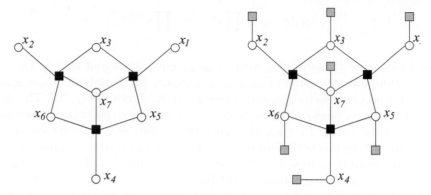

Fig. 9.6 *Left*: factor graph for the uniform distribution over the code defined in eqn (9.8). *Right*: factor graph for the distribution of the transmitted message conditional on the channel output. Grey function nodes encode the information carried by the channel output.

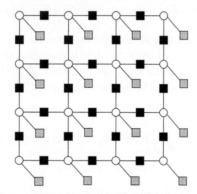

Fig. 9.7 Factor graph for an Edwards–Anderson model with size $L = 4$ in $d = 2$ dimensions. Filled squares correspond to pairwise interaction terms $-J_{ij}\sigma_i\sigma_j$. Grey squares denote magnetic-field terms $-B\sigma_i$.

Exercise 9.2 Suppose that a codeword in $\mathfrak{C}$ is transmitted through a binary memoryless channel, and that the message $(y_1, y_2, \ldots, y_7)$ is received. As argued in Chapter 6, in order to find the codeword which has been sent, one should consider the probability distribution of the transmitted message conditional on the channel output (see eqn (6.5)). Show that the factor graph representation for this distribution is the one given in Fig. 9.6, right-hand frame.

Let us now introduce an example from statistical physics. In Section 2.6 we introduced the Edwards–Anderson model, a statistical-mechanics model for spin glasses, whose energy function reads $E(\underline{\sigma}) = -\sum_{(ij)} J_{ij}\sigma_i\sigma_j - B\sum_i \sigma_i$. The Boltzmann distribution can be written as

$$\mu_\beta(\underline{\sigma}) = \frac{1}{Z} \prod_{(ij)} e^{\beta J_{ij}\sigma_i\sigma_j} \prod_i e^{\beta B\sigma_i} , \qquad (9.10)$$

where i runs over the sites of a d-dimensional cubic lattice of side L, i.e. $i \in [L]^d$, and (ij) runs over the pairs of nearest neighbours in the lattice. Once again, this distribution admits a factor graph representation, as shown in Fig. 9.7. This graph includes two types of function nodes. Nodes corresponding to pairwise interaction terms $-J_{ij}\sigma_i\sigma_j$ in the energy function are connected to two neighbouring variable nodes. Nodes representing magnetic-field terms $-B\sigma_i$ are connected to a unique variable.

The final example comes from combinatorial optimization. The satisfiability problem is a decision problem, introduced in Chapter 3. Given N Boolean variables $x_1, \dots, x_N \in \{T, F\}$ and M logical clauses containing them, one is asked to find a truth assignment that satisfies all of the clauses. The logical AND of the M clauses is usually called a formula. Consider, for instance, the following formula over $N = 7$ variables:

$$(x_1 \vee x_2 \vee \overline{x_4}) \wedge (x_2 \vee x_3 \vee x_5) \wedge (\overline{x_4} \vee \overline{x_5}) \wedge (x_5 \vee \overline{x_7} \vee \overline{x_6}) . \qquad (9.11)$$

For a given satisfiability formula, it is quite natural to consider the uniform probability distribution $\mu_{\text{sat}}(x_1, \dots, x_N)$ over the truth assignments which satisfy eqn (9.11) (whenever there exists at least one such assignment). A little thought shows that such a distribution can be written in the factorized form (9.2). For instance, the formula (9.11) yields

$$\mu_{\text{sat}}(x_1, \dots, x_7) = \frac{1}{Z_{\text{sat}}} \mathbb{I}(x_1 \vee x_2 \vee \overline{x_4}) \, \mathbb{I}(x_2 \vee x_3 \vee x_5)) \, \mathbb{I}(\overline{x_4} \vee \overline{x_5})$$
$$\times \mathbb{I}(x_5 \vee \overline{x_7} \vee \overline{x_6}) , \qquad (9.12)$$

where Z_{sat} is the number of distinct truth assignments which satisfy eqn (9.11). We invite the reader to draw the corresponding factor graph.

Exercise 9.3 Consider the problem of colouring a graph $\mathcal{G}$ with q colours, encountered earlier in Section 3.3. Build a factor graph representation for this problem, and write the associated compatibility functions.

[Hint: In the simplest such representation, the number of function nodes is equal to the number of edges of $\mathcal{G}$, and every function node has degree 2.]

9.2 Ensembles of factor graphs: Definitions

We shall be interested generically in understanding the properties of *ensembles* of probability distributions taking the factorized form (9.2). We introduce here a few useful ensembles of factor graphs. In the simple case, where every function node has degree 2, factor graphs are in one-to-one correspondence with the usual type of graphs, and we are just treating random graph ensembles, as first studied by Erdös and Renyi. The case of arbitrary factor graphs is a simple generalization. From the graph-theoretical point of view, they can be regarded either as **hypergraphs** (by associating a vertex with each variable node and a hyperedge with each function node) or as bipartite graphs (variable and function nodes are both associated with vertices in this case).

For any integer $k \geq 1$, the **random k-factor graph** with M function nodes and N variables nodes is denoted by $\mathbb{G}_N(k, M)$, and is defined as follows. For each function node $a \in \{1 \dots M\}$, the k-tuple ∂a is chosen uniformly at random from the $\binom{N}{k}$ k-tuples in $\{1 \dots N\}$.

Sometimes one encounters variations of this basic distribution. For instance, it can be useful to prevent any two function nodes from having the same neighbourhood, by imposing the condition $\partial a \neq \partial b$ for any $a \neq b$. This can be done in a natural way through the ensemble $\mathbb{G}_N(k, \alpha)$ defined as follows. For each of the $\binom{N}{k}$ k-tuples of variable nodes, a function node is added to the factor graph independently with probability $N\alpha / \binom{N}{k}$, and all of the variables in the k-tuple are connected to it. The total number M of function nodes in the graph is a random variable, with expectation $M_{\mathrm{av}} = \alpha N$.

In the following, we shall often be interested in large graphs ($N \to \infty$) with a finite density of function nodes. In $\mathbb{G}_N(k, M)$, this means that $M \to \infty$, with the ratio M/N kept fixed. In $\mathbb{G}_N(k, \alpha)$, the large-N limit is taken with α fixed. The exercises below suggest that, for some properties, the distinction between the two graph ensembles does not matter in this limit.

Exercise 9.4 Consider a factor graph from the ensemble $\mathbb{G}_N(k, M)$. What is the probability p_{dist} that for all pairs of function nodes, the corresponding neighbourhoods are distinct? Show that, in the limit $N \to \infty$, $M \to \infty$ with $M/N \equiv \alpha$ and k fixed,

$$
p_{\mathrm{dist}} = \begin{cases} O(e^{-\alpha^2 N/2}) & \text{if } k = 1, \\ e^{-\alpha^2}[1 + \Theta(N^{-1})] & \text{if } k = 2, \\ 1 + \Theta(N^{-k+2}) & \text{if } k \geq 3. \end{cases} \tag{9.13}
$$

Exercise 9.5 Consider a random factor graph from the ensemble $\mathbb{G}_N(k, \alpha)$, in the large-N limit. Show that the probability of getting a number of function nodes M different from its expectation αN by an 'extensive' number (i.e. a number of order N) is exponentially small. In mathematical terms, there exists a constant $A > 0$ such that, for any $\varepsilon > 0$,

$$
\mathbb{P}\left[|M - M_{\mathrm{av}}| > N\varepsilon \right] \leq 2\, e^{-AN\varepsilon^2}. \tag{9.14}
$$

Consider the distribution of a $\mathbb{G}_N(k, \alpha)$ random graph conditional on the number of function nodes being $\overline{M}$. Show that this is the same as the distribution of a $\mathbb{G}_N(k, \overline{M})$ random graph conditional on all the function nodes having distinct neighbourhoods.

An important local property of a factor graph is its **degree profile**. Given a graph, we denote by Λ_i and P_i the fractions of variable nodes and function nodes, respectively, of degree i. Note that $\Lambda \equiv \{\Lambda_n : n \geq 0\}$ and $P \equiv \{P_n : n \geq 0\}$ are in fact two distributions over the non-negative integers (they are both non-negative and normalized). Moreover, they have non-vanishing weight only at a finite number of degrees (at most N for Λ and M for P). The pair (Λ, P) is called the degree profile of the graph F. A practical representation of the degree profile is provided by the generating functions $\Lambda(x) = \sum_{n \geq 0} \Lambda_n x^n$ and $P(x) = \sum_{n \geq 0} P_n x^n$. Because of the above remarks, both $\Lambda(x)$ and $P(x)$ are in fact finite polynomials with non-negative coefficients. The average degree of a variable node and a function node are given by $\sum_{n \geq 0} \Lambda_n n = \Lambda'(1)$ and $\sum_{n \geq 0} P_n n = P'(1)$, respectively.

If the graph is randomly generated, its degree profile is a random variable. For instance, in the random k-factor graph ensemble $\mathbb{G}_N(k, M)$ defined above, the variable-node degree Λ depends upon the graph realization: we shall investigate some of its properties below. In contrast, the function node profile $P_n = \mathbb{I}(n = k)$ of this ensemble is deterministic.

It is convenient to consider *ensembles* of factor graphs with a prescribed degree profile. We therefore introduce the ensemble of **degree-constrained factor graphs** $\mathbb{D}_N(\Lambda, P)$ by endowing the set of graphs with degree profile (Λ, P) with the uniform probability distribution. Note that the number M of function nodes is fixed by the relation $MP'(1) = N\Lambda'(1)$. A special case which is important in this context is that of **random regular graphs**, in which the degree of variable nodes is fixed, as well as the degree of function nodes. In an (l, k) random regular graph, each variable node has degree l and each function node has degree k, corresponding to $\Lambda(x) = x^l$ and $P(x) = x^k$.

A degree-constrained factor graph ensemble is non-empty only if $N\Lambda_n$ and MP_n are integers for any $n \geq 0$. Even if these conditions are satisfied, it is not obvious how to construct a graph in $\mathbb{D}_N(\Lambda, P)$ efficiently. Since such ensembles play a crucial role in the theory of sparse-graph codes, we postpone this issue to Chapter 11.

9.3 Random factor graphs: Basic properties

For the sake of simplicity, we shall study here only the ensemble $\mathbb{G}_N(k, M)$ with $k \geq 2$. Generalizations to graphs in $\mathbb{D}_N(\Lambda, P)$ will be mentioned in Section 9.5.1 and developed further in Chapter 11. We study the asymptotic limit of large graphs, i.e. for $N \to \infty$ with k (the degree of function nodes) and $M/N = \alpha$ fixed.

9.3.1 Degree profile

The variable-node degree profile $\{\Lambda_n : n \geq 0\}$ is a random variable. By the linearity of expectation, $\mathbb{E}\,\Lambda_n = \mathbb{P}[\deg_i = n]$, where $\deg_i$ is the degree of the node i. Let p be the probability that a uniformly chosen k-tuple in $\{1, \ldots, N\}$ contains i. It is clear that $\deg_i$ is a binomial random variable (defined in Section A.3) with parameters M and p. Furthermore, since p does not depend upon the site i, it is equal to the probability that a randomly chosen site belongs to a fixed k-tuple. Expressed in formulae,

$$\mathbb{P}[\deg_i = n] = \binom{M}{n} p^n (1-p)^{M-n}, \qquad p = \frac{k}{N}. \tag{9.15}$$

If we consider the large-graph limit, with n fixed, we get

$$\lim_{N \to \infty} \mathbb{P} \left[\deg_i = n \right] = \lim_{N \to \infty} \mathbb{E} \, \Lambda_n = \mathrm{e}^{-k\alpha} \frac{(k\alpha)^n}{n!} \, . \qquad (9.16)$$

The degree of site i is asymptotically a Poisson random variable.

How correlated are the degrees of variable nodes? By a simple generalization of the above calculation, we can compute the joint probability distribution of $\deg_i$ and $\deg_j$ with $i \neq j$. We think of constructing the graph by choosing one k-tuple of variable nodes at a time and adding the corresponding function node to the graph. Each node can have one of four possible 'fates': it is connected to both of the nodes i and j (with probability p_2); it is connected only to i or only to j (each case has probability p_1); or it is connected neither to i nor to j (probability $p_0 \equiv 1 - 2p_1 - p_2$). A little thought shows that $p_2 = k(k-1)/N(N-1)$, $p_1 = k(N-k)/N(N-1)$, and

$$\mathbb{P}[\deg_i = n, \deg_j = m] = \sum_{l=0}^{\min(n,m)} \binom{M}{n-l, \, m-l, \, l} p_2^l p_1^{n+m-2l} p_0^{M-n-m+l}, \quad (9.17)$$

where l is the number of function nodes which are connected both to i and to j, and we have used the standard notation for multinomial coefficients (see Appendix A).

Once again, it is illuminating to look at the large-graph limit $N \to \infty$ with n and m fixed. It is clear that the $l = 0$ term dominates the sum in eqn (9.17). In fact, the multinomial coefficient is of order $\Theta(N^{n+m-l})$ and the various probabilities are of order $p_0 = \Theta(1)$, $p_1 = \Theta(N^{-1})$, and $p_2 = \Theta(N^{-2})$. Therefore the l-th term of the sum is of order $\Theta(N^{-l})$. Elementary calculus then shows that

$$\mathbb{P}[\deg_i = n, \deg_j = m] = \mathbb{P}[\deg_i = n] \, \mathbb{P}[\deg_j = m] + \Theta(N^{-1}) \, . \qquad (9.18)$$

This shows that, asymptotically, the degrees of the nodes are pairwise independent Poisson random variables. This fact can be used to show that the degree profile $\{\Lambda_n : n \geq 0\}$ is, for large graphs, close to its expectation. In fact,

$$\mathbb{E} \left[(\Lambda_n - \mathbb{E}\Lambda_n)^2 \right] = \frac{1}{N^2} \sum_{i,j=1}^{N} \left\{ \mathbb{P}[\deg_i = n, \deg_j = n] - \mathbb{P}[\deg_i = n] \mathbb{P}[\deg_j = n] \right\}$$

$$= \Theta(N^{-1}), \qquad (9.19)$$

which implies, via the Chebyshev inequality, $\mathbb{P}(|\Lambda_n - \mathbb{E}\Lambda_n| \geq \delta \, \mathbb{E}\Lambda_n) = \Theta(N^{-1})$ for any $\delta > 0$.

The pairwise independence expressed in eqn (9.18) is essentially a consequence of the fact that, given two distinct variable nodes i and j, the probability that they are connected to the same function node is of order $\Theta(N^{-1})$. It is easy to see that the same property holds when we consider any finite number of variable nodes. Suppose now that we look at a factor graph from the ensemble $\mathbb{G}_N(k, M)$ conditional on the function node a being connected to variable nodes $i_1, \ldots, i_k$. What is the distribution of the residual degrees $\deg'_{i_1}, \ldots, \deg'_{i_k}$? (By the residual degree $\deg'_i$, we mean the degree of node i once the function node a has been pruned from the graph.)?It is clear that the residual graph is distributed according to the ensemble $\mathbb{G}_N(k, M-1)$. Therefore the residual degrees are (in the large-graph limit) independent Poisson random variables with mean $k\alpha$. We can formalize these simple observations as follows.

Proposition 9.4 *Let $i_1, \ldots, i_n \in \{1, \ldots, N\}$ be n distinct variable nodes, and let G be a random graph from $\mathbb{G}_N(k, M)$ conditional on the neighbourhoods of m function nodes $a_1, \ldots, a_m$ being $\partial a_1, \ldots, \partial a_m$. Denote by $\deg_i'$ the degree of variable node i once $a_1, \ldots, a_m$ have been pruned from the graph. In the limit of large graphs $N \to \infty$ with $M/N \equiv \alpha$, k, n, and m fixed, the residual degrees $\deg_{i_1}', \ldots, \deg_{i_n}'$ converge in distribution to independent Poisson random variables with mean $k\alpha$.*

This property is particularly useful when one is investigating the local properties of a $\mathbb{G}_N(k, N\alpha)$ random graph. In particular, it suggests that such local properties are close to those of the ensemble $\mathbb{D}_N(\Lambda, P)$, where $P(x) = x^k$ and $\Lambda(x) = \exp[k\alpha(x-1)]$.

A remark: in the above discussion, we have focused on the probability of finding a node with some constant degree n in the asymptotic limit $N \to \infty$. One may wonder whether, in a typical graph $G \in \mathbb{G}_N(k, M)$, there might exist some variable nodes with exceptionally large degrees. The exercise below shows that this is not the case.

Exercise 9.6 We want to investigate the typical properties of the maximum variable-node degree $\Delta(G)$ in a random graph G from $\mathbb{G}_N(k, M)$.

(a) Let $\bar{n}_{\max}$ be the smallest value of $n > k\alpha$ such that $N\mathbb{P}[\deg_i = n] \leq 1$. Show that $\Delta(G) \leq \bar{n}_{\max}$ with probability approaching one in the large-graph limit.
 [Hint: Show that $N\mathbb{P}[\deg_i = \bar{n}_{\max} + 1] \to 0$ at large N.]

(b) Show that the following asymptotic form holds for $\bar{n}_{\max}$:

$$\frac{\bar{n}_{\max}}{k\alpha e} = \frac{z}{\log(z/\log z)} \left[1 + \Theta\left(\frac{\log \log z}{(\log z)^2} \right) \right], \tag{9.20}$$

 where $z \equiv (\log N)/(k\alpha e)$.

(c) Let $\underline{n}_{\max}$ be the largest value of n such that $N\mathbb{P}[\deg_i = n] \geq 1$. Show that $\Delta(G) \geq \underline{n}_{\max}$ with probability approaching one in the large-graph limit.
 [Hints: Show that $N\mathbb{P}[\deg_i = \underline{n}_{\max} - 1] \to \infty$ at large N. Apply the second-moment method to Z_l, the number of nodes of degree l.]

(d) What is the asymptotic behaviour of $\underline{n}_{\max}$? How does it compare with $\bar{n}_{\max}$?

9.3.2 Small subgraphs

The next simplest question one may ask, concerning a random graph, is the occurrence in it of a given small subgraph. We shall not give a general treatment of the problem here, but rather work out a few simple examples.

Let us begin by considering a fixed k-tuple of variable nodes $i_1, \ldots, i_k$ and ask for the probability p that they are connected by a function node in a graph $G \in \mathbb{G}_N(k, M)$. In fact, it is easier to compute the probability that they are *not* connected:

$$1 - p = \left[1 - \binom{N}{k}^{-1} \right]^M. \tag{9.21}$$

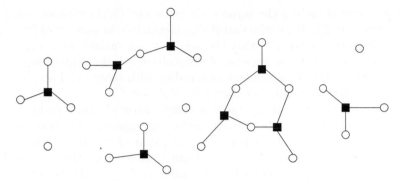

Fig. 9.8 A factor graph from the ensemble $\mathbb{G}_N(k, M)$ with $k = 3$, $N = 23$, and $M = 8$. It contains $Z_{\text{isol}} = 3$ isolated function nodes, $Z_{\text{isol},2} = 1$ isolated pairs of function nodes, and $Z_{\text{cycle},3} = 1$ cycle of length 3. The remaining three variable nodes have degree 0.

The quantity in brackets is the probability that a given function node *is not* a neighbour of $i_1, \ldots, i_k$. It is raised to the power M because the M function nodes are independent in the model $\mathbb{G}_N(k, M)$. In the large-graph limit, we get

$$p = \frac{\alpha\, k!}{N^{k-1}}[1 + \Theta(N^{-1})]. \tag{9.22}$$

This confirms an observation of the previous section: for any fixed set of nodes, the probability that a function node connects any two of them vanishes in the large graph limit.

As a first example, let us ask how many isolated function nodes appear in a graph $G \in \mathbb{G}_N(k, M)$. We say that a node is 'isolated' if all the neighbouring variable nodes have degree one. We denote the number of such function nodes by Z_{isol}. It is easy to compute the expectation of this quantity,

$$\mathbb{E}\, Z_{\text{isol}} = M \left[\binom{N}{k}^{-1} \binom{N-k}{k} \right]^{M-1}. \tag{9.23}$$

The factor M is due to the fact that each of the M function nodes can be isolated. Consider one such node a and its neighbours $i_1, \ldots, i_k$. The factor in $\binom{N}{k}^{-1}\binom{N-k}{k}$ is the probability that a function node $b \neq a$ is not incident on any of the variables $i_1, \ldots, i_k$. This must be counted for any $b \neq a$; hence the exponent $M - 1$. Once again, things become more transparent in the large-graph limit:

$$\mathbb{E}\, Z_{\text{isol}} = N\alpha e^{-k^2\alpha}[1 + \Theta(N^{-1})]. \tag{9.24}$$

So, there is a non-vanishing density of isolated function nodes, $\mathbb{E}\, Z_{\text{isol}}/N$. This density approaches 0 at small α (because there are few function nodes) and at large α (because function nodes are unlikely to be isolated). A more refined analysis shows that Z_{isol} is tightly concentrated around its expectation: the probability of an order-N fluctuation vanishes exponentially as $N \to \infty$.

There is a way of finding the asymptotic behaviour (9.24) without going through the exact formula (9.23). We notice that $\mathbb{E} \, Z_{\text{isol}}$ is equal to the number of function nodes ($M = N\alpha$) times the probability that the neighbouring variable nodes $i_1, \ldots, i_k$ have degree 0 in the residual graph. Because of Proposition 9.4, the degrees $\deg'_{i_1}, \ldots, \deg'_{i_k}$ are approximatively i.i.d. Poisson random variables with mean $k\alpha$. Therefore the probability for all of them to vanish is close to $(e^{-k\alpha})^k = e^{-k^2\alpha}$.

Of course, this last type of argument becomes extremely convenient when we are considering small structures which involve more than one function node. As a second example, let us compute the number $Z_{\text{isol},2}$ of pairss of function nodes which have exactly one variable node in common and are isolated from the rest of the factor graph (for instance, in the graph in Fig. 9.8, we have $Z_{\text{isol},2} = 1$). We obtain

$$
\mathbb{E} \, Z_{\text{isol},2} = \binom{N}{2k-1} \cdot \frac{k}{2} \binom{2k-1}{k} \cdot \left(\frac{\alpha k!}{N^{k-1}} \right)^2 \cdot (e^{-k\alpha})^{2k-1} \left[1 + \Theta \left(\frac{1}{N} \right) \right] . \quad (9.25)
$$

The first factor counts the ways of choosing the $2k - 1$ variable nodes which support the structure. Then we count the number of way of connecting two function nodes to $(2k - 1)$ variable nodes in such a way that they have only one variable in common. The third factor is the probability that the two function nodes are indeed present (see eqn (9.22)). Finally, we have to require that the residual graph of all the $(2k - 1)$ variable nodes is 0, which gives the factor $(e^{-k\alpha})^{2k-1}$. The above expression is easily rewritten as

$$
\mathbb{E} \, Z_{\text{isol},2} = N \cdot \frac{1}{2} (k\alpha)^2 \, e^{-k(2k-1)\alpha} \left[1 + \Theta(1/N) \right] . \quad (9.26)
$$

With some more work, one can prove again that $Z_{\text{isol},2}$ is in fact concentrated around its expected value: a random factor graph contains a finite density of isolated pairs of function nodes.

Let us consider, in general, the number of small subgraphs of some definite type. Its most important property is how it scales with N in the large N-limit. This is easily found. For instance, let us take another look at eqn (9.25): N enters only in counting the $(2k - 1)$-tuples of variable nodes which can support the chosen structure, and in the probability of having two function nodes in the desired positions. In general, if we consider a small subgraph with v variable nodes and f function nodes, the number $Z_{v,f}$ of such structures has an expectation which scales as

$$
\mathbb{E} \, Z_{v,f} \sim N^{v-(k-1)f} . \quad (9.27)
$$

This scaling has important consequences for the nature of small structures which appear in a large random graph. For discussing such structures, it is useful to introduce the notions of a 'connected (sub)graph', a 'tree', and a 'path' in a factor graph exactly in the same way as in the case of the usual type of graphs, whereby both variable and function nodes are viewed as vertices (see Chapter 3). We define, further, a **component** of a factor graph G as a subgraph C which is is connected and isolated, in the sense that there is no path between a node of C and a node of $G \backslash C$

Consider a connected factor graph with v variable nodes and f function nodes, all of them having degree k. This graph is a tree if and only if $v = (k-1)f + 1$. We denote by $Z_{\text{tree},v}$ the number of isolated trees over v variable nodes which are contained in a $\mathbb{G}_N(k, M)$ random graph. Because of eqn (9.27), we have $\mathbb{E} Z_{\text{tree},v} \sim N$: a random graph contains a finite density (when $N \to \infty$) of trees of any finite size. On the other hand, connected subgraphs which are not trees must have $v < (k-1)f + 1$, and eqn (9.27) shows that their number does not grow with N. In other words, most (more precisely, all but a vanishing fraction) of the *finite* components of a random factor graph are trees.

Exercise 9.7 Consider the largest component in the graph in Fig. 9.8 (the one with three function nodes), and let $Z_{\text{cycle},3}$ be the number of times it occurs as a component of a $\mathbb{G}_N(k, M)$ random graph. Compute $\mathbb{E} Z_{\text{cycle},3}$ in the large-graph limit.

Exercise 9.8 A factor graph is said to be **unicyclic** if it contains a unique (up to shifts) closed, self-avoiding path $\omega_0, \omega_1, \ldots, \omega_\ell = \omega_0$ ('self-avoiding' means that for any $t, s \in \{0 \ldots \ell - 1\}$ with $t \neq s$, one has $\omega_t \neq \omega_s$).

(a) Show that a connected factor graph with v variable nodes and f function nodes, all of them having degree k, is unicyclic if and only if $v = (k-1)f$.

(b) Let $Z_{\text{cycle},v}(N)$ be the number of unicyclic components over v nodes in a $\mathbb{G}_N(k, M)$ random graph. Use eqn (9.27) to show that $Z_{\text{cycle},v}$ is finite with high probability in the large-graph limit. More precisely, show that $\lim_{n \to \infty} \lim_{N \to \infty} \mathbb{P}_{\mathbb{G}_N}[Z_{\text{cycle},v} \geq n] = 0$.

9.4 Random factor graphs: The giant component

We have just argued that most finite-size components of a $\mathbb{G}_N(k, \alpha N)$ factor graph are trees in the large-N limit. However, finite-size trees do not always exhaust the graph. It turns out that when α becomes larger than a threshold value, a 'giant component' appears in the graph. This is a connected component containing an extensive (proportional to N) number of variable nodes, with many cycles.

9.4.1 Nodes in finite trees

We want to estimate what fraction of a random graph from the ensemble $\mathbb{G}_N(k, \alpha N)$ is covered by finite-size trees. This fraction is defined as

$$x_{\text{tr}}(\alpha, k) \equiv \lim_{s \to \infty} \lim_{N \to \infty} \frac{1}{N} \mathbb{E} N_{\text{trees},s}, \qquad (9.28)$$

where $N_{\text{trees},s}$ is the number of sites contained in trees of size not larger than s. In order to compute $\mathbb{E} N_{\text{trees},s}$, we use the number of trees of size equal to s, which we denote by $Z_{\text{trees},s}$. Using the approach discussed in the previous section, we get

$$\mathbb{E}\, N_{\text{trees},s} = \sum_{v=0}^{s} v \cdot \mathbb{E}\, Z_{\text{trees},v} \tag{9.29}$$

$$= \sum_{v=0}^{s} v \binom{N}{v} \cdot T_k(v) \cdot \left(\frac{\alpha k!}{N^{k-1}} \right)^{(v-1)/(k-1)} \cdot (\text{e}^{-k\alpha})^v \left[1 + \Theta \left(\frac{1}{N} \right) \right]$$

$$= N(\alpha k!)^{-1/(k-1)} \sum_{v=0}^{s} \frac{1}{(v-1)!} T_k(v) \left[(\alpha k!)^{1/(k-1)} \text{e}^{-k\alpha} \right]^v + \Theta(1) \,,$$

where $T_k(v)$ is the number of trees which can be built out of v distinct variable nodes and $f = (v-1)/(k-1)$ function nodes of degree k. The computation of $T_k(v)$ is a classical piece of enumerative combinatorics, which is described in Section 9.4.3 below. The result is

$$T_k(v) = \frac{(v-1)!\, v^{f-1}}{(k-1)!^f f!} \,, \tag{9.30}$$

and the generating function $\widehat{T}_k(z) = \sum_{v=1}^{\infty} T_k(v) z^v / (v-1)!$, which we need in order to compute $\mathbb{E} N_{\text{trees},s}$ from eqn (9.29), is found to satisfy the self-consistency equation

$$\widehat{T}_k(z) = z \exp \left\{ \frac{\widehat{T}_k(z)^{k-1}}{(k-1)!} \right\} . \tag{9.31}$$

It is a simple exercise to see that, for any $z \geq 0$, this equation has two solutions such that $\widehat{T}_k(z) \geq 0$, the relevant one being the smaller of the two (this is a consequence of the fact that $\widehat{T}_k(z)$ has a regular Taylor expansion around $z = 0$). Using this characterization of $\widehat{T}_k(z)$, one can show that $x_{\text{tr}}(\alpha, k)$ is the smallest positive solution of the equation

$$x_{\text{tr}} = \exp \left(-k\alpha + k\alpha\, x_{\text{tr}}^{k-1} \right) . \tag{9.32}$$

This equation is solved graphically in Fig. 9.9, left frame. In the range $\alpha \leq \alpha_{\text{p}} \equiv 1/(k(k-1))$, the only non-negative solution is $x_{\text{tr}} = 1$: all but a vanishing fraction of nodes belong to finite-size trees. When $\alpha > \alpha_{\text{p}}$, the solution has $0 < x_{\text{tr}} < 1$: the fraction of nodes in finite trees is strictly smaller than one.

9.4.2 Size of the giant component

This result is somewhat surprising. For $\alpha > \alpha_{\text{p}}$, a strictly positive fraction of variable nodes does not belong to any finite tree. On the other hand, we saw in the previous subsection that finite components with cycles contain a vanishing fraction of the nodes. Where are the other $N(1-x_{\text{tr}})$ nodes? It turns out that, roughly speaking, they belong to a unique connected component, called the giant component, which is not a tree. One basic result describing this phenomenon is the following.

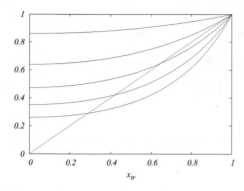 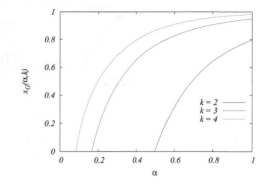

Fig. 9.9 *Left*: graphical representation of eqn (9.32) for the fraction of nodes of a $\mathbb{G}_N(k, M)$ random factor graph that belong to finite-size tree components. The curves refer to $k = 3$ and (from *top* to *bottom*) $\alpha = 0.05, 0.15, 0.25, 0.35, 0.45$. *Right*: typical size of the giant component.

Theorem 9.5 *Let X_1 be the size of the largest connected component in a $\mathbb{G}_N(k, M)$ random graph with $M = N[\alpha + o_N(1)]$, and let $x_{\mathrm{G}}(\alpha, k) = 1 - x_{\mathrm{tr}}(\alpha, k)$, where $x_{\mathrm{tr}}(\alpha, k)$ is defined as the smallest solution of eqn (9.32). Then, for any positive ε,*

$$|X_1 - N x_{\mathrm{G}}(\alpha, k)| \leq N\varepsilon, \tag{9.33}$$

with high probability.

Furthermore, the giant component contains many loops. We define the **cyclic number** c of a factor graph containing v vertices and f function nodes of degree k as $c = v - (k - 1)f - 1$. The cyclic number of the giant component is then $c = \Theta(N)$ with high probability.

Exercise 9.9 Convince yourself that there cannot be more than one component of size $\Theta(N)$. Here is a possible route. Consider the event of having two connected components of sizes $\lfloor N s_1 \rfloor$ and $\lfloor N s_2 \rfloor$, for two fixed positive numbers s_1 and s_2, in a $\mathbb{G}_N(k, M)$ random graph with $M = N[\alpha + o_N(1)]$ (with $\alpha \geq s_1 + s_2$). In order to estimate the probability of such an event, imagine constructing the $\mathbb{G}_N(k, M)$ graph by adding one function node at a time. What condition must hold when the number of function nodes is $M - \Delta M$? What can happen to the last ΔM nodes? Now take $\Delta M = \lfloor N^\delta \rfloor$ with $0 < \delta < 1$.

The appearance of a giant component is sometimes referred to as **percolation on the complete graph** and is one of the simplest instance of a phase transition. We shall now give a simple heuristic argument which predicts correctly the typical size of the giant component. This argument can be seen as the simplest example of the 'cavity method' that we shall develop in the following chapters. We first notice that, by the linearity of the expectation, $\mathbb{E}\, X_1 = N x_{\mathrm{G}}$, where x_{G} is the probability that a given variable node i belongs to the giant component. In the large-graph limit, site i is connected to $l(k - 1)$ distinct variable nodes, l being a Poisson random variable of

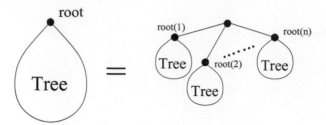

Fig. 9.10 A rooted tree G on $v+1$ vertices can be decomposed into a root and the union of n rooted trees $G_1, \ldots, G_n$ on $v_1, \ldots, v_n$ vertices, respectively.

mean $k\alpha$ (see Section 9.3.1). The node i belongs to the giant component if any of its $l(k-1)$ neighbours does. If we assume that the $l(k-1)$ neighbours belong to the giant component independently with probability x_G, then we get

$$x_\mathrm{G} = \mathbb{E}_l[1 - (1 - x_\mathrm{G})^{l(k-1)}], \qquad (9.34)$$

where l is Poisson distributed with mean $k\alpha$. Taking the expectation, we get

$$x_\mathrm{G} = 1 - \exp[-k\alpha + k\alpha(1 - x_\mathrm{G})^{k-1}], \qquad (9.35)$$

which coincides with eqn (9.32) if we set $x_\mathrm{G} = 1 - x_\mathrm{tr}$.

The above argument has several flaws, but only one of them is serious. In writing eqn (9.34), we assumed that the probability that none of l randomly chosen variable nodes belongs to the giant component is just the product of the probabilities that each of them does not. In the present case it is not difficult to fix the problem, but in subsequent chapters we shall see several examples of the same type of heuristic reasoning where the solution is less straightforward.

9.4.3   Appendix: Counting trees

This subsection is a technical appendix devoted to the computation of $T_k(v)$, the number of trees with v variable nodes, when the function nodes have degree k. Let us begin by considering the case $k = 2$. Note that if $k = 2$, we can uniquely associate with any factor graph F an ordinary graph G obtained by replacing each function node by an edge joining the neighbouring variables (for basic definitions concerning graphs, we refer readers to Chapter 3). In principle, G may contain multiple edges, but this does not concern us as long as we stick to the case where F is a tree. Therefore $T_2(v)$ is just the number of ordinary (non-factor) trees on v distinct vertices. Rather than computing $T_2(v)$, we shall compute the number $T_2^*(v)$ of **rooted** trees on v distinct vertices. Recall that a rooted graph is just a pair (G, i_*), where G is a graph and i_* is a distinguished node in G. Of course, we have the relation $T_2^*(v) = vT_2(v)$.

Consider now a rooted tree on $v+1$ vertices, and assume that the root has degree n (of course, $1 \le n \le v$). Erase the root together with its edges, and mark the n vertices that were connected to the root. We are left with n rooted trees of sizes $v_1, \ldots, v_n$ such that $v_1 + \cdots + v_n = v$. This naturally leads to the recursion

$$T_2^*(v+1) = (v+1) \sum_{n=1}^{v} \frac{1}{n!} \sum_{\substack{v_1 \ldots v_n > 0 \\ v_1 + \cdots + v_n = v}} \binom{v}{v_1, \cdots, v_n} T_2^*(v_1) \cdots T_2^*(v_n), \qquad (9.36)$$

which holds for any $v \geq 1$. Together with the initial condition $T_2^*(1) = 1$, this relation allows one to determine $T_2^*(v)$ recursively for any $v > 0$. This recursion is depicted in Fig. 9.10.

The recursion is most easily solved by introducing the generating function $\widehat{T}(z) = \sum_{v>0} T_2^*(v) z^v / v!$. Using this definition in eqn (9.36), we get

$$\widehat{T}(z) = z \exp\{\widehat{T}(z)\}, \qquad (9.37)$$

which is closely related to the definition of Lambert's W function (usually written as $W(z) \exp(W(z)) = z$). One has in fact the identity $\widehat{T}(z) = -W(-z)$. The expansion of $\widehat{T}(z)$ in powers of z can be obtained by the Lagrange inversion method (see the Exercise below). We get $T_2^*(v) = v^{v-1}$, and therefore $T_2(v) = v^{v-2}$. This result is known as the **Cayley formula** and is one of the most famous results in enumerative combinatorics.

Exercise 9.10 Assume that the generating function $A(z) = \sum_{n>0} A_n z^n$ is a solution of the equation $z = f(A(z))$, where f is an analytic function such that $f(0) = 0$ and $f'(0) = 1$. Use the Cauchy formula $A_n = \oint z^{-n-1} A(z) \, dz / (2\pi i)$ to show that

$$A_n = \operatorname{coeff} \left\{ f'(x) \left(x/f(x) \right)^{n+1}; x^{n-1} \right\}. \qquad (9.38)$$

Use this result, known as the 'Lagrange inversion method', to compute the power expansion of $\widehat{T}(z)$ and prove the Cayley formula $T_2(v) = v^{v-2}$.

Let us now return to the generic-k case. The reasoning is similar to the $k = 2$ case. One finds, after some work, that the generating function $\widehat{T}_k(z) \equiv \sum_{v>0} T_k^*(v) z^v / v!$ satisfies the equation

$$\widehat{T}_k(z) = z \exp \left\{ \frac{\widehat{T}_k(z)^{k-1}}{(k-1)!} \right\}, \qquad (9.39)$$

from which one can deduce the number of trees with v variable nodes,

$$T_k^*(v) = \frac{v! \, v^{f-1}}{(k-1)!^f f!}. \qquad (9.40)$$

In this expression, the number of function nodes f is fixed by $v = (k-1)f + 1$.

9.5 The locally tree-like structure of random graphs

9.5.1 Neighbourhood of a node

There exists a natural notion of a distance between variable nodes of a factor graph. Given a path $(\omega_0, \ldots, \omega_\ell)$ on the factor graph, we define its length as the number of function nodes on it. Then the **distance** between two variable nodes is defined as the length of the shortest path connecting them (by convention, it is set to $+\infty$ when the nodes belong to distinct connected components). We also define the **neighbourhood** of radius r of a variable node i, denoted by $\mathsf{B}_{i,r}(F)$, as the subgraph of F including all the variable nodes at a distance at most r from i, and all the function nodes connected only to those variable nodes.

What does the neighbourhood of a typical node look like in a random graph? It is convenient to step back for a moment from the ensemble $\mathbb{G}_N(k, M)$ and consider a degree-constrained factor graph $F \stackrel{\text{d}}{=} \mathbb{D}_N(\Lambda, P)$. Furthermore, we define the **edge-perspective** degree profiles as $\lambda(x) \equiv \Lambda'(x)/\Lambda'(1)$ and $\rho(x) \equiv P'(x)/P'(1)$. These are polynomials

$$\lambda(x) = \sum_{l=1}^{l_{\max}} \lambda_l \, x^{l-1}, \qquad \rho(x) = \sum_{k=1}^{k_{\max}} \rho_k \, x^{k-1}, \tag{9.41}$$

where λ_l and ρ_k are the probabilities that a randomly chosen edge in the graph is adjacent to a variable node and to a function node, respectively, of degree l or k. The explicit formulae

$$\lambda_l = \frac{l\Lambda_l}{\sum_{l'} l'\Lambda_{l'}}, \qquad \rho_k = \frac{kP_k}{\sum_{k'} k'P_{k'}}, \tag{9.42}$$

can be derived by noticing that the graph F contains $nl\Lambda_l$ and mkP_k edges adjacent to variable nodes of degree l and function nodes of degree k, respectively.

Imagine constructing the neighbourhoods of a node i of increasing radius r. Given $\mathsf{B}_{i,r}(F)$, let $i_1, \ldots, i_L$ be the nodes at distance r from i, and let $\deg'_{i_1}, \ldots, \deg'_{i_L}$ be their degrees in the residual graph $F \setminus \mathsf{B}_{i,r}(F)$. Arguments analogous to the ones leading to Proposition 9.4 imply that $\deg'_{i_1}, \ldots, \deg'_{i_L}$ are asymptotically i.i.d. random variables with $\deg'_{i_n} = l_n - 1$, and with l_n distributed according to λ_{l_n}. An analogous result holds for function nodes (we just invert the roles of variable and function nodes).

This motivates the following definition of an r-generation tree ensemble $\mathbb{T}_r(\Lambda, P)$. If $r = 0$, there is a unique element in the ensemble: a single isolated node, which is given the generation number 0. If $r > 0$, we first generate a tree from the ensemble $\mathbb{T}_{r-1}(\Lambda, P)$ ensemble. Then, for each variable node i of generation $r - 1$, we draw an independent integer $l_i \geq 1$ distributed according to λ_l and add to the graph $l_i - 1$ function nodes connected to the variable i (unless $r = 1$, in which case l_i function nodes are added, with l_i distributed according to Λ_{l_i}). Next, for each of the newly added function nodes $\{a\}$, we draw an independent integer $k_a \geq 1$ distributed according to ρ_k and add to the graph $k_a - 1$ variable nodes connected to the function a. Finally, the new variable nodes are given the generation number r. The case of uniformly chosen random graphs where function nodes have a fixed degree k corresponds to the tree ensemble $\mathbb{T}_r(e^{k\alpha(x-1)}, x^k)$. In this case, it is easy to check that the degrees in the residual graph have a Poisson distribution with mean $k\alpha$, in agreement with Proposition 9.4. With a slight abuse of notation, we shall use the shorthand $\mathbb{T}_r(k, \alpha)$ to denote this tree ensemble.

It is not unexpected that $\mathbb{T}_r(\Lambda, P)$ constitutes a good model for r-neighbourhoods in the degree-constrained ensemble. Analogously, $\mathbb{T}_r(k, \alpha)$ is a good model for r-neighbourhoods in the ensemble $\mathbb{G}_N(k, M)$ when $M \simeq N\alpha$. This is made more precise below.

Theorem 9.6 *Let F be a random factor graph in the ensemble $\mathbb{D}_N(\Lambda, P)$ or $\mathbb{G}_N(k, M)$, respectively, let i be a uniformly chosen random variable node in F, and let r be a non-negative integer. Then $\mathsf{B}_{i,r}(F)$ converges in distribution to $\mathbb{T}_r(\Lambda, P)$ or to $\mathbb{T}_r(k, \alpha)$ as $N \to \infty$ with Λ, P fixed or α, k fixed, respectively.*

In other words, the factor graph F looks locally like a random tree from the ensemble $\mathbb{T}_r(\Lambda, P)$.

9.5.2   Loops

We have seen that in the large-graph limit, a factor graph $F \overset{\mathrm{d}}{=} \mathbb{G}_N(k, M)$ converges locally to a tree. Furthermore, it has been shown in Section 9.3.2 that the number of 'small' cycles in such a graph is only $\Theta(1)$ as $N \to \infty$. It is therefore natural to ask at what distance from any given node loops start playing a role.

More precisely, let i be a uniformly chosen random node in F. We would like to know what the typical length of the shortest loop through i is. Of course, this question has a trivial answer if $k(k-1)\alpha < 1$, since in this case most of the variable nodes belong to small tree components (see Section 9.4). We shall therefore consider $k(k-1)\alpha > 1$ from now on.

A heuristic guess at the size of this loop can be obtained as follows. Assume that the neighbourhood $\mathsf{B}_{i,r}(F)$ is a tree. Each function node has $k-1$ adjacent variable nodes at the next generation. Each variable node has a Poisson-distributed number of adjacent function nodes at the next generation, with mean $k\alpha$. Therefore the average number of variable nodes at a given generation is $[k(k-1)\alpha]$ times the number at the previous generation. The total number of nodes in $\mathsf{B}_{i,r}(F)$ is about $[k(k-1)\alpha]^r$, and loops will appear when this quantity becomes comparable to the total number of nodes in the graph. This yields $[k(k-1)\alpha]^r = \Theta(N)$, or $r = \log N / \log[k(k-1)\alpha]$. This is of course a very crude argument, but it is also a very robust one: one can, for instance, change N to $N^{1\pm\varepsilon}$ and affect only the prefactor. It turns out that this result is correct, and can be generalized to the ensemble $\mathbb{D}_N(\Lambda, P)$.

Proposition 9.7 *Let F be a random factor graph in the ensemble $\mathbb{D}_N(\Lambda, P)$ or $\mathbb{G}_N(k, M)$, let i be a uniformly chosen random variable node in F, and let ℓ_i be the length of the shortest loop in F through i. Assume that $c = \lambda'(1)\rho'(1) > 1$ or $c = k(k-1)\alpha > 1$, respectively. Then, with high probability,*

$$\ell_i = \frac{\log N}{\log c}[1 + o(1)] . \tag{9.43}$$

We refer the reader to the literature for a proof. The following exercise gives a slightly more precise, but still heuristic, version of the previous argument.

Exercise 9.11 Assume that the neighbourhood $B_{i,r}(F)$ is a tree and that it includes n 'internal' variable nodes (i.e. nodes whose distance from i is smaller than r), n_l 'boundary' variable nodes (whose distance from i is equal to r), and m function nodes. Let F_r be the residual graph, i.e. F minus the subgraph $B_{i,r}(F)$. It is clear that $F_r \stackrel{\mathrm{d}}{=} \mathbb{G}_{N-n}(k, M - m)$. Show that the probability p_r that a function node of F_r connects two of the variable nodes on the boundary of $B_{i,r}(F)$ is

$$p_r = 1 - \left[(1 - q)^k + k\,(1 - q)^{k-1}\,q \right]^{M-m}, \tag{9.44}$$

where $q \equiv n_l/(N - n)$. As a first estimate of p_r, we can substitute n_l, n, and m by their expectations (in the tree ensemble) in this equation, and call the corresponding estimate $\bar{p}_r$. Assuming that $r = \rho \frac{\log N}{\log[k(k-1)\alpha]}$, show that

$$\bar{p}_r = 1 - \exp\left\{ -\frac{1}{2}k(k - 1)\alpha N^{2\rho-1} \right\} \left[1 + O(N^{-2+3\rho}) \right]. \tag{9.45}$$

If $\rho > 1/2$, this indicates that, under the assumption that there is no loop of length $2r$ or smaller through i, there is, with high probability, a loop of length $2r + 1$. If, on the other hand, $\rho < 1/2$, this indicates that there is no loop of length $2r + 1$ *or smaller* through i. This argument suggests that the length of the shortest loop through i is about $\log N / \log[k(k - 1)\alpha]$.

Notes

A nice introduction to factor graphs is provided by the paper by Kschischang *et al.* (2001); see also Aji and McEliece (2000). Factor graphs are related to graphical models (Jordan, 1998), Bayesian networks (Pearl, 1988), and to Tanner graphs in coding theory (Tanner, 1981). Among the alternatives to factor graphs, it is worth recalling the 'normal realizations' discussed by Forney (2001).

The proof of the Hammersley–Clifford theorem (initially motivated by the probabilistic modelling of some physical problems) goes back to 1971. A proof, more detailed references, and some historical comments can be found in Clifford (1990).

The theory of random graphs was pioneered by Erdös and Renyi (1960). The emergence of a giant component in a random graph is a classic result which goes back to their work. Two standard textbooks on random graphs, by Bollobás (2001) and Janson *et al.* (2000), provide, in particular, a detailed study of the phase transition. Graphs with constrained degree profiles were studied by Bender and Canfield (1978). A convenient 'configuration model' for analysing them was introduced by Bollobás (1980) and allowed for the location of the phase transition by Molloy and Reed (1995). Finally, Wormald (1999) provides a useful survey (including short-loop properties) of degree-constrained ensembles.

For general background on hypergraphs, see Duchet (1995). The threshold for the emergence of a giant component in a random hypergraph with edges of fixed size k

(corresponding to the factor graph ensemble $\mathbb{G}_N(k, M)$) was discussed by Schmidt-Pruzan and Shamir (1985). The neighbourhood of the threshold was analysed by Karónski and Luczak (2002) and in references therein.

In enumerating trees, we used generating functions. This approach to combinatorics is developed thoroughly in Flajolet and Sedgewick (2008).

Ensembles with hyperedges of different sizes have been considered recently in combinatorics (Darling and Norris, 2005), as well as in coding theory (as code ensembles). Our definitions and notation for degree profiles and degree-constrained ensembles follows the coding literature (Luby *et al.*, 1997; Richardson and Urbanke, 2001*a*).

The local structure of random graphs and of more complex random objects (in particular, random *labelled* graphs) is the object of the theory of *local weak convergence* (Aldous and Steele, 2003).

10

Satisfiability

Because of Cook's theorem (see Chapter 3) satisfiability lies at the heart of computational complexity theory: this fact has motivated intense research activity on this problem. This chapter will not be a comprehensive introduction to such a vast topic, but rather will present some selected research directions. In particular, we shall pay special attention to the definition and analysis of ensembles of random satisfiability instances. There are various motivations for studying random instances. In order to test and improve algorithms that are aimed at solving satisfiability, it is highly desirable to have an automatic generator of 'hard' instances at hand. As we shall see, properly 'tuned' ensembles provide such a generator. Also, the analysis of ensembles has revealed a rich structure and stimulated fruitful contacts with other disciplines. The present chapter focuses on 'standard' algorithmic and probabilistic approaches. We shall come back to satisfiability, using methods inspired by statistical physics, in Chapter 20.

Section 10.1 recalls the definition of satisfiability and introduces some standard terminology. A basic, widely adopted strategy for solving decision problems consists in exploring exhaustively the tree of possible assignments of the problem's variables. Section 10.2 presents a simple implementation of this strategy. In Section 10.3, we introduce some important ensembles of random instances. The hardness of the satisfiability problem depends on the maximum clause length. When clauses have length 2, the decision problem is solvable in polynomial time. This is the topic of Section 10.4. Finally, in Section 10.5, we discuss the existence of a phase transition for random K-satisfiability with $K \geq 3$, when the density of clauses is varied, and derive some rigorous bounds on the location of this transition.

10.1 The satisfiability problem

10.1.1 SAT and UNSAT formulae

An instance of the satisfiability problem is defined in terms of N Boolean variables and a set of M constraints between them, where each constraint takes the special form of a clause. A clause is a logical OR of some variables or their negations. Here we shall adopt the following representation: a variable x_i, with $i \in \{1, \ldots, N\}$, takes values in $\{0, 1\}$, where 1 corresponds to 'true' and 0 to 'false'; the negation of x_i is $\overline{x}_i \equiv 1 - x_i$. A variable or its negation is called a literal, and we shall denote a literal by z_i, with $i \in \{1, \ldots, N\}$ (therefore z_i denotes any of x_i, $\overline{x}_i$). A clause a, with $a \in \{1, \ldots, M\}$, involving K_a variables is a constraint which forbids exactly one among the 2^{K_a} possible assignments to these K_a variables. It is written as a logical OR

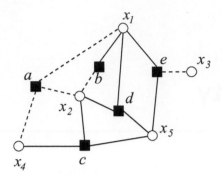

Fig.   10.1 Factor graph representation of the formula $(\bar{x}_1 \vee \bar{x}_2 \vee \bar{x}_4) \wedge (x_1 \vee \bar{x}_2)$ $\wedge (x_2 \vee x_4 \vee x_5) \wedge (x_1 \vee x_2 \vee x_5) \wedge (x_1 \vee \bar{x}_3 \vee x_5)$.

(denoted by $\vee$) function of some variables or their negations. For instance, the clause $x_2 \vee \bar{x}_{12} \vee x_{37} \vee \bar{x}_{41}$ is satisfied by all assignments of the variables except those where $x_2 = 0, x_{12} = 1, x_{37} = 0, x_{41} = 1$. When it is not satisfied, a clause is said to be violated.

We denote by ∂a the subset $\{i_1^a, \ldots, i_{K_a}^a\} \subseteq \{1, \ldots, N\}$ containing the indices of the $K_a = |\partial a|$ variables involved in clause a. Clause a can be written as $C_a = z_{i_1^a} \vee z_{i_2^a} \vee \cdots \vee z_{i_{K_a}^a}$. An instance of the satisfiability problem can be summarized as the following logical formula (in **conjunctive normal form** (CNF)):

$$F = C_1 \wedge C_2 \wedge \cdots \wedge C_M . \tag{10.1}$$

As we have seen in Section 9.1.2, there exists[1] a simple and natural representation of a satisfiability formula as a factor graph associated with the indicator function $\mathbb{I}(\underline{x} \text{ satisfies } F)$. Actually, it is often useful to use a slightly more elaborate factor graph with two types of edges: a full edge is drawn between a variable vertex i and a clause vertex a whenever x_i appears in a, and a dashed edge is drawn whenever $\bar{x}_i$ appears in a. In this way, there is a one-to-one correspondence between a CNF formula and its graph. An example is shown in Fig. 10.1.

Given the formula F, the question is whether there exists an assignment of the variables x_i to $\{0, 1\}$ (among the 2^N possible assignments) such that the formula F is true. An algorithm that solves the satisfiability problem must be able, given a formula F, either to answer 'yes' (the formula is then said to be **SAT**), and provide such an assignment, called a **SAT-assignment**, or to answer 'no', in which case the formula is said to be **UNSAT**. The restriction of the satisfiability problem obtained by requiring that all the clauses in F have the same length $K_a = K$ is called the K-**satisfiability** (or K-SAT) problem.

As usual, an optimization problem is naturally associated with the decision version of the satisfiability provlem: given a formula F, one is asked to find an assignment which violates the smallest number of clauses. This is called the **MAX-SAT** problem.

---

[1]It may happen that there does not exist any assignment satisfying F, so that one cannot use this indicator function to define a probability measure. However, one can still characterize the local structure of $\mathbb{I}(\underline{x} \text{ satisfies } F)$ by a factor graph.

Exercise 10.1 Consider the 2-SAT instance defined by the formula $F_1 = (x_1 \vee \overline{x}_2) \wedge (x_2 \vee \overline{x}_3) \wedge (\overline{x}_2 \vee x_4) \wedge (x_4 \vee \overline{x}_1) \wedge (\overline{x}_3 \vee \overline{x}_4) \wedge (\overline{x}_2 \vee x_3)$. Show that this formula is SAT and write a SAT-assignment.

[Hint: Assign, for instance, $x_1 = 1$; the clause $x_4 \vee \overline{x}_1$ is then reduced to x_4; this is a **unit clause** which fixes $x_4 = 1$; the chain of 'unit clause propagation' leads either to a SAT assignment or to a contradiction.]

Exercise 10.2 Consider the 2-SAT formula $F_2 = (x_1 \vee \overline{x}_2) \wedge (x_2 \vee \overline{x}_3) \wedge (\overline{x}_2 \vee x_4) \wedge (x_4 \vee \overline{x}_1) \wedge (\overline{x}_3 \vee \overline{x}_4) \wedge (\overline{x}_2 \vee \overline{x}_3)$. Show that this formula is UNSAT by using the same method as in the previous exercise.

Exercise 10.3 Consider the 3-SAT formula $F_3 = (x_1 \vee x_2 \vee \overline{x}_3) \wedge (x_1 \vee x_3 \vee \overline{x}_4) \wedge (x_2 \vee x_3 \vee x_4) \wedge (\overline{x}_1 \vee x_2 \vee \overline{x}_4) \wedge (x_1 \vee \overline{x}_2 \vee x_4) \wedge (\overline{x}_1 \vee \overline{x}_2 \vee x_4) \wedge (\overline{x}_2 \vee \overline{x}_3 \vee \overline{x}_4) \wedge (x_2 \vee \overline{x}_3 \vee x_4) \wedge (\overline{x}_1 \vee x_3 \vee \overline{x}_4)$. Show that it is UNSAT.

[Hint: Try to generalize the previous method by using a decision tree (see Section 10.2.2) or list the 16 possible assignments and cross out which one is eliminated by each clause.]

As we have already mentioned, satisfiability was the first problem to be proved to be NP-complete. The restriction defined by requiring that $K_a \leq 2$ for each clause a is polynomial. However, if one relaxes this condition to $K_a \leq K$, with $K = 3$ or more, the resulting problem is NP-complete. For instance, 3-SAT is NP-complete, while 2-SAT is polynomial. It is intuitively clear that MAX-SAT is 'at least as hard' as SAT: an instance is SAT if and only if the minimum number of violated clauses (that is, the output of MAX-SAT) vanishes. It is less obvious that MAX-SAT can be 'much harder' than SAT. For instance, MAX-2-SAT is NP-hard, while, as said above, its decision counterpart is in P.

The study of applications is not the aim of this book, but one should keep in mind that the satisfiability problem is related to a myriad of other problems, some of which have enormous practical relevance. It is a problem of direct relevance to the fields of mathematical logic, computation theory, and artificial intelligence. Applications range from integrated circuit design (modelling, placement, routing, testing, ...), through computer architecture design (compiler optimization, scheduling and task partitioning, ...), to computer graphics, image processing, etc...

10.2 Algorithms

10.2.1 A simple case: 2-SAT

The reader who has worked out Exercises 10.1 and 10.2 will already have a feeling that 2-SAT is an easy problem. The main tool for solving it is the **unit clause propagation** (UCP) procedure. If we start from a 2-clause $C = z_1 \vee z_2$ and fix the literal z_1, two things may happen:

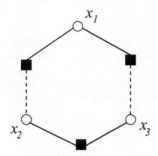 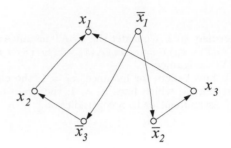

Fig. 10.2 Factor graph representation of the 2SAT formula $F = (x_1 \vee \overline{x}_2) \wedge (x_1 \vee \overline{x}_3) \wedge (x_2 \vee x_3)$ (left) and the corresponding directed graph $\mathcal{D}(F)$ (right).

- If we fix $z_1 = 1$, the clause is satisfied and disappears from the formula.
- If we fix $z_1 = 0$, the clause is transformed into the unit clause z_2, which implies that $z_2 = 1$.

Given a 2-SAT formula, one can start from a variable x_i, $i \in \{1, \ldots, N\}$, and fix $x_i = 0$, for instance. Then apply the reduction rule described above to all the clauses in which x_i or $\overline{x}_i$ appears. Finally, one fixes recursively, in the same way, all the literals which appear in unit clauses. This procedure may halt for one of the following reasons: (i) the formula does not contain any unit clause, or (ii) the formula contains the unit clause z_j together with its negation $\overline{z}_j$.

In the first case, a partial SAT assignment (i.e. an assignment of a subset of the variables such that no clause is violated) has been found. We shall prove below that such a partial assignment can be extended to a complete SAT assignment if and only if the formula is SAT. One therefore repeats the procedure by fixing a not-yet-assigned variable x_j.

In the second case, the partial assignment cannot be extended to a SAT assignment. One proceeds by changing the initial choice and setting $x_i = 1$. Once again, if the procedure stops because of reason (i), then the formula can be effectively reduced and the already-fixed variables do not need to be reconsidered in the following steps. If on the other hand, the choice $x_i = 1$ also leads to a contradiction (i.e. the procedure stops because of (ii)), then the formula is UNSAT.

It is clear that the algorithm defined in this way is very efficient. Its complexity can be measured by the number of variable-fixing operations that it involves. Since each variable is considered at most twice, this number is at most $2N$.

To prove the correctness of this algorithm, we still have to show the following fact: if the formula is SAT and UCP stops because of reason (i), then the resulting partial assignment can be extended to a global SAT assignment. (The implication in the reverse direction is obvious.) The key point is that the residual formula is formed by a subset $\mathcal{R}$ of the variables (the ones which have not yet been fixed) together with a *subset of the original clauses* (those which involve, uniquely, variables in $\mathcal{R}$). If a SAT assignment exists, its restriction to $\mathcal{R}$ satisfies the residual formula and constitutes an extension of the partial assignment generated by UCP.

Exercise 10.4 Write a code for solving 2-SAT using the algorithm described above.

Exercise 10.5 A nice way to understand UCP, and why it is so effective for 2-SAT, consists in associating with the formula F a directed graph $\mathcal{D}(F)$ (not to be confused with the factor graph!) as follows. We associate a vertex with each of the $2N$ literals (for instance we have one vertex for x_1 and one vertex for $\bar{x}_1$). Whenever a clause such as $\bar{x}_1 \vee x_2$ appears in the formula, we have two implications: if $x_1 = 1$, then $x_2 = 1$; if $x_2 = 0$, then $x_1 = 0$. We represent them graphically by drawing a directed edge from the vertex x_1 towards x_2, and an directed edge from $\bar{x}_2$ to $\bar{x}_1$.

Show that F is UNSAT if and only if there exists a variable index $i \in \{1, \ldots, N\}$ such that $\mathcal{D}(F)$ contains a directed path from x_i to $\bar{x}_i$, *and* a directed path from $\bar{x}_i$ to x_i.

[Hint: Consider the UCP procedure described above and rephrase it in terms of the directed graph $\mathcal{D}(F)$. Show that it can be regarded as an algorithm for finding a pair of paths from x_i to $\bar{x}_i$ and vice-versa in $\mathcal{D}(F)$.]

Let us note, finally, that the procedure described above does not give any clue about an efficient solution of MAX-2SAT, apart from determining whether the minimum number of violated clauses vanishes or not. As already mentioned, MAX-2SAT is NP-hard.

10.2.2 A general complete algorithm

As soon as we allow an unbounded number of clauses of length 3 or larger, satisfiability becomes an NP-complete problem. Exercise 10.3 shows how the UCP strategy fails: fixing a variable in a 3-clause may leave a 2-clause. As a consequence, UCP may halt without contradictions and produce a residual formula containing clauses which were not present in the original formula. Therefore, it can happen that the partial assignment produced by UCP cannot be extended to a global SAT assignment even if the original formula is SAT. Once a contradiction is found, it may be necessary to change any of the choices made so far in order to find a SAT assignment (in contrast to 2-SAT where only the last choice had to be changed). The exploration of all such possibilities is most conveniently described through a decision tree. Each time a contradiction is found, the search algorithm backtracks to the last choice for which one possibility was not explored.

The most widely used **complete algorithms** (i.e. algorithms which are able to either find a satisfying assignment or prove that there is no such assignment) rely on this idea. They are known under the name of **DPLL** search algorithms, from the initials of their inventors, Davis, Putnam, Logemann, and Loveland. The basic recursive process is best explained with an example, as in Fig. 10.3. Its structure can be summarized in a few lines, using the recursive procedure DPLL, described below, which takes as input a CNF formula F, a partial assignment of the variables A, and the list of indices of unassigned variables V, and returns either 'UNSAT' or a SAT assignment. To solve a problem given by the CNF formula F, written in terms of the N variables x_1, $dots, x_N$, the initial call to this procedure should be DPLL(F, $\emptyset$, $\{1, \ldots, N\}$).

DPLL (formula F, partial assignment A, unassigned variables V)
1:   if $V \neq \emptyset$:
2:       Choose an index $i \in V$;
3:       B=DPLL($F|\{x_i = 0\}$, $A \cup \{x_i = 0\}$, $V \setminus i$);
4:       if B=UNSAT B=DPLL($F|\{x_i = 1\}$, $A \cup \{x_i = 1\}$, $V \setminus i$);
5:           else **return** $A \cup \{x_i = 0\} \cup B$;
6:       if B=UNSAT **return** B;
7:           else **return** $A \cup \{x_i = 1\} \cup B$;
8:   else:
9:       if F has no clause **return** A;
10:          else **return** UNSAT;

The notation $F|\{x_i = 0\}$ refers to the formula obtained from F by assigning x_i to 0: all clauses of F which contain the literal $\overline{x}_i$ are eliminated, while clauses that contain x_i are shortened, namely $y \lor x_i$ is reduced to y. The reduced formula $F|\{x_i = 1\}$ is defined analogously.

As shown in Fig. 10.3, the algorithm can be represented as a walk in the decision tree. When it finds a contradiction, i.e. it reaches an 'UNSAT' leaf of the tree, it backtracks and searches a different branch.

In the above pseudocode, we did not specify how to select the next variable to be fixed in step 2. The various versions of the DPLL algorithm differ in the order in which the variables are taken into consideration and in which the branching process is performed. Unit clause propagation can be rephrased in the present setting as the following rule: whenever the formula F contains clauses of length 1, x_i must be chosen from the variables appearing in such clauses. In such a case, no branching takes place. For instance, if the literal x_i appears in a unit clause, setting $x_i = 0$ would produce an empty clause and therefore a contradiction: one is forced to set $x_i = 1$.

Apart from the case of unit clauses, deciding on which variable the next branching will be done on is an art, and can result in strongly varying performance. For instance, it is a good idea to branch on a variable which appears in many clauses, but other criteria, such as the number of unit clauses that a branching will generate, can also be used. It is customary to characterize the performances of this class of algorithms by the number of branching nodes that it generates. This does not correspond to the actual number of operations executed, which may depend on the heuristic. However, for many reasonable heuristics, the actual number of operations is within a polynomial factor (in the size of the instance) of the number of branchings, and such a factor does not affect the leading exponential behaviour.

Whenever the DPLL procedure does not return a SAT assignment, the formula is UNSAT: a representation of the search tree explored provides a proof of unsatisfiability. This is sometimes also called an UNSAT **certificate**. Note that the length of an UNSAT certificate is (in general) larger than polynomial in the size of the input. This

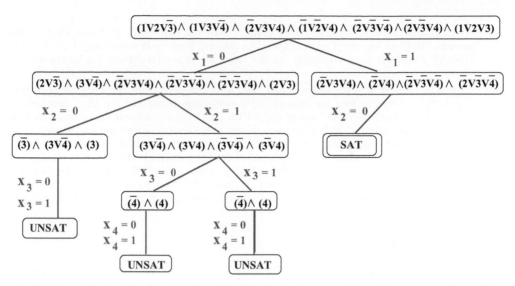

Fig. 10.3 A sketch of the DPLL algorithm, acting on the formula $(x_1 \vee x_2 \vee \overline{x}_3) \wedge (x_1 \vee x_3 \vee \overline{x}_4) \wedge (\overline{x}_2 \vee x_3 \vee x_4) \wedge (\overline{x}_1 \vee x_2 \vee x_4) \wedge (\overline{x}_2 \vee \overline{x}_3 \vee \overline{x}_4) \wedge$ $(\overline{x}_2 \vee \overline{x}_3 \vee x_4) \wedge (x_1 \vee x_2 \vee x_3) \wedge (\overline{x}_1 \vee x_2 \vee \overline{x}_4)$. In order to get a more readable figure, the notation has been simplified: a clause such as $(\overline{x}_1 \vee x_2 \vee x_4)$ is denoted here by $(\overline{1}\,2\,4)$. A first variable is fixed, here $x_1 = 0$. The problem is then reduced: clauses containing x_1 are eliminated, and clauses containing $\overline{x}_1$ are shortened by eliminating the literal $\overline{x}_1$. Then one proceeds by fixing a second variable, etc. At each step, if a unit clause is present, the next variable to be fixed is chosen from those appearing in unit clauses. This corresponds to the unit clause propagation rule. When the algorithm finds a contradiction (where two unit clauses fix a variable simultaneously to 0 and to 1), it backtracks to the last not-yet-explored branching node and explores another choice for the corresponding variable. In the present case, for instance, the algorithm first fixes $x_1 = 0$ and then it fixes $x_2 = 0$, which implies through UCP that $x_3 = 0$ and $x_3 = 1$. This is a contradiction, and therefore the algorithm backtracks to the last choice, which was $x_2 = 0$, and tries instead the other choice: $x_2 = 1$, etc. Here, branching follows the order of appearance of variables in the formula.

is at variance with a SAT certificate, which is provided, for instance, by a particular SAT assignment.

Exercise 10.6 Resolution and DPLL.

(a) A powerful approach to proving that a formula is UNSAT relies on the idea of a **resolution proof**. Imagine that F contains two clauses $x_j \vee A$ and $\overline{x}_j \vee B$, where A and B are subclauses. Show that these two clauses automatically imply the **resolvent on** x_j that is the clause $A \vee B$.

(b) A resolution proof is constructed by adding resolvent clauses to F. Show that if this process produces an empty clause, then the original formula is necessarily UNSAT. An UNSAT certificate is given simply by the sequence of resolvents leading to the empty clause.

(c) Although the use of a resolution proof may look different from the DPLL algorithm, any DPLL tree is in fact an example of a resolution proof. To see this, proceed as follows. Label each 'UNSAT' leaf of the DPLL tree by the resolvent of a pair of clauses of the original formula which have been shown to be contradictory on this branch (e.g. the leftmost such leaf in Fig. 10.3 corresponds to the pair of initial clauses $x_1 \vee x_2 \vee \bar{x}_3$ and $x_1 \vee x_2 \vee x_3$, so that it can be labelled by the resolvent of these two clauses on x_3, namely $x_1 \vee x_2$). Show that each branching node of the DPLL tree can be labelled by a clause which is a resolvent of the two clauses labeling its children, and that this process, when carried out on an UNSAT formula, produces a root (the top node of the tree) which is an empty clause.

10.2.3 Incomplete search

As we have seen above, proving that a formula is SAT is much easier than proving that it is UNSAT: one 'just' needs to exhibit an assignment that satisfies all the clauses. One can therefore relax the initial objective, and look for an algorithm that tries only to deal with the first task. This is often referred to as an **incomplete search** algorithm. Such an algorithm can either return a satisfying assignment or just say 'I do not know' whenever it is unable to find one (or to prove that the formula is UNSAT).

One basic algorithm for incomplete search, due to Schöning, is based on the following simple random-walk routine.

WALK (CNF formula F in N variables)

1:    for each variable i, set $x_i = 0$ or $x_i = 1$ with probability $1/2$;
2:    repeat $3N$ times:
3:       if the current assignment satisfies F, **return it** and **stop**;
4:       else:
5:          choose an unsatisfied clause a uniformly at random;
6:          choose a variable index i uniformly at random in ∂a;
7:          flip the variable i (i.e. $x_i \leftarrow 1 - x_i$);
8:    **end**

For this algorithm, one can obtain a guarantee of performance.

Proposition 10.1 *Denote by $p(F)$ the probability that the routine Walk, when executed on a formula F, returns a satisfying assignment. If F is SAT, then $p(F) \geq p_N$, where*

$$p_N = \frac{2}{3} \left(\frac{K}{2(K-1)} \right)^N . \tag{10.2}$$

One can therefore run the routine many times (with independent random numbers each time) in order to increase the probability of finding a solution. Suppose that the formula is SAT. If the routine is run $20/p_N$ times, the probability of not finding any solution is $(1 - p_N)^{20/p_N} \le e^{-20}$. While this is of course not a proof of unsatisfiability, it is very close to it. In general, the time required for this procedure to reduce the error probability below any fixed ε grows as

$$\tau_N \doteq \left(\frac{2(K-1)}{K} \right)^N . \tag{10.3}$$

This simple randomized algorithm achieves an exponential improvement over the naive exhaustive search, which takes about 2^N operations.

Proof We prove the lower bound (10.2) on the probability of finding a satisfying assignment during a single run of the routine Walk. Since, by assumption, F is SAT, we can consider a particular SAT assignment, say $\underline{x}_*$. Let $\underline{x}_t$ be the assignment produced by Walk(F) after t flips, and let d_t be the Hamming distance between $\underline{x}_*$ and $\underline{x}_t$. Obviously, at time 0 we have

$$\mathbb{P}\{d_0 \doteq d\} = \frac{1}{2^N} \binom{N}{d} . \tag{10.4}$$

Since $\underline{x}_*$ satisfies F, each clause is satisfied by at least one variable as assigned in $\underline{x}_*$. We mark *exactly* one such variable per clause. Each time Walk($\cdot$) chooses a violated clause, it flips a marked variable with probability $1/K$, reducing the Hamming distance by one. Of course, the Hamming distance can also decrease when another variable is flipped (if more than one variable in $\underline{x}_*$ satisfies this clause). In order to obtain a bound, we introduce an auxiliary integer variable $\hat{d}_t$ which decreases by one each time a marked variable is selected, and increases by one (the maximum possible increase in the Hamming distance due to a single flip) otherwise. If we choose the initial condition $\hat{d}_0 = d_0$, it follows from the previous observations that $d_t \le \hat{d}_t$ for any $t \ge 0$. We can therefore bound from below the probability that Walk finds a solution by the probability that $\hat{d}_t = 0$ for some $0 \le t \le 3N$. But the random process $\hat{d}_t = 0$ is simply a biased random walk on the a half-line with the initial condition (10.4): at each time step, it moves to the left with probability $1/K$ and to the right with probability $1 - 1/K$. The probability of hitting the origin can then be estimated as in Eq. (10.2), as shown in the following exercise.

Exercise 10.7 Analysis of the biased random walk $\hat{d}_t$.

(a) Show that the probability for $\hat{d}_t$ to start at position d at $t = 0$ and be at the origin at time t is

$$\mathbb{P}\{\hat{d}_0 = d ; \hat{d}_t = 0\} = \frac{1}{2^N} \binom{N}{d} \frac{1}{K^t} \binom{t}{(t-d)/2} (K-1)^{(t-d)/2} \tag{10.5}$$

for $t + d$ even, and vanishes otherwise.

(b) Use Stirling's formula to derive an approximation for this probability to the leading exponential order, $\mathbb{P}\{\hat{d}_0 = d; \ \hat{d}_t = 0\} \doteq \exp\{-N\Psi(\theta, \delta)\}$, where $\theta = t/N$ and $\delta = d/N$.

(c) Minimize $\Psi(\theta, \delta)$ with respect to $\theta \in [0,3]$ and $\delta \in [0,1]$, and show that the minimum value is $\Psi_* = \log[2(K-1)/K]$. Argue that $p_N \doteq \exp\{-N\Psi_*\}$ to the leading exponential order.

□

Notice that the above algorithm applies a very noisy strategy. Although it 'focuses' on unsatisfied clauses, it makes essentially random steps. The opposite philosophy would be that of making greedy steps. An example of a 'greedy' step is the following: flip a variable which will lead to the largest positive increase in the number of satisfied clauses.

There exist several refinements of the simple random-walk algorithm. One of the greatest improvement consists in applying a mixed strategy: with probability p, pick an unsatisfied clause, and flip a randomly chosen variable in this clause (as in Walk); with probability $1 - p$, perform a 'greedy' step as defined above.

The pseudocode of this 'Walksat' algorithm is given below, using the following notation: $E(\underline{x})$ is the number of clauses violated by the assignment $\underline{x} = (x_1, \ldots, x_N)$ and $\underline{x}^{(i)}$ is the assignment obtained from $\underline{x}$ by flipping $x_i \to 1 - x_i$.

WalkSAT (CNF formula F, number of flips f, mixing p)

1 : $t = 0$;
2 : Initialize $\underline{x}$ to a random assignment;
3 : While $t < f$ do
4 :   If $\underline{x}$ satisfies $\mathcal{F}$, return $\underline{x}$;
5 :   Let r be uniformly random in $[0, 1]$;
6 :   If $r < 1 - p$ then
7 :   For each $i \in V$, let $\Delta_i = E(\underline{x}^{(i)}) - E(\underline{x})$;
8 :   Flip a variable x_i for which Δ_i is minimal;
9 :   else
10:      Choose a violated clause a uniformly at random;
11:      Flip a uniformly random variable x_i, $i \in \partial a$;
12: End-While
13: Return 'Not found';

This strategy works reasonably well if p is properly optimized. The greedy steps drive the assignment towards 'quasi-solutions', while the noise term allows the algorithm to escape from local minima.

10.3 Random K-satisfiability ensembles

The satisfiability problem is NP-complete. One thus expects a complete algorithm to take exponential time in the worst case. However, empirical studies have shown that many formulae are very easy to solve. A natural research direction is therefore to characterize ensembles of problems which are easy, separating them from those which are hard. Such ensembles can be defined by introducing a probability measure over the space of instances.

One of the most interesting families of ensembles is **random K-SAT**. An instance of random K-SAT contains only clauses of length K. The ensemble is further characterized by the number of variables N and the number of clauses M, and is denoted by $\mathsf{SAT}_N(K, M)$. A formula in $\mathsf{SAT}_N(K, M)$ is generated by selecting M clauses of size K uniformly at random from the $\binom{N}{K}2^K$ such clauses. Note that the factor graph associated with a random K-SAT formula from the ensemble $\mathsf{SAT}_N(K, M)$ is in fact a random $\mathbb{G}_N(K, M)$ factor graph.

It turns out that a crucial parameter characterizing the random K-SAT ensemble is the **clause density** $\alpha \equiv M/N$. We shall define the 'thermodynamic' limit as $M \to \infty$, $N \to \infty$, with a fixed density α. In this limit, several important properties of random formulae concentrate in probability around their typical values.

As in the case of random graphs, it is sometimes useful to consider slight variants of the above definition. One such variant is the ensemble $\mathsf{SAT}_N(K, \alpha)$. A random instance from this ensemble is generated by including in the formula each of the $\binom{N}{K}2^K$ possible clauses independently with probability $\alpha N 2^{-K}/\binom{N}{K}$. Once again, the corresponding factor graph will be distributed according to the ensemble $\mathbb{G}_N(K, \alpha)$ introduced in Chapter 9. For many properties, the differences between such variants vanish in the thermodynamic limit (this is analogous to the equivalence of different factor graph ensembles).

10.3.1 Numerical experiments

Using the DPLL algorithm, one can investigate the properties of typical instances of the random K-SAT ensemble $\mathsf{SAT}_N(K, M)$. Figure 10.4 shows the probability $P_N(K, \alpha)$ that a randomly generated formula is satisfiable, for $K = 2$ and $K = 3$. For fixed K and N, this is a decreasing function of α, which goes to 1 in the $\alpha \to 0$ limit and goes to 0 in the $\alpha \to \infty$ limit. One interesting feature in these simulations is the fact that the crossover from high to low probability becomes sharper and sharper as N increases. This numerical result points to the existence of a phase transition at a finite value $\alpha_{\mathrm{s}}(K)$: for $\alpha < \alpha_{\mathrm{s}}(K)$ ($\alpha > \alpha_{\mathrm{s}}(K)$) a random K-SAT formula is SAT (respectively, UNSAT) with probability approaching 1 as $N \to \infty$. On the other hand, for $\alpha > \alpha_{\mathrm{s}}(K)$ a random K-SAT formula is UNSAT with probability approaching 1 as $N \to \infty$.

The conjectured phase transition in random satisfiability problems with $K \geq 3$ has drawn considerable attention. One important reason for this interest comes from the study of the computational effort needed to solve the problem. Figure 10.5 shows the typical number of branching nodes in the DPLL tree required to solve a typical random 3-SAT formula. One may notice two important features. For a given value of the number of variables N, the computational effort has a peak in the region of clause density where a phase transition seems to occur (compare Fig. 10.4). In this region, the computational effort also increases rapidly with N. Looking carefully at the data, one can distinguish qualitatively three different regions: at low α, the solution is 'easily' found and the computer time grows polynomially; at intermediate α, in the phase transition region, the problem becomes typically very hard and the computer

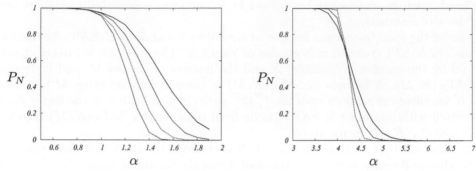

Fig. 10.4 The probability that a formula generated from the random K-SAT ensemble is satisfied, plotted versus the clause density α. *Left*, $K = 2$; *right*, $K = 3$. The curves were generated using a DPLL algorithm. Each point is the result of averaging over 10^4 random formulae. The curves for $K = 2$ correspond to formulae of size $N = 50$, 100, 200, and 400 (from *right* to *left*). In the case $K = 3$, the curves correspond to $N = 50$ (full line), $N = 100$ (dashed), and $N = 200$ (dotted). The transition between satisfiable and unsatisfiable formulae becomes sharper as N increases.

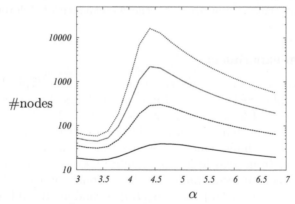

Fig. 10.5 Computational effort of our DPLL algorithm applied to random 3-SAT formulae. The logarithm of the number of branching nodes was averaged over 10^4 instances. From *bottom* to *top*: $N = 50$, 100, 150, 200.

time grows exponentially; and at larger α, in the region where a random formula is almost always UNSAT, the problem becomes easier, although the size of the DPLL tree still grows exponentially with N.

The hypothetical phase transition region is therefore the one where the hardest instances of random 3-SAT are located. This makes such a region particularly interesting, both from the point of view of computational complexity and from that of statistical physics.

10.4   Random 2-SAT

From the point of view of computational complexity, 2-SAT is polynomial, while K-SAT is NP-complete for $K \geq 3$. It turns out that random 2-SAT is also much simpler to analyse than the other cases. One important reason is the existence of the polynomial decision algorithm described in Section 10.2.1 (see, in particular, Exercise 10.5). This can be analysed in detail using the representation of a 2-SAT formula as a directed graph whose vertices are associated with literals. One can then use the mathematical theory of random directed graphs. In particular, the existence of a phase transition at a critical clause density $\alpha_s(2) = 1$ can be established.

Theorem 10.2 *Let $P_N(K = 2, \alpha)$ be the probability for a* SAT$_N(K = 2, M)$ *random formula to be SAT. Then*

$$\lim_{N \to \infty} P_N(K = 2, \alpha) = \begin{cases} 1 & \text{if } \alpha < 1 \text{ ,} \\ 0 & \text{if } \alpha > 1 \text{ .} \end{cases} \tag{10.6}$$

Proof Here we shall prove that a random formula is SAT with high proability for $\alpha < 1$. It follows from Theorem 10.5 below that it is, with high probability, UNSAT for $\alpha > 1$.

We use the directed-graph representation defined in Exercise 10.5. In this graph, we define a bicycle of length s as a path $(u, w_1, w_2, \ldots, w_s, v)$, where the w_i's are literals on s distinct variables, and $u, v \in \{w_1, \ldots, w_s, \overline{w}_1, \ldots, \overline{w}_s\}$. As we saw in Example 10.5, if a formula F is UNSAT, its directed graph $\mathcal{D}(F)$ has a cycle containing the two literals x_i and $\overline{x}_i$ for some i. From such a cycle, one can easily build a bicycle. The probability that a bicycle appears in $\mathcal{D}(F)$ is, in turn, bounded from above by the expected number of bicycles. Therefore,

$$\mathbb{P}(F \text{ is UNSAT}) \leq \mathbb{P}(\mathcal{D}(F) \text{ has a bicycle}) \leq \sum_{s=2}^{N} N^s 2^s (2s)^2 M^{s+1} \left(\frac{1}{4\binom{N}{2}} \right)^{s+1} . \tag{10.7}$$

The sum is over the size s of the bicycle; N^s is an upper bound on $\binom{N}{s}$, the number of ways one can choose the s variables; 2^s corresponds to the choice of literals, given the variables; $(2s)^2$ corresponds to the choice of u, v; M^{s+1} is an upper bound on $\binom{M}{s+1}$, the number of choices for the clauses involved in the bicycle; and the last factor is the probability that each of the chosen clauses of the bicycle appears in the random formula. A direct summation of the series in eqn (10.7) shows that, in the large-N limit, the result is $O(1/N)$ whenever $\alpha < 1$. $\square$

10.5   The phase transition in random $K (\geq 3)$-SAT

10.5.1   The satisfiability threshold conjecture

As noted above, numerical studies suggest that random K-SAT undergoes a phase transition between a SAT phase and an UNSAT phase, for any $K \geq 2$. There is a widespread belief that this is indeed true, as formalized by the following conjecture.

Conjecture 10.3 *For any $K \geq 2$, there exists a threshold $\alpha_s(K)$ such that*

$$\lim_{N \to \infty} P_N(K, \alpha) = \begin{cases} 1 & \text{if } \alpha < \alpha_s(K) \ , \\ 0 & \text{if } \alpha > \alpha_s(K) \ . \end{cases} \tag{10.8}$$

As discussed in the previous section, this conjecture has been proved in the case $K = 2$. The existence of a phase transition is still an open problem for larger K, although the following theorem gives some strong support.

Theorem 10.4. (Friedgut) *Let $P_N(K, \alpha)$ be the probability for a random formula from the ensemble $\mathsf{SAT}_N(K, M)$ to be satisfiable, and assume that $K \geq 2$. There then exists a sequence of $\alpha_s^{(N)}(K)$ such that, for any $\varepsilon > 0$,*

$$\lim_{N \to \infty} P_N(K, \alpha_N) = \begin{cases} 1 & \text{if } \alpha_N < \alpha_s^{(N)}(K) - \varepsilon \ , \\ 0 & \text{if } \alpha_N > \alpha_s^{(N)}(K) + \varepsilon \ , \end{cases} \tag{10.9}$$

In other words, the crossover from SAT to UNSAT becomes sharper and sharper as N increases. For N large enough, it takes place in a window smaller than any fixed width ε. The 'only' missing piece needed to prove the satisfiability threshold conjecture (10.3) is the convergence of $\alpha_s^{(N)}(K)$ to some value $\alpha_s(K)$ as $N \to \infty$.

10.5.2   Upper bounds

Rigorous studies have allowed to establish bounds on the satisfiability threshold $\alpha_s^{(N)}(K)$. Usually one focuses on the large-N limit of such bounds. Upper bounds are obtained by using the first moment method. The general strategy is to introduce a function $U(F)$ acting on formulae such that

$$U(F) = \begin{cases} 0 & \text{if } F \text{ is UNSAT,} \\ \geq 1 & \text{otherwise.} \end{cases} \tag{10.10}$$

Therefore, if F is a random K-SAT formula

$$\mathbb{P}\{F \text{ is SAT}\} \leq \mathbb{E}\, U(F) \ . \tag{10.11}$$

The inequality becomes an equality if $U(F) = \mathbb{I}(F \text{ is SAT})$. Of course, we do not know how to compute the expectation in this case. The idea is to find some function $U(F)$ which is simple enough that $\mathbb{E}\, U(F)$ can be computed, and which has an expectation value that vanishes as $N \to \infty$, for large enough α.

The simplest such choice is $U(F) = Z(F)$, the number of SAT assignments (this is the analogue of a 'zero-temperature' partition function). The expectation $\mathbb{E}\, Z(F)$ is equal to the number of assignments, 2^N, times the probability that an assignment is SAT (which does not depend on the assignment). Consider, for instance, the all-zero assignment $x_i = 0$, $i = 1, \ldots, N$. The probability that it is SAT is equal to the product of the probabilities that it satisfies each of the M clauses. The probability that the all-zero assignment satisfies a clause is $(1 - 2^{-K})$ because a K-clause excludes one of the 2^K assignments of variables which appear in it. Therefore

$$\mathbb{E}\, Z(F) = 2^N (1 - 2^{-K})^M = \exp\left[N\left(\log 2 + \alpha \log(1 - 2^{-K})\right)\right] . \qquad (10.12)$$

This result shows that, for $\alpha > \alpha_{\mathrm{UB},1}(K)$, where

$$\alpha_{\mathrm{UB},1}(K) \equiv -\frac{\log 2}{\log(1 - 2^{-K})} , \qquad (10.13)$$

$\mathbb{E}\, Z(F)$ is exponentially small at large N. Equation (10.11) implies that the probability of a formula being SAT also vanishes at large N for such an α.

Theorem 10.5 *If $\alpha > \alpha_{\mathrm{UB},1}(K)$, then $\lim_{N \to \infty} \mathbb{P}\{F \text{ is SAT}\} = 0$, whence $\alpha_{\mathrm{s}}^{(N)}(K) \leq \alpha_{\mathrm{UB},1}(K)$.*

One should not expect this bound to be tight. The reason is that, in the SAT phase, $Z(F)$ takes exponentially large values, and its fluctuations tend to be exponential in the number of variables.

Example 10.6 As a simple illustration, consider a toy example: the random 1-SAT ensemble $\mathrm{SAT}_N(1, \alpha)$. A formula is generated by including each of the $2N$ literals as a clause independently with probability $\alpha/2$ (we assume that $\alpha \leq 2$). In order for the formula to be SAT, for each of the N variables, at most one of the two corresponding literals must be included. We therefore have

$$P_N(K = 1, \alpha) = (1 - \alpha^2/4)^N . \qquad (10.14)$$

In other words, the probability for a random formula to be SAT goes exponentially fast to 0 for any $\alpha > 0$: $\alpha_{\mathrm{s}}(K = 1) = 0$. On the other hand, the upper bound deduced from $\mathbb{E}Z(F)$ is $\alpha_{\mathrm{UB},1}(K) = 1$. This is due to large fluctuations in the number of SAT assignments Z, as we shall see in the next exercise.

Exercise 10.8 Consider the distribution of $Z(F)$ in the random 1-SAT ensemble.

(a) Show that

$$\mathbb{P}\{Z(F) = 2^n\} = \binom{N}{n} \left(1 - \frac{\alpha}{2}\right)^{2n} \left[\alpha\left(1 - \frac{\alpha}{4}\right)\right]^{N-n}, \tag{10.15}$$

for any $n \geq 0$.

[Hint: If F is SAT, then $Z(F) = 2^n$, where n is the number of variables which do not appear in any clause.]

(b) From this expression, deduce the large-deviation principle

$$\mathbb{P}\left\{Z(F) = 2^{N\nu}\right\} \doteq \exp\{-N I_\alpha(\nu)\}, \tag{10.16}$$

where

$$I_\alpha(\nu) \equiv -\mathcal{H}(\nu) - 2\nu \log(1 - \alpha/2) - (1 - \nu) \log(\alpha(1 - \alpha/4)). \tag{10.17}$$

What is the most probable value of ν?

(c) Show that

$$\mathbb{E}\, Z(F) \doteq \exp\left\{N \max_\nu[-I_\alpha(\nu) + \nu \log 2]\right\}. \tag{10.18}$$

What is the value of ν where the maximum is achieved, ν^*? Show that $I_\alpha(\nu^*) > 0$: the probability of having $Z(F) \doteq 2^{N\nu^*}$ is exponentially small, and therefore $\mathbb{E}\, Z(F)$ is dominated by rare events.

Exercise 10.9 Repeat the derivation of Theorem 10.5 for the ensemble $\mathsf{SAT}_N(K, \alpha)$ (i.e. compute $\mathbb{E}\, Z(F)$ for this ensemble and find for what values of α this expectation is exponentially small). Show that the upper bound obtained in this case is $\alpha = 2^K \log 2$. This is worse than the previous upper bound $\alpha_{\mathrm{UB},1}(K)$, although one would expect the threshold to be the same. Why?

[Hint: The number of clauses M in a $\mathsf{SAT}_N(K, \alpha)$ formula has a binomial distribution with parameters N and α. What values of M provide the dominant contribution to $\mathbb{E}\, Z(F)$?]

In order to improve upon Theorem 10.5 using the first-moment method, one needs a better (but still simple) choice of the function $U(F)$. A possible strategy consists in defining some small subclass of 'special' SAT assignments, such that if a SAT assignment exists, then a special SAT assignment exists too. If the subclass is small enough, one can hope to reduce the fluctuations in $U(F)$ and sharpen the bound.

One choice of such a subclass is 'locally maximal' SAT assignments. Given a formula F, an assignment $\underline{x}$ for this formula is said to be a locally maximal SAT assignment if and only if (1) it is a SAT assignment, and (2) for any i such that x_i to 0, the assignment obtained by flipping it to $x_i = 1$ is UNSAT. We define $U(F)$ as the number of locally maximal SAT assignments and apply the first-moment method to this function. This gives the following result.

Theorem 10.7 *For any $K \geq 2$, let $\alpha_{\mathrm{UB},2}(K)$ be the unique positive solution of the equation*

$$\alpha \log(1 - 2^{-K}) + \log\left[2 - \exp\left(-\frac{K\alpha}{2^K - 1}\right)\right] = 0. \qquad (10.19)$$

Then $\alpha_{\mathrm{s}}^{(N)}(K) \leq \alpha_{\mathrm{UB},2}(K)$ for large enough N.

The proof is left as the following exercise.

Exercise 10.10 Consider an assignment $\underline{x}$ where exactly L variables are set to 0, the remaining $N - L$ variables being set to 1. Without loss of generality, assume $x_1, \ldots, x_L$ to be the variables set to zero.

(a) Let p be the probability that a clause constrains the variable x_1, *given that* the clause is satisfied by the assignment $\underline{x}$ (By a clause that constrains x_1, we mean that the clause becomes unsatisfied if x_1 is flipped from 0 to 1.) Show that $p = \binom{N-1}{K-1}[(2^K - 1)\binom{N}{K}]^{-1}$.

(b) Show that the probability that the variable x_1 is constrained by at least one of the M clauses, given that all these clauses are satisfied by the assignment $\underline{x}$, is equal to $q = 1 - (1 - p)^M$.

(c) Let $\mathcal{C}_i$ be the event that x_i is constrained by at least one of the M clauses. If $\mathcal{C}_1, \ldots,$ $\mathcal{C}_L$ were independent events, under the condition that $\underline{x}$ satisfies F, the probability that $x_1, \ldots x_L$ are constrained would be equal to q^L. Of course, $\mathcal{C}_1, \ldots, \mathcal{C}_L$ are not independent. Find a heuristic argument to show that they are anti-correlated and that their joint probability is *at most* q^L (consider, for instance the case $L = 2$).

(d) Assume the claim in (c) to be true. Show that $\mathbb{E}[U(F)] \leq (1 - 2^{-K})^M \sum_{L=0}^{N} \binom{N}{L} q^L = (1 - 2^{-K})^M [1 + q]^N$ and finish the proof by working out the large-N asymptotics of this formula (with $\alpha = M/N$ fixed).

In Table 10.1, we report the numerical values of the upper bounds $\alpha_{\mathrm{UB},1}(K)$ and $\alpha_{\mathrm{UB},2}(K)$ for a few values of K. These results can be slightly improved upon by pursuing the same strategy further. For instance, one may strengthen the condition of maximality to flipping two or more variables. However, the quantitative improvement in the bound is rather small.

10.5.3 Lower bounds

Two main strategies have been used to derive lower bounds on $\alpha_{\mathrm{c}}^{(N)}(K)$ in the large-N limit. In both cases, one takes advantage of Theorem 10.4: in order to show that $\alpha_{\mathrm{c}}^{(N)}(K) \geq \alpha^*$, it is sufficient to prove that a random $\mathsf{SAT}_N(K, M)$ formula, with $M = \alpha N$, is SAT with non-vanishing probability in the $N \to \infty$ limit.

The first approach consists in analysing explicit heuristic algorithms for finding SAT assignments. The idea is to prove that a particular algorithm finds a SAT assignment with positive probability as $N \to \infty$ when α is smaller than some value.

One of the simplest such bounds is obtained by considering unit clause propagation. Whenever there exists a unit clause, we assign the variables appearing in one such clause in such a way to satisfy it, and proceed recursively. Otherwise, we choose a variable uniformly at random from those which have not yet been fixed and assign it to 0 or 1 with probability 1/2. The algorithm halts if it finds a contradiction (i.e. a pair of opposite unit clauses) or if all the variables have been assigned. In the latter case, the assignment produced by the algorithm satisfies the formula.

This algorithm is then applied to a random K-SAT formula with clause density α. It can be shown that a SAT assignment will be found with positive probability for α small enough: this gives the lower bound $\alpha_c^{(N)}(K) \geq ((K-1)/(K-2))^{K-2} 2^{K-1}/K$ in the $N \to \infty$ limit. In the exercise below, we give the main steps of the reasoning for the case $K = 3$; we refer to the literature for more detailed proofs.

Exercise 10.11 After T iterations, the formula will contain 3-clauses, as well as 2-clauses and 1-clauses. Denote by $\mathcal{C}_s(T)$ the set of s-clauses, $s = 1, 2, 3$, and denote its size by $C_s(T) \equiv |\mathcal{C}_s(T)|$. Let $\mathcal{V}(T)$ be the set of variables which have not yet been fixed, and let $\mathcal{L}(T)$ be the set of literals on the variables of $\mathcal{V}(T)$ (obviously, we have $|\mathcal{L}(T)| = 2|\mathcal{V}(T)| = 2(N-T)$). Finally, if a contradiction is encountered after T_{halt} steps, we adopt the convention that the formula remains unchanged for all $T \in \{T_{\text{halt}}, \ldots, N\}$.

(a) Show that, for any $T \in \{1, \ldots, N\}$, each clause in $\mathcal{C}_s(T)$ is uniformly distributed among the s-clauses over the literals in $\mathcal{L}(T)$.

(b) Show that the expected changes in the numbere of 3- and 2-clauses are given by $\mathbb{E}[C_3(T+1) - C_3(T)] = -3C_3(T)/(N-T)$ and $\mathbb{E}[C_2(T+1) - C_2(T)] = 3C_3(T)/(2(N-T)) - 2C_2(T)/(N-T)$.

(c) Show that, conditional on $C_1(T)$, $C_2(T)$, and $C_3(T)$, the change in the number of 1-clauses is distributed as follows: $C_1(T+1) - C_1(T) \overset{\text{d}}{=} -\mathbb{I}(C_1(T) > 0) + B(C_2(T), 1/(N-T))$. (We recall that $B(n, p)$ denotes a binomial random variable with parameters n and p (see Appendix A)).

(d) It can be shown that, as $N \to \infty$ at fixed $t = T/N$, the variables $C_s(T)/N$ for $s \in \{2, 3\}$ concentrate around their expectation values, and these converge to smooth functions $c_s(t)$. Argue that these functions must solve the ordinary differential equations $(dc_3/dt) = -3c_3(t)/(1-t)$ and $(dc_2/dt) = 3c_3(t)/(2(1-t)) - 2c_2(t)/(1-t)$. Check that the solutions of these equations are $c_3(t) = \alpha(1-t)^3$ and $c_2(t) = (3\alpha/2)t(1-t)^2$.

(e) Show that the number of unit clauses follows a Markov process described by $C_1(0) = 0$, $C_1(T+1) - C_1(T) \overset{\text{d}}{=} -\mathbb{I}(C_1(T) > 0) + \eta(T)$, where $\eta(T)$ is a Poisson-distributed random variable with mean $c_2(t)/(1-t)$, and where $t = T/N$. Given C_1 and a time T, show that the probability that there is no contradiction generated by the unit clause algorithm up to time T is $\prod_{\tau=1}^{T} (1 - 1/(2(N-\tau)))^{[C_1(\tau)-1]\mathbb{I}(C_1(\tau)\geq 1)}$.

(f) Let $\rho(T)$ be the probability that there is no contradiction up to time T. Consider $T = N(1-\epsilon)$; show that $\rho(N(1-\epsilon)) \geq (1-1/(2N\epsilon))^{AN+B} \mathbb{P}(\sum_{\tau=1}^{N(1-\epsilon)} C_1(\tau) \leq AN+B)$. Assume that α is such that, $\forall t \in [0, 1-\epsilon]$, $c_2(t)/(1-t) < 1$. Show that there exist A, B such that $\lim_{N\to\infty} \mathbb{P}(\sum_{\tau=1}^{N(1-\epsilon)} C_1(\tau) \leq AN + B)$ is finite. Deduce that in the large-N limit, there is a finite probability that, at time $N(1-\epsilon)$, the unit clause algorithm has not produced any contradiction so far, and $C_1(N(1-\epsilon)) = 0$.

(g) Conditionnal on the fact that the algorithm has not produced any contradiction and $C_1(N(1-\epsilon)) = 0$, consider the residual formula at time $T = N(1-\epsilon)$. Transform each 3-clause into a 2-clause by removing a uniformly random variable from it. Show that one obtains, for ϵ small enough, a random 2-SAT problem with a small clause density $\leq 3\epsilon^2/2$, so that this is a satisfiable instance.

(h) Deduce that, for $\alpha < 8/3$, the unit clause propagation algorithm finds a solution with a finite probability

More refined heuristics have been analysed using this method and lead to better lower bounds on $\alpha_c^{(N)}(K)$. We shall not elaborate on this approach here, but rather present a second strategy, based on a structural analysis of the problem. The idea is to use the second-moment method. More precisely, we consider a function $U(F)$ of the SAT formula F, such that $U(F) = 0$ whenever F is UNSAT and $U(F) > 0$ otherwise. We then make use of the following inequality:

$$\mathbb{P}\{F \text{ is SAT}\} = \mathbb{P}\{U(F) > 0\} \geq \frac{[\mathbb{E}\,U(F)]^2}{\mathbb{E}[U(F)^2]} \,. \tag{10.20}$$

The present strategy is more delicate to implement than the first-moment method, which we used in Section 10.5.2 to derive upper bounds on $\alpha_c^{(N)}(K)$. For instance, the simple choice $U(F) = Z(F)$ does not give any result: it turns out that the ratio $[\mathbb{E}\,Z(F)]^2/\mathbb{E}[Z(F)^2]$ is exponentially small in N for any non-vanishing value of α, so that the inequality (10.20) is useless. Again, one needs to find a function $U(F)$ whose fluctuations are smaller than those of the number $Z(F)$ of SAT assignments. More precisely, one needs the ratio $[\mathbb{E}\,U(F)]^2/\mathbb{E}[U(F)^2]$ to be non-vanishing in the $N \to \infty$ limit.

One successful idea uses a weighted sum of SAT assignments,

$$U(F) = \sum_{\underline{x}} \prod_{a=1}^{M} W(\underline{x}, a) \,. \tag{10.21}$$

Here the sum is over all the 2^N assignments, and $W(\underline{x}, a)$ is a weight associated with clause a. This weight must be such that $W(\underline{x}, a) = 0$ when the assignment $\underline{x}$ does not satisfy clause a, and $W(\underline{x}, a) > 0$ otherwise. Let us choose a weight which depends on the number $r(\underline{x}, a)$ of variables which satisfy clause a in the assignment $\underline{x}$:

$$W(\underline{x}, a) = \begin{cases} \varphi(r(\underline{x}, a)) & \text{if } r(\underline{x}, a) \geq 1, \\ 0 & \text{otherwise.} \end{cases} \tag{10.22}$$

It is then easy to compute the first two moments of $U(F)$:

$$\mathbb{E}\,U(F) = 2^N \left[2^{-K} \sum_{r=1}^{K} \binom{K}{r} \varphi(r) \right]^M \,, \tag{10.23}$$

$$\mathbb{E}\left[U(F)^2\right] = 2^N \sum_{L=0}^{N} \binom{N}{L} \left[g_\varphi(N, L)\right]^M \,. \tag{10.24}$$

Here $g_\varphi(N, L)$ is the expectation value of the product $W(\underline{x}, a)W(\underline{y}, a)$ when a clause a is chosen uniformly at random, given that $\underline{x}$ and $\underline{y}$ are two assignments of N variables which agree on *exactly* L of them.

In order to compute $g_\varphi(N, L)$, it is convenient to introduce two binary vectors $\vec{u}, \vec{v} \in \{0, 1\}^K$. They encode the following information. Consider a clause a, and fix $u_s = 1$ if, in the assignment $\underline{x}$, the s-th variable of clause a satisfies the clause, and fix $u_s = 0$ otherwise. The components of $\vec{v}$ are defined similarly, but with the assignment $\underline{y}$. Furthermore, we denote the Hamming distance between these vectors by $d(\vec{u}, \vec{v})$, and their Hamming weights (number of non-zero components) by $w(\vec{u})$, $w(\vec{v})$. Then

$$g_\varphi(N, L) = 2^{-K} {\sum_{\vec{u}, \vec{v}}}' \varphi\left(w(\vec{u})\right) \varphi\left(w(\vec{v})\right) \left(\frac{L}{N}\right)^{d(\vec{u}, \vec{v})} \left(1 - \frac{L}{N}\right)^{K - d(\vec{u}, \vec{v})}. \quad (10.25)$$

Here the sum $\sum'$ runs over K-component vectors $\vec{u}$, $\vec{v}$ with at least one non-zero component. A particularly simple choice is $\varphi(r) = \lambda^r$. Denoting L/N by z, we find that

$$g_w(N, L) = 2^{-K} \left(\left[(\lambda^2 + 1)z + 2\lambda(1 - z)\right]^K - 2\left[z + \lambda(1 - z)\right]^K + z^k \right). \quad (10.26)$$

The first two moments can be evaluated from Eqs. (10.23) and (10.24):

$$\mathbb{E}\, U(F) \doteq \exp\{N h_1(\lambda, \alpha)\}, \quad \mathbb{E}\left[U(F)^2\right] \doteq \exp\{N \max_z h_2(\lambda, \alpha, z)\}, \quad (10.27)$$

where the maximum is taken over $z \in [0, 1]$, and

$$h_1(\lambda, \alpha) \equiv \log 2 - \alpha K \log 2 + \alpha \log\left[(1 + \lambda)^K - 1\right], \quad (10.28)$$

$$\begin{aligned}h_2(\lambda, \alpha, z) \equiv{}& \log 2 - z \log z - (1 - z)\log(1 - z) - \alpha K \log 2 && (10.29)\\&+ \alpha \log\left(\left[(\lambda^2 + 1)z + 2\lambda(1 - z)\right]^K - 2\left[z + \lambda(1 - z)\right]^K + z^k\right).\end{aligned}$$

Evaluating the above expression for $z = 1/2$, one finds $h_2(\lambda, \alpha, 1/2) = 2h_1(\lambda, \alpha)$. The interpretation is as follows. Setting $z = 1/2$ amounts to assuming that the second moment of $U(F)$ is dominated by completely uncorrelated assignments (two uniformly random assignments agree on about half of the variables). This results in the factorization of the expectation $\mathbb{E}\left[U(F)^2\right] \approx [\mathbb{E}\, U(F)]^2$.

Two cases are possible: either the maximum of $h_2(\lambda, \alpha, z)$ over $z \in [0, 1]$ is achieved only at $z = 1/2$, or it is not.

(i) In the latter case, $\max_z h_2(\lambda, \alpha, z) > 2h_1(\lambda, \alpha)$ strictly, and therefore the ratio $[\mathbb{E}\, U(F)]^2/\mathbb{E}[U(F)^2]$ is exponentially small in N, and the second-moment inequality (10.20) is useless.

(ii) If, on the other hand, the maximum of $h_2(\lambda, \alpha, z)$ is achieved only at $z = 1/2$, then the ratio $[\mathbb{E}\, U(F)]^2/\mathbb{E}[U(F)^2]$ is 1 to the leading exponential order. It is not difficult to work out the precise asymptotic behaviour (i.e. to compute the prefactor of the exponential). One finds that $[\mathbb{E}\, U(F)]^2/\mathbb{E}[U(F)^2]$ remains finite when $N \to \infty$. As a consequence, $\alpha \leq \alpha_c^{(N)}(K)$ for N large enough.

Table 10.1 Satisfiability thresholds for random K-SAT. We report the lower bound from Theorem 10.8 and the upper bounds obtained from Eqs. (10.13) and (10.19).

K	3	4	5	6	7	8	9	10
$\alpha_{\mathrm{LB}}(K)$	2.548	7.314	17.62	39.03	82.63	170.6	347.4	701.5
$\alpha_{\mathrm{UB},1}(K)$	5.191	10.74	21.83	44.01	88.38	177.1	354.5	709.4
$\alpha_{\mathrm{UB},2}(K)$	4.666	10.22	21.32	43.51	87.87	176.6	354.0	708.9

A necessary condition for the second case to occur is that $z = 1/2$ is a local maximum of $h_2(\lambda, \alpha, z)$. This implies that λ must be the (unique) strictly positive root of

$$(1 + \lambda)^{K-1} = \frac{1}{1 - \lambda}. \tag{10.30}$$

We have thus proved the following result.

Theorem 10.8 *Let λ be the positive root of Eq. (10.30), and let the function $h_2(\cdot)$ be defined as in Eq. (10.29). Assume that $h_2(\lambda, \alpha, z)$ achieves its maximum, as a function of $z \in [0,1]$ only at $z = 1/2$. Then a random $\mathsf{SAT}_N(K, \alpha)$ is SAT with probability approaching one as $N \to \infty$.*

Let $\alpha_{\mathrm{LB}}(K)$ be the largest value of α such that the hypotheses of this theorem are satisfied. The Theorem implies an explicit lower bound on the satisfiability threshold: $\alpha_s^{(N)}(K) \geq \alpha_{\mathrm{LB}}(K)$ in the $N \to \infty$ limit. Table 10.1 summarizes some of the values of the upper and lower bounds found in this section for a few values of K. In the large-K limit, the following asymptotic behaviours can be shown to hold:

$$\alpha_{\mathrm{LB}}(K) = 2^K \log 2 - 2(K + 1) \log 2 - 1 + o(1), \tag{10.31}$$

$$\alpha_{\mathrm{UB},1}(K) = 2^K \log 2 - \frac{1}{2} \log 2 + o(1). \tag{10.32}$$

In other words, the simple methods set out in this chapter allow one to determine the satisfiability threshold with a relative error that behaves as 2^{-K} in the large K-limit. More sophisticated tools, to be discussed in the following chapters, are necessary for obtaining sharp results at finite K.

Exercise 10.12 [Research problem] Show that the choice of weights $\varphi(r) = \lambda^r$ is optimal: all other choices for $\varphi(r)$ give a worse lower bound. What strategy could be followed to improve the bound further?

Notes

The review paper by Gu *et al.* (1996) is a rather comprehensive source of information on the algorithmic aspects of satisfiability. The reader interested in applications will also find a detailed and referenced list of applications there.

Davis and Putnam first studied an algorithm for satisfiability in Davis and Putnam (1960). This was based on a systematic application of the resolution rule. The backtracking algorithm discussed in this chapter was introduced in Davis *et al.* (1962).

Other ensembles of random CNF formulae have been studied, but it turns out that it is not so easy to find hard formulae. For instance, take N variables, and generate M clauses independently according to the following rule. In a clause a, each of the variables appears as either x_i or $\bar{x}_i$ with the same probability $p \leq 1/2$ in both cases, and does not appear with probability $1 - 2p$. The reader is invited to study this ensemble; an introduction and a guide to the corresponding literature can be found in Franco (2000). Another useful ensemble is the '$2+p$' SAT problem, which interpolates between $K = 2$ and $K = 3$ by picking pM 3-clauses and $(1 - p)M$ 2-clauses; see Monasson and Zecchina (1998), and Monasson *et al.* (1999).

The polynomial nature of 2-SAT is discussed by Cook (1971). MAX-2SAT was shown to be NP-complete by Garey *et al.* (1976).

Schöning's algorithm was introduced in Schöning (1999) and discussed further in Schöning (2002). More general random-walk strategies for SAT are treated in Papadimitriou (1991), Selman and Kautz (1993), and Selman *et al.* (1994).

The threshold $\alpha_s = 1$ for random 2-SAT was proved by Chvátal and Reed (1992), Goerdt (1996), and de la Vega (1992); see also de la Vega (2001). The scaling behaviour near to the threshold has been analysed using graph theoretical methods by Chayes *et al.* (2001).

The numerical identification of the phase transition in random 3-SAT, together with the observation that difficult formulae are found near the phase transition, were done by Kirkpatrick and Selman (1994) and Selman and Kirkpatrick (1996). See also Selman *et al.* 1996).

Friedgut's theorem was proved by Friedgut (1999).

Upper bounds on the threshold were discussed by Dubois and Boufkhad (1997) and Kirousis *et al.* (1998). Lower bounds for the threshold in random K-SAT, based on the analysis of search algorithms, were pioneered by Chao and Franco. The paper by Chao and Franco (1986) corresponds to Exercise 10.11, and a generalization can be found in Chao and Franco (1990). A review of this type of methods is provided by Achlioptas (2001). Backtracking algorithms were first analysed using an heuristic approach by Cocco and Monasson (2001*a*). Cocco *et al.* (2006) gives a survey of the analysis of algorithms based on statistical-physics methods.

The idea of deriving a lower bound by the weighted second-moment method was introduced by Achlioptas and Moore (2007). The lower bound which we discuss here is derived by Achlioptas and Peres (2004); this paper also solves the first question of Exercise 10.12. A simple introduction to the use of the second moment method in various constraint satisfaction problems is provided by Achlioptas *et al.* (2005); see also Gomes and Selman (2005).

11
Low-density parity-check codes

Low-density parity-check (LDPC) error-correcting codes were introduced in 1963 by Robert Gallager in his PhD thesis. The basic motivation came from the observation that random linear codes (see Section 6.6), had excellent theoretical performance (in terms of the number of channel errors they could correct) but were unpractical. In particular, no efficient algorithm was known for decoding. In retrospect, this is not surprising, since it was later shown that decoding for linear codes is an NP-hard problem.

The idea was then to restrict the random linear code ensemble, introducing some structure that could be exploited for more efficient decoding. Of course, the risk is that such a restriction of the ensemble might spoil its performance. Gallager's proposal was simple and successful (but ahead of its time): LDPC codes are among the most efficient codes around.

In this chapter, we introduce one of the most important families of LDPC ensembles and derive some of their basic properties. As for any code, one can take two quite different points of view. The first is to study the performance of the code with respect to an appropriate metric, under *optimal* decoding, in which no constraint is imposed on the computational complexity of the decoding procedure. For instance, decoding by a scan of the whole, exponentially large codebook is allowed. The second approach consists in analysing the performance of the code under some specific, efficient decoding algorithm. Depending on the specific application, one may be interested in algorithms of polynomial complexity, or even require the complexity to be linear in the block length.

Here we shall focus on performance under optimal decoding. We shall derive rigorous bounds, showing that appropriately chosen LDPC ensembles allow one to communicate reliably at rates close to Shannon's capacity. However, the main interest of LDPC codes is that they can be decoded efficiently, and we shall discuss a simple example of a decoding algorithm with linear time complexity. A more extensive study of LDPC codes under practical decoding algorithms is deferred to Chapter 15.

After defining LDPC codes and LDPC code ensembles in Section 11.1, we discuss some geometric properties of their codebooks in Section 11.2. In Section 11.3 we use these properties to derive a lower bound on the threshold for reliable communication. An upper bound follows from information-theoretic considerations. Section 11.4 discusses a simple decoding algorithm, which is shown to correct a finite fraction of errors.

11.1 Definitions

11.1.1 Linear algebra with binary variables

Recall that a code is characterized by its codebook $\mathfrak{C}$, which is a subset of $\{0,1\}^N$. LDPC codes are **linear codes**, which means that the codebook is a linear subspace of $\{0,1\}^N$. In practice, such a subspace can be specified through an $M \times N$ matrix $\mathbb{H}$, with binary entries $\mathbb{H}_{ij} \in \{0,1\}$, where $M < N$. The codebook is defined as the kernel of $\mathbb{H}$:

$$\mathfrak{C} = \{\underline{x} \in \{0,1\}^N \; : \; \mathbb{H}\underline{x} = \underline{0}\}. \tag{11.1}$$

Here and in all of this chapter, the multiplications and sums involved in $\mathbb{H}\underline{x}$ are understood as being computed modulo 2. The matrix $\mathbb{H}$ is called the **parity check matrix** of the code. The size of the codebook is $|\mathfrak{C}| = 2^{N-\mathrm{rank}(\mathbb{H})}$, where $\mathrm{rank}(\mathbb{H})$ denotes the rank of the matrix $\mathbb{H}$ (the number of linearly independent rows). As $\mathrm{rank}(\mathbb{H}) \leq M$, we have $|\mathfrak{C}| \geq 2^{N-M}$. With a slight modification with respect to the notation of Chapter 1, we let $L \equiv N - M$. The rate R of the code therefore satisfies $R \geq L/N$, equality being obtained when all of the rows of $\mathbb{H}$ are linearly independent.

Given such a code, encoding can always be implemented as a linear operation. There exists an $N \times L$ binary matrix $\mathbb{G}$, called the generator matrix, such that the codebook is the image of $\mathbb{G}$: $\mathfrak{C} = \{\underline{x} = \mathbb{G}\underline{z}, \text{ where } \underline{z} \in \{0,1\}^L\}$. Encoding is therefore realized as the mapping $\underline{z} \mapsto \underline{x} = \mathbb{G}\underline{z}$. (Note that the product $\mathbb{H}\mathbb{G}$ is an $M \times L$ 'null' matrix with all entries equal to zero.)

11.1.2 Factor graph

In Section 9.1.2, we described the factor graph associated with one particular linear code (the Hamming code of eqn (9.8)). The recipe to build the factor graph, knowing $\mathbb{H}$, is as follows. Let us denote by $i_1^a, \ldots, i_{k(a)}^a \in \{1, \ldots, N\}$ the column indices such that $\mathbb{H}$ has a matrix element equal to 1 at row a and column i_j^a. The a-th coordinate of the vector $\mathbb{H}\underline{x}$ is then equal to $x_{i_1^a} \oplus \cdots \oplus x_{i_{k(a)}^a}$. Let $\mu_{0,\mathbb{H}}(\underline{x})$ be the uniform distribution over all codewords of the code $\mathbb{H}$ (hereafter, we shall often identify a code with its parity check matrix). It is given by

$$\mu_{0,\mathbb{H}}(\underline{x}) = \frac{1}{Z} \prod_{a=1}^{M} \mathbb{I}(x_{i_1^a} \oplus \cdots \oplus x_{i_k^a} = 0). \tag{11.2}$$

Therefore, the factor graph associated with $\mu_{0,\mathbb{H}}(\underline{x})$ (or with $\mathbb{H}$) includes N variable nodes, one for each column of $\mathbb{H}$, and M function nodes (also called, in this context, **check nodes**), one for each row. A factor node and a variable node are joined by an edge if the corresponding entry in $\mathbb{H}$ is non-vanishing. Clearly, this procedure can be inverted: with any factor graph with N variable nodes and M function nodes, we can associate an $M \times N$ binary matrix $\mathbb{H}$, the **adjacency matrix** of the graph, whose non-zero entries correspond to the edges of the graph.

11.1.3 Ensembles with given degree profiles

In Chapter 9, we introduced the ensembles of factor graphs $\mathbb{D}_N(\Lambda, P)$. These have N variable nodes, and the two polynomials $\Lambda(x) = \sum_{n=0}^{\infty} \Lambda_n x^n$ and $P(x) = \sum_{n=0}^{\infty} P_n x^n$

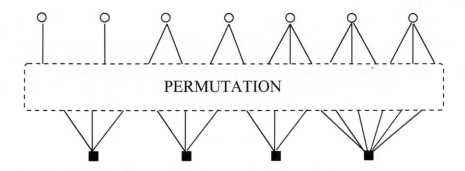

Fig. 11.1 Construction of a graph/code with given degree profiles. Here, a graph with $N = 7$, $M = 4$, $\Lambda(x) = \frac{1}{7}(2x + 2x^2 + 3x^3)$, and $P(x) = \frac{1}{4}(3x^3 + x^6)$ is shown. The 'sockets' from the variable nodes and those from the check nodes are connected through a uniformly random permutation.

define the degree profiles: Λ_n is the probability that a randomly chosen variable node has degree n, and P_n is the probability that a randomly chosen function node has degree n.

We define $\mathrm{LDPC}_N(\Lambda, P)$ to be the ensemble of linear codes whose parity check matrix is the adjacency matrix of a random graph from the ensemble $\mathbb{D}_N(\Lambda, P)$. We shall be interested in the limit $N \to \infty$ while keeping the degree profiles fixed. Therefore each vertex has bounded degree, and hence the parity check matrix has a 'low density.'

In order to eliminate trivial cases, we always assume that variable nodes have a degree at least 1, and function nodes at least 2. The numbers of parity check and variable nodes satisfy the relation $M = N\Lambda'(1)/P'(1)$. The ratio $L/N = (N-M)/N = 1 - \Lambda'(1)/P'(1)$, which is a lower bound on the actual rate R, is called the **design rate** R_{des} of the code (or of the ensemble). The actual rate of a code from the ensemble $\mathrm{LDPC}_N(\Lambda, P)$ is of course a random variable, but we shall see below that it is, in general, sharply concentrated 'near' R_{des}.

A special case which is often considered is that of 'regular' graphs: all variable nodes have degree l and all function nodes have degree k (i.e. $P(x) = x^k$ and $\Lambda(x) = x^l$). The corresponding code ensemble is usually denoted simply by $\mathrm{LDPC}_N(l, k)$, or, more synthetically, (l, k). It has a design rate $R_{\mathrm{des}} = 1 - (l/k)$.

Generating a uniformly random graph from the ensemble $\mathbb{D}_N(\Lambda, P)$ is not a trivial task. The simplest way to bypass this problem is to substitute the uniformly random ensemble by a slightly different one which has a simple algorithmic description. One can, for instance, proceed as follows. First, separate the set of variable nodes uniformly at random into subsets of sizes $N\Lambda_0$, $N\Lambda_1$, ..., $N\Lambda_{l_{\max}}$, and attribute zero 'sockets' to the nodes in the first subset, one socket to each of the nodes in the second, and so on. Analogously, separate the set of check nodes into subsets of size MP_0, MP_1, ..., $MP_{k_{\max}}$ and attribute $0, 1, \ldots, k_{\max}$ sockets to the nodes in each subset. At this point the variable nodes have $N\Lambda'(1)$ sockets, and so have the check nodes. Draw a uniformly random permutation from $N\Lambda'(1)$ objects and connect the sockets on the two sides accordingly (see Fig. 11.1).

Exercise 11.1 In order to sample a graph as described above, one needs two routines. The first one separates a set of N objects uniformly into subsets of prescribed sizes. The second one samples a random permutation of $N\Lambda'(1)$ objects. Show that both of these tasks can be accomplished with $O(N)$ operations (having at your disposal a random number generator).

This procedure a flaw: it may generate multiple edges joining the same pair of nodes in the graph.

In order to cure this problem, we shall agree that each time n edges join any two nodes, they must be erased if n is even, and they must be replaced by a single edge if n is odd. Of course, the resulting graph does not necessarily have the prescribed degree profile (Λ, P), and even if we condition on this to be the case, its distribution is not uniform. We shall nevertheless insist on denoting the ensemble by $\text{LDPC}_N(\Lambda, P)$. The intuition is that, for large N, the degree profile will be 'close' to the prescribed one and the distribution will be 'uniform enough' for our purposes. Moreover –and this is really important– this and similar graph generation techniques are used in practice.

Exercise 11.2 This exercise aims at proving that, for large N, the degree profile produced by the explicit construction described above is close to the prescribed degree profile.

(a) Let m be the number of multiple edges appearing in the graph; compute its expectation. Show that $\mathbb{E}\,m = O(1)$ as $N \to \infty$ with Λ and P fixed.

(b) Let (Λ', P') be the degree profile produced by the above procedure. Denote by

$$d \equiv \sum_l |\Lambda_l - \Lambda'_l| + \sum_k |P_k - P'_k| \tag{11.3}$$

the 'distance' between the prescribed and the actual degree profiles. Derive an upper bound on d in terms of m and show that it implies $\mathbb{E}\,d = O(1/N)$.

11.2 The geometry of the codebook

As we saw in Section 6.2, a classical approach to the analysis of error-correcting codes is to study the geometric properties of the corresponding codebooks. An important example of such properties is the distance enumerator $\mathcal{N}_{\underline{x}_0}(d)$, giving the number of codewords at a Hamming distance d from $\underline{x}_0$. In the case of linear codes, the distance enumerator does not depend upon the reference codeword $\underline{x}_0$ (the reader is invited to prove this statement). It is therefore customary to take the all-zeros codeword as the reference, and to use the term **weight enumerator**: $\mathcal{N}(w) = \mathcal{N}_{\underline{x}_0}(d = w)$ is the number of codewords having a **weight** (the number of ones in the codeword) equal to w.

In this section, we want to estimate the expected weight enumerator $\overline{\mathcal{N}}(w) \equiv \mathbb{E}\,\mathcal{N}(w)$ for a random code in the ensemble $\text{LDPC}_N(\Lambda, P)$. In the case of the random code ensemble of Section 6.2, the corresponding $\overline{\mathcal{N}}(w)$ grows exponentially in the block length N, and that most of the codewords have a weight $w = N\omega$ growing linearly with N. We shall in fact compute the exponential growth rate $\phi(\omega)$, defined by

$$\overline{\mathcal{N}}(w = N\omega) \doteq e^{N\phi(\omega)} . \tag{11.4}$$

In the jargon of statistical physics, $\overline{\mathcal{N}}(w)$ is an 'annealed average' and hence it may be dominated by rare instances in the ensemble. On the other hand, one expects $\log \mathcal{N}(w)$ to be tightly concentrated around its typical value $N\phi_{\mathrm{q}}(\omega)$. The typical exponent $\phi_{\mathrm{q}}(\omega)$ can be computed through a quenched calculation, for instance considering $\lim_{N \to \infty} N^{-1} \mathbb{E} \log \left[1 + \mathcal{N}(w)\right]$. Of course, $\phi_{\mathrm{q}}(\omega) \leq \phi(\omega)$ because of the concavity of the logarithm. In this chapter, we shall keep to the annealed calculation, which is much easier and gives an upper bound on the quenched result ϕ_{q}.

Let $\underline{x} \in \{0,1\}^N$ be a binary word of length N and weight w. Note that $\mathbb{H}\underline{x} = 0 \mod 2$ if and only if the corresponding factor graph has the following property. Consider all the w variable nodes i such that $x_i = 1$, and colour all edges incident on these nodes in red. Colour all of the other edges blue. All of the check nodes must then have an even number of incident red edges. A little thought shows that $\overline{\mathcal{N}}(w)$ is the number of 'coloured' factor graphs that have this property for some set of w variable nodes, divided by the total number of factor graphs in the ensemble. We shall compute this number first for a graph with fixed degrees, i.e. for codes in the ensemble $\mathrm{LDPC}_N(l,k)$, and then we shall generalize to arbitrary degree profiles.

11.2.1 Weight enumerator: Regular ensembles

In the fixed-degree case we have N variable nodes of degree l, and M function nodes of degree k. We denote by $F = Mk = Nl$ the total number of edges. A valid coloured graph must have $E = wl$ red edges. It can be constructed as follows. First, choose w variable nodes, which can be done in $\binom{N}{w}$ ways. Assign l red sockets to each node in this set, and l blue sockets to each node outside the set. Then, for each of the M function nodes, colour an even subset of its sockets red in such a way that the total number of red sockets is $E = wl$. Let m_r be the number of function nodes with r red sockets. The numbers m_r can be non-zero only when r is even, and they are constrained by $\sum_{r=0}^{k} m_r = M$ and $\sum_{r=0}^{k} r m_r = lw$. The number of ways one can colour the sockets of the function nodes is thus

$$\mathcal{C}(k, M, w) = \sum_{m_0, \ldots, m_k}^{(e)} \binom{M}{m_0, \ldots, m_k} \prod_r \binom{k}{r}^{m_r}$$

$$\times \mathbb{I}\left(\sum_{r=0}^{k} m_r = M\right) \mathbb{I}\left(\sum_{r=0}^{k} r m_r = lw\right), \tag{11.5}$$

where the sum $\sum^{(e)}$ means that non-zero m_r's appear only for r even. Finally, we join the variable-node and check node sockets in such a way that colours are matched. There are $(lw)!(F - lw)!$ such matchings out of the total number $F!$ corresponding to different elements in the ensemble. Putting everything together, we get the final formula,

$$\overline{\mathcal{N}}(w) = \frac{(lw)!(F - lw)!}{F!} \binom{N}{w} \mathcal{C}(k, M, w) . \tag{11.6}$$

In order to compute the function $\phi(\omega)$ in eqn (11.4), we need to work out the asymptotic behaviour of this formula when $N \to \infty$ at fixed $\omega = w/N$. Assuming that $m_r = x_r M = x_r N l/k$, we can expand the multinomial factors using Stirling's formula. This gives

$$\phi(\omega) = \max_{\{x_r\}}^* \left[(1-l)\mathcal{H}(\omega) + \frac{l}{k} \sum_r \left(-x_r \log x_r + x_r \log \binom{k}{r} \right) \right] , \qquad (11.7)$$

where the $\max^*$ is taken over all choices of $x_0, x_2, x_4, \ldots$ in $[0,1]$, subject to the two constraints $\sum_r x_r = 1$ and $\sum_r r x_r = k\omega$. The maximization can be done by imposing these constraints via two Lagrange multipliers. We obtain $x_r = C z^r \binom{k}{r} \mathbb{I}(r \text{ even})$, where C and z are two constants fixed by the equations

$$C = \frac{2}{(1+z)^k + (1-z)^k} , \qquad (11.8)$$

$$\omega = z \frac{(1+z)^{k-1} - (1-z)^{k-1}}{(1+z)^k + (1-z)^k} . \qquad (11.9)$$

Substituting the resulting x_r back into the equation (11.7) for ϕ, this gives, finally,

$$\phi(\omega) = (1-l)\mathcal{H}(\omega) + \frac{l}{k} \log \frac{(1+z)^k + (1-z)^k}{2} - \omega l \log z , \qquad (11.10)$$

where z is the function of ω defined in eqn (11.9).

We shall see in the next sections how to use this result, but let us first explain how it can be generalized.

11.2.2   Weight enumerator: General case

We want to compute the leading exponential behaviour $\overline{\mathcal{N}}(w) \doteq \exp[N\phi(\omega)]$ of the expected weight enumerator for a general $\text{LDPC}_N(\Lambda, P)$ code. The idea of the approach is the same as the one we have just used for the case of regular ensembles, but the computation becomes heavier. It is therefore useful to adopt a more powerful formalism. Altogether, this subsection is more technical than the others: the reader who is not interested in the details can skip it and go to the results.

We want to build a valid coloured graph; let us denote its number of red edges (which is no longer fixed by w) by E. There are $\text{coeff}[\prod_l (1 + xy^l)^{N\Lambda_l}, x^w y^E]$ ways of choosing the w variable nodes in such a way that their degrees add up to E.[1] As before, for each of the M function nodes, we colour an even subset of its sockets red in such a way that the total number of red sockets is E. This can be done in $\text{coeff}[\prod_k q_k(z)^{MP_k}, z^E]$ ways, where $q_k(z) \equiv \frac{1}{2}(1+z)^k + \frac{1}{2}(1-z)^k$. The numbers of ways one can match the red sockets in the variable and function nodes is still $E!(F-E)!$, where $F = N\Lambda'(1) = MP'(1)$ is the total number of edges in the graph. This gives the result

---

[1] We denote the coefficient of x^n in the formal power series $f(x)$ by $\text{coeff}[f(x), x^n]$.

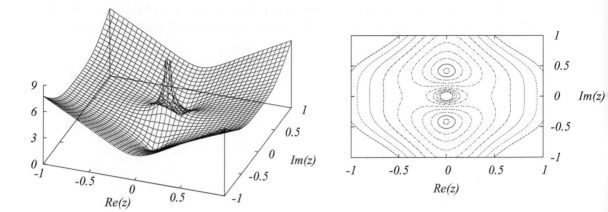

Fig. 11.2 Modulus of the function $z^{-3\xi}q_4(z)^{3/4}$ for $\xi = 1/3$.

$$\overline{\mathcal{N}}(w) = \sum_{E=0}^{F} \frac{E!(F-E)!}{F!}$$

$$\times \operatorname{coeff}\left[\prod_{l=1}^{l_{\max}}(1+xy^l)^{N\Lambda_l}, x^w y^E\right] \operatorname{coeff}\left[\prod_{k=2}^{k_{\max}} q_k(z)^{MP_k}, z^E\right]. \quad (11.11)$$

In order to estimate the leading exponential behaviour of $\overline{\mathcal{N}}(w)$ at large N, when $w = N\omega$, we set $E = F\xi = N\Lambda'(1)\xi$. The asymptotic behaviour of the $\operatorname{coeff}[\ldots,\ldots]$ terms can be estimated using the saddle point method. Here, we sketch the idea for the second of these terms. By the Cauchy theorem,

$$\operatorname{coeff}\left[\prod_{k=2}^{k_{\max}} q_k(z)^{MP_k}, z^E\right] = \oint \frac{1}{z^{N\Lambda'(1)\xi+1}} \prod_{k=2}^{k_{\max}} q_k(z)^{MP_k} \frac{\mathrm{d}z}{2\pi i} \equiv \oint \frac{f(z)^N}{z} \frac{\mathrm{d}z}{2\pi i},$$

$$(11.12)$$

where the integral runs over any path encircling the origin of the complex z plane in the anticlockwise direction, and

$$f(z) \equiv \frac{1}{z^{\Lambda'(1)\xi}} \prod_{k=2}^{k_{\max}} q_k(z)^{\Lambda'(1)P_k/P'(1)}. \quad (11.13)$$

In Fig. 11.2, we plot the modulus of the function $f(z)$ for degree distributions $\Lambda(x) = x^3$, $P(x) = x^4$, and $\xi = 1/3$. The function has two saddle points of $\pm z_*$, where $z_* = z_*(\xi) \in \mathbb{R}_+$ solves the equation $f'(z_*) = 0$, namely

$$\xi = \sum_{k=2}^{k_{\max}} \rho_k z_* \frac{(1+z_*)^{k-1} - (1-z_*)^{k-1}}{(1+z_*)^k + (1-z_*)^k}. \quad (11.14)$$

Here we have used the notation $\rho_k \equiv kP_k/P'(1)$ introduced in Section 9.5 (analogously, we shall write $\lambda_l \equiv l\Lambda_l/\Lambda'(1)$). This equation generalizes (11.9). If we take the integration contour in eqn (11.12) to be the circle of radius z_* centred at $z = 0$, the integral is dominated by the saddle point at z_* (together with the symmetric point $-z_*$). We therefore get

$$\text{coeff}\left[\prod_{k=2}^{k_{\max}} q_k(z)^{MP_k}, z^E\right] \doteq \exp\left\{N\left[-\Lambda'(1)\xi \log z_* + \frac{\Lambda'(1)}{P'(1)}\sum_{k=2}^{k_{\max}} P_k \log q_k(z_*)\right]\right\}.$$

Proceeding analogously with the second $\text{coeff}[\ldots,\ldots]$ term in eqn (11.11), we get $\overline{\mathcal{N}}(w = N\omega) \doteq \exp\{N\phi(\omega)\}$. The function ϕ is given by

$$\phi(\omega) = \sup_\xi \inf_{x,y,z}\left\{-\Lambda'(1)\mathcal{H}(\xi) - \omega \log x - \Lambda'(1)\xi \log(yz)\right.$$
$$\left. + \sum_{l=2}^{l_{\max}} \Lambda_l \log(1 + xy^l) + \frac{\Lambda'(1)}{P'(1)}\sum_{k=2}^{k_{\max}} P_k \log q_k(z)\right\}, \quad (11.15)$$

where the minimization over x, y, z is understood to be taken over the positive real axis while $\xi \in [0,1]$. The stationarity condition with respect to variations of z is given by eqn (11.14). Stationarity with respect to ξ, x, y yields

$$\xi = \frac{yz}{1+yz}, \qquad \omega = \sum_{l=1}^{l_{\max}} \Lambda_l \frac{xy^l}{1+xy^l}, \qquad \xi = \sum_{l=1}^{l_{\max}} \lambda_l \frac{xy^l}{1+xy^l}, \qquad (11.16)$$

respectively. If we use the first of these equations to eliminate ξ, we obtain the final parametric representation (in the parameter $x \in [0,\infty[$) of $\phi(\omega)$

$$\phi(\omega) = -\omega \log x - \Lambda'(1)\log(1+yz) + \sum_{l=1}^{l_{\max}} \Lambda_l \log(1+xy^l) \qquad (11.17)$$
$$+ \frac{\Lambda'(1)}{P'(1)}\sum_{k=2}^{k_{\max}} P_k \log q_k(z),$$

$$\omega = \sum_{l=1}^{l_{\max}} \Lambda_l \frac{xy^l}{1+xy^l}. \qquad (11.18)$$

The two functions $y = y(x)$ and $z = z(x)$ are solutions of the coupled equations

$$y = \frac{\sum_{k=2}^{k_{\max}} \rho_k\, p_k^-(z)}{\sum_{k=2}^{k_{\max}} \rho_k\, p_k^+(z)}, \qquad z = \frac{\sum_{l=1}^{l_{\max}} \lambda_l xy^{l-1}/(1+xy^l)}{\sum_{l=1}^{l_{\max}} \lambda_l/(1+xy^l)}, \qquad (11.19)$$

where we have defined $p_k^\pm(z) \equiv [(1+z)^{k-1} \pm (1-z)^{k-1}]/[(1+z)^k + (1-z)^k]$.

Exercise 11.3 The numerical solution of eqns (11.18) and (11.19) can be somewhat tricky. Here is a simple iterative procedure which usually works reasonably well. Readers are invited to try it with their favourite degree distributions Λ, P.

First, solve eqn (11.18) for x at given $y \in [0, \infty[$ and $\omega \in [0, 1]$, using a bisection method. Next, substitute this value of x into eqn (11.19), and write the resulting equations as $y = f(z)$ and $z = g(y, \omega)$. Define $F_\omega(y) \equiv f(g(y, \omega))$. Solve the equation $y = F_\omega(y)$ by iteration of the map $y_{n+1} = F_\omega(y_n)$ Once the fixed point y_* has been found, the other parameters are computed as follows $z_* = g(y_*, \omega)$, and x_* is the solution of eqn (11.18) for $y = y_*$. Finally, x_*, y_*, z_* are substituted into eqn (11.17) to obtain $\phi(\omega)$.

Examples of functions $\phi(\omega)$ are shown in Figs 11.3–11.5. We shall now discuss these results, paying special attention to the region of small ω.

11.2.3 Short-distance properties

In the low-noise limit, the performance of a code depends a lot on the existence of codewords at a short distance from the transmitted one. For linear codes and symmetric communication channels, we can assume without loss of generality that the all-zeros codeword has been transmitted. Here, we shall work out the short-distance (i.e. for a mall weight ω) behaviour of $\phi(\omega)$ for several LDPC ensembles. These properties will be used to characterize the performance of the code in Section 11.3.

As $\omega \to 0$, solving eqns (11.18) and (11.19) yields $y, z \to 0$. By Taylor expansion of these equations, we get

$$y \simeq \rho'(1)z, \quad z \simeq \lambda_{l_{\min}} x y^{l_{\min}-1}, \quad \omega \simeq \Lambda_{l_{\min}} x y^{l_{\min}}, \tag{11.20}$$

where we have neglected higher-order terms in y, z. At this point we must distinguish whether $l_{\min} = 1$, $l_{\min} = 2$, or $l_{\min} \geq 3$.

We start with the case $l_{\min} = 1$. In this case x, y, z all scale like $\sqrt{\omega}$, and a short computation shows that

$$\phi(\omega) = -\frac{1}{2} \omega \log\left(\frac{\omega}{\Lambda_1^2}\right) + O(\omega). \tag{11.21}$$

In particular, $\phi(\omega)$ is strictly positive for ω sufficiently small. The expected number of codewords within a small relative Hamming distance $w = N\omega$ from a given codeword is exponential in N. Furthermore, eqn (11.21) is reminiscent of the behaviour in the absence of any parity check, where one gets $\phi(\omega) = \mathcal{H}(\omega) \simeq -\omega \log \omega$.

Exercise 11.4 In order to check eqn (11.21), compute the weight enumerator for the regular ensemble LDPC$_N(l = 1, k)$. Note that, in this case, the weight enumerator does not depend on the realization of the code, and admits the simple representation $\mathcal{N}(w) = \text{coeff}[q_k(z)^{N/k}, z^w]$.

An example of a weight enumerator for an irregular code with $l_{\min} = 1$ is shown in Fig. 11.3. The behaviour (11.21) is quite bad for an error-correcting code. In order

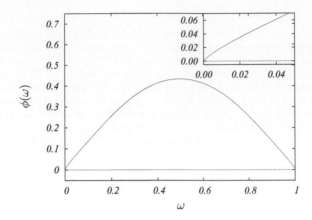

Fig. 11.3 Logarithm of the expected weight enumerator, $\phi(\omega)$, plotted versus the reduced weight $\omega = w/N$ for the ensemble $\text{LDPC}_N(\frac{1}{4}x + \frac{1}{4}x^2 + \frac{1}{2}x^3, x^6)$. *Inset*: small-weight region. $\phi(\omega)$ is positive near to the origin, and in fact its derivative diverges as $\omega \to 0$: each codeword is surrounded by a large number of very close other codewords. This makes it a bad error correcting code.

to understand why, let us for a moment forget that this result was obtained by taking $\omega \to 0$ *after* $N \to \infty$, and apply it in the regime $N \to \infty$ at $w = N\omega$ fixed. We get

$$\overline{\mathcal{N}}(w) \sim \left(\frac{N}{w}\right)^{w/2} . \tag{11.22}$$

It turns out that this result holds not only on average but for most codes in the ensemble. In other words, even at a Hamming distance of 2 from any given codeword, there are $\Theta(N)$ other codewords. It is intuitively clear that discriminating between two codewords at Hamming distance $\Theta(1)$, given a noisy observation, is in most cases impossible. Because of these remarks, one usually discards $l_{\min} = 1$ ensembles in error correction.

Consider now the case $l_{\min} = 2$. From eqn (11.20), we get

$$\phi(\omega) \simeq A\,\omega\,, \qquad A \equiv \log\left[\frac{P''(1)}{P'(1)} \frac{2\Lambda_2}{\Lambda'(1)}\right] = \log\left[\lambda'(0)\rho'(1)\right] . \tag{11.23}$$

As will be apparent in Chapter 15, the combination $\lambda'(0)\rho'(1)$ has an important concrete interpretation.

The code ensemble has significantly different properties depending on the sign of A. If $A > 0$, the expected number of codewords within a small (but of order $\Theta(N)$) Hamming distance from any given codeword is exponential in the block length. The situation seems similar to the $l_{\min} = 1$ case. Note, however, that $\phi(\omega)$ goes much more quickly to 0 as $\omega \to 0$ in the present case. Assuming again that eqn (11.23) holds beyond the asymptotic regime in which it was derived, we get

$$\mathcal{N}(w) \sim e^{Aw} . \tag{11.24}$$

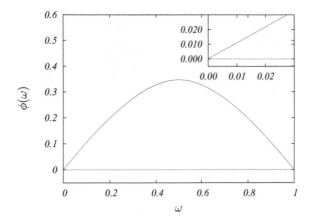

Fig. 11.4 Logarithm of the expected weight enumerator for the ensemble $\text{LDPC}_N(2,4)$. The degree profiles are $\Lambda(x) = x^2$, meaning that all variable nodes have degree 2, and $P(x) = x^4$, meaning that all function nodes have degree 4. *Inset*: small-weight region. The constant A is positive, so there exist codewords at short distances.

In other words, the number of codewords around any particular codeword is $o(N)$ until we reach a Hamming distance $d_* \simeq \log N/A$. For many purposes d_* plays the role of an 'effective' minimum distance. The example of the regular code $\text{LDPC}_N(2,4)$, for which $A = \log 3$, is shown in Fig. 11.4.

If, on the other hand $A < 0$, then $\phi(\omega) < 0$ in some interval $\omega \in]0, \omega_*[$. The first-moment method then shows that there are no codewords of weight 'close to' $N\omega$ for any ω in this range.

A similar conclusion is reached if $l_{\min} \geq 3$, where one finds

$$\phi(\omega) \simeq \left(\frac{l_{\min} - 2}{2}\right) \omega \log\left(\frac{\omega}{\Lambda_{l_{\min}}}\right). \tag{11.25}$$

An example of a weight enumerator exponent for a code with good short-distance properties, the code $\text{LDPC}_N(3,6)$, is given in Fig. 11.5.

This discussion can be summarized as follows.

Proposition 11.1 *Consider a random linear code from the ensemble $\text{LDPC}_N(\Lambda, P)$ with $l_{\min} \geq 3$. Let $\omega_* \in]0, 1/2[$ be the first non-trivial zero of $\phi(\omega)$, and consider any interval $[\omega_1, \omega_2] \subset]0, \omega_*[$. With high probability, there does not exist any pair of codewords with a distance belonging to this interval. The same result holds when $l_{\min} = 2$ and $\lambda'(0)\rho'(1) = (P''(1)/P'(1))(2\Lambda_2/\Lambda'(1)) < 1$.*

Note that our study deals only with weights $w = \omega N$ which grow linearly with N. The proposition excludes the existence of codewords of arbitrarily small ω, but it does not tell us anything about possible codewords of sublinear weight, i.e. $w = o(N)$ (for instance, with w finite as $N \to \infty$). It turns out that if $l_{\min} \geq 3$, the code has with high probability no such codewords, and its minimum distance is at least $N\omega_*$. If, on the other hand, $l_{\min} = 2$, the code typically has some codewords of finite weight

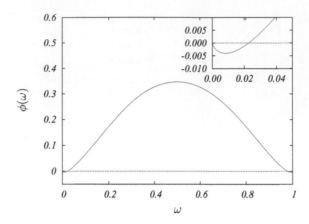

Fig. 11.5 Logarithm of the expected weight enumerator for the ensemble $\text{LDPC}_N(3,6)$. *Inset*: small-weight region; $\phi(\omega) < 0$ for $\omega < \omega_* \approx 0.02$. There are no codewords except for the 'all-zeros' one in the region $\omega < \omega_*$.

w. However (if $A < 0$), they can be eliminated without changing the code rate by an 'expurgation' procedure similar to that described in Section 6.5.1.

11.2.4 Rate

The weight enumerator can also be used to obtain a precise characterization of the rate of an $\text{LDPC}_N(\Lambda, P)$ code. For $\omega = 1/2$, $x = y = z = 1$ satisfy eqns (11.18) and (11.19). This gives

$$\phi\left(\omega = \frac{1}{2}\right) = \left(1 - \frac{\Lambda'(1)}{P'(1)}\right) \log 2 = R_{\text{des}} \log 2 . \tag{11.26}$$

It turns out that, in most cases of practical interest, the curve $\phi(\omega)$ has its maximum at $\omega = 1/2$ (see for instance Figs 11.3–11.5). In such cases the result (11.26) shows that the rate equals the design rate.

Proposition 11.2 *Let R be the rate of a code from the ensemble $\text{LDPC}_N(\Lambda, P)$, let $R_{\text{des}} = 1 - \Lambda'(1)/P'(1)$ be the associated design rate, and let $\phi(\omega)$ be the function defined in eqns (11.17)–(11.19). Assume that $\phi(\omega)$ achieves its absolute maximum over the interval $[0,1]$ at $\omega = 1/2$. Then, for any $\delta > 0$, there exists a positive N-independent constant $C_1(\delta)$ such that*

$$\mathbb{P}\{|R - R_{\text{des}}| \geq \delta\} \leq C_1(\delta)\, 2^{-N\delta/2} . \tag{11.27}$$

Proof Since we have already established that $R \geq R_{\text{des}}$, we only need to prove an upper bound on R. The rate is defined as $R \equiv (\log_2 \mathcal{N})/N$, where $\mathcal{N}$ is the total number of codewords. Markov's inequality gives:

$$\mathbb{P}\{R \geq R_{\text{des}} + \delta\} = \mathbb{P}\{\mathcal{N} \geq 2^{N(R_{\text{des}}+\delta)}\} \leq 2^{-N(R_{\text{des}}+\delta)}\, \mathbb{E}\mathcal{N} . \tag{11.28}$$

The expectation of the number of codewords is $\mathbb{E}\mathcal{N}(w) \doteq \exp\{N\phi(w/N)\}$, and there are only $N+1$ possible values of the weight w; therefore,

$$\mathbb{E}\mathcal{N} \doteq \exp\{N \sup_{\omega \in [0,1]} \phi(\omega)\}. \tag{11.29}$$

As $\sup \phi(\omega) = \phi(1/2) = R_{\text{des}} \log 2$ by hypothesis, there exists a constant $C_1(\delta)$ such that $\mathbb{E}\mathcal{N} \leq C_1(\delta) 2^{N(R_{\text{des}}+\delta/2)}$ for any N. Plugging this into eqn (11.28), we get

$$\mathbb{P}\{R \geq R_{\text{des}} + \delta\} \leq C_1(\delta) \, 2^{-N\delta/2}. \tag{11.30}$$

□

11.3 LDPC codes for the binary symmetric channel

Our study of the weight enumerator has shown that codes from the ensemble $\text{LDPC}_N(\Lambda, P)$ with $l_{\min} \geq 3$ have a good short-distance behaviour. The absence of codewords within an extensive distance $N\omega_*$ from the transmitted codeword guarantees that any error (even one introduced by an adversarial channel) that changes a fraction of the bits smaller than $\omega_*/2$ can be corrected. Here we want to study the performance of these codes in correcting *typical* errors introduced by a given (probabilistic) channel. We shall focus on the binary symmetric channel denoted by $\text{BSC}(p)$, which flips each bit independently with probability $p < 1/2$. Supposing as usual that the all-zero codeword $\underline{x}^{(0)} = \underline{0}$ has been transmitted, let us denote the received message by $\underline{y} = (y_1 \ldots y_N)$. Its components are i.i.d. random variables taking the value 0 with probability $1 - p$, and the value 1 with probability p. The decoding strategy which minimizes the block error rate is word MAP (or maximum-likelihood) decoding, which outputs the codeword closest to the channel output $\underline{y}$. As already mentioned, we shall not bother about the practical implementation of this strategy and its computational complexity.

The block error probability for a code $\mathfrak{C}$, denoted by $P_B(\mathfrak{C})$, is the probability that there exists a 'wrong' codeword, distinct from $\underline{0}$, whose distance from $\underline{y}$ is smaller than $d(\underline{0}, \underline{y})$. Its expectation value over the code ensemble, $P_B = \mathbb{E} P_B(\mathfrak{C})$, is an important indicator of the performance of the ensemble. We shall show that in the large-N limit, codes with $l_{\min} \geq 3$ undergo a phase transition, separating a low-noise phase, where $p < p_{\text{MAP}}$, in which $\lim_{N\to\infty} P_B$ is zero, from a high-noise phase, where $p > p_{\text{MAP}}$, in which the limit is not zero. Although the computation of p_{MAP} is deferred to Chapter 15, we derive rigorous bounds here which imply that appropriate LDPC codes have very good performance, close to Shannon's information-theoretic limit, under MAP decoding.

11.3.1 Lower bound on the threshold

We start by deriving a general bound on the block error probability $P_B(\mathfrak{C})$ for $\text{BSC}(p)$, valid for any linear code. Let $\mathcal{N} = 2^{NR}$ be the size of the codebook $\mathfrak{C}$. By the union bound,

$$P_B(\mathfrak{C}) = \mathbb{P}\left\{\exists \alpha \neq 0 \quad \text{such that} \quad d(\underline{x}^{(\alpha)}, \underline{y}) \leq d(\underline{0}, \underline{y})\right\}$$

$$\leq \sum_{\alpha=1}^{\mathcal{N}-1} \mathbb{P}\left\{d(\underline{x}^{(\alpha)}, \underline{y}) \leq d(\underline{0}, \underline{y})\right\}. \tag{11.31}$$

As the components of $\underline{y}$ are i.i.d. Bernoulli variables, the probability $\mathbb{P}\{d(\underline{x}^{(\alpha)}, \underline{y}) \leq d(\underline{0}, \underline{y})\}$ depends on the vector $\underline{x}^{(\alpha)}$ only through its weight w. Let $\underline{x}(w)$ be the vector formed by w ones followed by $N - w$ zeros, and denote by $\mathcal{N}(w)$ the weight enumerator of the code $\mathfrak{C}$. Then

$$P_B(\mathfrak{C}) \leq \sum_{w=1}^{N} \mathcal{N}(w) \, \mathbb{P}\left\{d(\underline{x}(w), \underline{y}) \leq d(\underline{0}, \underline{y})\right\} . \tag{11.32}$$

The probability $\mathbb{P}\left\{d(\underline{x}(w), \underline{y}) \leq d(\underline{0}, \underline{y})\right\}$ can be written as $\sum_u \binom{w}{u} p^u (1-p)^{w-u} \, \mathbb{I}(u \geq w/2)$, where u is the number of sites $i \in \{1, \dots, w\}$ such that $y_i = 1$. A good bound is provided by a standard method known as the **Chernoff bound**.

Exercise 11.5 Let X be a random variable. Show that, for any a and any $\lambda > 0$,

$$\mathbb{P}(X \geq a) \leq e^{-\lambda a} \, \mathbb{E}\left(e^{\lambda X}\right) . \tag{11.33}$$

In our case this gives

$$\mathbb{P}\left\{d(\underline{x}(w), \underline{y}) \leq d(\underline{0}, \underline{y})\right\} \leq \mathbb{E}e^{\lambda[d(\underline{0}, \underline{y}) - d(\underline{x}(w), \underline{y})]} = [(1-p)\,e^{-\lambda} + p\,e^{\lambda}]^w .$$

The bound is optimized for $\lambda = \frac{1}{2} \log\left((1-p)/p\right) > 0$, and gives

$$P_B(\mathfrak{C}) \leq \sum_{w=1}^{N} \mathcal{N}(w) \, e^{-\gamma w} , \tag{11.34}$$

where $\gamma \equiv -\log \sqrt{4p(1-p)} \geq 0$. The quantity $\sqrt{4p(1-p)}$ is sometimes referred to as the **Bhattacharya parameter** of the channel BSC(p).

Exercise 11.6 Consider the case of a general binary memoryless symmetric channel with a transition probability $Q(y|x)$, $x \in \{0, 1\}$, $y \in \mathcal{Y} \subseteq \mathbb{R}$. First, show that eqn (11.31) remains valid if the Hamming distance $d(\underline{x}, \underline{y})$ is replaced by the log-likelihood

$$d_Q(\underline{x}|\underline{y}) = -\sum_{i=1}^{N} \log Q(y_i|x_i) . \tag{11.35}$$

[Hint: Remember the general expressions (6.5) for the probability $\mu_{\underline{y}}(\underline{x}) = \mathbb{P}(\underline{x}|\underline{y})$ that the transmitted codeword was $\underline{x}$, given that the received message is $\underline{y}$.] Then repeat the derivation from eqn (11.31) to eqn (11.34). The final expression involves $\gamma = -\log B_Q$, where the Bhattacharya parameter is defined as $B_Q = \sum_y \sqrt{Q(y|1)Q(y|0)}$.

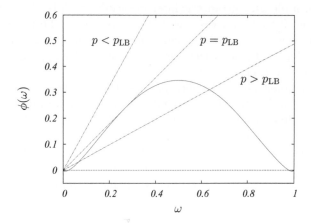

Fig. 11.6 Geometric construction yielding a lower bound on the threshold for reliable communication for the ensemble LDPC$_N(3, 6)$ used over the binary symmetric channel. In this case $p_{\mathrm{LB}} \approx 0.0438737$, shown by the line labelled '$p = p_{\mathrm{LB}}$'. The other two lines refer to $p = 0.01 < p_{\mathrm{LB}}$ and $p = 0.10 > p_{\mathrm{LB}}$.

Equation (11.34) shows that the block error probability depends on two factors: the first is the weight enumerator, and the second, $\exp(-\gamma w)$, is a channel-dependent term: as the weight of the codewords increases, their contribution is scaled down by an exponential factor because it is less likely that the received message $\underline{y}$ will be closer to a codeword of large weight than to the all-zero codeword.

So far, the discussion has been valid for any given code. Let us now consider the average over LDPC$_N(\Lambda, P)$ code ensembles. A direct averaging gives the bound

$$\mathrm{P_B} \equiv \mathbb{E}_{\mathfrak{C}} \mathrm{P_B}(\mathfrak{C}) \leq \sum_{w=1}^{N} \overline{\mathcal{N}}(w)\, \mathrm{e}^{-\gamma w} \doteq \exp\left\{ N \sup_{\omega \in]0,1]} [\phi(\omega) - \gamma \omega] \right\}. \qquad (11.36)$$

As such, this expression is useless, because the $\sup_\omega[\phi(\omega) - \gamma\omega]$, being greater than or equal to the value at $\omega = 0$, is always positive. However, if we restrict to ensembles with $l_{\min} \geq 3$, we know that, with a probability going to one in the large-N limit, there exists no codeword in the ω interval $]0, \omega_*[$. In such cases, the maximization over ω in eqn (11.36) can be performed in the interval $[\omega_*, 1]$ instead of $]0, 1]$. (By the Markov inequality, this is true whenever $N \sum_{w=1}^{N\omega_* - 1} \overline{\mathcal{N}}(w) \to 0$ as $N \to \infty$.) The bound becomes useful whenever the supremum $\sup_{\omega \in [\omega_*, 1]} [\phi(\omega) - \gamma\omega]$ is less than 0: then $\mathrm{P_B}$ vanishes in the large-N limit. We have thus obtained the following result.

Proposition 11.3 *Consider the average block error rate* $\mathrm{P_B}$ *for a random code in the ensemble* LDPC$_N(\Lambda, P)$, *with* $l_{\min} \geq 3$, *used over a channel* BSC(p), *with* $p < 1/2$. *Let* $\gamma \equiv -\log\sqrt{4p(1-p)}$, *and let* $\phi(\omega)$ *be the the weight enumerator exponent defined in eqn (11.4) ($\phi(\omega)$ can be computed using eqns (11.17)–(11.19)). If* $\phi(\omega) < \gamma\omega$ *for any* $\omega \in (0, 1]$ *such that* $\phi(\omega) \geq 0$, *then* $\mathrm{P_B} \to 0$ *in the large-block-length limit.*

This result has a pleasing geometric interpretation which is illustrated in Fig. 11.6. As p increases from 0 to $1/2$, γ decreases from $+\infty$ to 0. The condition $\phi(\omega) < \gamma\omega$ can be rephrased by saying that the weight enumerator exponent $\phi(\omega)$ must lie below a straight line of slope γ through the origin. We denote by p_{LB} the smallest value of p such that the line $\gamma\omega$ touches $\phi(\omega)$.

This geometric construction implies $p_{\mathrm{LB}} > 0$. Furthermore, for p large enough Shannon's theorem implies that P_{B} is bounded away from 0 for any non-vanishing rate $R > 0$. The **MAP threshold** p_{MAP} for the ensemble $\mathrm{LDPC}_N(\Lambda, P)$ can be defined as the largest (or, more precisely, the supremum) value of p such that $\lim_{N \to \infty} P_{\mathrm{B}} = 0$. This definition has a very concrete practical meaning: for any $p < p_{\mathrm{MAP}}$ one can communicate with an arbitrarily small error probability by using a code from the $\mathrm{LDPC}_N(\Lambda, P)$ ensemble, provided N is large enough. Proposition 11.3 then implies that

$$p_{\mathrm{MAP}} \ge p_{\mathrm{LB}} \,. \tag{11.37}$$

In general, one expects $\lim_{N \to \infty} P_{\mathrm{B}}$ to exist (and to be strictly positive) for $p > p_{\mathrm{MAP}}$. However, there exists no proof of this statement.

It is interesting to note that, at $p = p_{\mathrm{LB}}$, our upper bound on P_{B} is dominated by codewords of weight $w \approx N\tilde{\omega}$, where $\tilde{\omega} > 0$ is the value where $\phi(\omega) - \gamma\omega$ is maximum. This suggests that each time an error occurs, a finite fraction of the bits are decoded incorrectly and this fraction fluctuates little from transmission to transmission (or from code to code in the ensemble). The geometric construction also suggests the less obvious (but essentially correct) guess that this fraction jumps discontinuously from 0 to a finite value when p crosses the critical value p_{MAP}. Finally, $\tilde{\omega} > \omega_*$ strictly: dominant error events are not triggered by the closest codewords!

Exercise 11.7 Let us study the case $l_{\min} = 2$. Proposition 11.3 is no longer valid, but we can still apply eqn (11.36).

(a) Consider the ensemble $(2,4)$ whose weight enumerator exponent is plotted in Fig. 11.4, the small-weight behaviour being given by eqn (11.24). At small enough p, it is reasonable to assume that the block error rate is dominated by small weight codewords. Estimate P_{B} using eqn (11.36) under this assumption.

(b) Show that the above assumption breaks down for $p \ge p_{\mathrm{loc}}$, where $p_{\mathrm{loc}} \le 1/2$ solves the equation $3\sqrt{4p(1-p)} = 1$.

(c) Discuss the case of a general code ensemble with $l_{\min} = 2$, and $\phi(\omega)$ concave for $\omega \in [0,1]$. Draw a weight enumerator exponent $\phi(\omega)$ such that the assumption of dominance by low-weight codewords breaks down before p_{loc}. What do you expect for the average bit error rate P_{b} for $p < p_{\mathrm{loc}}$? And for $p > p_{\mathrm{loc}}$?

Exercise 11.8 Discuss the qualitative behaviour of the block error rate for cases where $l_{\min} = 1$.

11.3.2 Upper bound on the threshold

Let us consider as before communication over BSC(p), keeping for simplicity to regular codes LDPC$_N(l, k)$. Gallager has proved the following bound.

Theorem 11.4 *Let p_{MAP} be the threshold for reliable communication over a binary symmetric channel using codes from the ensemble LDPC$_N(l, k)$, with design rate $R_{\mathrm{des}} = 1 - k/l$. Then $p_{\mathrm{MAP}} \leq p_{\mathrm{UB}}$, where $p_{\mathrm{UB}} \leq 1/2$ is the solution of*

$$\mathcal{H}(p) = (1 - R_{\mathrm{des}})\mathcal{H}\left(\frac{1 - (1 - 2p)^k}{2}\right). \tag{11.38}$$

We shall not give a full proof of this result, but we shall show in this section a sequence of heuristic arguments which can be turned into a proof. The details can be found in the original literature.

Assume that the all-zero codeword $\underline{0}$ has been transmitted and that a noisy vector $\underline{y}$ has been received. The receiver will look for a vector $\underline{x}$ at a Hamming distance of about Np from $\underline{y}$, satisfying all the parity check equations. In other words, let us denote the **syndrome** by $\underline{z} = \mathbb{H}\underline{x}$, $\underline{z} \in \{0,1\}^M$ (here $\mathbb{H}$ is the parity check matrix and multiplication is performed modulo 2). This is a vector with M components. If $\underline{x}$ is a codeword, all parity checks are satisfied, and we have $\underline{z} = \underline{0}$. There is at least one vector $\underline{x}$ fulfilling the conditions $d(\underline{x}, \underline{y}) \approx Np$ and $\underline{z} = \underline{0}$: the transmitted codeword $\underline{0}$. Decoding is successful only if this is the unique such vector.

The number of vectors $\underline{x}$ whose Hamming distance from $\underline{y}$ is close to Np is approximatively $2^{N\mathcal{H}(p)}$. Let us now estimate the number of distinct syndromes $\underline{z} = \mathbb{H}\underline{x}$ when $\underline{x}$ is on the sphere $d(\underline{x}, \underline{y}) \approx Np$. Writing $\underline{x} = \underline{y} \oplus \underline{x}'$, this is equivalent to counting the number of distinct vectors $\underline{z}' = \mathbb{H}\underline{x}'$ when the weight of $\underline{x}'$ is about Np. It is convenient to think of $\underline{x}'$ as a vector of N i.i.d. Bernoulli variables of mean p: we are then interested in the number of distinct *typical* vectors $\underline{z}'$. Note that, since the code is regular, each entry z_i' is a Bernoulli variable of parameter

$$p_k = \sum_{n \text{ odd}}^{k} \binom{k}{n} p^n (1-p)^{k-n} = \frac{1 - (1 - 2p)^k}{2}. \tag{11.39}$$

If the bits of $\underline{z}'$ were independent, the number of typical vectors $\underline{z}'$ would be $2^{N(1-R_{\mathrm{des}})\mathcal{H}(p_k)}$ (the dimension of $\underline{z}'$ being $M = N(1 - R_{\mathrm{des}})$). It turns out that correlations between the bits decrease this number, so we can use the i.i.d. estimate to get an upper bound.

Let us now assume that for each $\underline{z}$ in this set, the number of reciprocal images (i.e. of vectors $\underline{x}$ such that $\underline{z} = \mathbb{H}\underline{x}$) is approximatively the same. If $2^{N\mathcal{H}(p)} \gg 2^{N(1-R_{\mathrm{des}})\mathcal{H}(p_k)}$, for each $\underline{z}$ there is an exponential number of vectors $\underline{x}$ such that $\underline{z} = \mathbb{H}\underline{x}$. This will be true, in particular, for $\underline{z} = \underline{0}$: the received message is therefore not uniquely decodable. In the alternative situation most of the vectors $\underline{z}$ correspond to (at most) a single $\underline{x}$. This will be the case for $\underline{z} = \underline{0}$: decoding can be successful.

11.3.3 Summary of the bounds

In Table 11.1 we consider a few regular LDPC$_N(\Lambda, P)$ ensembles over the channel BSC(p). We show the window of possible values of the noise threshold p_{MAP}, using

Table 11.1 Bounds on the threshold for reliable communication over BSC(p) using LDPC$_N(l,k)$ ensembles with MAP decoding. The fourth and fifth columns list the lower bound (LB) of Proposition 11.3 and the upper bound (UB) of Theorem 11.4. The sixth column lists an improved lower bound obtained by Gallager.

l	k	R_{des}	LB of Section 11.3.1	Gallager UB	Gallager LB	Shannon limit
3	4	1/4	0.1333161	0.2109164	0.2050273	0.2145018
3	5	2/5	0.0704762	0.1397479	0.1298318	0.1461024
3	6	1/2	0.0438737	0.1024544	0.0914755	0.1100279
4	6	1/3	0.1642459	0.1726268	0.1709876	0.1739524
5	10	1/2	0.0448857	0.1091612	0.1081884	0.1100279

the lower bound of Proposition 11.3 and the upper bound of Theorem 11.4. In most cases, the comparison is not satisfactory (the gap between the upper and lower bounds is close to a factor of 2). A much smaller uncertainty is achieved using an improved lower bound again derived by Gallager, based on a refinement of the arguments in the previous section. As we shall see in Chapter 15 by computing p_{MAP}, neither of the bounds is tight. On the other hand, they are sufficiently good to show that, for large k, l the MAP threshold of these ensembles is close to Shannon, capacity (although bounded away from it). Indeed, by studying the asymptotic behaviour of these bounds, one can show that the MAP threshold of the (k, l) ensemble converges to p_{Sh} as $k, l \to \infty$ with a fixed ratio l/k.

Exercise 11.9 Let p_{Sh} be the upper bound on p_{MAP} provided by the Shannon channel coding theorem. Explicitly, $p_{\text{Sh}} \leq 1/2$ is the solution of $\mathcal{H}(p) = 1-R$. Prove that if $R = R_{\text{des}}$ (as is the case with high probability for LDPC$_N(l,k)$ ensembles), then $p_{\text{UB}} < p_{\text{Sh}}$.

11.4 A simple decoder: Bit flipping

So far, we have analysed the behaviour of LDPC ensembles under the optimal (word MAP) decoding strategy. However, there is no known way of implementing this decoder with an efficient algorithm. The naive algorithm goes through each codeword $\underline{x}^{(\alpha)}$, $\alpha = 0, \ldots, 2^{NR} - 1$, and outputs the codeword of greatest likelihood $Q(\underline{y}|\underline{x}^{(\alpha)})$. However, this approach takes a time which grows exponentially with the block length N. For large N (which is the regime where the error rate becomes close to optimal), this is unpractical.

LDPC codes are interesting because there exist fast suboptimal decoding algorithms with performances close to the theoretical optimal performance, and therefore close to Shannon's limit. Here we show one example of a very simple decoding method, called the **bit flipping** algorithm. After transmission through a BSC, we have received the message $\underline{y}$, and we try to find the sent codeword $\underline{x}$ by use of the following algorithm.

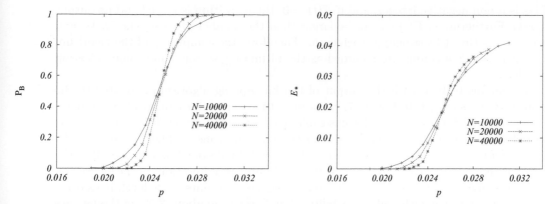

Fig. 11.7 Performance of the bit-flipping decoding algorithm on random codes from the $(5, 10)$ regular LDPC ensemble, used over the channel BSC(p). *Left*: block error rate. *Right*: residual number of unsatisfied parity checks after the algorithm has halted. The statistical error bars are smaller than the symbols.

BIT-FLIPPING DECODER (received message $\underline{y}$)

1:   Set $\underline{x}(0) = \underline{y}$.
2:   **for** $t = 1, \ldots, N$:
3:       find a bit belonging to more unsatisfied than satisfied parity checks;
4:       if such a bit exists, flip it: $x_i(t + 1) = x_i(t) \oplus 1$,
           and keep the other bits: $x_j(t + 1) = x_j(t)$ for all $j \neq i$;
5:       if there is no such bit, **return** $\underline{x}(t)$ and halt.

The bit to be flipped is usually chosen uniformly at random from the bits satisfying the condition at step 3. However, this is irrelevant in the analysis below.

Exercise 11.10 Consider a code from the (l, k) regular LDPC ensemble (with $l \geq 3$). Assume that the received message differs from the transmitted message in only one position. Show that the bit-flipping algorithm always corrects such an error.

Exercise 11.11 Assume now that the channel has introduced two errors. Draw the factor graph of a regular (l, k) code for which the bit-flipping algorithm is unable to recover from such an error event. What can you say of the probability of this type of graph in the ensemble?

In order to monitor the bit-flipping algorithm, it is useful to introduce the 'energy.'

$$E(t) \equiv \text{Number of parity check equations not satisfied by } \underline{x}(t). \tag{11.40}$$

This is a non-negative integer, and if $E(t) = 0$, the algorithm is halted and its output is $\underline{x}(t)$. Furthermore, $E(t)$ cannot be larger than the number of parity checks M, and it decreases (by at least one) at each cycle. Therefore, the complexity of the algorithm is $O(N)$ (this is a commonly regarded as the ultimate goal for many communication problems).

It remains to be seen if the output of the bit-flipping algorithm is related to the transmitted codeword. In Fig. 11.7, we present the results of a numerical experiment. We considered the $(5, 10)$ regular ensemble and generated about 1000 random code and channel realizations for each value of the noise in a mesh. Then we applied the above algorithm and plotted the fraction of successfully decoded blocks, as well as the residual energy $E_* = E(t_*)$, where t_* is the total number of iterations of the algorithm. The data suggests that bit-flipping is able to overcome a finite noise level: it recovers the original message with high probability when fewer than about 2.5% of the bits are corrupted by the channel. Furthermore, the curves for $P_{\mathrm{B}}^{\mathrm{bf}}$ under bit-flipping decoding become steeper and steeper as the size of the system is increased. It is natural to conjecture that asymptotically, a phase transition takes place at a well-defined noise level p_{bf}: $P_{\mathrm{B}}^{\mathrm{bf}} \rightarrow 0$ for $p < p_{\mathrm{bf}}$ and $P_{\mathrm{B}}^{\mathrm{bf}} \rightarrow 1$ for $p > p_{\mathrm{bf}}$. Numerically, $p_{\mathrm{bf}} = 0.025 \pm 0.005$.

This threshold can be compared with that for MAP decoding. The results in Table 11.1 imply $0.108188 \leq p_{\mathrm{MAP}} \leq 0.109161$ for the $(5, 10)$ ensemble. Bit-flipping is significantly suboptimal, but is still surprisingly good, given the extreme simplicity of the algorithm.

Can we provide any *guarantee* on the performance of the bit-flipping decoder? One possible approach consists in using the expansion properties of the underlying factor graph. Consider a graph from the (l, k) ensemble. We say that it is an (ε, δ) **expander** if, for any set U of variable nodes such that $|U| \leq N\varepsilon$, the set $|D|$ of neighbouring check nodes has a size $|D| \geq \delta|U|$. Roughly speaking, if the factor graph is an expander with a large **expansion constant** δ, any small set of corrupted bits induces a large number of unsatisfied parity checks. The bit-flipping algorithm can exploit these checks to successfully correct the errors.

It turns out that random graphs are very good expanders. This can be understood as follows. Consider a fixed subset U. As long as U is small, the subgraph induced by U and the neighbouring factor nodes D is a tree with high probability. If this is the case, elementary counting shows that $|D| = (l-1)|U| + 1$. This would suggest that one can achieve an expansion factor of $l - 1$ (or close to it), for small enough ε. Of course, this argument has several flaws. First of all, the subgraph induced by U is a tree only if U has a sublinear size, but we are interested in all subsets U with $|U| \leq \varepsilon N$ for some fixed N. Then, while most of the small subsets U are trees, we need to be sure that *all* subsets expand well. Nevertheless, one can prove that the heuristic expansion factor is essentially correct.

Proposition 11.5 *Consider a random factor graph $\mathcal{F}$ from the (l, k) ensemble. Then, for any $\delta < l - 1$, there exists a constant $\varepsilon = \varepsilon(\delta; l, k) > 0$ such that $\mathcal{F}$ is an (ε, δ) expander with probability approaching 1 as $N \rightarrow \infty$.*

In particular, this implies that, for $l \geq 5$, a random (l, k) regular factor graph is, with high probability an $(\varepsilon, \frac{3}{4}l)$ expander. In fact, this is enough to ensure that the code will perform well at a low noise level.

Theorem 11.6 *Consider a regular (l, k) LDPC code $\mathfrak{C}$, and assume that the corresponding factor graph is an $(\varepsilon, \frac{3}{4} l)$ expander. Then, the bit-flipping algorithm is able to correct any pattern of fewer than $N\varepsilon/2$ errors produced by a binary symmetric channel. In particular, $\mathrm{P}_{\mathrm{B}}(\mathfrak{C}) \to 0$ for communication over $\mathrm{BSC}(p)$ with $p < \varepsilon/2$.*

Proof As usual, we assume the channel input to be the all-zeros codeword $\underline{0}$. We denote by $w = w(t)$ the weight of $\underline{x}(t)$ (the current configuration of the bit-flipping algorithm), and by $E = E(t)$ the number of unsatisfied parity checks, as in eqn (11.40). Finally, we denote by F the number of *satisfied* parity checks among those which are neighbours of at least one corrupted bit in $\underline{x}(t)$ (a bit is 'corrupted' if it takes the value 1).

Assume, first, that $0 < w(t) \leq N\varepsilon$ at some time t. Because of the expansion property of the factor graph, we have $E + F > \frac{3}{4} lw$. On the other hand, every unsatisfied parity check is a neighbour of at least one corrupted bit, and every satisfied check which is a neighbour of a corrupted bit must involve at least two of them. Therefore $E + 2F \leq lw$. Eliminating F from the above inequalities, we deduce that $E(t) > \frac{1}{2} lw(t)$. Let $E_i(t)$ be the number of unsatisfied checks involving the bit x_i. Then,

$$\sum_{i:x_i(t)=1} E_i(t) \geq E(t) > \frac{1}{2} lw(t). \tag{11.41}$$

Therefore, there must be at least one bit that has more unsatisfied than satisfied neighbours, and the algorithm does not halt.

Let us now suppose that we start the algorithm with $w(0) \leq N\varepsilon/2$. It must halt at some time t_*, either with $E(t_*) = w(t_*) = 0$ (and therefore decoding is successful) or with $w(t_*) \geq N\varepsilon$. In the second case, as the weight of $\underline{x}(t)$ changes by one at each step, we have $w(t_*) = N\varepsilon$. The above inequalities imply $E(t_*) > Nl\varepsilon/2$ and $E(0) \leq lw(0) \leq Nl\varepsilon/2$. This contradicts the fact that $E(t)$ is a strictly decreasing function of t. Therefore the algorithm, when started with $w(0) \leq N\varepsilon/2$, ends up in the state $w = 0$, $E = 0$. $\square$

The approach based on expansion of the graph has the virtue of pointing out one important mechanism for the good performance of random LDPC codes, namely the local tree-like structure of the factor graph. It also provides explicit lower bounds on the critical noise level p_{bf} for bit-flipping. However, these bounds turn out to be quite pessimistic. For instance, in the case of the $(5, 10)$ ensemble, it has been proved that a typical factor graph is an $(\varepsilon, \frac{3}{4}l) = (\varepsilon, \frac{15}{4})$ expander for $\varepsilon < \varepsilon_* \approx 10^{-12}$. On the other hand, numerical simulations (see Fig. 11.7) show that the bit-flipping algorithm performs well up to noise levels much larger than $\varepsilon_*/2$.

Notes

Modern (post-Cook's Theorem) complexity theory was first applied to coding by Berlekamp *et al.* (1978), who showed that maximum-likelihood decoding of linear codes is NP-hard.

LDPC codes were first introduced by Gallager in his PhD thesis (Gallager, 1963; Gallager, 1962), which is in fact older than these complexity results. An excellent, detailed account of modern developments is provided by Richardson and Urbanke (2008).

Gallager's proposal did not receive enough consideration at the time. One possible explanation is the lack of computational power for simulating large codes in the 1960s. The rediscovery of LDPC codes in the 1990s (MacKay, 1999) was (at least in part) a consequence of the invention of turbo codes by Berrou and Glavieux (1996). Both of these classes of codes were soon recognized to be prototypes of a larger family: codes on sparse graphs.

The major technical advance after this rediscovery was the introduction of irregular ensembles (Luby *et al.* 1997; Luby *et al.* 1998). There exists no formal proof of the 'equivalence' (whatever this means) of the various possible definitions of LDPC ensembles in the large-block-length limit. But, as we shall see in Chapter 15, the main property that enters in the analysis of LDPC ensembles is the local tree-like structure of the factor graph described in Section 9.5.1; and this property is rather robust with respect to a change of the ensemble.

Gallager (1963) was the first to compute the expected weight enumerator for regular ensembles, and to use it in order to bound the threshold for reliable communication. General ensembles were considered by Litsyn and Shevelev (2003), Burshtein and Miller (2004), and Di *et al.* (2006). It turns out that the expected weight enumerator coincides with the typical (most likely) weight enumerator to leading exponential order for regular ensembles (in statistical-physics jargon, the annealed computation coincides with the quenched one). This is not the case for irregular ensembles, as pointed out by Di *et al.* (2004).

Proposition 11.2 has, essentially, been known since Gallager (1963). The formulation quoted here is from Méasson *et al.* (2008). That paper contains some examples of 'exotic' LDPC ensembles such that the maximum of the expected weight enumerator is at a weight $w = N\omega_*$, with $\omega_* \neq 1/2$.

A proof of the upper bound in Theorem 11.4 can be found in Gallager (1963). For some recent refinements, see Burshtein *et al.* (2002).

Bit-flipping algorithms played an important role in the revival of LDPC codes, especially following the work of Sipser and Spielman (1996). These authors focused on explicit code construction based on expander graphs. They also provided bounds on the expansion of random $LDPC_N(l, k)$ codes. The lower bound on the expansion mentioned in Section 11.4 is taken from Richardson and Urbanke (2008).

12
Spin glasses

We have already encountered several examples of spin glasses in Chapters 2 and 8. Like most problems in equilibrium statistical physics, they can be formulated in the general framework of factor graphs. Spin glasses are disordered systems, whose magnetic properties are dominated by randomly placed impurities. The theory aims at describing the behaviour of a typical sample of such materials. This motivates the definition and study of spin glass ensembles.

In this chapter, we shall explore the glass phase of these models. It is useful to have a good understanding of glass phases, as we shall see them appearing in various problems from optimization and coding theory. In general, the occurrence of a glass phase is described physically in terms of a dramatic slowdown in a dynamical relaxation process. Here, we shall focus instead on purely static characterizations of glass phases, which can be applied to a broad class of problems. The focus of our presentation is on 'mean-field models', for at least two reasons: (*i*) a deep mathematical theory (still under developement) can provide a precise understanding of their behaviour; (*ii*) the ensembles of combinatorial optimization and coding problems to be considered in the following fall naturally into this class. We shall discuss the two types of spin glass transition that have been encountered in such models.

In contrast to these 'soluble' cases, it must be stressed that very little is known (let alone proven) for realistic models of real spin glass materials. Even the existence of a spin glass phase has not been established rigorously in the latter case.

We first discuss, in Section 12.1, how Ising models and their generalizations can be formulated in terms of factor graphs, and introduce several ensembles of these models. Frustration is a crucial feature of spin glasses; in Section 12.2, we discuss it in conjunction with gauge transformations. This section also explains how to derive some exact results solely with the use of gauge transformations. Section 12.3 describes the spin glass phase and the main approaches to its characterization. Finally, the phase diagram of a spin glass model with several glassy phases is traced in Section 12.4.

12.1 Spin glasses and factor graphs

12.1.1 Generalized Ising models

Let us recall the main ingredients of magnetic systems with interacting Ising spins. The variables are N Ising spins $\underline{\sigma} = \{\sigma_1, \ldots, \sigma_N\}$, taking values in $\{+1, -1\}$. These are jointly distributed according to Boltzmann's law for the energy function

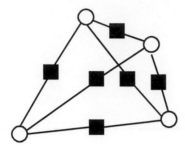 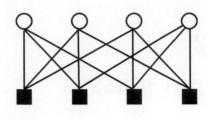

Fig. 12.1 Factor graph representation of the SK model with $N = 4$ (*left*), and of the fully connected 3-spin model with $N = 4$ (*right*). The squares denote the interactions between the spins.

$$E(\underline{\sigma}) = -\sum_{p=1}^{p_{\max}} \sum_{i_1 < \cdots < i_p} J_{i_1 \ldots i_p} \sigma_{i_1} \cdots \sigma_{i_p} \, . \tag{12.1}$$

The index p gives the order of the interaction. One-body terms ($p = 1$) are also referred to as external-field interactions, and will be sometimes written as $-B_i \sigma_i$. If $J_{i_1 \ldots i_p} \geq 0$ for any $i_1 \ldots i_p$, and $p \geq 2$, the model is said to be a ferromagnet. If $J_{i_1 \ldots i_p} \leq 0$, it is an **antiferromagnet**. Finally, if both positive and negative couplings are present for $p \geq 2$, the model is a spin glass.

The energy function can be rewritten as $E(\underline{\sigma}) = \sum_a E_a(\underline{\sigma}_{\partial a})$, where $E_a(\underline{\sigma}_{\partial a}) \equiv -J_a \sigma_{i_1^a} \cdots \sigma_{i_{p_a}^a}$. Each interaction term a involves the spins contained in a subset $\underline{\sigma}_{\partial a} = \{\sigma_{i_1^a}, \ldots, \sigma_{i_{p_a}^a}\}$, of size p_a. We then introduce a factor graph, in which each interaction term is represented by a square vertex and each spin is represented by a circular vertex. Edges are drawn between an interaction vertex a and a variable vertex i whenever the spin σ_i appears in $\underline{\sigma}_{\partial a}$. We have already seen, in Fig. 9.7, the factor graph of a two-dimensional Edwards-Anderson spin glass, where the energy contains terms with $p = 1$ and $p = 2$. Figure 12.1 shows the factor graphs of some small samples of the SK model in zero magnetic field ($p = 2$ only), and the '3-spin model', in which terms with $p = 3$ appear in the energy function.

The energy function (12.1) can be straightforwardly interpreted as a model of a magnetic system. We have used, so far, the language inherited from this application: the spins $\{\sigma_i\}$ are 'rotational' degrees of freedom associated with magnetic particles, their average is the magnetization, etc. In this context, the most relevant interaction between distinct degrees of freedom is pairwise: $-J_{ij}\sigma_i\sigma_j$.

Higher-order terms naturally arise in other applications, one of the simplest being lattice particle systems. These are used to model liquid-to-gas, liquid-to-solid, and similar phase transitions. One normally starts by considering some base graph $\mathcal{G}$ with N vertices, which is often taken to be a portion of $\mathbb{Z}^d$ (to model a real physical system, the dimension of choice is of course $d = 3$). Each vertex in the graph can be either occupied by a particle or empty. The particles are assumed to be indistinguishable from each other, and a configuration is characterized by occupation variables $n_i = \{0, 1\}$. The energy is a function $E(\underline{n})$ of the occupancies $\underline{n} = \{n_1, \ldots, n_N\}$, which takes into account local interactions between neighbouring particles. Usually, it can be rewritten in the form (12.1), using the mapping $\sigma_i = 1 - 2n_i$. We give a few examples in the exercises below.

Exercise 12.1 Consider an empty box which is free to exchange particles with a reservoir, and assume that particles do not interact with each other (except for the fact that they cannot be superimposed). This can be modelled by taking $\mathcal{G}$ to be a cube of side L in $\mathbb{Z}^d$, and specifying that each particle in the system contributes a constant amount $-\mu$ to the energy: $E(\underline{n}) = -\mu \sum_i n_i$. This is a model for what is usually called an **ideal gas**.

Compute the partition function. Rewrite the energy function in terms of spin variables and draw the corresponding factor graph.

Exercise 12.2 In the same problem, imagine that particles attract each other at short distances: whenever two neighbouring vertices i and j are occupied, the system gains an energy $-\epsilon$. This is a model for the liquid–gas phase transition.

Write the corresponding energy function in terms both of occupancy variables $\{n_i\}$ and spin variables $\{\sigma_i\}$. Draw the corresponding factor graph. Based on the phase diagram of the Ising model (see Section 2.5) discuss the behaviour of this particle system. What physical quantity corresponds to the magnetization of the Ising model?

Exercise 12.3 In some materials, molecules cannot be packed in a regular lattice at high density, and this may result in an amorphous solid material. In order to model this phenomenon, one can modify the energy function of the previous exercises as follows. Each time that a particle (i.e. an occupied vertex) is surrounded by more than k other particles in the neighbouring vertices, a penalty $+\delta$ is added to the energy.

Write the corresponding energy function (in terms both of $\{n_i\}$ and $\{\sigma_i\}$), and draw the factor graph associated with it.

12.1.2 Spin glass ensembles

A sample (or an instance) of a spin glass is defined by:

- its factor graph, which specifies the subsets of spins which interact;
- the value of the coupling constant $J_a \in \mathbb{R}$ for each function node in the factor graph.

An ensemble is defined by a probability distribution over the space of samples. In all of the cases which we shall consider, the couplings are assumed to be i.i.d. random variables, independent of the factor graph. The most studied cases are those with Gaussian J_a's or those with J_a taking values $\{+1, -1\}$ with equal probability (in the jargon, this is called the $\pm J$ model). More generally, we shall denote the pdf of J_a by $\mathcal{P}(J)$.

One can distinguish two large families of spin glass ensembles which have attracted the attention of physicists: 'realistic' and 'mean-field' ones. While in the first case the focus is on modelling actual physical systems, mean-field models have proved to be analytically tractable, and have revealed a rich mathematical structure. The relation between these two classes is a fascinating open problem that we shall not try to address.

Physical spin glasses are mostly three-dimensional systems, but in some cases they can be two-dimensional. The main feature of realistic ensembles is that they retain this geometric structure: a position x in d dimensions can be associated with each spin. The interaction strength (the absolute value of the coupling J) decays rapidly with the distance between the positions of the associated spins. The Edwards–Anderson model is the most studied example in this family. Here, the spins are located on the vertices of a d-dimensional hypercubic lattice. Neighbouring spins interact, through two-body interactions (i.e. $p_{\max} = 2$ in eqn (12.1)). The corresponding factor graph is not random, as can be seen for the two-dimensional example of Fig. 9.7. The only source of disorder is the random couplings J_{ij}, distributed according to $\mathcal{P}(J)$. It is customary to add a uniform magnetic field B, which is written as a $p = 1$ term with $J_i = B$. Very little is known about these models when $d \geq 2$, and most of our knowledge comes from numerical simulations. These suggest the existence of a glass phase when $d \geq 3$, but this has not been proven yet.

There exists no general mathematical definition of a mean-field model. From a technical point of view, mean-field models admit exact expressions for the asymptotic ($N \to \infty$) free-energy density, as the optimum of some sort of large-deviation rate function. The distinctive feature that allows a solution in this form is the lack of any finite-dimensional geometrical structure.

The p-spin glass model discussed in Section 8.2 (and, in particular, the $p = 2$ case, which is the SK model) is a mean-field model. In this case also, the factor graph is non-random, and the disorder enters only into the random couplings. The factor graph is a regular bipartite graph. It contains $\binom{N}{p}$ function nodes, one for each p-tuple of spins; for this reason, it is called **fully connected**. Each function node has degree p, and each variable node has degree $\binom{N-1}{p-1}$. Since the degree diverges with N, the coupling distribution $\mathcal{P}(J)$ must be scaled appropriately with N (see eqn (8.25)).

Fully connected models are among the best understood in the mean-field family. They can be studied either via the replica method, as in Chapter 8, or via the cavity method, which we shall develop in the following chapters. Some of the predictions from these two heuristic approaches have been confirmed rigorously.

One unrealistic feature of fully connected models is that each spin interacts with a diverging number of other spins (the degree of a spin variable in the factor graph diverges in the thermodynamic limit). In order to eliminate this feature, one can study spin glass models on Erdös–Rényi random graphs with a finite average degree. Spins are associated with vertices in the graph, and $p = 2$ interactions (with couplings that are i.i.d. random variables drawn from $\mathcal{P}(J)$) are associated with edges in the graph. The generalization to p-spin interactions is immediate. The corresponding spin glass models are called **diluted spin glasses** (DSGd). We define the ensemble $\mathsf{DSG}_N(p, M, \mathcal{P})$ as follows:

- Generate a factor graph from the ensemble $\mathbb{G}_N(p, M)$ (the graph therefore has M function nodes, all of degree p).

- For every function node a in the graph, connecting spins $i_1^a, \dots, i_p^a$, draw a random coupling $J_{i_1^a, \dots, i_p^a}$ from the distribution $\mathcal{P}(J)$, and introduce an energy term

$$E_a(\underline{\sigma}_{\partial a}) = -J_{i_1^a, \dots, i_p^a} \sigma_{i_1^a} \cdots \sigma_{i_p^a} \ . \tag{12.2}$$

- The final energy is $E(\underline{\sigma}) = \sum_{a=1}^{M} E_a(\underline{\sigma}_{\partial a})$.

The thermodynamic limit is taken by letting $N \to \infty$ at fixed $\alpha = M/N$.

As in the case of random graphs, one can introduce some variants of this definition. In the ensemble $\mathsf{DSG}(p, \alpha, \mathcal{P})$, the factor graph is drawn from $\mathbb{G}_N(p, \alpha)$: each p-tuple of variable nodes is connected by a function node independently, with probability $\alpha/\binom{N}{p}$. As we shall see, the ensembles $\mathsf{DSG}_N(p, M, \mathcal{P})$ and $\mathsf{DSG}_N(p, \alpha, \mathcal{P})$ have the same free energy per spin in the thermodynamic limit, and many of their thermodynamic properties are identical. One basic reason for this phenomenon is that any finite neighbourhood of a random site i has the same asymptotic distribution in the two ensembles.

Obviously, any ensemble of random graphs can be turned into an ensemble of spin glasses by the same procedure. Some of these ensembles have been considered in the literature. Mimicking the notation defined in Section 9.2, we shall introduce general diluted spin glasses with constrained degree profiles, denoted by $\mathsf{DSG}_N(\Lambda, P, \mathcal{P})$, which indicates the ensemble derived from the random graphs in $\mathbb{D}_N(\Lambda, P)$.

Diluted spin glasses are a very interesting class of models, which are intimately related to sparse graph codes and to random satisfiability problems, among other things. Our understanding of DSGs is at an intermediate level between fully connected models and realistic models. It is believed that both the replica and the cavity method should allow one to compute many thermodynamic properties exactly for most of these models. However, the number of such exact results is still rather small, and only a fraction of these have been proved rigorously.

12.2 Spin glasses: Constraints and frustration

Spin glasses at zero temperature can be seen as constraint satisfaction problems. Consider, for instance, a model with two-body interactions

$$E(\underline{\sigma}) = - \sum_{(i,j) \in \mathcal{E}} J_{ij} \sigma_i \sigma_j \ , \tag{12.3}$$

where the sum is over the edge set $\mathcal{E}$ of a graph $\mathcal{G}$ (the corresponding factor graph is obtained by associating a function node a with each edge $(ij) \in \mathcal{E}$). At zero temperature the Boltzmann distribution is concentrated on those configurations which minimize the energy. Each edge (i, j) therefore induces a constraint between the spins σ_i and σ_j: they should be aligned if $J_{ij} > 0$, or anti-aligned if $J_{ij} < 0$. If there exists a spin configuration which satisfies all the constraints, the ground state energy is $E_{\mathrm{gs}} = -\sum_{(i,j) \in \mathcal{E}} |J_{ij}|$ and the sample is said to be **unfrustrated** (see Section 2.6). Otherwise, it is **frustrated**. In this case one defines a ground state as a spin configuration which violates the minimum possible number of constraints.

As shown in the exercise below, there are several methods to check whether an energy function of the form (12.3) is frustrated.

Exercise 12.4 We define a 'plaquette' of a graph as a circuit $i_1, i_2, \ldots, i_L, i_1$ such that no shortcut exists: $\forall r, s \in \{1, \ldots, L\}$, the edge (i_r, i_s) is absent from the graph whenever $r \neq s \pm 1 \pmod{L}$. Show that a spin glass sample is unfrustrated if and only if the product of the couplings along every plaquette of the graph is positive.

Exercise 12.5 Consider a spin glass of the form (12.3), and define the Boolean variables $x_i = (1 - \sigma_i)/2$. Show that the spin glass constraint satisfaction problem can be transformed into an instance of the 2-satisfiability problem.
    [Hint: Write the constraint $J_{ij}\sigma_i\sigma_j > 0$ in Boolean form using x_i and x_j.]

Since 2-SAT is in P, and because of the equivalence demonstrated in the last exercise, one can check in polynomial time whether the energy function (12.3) is frustrated or not. This approach does not work when $p \geq 3$, because K-SAT is NP-complete for $K \geq 3$. However, as we shall see in Chapter 18, checking whether a spin glass energy function is frustrated remains a polynomial problem for any p.

12.2.1 Gauge transformation

When a spin glass sample has some negative couplings but is unfrustrated, one is in fact dealing with a 'disguised ferromagnet'. By this we mean that, through a change of variables, the problem of computing the partition function for such a system can be reduced to the problem of computing the partition function of a ferromagnet. Indeed, by assumption, there exists a ground state spin configuration $\sigma_i^* \in \{\pm 1\}$ such that, $\forall (i,j) \in \mathcal{E}$, $J_{ij}\sigma_i^*\sigma_j^* > 0$. Given a configuration $\underline{\sigma}$, we define $\tau_i = \sigma_i\sigma_i^*$, and notice that $\tau_i \in \{+1, -1\}$. Then the energy of the configuration is $E(\underline{\sigma}) = E_*(\underline{\tau}) \equiv -\sum_{(i,j) \in \mathcal{E}} |J_{ij}|\tau_i\tau_j$. Obviously, the partition function for a system with the energy function $E_*(\cdot)$ (which is a ferromagnet, since $|J_{ij}| > 0$) is the same as for the original system.

This change of variables is an example of a **gauge transformation**. In general, such a transformation amounts to changing all spins and, simultaneously, all couplings according to:

$$\sigma_i \mapsto \sigma_i^{(s)} = \sigma_i s_i , \qquad J_{ij} \mapsto J_{ij}^{(s)} = J_{ij} s_i s_j , \tag{12.4}$$

where $\underline{s} = \{s_1, \ldots, s_N\}$ is an arbitrary configuration in $\{-1, 1\}^N$. If we regard the partition function as a function of the coupling constants $\underline{J} = \{J_{ij} : (ij) \in \mathcal{E}\}$,

$$Z[\underline{J}] = \sum_{\{\sigma_i\}} \exp\left(\beta \sum_{(ij) \in \mathcal{E}} J_{ij}\sigma_i\sigma_j\right) , \tag{12.5}$$

then we have

$$Z[\underline{J}] = Z[\underline{J}^{(s)}] . \tag{12.6}$$

The system with coupling constants $\underline{J}^{(s)}$ is sometimes called the 'gauge-transformed system'.

Exercise 12.6 Consider adding a uniform magnetic field (i.e. a linear term of the form $-B \sum_i \sigma_i$) to the energy function (12.3), and apply a generic gauge transformation to such a system. How must the uniform magnetic field be changed in order to keep the partition function unchanged? Is the new magnetic-field term still uniform?

Exercise 12.7 Generalize the above discussion of frustration and gauge transformations to the $\pm J$ 3-spin glass (i.e. a model of the type (12.1) involving only terms with $p = 3$).

12.2.2 The Nishimori temperature...

In many spin glass ensembles, there exists a special temperature (called the **Nishimori temperature**) at which some thermodynamic quantities, such as the internal energy, can be computed exactly. This nice property is particularly useful in the study of inference problems (a particular instance being symbol MAP decoding of error-correcting codes), since the Nishimori temperature arises naturally in these contexts. There are in fact two ways of deriving it: either as an application of gauge transformations (this is how it was discovered in physics), or by mapping the system onto an inference problem.

Let us begin by taking the first point of view. Consider, for the sake of simplicity, the model (12.3). The underlying graph $\mathcal{G} = (\mathcal{V}, \mathcal{E})$ can be arbitrary, but we assume that the couplings J_{ij} on all the edges $(ij) \in \mathcal{E}$ are i.i.d. random variables taking values $J_{ij} = +1$ with probability $1 - p$ and $J_{ij} = -1$ with probability p. We denote by $\mathbb{E}$ the expectation with respect to this distribution.

The Nishimori temperature for this system is given by $T_{\mathrm{N}} = 1/\beta_{\mathrm{N}}$, where $\beta_{\mathrm{N}} = \frac{1}{2} \log \left((1 - p)/p \right)$. It is chosen in such a way that the coupling-constant distribution $\mathcal{P}(J)$ satisfies the condition

$$\mathcal{P}(J) = \mathrm{e}^{-2\beta_{\mathrm{N}} J} \, \mathcal{P}(-J) \,. \tag{12.7}$$

An equivalent way of stating the same condition is to write

$$\mathcal{P}(J) = \frac{\mathrm{e}^{\beta_{\mathrm{N}} J}}{2 \cosh(\beta_{\mathrm{N}} J)} \, \mathcal{Q}(|J|) \,, \tag{12.8}$$

where $\mathcal{Q}(|J|)$ denotes the distribution of the absolute values of the couplings (in the present example, this is a Dirac delta on $|J| = 1$).

Let us now turn to the computation of the average, over the distribution of couplings, of the internal energy[1] $U \equiv \mathbb{E}\langle E(\underline{\sigma}) \rangle$. More explicitly,

---

[1]The same symbol U was used in Chapter 2 to denote the internal energy $\langle E(\underline{\sigma}) \rangle$ (instead of its average). There should be no confusion with the present use.

$$U = \mathbb{E}\left\{\frac{1}{Z[\underline{J}]} \sum_{\underline{\sigma}} \left(-\sum_{(kl)} J_{kl}\sigma_k\sigma_l\right) e^{\beta \sum_{(ij)} J_{ij}\sigma_i\sigma_j}\right\}. \tag{12.9}$$

In general, it is very difficult to compute U. It turns out, however, that at the Nishimori temperature, the gauge invariance allows an easy computation. The average internal energy U can be expressed as $U = \mathbb{E}\{Z_U[\underline{J}]/Z[\underline{J}]\}$, where $Z_U[\underline{J}] = -\sum_{\underline{\sigma}} \sum_{(kl)} J_{kl}\sigma_k\sigma_l \prod_{(ij)} e^{\beta_{\mathrm{N}} J_{ij}\sigma_i\sigma_j}$.

Let $\underline{s} \in \{-1, 1\}^N$. By an obvious generalization of eqn (12.6), we have $Z_U[\underline{J}^{(\underline{s})}] = Z_U[\underline{J}]$, and therefore

$$U = 2^{-N} \sum_{\underline{s}} \mathbb{E}\{Z_U[\underline{J}^{(\underline{s})}]/Z[\underline{J}^{(\underline{s})}]\}. \tag{12.10}$$

If the coupling constants J_{ij} are i.i.d. with the distribution (12.8), then the gauge-transformed constants $J'_{ij} = J^{(\underline{s})}_{ij}$ are equally independent but with a distribution

$$\mathcal{P}_{\underline{s}}(J_{ij}) = \frac{e^{\beta_{\mathrm{N}} J_{ij} s_i s_j}}{2\cosh\beta_{\mathrm{N}}}. \tag{12.11}$$

Equation (12.10) can therefore be written as $U = 2^{-N} \sum_{\underline{s}} \mathbb{E}_{\underline{s}}\{Z_U[\underline{J}]/Z[\underline{J}]\}$, where $\mathbb{E}_{\underline{s}}$ denotes the expectation with respect to the modified measure $\mathcal{P}_{\underline{s}}(J_{ij})$. Using eqn (12.11), and denoting by $\mathbb{E}_0$ the expectation with respect to the uniform measure over $J_{ij} \in \{\pm 1\}$, we get

$$U = 2^{-N} \sum_{\underline{s}} \mathbb{E}_0\left\{\prod_{(ij)} \frac{e^{\beta_{\mathrm{N}} J_{ij} s_i s_j}}{\cosh\beta_{\mathrm{N}}} \frac{Z_U[\underline{J}]}{Z[\underline{J}]}\right\} \tag{12.12}$$

$$= 2^{-N}(\cosh\beta_{\mathrm{N}})^{-|\mathcal{E}|} \mathbb{E}_0\left\{\sum_{\underline{s}} e^{\beta_{\mathrm{N}} \sum_{(ij)} J_{ij} s_i s_j} \frac{Z_U[\underline{J}]}{Z[\underline{J}]}\right\} \tag{12.13}$$

$$= 2^{-N}(\cosh\beta_{\mathrm{N}})^{-|\mathcal{E}|} \mathbb{E}_0\{Z_U[\underline{J}]\}. \tag{12.14}$$

It is easy to compute $\mathbb{E}_0 Z_U[\underline{J}] = -2^N(\cosh\beta_{\mathrm{N}})^{|\mathcal{E}|-1}\sinh\beta_{\mathrm{N}}$. This implies our final result for the average energy at the Nishimori temperature:

$$U = -|\mathcal{E}|\tanh(\beta_{\mathrm{N}}). \tag{12.15}$$

Note that this simple result holds for any choice of the underlying graph. Furthermore, it is easy to generalize it to other choices of the coupling distribution that satisfy eqn (12.8) and to models with multispin interactions of the form (12.1). An even wider generalization is treated below.

12.2.3   . . . and its relation to probability

The calculation of the internal energy in the previous subsection is straightforward but, in a sense, mysterious. It is hard to grasp what the fundamental reason is that makes things simpler at the Nishimori temperature. Here we discuss a more general derivation, in a slightly more abstract setting, which is related to the connection with inference problems mentioned above.

Consider the following process. A configuration $\underline{\sigma} \in \{\pm 1\}$ is chosen uniformly at random; we call the corresponding distribution $\mathbb{P}_0(\underline{\sigma})$. Next, a set of coupling constants $\underline{J} = \{J_a\}$ is chosen according to the conditional distribution

$$\mathbb{P}(\underline{J}|\underline{\sigma}) = \mathrm{e}^{-\beta E_{\underline{J}}(\underline{\sigma})} \, \mathbb{Q}_0(\underline{J}) \,. \tag{12.16}$$

Here $E_{\underline{J}}(\underline{\sigma})$ is an energy function with coupling constants $\underline{J}$, and $\mathbb{Q}_0(\underline{J})$ is some reference measure (which can be chosen in such a way that the resulting $\mathbb{P}(\underline{J}|\underline{\sigma})$ is normalized). This can be interpreted as a communication process. The information source produces the message $\underline{\sigma}$ uniformly at random, and the receiver observes the couplings $\underline{J}$.

The joint distribution of $\underline{J}$ and $\underline{\sigma}$ is $\mathbb{P}(\underline{J}, \underline{\sigma}) = \mathrm{e}^{-\beta E_{\underline{J}}(\underline{\sigma})} \, \mathbb{Q}_0(\underline{J}) \mathbb{P}_0(\underline{\sigma})$. We shall denote the expectation with respect to this joint distribution by 'Av' in order to distinguish it from the thermal average (that over the Boltzmann measure, denoted by $\langle \, . \, \rangle$) and from the quenched average over the couplings, denoted by $\mathbb{E}$.

We assume that this process possesses a gauge symmetry: this assumption defines the Nishimori temperature. By this we mean that, given $\underline{s} \in \{\pm 1\}^N$, there exists an invertible mapping $\underline{J} \to \underline{J}^{(\underline{s})}$ such that $\mathbb{Q}_0(\underline{J}^{(\underline{s})}) = \mathbb{Q}_0(\underline{J})$ and $E_{\underline{J}^{(\underline{s})}}(\underline{\sigma}^{(\underline{s})}) = E_{\underline{J}}(\underline{\sigma})$. It is then clear that the joint probability distribution of the coupling and the spins and the conditional distribution possess the same symmetry:

$$\mathbb{P}(\underline{\sigma}^{(\underline{s})}, \underline{J}^{(\underline{s})}) = \mathbb{P}(\underline{\sigma}, \underline{J}) \,, \qquad \mathbb{P}(\underline{J}^{(\underline{s})}|\underline{\sigma}^{(\underline{s})}) = \mathbb{P}(\underline{J}|\underline{\sigma}) \,. \tag{12.17}$$

Let us introduce the quantity

$$\mathcal{U}(\underline{J}) = \mathrm{Av}(E_{\underline{J}}(\underline{\sigma})|\underline{J}) = \sum_{\underline{\sigma}} \mathbb{P}(\underline{\sigma}|\underline{J}) E_{\underline{J}}(\underline{\sigma}) \,, \tag{12.18}$$

and denote $\sum_{\underline{J}} \mathbb{P}(\underline{J}|\underline{\sigma}_0) \mathcal{U}(\underline{J})$ by $U(\underline{\sigma}_0)$. This is nothing but the average internal energy for a disordered system with an energy function $E_{\underline{J}}(\underline{\sigma})$ and coupling distribution $\mathbb{P}(\underline{J}|\underline{\sigma}_0)$. For instance, if we take $\underline{\sigma}_0$ as the 'all-plus' configuration, $\mathbb{Q}_0(\underline{J})$ to be proportional to the uniform measure over $\{\pm 1\}^{\mathcal{E}}$, and $E_{\underline{J}}(\underline{\sigma})$ as given by eqn (12.3), then $U(\underline{\sigma}_0)$ is exactly the quantity U that we computed in the previous section.

Gauge invariance implies that $\mathcal{U}(\underline{J}) = \mathcal{U}(\underline{J}^{(\underline{s})})$ for any $\underline{s}$, and $U(\underline{\sigma}_0)$ does not depend upon $\underline{\sigma}_0$. We can therefore compute $U = U(\underline{\sigma}_0)$ by averaging over $\underline{\sigma}_0$. We obtain

$$\begin{aligned} U &= \sum_{\underline{\sigma}_0} \mathbb{P}_0(\underline{\sigma}_0) \sum_{\underline{J}} \mathbb{P}(\underline{J}|\underline{\sigma}_0) \sum_{\underline{\sigma}} \mathbb{P}(\underline{\sigma}|\underline{J}) E_{\underline{J}}(\underline{\sigma}) \\ &= \sum_{\underline{\sigma}, \underline{J}} \mathbb{P}(\underline{\sigma}, \underline{J}) E_{\underline{J}}(\underline{\sigma}) = \sum_{\underline{J}} \mathbb{P}(\underline{J}|\underline{\sigma}_0) E_{\underline{J}}(\underline{\sigma}) \,, \end{aligned} \tag{12.19}$$

where we have used gauge invariance once more in the last step. The final expression is generally easy to evaluate, since the coublings J_a are generically independent under $\mathbb{P}(\underline{J}|\underline{\sigma}_0)$ In particular, it is straightforward to recover eqn (12.15) for the case treated in the last subsection.

Exercise 12.8 Consider a spin glass model on an arbitrary graph, with an energy given by eqn (12.3), and i.i.d. random couplings on the edges, drawn from the distribution $\mathcal{P}(J) = \mathcal{P}_0(|J|)e^{aJ}$. Show that the Nishimori inverse temperature is $\beta_N = a$, and that the internal energy at this point is given by $U = -|\mathcal{E}| \sum_J \mathcal{P}_0(|J|) J \sinh(\beta_N J)$. In the case where $\mathcal{P}$ is a Gaussian distribution of mean J_0, show that $U = -|\mathcal{E}| J_0$.

12.3 What is a glass phase?

12.3.1 Spontaneous local magnetizations

In physics, a 'glass' is defined through its dynamical properties. For classical spin models such as the ones we are considering here, one can define several types of physically meaningful dynamics. For definiteness, we shall use the single-spin-flip Glauber dynamics defined in Section 4.5. The main features of our discussion should remain unchanged as long as we keep to local dynamics (i.e. a bounded number of spins is flipped at each step), which obey detailed balance.

Consider a system at equilibrium at time 0 (i.e. assume $\underline{\sigma}(0)$ to be distributed according to the Boltzmann distribution), and denote by $\langle \cdot \rangle_{\underline{\sigma}(0)}$ the expectation with respect to Glauber dynamics *conditional* on the initial configuration. Within a 'solid'[2] phase, spins are correlated with their initial values on long time scales:

$$\lim_{t \to \infty} \lim_{N \to \infty} \langle \sigma_i(t) \rangle_{\underline{\sigma}(0)} \equiv m_{i,\underline{\sigma}(0)} \neq \langle \sigma_i \rangle. \tag{12.20}$$

In other words, on arbitrary long but finite (in terms of the system size) time scales, the system converges to a 'quasi-equilibrium' state, which we shall call for brevity a 'quasi-state', with local magnetizations $m_{i,\underline{\sigma}(0)}$ depending on the initial condition.

The condition (12.20) is, for instance, satisfied by a $d \geq 2$ Ising ferromagnet in zero external field, at temperatures below the ferromagnetic phase transition. In this case we have either $m_{i,\underline{\sigma}(0)} = M(\beta)$ or $m_{i,\underline{\sigma}(0)} = -M(\beta)$, depending on the initial condition, where $M(\beta)$ is the spontaneous magnetization of the system. There are two quasi-states, invariant under translation and related by a simple symmetry transformation. If the different quasi-states are neither periodic nor related by any symmetry, one may speak of a glass phase.

It is also very important to characterize the glass phase at the level of equilibrium statistical mechanics, without introducing a specific dynamics. For the case of ferromagnets we have already seen the solution to this problem in Chapter 2. Let $\langle \cdot \rangle_B$ denote the expectation with respect to the Boltzmann measure for the energy function (12.1), after a uniform magnetic field has been added. We then define the two quasi-states by

$$m_{i,\pm} \equiv \lim_{B \to 0\pm} \lim_{N \to \infty} \langle \sigma_i \rangle_B. \tag{12.21}$$

A natural generalization to glasses consists in adding a small magnetic field which is not uniform. Let us add to the energy function (12.1) a term of the form $-\epsilon \sum_i s_i \sigma_i$,

---

[2]The name comes from the fact that in a solid, the preferred positions of the atoms are time-independent; for instance, in a crystal, they are the vertices of a periodic lattice.

where $\underline{s} \in \{\pm 1\}^N$ is an arbitrary configuration. We denote by $\langle \cdot \rangle_{\epsilon,\underline{s}}$ the expectation with respect to the corresponding Boltzmann distribution, and let

$$m_{i,\underline{s}} \equiv \lim_{\epsilon \to 0\pm} \lim_{N \to \infty} \langle \sigma_i \rangle_{\epsilon,\underline{s}} . \qquad (12.22)$$

The **Edwards–Anderson order parameter**, defined as

$$q_{\mathrm{EA}} \equiv \lim_{\epsilon \to 0\pm} \lim_{N \to \infty} \frac{1}{N} \sum_i \langle \sigma_i \rangle_{\epsilon,\underline{s}}^2 , \qquad (12.23)$$

where $\underline{s}$ is an equilibrium configuration sampled from the Boltzmann distribution, then signals the onset of the spin glass phase.

The careful reader will notice that eqn (12.20) is not really completely defined. How should we take the $N \to \infty$ limit? Do the limits exist, and how does the result depend on $\underline{s}$? These are subtle questions. They underlie the problem of defining properly the pure states (extremal Gibbs states) in disordered systems. We shall come back to these issues in Chapter 22.

An extremely fruitful idea is, instead, to study glassy phases by comparing several equilibrated (i.e. drawn from the Boltzmann distribution) configurations of the system: one can then use one configuration to define the direction of the polarizing field, as we just did for the Edwards–Anderson order parameter. Remarkably, this idea underlies the formal manipulations in the replica method.

We shall explain below in greater detail two distinct criteria, based on this idea, which can be used to define a glass phase. Before this, let us discuss a criterion for stability of the high-temperature phase.

12.3.2 The spin glass susceptibility

We take a spin glass sample, with energy (12.1), and add to it a local magnetic field on site i, B_i. The magnetic susceptibility of spin j with respect to the field B_i is defined as the rate of change of $m_j = \langle \sigma_j \rangle_{B_i}$ with respect to B_i,

$$\chi_{ji} \equiv \left. \frac{\mathrm{d}m_j}{\mathrm{d}B_i} \right|_{B_i=0} = \beta(\langle \sigma_i \sigma_j \rangle - \langle \sigma_i \rangle \langle \sigma_j \rangle) , \qquad (12.24)$$

where we have used the fluctuation–dissipation relation (2.44).

The uniform (ferromagnetic) susceptibility defined in Section 2.5.1 gives the rate of change of the average magnetization with respect to an infinitesimal global uniform field, i.e. $\chi = (1/N) \sum_{i,j} \chi_{ji}$. Consider a ferromagnetic Ising model, as introduced in Section 2.5. Within the ferromagnetic phase (i.e. at zero external field and below the critical temperature), χ diverges with the system size N. One way to understand this divergence is the following. If we denote by $m(B)$ the magnetization in infinite volume in a magnetic field B, and denote the spontaneous magnetization by $M(\beta)$, we have

$$\chi = \lim_{B \to 0} \frac{1}{2B} [m(B) - m(-B)] = \lim_{B \to 0+} \frac{M(\beta)}{B} = \infty , \qquad (12.25)$$

within the ferromagnetic phase.

The above argument relates the divergence of the susceptibility to the existence of two distinct pure states of the system ('plus' and 'minus'). What is the appropriate susceptibility for detecting a spin glass ordering? Following our previous discussion, we should consider the addition of a small non-uniform field $B_i = s_i \epsilon$. The local magnetizations are given by

$$\langle \sigma_i \rangle_{\epsilon, \underline{s}} = \langle \sigma_i \rangle_0 + \epsilon \sum_j \chi_{ij} s_j + O(\epsilon^2) \,. \tag{12.26}$$

As suggested by eqn (12.25), we compare the local magnetizations obtained by perturbing the system in two different directions $\underline{s}$ and $\underline{s}'$:

$$\langle \sigma_i \rangle_{\epsilon, \underline{s}} - \langle \sigma_i \rangle_{\epsilon, \underline{s}'} = \epsilon \sum_j \chi_{ij} (s_j - s'_j) + O(\epsilon^2) \,. \tag{12.27}$$

How should we choose $\underline{s}$ and $\underline{s}'$? A simple choice takes them as independent and uniformly random in $\{\pm 1\}^N$; let us denote the expectation with respect to this distribution by $\mathbb{E}_s$. The above difference therefore becomes a random variable with zero mean. Its second moment allows us to define the **spin glass susceptibility** (sometimes called the **non-linear susceptibility**),

$$\chi_{\mathrm{SG}} \equiv \lim_{\epsilon \to 0} \frac{1}{2N\epsilon^2} \sum_i \mathbb{E}_s \left(\langle \sigma_i \rangle_{\epsilon, \underline{s}} - \langle \sigma_i \rangle_{\epsilon, \underline{s}'} \right)^2 \,. \tag{12.28}$$

This is in a sense the equivalent of eqn (12.25) for the spin glass case. Using eqn (12.27), one gets the expression $\chi_{\mathrm{SG}} = (1/N) \sum_{ij} (\chi_{ij})^2$, which can also be written, using the fluctuation–dissipation relation, as

$$\chi_{\mathrm{SG}} = \frac{\beta^2}{N} \sum_{i,j} [\langle \sigma_i \sigma_j \rangle - \langle \sigma_i \rangle \langle \sigma_j \rangle]^2 \,. \tag{12.29}$$

Usually, a necessary condition for the system to be in a paramagnetic, non-solid phase is that χ_{SG} remains finite when $N \to \infty$. We shall see below that this necessary condition of local stability is not always sufficient.

Exercise 12.9 Another natural choice would consist in choosing $\underline{s}$ and $\underline{s}'$ as independent configurations drawn from the Boltzmann distribution. Show that with such a choice one would get $\chi_{\mathrm{SG}} = (1/N) \sum_{i,j,k} \chi_{ij} \chi_{jk} \chi_{ki}$. This susceptibility has not been studied in the literature, but it is reasonable to expect that it will lead generically to the same criterion of stability as the usual one given in eqn (12.29).

12.3.3 The overlap distribution function $P(q)$

One of the main indicators of a glass phase is the overlap distribution, which we defined in Section 8.2.2. Given a general magnetic model of the type (12.1), one generates two independent configurations $\underline{\sigma}$ and $\underline{\sigma}'$ from the associated Boltzmann distribution and considers their overlap $q_{\underline{\sigma},\underline{\sigma}'} = N^{-1}\sum_i \sigma_i \sigma_i'$. The overlap distribution $P_N(q)$ is the distribution of $q_{\underline{\sigma},\underline{\sigma}'}$ when the couplings and the underlying factor graph are taken randomly from an ensemble. Its infinite-N limit is denoted by $P(q)$. Its moments are given by

$$\int P_N(q) q^r \, \mathrm{d}q = \mathbb{E}\Big\{ \frac{1}{N^r} \sum_{i_1,\dots,i_r} \langle \sigma_{i_1} \dots \sigma_{i_r} \rangle^2 \Big\}. \tag{12.30}$$

In particular, the first moment $\int P_N(q)\, q \, \mathrm{d}q = N^{-1}\sum_i m_i^2$ is the expected overlap, and the variance $\mathrm{Var}(q) \equiv \int P_N(q)\, q^2 \, \mathrm{d}q - \left[\int P_N(q)\, q \, \mathrm{d}q\right]^2$ is related to the spin glass susceptibility:

$$\mathrm{Var}(q) = \mathbb{E}\Big\{ \frac{1}{N^2} \sum_{i,j} [\langle \sigma_i \sigma_j \rangle - \langle \sigma_i \rangle \langle \sigma_j \rangle]^2 \Big\} = \frac{1}{N}\chi_{\mathrm{SG}}. \tag{12.31}$$

How is a glass phase detected through the behaviour of the overlap distribution $P(q)$? We shall discuss here two scenarios that appear to be remarkably universal within mean-field models. In the next subsection, we shall see that the overlap distribution is in fact related to the idea, discussed in Section 12.3.1, of perturbing the system in order to explore its quasi-states.

Generically, at small β, a system of the type (12.1) is found in a 'paramagnetic', or 'liquid', phase. In this regime, $P_N(q)$ becomes concentrated as $N \to \infty$ on a single (deterministic) value $q(\beta)$: with high probability, two independent configurations $\underline{\sigma}$ and $\underline{\sigma}'$ have an overlap close to $q(\beta)$. In fact, in such a phase, the spin glass susceptibility χ_{SG} is finite, and the variance of $P_N(q)$ therefore vanishes as $1/N$.

For β larger than a critical value β_c, the distribution $P(q)$ may acquire some structure, in the sense that several values of the overlap have non-zero probability in the $N \to \infty$ limit. The temperature $T_c = 1/\beta_c$ is called the **static** (or **equilibrium**) **glass transition temperature**. For $\beta > \beta_c$, the system is in an equilibrium glass phase.

What does $P(q)$ look like at $\beta > \beta_c$? Generically, the transition falls into one of the following two categories, the names of which come from the corresponding replica-symmetry-breaking patterns found in the replica approach:

(*i*) **Continuous** ('full replica-symmetry-breaking', 'FRSB') glass transition. In Fig. 12.2, we sketch the behaviour of the thermodynamic limit of $P(q)$ in this case. The delta function present at $\beta < \beta_c$ 'broadens' for $\beta > \beta_c$, giving rise to a distribution with support in some interval $[q_0(\beta), q_1(\beta)]$. The width $q_1(\beta) - q_0(\beta)$ vanishes continuously when $\beta \downarrow \beta_c$. Furthermore, the asymptotic distribution has a continuous density which is strictly positive in $]q_0(\beta), q_1(\beta)[$, and two discrete (delta) contributions at $q_0(\beta)$ and $q_1(\beta)$. This type of transition has a 'precursor'. If we consider the $N \to \infty$ limit of the spin glass susceptibility, this diverges as $\beta \uparrow \beta_c$. This phenomenon is quite important for identifying the critical temperature experimentally, numerically, and analytically.

(*ii*) **Discontinuous** ('1RSB') glass transition (see Fig. 12.3). Again, $P(q)$ acquires a non-trivial structure in the glass phase, but the scenario is different. When β increases above β_c, the δ-peak at $q(\beta)$, which had unit mass at $\beta \leq \beta_c$, becomes a

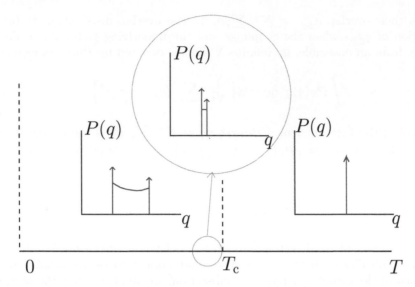

Fig. 12.2 Typical behaviour of the order parameter $P(q)$ (the asymptotic overlap distribution) at a continuous (FRSB) glass transition. Vertical arrows denote Dirac delta functions.

peak at $q_0(\beta)$, with a mass $1 - \mathbf{x}(\beta) < 1$. Simultaneously, a second δ-peak appears at a value of the overlap $q_1(\beta) > q_0(\beta)$ with mass $\mathbf{x}(\beta)$. As $\beta \downarrow \beta_c$, $q_0(\beta) \to q(\beta_c)$ and $\mathbf{x}(\beta) \to 0$. Unlike in a continuous transition, the width $q_1(\beta) - q_0(\beta)$ does not vanish as $\beta \downarrow \beta_c$, and the open interval $]q_0(\beta), q_1(\beta)[$ has vanishing probability in the $N \to \infty$ limit. Furthermore, the thermodynamic limit of the spin glass susceptibility χ_{SG} has a finite limit as $\beta \uparrow \beta_c$. This type of transition has no 'simple' precursor (but we shall describe below a more subtle indicator).

The two-peak structure of $P(q)$ in a discontinuous transition has a particularly simple geometrical interpretation. When two configurations $\underline{\sigma}$ and $\underline{\sigma}'$ are chosen independently with the Boltzmann measure, their overlap is (with high probability) approximately equal to either q_0 or q_1. In other words, their Hamming distance is either $N(1 - q_1)/2$ or $N(1 - q_0)/2$. This means that the Boltzmann measure $\mu(\underline{\sigma})$ is concentrated in some regions of the Hamming space $\{-1, 1\}^N$, called **clusters**. With high probability, two independent random configurations in the same cluster have a distance (close to) $N(1 - q_1)/2$, and two configurations in distinct clusters have a distance (close to) $N(1 - q_0)/2$. In other words, while the overlap does not concentrate in probability when $\underline{\sigma}$ and $\underline{\sigma}'$ are drawn from the Boltzmann measure, it does when this measure is restricted to one cluster. In a more formal (but still imprecise) way, we might write

$$\mu(\underline{\sigma}) \approx \sum_\alpha w_\alpha \mu_\alpha(\underline{\sigma}), \qquad (12.32)$$

where the $\mu_\alpha(\cdot)$ are probability distributions concentrated on a single cluster, and the w_α are the weights attributed by the Boltzmann distribution to each cluster.

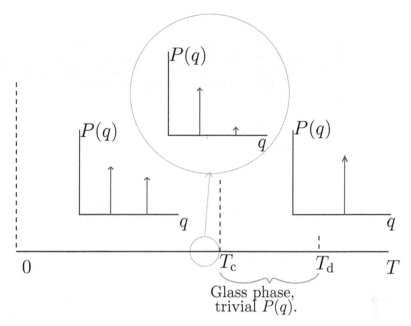

Fig. 12.3 Typical behaviour of the order parameter $P(q)$ (the asymptotic overlap distribution) in a discontinuous (1RSB) glass transition. Vertical arrows denote Dirac delta functions.

According to this interpretation, $\mathbf{x}(\beta) = \mathbb{E}\left\{\sum_\alpha w_\alpha^2\right\}$. Note that, since $\mathbf{x}(\beta) > 0$ for $\beta > \beta_c$, the weights are sizeable only for a finite number of clusters (if there were R clusters, all with the same weight $w_\alpha = 1/R$, one would have $\mathbf{x}(\beta) = 1/R$). This is what we have found already in the REM, as well as in the replica solution of the completely connected p-spin model (see. Section 8.2).

Generically, clusters already exist in some region of temperature above T_c, but the measure is not yet condensed on a finite number of them. The existence of clusters in this intermediate-temperature region can be detected instead using the tools described below.

There is no clear criterion that allows one to distinguish a priori between systems that undergo one or the other type of transition. Experience gained with models solved via the replica and cavity methods indicates that a continuous transition typically occurs in standard spin glasses with $p = 2$-body interactions, but also, for instance, in the vertex-covering problem. A discontinuous transition is instead found in structural glasses, generalized spin glasses with $p \geq 3$, random satisfiability, and coloring. To complicate things, both types of transition may occur in the same system at different temperatures (or when some other parameter is varied). This may lead to a rich phase diagram with several glass phases of different nature.

It is natural to wonder whether gauge transformations might give some information about $P(q)$. Unfortunately, it turns out that the Nishimori temperature never enters a spin glass phase: the overlap distribution at T_N is concentrated on a single value, as suggested by the next exercise.

Exercise 12.10 Using the gauge transformation of Section 12.2.1, show that, at the Nishimori temperature, the overlap distribution $P_N(q)$ is equal to the distribution of the magnetization per spin $m(\underline{\sigma}) \equiv N^{-1}\sum_i \sigma_i$. (In many spin glass models, one expects that this distribution of magnetization per spin will obey a large-deviation principle, and become concentrated on a single value as $N \to \infty$.)

12.3.4 The ϵ-coupling method

The overlap distribution is in fact related to the idea of quasi-states introduced in Section 12.3.1. Let us again consider a perturbation of the Boltzmann distribution defined by adding to the energy a magnetic-field term $-\epsilon \sum_i s_i \sigma_i$, where $\underline{s} = (s_1, \ldots, s_N)$ is a generic configuration. We introduce the ϵ-perturbed energy of a configuration $\underline{\sigma}$ as

$$E_{\epsilon,\underline{s}}(\underline{\sigma}) = E(\underline{\sigma}) - \epsilon \sum_{i=1}^{N} s_i \sigma_i \ . \tag{12.33}$$

It is important to realize that both the original energy $E(\underline{\sigma})$ and the new term $-\epsilon \sum_i s_i \sigma_i$ are extensive, i.e. they grow in proportion to N as $N \to \infty$. Therefore, in this limit the presence of the perturbation can be relevant. The ϵ-perturbed Boltzmann measure is

$$\mu_{\epsilon,\underline{s}}(\underline{\sigma}) = \frac{1}{Z_{\epsilon,\underline{s}}} \, \mathrm{e}^{-\beta E_{\epsilon,\underline{s}}(\underline{\sigma})} \ . \tag{12.34}$$

In order to quantify the effect of the perturbation, let us measure the expected distance between $\underline{\sigma}$ and $\underline{s}$,

$$d(\underline{s}, \epsilon) \equiv \frac{1}{N} \sum_{i=1}^{N} \frac{1}{2}(1 - s_i \langle \sigma_i \rangle_{\underline{s},\epsilon}) \tag{12.35}$$

(note that $\sum_i(1 - s_i \sigma_i)/2$ is just the number of positions in which $\underline{\sigma}$ and $\underline{s}$ differ). For $\epsilon > 0$, the coupling between $\underline{\sigma}$ and $\underline{s}$ is attractive; for $\epsilon < 0$, it is repulsive. In fact, it is easy to show that $d(\underline{s}, \epsilon)$ is a decreasing function of ϵ.

In the ϵ-**coupling method**, $\underline{s}$ is taken as a random variable, drawn from the (unperturbed) Boltzmann distribution. The rationale for this choice is that, in this way, $\underline{s}$ will point in the directions corresponding to quasi-states. The average distance induced by the ϵ-perturbation is then obtained, after averaging over $\underline{s}$ and over the choice of sample:

$$d(\epsilon) \equiv \mathbb{E}\left\{ \sum_{\underline{s}} \frac{1}{Z} \, \mathrm{e}^{-\beta E(\underline{s})} \, d(\underline{s}, \epsilon) \right\}. \tag{12.36}$$

There are two important differences between the ϵ-coupling method and the computation of the overlap distribution $P_N(q)$. (*i*) When $P_N(q)$ is computed, the two copies of the system are treated on an equal footing: they are independent and distributed according to Boltzmann law. In the ϵ-coupling method, one of the copies is distributed according to Boltzmann's law, while the other follows a perturbed distribution depending on the first copy. (*ii*) In the ϵ-coupling method, the $N \to \infty$ limit is taken *at fixed ϵ*. Therefore, the sum in eqn (12.36) can be dominated by values of the overlap $q(\underline{s}, \underline{\sigma})$ which would have been exponentially unlikely for the original (unperturbed) measure. When the $N \to \infty$ limit $P(q)$ is computed, such values of the overlap have a vanishing weight. The two approaches provide complementary information.

Within a paramagnetic phase, $d(\epsilon)$ remains a smooth function of ϵ in the neighbourhood of $\epsilon = 0$, even after the $N \to \infty$ limit has been taken: perturbing the system does not have any dramatic effect. But in a glass phase, $d(\epsilon)$ becomes singular: it develops a discontinuity at $\epsilon = 0$, which can be detected by defining

$$\Delta = \lim_{\epsilon \to 0+} \lim_{N \to \infty} d(\epsilon) - \lim_{\epsilon \to 0-} \lim_{N \to \infty} d(\epsilon) \,. \tag{12.37}$$

Notice that the limit $N \to \infty$ is taken first: for finite N, there cannot be any discontinuity.

One expects Δ to be non-zero if and only if the system is in a 'solid' phase. In order to get an intuitive understanding of this, one can think of the process of adding a positive ϵ-coupling and then letting it tend to 0 as of a physical process. The system is first forced into an energetically favourable configuration (given by $\underline{s}$). The forcing is then gradually removed and one checks whether any memory of the preparation is retained ($\Delta > 0$) or whether the system 'liquefies' ($\Delta = 0$).

The advantage of the ϵ-coupling method with respect to the overlap distribution $P(q)$ is twofold:

- In some cases the dominant contribution to the Boltzmann measure comes from several distinct clusters, but a single cluster dominates over the others. More precisely, it may happen that the weights for subdominant clusters scale as $w_\alpha = \exp[-\Theta(N^\theta)]$, with $\theta \in]0, 1[$. In this case, $P(q)$ is a delta function and does not allow one to distinguish the state of the system from a purely paramagnetic phase. However, the ϵ-coupling method identifies the phase transition through a singularity in $d(\epsilon)$ at $\epsilon = 0$.

- One can use the ϵ-coupling method to analyse a system undergoing a discontinuous transition when it is in a glass phase but in the $T > T_c$ regime. In this case, the existence of clusters cannot be detected from $P(q)$ because the Boltzmann measure is spread over an exponential number of them. This situation will be the subject of the next subsection.

12.3.5   1RSB clusters and the potential method

The 1RSB equilibrium glass phase corresponds to a condensation of the measure onto a small number of clusters of configurations. However, the most striking phenomenon is the appearance of clusters themselves. In the following chapters, we shall argue that this has important consequences for Monte Carlo dynamics, as well as for other algorithmic approaches to these systems. It turns out that the Boltzmann measure splits into clusters at a distinct temperature $T_d > T_c$. In the region of temperatures $[T_c, T_d]$, we shall say that the system is in a **clustered phase, dynamical glass phase**, or **dynamical 1RSB phase**. The phase transition at T_d will be referred to as the **clustering** or **dynamical** (glass) *transition.* In this regime, an exponential number of clusters $\mathcal{N} \doteq e^{N\Sigma}$ carry a roughly equal weight. The rate of growth Σ is called the **complexity**[3] or **configurational entropy**.

The thermodynamic limit of the overlap distribution, $P(q)$, does not show any signature of the clustered phase. In order to understand this point, it is useful to work through a toy example. Assume that the Boltzmann measure is entirely supported on *exactly* $e^{N\Sigma}$ sets of configurations in $\{\pm 1\}^N$ (each set is a cluster), denoted by $\alpha = 1, \ldots, e^{N\Sigma}$, and that the Boltzmann probability of each of these sets is $w = e^{-N\Sigma}$. Assume furthermore that, for any two configurations belonging to the same cluster $\underline{\sigma}, \underline{\sigma}' \in \alpha$, their overlap is $q_{\underline{\sigma}, \underline{\sigma}'} = q_1$, whereas, if they belong to different clusters $\underline{\sigma} \in \alpha$, $\underline{\sigma}' \in \alpha'$, $\alpha \neq \alpha'$, their overlap is $q_{\underline{\sigma}, \underline{\sigma}'} = q_0 < q_1$. Although it might be difficult to actually construct such a measure, we shall neglect this problem for a moment, and compute the overlap distribution. The probability that two independent configurations fall into the same cluster is $e^{N\Sigma} w^2 = e^{-N\Sigma}$. Therefore, we have

$$P_N(q) = (1 - e^{-N\Sigma}) \, \delta(q - q_0) + e^{-N\Sigma} \, \delta(q - q_1), \qquad (12.38)$$

which converges to $\delta(q - q_0)$ as $N \to \infty$: $P(q)$ has a single delta function, as in the paramagnetic phase.

A first signature of the clustered phase is provided by the ϵ-coupling method described in the previous subsection. The reason is very clear if we look at eqn (12.33): the ϵ-coupling 'tilts' the Boltzmann distribution in such a way that unlikely values of the overlap acquire a strictly positive probability. It is easy to compute the thermodynamic limit $d_*(\epsilon) \equiv \lim_{N \to \infty} d(\epsilon)$. We get

$$d_*(\epsilon) = \begin{cases} (1 - q_0)/2 & \text{for } \epsilon < \epsilon_c, \\ (1 - q_1)/2 & \text{for } \epsilon > \epsilon_c, \end{cases} \qquad (12.39)$$

where $\epsilon_c = \Sigma/\beta(q_1 - q_0)$. As $T \downarrow T_c$, clusters becomes less and less numerous and $\Sigma \to 0$. Correspondingly, $\epsilon_c \downarrow 0$ as the equilibrium glass transition is approached.

The picture provided by this toy example is essentially correct, with the caveats that the properties of clusters will hold only within some accuracy and with high probability. Nevertheless, one expects $d_*(\epsilon)$ to have a discontinuity at some $\epsilon_c > 0$ for all temperatures in an interval $]T_c, T_d']$. Furthermore, $\epsilon_c \downarrow 0$ as $T \downarrow T_c$.

In general, the temperature T_d' computed through the ϵ-coupling method does not coincide with that of the clustering transition. The reason is easily understood. As illustrated by the above example, we are estimating the exponentially small probability $\mathbb{P}(q|\underline{s}, \underline{J})$ that an equilibrated configuration $\underline{\sigma}$ has an overlap q with the reference configuration $\underline{s}$, in a sample $\underline{J}$. In order to do this, we compute the distance $d(\epsilon)$ in a problem with a tilted measure. As we have seen already several times since Chapter 5, exponentially small (or large) quantities usually do not concentrate in probability, and $d(\epsilon)$ may be dominated by exponentially rare samples. We have also learnt the cure for this problem: take logarithms! We therefore define[4] the **glass potential**

---

[3]This use of the term 'complexity' here, which is customary in statistical physics, should not be confused with its use in theoretical computer science.

[4]One should introduce a resolution here, so that the overlap is actually constrained to be in some window around q. The width of this window can be let tend to 0 *after* $N \to \infty$.

$$V(q) = -\lim_{N\to\infty} \frac{1}{N\beta} \mathbb{E}_{\underline{s},\underline{J}} \{ \log \mathbb{P}(q|\underline{s},\underline{J}) \} . \tag{12.40}$$

Here (as in the ϵ-coupling method), the reference configuration is drawn from the Boltzmann distribution. In other words,

$$\mathbb{E}_{\underline{s},\underline{J}}(\cdots) = \mathbb{E}_{\underline{J}} \left\{ \frac{1}{Z_{\underline{J}}} \sum_{\underline{s}} e^{-\beta E_{\underline{J}}(\underline{s})}(\cdots) \right\} . \tag{12.41}$$

If, as expected, $\log \mathbb{P}(q|\underline{s},\underline{J})$ concentrates in probability, one has $\mathbb{P}(q|\underline{s},\underline{J}) \doteq e^{-NV(q)}$ with high probability.

Exercise 12.11 Consider the following refined version of the toy model (12.38): $\mathbb{P}(q|\underline{s},\underline{J}) = (1 - e^{-N\Sigma(\underline{s},\underline{J})}) G_{q_0(\underline{s},\underline{J});(b_0/N\beta)}(q) + e^{-N\Sigma(\underline{s},\underline{J})} G_{q_1(\underline{s},\underline{J});(b_1/N\beta)}(q)$, where $G_{a,b}$ is a Gaussian distribution of mean a and variance b. We suppose that b_0, b_1 are constants, but that $\Sigma(\underline{s},\underline{J}), q_0(\underline{s},\underline{J}), q_1(\underline{s},\underline{J})$ fluctuate as follows: when $\underline{J}$ and $\underline{s}$ are distributed according to the correct joint distribution (12.41), then $\Sigma(\underline{s},\underline{J}), q_0(\underline{s},\underline{J}), q_1(\underline{s},\underline{J})$ are independent Gaussian random variables of means $\overline{\Sigma}, \overline{q}_0, \overline{q}_1$, respectively, and variances $\delta\Sigma^2/N, \delta q_0^2/N, \delta q_1^2/N$.

Assuming for simplicity that $\delta\Sigma^2 < 2\overline{\Sigma}$, compute $P(q)$ and $d(\epsilon)$ for this model. Show that the glass potential $V(q)$ is given by two arcs of parabolas as follows:

$$V(q) = \min \left\{ \frac{(q - \overline{q}_0)^2}{2b_0} , \frac{(q - \overline{q}_1)^2}{2b_1} + \frac{1}{\beta}\overline{\Sigma} \right\} . \tag{12.42}$$

The glass potential $V(q)$ has been computed using the replica method in only a small number of cases, mainly fully connected p-spin glasses. Here we shall just mention the qualitative behaviour that is expected on the basis of these computations. The result is summarized in Fig. 12.4. At small enough β, the glass potential is convex. When β is increased, one first encounters a value β_* where $V(q)$ ceases to be convex. When $\beta > \beta_d = 1/T_d$, $V(q)$ develops a secondary minimum, at $q = q_1(\beta) > q_0(\beta)$. This secondary minimum is in fact an indication of the existence of an exponential number of clusters, such that two configurations in the same cluster typically have an overlap q_1, while two configurations in distinct clusters have an overlap q_0. A little thought shows that the difference between the values of the glass potential at the two minima gives the complexity: $V(q_1) - V(q_0) = T\Sigma$.

In models in which the glass potential has been computed exactly, the temperature T_d computed in this way coincides with a dramatic slowing down of the relaxational dynamics. More precisely, a properly defined relaxation time for Glauber-type dynamics is finite for $T > T_d$ and diverges exponentially in the system size for $T < T_d$.

12.3.6 Cloning and the complexity function

When the various clusters do not all have the same weight, the system is most appropriately described through a **complexity function**. Consider a cluster of configurations, called α. Its free energy F_α can be defined by restricting the partition function

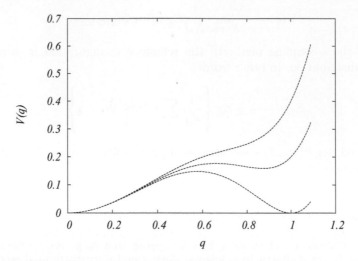

Fig. 12.4 Qualitative shapes of the glass potential $V(q)$ at various temperatures. When the temperature is very high (not shown), $V(q)$ is convex. Below $T = T_{\mathrm{d}}$, it develops a secondary minimum. The height difference between the two minima is $V(q_1) - V(q_0) = T\Sigma$. In the case shown here, $q_0 = 0$ is independent of the temperature.

to configurations in the cluster α. One way of imposing this restriction is to choose a reference configuration $\underline{\sigma}_0 \in \alpha$, and restrict the Boltzmann sum to those configurations $\underline{\sigma}$ whose distance from $\underline{\sigma}_0$ is smaller than $N\delta$. In order to correctly identify clusters, one has to take $(1 - q_1)/2 < \delta < (1 - q_*)/2$, where $q_* > q_1$ is such that $V(q_*) > V(q_1)$.

Let $\mathcal{N}_\beta(f)$ be the number of clusters such that $F_\alpha = Nf$ (more precisely, this is an unnormalized measure attributing unit weight to the points F_α/N). We expect it to satisfy a large-deviation principle of the form

$$\mathcal{N}_\beta(f) \doteq \exp\{N\Sigma(\beta, f)\}. \tag{12.43}$$

The rate function $\Sigma(\beta, f)$ is the complexity function. If clusters are defined as above, with the cut-off δ in the appropriate interval, they are expected to be disjoint up to a subset of configurations of exponentially small Boltzmann weight. Therefore the total partition function is given by

$$Z = \sum_\alpha \mathrm{e}^{-\beta F_\alpha} \doteq \int \mathrm{e}^{N[\Sigma(\beta,f)-\beta f]} \, \mathrm{d}f \doteq \mathrm{e}^{N[\Sigma(\beta,f_*)-\beta f_*]}, \tag{12.44}$$

where we have applied the saddle point method as in standard statistical-mechanics calculations (see Section 2.4). Here $f_* = f_*(\beta)$ solves the saddle point equation $\partial\Sigma/\partial f = \beta$.

For several reasons, it is interesting to determine the full complexity function $\Sigma(\beta, f)$ as a function of f for a given inverse temperature β. The **cloning method** is a particularly efficient (although non-rigorous) way to do this computation. Here we sketch the basic idea: several applications will be discussed in the following chapters.

One begins by introducing m identical 'clones' of the initial system. These are non-interacting except for the fact that they are constrained to be in the same cluster. In practice, one can constrain all their pairwise Hamming distances to be smaller than $N\delta$, where $(1-q_1)/2 < \delta < (1-q_*)/2$. The partition function for the systems of m clones is therefore

$$Z_m = \sum_{\underline{\sigma}^{(1)},\dots,\underline{\sigma}^{(m)}}{}' \exp\left\{-\beta E(\underline{\sigma}^{(1)})\cdots - \beta E(\underline{\sigma}^{(m)})\right\}. \tag{12.45}$$

where the prime reminds us that $\underline{\sigma}^{(1)}, \dots \underline{\sigma}^{(m)}$ stay in the same cluster. By splitting the sum over the various clusters, we have

$$Z_m = \sum_{\alpha} \sum_{\underline{\sigma}^{(1)}\dots\underline{\sigma}^{(m)}\in\alpha} e^{-\beta E(\underline{\sigma}^{(1)})\cdots-\beta E(\underline{\sigma}^{(m)})} = \sum_{\alpha}\left(\sum_{\underline{\sigma}\in\alpha}e^{-\beta E(\underline{\sigma})}\right)^m. \tag{12.46}$$

At this point we can proceed as for the calculation of the usual partition function, and obtain

$$Z_m = \sum_{\alpha} e^{-\beta m F_\alpha} \doteq \int e^{N[\Sigma(\beta,f)-\beta m f]}\, df \doteq e^{N[\Sigma(\beta,\hat{f})-\beta m \hat{f}]}, \tag{12.47}$$

where $f_* = f_*(\beta, m)$ solves the saddle point equation $\partial\Sigma/\partial f = \beta m$.

The free-energy density per clone of the cloned system is defined as

$$\Phi(\beta, m) = -\lim_{N\to\infty} \frac{1}{\beta m N} \log Z_m. \tag{12.48}$$

The saddle point estimate (12.47) implies that $\Phi(\beta, m)$ is related to $\Sigma(\beta, f)$ through Legendre transform:

$$\Phi(\beta, m) = f - \frac{1}{\beta m}\Sigma(\beta, f), \qquad \frac{\partial\Sigma}{\partial f} = \beta m. \tag{12.49}$$

If we forget that m is an integer, and admit that $\Phi(\beta, m)$ can be 'continued' to non-integer m, the complexity $\Sigma(\beta, f)$ can be computed from $\Phi(\beta, m)$ by inverting this Legendre transform. The similarity to the procedure used in the replica method is not fortuitous. Note, however, that replicas were introduced to deal with quenched disorder, while cloning is more general: it also applies to systems without disorder.

Exercise 12.12 In the REM, the natural definition of the overlap between two configurations $i, j \in \{1, \dots, 2^N\}$ is $q_{i,j} = \mathbb{I}(i = j)$. Taking a configuration j_0 as the reference, the ϵ-perturbed energy of a configuration j is $E'(\epsilon, j) = E_j - N\epsilon\mathbb{I}(j = j_0)$. (Note the extra

N multiplying ϵ, introduced in order to ensure that the new ϵ-coupling term is typically extensive.)

(a)   Consider the high-temperature phase where $\beta < \beta_c = 2\sqrt{\log 2}$. Show that the ϵ-perturbed system has a phase transition at $\epsilon = (\log 2)/\beta - \beta/4$.

(b)   In the low temperature phase where $\beta > \beta_c$, show that the phase transition takes place at $\epsilon = 0$.

Therefore, in the REM, clusters exist at any β, and every cluster is reduced to one single configuration: one must have $\Sigma(\beta, f) = \log 2 - f^2$ independently of β. Show that this is compatible with the cloning approach, through a computation of the potential $\Phi(\beta, m)$:

$$\Phi(\beta, m) = \begin{cases} -\frac{\log 2}{\beta m} - \frac{\beta m}{4} & \text{for } m < \frac{\beta_c}{\beta}, \\ -\sqrt{\log 2} & \text{for } m > \frac{\beta_c}{\beta}. \end{cases} \tag{12.50}$$

12.4   An example: The phase diagram of the SK model

Several mean-field models have been solved using the replica method. Sometimes a model may present two or more glass phases with different properties. Determining the phase diagram can be particularly challenging in those cases.

A classical example is provided by the Sherrington–Kirkpatrick model with ferromagnetically biased couplings. As in the other examples in this chapter, this is a model for N Ising spins $\underline{\sigma} = (\sigma_1, \ldots, \sigma_N)$. The energy function is

$$E(\underline{\sigma}) = -\sum_{(i,j)} J_{ij} \sigma_i \sigma_j , \tag{12.51}$$

where (i, j) are unordered pairs, and the couplings J_{ij} are i.i.d. Gaussian random variables with mean J_0/N and variance $1/N$. In a sense, this model interpolates between the Curie–Weiss model treated in Section 2.5.2, corresponding to $J_0 \to \infty$, and the unbiased SK model considered in Chapter 8, corresponding to $J_0 = 0$.

The phase diagram can be plotted in terms of two parameters: the ferromagnetic bias J_0, and the temperature T. Depending on their values, the system can be found in one of four phases (see Fig. 12.5): the paramagnetic (P), ferromagnetic (F), symmetric spin glass (SG), or mixed ferromagnetic–spin glass (F-SG) phase. A simple characterization of these four phases can be obtained in terms of two quantities: the average magnetization and the overlap. In order to define these quantities, we must first observe that, since $E(\underline{\sigma}) = E(-\underline{\sigma})$, in the present model $\langle \sigma_i \rangle = 0$ identically for all values of J_0, and T. In order to break this symmetry, we may add a magnetic-field term $-B \sum_i \sigma_i$ and let $B \to 0$ after taking the thermodynamic limit. We then define

$$m = \lim_{B \to 0+} \lim_{N \to \infty} \mathbb{E}\langle \sigma_i \rangle_B , \qquad \bar{q} = \lim_{B \to 0+} \lim_{N \to \infty} \mathbb{E}(\langle \sigma_i \rangle_B^2) , \tag{12.52}$$

(which do not depend on i, because the coupling distribution is invariant under a permutation of the sites). In the P phase, one has $m = 0, \bar{q} = 0$; in the SG phase, $m = 0, \bar{q} > 0$, and in the F and F-SG phases, $m > 0, \bar{q} > 0$.

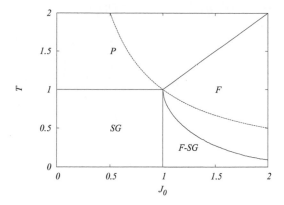

Fig. 12.5 Phase diagram of the SK model in zero magnetic field. When the temperature T and the ferromagnetic bias J_0 are varied, four distinct phases are encountered: paramagnetic (P), ferromagnetic (F), spin glass (SG) and mixed ferromagnetic–spin glass (F-SG). The full lines separate these various phases. The dashed line is the location of the Nishimori temperature.

A more complete description can be obtained in terms of the overlap distribution $P(q)$. Because of the symmetry under spin inversion mentioned above, $P(q) = P(-q)$ identically. The qualitative shape of $P(q)$ in the thermodynamic limit is shown in Fig. 12.6. In the P phase, it consists of a single delta function with unit weight at $q = 0$: two independent configurations drawn from the Boltzmann distribution have, with high probability, an overlap close to 0. In the F phase, it is concentrated on two symmetric values $q(J_0, T) > 0$ and $-q(J_0, T) < 0$, each carrying a weight of one-half. We can summarize this behaviour by saying that a random configuration drawn from the Boltzmann distribution is found, with equal probability, in one of two different states. In the first one the local magnetizations are $\{m_i\}$, and in the second one they are $\{-m_i\}$. If one draws two independent configurations, they fall into the same state (corresponding to the overlap value $q(J_0, T) = N^{-1} \sum_i m_i^2$) or into opposite states (overlap $-q(J_0, T)$) with probability $1/2$. In the SG phase, the support of $P(q)$ is a symmetric interval $[-q_{\max}, q_{\max}]$, with $q_{\max} = q_{\max}(J_0, T)$. Finally, in the F-SG phase, the support is the union of two intervals $[-q_{\max}, -q_{\min}]$ and $[q_{\min}, q_{\max}]$. In both the SG and the F-SG phase, the presence of a whole range of overlap values carrying non-vanishing probability, suggests the existence of a multitude of quasi-states (in the sense discussed in the previous section).

In order to remove the degeneracy due to the symmetry under spin inversion, one sometimes defines an asymmetric overlap distribution by adding a magnetic-field term, and taking the thermodynamic limit as in eqn (12.52):

$$P_+(q) = \lim_{B \to 0+} \lim_{N \to \infty} P_B(q). \qquad (12.53)$$

Somewhat surprisingly, it turns out that $P_+(q) = 0$ for $q < 0$, while $P_+(q) = 2P(q)$ for $q > 0$. In other words, $P_+(q)$ is equal to the distribution of the *absolute value* of the overlap.

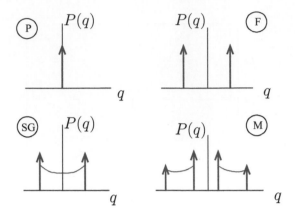

Fig. 12.6 The typical shape of the distribution $P(q)$ in each of the four phases of the SK model with ferromagnetically biased couplings.

Exercise 12.13 Consider the Curie–Weiss model in a magnetic field, (see Section 2.5.2). Draw the phase diagram and compute the asymptotic overlap distribution. Discuss its qualitative features for different values of the temperature and magnetic field.

We shall now add a few words for the reader interested in how one derives this diagram: Some of the phase boundaries have already been derived using the replica method in Exercise 8.12. The boundary P–F is obtained by solving the RS equation (8.68) for q, ω, x. The P–SG and F–M lines are obtained by the AT stability condition (8.69). Deriving the phase boundary between the SG and F–SG phases is much more challenging, because it separates glassy phases, and therefore it cannot be derived within the RS solution. It is known to be approximately vertical, but there is no simple expression for it. The Nishimori temperature is deduced from the condition (12.7), which gives $T_N = 1/J_0$, and the line $T = 1/J_0$ is usually called the 'Nishimori line'. The internal energy per spin on this line is $U/N = -J_0/2$. Notice that the line does not enter any of the glass phases, as we know from general arguments.

An important aspect of the SK model is that the appearance of the glass phase on the lines separating P from SG and on the line separating F from F–SG is a continuous transition. Therefore it is associated with a divergence of the non-linear susceptibility χ_{SG}. The following exercise, reserved for replica aficionados, sketches the main lines of the argument showing this.

Exercise 12.14 Let us see how to compute the non-linear susceptibility of the SK model, $\chi_{SG} = (\beta^2/N)\sum_{i\neq j}(\langle\sigma_i\sigma_j\rangle - \langle\sigma_i\rangle\langle\sigma_j\rangle)^2$, by the replica method. Show that

$$\chi_{SG} = \lim_{n\to 0}\frac{\beta^2}{N}\sum_{i\neq j}\left(\binom{n}{2}^{-1}\sum_{(ab)}\langle\sigma_i^a\sigma_i^b\sigma_j^a\sigma_j^b\rangle - \binom{n}{3}^{-1}\sum_{(abc)}\langle\sigma_i^a\sigma_i^b\sigma_j^a\sigma_j^c\rangle\right.$$
$$\left. + \binom{n}{4}^{-1}\sum_{(abcd)}\langle\sigma_i^a\sigma_i^b\sigma_j^c\sigma_j^d\rangle\right)$$
$$= N\lim_{n\to 0}\int e^{-NG(Q,\lambda)}A(Q)\prod_{(ab)}(\mathrm{d}Q_{ab}\mathrm{d}\lambda_{ab}) , \tag{12.54}$$

where we follow the notation of eqn (8.30), the sum over $(a_1a_2\ldots a_k)$ is understood to run over all the k-tuples of distinct replica indices, and

$$A(Q) \equiv \binom{n}{2}^{-1}\sum_{(ab)}Q_{ab}^2 - \binom{n}{3}^{-1}\sum_{(abc)}Q_{ab}Q_{ac} + \binom{n}{4}^{-1}\sum_{(abcd)}Q_{ab}Q_{cd} . \tag{12.55}$$

Analyse the divergence of χ_{SG} along the following lines. The leading contribution to eqn (12.54) should come from the saddle point and be given, in the high-temperature phase, by $A(Q_{ab} = q)$, where $Q_{ab} = q$ is the RS saddle point. However, this contribution clearly vanishes when one takes the $n \to 0$ limit. One must thus consider the fluctuations around the saddle point. Each of the terms of the type $Q_{ab}Q_{cd}$ in $A(Q)$ gives a factor of $1/N$ times the appropriate matrix element of the inverse of the Hessian matrix. When this Hessian matrix is non-singular, these elements are all finite and one obtains a finite result in the $N \to \infty$ limit. (The $1/N$ cancels the factor N in eqn (12.54).) When one reaches the AT instability line, the elements of the inverse of the Hessian matrix diverge, and therefore χ_{SG} also diverges.

Notes

Lattice gas models of atomic systems, such as those discussed in the first two exercises, are discussed in statistical-physics textbooks; see, for instance, Ma (1985). The simple model of a glass in Exercise 12.3 was introduced and solved on sparse random graphs using the cavity method by Biroli and Mézard (2002).

The order parameter for spin glasses defined by Edwards and Anderson (1975) is a dynamic order parameter which captures the long-time persistence of the spins. The static definition that we have introduced here should give the same result as the original dynamical definition (although of course we have no proof of this statement in general). A review of simulations of the Edwards–Anderson model can be found in Marinari *et al.* (1997).

Mathematical results on mean-field spin glasses can be found in the book by Tanagrand (2003). A short recent survey is provided by Guerra (2005).

Diluted spin glasses were introduced by Viana and Bray (1985).

The implications of the gauge transformation were derived by Hidetoshi Nishimori and his coworkers, and are explained in detail in his book (Nishimori, 2001).

The notion of pure states in phase transitions, and the decomposition of Gibbs measures into pure states, is discussed in the book by Georgii (1988). We shall discuss this topic further in Chapter 22.

The divergence of the spin glass susceptibility is especially relevant because this susceptibility can be measured in zero field. The experiments of Monod and Bouchiat (1982) present evidence of a divergence. This supports the existence of a spin glass transition in real (three-dimensional) spin glasses in zero magnetic field, at non-zero temperature.

The existence of two transition temperatures $T_c < T_d$ was first discussed by Kirkpatrick and Wolynes (1987) and Kirkpatrick and Thirumalai (1987), who pointed out the relevance to the theory of structural glasses. In particular, Kirkpatrick and Thirumalai (1987) discussed the case of the p-spin glass. A review of this line of approach to structural glasses, and particularly its relevance to dynamical effects, is provided by Bouchaud *et al.* (1997).

The ϵ-coupling method was introduced by Caracciolo *et al.* (1990). The idea of cloning in order to study the complexity function is due to Monasson (1995). Its application to studies of the glass transition without quenched disorder was developed by Mézard and Parisi (1999).

The glass potential method was introduced by Franz and Parisi (1995).

13

Bridges: Inference and the Monte Carlo method

We have seen in the last three chapters how problems with very different origins can be cast into the unifying framework of factor graph representations. The underlying mathematical structure, namely the locality of probabilistic dependencies between variables, is also present in many problems of probabilistic inference, which provides another unifying view of the field. We shall see through examples that several fundamental questions in physics, coding, and constraint satisfaction problems can be formulated as inference problems.

On the other hand, locality is also an important ingredient that allows sampling from complex distributions using the Monte Carlo technique. It is thus very natural to apply this technique to our problems, and to study its limitations. We shall see here that the existence of large free-energy barriers, in particular in low-temperature regimes, considerably slows down the Monte Carlo sampling procedure.

In Section 13.1, we present some basic terminology and simple examples of statistical inference problems. Statistical inference is an interesting field in itself, with many important applications (ranging from artificial intelligence to modelling and statistics). Here we emphasize the possibility of considering coding theory, statistical mechanics, and combinatorial optimization as inference problems.

Section 13.2 develops a very general tool, the Markov chain Monte Carlo (MCMC) technique, already introduced in Section 4.5. This is often a very powerful approach. Furthermore, Monte Carlo sampling can be regarded as a statistical inference method, and Monte Carlo dynamics is a simple prototype of the local search strategies introduced in Sections 10.2.3 and 11.4. Many of the difficulties encountered in decoding, in constraint satisfaction problems, and in the study of glassy phases are connected to a dramatic slowing down of MCMC dynamics. We present the results of simple numerical experiments on some examples, and identify regions in the phase diagram where the MCMC slowdown implies poor performance as a sampling/inference algorithm. Finally, in Section 13.3, we explain a rather general argument to estimate the amount of time an MCMC algorithm as to be run in order to produce roughly independent samples with the desired distribution.

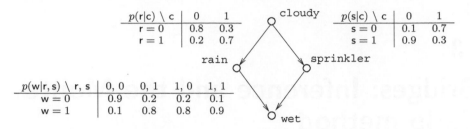

Fig. 13.1 The rain–sprinkler Bayesian network.

13.1 Statistical inference

13.1.1 Bayesian networks

It is common practice in artificial intelligence and statistics, to formulate inference problems in terms of Bayesian networks. Although any such problem can also be represented in terms of a factor graph, it is worth to briefly introduce this alternative language. A famous toy example is the 'rain–sprinkler' network.

Example 13.1 During a walk to the park, a statistician notices that the grass is wet. There are two possible reasons for this: either it rained during the night, or the sprinkler was activated in the morning to irrigate the lawn. Both events are, in turn, correlated with the weather condition in the last 24 hours.

After a little thought, the statistician formalizes these considerations as the probabilistic model depicted in Fig. 13.1. The model includes four random variables: cloudy, rain, sprinkler, and wet, taking values in $\{0, 1\}$ (which correspond, respectively, to false and true). The variables are organized as the vertices of a directed graph. A directed edge corresponds, intuitively, to a relation of causality. The joint probability distribution of the four variables is given in terms of conditional probabilities associated with the edges. Explicitly (variables are indicated by their initials),

$$p(\mathsf{c}, \mathsf{s}, \mathsf{r}, \mathsf{w}) = p(\mathsf{c})\, p(\mathsf{s}|\mathsf{c})\, p(\mathsf{r}|\mathsf{c})\, p(\mathsf{w}|\mathsf{s}, \mathsf{r})\,. \tag{13.1}$$

The three conditional probabilities in this formula are given by the tables in Fig. 13.1. A 'uniform prior' is assumed on the event that the day was cloudy: $p(\mathsf{c} = 0) = p(\mathsf{c} = 1) = 1/2$.

Given that wet grass was observed, we may want to know whether the most likely cause was the rain or the sprinkler. This amounts to computing the marginal probabilities

$$p(\mathsf{s}|\mathsf{w} = 1) = \frac{\sum_{\mathsf{c}, \mathsf{r}} p(\mathsf{c}, \mathsf{s}, \mathsf{r}, \mathsf{w} = 1)}{\sum_{\mathsf{c}, \mathsf{r}, \mathsf{s}'} p(\mathsf{c}, \mathsf{s}', \mathsf{r}, \mathsf{w} = 1)}\,, \tag{13.2}$$

$$p(\mathsf{r}|\mathsf{w} = 1) = \frac{\sum_{\mathsf{c}, \mathsf{s}} p(\mathsf{c}, \mathsf{s}, \mathsf{r}, \mathsf{w} = 1)}{\sum_{\mathsf{c}, \mathsf{r}, \mathsf{s}'} p(\mathsf{c}, \mathsf{s}', \mathsf{r}, \mathsf{w} = 1)}\,. \tag{13.3}$$

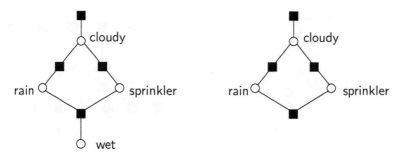

Fig. 13.2 *Left*: factor graph corresponding to the rain–sprinkler Bayesian network represented in Fig. 13.1. *Right*: factor graph for the same network under the observation of the variable w.

Using the numbers in Fig. 13.1, we get $p(\mathsf{s} = 1|\mathsf{w} = 1) \approx 0.63$ and $p(\mathsf{r} = 1|\mathsf{w} = 1) \approx 0.48$: the most likely cause of the wet grass is the sprinkler.

In Fig. 13.2 we show the factor graph representation of eqn (13.1), and that corresponding to the conditional distribution $p(\mathsf{c}, \mathsf{s}, \mathsf{r}|\mathsf{w} = 1)$. As is clear from the factor graph representation, the observation $\mathsf{w} = 1$ induces a further dependency between the variables s and r, beyond that induced by their relation to c. The reader is invited to draw the factor graph associated with the *marginal* distribution $p(\mathsf{c}, \mathsf{s}, \mathsf{r})$.

In general, a **Bayesian network** describes the joint distributions of variables associated with the vertices of a directed acyclic graph $G = (V, E)$. A directed graph is an ordinary graph with a direction (i.e. an ordering of the adjacent vertices) chosen on each of its edges, and no cycles. For such a graph, we say that a vertex $u \in V$ is a **parent** of v, and write $u \in \pi(v)$, if (u, v) is a (directed) edge of G. A random variable X_v is associated with each vertex v of the graph (for simplicity, we assume all the variables to take values in the same finite set $\mathcal{X}$). The joint distribution of $\{X_v, \ v \in V\}$ is determined by the conditional probability distributions $\{p(x_v|\underline{x}_{\pi(v)})\}$, where $\pi(v)$ denotes the set of parents of the vertex v, and $\underline{x}_{\pi(v)} = \{x_u : u \in \pi(v)\}$:

$$p(\underline{x}) = \prod_{v \in \pi(G)} p(x_v) \prod_{v \in G \backslash \pi(G)} p(x_v|\underline{x}_{\pi(v)}), \qquad (13.4)$$

where $\pi(G)$ denotes the set of vertices that have no parent in G.

A general class of statistical inference problems can be formulated as follows. We are given a Bayesian network, i.e. a directed graph G plus the associated conditional probability distributions $\{p(x_v|\underline{x}_{\pi(v)})\}$. A subset $O \subseteq V$ of the variables is observed and takes values $\underline{x}_O$. The problem is to compute marginals of the conditional distribution $p(\underline{x}_{V \backslash O}|\underline{x}_O)$.

Given a Bayesian network G and a set of observed variable O, it is easy to obtain a factor graph representation of the conditional distribution $p(\underline{x}_{V \backslash O}|\underline{x}_O)$, by a generalization of the procedure that we applied in Fig. 13.2. The general rule is as follows: (i) associate a variable node with each non-observed variable (i.e. each variable in $\underline{x}_{V \backslash O}$); ($ii$) for each variable in $\pi(G) \backslash O$, add a degree-1 function node connected uniquely to

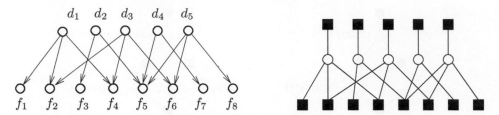

Fig. 13.3 *Left*: toy version of the QMR-DT Bayesian network. *Right*: factor graph representation of the conditional distribution of the diseases d_1, ..., d_5, given the findings f_1, ...,f_8.

that variable; (*iii*) for each non-observed vertex v which is not in $\pi(G)$, add a function node and connect it to v and to all the parents of v; (*iv*) finally, for each observed variable u, add a function node and connect it to all the parents of u.

Here is an example showing the practical utility of Bayesian networks.

Example 13.2 The Quick Medical Reference, Decision Theoretic (QMR-DT) network is a two-level Bayesian network that was developed for automatic medical diagnosis. A schematic example is shown in Fig. 13.3. Variables in the top level, denoted by $d_1, \dots, d_N$, are associated with *diseases*. Variables in the bottom level, denoted by $f_1, \dots, f_M$, are associated with symptoms, or *findings*. Both diseases and findings are described by binary variables. An edge connects a disease d_i to a finding f_a whenever that disease may be a cause for that finding. Such networks of implications are constructed on the basis of accumulated medical experience.

The network is completed by two types of probability distribution. For each disease d_i, we are given an *a priori* occurrence probability $p(d_i)$. Furthermore, for each finding, we have a conditional probability distribution for that finding given a certain disease pattern. This usually takes the 'noisy-OR' form

$$p(f_a = 0|d) = \frac{1}{z_a} \exp\left\{-\sum_{i=1}^{N} \theta_{ia} d_i\right\}. \qquad (13.5)$$

This network is used for diagnostic purposes. The findings are set to values determined by observation of a patient. Given this pattern of symptoms, one aims to compute the marginal probability that any given disease is indeed present.

13.1.2   Inference in coding, physics, and optimization

Several of the problems encountered so far in this book can be recast in the language of statistical inference.

Let us start with the decoding of error-correcting codes. As discussed in Chapter 6, in order to implement symbol MAP decoding, one has to compute the marginal distribution of input symbols, given the channel output. For instance, in the case of an LDPC code ensemble, dependencies between input symbols are induced by the parity check constraints. The joint probability distribution to be marginalized has a natural graphical representation (although we have used factor graphs rather than Bayesian networks). The introduction of finite-temperature decoding allows us to view word MAP decoding as the zero-temperature limit case of a one-parameter family of inference problems.

In statistical-mechanics models, one is mainly interested in the expectations and covariances of local observables with respect to the Boltzmann measure. For instance, the paramagnetic-to-ferromagnetic transition in an Ising ferromagnet, (see Section 2.5) can be located using the magnetization $M_N(\beta, B) = \langle \sigma_i \rangle_{\beta,B}$. The computation of co-variances, such as the correlation function $C_{ij}(\beta, B) = \langle \sigma_i; \sigma_j \rangle_{\beta,B}$, is a natural gener-alization of the simple inference problem discussed so far.

Let us consider, finally, consider the case of combinatorial optimization. Assume, for the sake of definiteness, that a feasible solution is an assignment of the variables $\underline{x} = (x_1, x_2, \ldots, x_N) \in \mathcal{X}^N$ and that its cost $E(\underline{x})$ can be written as a sum of 'local' terms,

$$E(\underline{x}) = \sum_a E_a(\underline{x}_a). \tag{13.6}$$

Here $\underline{x}_a$ denotes a subset of the variables $(x_1, x_2, \ldots, x_N)$. Let $\mu_*(\underline{x})$ denote the uniform distribution over the optimal solutions. The minimum energy can be computed as a sum of expectations with respect to this distribution: $E_* = \sum_a [\sum_{\underline{x}} \mu_*(\underline{x}) E_a(\underline{x}_a)]$. Of course, the distribution $\mu_*(\underline{x})$ does not necessarily have a simple representation, and therefore the computation of E_* can be significantly harder than simple inference.[1]

This problem can be overcome by 'softening' the distribution $\mu_*(\underline{x})$. One possibility is to introduce a finite temperature and define $\mu_\beta(\underline{x}) = \exp[-\beta E(\underline{x})]/Z$, as already done in Section 4.6: if β is large enough, this distribution is concentrated on optimal solutions. At the same time, it has an explicit representation (apart from the value of the normalization constant Z) at any value of β.

How large should β be in order to get a good estimate of E_*? The exercise below gives the answer under some rather general assumptions.

Exercise 13.1 Assume that the cost function $E(\underline{x})$ takes integer values, and let $U(\beta) = \langle E(\underline{x}) \rangle_\beta$. Owing to the form of eqn (13.6), the computation of $U(\beta)$ is essentially equivalent to statistical inference. We assume, furthermore that $\Delta_{\max} = \max[E(\underline{x}) - E_*]$ is bounded by a polynomial in N. Show that

$$0 \le \frac{\partial U}{\partial T} \le \frac{1}{T^2} \Delta_{\max}^2 |\mathcal{X}|^N e^{-1/T}, \tag{13.7}$$

where $T = 1/\beta$. Deduce that by taking $T = \Theta(1/N)$, one can obtain $|U(\beta) - E_*| \le \varepsilon$ for any fixed $\varepsilon > 0$.

---

[1]Consider, for instance, the MAX-SAT problem, and let $E(\underline{x})$ be the number of unsatisfied clauses under the assignment $\underline{x}$. If the formula under study is satisfiable, then $\mu_*(\underline{x})$ is proportional to the product of the characteristic functions associated with the clauses. In the opposite case, no explicit representation is known.

In fact, a much larger temperature (smaller β) can be used in many important cases. We refer to Chapter 2 for examples in which $U(\beta) = E_* + E_1(N) \, e^{-\beta} + O(e^{-2\beta})$, with $E_1(N)$ growing polynomially in N. In such cases one expects $\beta = \Theta(\log N)$ to be large enough.

13.2 The Monte Carlo method: Inference via sampling

Consider the statistical inference problem of computing the marginal probability $\mu(x_i = x)$ from a joint distribution $\mu(\underline{x})$, $\underline{x} = (x_1, x_2, \ldots, x_N) \in \mathcal{X}^N$. Given L i.i.d. samples $\{\underline{x}^{(1)}, \ldots, \underline{x}^{(L)}\}$ drawn from the distribution $\mu(\underline{x})$, the desired marginal $\mu(x_i = x)$ can be estimated as the the fraction of such samples for which $x_i = x$.

'Almost i.i.d.' samples from $p(\underline{x})$ can be produced, in principle, using the Markov chain Monte Carlo (MCMC) method of Section 4.5. Therefore the MCMC method can be viewed as a general-purpose inference strategy which can be applied in a variety of contexts.

Notice that the locality of the interactions, expressed by the factor graph, is very useful since it allows one to generate 'local' changes in $\underline{x}$ (e.g. changing only one x_i or a small number of them) easily. This will typically change the value of a few compatibility functions and hence produce only a small change in $p(\underline{x})$ (i.e., in physical terms, in the energy of $\underline{x}$). The possibility of generating, given $\underline{x}$, a new configuration close in energy is important for the MCMC method to work. In fact, moves that increase the system energy by a large amount are typically rejected within MCMC rules.

One should also be aware that sampling, for instance by the MCMC method, only allows one to estimate marginals or expectations which involve a small subset of variables. It would be very hard, for instance, to estimate the probability of a particular configuration $\underline{x}$ through the number $L(\underline{x})$ of its occurrences in the samples. The reason is that at least $1/\mu(\underline{x})$ samples would be required for us to obtain any accuracy, and this is typically a number exponentially large in N.

13.2.1 LDPC codes

Consider a code $\mathfrak{C}$ from one of the LDPC ensembles introduced in Chapter 11, and assume that it has been used to communicate over a binary-input memoryless symmetric channel with transition probability $Q(y|x)$. As shown in Chapter 6 (see eqn (6.5)), the conditional distribution of the channel input $\underline{x}$, given the output $\underline{y}$, reads

$$\mu_y(\underline{x}) \equiv \mathbb{P}(\underline{x}|\underline{y}) = \frac{1}{Z(\underline{y})} \, \mathbb{I}(\underline{x} \in \mathfrak{C}) \prod_{i=1}^{N} Q(y_i|x_i) \, . \tag{13.8}$$

We can use the explicit representation of the code membership function to write

$$\mu_{y,\beta}(\underline{x}) = \frac{1}{Z(\underline{y})} \prod_{a=1}^{M} \mathbb{I}(x_{i_1^a} \oplus \cdots \oplus x_{i_k^a} = 0) \prod_{i=1}^{N} Q(y_i|x_i) \, . \tag{13.9}$$

in order to implement symbol MAP decoding, we must compute the marginals $\mu_y^{(i)}(x_i)$ of this distribution. Let us see how this can be done in an approximate way via MCMC sampling.

Unfortunately, the simple MCMC algorithms introduced in Section 4.5 (single bit flip with acceptance test satisfying detailed balance) cannot be applied in the present case. In any reasonable LDPC code, each variable x_i is involved in at least one parity check constraint. Suppose that we start the MCMC algorithm from a random configuration $\underline{x}$ distributed according to eqn (13.9). Since $\underline{x}$ has a non-vanishing probability, it satisfies all the parity check constraints. If we propose a new configuration where bit x_i is flipped, this configuration will violate all the parity check constraints involving x_i. As a consequence, such a move will be rejected by any rule satisfying detailed balance. The Markov chain is therefore reducible (each codeword forms a separate ergodic component), and useless for sampling purposes.

In good codes, this problem is not easily cured by allowing for moves that flip more than a single bit. The reason is that a number of bits greater than or equal to the minimum distance must be flipped simultaneously. But as we saw in Section 11.2, if $\mathfrak{C}$ is drawn from an LDPC ensemble with variable nodes of degree greater than or equal to 2, its minimum distance diverges with the block length (logarithmically if the minimum degree is 2, linearly if this is at least 3). Large moves of this type are likely to be rejected, because they imply a large, uncontrolled variation in the likelihood $\prod_{i=1}^{N} Q(y_i|x_i)$.

A way out of this dilemma consists in 'softening' the parity check constraint by introducing a 'parity check temperature' β and the associated distribution

$$\mu_{y,\beta}(\underline{x}) = \frac{1}{Z(\underline{y},\beta)} \prod_{a=1}^{M} e^{-\beta E_a(x_{i_1^a}\ldots x_{i_k^a})} \prod_{i=1}^{N} Q(y_i|x_i). \tag{13.10}$$

Here the energy term $E_a(x_{i_1^a}\ldots x_{i_k^a})$ takes values 0 if $x_{i_1^a} \oplus \cdots \oplus x_{i_k^a} = 0$, and 1 otherwise. In the limit $\beta \to \infty$, the distribution (13.10) reduces to eqn (13.9). The idea is now to estimate the marginals of eqn (13.10), $\mu_{y,\beta}^{(i)}(x_i)$, via MCMC sampling and then to use the decoding rule

$$x_i^{(\beta)}(\underline{y}) \equiv \arg\max_{x_i} \mu_{y,\beta}^{(i)}(x_i). \tag{13.11}$$

For any finite β, this prescription is certainly suboptimal with respect to symbol MAP decoding. In particular, the distribution (13.10) gives non-zero weight to words $\underline{x}$ which do not belong to the codebook $\mathfrak{C}$. On the other hand, one may hope that for β large enough, the above prescription achieves a close-to-optimal bit error rate.

We can simplify the above strategy further by giving up the objective of approximating the marginal $\mu_{y,\beta}^{(i)}(x_i)$ within any prescribed accuracy. Instead, we run the Glauber single-bit-flip MCMC algorithm for a fixed computer time and extract an estimate of $\mu_{y,\beta}^{(i)}(x_i)$ from this run. Figure 13.4 shows the results of Glauber dynamics executed for $2LN$ steps starting from a uniformly random configuration. At each step, a bit i is chosen uniformly at random and flipped with a probability

$$w_i(\underline{x}) = \frac{\mu_{y,\beta}(\underline{x}^{(i)})}{\mu_{y,\beta}(\underline{x}^{(i)}) + \mu_{y,\beta}(\underline{x})}, \tag{13.12}$$

where $\underline{x}$ is the current configuration and $\underline{x}^{(i)}$ is the configuration obtained from $\underline{x}$ by flipping the i-th bit.

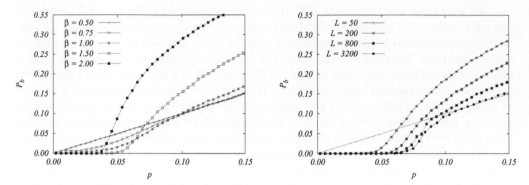

Fig. 13.4 Decoding LDPC codes from the $(3, 6)$ ensemble, used over a bynary symmetric channel with flip probability p, using MCMC sampling. The bit error rate is plotted versus p. The block length was fixed to $N = 2000$, and the number of sweeps was $2L$. *Left*: for $L = 100$ and several values of the effective inverse temperature β. *Right*: improvement of the performance as the number of sweeps increases at fixed $\beta = 1.5$.

Exercise 13.2 Derive an explicit expression for $w_i(\underline{x})$, and show that this probability can be computed with a number of operations independent of the block length.

In this context, one often refers to a sequence of N successive updates as a **sweep** (on average, one flip is proposed for each bit in a sweep). The value of x_i is recorded in each of the last L sweeps, and the decoder output is $x_i = 0$ or $x_i = 1$ depending on which value occurs more often in this record.

The data in Fig. 13.4 refers to communication over a binary symmetric channel with flip probability p. In the left frame, we have fixed $L = 100$ and used several values of β. At small β, the resulting bit error rate is almost indistinguishable from that in the absence of coding, namely $\mathrm{P_b} = p$. As β increases, parity checks are enforced more and more strictly and the error-correcting capabilities improve at low noise. The behaviour is qualitatively different for larger noise levels: for $p \gtrsim 0.05$, the bit error rate increases with β. The reason for this change is essentially dynamical. The Markov chain used for sampling from the distribution (13.10) decorrelates more and more slowly from its initial condition. Since the initial condition is uniformly random, thus yielding $\mathrm{P_b} = 1/2$, the bit error rate obtained through our algorithm approaches $1/2$ at large β (and above a certain threshold of p). This interpretation is confirmed by the data in the right frame of the figure.

We shall see in Chapter 15 that in the large-block-length limit, the threshold for errorless bit MAP decoding in this case is predicted to be $p_c \approx 0.101$. Unfortunately, because of its slow dynamics, our MCMC decoder cannot be used in practice if the channel noise is close to this threshold.

The sluggish dynamics of the single spin-flip MCMC for the distribution (13.10) is partially related to its reducibility for the model with hard constraints (13.9). An initial intuitive picture is as follows. At large β, codewords correspond to isolated 'lumps' of probability with $\mu_{y,\beta}(\underline{x}) = \Theta(1)$, separated by improbable regions such that $\mu_{y,\beta}(\underline{x}) = \Theta(e^{-2\beta})$ or smaller. In order to decorrelate, the Markov chain must spend a long time (at least of the order of the minimum distance of the code) in an improbable region, and this happens only very rarely. This rough explanation is neither complete nor entirely correct, but we shall refine it in Chapters 15 and 21.

13.2.2 Ising model

Some of the basic mechanisms responsible for the slowing down of Glauber dynamics can be understood with simple statistical-mechanics models. In this subsection, we consider a ferromagnetic Ising model with an energy function

$$E(\underline{\sigma}) = -\sum_{(ij) \in G} \sigma_i \sigma_j \,. \tag{13.13}$$

Here G is an ordinary graph with N vertices, whose precise structure will depend on the particular example. The Monte Carlo method is applied to the problem of sampling from the Boltzmann distribution $\mu_\beta(\underline{\sigma})$ at an inverse temperature β.

As in the previous subsection, we focus on Glauber (or heat bath) dynamics, but rescale the time: in an infinitesimal interval dt, a flip is proposed with probability $N dt$ at a uniformly random site i. The flip is accepted with the usual heat bath probability (here, $\underline{\sigma}$ is the current configuration and $\underline{\sigma}^{(i)}$ is the configuration obtained by flipping the spin σ_i),

$$w_i(\underline{\sigma}) = \frac{\mu_\beta(\underline{\sigma}^{(i)})}{\mu_\beta(\underline{\sigma}) + \mu_\beta(\underline{\sigma}^{(i)})} \,. \tag{13.14}$$

Let us consider equilibrium dynamics first. We assume, therefore, that the initial configuration $\underline{\sigma}(0)$ is sampled from the equilibrium distribution $\mu_\beta(\cdot)$ and ask how many Monte Carlo steps must be performed in order to obtain an effectively independent random configuration. A convenient way of monitoring the equilibrium dynamics consists in computing the time correlation function

$$C_N(t) \equiv \frac{1}{N} \sum_{i=1}^{N} \langle \sigma_i(0) \sigma_i(t) \rangle \,. \tag{13.15}$$

Here the average $\langle \cdot \rangle$ is taken with respect to the realization of the Monte Carlo dynamics, as well as with respect to the initial state $\underline{\sigma}(0)$. Note that $(1 - C(t))/2$ is the average fraction of spins which differ in the configurations $\underline{\sigma}(0)$ and $\underline{\sigma}(t)$. One therefore expects $C(t)$ to decrease with t, asymptotically reaching 0 when $\underline{\sigma}(0)$ and $\underline{\sigma}(t)$ are well decorrelated (in this model, each spin is equally likely to take a value $+1$ or -1).

The reader may wonder how one can sample $\underline{\sigma}(0)$ from the equilibrium (Boltzmann) distribution. As already suggested in Section 4.5, within the Monte Carlo approach one can obtain an 'almost' equilibrium configuration by starting from an arbitrary configuration and running the Markov chain for sufficiently many steps. In practice, we initialize our chain from a uniformly random configuration (i.e. a $\beta = 0$, or infinite-temperature, equilibrium configuration) and run the dynamics for t_w sweeps. We call

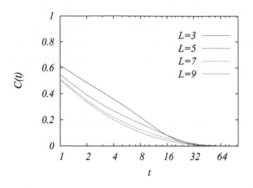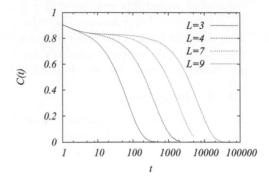

Fig. 13.5 Equilibrium correlation function for an Ising model on a two-dimensional grid of side L. *Left*: high temperature, $T = 3$. *Right*: low temperature, $T = 2$.

the configuration obtained after this process $\underline{\sigma}(0)$ and we then run for t more sweeps in order to measure $C(t)$. The choice of t_w is, of course, crucial: in general, the above procedure will produce a configuration $\underline{\sigma}(0)$ whose distribution is not the equilibrium one, and depends on t_w. The measured correlation function will also depend on t_w. Determining how large t_w must be in order to obtain a good enough approximation to $C(t)$ is a subject of intense theoretical work. A simple empirical rule is to measure $C(t)$ for a given large t_w, then double t_w and check that nothing has changed.

Exercise 13.3 Using the instructions above, the reader is invited to write an MCMC code for the Ising model on a general graph and reproduce the data in the following examples.

Example 13.3 We begin by considering an Ising model on a two-dimensional grid of side L, with periodic boundary conditions. The vertex set is $\{(x_1, x_2) : 1 \leq x_a \leq L\}$. Edges join any two vertices at a (Euclidean) distance one, and also join the vertices (L, x_2) to $(1, x_2)$ and (x_1, L) to $(x_1, 1)$. We denote by $C_L(t)$ the correlation function for such a graph.

In Chapter 2, we saw that this model undergoes a phase transition at a critical temperature $1/\beta_c = T_c = 2/\log(1 + \sqrt{2}) \approx 2.269185$. The correlation functions plotted in Fig. 13.5 are representative of the qualitative behaviour in the high-temperature (left) and low-temperature (right) phases. At high temperature, $C_L(t)$ depends only mildly on the linear size L of the system. As L increases, the correlation function rapidly approaches a limit curve $C(t)$ which decreases from 1 to 0 on a finite time scale (more precisely, for any $\delta > 0$, $C_L(t)$ decreases below δ in a time which is bounded as $L \to \infty$).

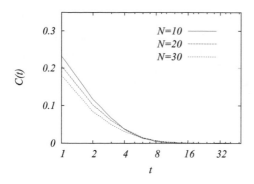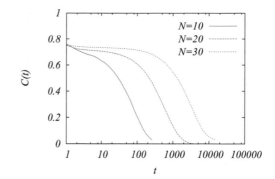

Fig. 13.6 Equilibrium correlation function for an Ising model on random graphs from the ensemble $\mathbb{G}_N(2, M)$, with $M = 2N$. *Left*: high temperature, $T = 5$. *Right*: low temperature, $T = 2$.

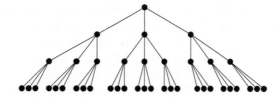

Fig. 13.7 A rooted ternary tree with $n = 4$ generations and $N = 40$ vertices.

At low temperature, there exists no limiting curve $C(t)$ decreasing from 1 to 0 such that $C_L(t) \to C(t)$ as $L \to \infty$. The time required for the correlation function $C_L(t)$ to get close to 0 is much larger than in the high-temperature phase. More importantly, it depends strongly on the system size. This suggests that strong cooperative effects are responsible for the slowing down of the dynamics.

Example 13.4 We take G as a random graph from the ensemble $\mathbb{G}_N(2, M)$, with $M = N\alpha$. As we shall see in Chapter 17, this model undergoes a phase transition when $N \to \infty$ at a critical temperature β_c, satisfying the equation $2\alpha \tanh \beta = 1$. In Fig. 13.6, we present numerical data for a few values of N, and $\alpha = 2$ (corresponding to a critical temperature $T_c \approx 3.915230$).

The curves presented here are representative of the high-temperature and low-temperature phases. As in the previous example, the relaxation time scale depends strongly on the system size at low temperature.

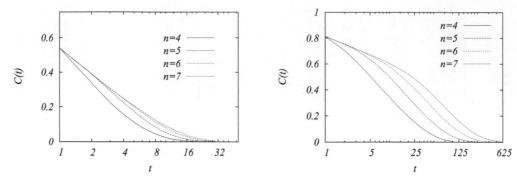

Fig. 13.8 Equilibrium correlation function for a ferromagnetic Ising model on a regular ternary tree. *Left*: high temperature, $T = 2$. *Right*: low temperature, $T = 1.25$.

Example 13.5 We take G as a rooted ternary tree with n generations (see Fig. 13.7). This tree contains $N = (3^n - 1)/(3 - 1)$ vertices and $N - 1$ edges. Using the methods of Chapter 17, one can show that this model undergoes a phase transition at a critical temperature β_c, which satisfies the equation $3(\tanh \beta)^2 = 1$. We therefore get $T_c \approx 1.528651$. In this case the dynamics of a spin depends strongly upon its distance from the root. In particular, leaf spins are much less constrained than the others. In order to single out the 'bulk' behaviour, we modify the definition of the correlation function (13.15) by averaging only over the spins σ_i in the first $\underline{n} = 3$ generations. We keep $\underline{n}$ fixed as $n \to \infty$.

As in the previous examples, $C_N(t)$ (see Fig. 13.8) has a well-defined $N \to \infty$ limit in the high-temperature phase, and is strongly size-dependent at low temperature.

We can summarize the last three examples by comparing the size dependence of the relaxation times scale in the respective low-temperature phases. A simple way to define such a time scale is to look for the smallest time such that $C(t)$ decreases below some given threshold:

$$\tau(\delta; N) = \min\{ t > 0 \ \text{such that} \ \ C_N(t) \le \delta \} . \tag{13.16}$$

In Fig. 13.9, we plot estimates obtained from the data presented in the previous examples, using $\delta = 0.2$, and restricting ourselves to the data in the low-temperature (ferromagnetic) phase. The size dependence of $\tau(\delta; N)$ is very clear. However, it is much stronger for the random-graph and square-grid cases (and, in particular, in the former of these) than for the tree case. In fact, it can be shown that, in the ferromagnetic phase,

$$\tau(\delta; N) = \begin{cases} \exp\{\Theta(N)\} & \text{random graph,} \\ \exp\{\Theta(\sqrt{N})\} & \text{square lattice,} \\ \exp\{\Theta(\log N)\} & \text{tree.} \end{cases} \tag{13.17}$$

Section 13.3 will explain the origin of these different behaviours.

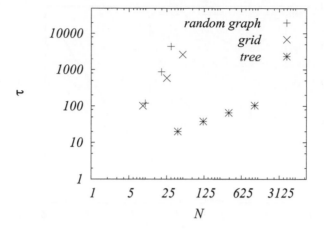

Fig. 13.9 Size dependence of the relaxation time in ferromagnetic Ising models in the low-temperature phase. The various symbols refer to the various families of graphs considered in Examples 13.3–13.5.

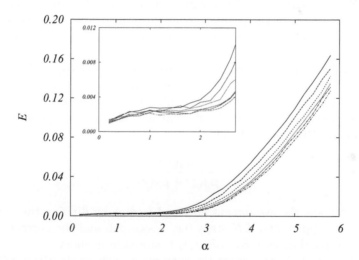

Fig. 13.10 Minimization of the number of unsatisfied clauses in random 3-SAT formulae via Glauber dynamics. The number of variables $N = 1000$ was kept fixed. Here we take $T = 0.25$ and, from *top* to *bottom* $L = 2.5 \times 10^3$, 5×10^3, 10^4, 2×10^4, 4×10^4, and 8×10^4 iterations. The *inset* show the small-α regime in greater detail.

13.2.3 MAX-SAT

Given a satisfiability formula over N Boolean variables $(x_1, \ldots, x_N) = \underline{x}$, $x_i \in \{0, 1\}$, the MAX-SAT optimization problem requires one to find a truth assignment which satisfies the largest number of clauses. We denote by $\underline{x}_a$ the set of variables involved in the a-th clause, and by $E_a(\underline{x}_a)$ a function of the truth assignment that takes the

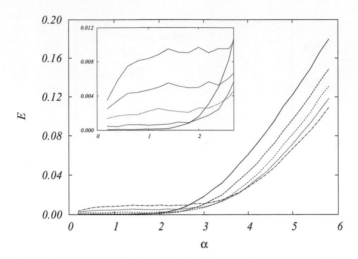

Fig. 13.11 Same numerical simulation as in Fig. 13.10. Here we fix $L = 4 \cdot 10^4$ and take, from *top* to *bottom* at large α, $T = 0.15, 0.20, 0.25, 0.30, 0.35$.

value 0 if the clause is satisfied and 1 otherwise. With these definitions, the MAX-SAT problem can be rephrased as the problem of minimizing an energy function of the form $E(\underline{x}) = \sum_a E_a(\underline{x}_a)$, and we can therefore apply the general approach discussed after eqn (13.6).

We thus consider the Boltzmann distribution $\mu_\beta(\underline{x}) = \exp[-\beta E(\underline{x})]/Z$ and try to sample a configuration from $\mu_\beta(\underline{x})$ at large enough β using the MCMC method. The assignment $\underline{x} \in \{0, 1\}^N$ is initialized uniformly at random. At each time step, a variable index i is chosen uniformly at random and the corresponding variable is flipped according to the heat bath rule

$$w_i(\underline{x}) = \frac{\mu_\beta(\underline{x}^{(i)})}{\mu_\beta(\underline{x}) + \mu_\beta(\underline{x}^{(i)})} \, . \tag{13.18}$$

As above, $\underline{x}^{(i)}$ denotes the assignment obtained from $\underline{x}$ by flipping the i-th variable. The algorithm is stopped after LN steps (i.e. L sweeps), and the current assignment $\underline{x}_*$ (and the corresponding cost $E_* = E(\underline{x}_*)$) is stored in memory.

In Fig. 13.10 and 13.11, we present the outcome of such an algorithm when applied to random 3-SAT formulae from the ensemble $\mathsf{SAT}_N(3, M)$ with $\alpha = M/N$. Here we focus on the mean cost $\langle E_* \rangle$ of the returned assignment. One expects that, as $N \to \infty$ with fixed L, the cost scales as $\langle E_* \rangle = \Theta(N)$, and order-$N$ fluctuations of E_* away from the mean become exponentially unlikely. At low enough temperature, the behaviour depends dramatically on the value of α. For small α, E_*/N is small and grows rather slowly with α. Furthermore, it seems to decrease to 0 as β increases. Our strategy is essentially successful and finds an (almost) satisfying assignment. Above $\alpha \approx 2$ to 3, E_*/N starts to grow more rapidly with α, and does not seem to vanish as $\beta \to \infty$. Even more striking is the behaviour as the number of sweeps L increases. In the small-α regime, E_*/N decreases rapidly to some roughly L-independent saturation

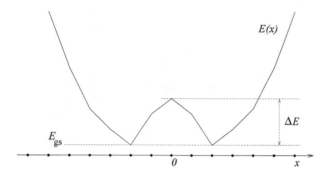

Fig. 13.12 Random walk in a double-well energy landscape. After how many steps is the walker distributed (approximately) according to the equilibrium distribution?

value, already reached after about 10^3 sweeps. At large α there seems also to be an asymptotic value, but this is reached much more slowly, and even after 10^5 sweeps there is still space for improvement.

13.3   Free-energy barriers

The examples above show that the time scale required for a Monte Carlo algorithm to produce (approximately) statistically independent configurations may vary wildly depending on the particular problem at hand. The same is true if we consider the time required in order to generate a configuration distributed approximately according to the equilibrium distribution, starting from an arbitrary initial condition.

There exist various sophisticated techniques for estimating these time scales analytically, at least in the case of unfrustrated problems. In this section, we discuss a simple argument which is widely used in statistical physics as well as in probability theory, that of free-energy barriers. The basic intuition can be conveyed by simple examples.

Example 13.6 Consider a particle moving on the line of integers, and denote its position by $x \in \mathbb{Z}$. Each point x on the line has an energy $E(x) \geq E_{\text{gs}}$ associated with it, as depicted in Fig. 13.12. At each time step, the particle attempts to move to one of the adjacent positions (either to the right or to the left) with probability $1/2$. If we denote by x' the position that the particle is trying to move to, the move is accepted according to the Metropolis rule

$$w(x \to x') = \min\left\{ e^{-\beta[E(x')-E(x)]}, 1 \right\}. \qquad (13.19)$$

The equilibrium distribution is of course given by Boltzmann's law $\mu_\beta(x) = \exp[-\beta E(x)]/Z(\beta)$.

Suppose we start with, say, $x = 10$. How many steps must we wait for in order that the random variable x will be distributed according to $\mu_\beta(x)$? It is intuitively

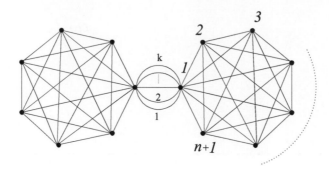

Fig. 13.13 How much time does a random walk need to explore this graph?

clear that, in order to equilibrate, the particle must spend some amount of time *both* in the right *and* in the left well, and therefore it must visit the site $x = 0$. At equilibrium, this is visited on average a fraction $\mu_\beta(0)$ of the time. This gives an estimate of the time needed to see the system change from one well to the other:

$$\tau \approx \frac{1}{\mu_\beta(0)}.\qquad(13.20)$$

One is often interested in the low-temperature limit of τ. Assuming that $E(x)$ diverges fast enough as $|x| \to \infty$, the leading exponential behaviour of Z is $Z(\beta) \doteq e^{-\beta E_{gs}}$, and therefore $\tau \doteq \exp\{\beta\Delta E\}$, where $\Delta E = E(0) - E_{gs}$ is the energy barrier to be crossed in order to pass from one well to the other. A low-temperature asymptotics of the type $\tau \doteq \exp\{\beta\Delta E\}$ is referred to as an **Arrhenius law**.

Exercise 13.4 Consider a random walk on the graph in Fig. 13.13 (two cliques with $n+1$ vertices, joined by a k-fold degenerate edge). At each time step, the walker chooses one of the adjacent edges uniformly at random and moves along it to the next node. What is the stationary distribution $\mu_{eq}(x)$, $x \in \{1, \ldots, 2n\}$? Show that the probability to be at node 1 is $\frac{1}{2}(k+n-1)/(n^2+k-n)$.

Suppose we start with a walker distributed according to $\mu_{eq}(x)$. Using an argument similar to that in the previous example, estimate the number of time steps τ that one must wait for in order to obtain an approximatively independent value of x. Show that $\tau \simeq 2n$ when $n \gg k$, and interpret this result. In this case the k-fold degenerate edge joining the two cliques is called a bottleneck, and one speaks of an entropy barrier.

In order to obtain a precise mathematical formulation of the intuition gained in the preceding examples, we must define what we mean by 'relaxation time'. We shall focus here on ergodic continuous-time Markov chains on a finite state space $\mathcal{X}$. Such a chain is described by its transition rates $w(x \to y)$. If, at time t, the chain is in a state $x(t) = x \in \mathcal{X}$, then, for any $y \neq x$, the probability that the chain is in a state y, 'just after' time t, is

$$\mathbb{P}\left\{x(t+\mathrm{d}t)=y \mid x(t)=x\right\}=w(x\to y)\mathrm{d}t\,. \tag{13.21}$$

Such a chain is said to satisfy the detailed balance condition (or to be reversible) if, for any $x\neq y$, $\mu(x)w(x\to y)=\mu(y)w(y\to x)$.

Exercise 13.5 Consider a discrete-time Markov chain and modify it as follows. Instead of waiting a unit time Δt between successive steps, wait an exponentially distributed random time (i.e. Δt is a random variable with pdf $p(\Delta t)=\exp(-\Delta t)$). Show that the resulting process is a continuous-time Markov chain. What are the corresponding transition rates?

Let $x\mapsto\mathcal{O}(x)$ be an observable (a function of the state), define the shorthand $\mathcal{O}(t)=\mathcal{O}(x(t))$, and assume $x(0)$ to be drawn from the stationary distribution. If the chain satisfies the detailed balance condition, one can show that the correlation function $C_{\mathcal{O}}(t)=\langle\mathcal{O}(0)\mathcal{O}(t)\rangle-\langle\mathcal{O}(0)\rangle\langle\mathcal{O}(t)\rangle$ is non-negative and monotonically decreasing, and that $C_{\mathcal{O}}(t)\to 0$ as $t\to\infty$. The exponential autocorrelation time for the observable $\mathcal{O}$, denoted by $\tau_{\mathcal{O},\mathrm{exp}}$, is defined by

$$\frac{1}{\tau_{\mathcal{O},\mathrm{exp}}}=-\lim_{t\to\infty}\frac{1}{t}\log C_{\mathcal{O}}(t)\,. \tag{13.22}$$

The time $\tau_{\mathcal{O},\mathrm{exp}}$ depends on the observable and tells us how fast its autocorrelation function decays to 0: $C_{\mathcal{O}}(t)\sim\exp(-t/\tau_{\mathcal{O},\mathrm{exp}})$. It is meaningful to look for the 'slowest' observable and to define the **exponential autocorrelation time** (also called the **inverse spectral gap**, or, for brevity, the **relaxation time**) of the Markov chain as

$$\tau_{\mathrm{exp}}=\sup_{\mathcal{O}}\left\{\tau_{\mathcal{O},\mathrm{exp}}\right\}\,. \tag{13.23}$$

The idea of a bottleneck, and its relationship to the relaxation time, is clarified by the following theorem.

Theorem 13.7 *Consider an ergodic continuous-time Markov chain with state space $\mathcal{X}$, and with transition rates $\{w(x\to y)\}$ satisfying detailed balance with respect to the stationary distribution $\mu(x)$. Given any two disjoint sets of states $\mathcal{A},\mathcal{B}\subset\mathcal{X}$, define the probability flux between them as $W(\mathcal{A}\to\mathcal{B})=\sum_{x\in\mathcal{A},\,y\in\mathcal{B}}\mu(x)\,w(x\to y)$. Then*

$$\tau_{\mathrm{exp}}\geq 2\,\frac{\mu(x\in\mathcal{A})\,\mu(x\notin\mathcal{A})}{W(\mathcal{A}\to\mathcal{X}\backslash\mathcal{A})}\,. \tag{13.24}$$

In other words, a lower bound on the correlation time can be constructed by looking for 'bottlenecks' in the Markov chain, i.e. partitions of the configuration space into two subsets. The lower bound will be good (and the Markov chain will be slow) if each of the subsets carries a reasonably large probability at equilibrium, but jumping from one to the other is unlikely.

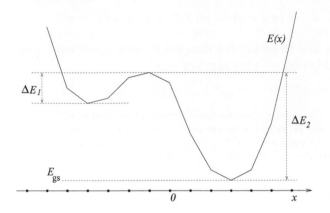

Fig. 13.14 Random walk in an asymmetric double well.

Example 13.8 Consider a random walk in the double-well energy landscape of Fig. 13.12, where we confine the random walk to some large interval $[-M : M]$ in order to have a finite state space. Let us apply Theorem 13.7, by taking $\mathcal{A} = \{x \geq 0\}$. We have $W(\mathcal{A} \to \mathcal{X}\backslash\mathcal{A}) = \mu_\beta(0)/2$ and, by symmetry, $\mu_\beta(x \in \mathcal{A}) = \frac{1}{2}(1 + \mu_\beta(0))$. The inequality (13.24) yields

$$\tau_{\exp} \geq \frac{1 - \mu_\beta(0)^2}{\mu_\beta(0)}. \tag{13.25}$$

Expanding the right-hand side in the low-temperature limit, we get $\tau_{\exp} \geq 2\, e^{\beta\Delta E}\, (1 + \Theta(e^{-c\beta}))$.

Exercise 13.6 Apply Theorem 13.7 to a random walk in the asymmetric double-well energy landscape of Fig. 13.14. Does the Arrhenius law $\tau_{\exp} \sim \exp(\beta\Delta E)$ apply to this case? What is the relevant energy barrier ΔE?

Exercise 13.7 Apply Theorem 13.7 to estimate the relaxation time of a random walk on the graph in Exercise 13.4.

Example 13.9 Consider Glauber dynamics for an Ising model on a two-dimensional $L \times L$ grid, with periodic boundary conditions, discussed earlier in Example 13.3. In the ferromagnetic phase, the distribution of the total magnetization $\mathcal{M}(\underline{\sigma}) \equiv \sum_i \sigma_i$, $N = L^2$, is concentrated around the values $\pm N M_+(\beta)$, where $M_+(\beta)$ is the spontaneous magnetization. It is natural to expect that the bottleneck will correspond to the global magnetization changing sign. Assuming, for instance, that L is odd, let us define

$$\mathcal{A} = \{\underline{\sigma} : \mathcal{M}(\underline{\sigma}) \geq 1\}, \quad \bar{\mathcal{A}} = \mathcal{X} \backslash \mathcal{A} = \{\underline{\sigma} : \mathcal{M}(\underline{\sigma}) \leq -1\}. \quad (13.26)$$

Using the symmetry under spin reversal, Theorem 13.7 yields

$$\tau_{\exp} \geq \left\{ 2 \sum_{\underline{\sigma} : \mathcal{M}(\underline{\sigma})=1} \sum_{i : \sigma_i = 1} \mu_\beta(\underline{\sigma}) \, w(\underline{\sigma} \to \underline{\sigma}^{(i)}) \right\}^{-1}. \quad (13.27)$$

A good estimate of this sum can be obtained by noticing that, for any $\underline{\sigma}$, $w(\underline{\sigma} \to \underline{\sigma}^{(i)}) \geq w(\beta) \equiv \frac{1}{2}(1 - \tanh 4\beta)$. Moreover, for any $\underline{\sigma}$ entering the sum, there are exactly $(L^2 + 1)/2$ sites i such that $\sigma_i = +1$. We therefore get $\tau_{\exp} \geq (L^2 w(\beta) \sum_{\underline{\sigma} : \mathcal{M}(\underline{\sigma})=1} \mu_\beta(\underline{\sigma}))^{-1}$ One suggestive way of writing this lower bound consists in defining a constrained free energy as follows:

$$F_L(m; \beta) \equiv -\frac{1}{\beta} \log \left\{ \sum_{\underline{\sigma} : \mathcal{M}(\underline{\sigma})=m} \exp[-\beta E(\underline{\sigma})] \right\}. \quad (13.28)$$

If we denote the usual (unconstrained) free energy by $F_L(\beta)$, our lower bound can be written as

$$\tau_{\exp} \geq \frac{1}{L^2 w(\beta)} \exp\{\beta[F_L(1; \beta) - F_L(\beta)]\}. \quad (13.29)$$

Apart from the pre-exponential factors, this expression has the same form as the Arrhenius law, the energy barrier ΔE being replaced by a 'free-energy barrier' $\Delta F_L(\beta) \equiv F_L(1; \beta) - F_L(\beta)$.

We are left with the task of estimating $\Delta F_L(\beta)$. Let us start by considering the $\beta \to \infty$ limit. In this regime, $F_L(\beta)$ is dominated by the all-plus and all-minus configurations, with energy $E_{\rm gs} = -2L^2$. Analogously, $F_L(1; \beta)$ is dominated by the lowest-energy configurations satisfying the constraint $\mathcal{M}(\underline{\sigma}) = 1$. An example of such a configuration is that shown in Fig. 13.15, whose energy is $E(\underline{\sigma}) = -2L^2 + 2(2L + 2)$. Of course, all configurations obtained from that in Fig. 13.15, through a translation, rotation, or spin inversion have the same energy. We therefore find $\Delta F_L(\beta) = 2(2L + 2) + \Theta(1/\beta)$.

It is reasonable to guess (and it can be proved rigorously) that the size dependence of $\Delta F_L(\beta)$ remains unchanged through the whole low-temperature phase:

$$\Delta F_L(\beta) \simeq 2\gamma(\beta)L, \quad (13.30)$$

where the **surface tension** $\gamma(\beta)$ is strictly positive at any $\beta > \beta_{\rm c}$, and vanishes as $\beta \downarrow \beta_{\rm c}$. This, in turn, implies the following lower bound on the correlation time:

$$\tau_{\exp} \geq \exp\{2\beta\gamma(\beta)L + o(L)\}. \quad (13.31)$$

This bound matches the numerical simulations in the previous section and can be proved to give the correct asymptotic size dependence.

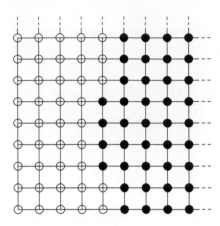

Fig. 13.15 Ferromagnetic Ising model on a 9×9 grid with periodic boundary conditions. Open circles correspond to $\sigma_i = +1$, and filled circles to $\sigma_i = -1$. The configuration shown here has an energy $E(\underline{\sigma}) = -122$ and a magnetization $\mathcal{M}(\underline{\sigma}) = +1$.

Exercise 13.8 Consider the ferromagnetic Ising model on a random graph from $\mathbb{G}_N(2, M)$ that we studied in Example 13.4, and assume, for definiteness, N even. Arguing as above, show that

$$\tau_{\exp} \geq C_N(\beta) \exp\{\beta[F_N(0; \beta) - F_N(\beta)]\}. \tag{13.32}$$

Here $F_N(m; \beta)$ is the free energy of the model constrained to $\mathcal{M}(\underline{\sigma}) = m$, $F_N(\beta)$ is the unconstrained free energy, and $C_N(\beta)$ is a constant which grows, with high probability, slower than exponentially with N.

For a graph G, let $\delta(G)$ be the minimum number of bicoloured edges if we colour half of the vertices red and half blue. Show that

$$F_N(0; \beta) - F_N(\beta) = 2\delta(G_N) + \Theta(1/\beta). \tag{13.33}$$

The problem of computing $\delta(G)$ for a given graph G is referred to as the **balanced minimum cut** (or **graph-partitioning**) problem, and is known to be NP-complete. For a random graph in $\mathbb{G}_N(2, M)$, it is known that $\delta(G_N) = \Theta(N)$ with high probability in the limit $N \to \infty$, $M \to \infty$, with $\alpha = M/N$ fixed and $\alpha > 1/2$. (If $\alpha < 1/2$, the graph does not contain a giant component and, obviously, $\delta(G) = o(N)$.)

This claim can be substantiated through the following calculation. Given a spin configuration $\underline{\sigma} = (\sigma_1, \ldots, \sigma_N)$ with $\sum_i \sigma_i = 0$, let $\Delta_G(\underline{\sigma})$ be the number of edges (i, j) in G such that $\sigma_i \neq \sigma_j$. Then

$$\mathbb{P}\{\delta(G) \leq n\} = \mathbb{P}\{\exists \underline{\sigma} \text{ such that } \Delta_G(\underline{\sigma}) \leq n\} \leq \sum_{m=0}^{n} \mathbb{E}\mathcal{N}_{G,m}, \tag{13.34}$$

where $\mathcal{N}_{G,m}$ denotes the number of spin configurations with $\Delta_G(\underline{\sigma}) = m$. Show that

$$\mathbb{E}\mathcal{N}_{G,m} = \binom{N}{N/2} \binom{N}{2}^{-M} \binom{M}{m} \left(\frac{N^2}{4}\right)^m \left[\binom{N}{2} - \frac{N^2}{4}\right]^{M-m}. \tag{13.35}$$

Estimate this expression for large N and M with $\alpha = M/N$ fixed, and show that it implies $\delta(G) \geq c(\alpha)N$ with high probability, where $c(\alpha) > 0$ for $\alpha > 1/2$.

Exercise 13.9 Repeat the same arguments as above for the case of a regular ternary tree described in Example 13.5, and derive a bound of the form (13.32). Show that, at low temperature, the Arrhenius law holds, i.e. $\tau_{\exp} \geq \exp\{\beta\Delta E_N + o(\beta)\}$. How does ΔE_N behave for large N?

[Hint: An upper bound can be obtained by constructing a sequence of configurations from the all-plus to the all-minus ground state such that any two consecutive configurations differ by a single spin flip.]

Notes

For introductions to Bayesian networks, see Jordan (1998) and Jensen (1996). Bayesian inference was proved to be NP-hard by Cooper (1990). In Dagm and Luby (1993), it was shown that approximate Bayesian inference remains NP-hard.

Decoding of LDPC codes via Glauber dynamics was considered by Franz *et al.* (2002). Satisfiability problems were considered by Svenson and Nordahl (1999).

The Arrhenius law and the concept of an energy barrier (or 'activation energy') were discovered by the Swedish chemist Svante Arrhenius in 1889, in his study of chemical kinetics. An introduction to the analysis of Markov chain Monte Carlo methods (with special emphasis on enumeration problems) and their equilibration time can be found in Jerrum and Sinclair (1996) and Sinclair (1998). Geometric techniques for bounding the spectral gap are also discussed in Diaconis and Stroock (1991) and Diaconis and Saloff-Coste (1993). A book in preparation (Aldous and Fill, 2008) provides a complete exposition of the subject from a probabilistic point of view. For a mathematical-physics perspective, we refer to the lectures by Martinelli (1999). The rather basic Theorem 13.7 can be found in any of these references.

For an early treatment of the Glauber dynamics of an Ising model on a tree, see Henley (1986). This paper contains a partial answer to Exercise 13.9. Rigorous estimates for this problem have been proved by Berger *et al.* (2005) and Martinelli *et al.* (2004).

Part IV

Short-range correlations

Part IV

Short-range correlations

14

Belief propagation

Consider the ubiquitous problem of computing marginals of a graphical model with N variables $\underline{x} = (x_1, \ldots, x_N)$ taking values in a finite alphabet $\mathcal{X}$. The naive algorithm, which sums over all configurations, takes a time of order $|\mathcal{X}|^N$. The complexity can be reduced dramatically when the underlying factor graph has some special structure. One extreme case is that of tree factor graphs. On trees, marginals can be computed in a number of operations which grows linearly with N. This can be done through a 'dynamic programming' procedure that recursively sums over all variables, starting from the leaves and progressing towards the 'centre' of the tree.

Remarkably, such a recursive procedure can be recast as a distributed 'message-passing' algorithm. Message-passing algorithms operate on 'messages' associated with edges of the factor graph, and update them recursively through local computations done at the vertices of the graph. The update rules that yield exact marginals on trees have been discovered independently in several different contexts: statistical physics (under the name 'Bethe–Peierls approximation'), coding theory (the 'sum–product' algorithm), and artificial intelligence ('belief propagation', BP). Here we shall adopt the artificial-intelligence terminology.

This chapter gives a detailed presentation of BP and, more generally, message-passing procedures, which provide one of the main building blocks that we shall use throughout the rest of the book. It is therefore important that the reader has a good understanding of BP.

It is straightforward to prove that BP computes marginals exactly on tree factor graphs. However, it was found only recently that it can be extremely effective on loopy graphs as well. One of the basic intuitions behind this success is that BP, being a local algorithm, should be successful whenever the underlying graph is 'locally' a tree. Such factor graphs appear frequently, for instance in error-correcting codes, and BP turns out to be very powerful in this context. However, even in such cases, its application is limited to distributions such that far-apart variables become approximately uncorrelated. The onset of long-range correlations, typical of the occurrence of a phase transition, leads generically to poor performance of BP. We shall see several applications of this idea in the following chapters.

We introduce the basic ideas in Section 14.1 by working out two simple examples. The general BP equations are stated in Section 14.2, which also shows how they provide exact results on tree factor graphs. Section 14.3 describes an alternative message-passing procedure, the max-product (or, equivalently, min-sum) algorithm, which can be used in optimization problems. In Section 14.4, we discuss the use of BP in graphs with loops. In the study of random constraint satisfaction problems, BP messages

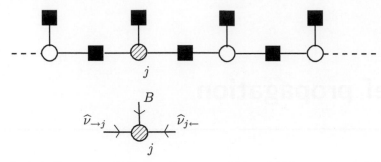

Fig. 14.1 *Top*: the factor graph of a one-dimensional Ising model in an external field. *Bottom*: the three messages arriving at site j describe the contributions to the probability distribution of σ_j due to the left chain ($\widehat{\nu}_{\to j}$), the right chain ($\widehat{\nu}_{j\leftarrow}$), and the external field B.

become random variables. The study of their distribution provides a large amount of information about such instances and can be used to characterize the corresponding phase diagram. The time evolution of these distributions is known under the name of 'density evolution', and the fixed-point analysis of them is done by the replica-symmetric cavity method. Both are explained in Section 14.6.

14.1 Two examples

14.1.1 Example 1: Ising chain

Consider the ferromagnetic Ising model on a line. The variables are Ising spins $(\sigma_1, \ldots, \sigma_N) = \underline{\sigma}$, with $\sigma_i \in \{+1, -1\}$, and their joint distribution takes the Boltzmann form

$$\mu_\beta(\underline{\sigma}) = \frac{1}{Z} e^{-\beta E(\underline{\sigma})}, \qquad E(\underline{\sigma}) = -\sum_{i=1}^{N-1} \sigma_i \sigma_{i+1} - B \sum_{i=1}^{N} \sigma_i. \tag{14.1}$$

The corresponding factor graph is shown in Figure 14.1.

Let us now compute the marginal probability distribution $\mu(\sigma_j)$ of spin σ_j. We shall introduce three 'messages' arriving at spin j, representing the contributions to $\mu(\sigma_j)$ from each of the function nodes which are connected to i. More precisely, we define

$$\widehat{\nu}_{\to j}(\sigma_j) = \frac{1}{Z_{\to j}} \sum_{\sigma_1 \ldots \sigma_{j-1}} \exp\left\{ \beta \sum_{i=1}^{j-1} \sigma_i \sigma_{i+1} + \beta B \sum_{i=1}^{j-1} \sigma_i \right\},$$

$$\widehat{\nu}_{j\leftarrow}(\sigma_j) = \frac{1}{Z_{j\leftarrow}} \sum_{\sigma_{j+1} \ldots \sigma_N} \exp\left\{ \beta \sum_{i=j}^{N-1} \sigma_i \sigma_{i+1} + \beta B \sum_{i=j+1}^{N} \sigma_i \right\}. \tag{14.2}$$

Messages are understood to be probability distributions and thus to be normalized. In the present case, the constants $Z_{\to j}$, $Z_{j\leftarrow}$ are set by the conditions $\widehat{\nu}_{\to j}(+1) + \widehat{\nu}_{\to j}(-1) = 1$, and $\widehat{\nu}_{j\leftarrow}(+1) + \widehat{\nu}_{j\leftarrow}(-1) = 1$. In the following, when dealing with normalized distributions, we shall avoid writing the normalization constants explicitly and instead use the symbol $\cong$ to denote 'equality up to a normalization'. With this notation, the first of the above equations can be rewritten as

$$\widehat{\nu}_{\to j}(\sigma_j) \cong \sum_{\sigma_1 \ldots \sigma_{j-1}} \exp\left\{ \beta \sum_{i=1}^{j-1} \sigma_i \sigma_{i+1} + \beta B \sum_{i=1}^{j-1} \sigma_i \right\} . \tag{14.3}$$

By rearranging the summation over spins σ_i, $i \neq j$, the marginal $\mu(\sigma_j)$ can be written as

$$\mu(\sigma_j) \cong \widehat{\nu}_{\to j}(\sigma_j)\, e^{\beta B \sigma_j}\, \widehat{\nu}_{j \leftarrow}(\sigma_j) . \tag{14.4}$$

In this expression, we can interpret each of the three factors as a 'message' sent to j from one of the three function nodes connected to the variable j. Each message coincides with the marginal distribution of σ_j in a modified graphical model. For instance, $\widehat{\nu}_{\to j}(\sigma_j)$ is the distribution of σ_j in the graphical model obtained by removing all of the factor nodes adjacent to j except for the one on its left (see Fig. 14.1).

This decomposition is interesting because the various messages can be computed iteratively. Consider, for instance, $\widehat{\nu}_{\to i+1}$. It is expressed in terms of $\widehat{\nu}_{\to i}$ as

$$\widehat{\nu}_{\to i+1}(\sigma) \cong \sum_{\sigma'} \widehat{\nu}_{\to i}(\sigma')\, e^{\beta \sigma' \sigma + \beta B \sigma'} . \tag{14.5}$$

Furthermore, $\widehat{\nu}_{\to 1}$ is the uniform distribution over $\{+1, -1\}$: $\widehat{\nu}_{\to 1}(\sigma) = \frac{1}{2}$ for $\sigma = \pm 1$. Equation (14.5) allows one to compute all of the messages $\widehat{\nu}_{\to i}$, $i \in \{1, \ldots, N\}$, in $O(N)$ operations. A similar procedure yields $\widehat{\nu}_{i \leftarrow}$, by starting from the uniform distribution $\widehat{\nu}_{N \leftarrow}$ and computing $\widehat{\nu}_{i-1 \leftarrow}$ from $\widehat{\nu}_{i \leftarrow}$ recursively. Finally, eqn (14.4) can be used to compute all of the marginals $\mu(\sigma_j)$ in linear time.

All of the messages are distributions over binary variables and can thus be parameterized by a single real number. One popular choice for such a parameterization is to use the log-likelihood ratio[1]

$$u_{\to i} \equiv \frac{1}{2\beta} \log \frac{\widehat{\nu}_{\to i}(+1)}{\widehat{\nu}_{\to i}(-1)} . \tag{14.6}$$

In statistical-physics terms, $u_{\to i}$ is an 'effective (or local) magnetic field': $\widehat{\nu}_{\to i}(\sigma) \cong e^{\beta u_{\to i} \sigma}$. Using this definition (and noticing that it implies $\widehat{\nu}_{\to i}(\sigma) = \frac{1}{2}(1 + \sigma \tanh(\beta u_{\to i})))$, eqn (14.5) becomes

$$u_{\to i+1} = f(u_{\to i} + B) , \tag{14.7}$$

where the function $f(x)$ is defined as

$$f(x) = \frac{1}{\beta} \operatorname{atanh}\left[\tanh(\beta) \tanh(\beta x)\right] . \tag{14.8}$$

The mapping $u \mapsto f(u + B)$ is differentiable, with its derivative bounded by $\tanh \beta < 1$. Therefore the fixed-point equation $u = f(u + B)$ has a unique solution u_*, and $u_{\to i}$ goes to u_* when $i \to \infty$. Consider a very long chain, and a node

---

[1]Note that our definition differs by a factor $1/2\beta$ from the standard definition of the log-likelihood in statistics. This factor is introduced to make contact with statistical-physics definitions.

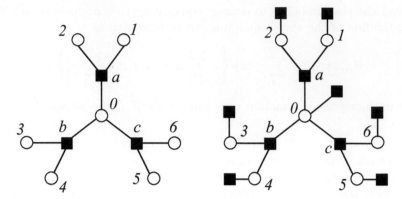

Fig. 14.2 *Left*: a simple parity check code with seven variables and three checks. *Right*: the factor graph corresponding to the problem of finding the sent codeword, given a received message.

in the bulk $j \in [\varepsilon N, (1 - \varepsilon)N]$. Then, as $N \to \infty$, both $u_{\to j}$ and $u_{j \leftarrow}$ converge to u^*, so that $\langle \sigma_j \rangle \to \tanh[\beta(2u^* + B)]$. This is the bulk magnetization. If, on the other hand, we consider a spin on the boundary, we get a smaller magnetization $\langle \sigma_1 \rangle = \langle \sigma_N \rangle \to \tanh[\beta(u^* + B)]$.

Exercise 14.1 Use the recursion (14.7) to show that, when N and j go to infinity, $\langle \sigma_j \rangle = M + O(\lambda^j, \lambda^{N-j})$, where $M = \tanh(2u_* + B)$ and $\lambda = f'(u_* + B)$. Compare this with the treatment of the one-dimensional Ising model in Section 2.5.

The above method can be generalized to the computation of joint distributions of two or more variables. Consider, for instance, the joint distribution $\mu(\sigma_j, \sigma_k)$, for $k > j$. Since we already know how to compute the marginal $\mu(\sigma_j)$, it is sufficient to consider the conditional distribution $\mu(\sigma_k | \sigma_j)$. For each of the two values of σ_j, the conditional distribution of $\sigma_{j+1}, \ldots, \sigma_N$ takes a form analogous to eqn (14.1) but with σ_j fixed. Therefore, the marginal $\mu(\sigma_k | \sigma_j)$ can be computed through the same algorithm as before. The only difference is in the initial condition, which becomes $\widehat{\nu}_{\to j}(+1) = 1$, $\widehat{\nu}_{\to j}(-1) = 0$ (if we condition on $\sigma_j = +1$) and $\widehat{\nu}_{\to j}(+1) = 0$, $\widehat{\nu}_{\to j}(-1) = 1$ (if we condition on $\sigma_j = -1$).

Exercise 14.2 Compute the correlation function $\langle \sigma_j \sigma_k \rangle$, when $j, k \in [N\varepsilon, N(1 - \varepsilon)]$ and $N \to \infty$. Check that when $B = 0$, $\langle \sigma_j \sigma_k \rangle = (\tanh \beta)^{|j-k|}$. Find a simpler derivation of this last result.

14.1.2 Example 2: A tree-parity-check code

Our second example deals with a decoding problem. Consider the simple linear code whose factor graph is reproduced in the left frame of Fig. 14.2. It has a block length $N = 7$, and the codewords satisfy the three parity check equations

$$x_0 \oplus x_1 \oplus x_2 = 0 \,, \tag{14.9}$$
$$x_0 \oplus x_3 \oplus x_4 = 0 \,, \tag{14.10}$$
$$x_0 \oplus x_5 \oplus x_6 = 0 \,. \tag{14.11}$$

One of the codewords is sent through a channel of the type BSC(p), defined earlier. Assume that the received message is $\underline{y} = (1, 0, 0, 0, 0, 1, 0)$. The conditional distribution for $\underline{x}$ to be the transmitted codeword, given the received message $\underline{y}$, takes the usual form $\mu_y(\underline{x}) = \mathbb{P}(\underline{x}|\underline{y})$:

$$\mu_y(\underline{x}) \cong \mathbb{I}(x_0 \oplus x_1 \oplus x_2 = 0)\mathbb{I}(x_0 \oplus x_3 \oplus x_4 = 0)\mathbb{I}(x_0 \oplus x_5 \oplus x_6 = 0) \prod_{i=0}^{6} Q(y_i|x_i) \,,$$

where $Q(0|0) = Q(1|1) = 1 - p$ and $Q(1|0) = Q(0|1) = p$. The corresponding factor graph is drawn in the right frame of Fig. 14.2.

In order to implement symbol MAP decoding, (see Chapter 6), we need to compute the marginal distribution of each bit. The computation is straightforward, but it is illuminating to recast it as a message-passing procedure similar to that in the Ising chain example. Consider, for instance, bit x_0. We start from the boundary. In the absence of the check a, the marginal of x_1 would be $\nu_{1 \to a} = (1 - p, p)$ (we use here the convention of writing distributions $\nu(x)$ over a binary variable as two-dimensional vectors $(\nu(0), \nu(1))$). This is interpreted as a message sent from variable 1 to check a.

Variable 2 sends an analogous message $\nu_{2 \to a}$ to a (in the present example, this happens to be equal to $\nu_{1 \to a}$). Knowing these two messages, we can compute the contribution to the marginal probability distribution of variable x_0 arising from the part of the factor graph containing the whole branch connected to x_0 through the check a:

$$\widehat{\nu}_{a \to 0}(x_0) \cong \sum_{x_1, x_2} \mathbb{I}(x_0 \oplus x_1 \oplus x_2 = 0) \, \nu_{1 \to a}(x_1)\nu_{2 \to a}(x_2) \,. \tag{14.12}$$

Clearly, $\widehat{\nu}_{a \to 0}(x_0)$ is the marginal distribution of x_0 in a modified factor graph that does not include either of the factor nodes b or c, and in which the received symbol y_0 has been erased. This is analogous to the messages $\widehat{\nu}_{\to j}(\sigma_j)$ used in the Ising chain example. The main difference is that the underlying factor graph is no longer a line, but a tree. As a consequence, the recursion (14.12) is no longer linear in the incoming messages. Using the rule (14.12), and analogous ones for $\widehat{\nu}_{b \to 0}(x_0)$ and $\widehat{\nu}_{c \to 0}(x_0)$, we obtain

$$\widehat{\nu}_{a \to 0} = (p^2 + (1 - p)^2, \; 2p(1 - p)) \,,$$
$$\widehat{\nu}_{b \to 0} = (p^2 + (1 - p)^2, \; 2p(1 - p)) \,,$$
$$\widehat{\nu}_{c \to 0} = (2p(1 - p), \; p^2 + (1 - p)^2) \,.$$

The marginal probability distribution of the variable x_0 is finally obtained by taking into account the contributions of each subtree, together with the channel output for bit x_0:

$$\mu(x_0) \cong Q(y_0|x_0)\,\widehat{\nu}_{a\to 0}(x_0)\widehat{\nu}_{b\to 0}(x_0)\widehat{\nu}_{c\to 0}(x_0)$$
$$\cong \left(2p^2(1-p)[p^2+(1-p)^2]^2,\ \ 4p^2(1-p)^3[p^2+(1-p)^2]\right).$$

In particular, the MAP decoding of the symbol x_0 is always $x_0 = 0$ in this case, for any $p < 1/2$.

An important fact emerges from this simple calculation. Instead of performing a summation over $2^7 = 128$ configurations, we were able to compute the marginal at x_0 by doing six summations (one for every factor node a, b, c and for every value of x_0), each one over two summands (see eqn (14.12)). Such complexity reduction was achieved by merely rearranging the order of sums and multiplications in the computation of the marginal.

Exercise 14.3 Show that the message $\nu_{0\to a}(x_0)$ is equal to $(1/2, 1/2)$, and deduce that $\mu(x_1) \cong ((1-p), p)$.

14.2 Belief propagation on tree graphs

We shall now define belief propagation and analyse it in the simplest possible setting: tree-graphical models. In this case, it solves several computational problems in an efficient and distributed fashion.

14.2.1 Three problems

Let us consider a graphical model such that the associated factor graph is a tree (we call this model a **tree-graphical model**). We use the same notation as in Section 9.1.1. The model describes N random variables $(x_1, \ldots, x_N) \equiv \underline{x}$ taking values in a finite alphabet $\mathcal{X}$, whose joint probability distribution has the form

$$\mu(\underline{x}) = \frac{1}{Z} \prod_{a=1}^{M} \psi_a(\underline{x}_{\partial a}), \tag{14.13}$$

where $\underline{x}_{\partial a} \equiv \{x_i \,|\, i \in \partial a\}$. The set $\partial a \subseteq [N]$, of size $|\partial a|$, contains all variables involved in constraint a. We shall always use indices $i, j, k, \ldots$ for the variables and $a, b, c, \ldots$ for the function nodes. The set of indices ∂i involves all function nodes a connected to i.

When the factor graph has no loops, the following are among the basic problems that can be solved efficiently with a message-passing procedure:

1. Compute the marginal distributions of one variable, $\mu(x_i)$, or the joint distribution of a small number of variables.
2. Sample from $\mu(\underline{x})$, i.e. draw independent random configurations $\underline{x}$ with a distribution $\mu(\underline{x})$.

3. Compute the partition function Z or, equivalently, in statistical-physics language, the free entropy $\log Z$.

These three tasks can be accomplished using belief propagation, which is an obvious generalization of the procedure exemplified in the previous section.

14.2.2 The BP equations

Belief propagation is an iterative 'message-passing' algorithm. The basic variables on which it acts are messages associated with directed edges on the factor graph. For each edge (i, a) (where i is a variable node and a a function node) there exist, at the t-th iteration, two messages $\nu_{i \to a}^{(t)}$ and $\hat{\nu}_{a \to i}^{(t)}$. Messages take values in the space of probability distributions over the single-variable space $\mathcal{X}$. For instance, $\nu_{i \to a}^{(t)} = \{\nu_{i \to a}^{(t)}(x_i) : x_i \in \mathcal{X}\}$, with $\nu_{i \to a}^{(t)}(x_i) \geq 0$ and $\sum_{x_i} \nu_{i \to a}^{(t)}(x_i) = 1$.

In tree-graphical models, the messages converge when $t \to \infty$ to fixed-point values (see Theorem 14.1). These coincide with single-variable marginals in modified graphical models, as we saw in the two examples in the previous section. More precisely, $\nu_{i \to a}^{(\infty)}(x_i)$ is the marginal distribution of variable x_i in a modified graphical model which does not include the factor a (i.e. the product in eqn (14.13) does not include a). Analogously, $\hat{\nu}_{a \to i}^{(\infty)}(x_i)$ is the distribution of x_i in a graphical model where all factors in ∂i except a have been erased.

Messages are updated through local computations at the nodes of the factor graph. By *local* we mean that a given node updates the outgoing messages on the basis of incoming ones at previous iterations. This is a characteristic feature of message-passing algorithms; the various algorithms in this family differ in the precise form of the update equations. The **belief propagation** (BP), or **sum–product**, update rules are

$$\nu_{j \to a}^{(t+1)}(x_j) \cong \prod_{b \in \partial j \backslash a} \hat{\nu}_{b \to j}^{(t)}(x_j) \,, \tag{14.14}$$

$$\hat{\nu}_{a \to j}^{(t)}(x_j) \cong \sum_{\underline{x}_{\partial a \backslash j}} \psi_a(\underline{x}_{\partial a}) \prod_{k \in \partial a \backslash j} \nu_{k \to a}^{(t)}(x_k) \,. \tag{14.15}$$

It is understood that, when $\partial j \backslash a$ is an empty set, $\nu_{j \to a}(x_j)$ is the uniform distribution. Similarly, if $\partial a \backslash j$ is empty, then $\hat{\nu}_{a \to j}(x_j) = \psi_a(x_j)$. A pictorial illustration of these rules is provided in Fig. 14.3. A BP fixed point is a set of t-independent messages $\nu_{i \to a}^{(t)} = \nu_{i \to a}$, $\hat{\nu}_{a \to i}^{(t)} = \hat{\nu}_{a \to i}$ which satisfy eqns (14.14) and (14.15). From these, one obtains $2|\mathcal{E}|$ equations (one equation for each directed edge of the factor graph) relating $2|\mathcal{E}|$ messages. We shall often refer to these fixed-point conditions as the **BP equations**.

After t iterations, one can estimate the marginal distribution $\mu(x_i)$ of variable i using the set of *all* incoming messages. The BP estimate is:

$$\nu_i^{(t)}(x_i) \cong \prod_{a \in \partial i} \hat{\nu}_{a \to i}^{(t-1)}(x_i) \,. \tag{14.16}$$

In writing the update rules, we have assumed that the update is done in parallel at all the variable nodes, then in parallel at all function nodes, and so on. Clearly, in this

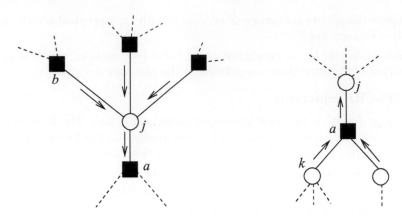

Fig. 14.3 *Left*: the portion of the factor graph involved in the computation of $\nu_{j\to a}^{(t+1)}(x_j)$. This message is a function of the 'incoming messages' $\widehat{\nu}_{b\to j}^{(t)}(x_j)$, with $b \neq a$. *Right*: the portion of the factor graph involved in the computation of $\widehat{\nu}_{a\to j}^{(t)}(x_j)$. This message is a function of the 'incoming messages' $\nu_{k\to a}^{(t)}(x_k)$, with $k \neq j$.

case, the iteration number must be incremented either at variable nodes or at factor nodes, but not necessarily at both. This is what happens in eqns (14.14) and (14.15). Other update schedules are possible and sometimes useful. For the sake of simplicity, however, we shall stick to the parallel schedule.

In order to fully define the algorithm, we need to specify an initial condition. It is a widespread practice to set initial messages to the uniform distribution over $\mathcal{X}$ (i.e. $\nu_{i\to a}^{(0)}(x_i) = 1/|\mathcal{X}|$). On the other hand, it can be useful to explore several distinct (random) initial conditions. This can be done by defining some probability measure P over the space $\mathfrak{M}(\mathcal{X})$ of distributions over $\mathcal{X}$ (i.e. the $|\mathcal{X}|$-dimensional simplex) and taking $\nu_{i\to a}^{(0)}(\cdot)$ as i.i.d. random variables with distribution P.

The BP algorithm can be applied to any graphical model, irrespective of whether the factor graph is a tree or not. One possible version of the algorithm is as follows.

	BP (graphical model (G, ψ), accuracy ϵ, iterations $t_{\max}$)
1:	Initialize BP messages as i.i.d. random variables with distribution P;
2:	For $t \in \{0, \dots, t_{\max}\}$
3:	For each $(j, a) \in E$
4:	Compute the new value of $\widehat{\nu}_{a\to j}$ using eqn (14.15);
5:	For each $(j, a) \in E$
6:	Compute the new value of $\nu_{j\to a}$ using eqn (14.14);
7:	Let Δ be the maximum message change;
8:	If $\Delta < \epsilon$ return current messages;
9:	End-For;
10:	Return 'Not Converged';

Among all message-passing algorithms, BP is uniquely characterized by the property of computing exact marginals on tree-graphical models.

Theorem 14.1. (BP is exact on trees) *Consider a tree-graphical model with diameter t_* (which means that t_* is the maximum distance between any two variable nodes). Then:*

1. *Irrespective of the initial condition, the BP update equations (14.14) and (14.15) converge after at most t_* iterations. In other words, for any edge (ia), and any $t > t_*$, $\nu_{i \to a}^{(t)} = \nu_{i \to a}^*$, $\widehat{\nu}_{a \to i}^{(t)} = \widehat{\nu}_{a \to i}^*$.*
2. *The fixed-point messages provide the exact marginals: for any variable node i, and any $t > t_*$, $\nu_i^{(t)}(x_i) = \mu(x_i)$.*

Proof As exemplified in the previous section, on tree factor graphs BP is just a clever way to organize the sum over configurations to compute marginals. In this sense, the theorem is obvious.

We shall sketch a formal proof here, leaving a few details to the reader. Given a directed edge $i \to a$ between a variable i and a factor node a, we define $\mathbb{T}(i \to a)$ as the subtree rooted on this edge. This is the subtree containing all nodes w which can be connected to i by a non-reversing path[2] which does not include the edge (i, a). Let $t_*(i \to a)$ be the *depth* of $\mathbb{T}(i \to a)$ (the maximal distance from a leaf to i).

We can show that, for any number of iterations $t > t_*(i \to a)$, the message $\nu_{i \to a}^{(t)}$ coincides with the marginal distribution of the root variable with respect to the graphical model $\mathbb{T}(i \to a)$. In other words, for tree graphs, the interpretation of BP messages in terms of modified marginals is correct.

This claim is proved by induction on the tree depth $t_*(i \to a)$. The base step of the induction is trivial: $\mathbb{T}(i \to a)$ is the graph formed by the unique node i. By definition, for any $t \geq 1$, $\nu_{i \to a}^{(t)}(x_i) = 1/|\mathcal{X}|$ is the uniform distribution, which coincides with the marginal of the trivial graphical model associated with $\mathbb{T}(i \to a)$.

The induction step is easy as well. Assuming the claim to be true for $t_*(i \to a) \leq \tau$, we have to show that it holds when $t_*(i \to a) = \tau + 1$. To this end, take any $t > \tau + 1$ and compute $\nu_{i \to a}^{(t+1)}(x_i)$ using eqns (14.14) and (14.15) in terms of messages $\nu_{j \to b}^{(t)}(x_j)$ in the subtrees for $b \in \partial i \setminus a$ and $j \in \partial b \setminus i$. By the induction hypothesis, and since the depth of the subtree $T(j \to b)$ is at most τ, $\nu_{j \to b}^{(t)}(x_j)$ is the root marginal in such a subtree. It turns out that by combining the marginals at the roots of the subtrees $\mathbb{T}(j \to b)$ using eqns (14.14) and (14.15), we can obtain the marginal at the root of $\mathbb{T}(i \to a)$. This proves the claim. $\square$

14.2.3 Correlations and energy

The use of BP is not limited to computing one-variable marginals. Suppose we want to compute the joint probability distribution $\mu(x_i, x_j)$ of two variables x_i and x_j. Since BP already enables to compute $\mu(x_i)$, this task is equivalent to computing the

---

[2]A **non-reversing path** on a graph $\mathcal{G}$ is a sequence of vertices $\omega = (j_0, j_1, \ldots, j_n)$ such that (j_s, j_{s+1}) is an edge for any $s \in \{0, \ldots, n-1\}$, and $j_{s-1} \neq j_{s+1}$ for $s \in \{1, \ldots, n-1\}$.

conditional distribution $\mu(x_j \mid x_i)$. Given a model that factorizes as in eqn (14.13), the conditional distribution of $\underline{x} = (x_1, \ldots, x_N)$ given $x_i = x$ takes the form

$$\mu(\underline{x}|x_i = x) \cong \prod_{a=1}^{M} \psi_a(\underline{x}_{\partial a}) \, \mathbb{I}(x_i = x) \,. \tag{14.17}$$

In other words, it is sufficient to add to the original graph a new function node of degree 1 connected to variable node i, which fixes $x_i = x$. One can then run BP on the modified factor graph and obtain estimates $\nu_j^{(t)}(x_j|x_i = x)$ for the conditional marginal of x_j.

This strategy is easily generalized to the joint distribution of any number m of variables. The complexity, however, grows exponentially in the number of variables involved, since we have to condition over $|\mathcal{X}|^{m-1}$ possible assignments.

Happily, for tree-graphical models, the marginal distribution of any number of variables admits an explicit expression in terms of messages. Let F_R be a subset of function nodes, let V_R be the subset of variable nodes adjacent to F_R, let R be the induced subgraph, and let $\underline{x}_R$ be the corresponding variables. Without loss of generality, we shall assume R to be connected. Further, we denote by ∂R the subset of function nodes that are not in F_R but are adjacent to a variable node in V_R.

Then, for $a \in \partial R$, there exists a unique $i \in \partial a \cap V_R$, which we denote by $i(a)$. It then follows immediately from Theorem 14.1, and its characterization of messages, that the joint distribution of variables in R is

$$\mu(\underline{x}_R) = \frac{1}{Z_R} \prod_{a \in F_R} \psi_a(\underline{x}_{\partial a}) \prod_{a \in \partial R} \widehat{\nu}^*_{a \to i(a)}(x_{i(a)}) \,, \tag{14.18}$$

where $\widehat{\nu}^*_{a \to i}(\,\cdot\,)$ are the fixed-point BP messages.

Exercise 14.4 Let us use the above result to write the joint distribution of the variables along a path in a tree factor graph. Consider two variable nodes i, j, and let $R = (V_R, F_R, E_R)$ be the subgraph induced by the nodes along the path between i and j. For any function node $a \in R$, denote by $i(a)$ and $j(a)$ the variable nodes in R that are adjacent to a. Show that the joint distribution of the variables along this path, $\underline{x}_R = \{x_l : l \in V_R\}$, takes the form

$$\mu(\underline{x}_R) = \frac{1}{Z_R} \prod_{a \in F_R} \tilde{\psi}_a(x_{i(a)}, x_{j(a)}) \prod_{l \in V_R} \tilde{\psi}_l(x_l) \,. \tag{14.19}$$

In other words, $\mu(\underline{x}_R)$ factorizes according to the subgraph R. Write expressions for the compatibility functions $\tilde{\psi}_a(\,\cdot\,,\,\cdot\,)$, $\tilde{\psi}_l(\,\cdot\,)$ in terms of the original compatibility functions and the messages going from ∂R to V_R.

A particularly useful case arises in the computation of the internal energy. In physics problems, the compatibility functions in eqn (14.13) take the form $\psi_a(\underline{x}_{\partial a}) = \mathrm{e}^{-\beta E_a(\underline{x}_{\partial a})}$, where β is the inverse temperature and $E_a(\underline{x}_{\partial a})$ is the energy function characterizing constraint a. Of course, any graphical model can be written in this form (allowing for the possibility of $E_a(\underline{x}_{\partial a}) = +\infty$ in the case of hard constraints), adopting for instance the convention $\beta = 1$, which we shall use from now on. The internal energy U is the expectation value of the total energy:

$$U = -\sum_{\underline{x}} \mu(\underline{x}) \sum_{a=1}^{M} \log \psi_a(\underline{x}_{\partial a}) . \tag{14.20}$$

This can be computed in terms of BP messages using eqn (14.18) with $F_R = \{a\}$. If, further, we use eqn (14.14) to express products of check-to-variable messages in terms of variable-to-check ones, we get

$$U = -\sum_{a=1}^{M} \frac{1}{Z_a} \sum_{\underline{x}_{\partial a}} \left(\psi_a(\underline{x}_{\partial a}) \log \psi_a(\underline{x}_{\partial a}) \prod_{i \in \partial a} \nu_{i \to a}^*(x_j) \right) , \tag{14.21}$$

where $Z_a \equiv \sum_{\underline{x}_{\partial a}} \psi_a(\underline{x}_{\partial a}) \prod_{i \in \partial a} \nu_{i \to a}^*(x_j)$. Notice that in this expression the internal energy is a sum of 'local' terms, one for each compatibility function.

On a loopy graph, eqns (14.18) and (14.21) are no longer valid, and, indeed, BP does not necessarily converge to fixed-point messages $\{\nu_{i \to a}^*, \widehat{\nu}_{a \to i}^*\}$. However, one can replace fixed-point messages with BP messages after any number t of iterations and take these as *definitions* of the BP estimates of the corresponding quantities. From eqn (14.18), one obtains an estimate of the joint distribution of a subset of variables, which we shall call $\nu^{(t)}(\underline{x}_R)$, and from (14.21), an estimate of the internal energy.

14.2.4 Entropy

Remember that the entropy of a distribution μ over $\mathcal{X}^V$ is defined as $H[\mu] = -\sum_{\underline{x}} \mu(\underline{x}) \log \mu(\underline{x})$. In a tree-graphical model, the entropy, like the internal energy, has a simple expression in terms of local quantities. This follows from an important decomposition property. Let us denote by $\mu_a(\underline{x}_{\partial a})$ the marginal probability distribution of all the variables involved in the compatibility function a, and by $\mu_i(x_i)$ the marginal probability distribution of variable x_i.

Theorem 14.2 *In a tree-graphical model, the joint probability distribution $\mu(\underline{x})$ of all of the variables can be written in terms of the marginals $\mu_a(\underline{x}_{\partial a})$ and $\mu_i(x_i)$ as*

$$\mu(\underline{x}) = \prod_{a \in F} \mu_a(\underline{x}_{\partial a}) \prod_{i \in V} \mu_i(x_i)^{1 - |\partial i|} . \tag{14.22}$$

Proof The proof is by induction on the number M of factors. Equation (14.22) holds for $M = 1$ (since the degrees $|\partial i|$ are all equal to 1). Assume that it is valid for any factor graph with up to M factors, and consider a specific factor graph G with $M+1$ factors. Since G is a tree, it contains at least one factor node such that all its adjacent variable nodes have degree 1, except for at most one of them. Call such a factor node a, and let i be the only neighbour with degree larger than one (the case in which no such neighbour exists is treated analogously). Further, let $\underline{x}_\sim$ be the vector of variables in

G that are not in $\partial a \setminus i$. Then (writing $\mathbb{P}_\mu(\,\cdot\,)$ for a probability under the distribution μ), the Markov property together with the Bayes rule yields

$$\mathbb{P}_\mu(\underline{x}) = \mathbb{P}_\mu(\underline{x}_\sim)\mathbb{P}_\mu(\underline{x}|\underline{x}_\sim) = \mathbb{P}_\mu(\underline{x}_\sim)\mathbb{P}_\mu(\underline{x}_{\partial a \setminus i}|x_i) = \mathbb{P}_\mu(\underline{x}_\sim)\mu_a(\underline{x}_{\partial a})\mu_i(x_i)^{-1}\,. \tag{14.23}$$

The probability $\mathbb{P}_\mu(\underline{x}_\sim)$ can be written as $\mathbb{P}(\underline{x}_\sim) \cong \tilde{\psi}_a(x_i)\prod_{b\in F\setminus a}\psi_b(\underline{x}_{\partial b})$, where $\tilde{\psi}_a(x_i) = \sum_{\underline{x}_{\partial a\setminus i}}\psi_a(\underline{x}_{\partial a})$. As the factor $\tilde{\psi}_a$ has degree one, it can be erased and incorporated into another factor as follows: take one of the other factors connected to i, $c \in \partial i \setminus a$, and change it to $\tilde{\psi}_c(\underline{x}_{\partial c}) = \psi_c(\underline{x}_{\partial c})\tilde{\psi}_a(x_i)$. In the reduced factor graph, the degree of i is smaller by one and the number of factors is M. Using the induction hypothesis, we get

$$\mathbb{P}_\mu(\underline{x}_\sim) = \mu_i(x_i)^{2-|\partial i|}\prod_{b\in F\setminus a}\mu_b(\underline{x}_{\partial b})\prod_{j\in V\setminus i}\mu_j(x_j)^{1-|\partial j|}\,. \tag{14.24}$$

The proof is completed by putting together eqns (14.23) and (14.24). $\square$

As an immediate consequence of eqn (14.22), the entropy of a tree-graphical model can be expressed as sums of local terms:

$$H[\mu] = -\sum_{a\in F}\mu_a(\underline{x}_{\partial a})\log\mu_a(\underline{x}_{\partial a}) - \sum_{i\in V}(1-|\partial i|)\,\mu_i(x_i)\log\mu_i(x_i)\,. \tag{14.25}$$

It is also easy to express the free entropy $\Phi = \log Z$ in terms of *local* quantities. Recalling that $\Phi = H[\mu]-U[\mu]$ (where $U[\mu]$ is the internal energy given by eqn (14.21)), we get $\Phi = \mathbb{F}[\mu]$, where

$$\mathbb{F}[\mu] = -\sum_{a\in F}\mu_a(\underline{x}_{\partial a})\log\left\{\frac{\mu_a(\underline{x}_{\partial a})}{\psi_a(\underline{x}_{\partial a})}\right\} - \sum_{i\in V}(1-|\partial i|)\mu_i(x_i)\log\mu_i(x_i)\,. \tag{14.26}$$

Expressing local marginals in terms of messages, via eqn (14.18), we can in turn write the free entropy as a function of the fixed-point messages. We introduce the function $\mathbb{F}_*(\underline{\nu})$, which yields the free entropy in terms of $2|E|$ messages $\underline{\nu} = \{\nu_{i\to a}(\,\cdot\,),\widehat{\nu}_{a\to i}(\,\cdot\,)\}$:

$$\mathbb{F}_*(\underline{\nu}) = \sum_{a\in F}\mathbb{F}_a(\underline{\nu}) + \sum_{i\in V}\mathbb{F}_i(\underline{\nu}) - \sum_{(ia)\in E}\mathbb{F}_{ia}(\underline{\nu})\,, \tag{14.27}$$

where

$$\mathbb{F}_a(\underline{\nu}) = \log\left[\sum_{\underline{x}_{\partial a}}\psi_a(\underline{x}_{\partial a})\prod_{i\in\partial a}\nu_{i\to a}(x_i)\right]\,, \quad \mathbb{F}_i(\underline{\nu}) = \log\left[\sum_{x_i}\prod_{b\in\partial i}\widehat{\nu}_{b\to i}(x_i)\right]\,,$$

$$\mathbb{F}_{ai}(\underline{\nu}) = \log\left[\sum_{x_i}\nu_{i\to a}(x_i)\widehat{\nu}_{a\to i}(x_i)\right]\,. \tag{14.28}$$

It is not hard to show that, by evaluating this functional at the BP fixed point $\underline{\nu}^*$, one gets $\mathbb{F}_*(\underline{\nu}^*) = \mathbb{F}[\mu] = \Phi$, thus recovering the correct free entropy. The function

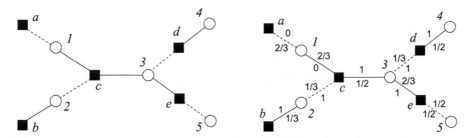

Fig. 14.4 *Left*: the factor graph of a small instance of the satisfiability problem with five variables and five clauses. A dashed line means that the variable appears negated in the adjacent clause. *Right*: the set of fixed-point BP messages for the uniform measure over solutions of this instance. All messages are normalized, and we show their weights for the value True. For any edge (a, i) (a being the clause and i the variable), the weight corresponding to the message $\widehat{\nu}_{a \to i}$ is shown above the edge, and the weight corresponding to $\nu_{i \to a}$ below the edge.

$\mathbb{F}_*(\underline{\nu})$ defined in eqn (14.27) is known as the **Bethe free entropy** (when multiplied by a factor $-1/\beta$, it is called the **Bethe free energy**). The above observations are important enough to be highlighted in a theorem.

Theorem 14.3. (the Bethe free entropy is exact on trees) *Consider a tree-graphical model. Let $\{\mu_a, \mu_i\}$ denote its local marginals, and let $\underline{\nu}^* = \{\nu^*_{i \to a}, \widehat{\nu}^*_{a \to i}\}$ be the fixed-point BP messages. Then $\Phi = \log Z = \mathbb{F}[\mu] = \mathbb{F}_*(\underline{\nu}^*)$.*

Notice that in the above statement, we have used the correct local marginals in $\mathbb{F}[\cdot]$ and the fixed-point messages in $\mathbb{F}_*(\cdot)$. In Section 14.4 we shall reconsider the Bethe free entropy for more general graphical models, and regard it as a function over the space of all 'possible' marginals/messages.

Exercise 14.5 Consider the instance of the satisfiability problem shown in Fig. 14.4, left. Show by exhaustive enumeration that it has only two satisfying assignments, $\underline{x} = (0, 1, 1, 1, 0)$ and $(0, 1, 1, 1, 1)$. Rederive this result using BP. Namely, compute the entropy of the uniform measure over satisfying assignments, and check that its value is indeed $\log 2$. The BP fixed point is shown in Fig. 14.4, right.

Exercise 14.6 In many systems some of the function nodes have degree 1 and amount to a local redefinition of the reference measure over $\mathcal{X}$. It is then convenient to single out these factors. Let us write $\mu(\underline{x}) \cong \prod_{a \in F} \psi_a(\underline{x}_{\partial a}) \prod_{i \in V} \psi_i(x_i)$, where the second product runs over degree-1 function nodes (indexed by the adjacent variable node), and the factors ψ_a have degree at least 2. In the computation of $\mathbb{F}_*$, the introduction of ψ_i adds N extra factor nodes and subtracts N extra 'edge' terms corresponding to the edge between the variable node i and the function node corresponding to ψ_i.

Show that these two effects cancel, and that the net effect is to replace the variable-node contribution in eqn (14.27) with

$$\mathbb{F}_i(\underline{\nu}) = \log \left[\sum_{x_i} \psi_i(x_i) \prod_{a \in \partial i} \widehat{\nu}_{a \to i}(x_i) \right]. \tag{14.29}$$

The problem of sampling from the distribution $\mu(\underline{x})$ over the large-dimensional space $\mathcal{X}^N$ reduces to that of computing one-variable marginals of $\mu(\underline{x})$, conditional on a subset of the other variables. In other words, if we have a black box that computes $\mu(x_i | \underline{x}_U)$ for any subset $U \subseteq V$, it can be used to sample a random configuration $\underline{x}$. The standard procedure for doing this is called **sequential importance sampling**. We can describe this procedure by the following algorithm in the case of tree-graphical models, using BP to implement such a 'black box'.

BP-GUIDED SAMPLING (fraphical model (G, ψ))

1:     initialize BP messages;
2:     initialize $U = \emptyset$;
3:     **for** $t = 1, \ldots, N$:
4:         run BP until convergence;
5:         choose $i \in V \setminus U$;
6:         compute the BP marginal $\nu_i(x_i)$;
7:         choose x_i^* distributed according to ν_i;
8:         fix $x_i = x_i^*$ and set $U \leftarrow U \cup \{i\}$;
9:         add a factor $\mathbb{I}(x_i = x_i^*)$ to the graphical model;
10:   **end**
11:   **return** $\underline{x}^*$.

14.2.5   Pairwise models

Pairwise graphical models, i.e. graphical models such that all factor nodes have degree 2, form an important class. A pairwise model can be conveniently represented as an ordinary graph $G = (V, E)$ over variable nodes. An edge joins two variables each time they are the arguments of the same compatibility function. The corresponding probability distribution reads

$$\mu(\underline{x}) = \frac{1}{Z} \prod_{(ij) \in E} \psi_{ij}(x_i, x_j). \tag{14.30}$$

Function nodes can be identified with edges $(ij) \in E$.

In this case belief propagation can be described as operating directly on G. Further, one of the two types of messages can be easily eliminated: here we shall work uniquely with variable-to-function messages, which we will denote by $\nu_{i \to j}^{(t)}(x_i)$, a shortcut for $\nu_{i \to (ij)}^{(t)}(x_i)$. The BP updates then read

$$\nu_{i \to j}^{(t+1)}(x_i) \cong \prod_{l \in \partial i \setminus j} \sum_{x_l} \psi_{il}(x_i, x_l) \nu_{l \to i}^{(t)}(x_l) \,. \tag{14.31}$$

Simplified expressions can be derived in this case for the joint distribution of several variables (see eqn (14.18)), as well as for the free entropy.

Exercise 14.7 Show that, for pairwise models, the free entropy given in eqn (14.27) can be written as $\mathbb{F}_*(\underline{\nu}) = \sum_{i \in V} \mathbb{F}_i(\underline{\nu}) - \sum_{(ij) \in E} \mathbb{F}_{(ij)}(\underline{\nu})$, where

$$\mathbb{F}_i(\underline{\nu}) = \log \left[\sum_{x_i} \prod_{j \in \partial i} \left(\sum_{x_j} \psi_{ij}(x_i, x_j) \nu_{j \to i}(x_j) \right) \right] \,,$$

$$\mathbb{F}_{(ij)}(\underline{\nu}) = \log \left[\sum_{x_i, x_j} \nu_{i \to j}(x_i) \psi_{ij}(x_i, x_j) \nu_{j \to i}(x_j) \right] \,. \tag{14.32}$$

14.3 Optimization: Max-product and min-sum

Message-passing algorithms are not limited to computing marginals. Imagine that you are given a probability distribution $\mu(\cdot)$ as in eqn (14.13), and you are asked to find a configuration $\underline{x}$ which maximizes the probability $\mu(\underline{x})$. Such a configuration is called a **mode** of $\mu(\cdot)$. This task is important in many applications, ranging from MAP estimation (e.g. in image reconstruction) to word MAP decoding.

It is not hard to devise a message-passing algorithm adapted to this task, which correctly solves the problem on trees.

14.3.1 Max-marginals

The role of marginal probabilities is played here by the **max-marginals**

$$M_i(x_i^*) = \max_{\underline{x}} \{ \mu(\underline{x}) : x_i = x_i^* \} \,. \tag{14.33}$$

In the same way as the tasks of sampling and of computing partition functions can be reduced to computing marginals, optimization can be reduced to computing max-marginals. In other words, given a black box that computes max-marginals, optimization can be performed efficiently.

Consider first the simpler case in which the max-marginals are non-degenerate, i.e., for each $i \in V$, there exists an x_i^* such that $M_i(x_i^*) > M_i(x_i)$ (strictly) for any $x_i \neq x_i^*$. The unique maximizing configuration is then given by $\underline{x}^* = (x_1^*, \ldots, x_N^*)$.

In the general case, the following 'decimation' procedure, which is closely related to the BP-guided sampling algorithm of Section 14.2.4, returns one of the maximizing configurations. Choose an ordering of the variables, say $(1, \ldots, N)$. Compute $M_1(x_1)$, and let x_1^* be one of the values maximizing it: $x^* \in \arg\max M_1(x_1)$. Fix x_1 to take this value, i.e. modify the graphical model by introducing the factor $\mathbb{I}(x_1 = x_1^*)$ (this corresponds to considering the conditional distribution $\mu(\underline{x}|x_1 = x_1^*)$). Compute $M_2(x_2)$ for the new model, fix x_2 to one value $x_2^* \in \arg\max M_2(x_2)$, and iterate this procedure, fixing all the x_i's sequentially.

14.3.2   Message passing

It is clear from the above that max-marginals need only to be computed up to a multiplicative normalization. We shall therefore stick to our convention of denoting equality between max-marginals up to an overall normalization by $\cong$. Adapting the message-passing update rules to the computation of max-marginals is not hard: it is sufficient to replace sums with maximizations. This yields the following **max-product** update rules:

$$\nu_{i \to a}^{(t+1)}(x_i) \cong \prod_{b \in \partial i \setminus a} \widehat{\nu}_{b \to i}^{(t)}(x_i) , \tag{14.34}$$

$$\widehat{\nu}_{a \to i}^{(t)}(x_i) \cong \max_{\underline{x}_{\partial a \setminus i}} \left\{ \psi_a(\underline{x}_{\partial a}) \prod_{j \in \partial a \setminus i} \nu_{j \to a}^{(t)}(x_j) \right\} . \tag{14.35}$$

The fixed-point conditions for this recursion are called the **max-product equations**. As in BP, it is understood that, when $\partial j \setminus a$ is an empty set, $\nu_{j \to a}(x_j) \cong 1$ is the uniform distribution. Similarly, if $\partial a \setminus j$ is empty, then $\widehat{\nu}_{a \to j}(x_j) \cong \psi_a(x_j)$. After any number of iterations, an estimate of the max-marginals is obtained as follows:

$$\nu_i^{(t)}(x_i) \cong \prod_{a \in \partial i} \widehat{\nu}_{a \to i}^{(t-1)}(x_i) . \tag{14.36}$$

As in the case of BP, the main motivation for the above updates comes from the analysis of graphical models on trees.

Theorem 14.4. (the max-product algorithm is exact on trees) *Consider a tree-graphical model with diameter t_*. Then:*

1. *Irrespective of the initialization, the max-product updates (14.34) and (14.35) converge after at most t_* iterations. In other words, for any edge (i, a) and any $t > t_*$, $\nu_{i \to a}^{(t)} = \nu_{i \to a}^*$ and $\widehat{\nu}_{a \to i}^{(t)} = \widehat{\nu}_{a \to i}^*$.*

2. *The max-marginals are estimated correctly, i.e., for any variable node i and any $t > t_*$, $\nu_i^{(t)}(x_i) = M_i(x_i)$.*

The proof follows closely that of Theorem 14.1, and is left as an exercise for the reader.

Exercise 14.8 The crucial property used in both Theorem 14.1 and Theorem 14.4 is the distributive property of the sum and the maximum with respect to the product. Consider, for instance, a function of the form $f(x_1, x_2, x_3) = \psi_1(x_1, x_2)\psi_2(x_1, x_3)$. Then one can decompose the sum and maximum as follows:

$$\sum_{x_1, x_2, x_3} f(x_1, x_2, x_3) = \sum_{x_1}\left[\left(\sum_{x_2}\psi_1(x_1, x_2)\right)\left(\sum_{x_3}\psi_2(x_1, x_3)\right)\right], \qquad (14.37)$$

$$\max_{x_1, x_2, x_3} f(x_1, x_2, x_3) = \max_{x_1}\left[\left(\max_{x_2}\psi_1(x_1, x_2)\right)\left(\max_{x_3}\psi_2(x_1, x_3)\right)\right]. \qquad (14.38)$$

Formulate a general 'marginalization' problem (with the ordinary sum and product substituted by general operations with a distributive property) and describe a message-passing algorithm that solves it on trees.

The max-product messages $\nu_{i\to a}^{(t)}(\cdot)$ and $\hat{\nu}_{a\to i}^{(t)}(\cdot)$ admit an interpretation which is analogous to that of sum–product messages. For instance, $\nu_{i\to a}^{(t)}(\cdot)$ is an estimate of the max-marginal of variable x_i with respect to the modified graphical model in which factor node a is removed from the graph. Along with the proof of Theorem 14.4, it is easy to show that, in a tree-graphical model, fixed-point messages do indeed coincide with the max-marginals of such modified graphical models.

The problem of finding the mode of a distribution that factorizes as in eqn (14.13) has an alternative formulation, namely as minimizing a cost (energy) function that can be written as a sum of local terms:

$$E(\underline{x}) = \sum_{a\in F} E_a(\underline{x}_{\partial a}). \qquad (14.39)$$

The problems are mapped onto each other by writing $\psi_a(\underline{x}_{\partial a}) = \mathrm{e}^{-\beta E_a(\underline{x}_{\partial a})}$ (with β some positive constant). A set of message-passing rules that is better adapted to the latter formulation is obtained by taking the logarithm of eqns (14.34) and (14.35). This version of the algorithm is known as the **min-sum** algorithm:

$$E_{i\to a}^{(t+1)}(x_i) = \sum_{b\in\partial i\setminus a} \hat{E}_{b\to i}^{(t)}(x_i) + C_{i\to a}^{(t)}, \qquad (14.40)$$

$$\hat{E}_{a\to i}^{(t)}(x_i) = \min_{\underline{x}_{\partial a\setminus i}}\left[E_a(\underline{x}_{\partial a}) + \sum_{j\in\partial a\setminus i} E_{j\to a}^{(t)}(x_j)\right] + \hat{C}_{a\to i}^{(t)}. \qquad (14.41)$$

The corresponding fixed-point equations are also known in statistical physics as the **energetic cavity equations**. Notice that, since the max-product marginals are relevant only up to a multiplicative constant, the min-sum messages are defined up to an overall additive constant. In the following, we shall choose the constants $C_{i\to a}^{(t)}$ and $\hat{C}_{a\to i}^{(t)}$ such that $\min_{x_i} E_{i\to a}^{(t+1)}(x_i) = 0$ and $\min_{x_i} \hat{E}_{a\to i}^{(t)}(x_i) = 0$, respectively. The analogue of the max-marginal estimate in eqn (14.36) is provided by the following log-max-marginal:

$$E_i^{(t)}(x_i) = \sum_{a \in \partial i} \widehat{E}_{a \to i}^{(t-1)}(x_i) + C_i^{(t)} . \tag{14.42}$$

In the case of tree-graphical models, the minimum energy $U_* = \min_{\underline{x}} E(\underline{x})$ can be immediately written in terms of the fixed-point messages $\{E_{i \to a}^*, \widehat{E}_{i \to a}^*\}$. We obtain, in fact,

$$U_* = \sum_a E_a(\underline{x}_{\partial a}^*) , \tag{14.43}$$

$$\underline{x}_{\partial a}^* = \arg \min_{\underline{x}_{\partial a}} \left\{ E_a(\underline{x}_{\partial a}) + \sum_{i \in \partial a} \widehat{E}_{i \to a}^*(x_i) \right\} . \tag{14.44}$$

In the case of non-tree graphs, this can be taken as a prescription to obtain a max-product estimate $U_*^{(t)}$ of the minimum energy. One just needs to replace the fixed-point messages in eqn (14.44) with the messages obtained after t iterations. Finally, a minimizing configuration $\underline{x}^*$ can be obtained through the decimation procedure described in the previous subsection.

Exercise 14.9 Show that U_* is also given by $U_* = \sum_{a \in F} \epsilon_a + \sum_{i \in V} \epsilon_i - \sum_{(ia) \in E} \epsilon_{ia}$, where

$$\epsilon_a = \min_{\underline{x}_{\partial a}} \left[E_a(\underline{x}_{\partial a}) + \sum_{j \in \partial a} E_{j \to a}^*(x_j) \right] , \qquad \epsilon_i = \min_{x_i} \left[\sum_{a \in \partial i} \widehat{E}_{a \to i}^*(x_i) \right] ,$$

$$\epsilon_{ia} = \min_{x_i} \left[E_{i \to a}^*(x_i) + \widehat{E}_{a \to i}^*(x_i) \right] . \tag{14.45}$$

[Hints: (*i*) Define $x_i^*(a) = \arg \min \left[\widehat{E}_{a \to i}^*(x_i) + E_{i \to a}^*(x_i) \right]$, and show that the minima in eqn (14.45) are achieved at $x_i = x_i^*(a)$ (for ϵ_i and ϵ_{ai}) and at $\underline{x}_{\partial a}^* = \{x_i^*(a)\}_{i \in \partial a}$ (for ϵ_a). (*ii*) Show that $\sum_{(ia)} \widehat{E}_{a \to i}^*(x_i^*(a)) = \sum_i \epsilon_i$.]

14.3.3 Warning propagation

A frequently encountered case is that of constraint satisfaction problems, where the energy function just counts the number of violated constraints:

$$E_a(\underline{x}_{\partial a}) = \begin{cases} 0 & \text{if constraint } a \text{ is satisfied}, \\ 1 & \text{otherwise}. \end{cases} \tag{14.46}$$

The structure of messages can be simplified considerably in this case. More precisely, if the messages are initialized in such a way that $\widehat{E}_{a \to i}^{(0)} \in \{0, 1\}$, this condition is preserved by the min-sum updates (14.40) and (14.41) at any subsequent time. Let us prove this statement by induction. Suppose it holds up to time $t-1$. From eqn (14.40), it follows that $E_{i \to a}^{(t)}(x_i)$ is a non-negative integer. Now consider eqn (14.41). Since both $E_{j \to a}^{(t)}(x_j)$ and $E_a(\underline{x}_{\partial a})$ are integers, it follows that $\widehat{E}_{a \to i}^{(t)}(x_i)$, the minimum of the right-hand side, is a non-negative integer as well. Further, since for each $j \in \partial a \setminus i$ there exists an x_j^* such that $E_{j \to a}^{(t)}(x_j^*) = 0$, the minimum in eqn (14.41) is at most 1, which proves our claim.

This argument also shows that the outcome of the minimization in eqn (14.41) depends only on which entries of the messages $E_{j \to a}^{(t)}(\,\cdot\,)$ vanish. If there exists an assignment x_j^* such that $E_{j \to a}^{(t)}(x_j^*) = 0$ for each $j \in \partial a \setminus i$, and $E_a(x_i, \underline{x}_{\partial a \setminus i}^*) = 0$, then the value of the minimum is 0. Otherwise, it is 1.

In other words, instead of keeping track of the messages $E_{i \to a}(\,\cdot\,)$, one can use their 'projections'

$$\mathrm{E}_{i \to a}(x_i) = \min\{1, E_{i \to a}(x_i)\} \ . \tag{14.47}$$

Proposition 14.5 *Consider an optimization problem with a cost function of the form (14.39) with $E_a(\underline{x}_{\partial a}) \in \{0,1\}$, and assume the min-sum algorithm to be initialized with $\widehat{E}_{a \to i}(x_i) \in \{0,1\}$ for all edges (i,a). Then, after any number of iterations, the function-node-to-variable-node messages coincide with those computed using the following update rules:*

$$\mathrm{E}_{i \to a}^{(t+1)}(x_i) = \min\left\{1, \sum_{b \in \partial i \setminus a} \widehat{E}_{b \to i}^{(t)}(x_i) + C_{i \to a}^{(t)}\right\}, \tag{14.48}$$

$$\widehat{E}_{a \to i}^{(t)}(x_i) = \min_{\underline{x}_{\partial a \setminus i}}\left\{E_a(\underline{x}_{\partial a}) + \sum_{j \in \partial a \setminus i} \mathrm{E}_{j \to a}^{(t)}(x_j)\right\} + \widehat{C}_{a \to i}^{(t)}, \tag{14.49}$$

where $C_{i \to a}^{(t)}$, $\widehat{C}_{a \to i}^{(t)}$ are normalization constants determined by $\min_{x_i} \widehat{E}_{a \to i}(x_i) = 0$ and $\min_{x_i} \mathrm{E}_{i \to a}(x_i) = 0$.

Finally, the ground state energy takes the same form as eqn. (14.45), with $\mathrm{E}_{i \to a}(\,\cdot\,)$ replacing $E_{i \to a}(\,\cdot\,)$.

We call the simplified min-sum algorithm with the update equations (14.49) and (14.48) the **warning propagation** algorithm.

The name is due to the fact that the messages $\mathrm{E}_{i \to a}(\,\cdot\,)$ can be interpreted as the following warnings:

> $\mathrm{E}_{i \to a}(x_i) = 1 \ \to$ *'according to the set of constraints $b \in \partial i \setminus a$, the i-th variable should not take the value x_i'.*
> $\mathrm{E}_{i \to a}(x_i) = 0 \ \to$ *'according to the set of constraints $b \in \partial i \setminus a$, the i-th variable can take the value x_i'.*

Warning propagation provides a procedure for finding all direct implications of a partial assignment of the variables in a constraint satisfaction problem. For instance, in the case of the satisfiability problem, it finds all implications found by unit clause propagation (see Section 10.2).

14.4 Loopy BP

We have seen how message-passing algorithms can be used efficiently in tree-graphical models. In particular, they allow one to exactly sample distributions that factorize according to tree factor graphs and to compute marginals, partition functions, and modes of such distributions. It would be very useful in a number of applications to be able to accomplish the same tasks when the underlying factor graph is no longer a tree.

It is tempting to use the BP equations in this more general context, hoping to get approximate results for large graphical models. Often, we shall be dealing with problems that are NP-hard even to approximate, and it is difficult to provide general guarantees of performance. Indeed, an important unsolved challenge is to identify classes of graphical models where the following questions can be answered:

1. Is there any set of messages $\{\nu^*_{i\to a}, \widehat{\nu}^*_{a\to i}\}$ that reproduces the local marginals of $\mu(\cdot)$ by use of eqn (14.18), within some prescribed accuracy?
2. Do such messages correspond to an (approximate) fixed point of the BP update rules (14.14) and (14.15)?
3. Do the BP update rules have at least one (approximate) fixed point? Is it unique?
4. Does such a fixed point have a non-empty 'basin of attraction' with respect to eqns (14.14) and (14.15)? Does this basin of attraction include all possible (or all 'reasonable') initializations?

We shall not treat these questions in depth, as a general theory is lacking. We shall, rather, describe the sophisticated picture that has emerged, building on a mixture of physical intuition, physical methods, empirical observations, and rigorous proofs.

Exercise 14.10 Consider a ferromagnetic Ising model on a two-dimensional grid with periodic boundary conditions (i.e. 'wrapped' around a torus), as defined in Section 9.1.2 (see Fig. 9.7). Ising spins σ_i, $i \in V$, are associated with the vertices of the grid, and interact along the edges:

$$\mu(\underline{\sigma}) = \frac{1}{Z} e^{\beta \sum_{(ij)\in E} \sigma_i \sigma_j} . \tag{14.50}$$

(*a*) Describe the associated factor graph.

(*b*) Write the BP equations.

(*c*) Look for a solution that is invariant under translation, i.e. $\nu_{i\to a}(\sigma_i) = \nu(\sigma_i)$, $\widehat{\nu}_{a\to i}(\sigma_i) = \widehat{\nu}(\sigma_i)$: write down the equations satisfied by $\nu(\cdot)$, $\widehat{\nu}(\cdot)$.

(*d*) Parameterize $\nu(\sigma)$ in terms of the log-likelihood $h = (1/2\beta)\log(\nu(+1)/\nu(-1))$ and show that h satisfies the equation $\tanh(\beta h) = \tanh(\beta)\tanh(3\beta h)$.

(*e*) Study this equation and show that, for $3\tanh\beta > 1$, it has three distinct solutions corresponding to three BP fixed points.

(*f*) Consider the iteration of the BP updates starting from a translation-invariant initial condition. Does the iteration converge to a fixed point? Which one?

(*g*) Discuss the appearance of three BP fixed points in relation to the structure of the distribution $\mu(\underline{\sigma})$ and the paramagnetic–ferromagnetic transition. What is the approximate value of the critical temperature obtained from BP? Compare with the exact value $\beta_c = \frac{1}{2}\log(1+\sqrt{2})$.

(*h*) What results does one obtain for an Ising model on a d-dimensional (instead of two-dimensional) grid?

14.4.1 The Bethe free entropy

As we saw in Section 14.2.4, the free entropy of a tree-graphical model has a simple expression in terms of local marginals (see eqn (14.26)). We can use it in graphs with loops with the hope that it provides a good estimate of the actual free entropy. In spirit, this approach is similar to the 'mean-field' free entropy introduced in Chapter 2, although it differs from it in several respects.

In order to define precisely the Bethe free entropy, we must first describe a space of 'possible' local marginals. A minimalistic approach is to restrict ourselves to the 'locally consistent marginals'. A set of **locally consistent marginals** is a collection of distributions $b_i(\cdot)$ over $\mathcal{X}$ for each $i \in V$, and $b_a(\cdot)$ over $\mathcal{X}^{|\partial a|}$ for each $a \in F$. Being distributions, they must be non-negative, i.e. $b_i(x_i) \geq 0$ and $b_a(\underline{x}_{\partial a}) \geq 0$, and they must satisfy the normalization conditions

$$\sum_{x_i} b_i(x_i) = 1 \quad \forall i \in V, \qquad \sum_{\underline{x}_{\partial a}} b_a(\underline{x}_{\partial a}) = 1 \quad \forall a \in F. \tag{14.51}$$

To be 'locally consistent', they must satisfy the marginalization condition

$$\sum_{\underline{x}_{\partial a \setminus i}} b_a(\underline{x}_{\partial a}) = b_i(x_i) \quad \forall a \in F, \ \forall i \in \partial a. \tag{14.52}$$

Given a factor graph G, we shall denote the set of locally consistent marginals by $\mathrm{LOC}(G)$, and the Bethe free entropy will be defined as a real-valued function on this space.

It is important to stress that, although the marginals of any probability distribution $\mu(\underline{x})$ over $\underline{x} = (x_1, \ldots, x_N)$ must be locally consistent, the converse is not true: one can find sets of locally consistent marginals that do not correspond to any distribution. In order to emphasize this point, locally consistent marginals are sometimes called '**beliefs**'.

Exercise 14.11 Consider the graphical model shown in Fig. 14.5, on binary variables (x_1, x_2, x_3), $x_i \in \{0, 1\}$. The figure also gives a set of beliefs in the vector/matrix form

$$b_i = \begin{bmatrix} b_i(0) \\ b_i(1) \end{bmatrix}, \quad b_{ij} = \begin{bmatrix} b_{ij}(00) & b_{ij}(01) \\ b_{ij}(10) & b_{ij}(11) \end{bmatrix}. \tag{14.53}$$

Check that this set of beliefs is locally consistent, but that they cannot be the marginals of any distribution $\mu(x_1, x_2, x_3)$.

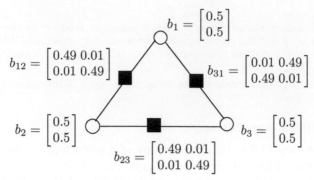

Fig. 14.5 A set of locally consistent marginals, 'beliefs', that cannot arise as the marginals of any global distribution.

Given a set of locally consistent marginals $\underline{b} = \{b_a, b_i\}$, we associate a **Bethe free entropy** with it exactly as in eqn (14.26):

$$\mathbb{F}[\underline{b}] = -\sum_{a \in F} b_a(\underline{x}_{\partial a}) \log \left\{ \frac{b_a(\underline{x}_{\partial a})}{\psi_a(\underline{x}_{\partial a})} \right\} - \sum_{i \in V} (1 - |\partial i|) b_i(x_i) \log b_i(x_i). \quad (14.54)$$

The analogy with the naive mean-field approach suggests that stationary points (and, in particular, maxima) of the Bethe free entropy should play an important role. This is partially confirmed by the following result.

Proposition 14.6 *Assume $\psi_a(\underline{x}_{\partial a}) > 0$ for every a and $\underline{x}_{\partial a}$. Then the stationary points of the Bethe free entropy $\mathbb{F}[\underline{b}]$ are in one-to-one correspondence with the fixed points of the BP algorithm.*

As will become apparent from the proof, the correspondence between BP fixed points and stationary points of $\mathbb{F}[\underline{b}]$ is completely explicit.

Proof We want to check stationarity with respect to variations of $\underline{b}$ within the set $\mathsf{LOC}(G)$, which is defined by the constraints (14.51) and (14.52), as well as $b_a(\underline{x}_{\partial a}) \geq 0$, $b_i(x_i) \geq 0$. We thus introduce a set of Lagrange multipliers $\underline{\lambda} = \{\lambda_i, i \in V;$ $\lambda_{ai}(x_i), (a,i) \in E, x_i \in \mathcal{X}\}$, where λ_i corresponds to the normalization of $b_i(\,\cdot\,)$ and $\lambda_{ai}(x_i)$ corresponds to the marginal of b_a coinciding with b_i. We then define the Lagrangian

$$\mathcal{L}(\underline{b}, \underline{\lambda}) = \mathbb{F}[\underline{b}] - \sum_{a \in F} \lambda_i \left[\sum_{x_i} b_i(x_i) - 1 \right] - \sum_{(ia), x_i} \lambda_{ai}(x_i) \left[\sum_{\underline{x}_{\partial a \setminus i}} b_a(\underline{x}_{\partial a}) - b_i(x_i) \right].$$

$$(14.55)$$

Notice that we have not introduced a Lagrange multiplier for the normalization of $b_a(\underline{x}_{\partial a})$, as this follows from the two constraints already enforced. The stationarity conditions with respect to b_i and b_a imply

$$b_i(x_i) \cong e^{-1/(|\partial i|-1)} \sum_{a \in \partial i} \lambda_{ai}(x_i) , \quad b_a(\underline{x}_{\partial a}) \cong \psi_a(\underline{x}_{\partial a}) \, e^{-\sum_{i \in \partial a} \lambda_{ai}(x_i)} . \quad (14.56)$$

The Lagrange multipliers must be chosen in such a way that eqn (14.52) is fulfilled. Any such set of Lagrange multipliers yields a stationary point of $\mathbb{F}[\underline{b}]$. Once the $\lambda_{ai}(x_j)$ have been found, the computation of the normalization constants in these expressions fixes λ_i. Conversely, any stationary point corresponds to a set of Lagrange multipliers satisfying the stated condition.

It remains to show that sets of Lagrange multipliers such that $\sum_{\underline{x}_{\partial a\setminus i}} b_a(\underline{x}_{\partial a}) = b_i(x_i)$ are in one-to-one correspondence with BP fixed points. In order to see this, we define the messages

$$\nu_{i\to a}(x_i) \cong e^{-\lambda_{ai}(x_i)} , \quad \widehat{\nu}_{a\to i}(x_i) \cong \sum_{\underline{x}_{\partial a\setminus i}} \psi_a(\underline{x}_{\partial a}) e^{-\sum_{j\in\partial a\setminus i}\lambda_{aj}(x_j)} . \quad (14.57)$$

It is clear from the definition that such messages satisfy

$$\widehat{\nu}_{a\to i}(x_i) \cong \sum_{\underline{x}_{\partial a\setminus i}} \psi_a(\underline{x}_{\partial a}) \prod_{j\in\partial a\setminus i} \nu_{i\to a}(x_i) . \quad (14.58)$$

Further, using the second equation of eqns (14.56) together with eqn. (14.57), we get $\sum_{\underline{x}_{\partial a\setminus i}} b_a(\underline{x}_{\partial a}) \cong \nu_{i\to a}(x_i)\widehat{\nu}_{a\to i}(x_i)$. On the other hand, from the first of eqns (14.56) together with eqn (14.57), we get $b_i(x_i) \cong \prod_b \nu_{i\to b}(x_i)^{1/(|\partial i|-1)}$. The marginalization condition thus implies

$$\prod_{b\in\partial i} \nu_{i\to b}(x_i)^{1/(|\partial i|-1)} \cong \nu_{i\to a}(x_i)\widehat{\nu}_{a\to i}(x_i) . \quad (14.59)$$

Taking the product of these equalities for $a\in\partial i\setminus b$, and eliminating $\prod_{a\in\partial i\setminus b}\nu_{i\to a}(x_i)$ from the resulting equation (which is possible if $\psi_a(\underline{x}_{\partial a}) > 0$), we get

$$\nu_{i\to b}(x_i) \cong \prod_{a\in\partial i\setminus b} \widehat{\nu}_{a\to i}(x_i) . \quad (14.60)$$

At this point we recognize in eqns (14.58) and (14.60) the fixed-point condition for BP (see eqns (14.14) and (14.15)). Conversely, given any solution of eqns (14.58) and (14.60), one can define a set of Lagrange multipliers using the first of eqns (14.57). It follows from the fixed point condition that the second of eqns (14.57) is fulfilled as well, and that the marginalization condition holds. $\square$

An important consequence of this proposition is the existence of BP fixed points.

Corollary 14.7 *Assume $\psi_a(\underline{x}_a) > 0$ for every a and $\underline{x}_{\partial a}$. The BP algorithm then has at least one fixed point.*

Proof Since $\mathbb{F}[\underline{b}]$ is bounded and continuous in $\mathsf{LOC}(G)$ (which is closed), it takes its maximum at some point $\underline{b}^* \in \mathsf{LOC}(G)$. Using the condition $\psi_a(\underline{x}_a) > 0$, it is easy to see that such a maximum is reached in the relative interior of $\mathsf{LOC}(G)$, i.e. that $b_a^*(\underline{x}_{\partial a}) > 0$, $b_i^*(x_i) > 0$ strictly. As a consequence, $\underline{b}^*$ must be a stationary point and therefore, by Proposition 14.6, there is a BP fixed point associated with it. $\square$

The 'variational principle' provided by Proposition 14.6 is particularly suggestive as it is analogous to naive mean-field bounds. For practical applications, it is sometimes

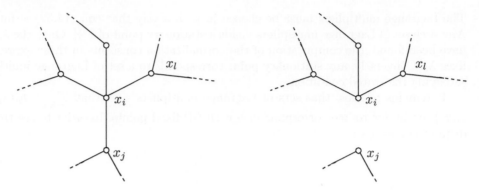

Fig. 14.6 *Left*: neighbourhood of a node i in a pairwise graphical model. *Right*: the modified graphical model used to define the message $\nu_{i\to j}(x_i)$.

more convenient to use the free-entropy functional $\mathbb{F}_*(\underline{\nu})$ of eqn (14.27). This can be regarded as a function from the space of messages to reals $\mathbb{F} : \mathfrak{M}(\mathcal{X})^{|\vec{E}|} \to \mathbb{R}$ (remember that $\mathfrak{M}(\mathcal{X})$ denotes the set of measures over $\mathcal{X}$, and $\vec{E}$ is the set of directed edges in the factor graph).[3] It satisfies the following variational principle.

Proposition 14.8 *The stationary points of the Bethe free entropy $\mathbb{F}_*(\underline{\nu})$ are fixed points of belief propagation. Conversely, any fixed point $\underline{\nu}$ of belief propagation such that $\mathbb{F}_*(\underline{\nu})$ is finite, is also a stationary point of $\mathbb{F}_*(\underline{\nu})$.*

The proof is simple calculus and is left to the reader.

It turns out that for tree graphs and unicyclic graphs, $\mathbb{F}[\underline{b}]$ is convex, and the above results then prove the existence and uniqueness of BP fixed points. But, for general graphs, $\mathbb{F}[\underline{b}]$ is non-convex and may have multiple stationary points.

14.4.2 Correlations

What is the origin of the error made when BP is used in an arbitrary graph with loops, and under what conditions can it be small? In order to understand this point, let us consider for notational simplicity a pairwise graphical model (see eqn (14.2.5)). The generalization to other models is straightforward. Taking seriously the probabilistic interpretation of messages, we want to compute the marginal distribution $\nu_{i\to j}(x_i)$ of x_i in a modified graphical model that does not include the factor $\psi_{ij}(x_i, x_j)$ (see Fig. 14.6). We denote by $\mu_{\partial i\backslash j}(\underline{x}_{\partial i\backslash j})$ the joint distribution of all variables in $\partial i \backslash j$ in the model where all the factors $\psi_{il}(x_i, x_l)$, $l \in \partial i$, have been removed. Then,

$$\nu_{i\to j}(x_i) \cong \sum_{\underline{x}_{\partial i\backslash j}} \prod_{l\in\partial i\backslash j} \psi_{il}(x_i, x_l)\mu_{\partial i\backslash j}(\underline{x}_{\partial i\backslash j}). \tag{14.61}$$

Comparing this expression with the BP equations (see eqn (14.31)), we deduce that the messages $\{\nu_{i\to j}\}$ solve these equations if

[3]On a tree, $\mathbb{F}_*(\underline{\nu})$ is (up to a change of variables) the Lagrangian dual of $\mathbb{F}(\underline{b})$.

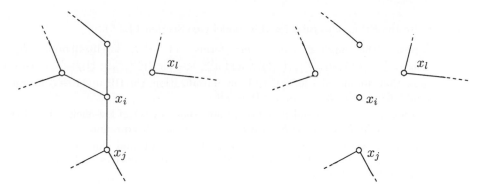

Fig. 14.7 *Left*: modified graphical model used to define $\nu_{l\to i}(x_l)$. *Right*: modified graphical model corresponding to the cavity distribution of the neighbours of i, $\mu_{\partial i\setminus j}(\underline{x}_{\partial i\setminus j})$.

$$\mu_{\partial i\setminus j}(\underline{x}_{\partial i\setminus j}) = \prod_{l\in\partial i\setminus j} \nu_{l\to i}(x_l)\,. \tag{14.62}$$

We can expect this to happen when two conditions are fulfilled:

1. Under $\mu_{\partial i\setminus j}(\,\cdot\,)$, the variables $\{x_l : l \in \partial i \setminus j\}$ are independent: $\mu_{\partial i\setminus j}(\underline{x}_{\partial i\setminus j}) = \prod_{l\in\partial i\setminus j} \mu_{\partial i\setminus j}(x_l)$.

2. The marginal of each of these variables under $\mu_{\partial i\setminus j}(\,\cdot\,)$ is equal to the corresponding message $\nu_{l\to i}(x_l)$. In other words, the two graphical models obtained by removing all the compatibility functions that involve x_i (namely, the model $\mu_{\partial i\setminus j}(\,\cdot\,)$) and by removing only $\psi_{il}(x_i, x_l)$ must have the same marginal for the variable x_l; see Fig. 14.7.

These two conditions are obviously fulfilled for tree-graphical models. They are also approximately fulfilled if the correlations among the variables $\{x_l : l \in \partial i\}$ are 'small' under $\mu_{\partial i\setminus j}(\,\cdot\,)$. As we have seen, in many cases of practical interest (LDPC codes, random K-SAT, etc.) the factor graph is locally tree-like. In other words, when node i is removed, the variables $\{x_l : l \in \partial i\}$ are, with high probability, far apart from each other. This suggests that, in such models, the two conditions above may indeed hold in the large-size limit, provided far-apart variables are weakly correlated. A simple illustration of this phenomenon is provided in the exercises below. The following chapters will investigate this property further and discuss how to cope with cases in which it does not hold.

Exercise 14.12 Consider an antiferromagnetic Ising model on a ring, with variables $(\sigma_1, \ldots, \sigma_N) \equiv \underline{\sigma}$, $\sigma_i \in \{+1, -1\}$ and distribution

$$\mu(\underline{\sigma}) = \frac{1}{Z}\, e^{-\beta \sum_{i=1}^{N} \sigma_i \sigma_{i+1}}\,, \tag{14.63}$$

where $\sigma_{N+1} \equiv \sigma_1$. This is a pairwise graphical model whose graph G is a ring over N vertices.

(a) Write the BP update rules for this model (see Section 14.2.5).

(b) Express the update rules in terms of the log-likelihoods $h^{(t)}_{i\rightarrow}$ $\equiv$ $\frac{1}{2}\log((\nu^{(t)}_{i\rightarrow i+1}(+1))/(\nu^{(t)}_{i\rightarrow i+1}(-1)))$, and $h^{(t)}_{\leftarrow i} \equiv \frac{1}{2}\log((\nu^{(t)}_{i\rightarrow i-1}(+1))/(\nu^{(t)}_{i\rightarrow i-1}(-1)))$.

(c) Show that, for any $\beta \in [0,\infty)$, and any initialization, the BP updates converge to the unique fixed point $h_{\leftarrow i} = h_{i\rightarrow} = 0$ for all i.

(d) Assume that $\beta = +\infty$ and N is even. Show that any set of log-likelihoods of the form $h_{i\rightarrow} = (-1)^i a$, $h_{\leftarrow i} = (-1)^i b$, with $a,b \in [-1,1]$, is a fixed point.

(e) Consider now the case where $\beta = \infty$ and N is odd, and show that the only fixed point is $h_{\leftarrow i} = h_{i\rightarrow} = 0$. Find an initialization of the messages such that BP does not converge to this fixed point.

Exercise 14.13 Consider a ferromagnetic Ising model on a ring with a magnetic field. This is defined through the distribution

$$\mu(\underline{\sigma}) = \frac{1}{Z} e^{\beta \sum_{i=1}^{N} \sigma_i \sigma_{i+1} + B \sum_{i=1}^{N} \sigma_i} , \qquad (14.64)$$

where $\sigma_{N+1} \equiv \sigma_1$. Notice that, with respect to the previous exercise, we have changed a sign in the exponent.

(a, b) As in the previous exercise.

(c) Show that, for any $\beta \in [0,\infty)$, and any initialization, the BP updates converge to the unique fixed point $h_{\leftarrow i} = h_{i\rightarrow} = h_*(\beta, B)$ for all i.

(d) Let $\langle \sigma_i \rangle$ be the expectation of spin σ_i with respect to the measure $\mu(\cdot)$, and let $\langle \sigma_i \rangle_{\mathrm{BP}}$ be the corresponding BP estimate. Show that $|\langle \sigma_i \rangle - \langle \sigma_i \rangle_{\mathrm{BP}}| = O(\lambda^N)$ for some $\lambda \in (0,1)$.

14.5 General message-passing algorithms

Both the sum–product and the max-product (or min-sum) algorithm are instances of a more general class of **message-passing** algorithms. All of the algorithms in this family share some common features, which we now highlight.

Given a factor graph, a message-passing algorithm is defined by the following ingredients:

1. An alphabet of messages M. This can be either continuous or discrete. The algorithm operates on messages $\nu^{(t)}_{i\rightarrow a}, \widehat{\nu}^{(t)}_{a\rightarrow i} \in$ M associated with the directed edges in the factor graph.

2. Update functions $\Psi_{i\rightarrow a} : \mathsf{M}^{|\partial i \setminus a|} \rightarrow \mathsf{M}$ and $\Phi_{a\rightarrow i} : \mathsf{M}^{|\partial a \setminus i|} \rightarrow \mathsf{M}$ that describe how to update messages.

3. An initialization, i.e. a mapping from the directed edges in the factor graph to M (this can be a random mapping). We shall denote by $\nu^{(0)}_{i\rightarrow a}, \widehat{\nu}^{(0)}_{a\rightarrow i}$ the image of such a mapping.

4. A decision rule, i.e. a local function from messages to a space of 'decisions' from which we are interested in making a choice. Since we shall be interested mostly

in computing marginals (or max-marginals), we shall assume the decision rule to be given by a family of functions $\widehat{\Psi}_i : \mathsf{M}^{|\partial i|} \to \mathfrak{M}(\mathcal{X})$.

Notice the characteristic feature of message-passing algorithms: messages going out from a node are functions of messages coming into the same node through the other edges.

Given these ingredients, a message-passing algorithm with parallel updating may be defined as follows. Assign the values of initial messages $\nu_{i\to a}^{(0)}, \widehat{\nu}_{a\to i}^{(0)}$ according to an initialization rule. Then, for any $t \geq 0$, update the messages through local operations at variable/check nodes as follows:

$$\nu_{i\to a}^{(t+1)} = \Psi_{i\to a}(\{\widehat{\nu}_{b\to i}^{(t)} : b \in \partial i \setminus a\})\,, \tag{14.65}$$

$$\widehat{\nu}_{a\to i}^{(t)} = \Phi_{a\to i}(\{\nu_{j\to a}^{(t)} : j \in \partial a \setminus i\})\,. \tag{14.66}$$

Finally, after a pre-established number of iterations t, take the decision using the rules $\widehat{\Psi}_i$; namely, return

$$\nu_i^{(t)}(x_i) = \widehat{\Psi}_i(\{\widehat{\nu}_{b\to i}^{(t-1)} : b \in \partial i\})(x_i)\,. \tag{14.67}$$

Many variants are possible concerning the update schedule. For instance, in the case of sequential updating one can pick out a directed edge uniformly at random and compute the corresponding message. Another possibility is to generate a random permutation of the edges and update the messages according to this permutation. We shall not discuss these 'details', but the reader should be aware that they can be important in practice: some update schemes may converge better than others.

Exercise 14.14 Recast the sum–product and min-sum algorithms in the general message-passing framework. In particular, specify the alphabet of the messages, and the update and decision rules.

14.6 Probabilistic analysis

In the following chapters, we shall repeatedly be concerned with the analysis of message-passing algorithms on random graphical models. In this context, messages become random variables, and their distribution can be characterized in the large-system limit, as we shall now see.

14.6.1 Assumptions

Before proceeding, it is necessary to formulate a few technical assumptions under which our approach works. The basic idea is that, in a 'random graphical model', distinct nodes should be essentially independent. Specifically, we shall consider below a setting which already includes many cases of interest; it is easy to extend our analysis to even more general situations.

A **random graphical model** is a (random) probability distribution on $\underline{x} = (x_1, \ldots, x_N)$ of the form[4]

$$\mu(\underline{x}) \cong \prod_{a \in F} \psi_a(\underline{x}_{\partial a}) \prod_{i \in V} \psi_i(x_i) \,, \tag{14.68}$$

where the factor graph $G = (V, F, E)$ (with variable nodes V, factor nodes F, and edges E) and the various factors ψ_a, ψ_i are independent random variables. More precisely, we assume that the factor graph is distributed according to one of the ensembles $\mathbb{G}_N(K, \alpha)$ or $\mathbb{D}_N(\Lambda, P)$ (see Chapter 9).

The random factors are assumed to be distributed as follows. For any given degree k, we are given a list of possible factors $\psi^{(k)}(x_1, \ldots, x_k; \widehat{J})$, indexed by a 'label' $\widehat{J} \in \mathsf{J}$, and a distribution $P_{\widehat{J}}^{(k)}$ over the set of possible labels J. For each function node $a \in F$ of degree $|\partial a| = k$, a label $\widehat{J}_a$ is drawn with distribution $P_{\widehat{J}}^{(k)}$, and the function $\psi_a(\cdot)$ is taken to be equal to $\psi^{(k)}(\cdot; \widehat{J}_a)$. Analogously, the factors ψ_i are drawn from a list of possible $\{\psi(\cdot; J)\}$, indexed by a label J which is drawn from a distribution P_J. The random graphical model is fully characterized by the graph ensemble, the set of distributions $P_{\widehat{J}}^{(k)}$, P_J, and the lists of factors $\{\psi^{(k)}(\cdot; \widehat{J})\}$, $\{\psi(\cdot; J)\}$.

We need to make some assumptions about the message update rules. Specifically, we assume that the variable-to-function-node update rules $\Psi_{i \to a}$ depend on $i \to a$ only through $|\partial i|$ and J_i, and the function-to-variable-node update rules $\Phi_{a \to i}$ depend on $a \to i$ only through $|\partial a|$ and $\widehat{J}_a$. With a slight misuse of notation, we shall denote the update functions by

$$\Psi_{i \to a}(\{\widehat{\nu}_{b \to i} : b \in \partial i \setminus a\}) = \Psi_l(\widehat{\nu}_1, \ldots, \widehat{\nu}_l; J_i) \,, \tag{14.69}$$

$$\Phi_{a \to i}(\{\nu_{j \to a} : j \in \partial a \setminus i\}) = \Phi_k(\nu_1, \ldots, \nu_k; \widehat{J}_a) \,, \tag{14.70}$$

where $l \equiv |\partial i| - 1$, $k \equiv |\partial a| - 1$, $\{\widehat{\nu}_1, \ldots, \widehat{\nu}_l\} \equiv \{\widehat{\nu}_{b \to i} : b \in \partial i \setminus a\}$, and $\{\nu_1, \ldots, \nu_k\} \equiv \{\nu_{j \to a} : j \in \partial a \setminus i\}$. A similar notation will be used for the decision rule $\widehat{\Psi}$.

Exercise 14.15 Let $G = (V, E)$ be a uniformly random graph with $M = N\alpha$ edges over N vertices, and let λ_i, $i \in V$, be i.i.d. random variables uniform in $[0, \bar{\lambda}]$. Recall that an independent set for G is a subset of the vertices $S \subseteq V$ such that if $i, j \in S$, then (ij) is not an edge. Consider the following weighted measure over independent sets:

$$\mu(S) = \frac{1}{Z} \, \mathbb{I}(S \text{ is an independent set}) \prod_{i \in S} \lambda_i \,. \tag{14.71}$$

---

[4]Note that the factors ψ_i, $i \in V$, could have been included as degree-1 function nodes, as we did in eqn (14.13); including them explicitly yields a description of density evolution which is more symmetric between variables and factors, and applies more directly to decoding.

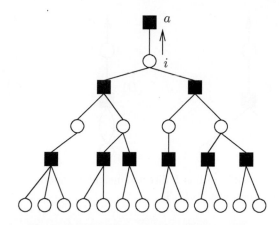

Fig. 14.8 A radius-2 directed neighbourhood $B_{i \to a, 2}(F)$.

(a) Write the distribution $\mu(\underline{S})$ as a graphical model with binary variables, and define the corresponding factor graph.

(b) Describe the BP algorithm to compute its marginals.

(c) Show that this model is a random graphical model in the sense defined above.

14.6.2 Density evolution equations

Consider a random graphical model, with factor graph $G = (V, F, E)$, and let (i, a) be a uniformly random edge in G. Let $\nu_{i \to a}^{(t)}$ be the message sent by the BP algorithm in iteration t along edge (i, a). We assume that the initial messages $\nu_{i \to a}^{(0)}$, $\widehat{\nu}_{a \to i}^{(0)}$ are i.i.d. random variables, with distributions independent of N. A considerable amount of information is contained in the distributions of $\nu_{i \to a}^{(t)}$ and $\widehat{\nu}_{a \to i}^{(t)}$ with respect to the realization of the model. We are interested in characterizing these distributions in the large-system limit $N \to \infty$. Our analysis will assume that both the message alphabet M and the node label alphabet J are subsets of $\mathbb{R}^d$ for some fixed d, and that the update functions $\Psi_{i \to a}$, $\Phi_{a \to i}$ are continuous with respect to the usual topology of $\mathbb{R}^d$.

It is convenient to introduce the **directed neighbourhood** of radius t of a directed edge $i \to a$, denoted by: $B_{i \to a, t}(G)$. This is defined as the subgraph of G that includes all of the variable nodes which can be reached from i by a non-reversing path of length at most t, whose first step *is not* the edge (i, a). It includes, as well, all of the function nodes connected only to those variable nodes; see Fig. 14.8. For illustrative reasons, we shall occasionally add a 'root edge', such as $i \to a$ in Fig. 14.8. Let us consider, to be definite, the case where G is a random factor graph from the ensemble $\mathbb{D}_N(\Lambda, P)$. In this case, $B_{i \to a, t}(F)$ converges in distribution, when $N \to \infty$, to the random tree ensemble $\mathbb{T}_t(\Lambda, P)$ defined in Section 9.5.1.

Exercise 14.16 Consider a random graph from the regular ensemble $\mathbb{D}_N(\Lambda, P)$

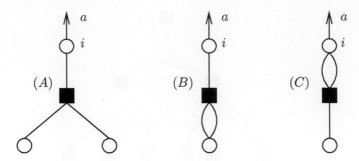

Fig. 14.9 The three possible radius-1 directed neighbourhoods in a random factor graph from the regular graph ensemble $\mathbb{D}_N(2,3)$.

with $\Lambda_2 = 1$ and $P_3 = 1$ (each variable node has degree 2 and each function node degree 3). The three possible radius-1 directed neighbourhoods appearing in such factor graphs are depicted in Fig. 14.9.

(a) Show that the probability that a given edge (i,a) has neighbourhoods as in (B) or (C) in the figure is $O(1/N)$.

(b) Deduce that $\mathsf{B}_{i \to a,1}(F) \overset{\mathrm{d}}{\to} \mathsf{T}_1$, where T_1 is distributed according to the tree model $\mathbb{T}_1(2,3)$ (i.e. it is the tree in Fig. 14.9, labelled (A)).

(c) Discuss the case of a radius-t neighbourhood.

For our purposes, it is necessary to include in the description of the neighbourhood $\mathsf{B}_{i \to a,t}(F)$ the value of the labels $J_i, \widehat{J}_b$ for function nodes b in this neighbourhood. It is understood that the tree model $\mathbb{T}_t(\Lambda, P)$ includes labels as well: these have to be drawn as i.i.d. random variables independent of the tree and with the same distribution as in the original graphical model.

Now consider the message $\nu_{i \to a}^{(t)}$. This is a function of the factor graph G, of the labels $\{J_j\}$, $\{\widehat{J}_b\}$, and of the initial condition $\{\nu_{j \to b}^{(0)}\}$. However, a moment of thought shows that its dependence on G and on the labels occurs only through the radius-$(t+1)$ directed neighbourhood $\mathsf{B}_{i \to a,t+1}(F)$. Its dependence on the initial condition is only through the messages $\nu_{j \to b}^{(0)}$ for $j, b \in \mathsf{B}_{i \to a,t}(F)$.

In view of the above discussion, let us pretend for a moment that the neighbourhood of (i,a) *is* a random tree T_{t+1} with distribution $\mathbb{T}_{t+1}(\Lambda, P)$. We define $\nu^{(t)}$ to be the message passed through the root edge of such a random neighbourhood after t message-passing iterations. Since $\mathsf{B}_{i \to a,t+1}(F)$ converges in distribution to the tree T_{t+1}, we find that[5] $\nu_{i \to a}^{(t)} \overset{\mathrm{d}}{\to} \nu^{(t)}$ as $N \to \infty$.

We have shown that, as $N \to \infty$, the distribution of $\nu_{i \to a}^{(t)}$ converges to that of a well-defined (N-independent) random variable $\nu^{(t)}$. The next step is to find a recursive characterization of $\nu^{(t)}$. Consider a random tree from the ensemble $\mathbb{T}_r(\Lambda, P)$ and let

---

[5]The mathematically suspicious reader may wonder about the topology we are assuming for the message space. In fact, no assumption is necessary if the distribution of labels $J_i, \widehat{J}_a$ is independent of N. If it is N-dependent but converges, then the topology must be such that the message updates are continuous with respect to it.

$j \rightarrow b$ be an edge directed towards the root, at a distance d from it. The directed subtree rooted at $j \rightarrow b$ is distributed according to $\mathbb{T}_{r-d}(\Lambda, P)$. Therefore the message passed through it after $r - d - 1$ (or more) iterations is distributed as $\nu^{(r-d-1)}$. The degree of the root variable node i (including the root edge) has a distribution λ_l. Each check node connected to i has a number of other neighbours (distinct from i) which is a random variable distributed according to ρ_k. These facts imply the following distributional equations for $\nu^{(t)}$ and $\widehat{\nu}^{(t)}$:

$$\nu^{(t+1)} \stackrel{\mathrm{d}}{=} \Psi_l(\widehat{\nu}_1^{(t)}, \dots, \widehat{\nu}_l^{(t)}; J), \qquad \widehat{\nu}^{(t)} \stackrel{\mathrm{d}}{=} \Phi_k(\nu_1^{(t)}, \dots, \nu_k^{(t)}; \widehat{J}). \qquad (14.72)$$

Here $\widehat{\nu}_b^{(t)}$, $b \in \{1, \dots, l-1\}$, are independent copies of $\widehat{\nu}^{(t)}$, and $\nu_j^{(t)}$, $j \in \{1, \dots, k-1\}$, are independent copies of $\nu^{(t)}$. As for l and k, these are independent random integers distributed according to λ_l and ρ_k, respectively; $\widehat{J}$ is distributed as $P_{\widehat{J}}^{(k)}$, and J is distributed as P_J. It is understood that the recursion is initiated with $\nu^{(0)} \stackrel{\mathrm{d}}{=} \nu_{i \rightarrow a}^{(0)}$, $\widehat{\nu}^{(0)} \stackrel{\mathrm{d}}{=} \widehat{\nu}_{a \rightarrow i}^{(0)}$.

In coding theory, the equations (14.72) are referred to as **density evolution**; sometimes, this term is also applied to the sequence of random variables $\{\nu^{(t)}, \widehat{\nu}^{(t)}\}$. In probabilistic combinatorics, they are also called **recursive distributional equations**. We have proved the following characterization of the distribution of messages.

Proposition 14.9 *Consider a random graphical model satisfying the assumptions in Section 14.6.1. Let $t \geq 0$ and let (ia) be a uniformly random edge in the factor graph. Then, as $N \rightarrow \infty$, the messages $\nu_{i \rightarrow a}^{(t)}$ and $\widehat{\nu}_{i \rightarrow a}^{(t)}$ converge in distribution to the random variables $\nu^{(t)}$ and $\widehat{\nu}^{(t)}$, respectively, defined through the density evolution equations (14.72).*

We shall discuss several applications of the idea of density evolution in the following chapters. Here we shall just mention that it allows one to compute the asymptotic distribution of message-passing decisions at a uniformly random site i. Recall that the general message-passing decision after t iterations is taken using the rule (14.67), with $\widehat{\Psi}_i(\{\widehat{\nu}_b\}) = \widehat{\Psi}_l(\widehat{\nu}_1, \dots, \widehat{\nu}_l; J_i)$ (where $l \equiv |\partial i|$). Arguing as in the previous paragraphs, it is easy to show that in the large-N limit, $\nu_i^{(t)} \stackrel{\mathrm{d}}{\rightarrow} \nu^{(t)}$, where the random variable $\nu^{(t)}$ is distributed according to

$$\nu^{(t)} \stackrel{\mathrm{d}}{=} \widehat{\Psi}_l(\widehat{\nu}_1^{(t-1)}, \dots, \widehat{\nu}_l^{(t-1)}; J). \qquad (14.73)$$

As above, $\widehat{\nu}_1^{(t-1)}, \dots, \widehat{\nu}_l^{(t-1)}$ are i.i.d. copies of $\widehat{\nu}^{(t-1)}$, J is an independent copy of the variable-node label J_i, and l is a random integer distributed according to Λ_l.

14.6.3 The replica-symmetric cavity method

The replica-symmetric (RS) cavity method of statistical mechanics adopts a point of view which is very close to the previous one, but less algorithmic. Instead of considering the BP update rules as an iterative message-passing rule, it focuses on the fixed-point BP equations themselves.

The idea is to compute the partition function recursively, by adding one variable node at a time. Equivalently, one may think of taking one variable node out of the system and computing the change in the partition function. The name of the method comes exactly from this image: one digs a 'cavity' in the system.

As an example, take the original factor graph, and delete the factor node a and all the edges incident on it. If the graph is a tree, this procedure separates it into $|\partial a|$ disconnected trees. Consider now the tree-graphical model described by the connected component containing the variable $j \in \partial a$. Denote the corresponding partition function, when the variable j is fixed to the value x_j, by $Z_{j \to a}(x_j)$. This partial partition function can be computed iteratively as

$$Z_{j \to a}(x_j) = \prod_{b \in \partial j \setminus a} \left[\sum_{\underline{x}_{\partial b \setminus j}} \psi_b(\underline{x}_{\partial b}) \prod_{k \in \partial b \setminus j} Z_{k \to b}(x_k) \right]. \tag{14.74}$$

The equations obtained by letting $j \to b$ be a generic directed edge in G are called the **cavity equations**, or **Bethe equations**.

The cavity equations are mathematically identical to the BP equations, but with two important conceptual differences: (i) one is naturally led to think that the equations (14.74) must have a fixed point, and to give special importance to it; (ii) the partial partition functions are unnormalized messages, and, as we shall see in Chapter 19, their normalization provides useful information. The relation between BP messages and partial partition functions is

$$\nu_{j \to a}(x_j) = \frac{Z_{j \to a}(x_j)}{\sum_y Z_{j \to a}(y)}. \tag{14.75}$$

In the cavity approach, the **replica symmetry assumption** consists in pretending that, for random graphical models of the kind introduced above, and in the large-N limit, the following conditions apply:

1. There exists a solution (or quasi-solution[6]) to these equations.
2. This solution provides good approximations to the marginals of the graphical model.
3. The messages in this solution are distributed according to a density evolution fixed point.

The last statement amounts to assuming that the normalized variable-to-factor messages $\nu_{i \to a}$ (see eqn (14.75)), converge in distribution to a random variable ν that solves the following distributional equations:

$$\nu \overset{\mathrm{d}}{=} \Psi(\widehat{\nu}_1, \dots, \widehat{\nu}_{k-1}; J), \qquad \widehat{\nu} \overset{\mathrm{d}}{=} \Phi(\nu_1, \dots, \nu_{l-1}; \widehat{J}). \tag{14.76}$$

Here we have used the same notation as in eqn (14.72): $\widehat{\nu}_b$, $b \in \{1, \dots, l-1\}$, are independent copies of $\widehat{\nu}^{(t)}$; $\nu_j^{(t)}$, $j \in \{1, \dots, k-1\}$, are independent copies of $\nu^{(t)}$; l

---

[6]A quasi-solution is a set of messages $\nu_{j \to a}$ such that the average difference between the left- and right-hand sides of the BP equations goes to zero in the large-N limit.

and k are independent random integers distributed according to λ_l and ρ_k respectively; and J and $\widehat{J}$ are distributed as the variable and function node labels J_i and $\widehat{J}_a$.

Using the distributions of ν and $\widehat{\nu}$, the expected Bethe free entropy per variable $\mathbb{F}/N$ can be computed by taking the expectation of eqn (14.27). The result is

$$\mathrm{f}^{\mathrm{RS}} = \mathrm{f}_{\mathrm{v}}^{\mathrm{RS}} + n_{\mathrm{f}}\mathrm{f}_{\mathrm{f}}^{\mathrm{RS}} - n_{\mathrm{e}}\mathrm{f}_{\mathrm{e}}^{\mathrm{RS}} , \tag{14.77}$$

where n_{f} is the average number of function nodes per variable, and n_{e} is the average number of edges per variable. In the ensemble $\mathbb{D}_N(\Lambda, P)$ we have $n_{\mathrm{f}} = \Lambda'(1)/P'(1)$ and $n_{\mathrm{e}} = \Lambda'(1)$; in the ensemble $\mathbb{G}_N(K, \alpha)$, $n_{\mathrm{f}} = \alpha$ and $n_{\mathrm{e}} = K\alpha$. The contributions of the variable nodes $\mathrm{f}_{\mathrm{v}}^{\mathrm{RS}}$, function nodes $\mathrm{f}_{\mathrm{f}}^{\mathrm{RS}}$, and edges $\mathrm{f}_{\mathrm{e}}^{\mathrm{RS}}$ are

$$\mathrm{f}_{\mathrm{v}}^{\mathrm{RS}} = \mathbb{E}_{l,J,\{\widehat{\nu}\}} \log \left[\sum_x \psi(x; J)\, \widehat{\nu}_1(x)\cdots\widehat{\nu}_l(x) \right] ,$$

$$\mathrm{f}_{\mathrm{f}}^{\mathrm{RS}} = \mathbb{E}_{k,\widehat{J},\{\nu\}} \log \left[\sum_{x_1,\ldots,x_k} \psi^{(k)}(x_1,\ldots,x_k; \widehat{J})\, \nu_1(x_1)\cdots\nu_k(x_k) \right] ,$$

$$\mathrm{f}_{\mathrm{e}}^{\mathrm{RS}} = \mathbb{E}_{\nu,\widehat{\nu}} \log \left[\sum_x \nu(x)\widehat{\nu}(x) \right] . \tag{14.78}$$

In these expressions, $\mathbb{E}$ denotes the expectation with respect to the random variables given in subscript. For instance, if G is distributed according to the ensemble $\mathbb{D}_N(\Lambda, P)$, $\mathbb{E}_{l,J,\{\widehat{\nu}\}}$ implies that l is drawn from the distribution Λ, J is drawn from P_J, and $\widehat{\nu}_1,\ldots,\widehat{\nu}_l$ are l independent copies of the random variable $\widehat{\nu}$.

Instead of estimating the partition function, the cavity method can be used to compute the ground state energy. One then uses min-sum-like messages instead of those in eqn (14.74). The method is then called the 'energetic cavity method'; we leave to the reader the task of writing the corresponding average ground state energy per variable.

14.6.4 Numerical methods

Generically, the RS cavity equations (14.76), as well as the density evolution equations (14.72), cannot be solved in closed form, and one must use numerical methods to estimate the distribution of the random variables ν, $\widehat{\nu}$. Here we limit ourselves to describing a stochastic approach that has the advantage of being extremely versatile and simple to implement. It has been used in coding theory under the name of 'sampled density evolution' or the 'Monte Carlo method', and is known in statistical physics as **population dynamics**, a name which we shall adopt in the following.

The idea is to approximate the distribution of ν (or $\widehat{\nu}$) through a sample of (ideally) N i.i.d. copies of ν (or $\widehat{\nu}$, respectively). As N becomes large, the empirical distribution of such a sample should converge to the actual distribution of ν (or $\widehat{\nu}$). We shall call the sample $\{\nu_i\} \equiv \{\nu_1,\ldots,\nu_N\}$ (or $\{\widehat{\nu}_i\} \equiv \{\widehat{\nu}_1,\ldots,\widehat{\nu}_N\}$) a **population**.

The algorithm is described by the pseudocode below. As inputs, it requires the population size N, the maximum number of iterations T, and a specification of the ensemble of (random) graphical models. The latter is a description of the (edge-perspective) degree distributions λ and ρ, the variable node labels P_J, and the factor node labels $P_{\widehat{J}}^{(k)}$.

POPULATION DYNAMICS (model ensemble, size N, iterations T)

1:    Initialize $\{\nu_i^{(0)}\}$;
2:    **for** $t = 1, \ldots, T$:
3:        **for** $i = 1, \ldots, N$:
4:            Draw an integer k with distribution ρ;
5:            Draw $i(1), \ldots, i(k-1)$ uniformly in $\{1, \ldots, N\}$;
6:            Draw $\widehat{J}$ with distribution $P_{\widehat{J}}^{(k)}$;
7:            Set $\widehat{\nu}_i^{(t)} = \Phi_k(\nu_{i(1)}^{(t-1)}, \ldots, \nu_{i(k-1)}^{(t-1)}; \widehat{J})$;
8:        **end;**
9:        **for** $i = 1, \ldots, N$:
10:           Draw an integer l with distribution λ;
11:           Draw $i(1), \ldots, i(l-1)$ uniformly in $\{1, \ldots, N\}$;
12:           Draw J with distribution P_J;
13:           Set $\nu_i^{(t)} = \Psi_l(\widehat{\nu}_{i(1)}^{(t)}, \ldots, \widehat{\nu}_{i(l-1)}^{(t)}; J)$;
14:       **end;**
15:   **end;**
16:   **return** $\{\nu_i^{(T)}\}$ and $\{\widehat{\nu}_i^{(T)}\}$.

In step 1, the initialization is done by drawing $\nu_1^{(0)}, \ldots, \nu_N^{(0)}$ independently with the same distribution P that was used for the initialization of the BP algorithm.

It is not hard to show that, for any fixed T, the empirical distribution of $\{\nu_i^{(T)}\}$ (or $\{\widehat{\nu}_i^{(T)}\}$) converges, as $N \to \infty$, to the distribution of the density evolution random variable $\nu^{(t)}$ (or $\widehat{\nu}^{(t)}$). The limit $T \to \infty$ is trickier. Let us assume first that the density evolution has a unique fixed point, and $\nu^{(t)}$, $\widehat{\nu}^{(t)}$ converge to this fixed point. We then expect the empirical distribution of $\{\nu_i^{(T)}\}$ also to converge to this fixed point if the $N \to \infty$ limit is taken after $T \to \infty$. When the density evolution has more than one fixed point, which is probably the most interesting case, the situation is more subtle. The population $\{\nu_i^{(T)}\}$ evolves according to a large but finite-dimensional Markov chain. Therefore (under some technical conditions) the distribution of the population is expected to converge to the unique fixed point of this Markov chain. This seems to imply that population dynamics cannot describe the multiple fixed points of density evolution. Luckily, the convergence of the population dynamics algorithm to its unique fixed point appears to happen on a time scale that increases very rapidly with N. For large N and on moderate time scales T, it converges instead to one of several 'quasi-fixed points' that correspond to the fixed points of the density evolution algorithm.

In practice, one can monitor the effective convergence of the algorithm by computing, after any number of iterations t, averages of the form

$$\langle\varphi\rangle_t \equiv \frac{1}{N} \sum_{i=1}^{N} \varphi(\nu_i^{(t)}),$$ (14.79)

for a smooth function $\varphi : \mathfrak{M}(\mathcal{X}) \to \mathbb{R}$. If these averages are well settled (up to statistical fluctuations of order $1/\sqrt{N}$), this is interpreted as a signal that the iteration has converged to a 'quasi-fixed point.'

The populations produced by the above algorithm can be used to to estimate expectations with respect to the density-evolution random variables $\nu, \widehat{\nu}$. For instance, the expression in eqn (14.79) is an estimate for $\mathbb{E}\{\varphi(\nu)\}$. When $\varphi = \varphi(\nu_1, \dots, \nu_l)$ is a function of l i.i.d. copies of ν, the above formula is modified to

$$\langle\varphi\rangle_t \equiv \frac{1}{R} \sum_{n=1}^{R} \varphi(\nu_{i_n(1)}^{(t)}, \dots, \nu_{i_n(l)}^{(t)}).$$ (14.80)

Here R is a large number (typically of the same order as N), and $i_n(1), \dots, i_n(l)$ are i.i.d. indices in $\{1, \dots, N\}$. Of course such estimates will be reasonable only if $l \ll N$.

A particularly important example is the computation of the free entropy (14.77). Each of the terms f_v^{RS}, f_f^{RS} and f_e^{RS} can be estimated as in eqn (14.80). The precision of these estimates can be improved by repeating the computation for several iterations and averaging the result.

Notes

The belief propagation equations have been rediscovered several times. They were developed by Pearl (1988) as an exact algorithm for probabilistic inference in acyclic Bayesian networks. In the early 1960s, Gallager had introduced them as an iterative procedure for decoding low-density-parity-check codes (Gallager, 1963). Gallager described several message-passing procedures, among them being the sum–product algorithm. In the field of coding theory, the basic idea of this algorithm was rediscovered in several works in the 1990s, in particular by Berrou and Glavieux (1996). In the physics context, the history is even longer. In 1935, Bethe used a free-energy functional written in terms of pseudo-marginals to approximate the partition function of the ferromagnetic Ising model (Bethe, 1935). Bethe's equations were of the simple form discussed in Exercise 14.10, because of the homogeneity (translation invariance) of the underlying model. Their generalization to inhomogeneous systems, which has a natural algorithmic interpretation, waited until the application of Bethe's method to spin glasses (Thouless *et al.*, 1977; Klein *et al.*, 1979; Katsura *et al.*, 1979; Morita, 1979; Nakanishi, 1981).

The review paper by Kschischang *et al.* (2001) gives a general overview of belief propagation in the framework of factor graphs. The role of the distributive property, mentioned in Exercise 14.8, was emphasized by Aji and McEliece (2000). On tree graphs, belief propagation can be regarded as an instance of the junction–tree algorithm (Lauritzen, 1996). This algorithm constructs a tree from the graphical model under study by grouping some of its variables. Belief propagation is then applied to this tree.

Although implicit in these earlier works, the equivalence between BP, the Bethe approximation, and the sum–product algorithm was only recognized in the 1990s. The turbodecoding and the sum–product algorithms were shown to be instances of BP by McEliece *et al.* (1998). A variational derivation of the turbo decoding algorithm was proposed by Montanari and Sourlas (2000). The equivalence between BP and the Bethe approximation was first put forward by Kabashima and Saad (1998) and, in a more general setting, by Yedidia *et al.* (2001) and Yedidia *et al.* (2005).

The last of these papers proved, in particular, the variational formulation in Proposition 14.8. This suggests that one should look for fixed points of BP by seeking stationary points of the Bethe free entropy directly, without iterating the BP equations. An efficient such procedure, based on the observation that the Bethe free entropy can be written as a difference between a convex and a concave function, was proposed by Yuille (2002). An alternative approach consists in constructing convex surrogates of the Bethe free energy (Wainwright *et al.*, 2005 *a,b*) which allow one to define provably convergent message-passing procedures.

The Bethe approximation can also be regarded as the first step in a hierarchy of variational methods describing larger and larger clusters of variables exactly. This point of view was first developed by Kikuchi (1951), leading to the 'cluster variational method' in physics. The algorithmic version of this approach is referred to as 'generalized BP', and is described in detail by Yedidia *et al.* (2005).

The analysis of iterative message-passing algorithms on random graphical models dates back to Gallager (1963). These ideas were developed into a systematic method, thanks also to efficient numerical techniques, by Richardson and Urbanke (2001 *b*), who coined the name 'density evolution'. The point of view taken in this book, however, is closer to that of 'local weak convergence' (Aldous and Steele, 2003).

In physics, the replica-symmetric cavity method for sparse random graphical models was first discussed by Mézard and Parisi (1987). The use of population dynamics first appeared in Abou-Chacra *et al.* (1973) and was developed further for spin glasses by Mézard and Parisi (2001), but that paper deals mainly with RSB effects, which will be the subject of Chapter 19.

15

Decoding with belief propagation

As we have already seen, symbol MAP decoding of error-correcting codes can be regarded as a statistical inference problem. It is a very natural idea to accomplish this task using belief propagation. For properly constructed codes (in particular, LDPC ensembles), this approach has low complexity while achieving very good performance.

However, it is clear that an error-correcting code cannot achieve good performance unless the associated factor graph has loops. As a consequence, belief propagation has to be regarded only as an approximate inference algorithm in this context. A major concern of the theory is to establish conditions for its optimality, and, more generally, the relation between message passing and optimal (exact symbol MAP) decoding.

In this chapter, we discuss belief propagation decoding of the LDPC ensembles introduced in Chapter 11. The message-passing approach can be generalized to several other applications within information and communication theory: other code ensembles, source coding, channels with memory, etc. Here we shall keep to the 'canonical' example of channel coding, as most of the theory has been developed in this context.

BP decoding is defined in Section 15.1. One of the main tools in the analysis is the 'density evolution' method, which we discuss in Section 15.2. This allows one to determine the threshold for reliable communication under BP decoding, and to optimize the code ensemble accordingly. The whole process is considerably simpler for an erasure channel, which is treated in Section 15.3. Finally, Section 15.4 explains the relation between optimal (MAP) decoding and BP decoding in the large-block-length limit: the two approaches can studied within a unified framework based on the Bethe free energy.

15.1 BP decoding: The algorithm

In this chapter, we shall consider communication over a **binary-input, output-symmetric, memoryless (BMS)** channel. This is a channel in which the transmitted codeword is binary, $\underline{x} \in \{0, 1\}^N$, and the output $\underline{y}$ is a sequence of N letters y_i from an alphabet $\mathcal{Y} \subset \mathbb{R}$. The probability of receiving letter y when bit x is sent, $Q(y|x)$, possesses the symmetry property $Q(y|0) = Q(-y|1)$.

Let us suppose that an LDPC error-correcting code is used in this communication. The conditional probability for the channel input to be $\underline{x} \in \{0, 1\}^N$ given the output $\underline{y}$ is $\mathbb{P}(\underline{x}|\underline{y}) = \mu_y(\underline{x})$, where

$$\mu_y(\underline{x}) = \frac{1}{Z(\underline{y})} \prod_{i=1}^{N} Q(y_i|x_i) \prod_{a=1}^{M} \mathbb{I}(x_{i_1^a} \oplus \cdots \oplus x_{i_{k(a)}^a} = 0). \tag{15.1}$$

The factor graph associated with this distribution is the usual one: an edge joins a variable node i to a check node a whenever the variable x_i appears in the a-th parity check equation.

Messages $\nu_{i \to a}(x_i)$, $\widehat{\nu}_{a \to i}(x_i)$ are exchanged along the edges. We shall assume a parallel updating of BP messages, as introduced in Section 14.2:

$$\nu_{i \to a}^{(t+1)}(x_i) \cong Q(y_i|x_i) \prod_{b \in \partial i \setminus a} \widehat{\nu}_{b \to i}^{(t)}(x_i) , \tag{15.2}$$

$$\widehat{\nu}_{a \to i}^{(t)}(x_i) \cong \sum_{\{x_j\}} \mathbb{I}(x_i \oplus x_{j_1} \oplus \cdots \oplus x_{j_{k-1}} = 0) \prod_{j \in \partial a \setminus i} \nu_{j \to a}^{(t)}(x_j) , \tag{15.3}$$

where we have used the notation $\partial a \equiv \{i, j_1, \dots, j_{k-1}\}$, and the symbol $\cong$ denotes, as before, 'equality up to a normalization constant'. We expect that the asymptotic performance at large t and large N of such BP decoding, for instance its asymptotic bit error rate, should be insensitive to the precise update schedule. On the other hand, this schedule can have an important influence on the speed of convergence, and on performance at moderate N. Here we shall not address these issues.

The BP estimate for the marginal distribution at node i and time t, also called the 'belief' or **soft decision**, is

$$\nu_i^{(t)}(x_i) \cong Q(y_i|x_i) \prod_{b \in \partial i} \widehat{\nu}_{b \to i}^{(t-1)}(x_i) . \tag{15.4}$$

Based on this estimate, the optimal BP decision for bit i at time t, sometimes called the **hard decision**, is

$$\widehat{x}_i^{(t)} = \arg \max_{x_i} \nu_i^{(t)}(x_i) . \tag{15.5}$$

In order to fully specify the algorithm, one must address two more issues: (1) How are the messages initialized? (2) After how many iterations t is the hard decision (15.5) taken?

In practice, one usually initializes the messages to $\nu_{i \to a}^{(0)}(0) = \nu_{i \to a}^{(0)}(1) = 1/2$. One alternative choice, which is sometimes useful for theoretical reasons, is to take the messages $\nu_{i \to a}^{(0)}(\cdot)$ as independent random variables, for instance by choosing $\nu_{i \to a}^{(0)}(0)$ uniformly on $[0, 1]$.

In relation to the number of iterations, one would like to have a stopping criterion. In practice, a convenient criterion is to check whether $\widehat{x}^{(t)}$ is a codeword, and to stop if this is the case. If this condition is not fulfilled, the algorithm is stopped after a fixed number of iterations $t_{\max}$. On the other hand, for the purpose of performance analysis, we shall instead fix $t_{\max}$ and assume that the belief propagation algorithm is always run for $t_{\max}$ iterations, regardless of whether a valid codeword is reached at an earlier stage.

Since the messages are distributions over binary-valued variables, we parameterize them by the log-likelihoods

$$h_{i \to a} = \frac{1}{2} \log \frac{\nu_{i \to a}(0)}{\nu_{i \to a}(1)} , \qquad u_{a \to i} = \frac{1}{2} \log \frac{\widehat{\nu}_{a \to i}(0)}{\widehat{\nu}_{a \to i}(1)} . \tag{15.6}$$

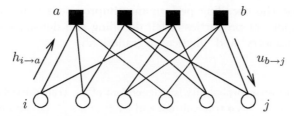

Fig. 15.1 Factor graph of a (2,3) regular LDPC code, and notation for the belief propagation messages.

Further, we introduce the a priori log-likelihood for bit i, given the received message y_i:

$$B_i = \frac{1}{2} \log \frac{Q(y_i|0)}{Q(y_i|1)} \,. \tag{15.7}$$

For instance, if communication takes place over a binary symmetric channel with flip probability p, one has $B_i = \frac{1}{2} \log((1-p)/p)$ on variable nodes which have received $y_i = 0$, and $B_i = -\frac{1}{2} \log((1-p)/p)$ on those with $y_i = 1$. The BP update equations (15.2) and (15.3) read in this notation (see Fig. 15.1)

$$h_{i \to a}^{(t+1)} = B_i + \sum_{b \in \partial i \backslash a} u_{b \to i}^{(t)} \,, \qquad u_{a \to i}^{(t)} = \mathrm{atanh}\Big\{ \prod_{j \in \partial a \backslash i} \tanh h_{j \to a}^{(t)} \Big\} \,. \tag{15.8}$$

The hard-decision decoding rule depends on the overall BP log-likelihood

$$h_i^{(t+1)} = B_i + \sum_{b \in \partial i} u_{b \to i}^{(t)} \,, \tag{15.9}$$

and is given by (using, for definiteness, a fair-coin outcome in the case of a tie)

$$\widehat{x}_i^{(t)}(\underline{y}) = \begin{cases} 0 & \text{if } h_i^{(t)} > 0, \\ 1 & \text{if } h_i^{(t)} < 0, \\ 0 \text{ or } 1 & \text{with probability } 1/2 \text{ if } h_i^{(t)} = 0. \end{cases} \tag{15.10}$$

15.2 Analysis: Density evolution

Let us study BP decoding of random codes from the ensemble $\mathrm{LDPC}_N(\Lambda, P)$ in the large-block-length limit. The code ensemble is specified by the degree distributions of the variable nodes $\Lambda = \{\Lambda_l\}$ and of the check nodes $P = \{P_k\}$. We assume for simplicity that messages are initialized to $u_{a \to i}^{(0)} = 0$.

Because of the symmetry of the channel, under the above hypotheses, the bit (or block) error probability is independent of the transmitted codeword. An explicit derivation of this fact is outlined in Exercise 15.1 below. Thanks to this freedom, we can assume that the all-zero codeword has been transmitted. We shall first write the density evolution recursion as a special case of the recursion given in Section 14.6.2. It turns out that this recursion can be analysed in quite some detail, and, in particular, one can show that the decoding performance improves as t increases. The analysis hinges on two important properties of BP decoding and density evolution, related to the notions of 'symmetry' and 'physical degradation'.

Exercise 15.1 The error probability is independent of the transmitted codeword. Assume that the codeword $\underline{x}$ has been transmitted, and let $B_i(\underline{x})$, $u_{a \to i}^{(t)}(\underline{x})$ and $h_{i \to a}^{(t)}(\underline{x})$ be the corresponding channel log-likelihoods and messages. Because of the randomness in the channel realization, they are random variables. Furthermore, let $\sigma_i = \sigma_i(\underline{x}) = +1$ if $x_i = 0$, and $= -1$ otherwise.

(a) Prove that the distribution of $\sigma_i B_i$ is independent of $\underline{x}$.

(b) Use the equations (15.8) to prove, by induction over t, that the (joint) distribution of $\{\sigma_i h_{i \to a}^{(t)}, \sigma_i u_{a \to i}^{(t)}\}$ is independent of $\underline{x}$.

(c) Use eqn (15.9) to show that the distribution of $\{\sigma_i h_i^{(t)}\}$ is independent of $\underline{x}$ for any $t \geq 0$. Finally, prove that the distribution of the 'error vector' $\underline{z}^{(t)} \equiv \underline{x} \oplus \widehat{\underline{x}}^{(t)}(\underline{y})$ is independent of $\underline{x}$ as well. Write the bit and block error rates in terms of the distribution of $\underline{z}^{(t)}$.

15.2.1 Density evolution equations

Let us consider the distribution of messages after a fixed number t of iterations. As we saw in Section 14.6.2, in the large-N limit, the directed neighbourhood of any given edge is with high probability a tree, whose distribution converges to the model $\mathbb{T}_t(\Lambda, P)$. This implies the following recursive distributional characterization for $h^{(t)}$ and $u^{(t)}$:

$$h^{(t+1)} \stackrel{\mathrm{d}}{=} B + \sum_{b=1}^{l-1} u_b^{(t)}, \qquad u^{(t)} \stackrel{\mathrm{d}}{=} \operatorname{atanh}\left\{ \prod_{j=1}^{k-1} \tanh h_j^{(t)} \right\}. \tag{15.11}$$

Here $u_b^{(t)}$, $b \in \{1, \ldots, l-1\}$, are independent copies of $u^{(t)}$; $h_j^{(t)}$, $j \in \{1, \ldots, k-1\}$, are independent copies of $h^{(t)}$; and l and k are independent random integers distributed according to λ_l and ρ_k, respectively. Finally, $B = \frac{1}{2}\log(Q(y|0)/Q(y|1))$, where y is independently distributed according to $Q(y|0)$. The recursion is initialized with $u^{(0)} = 0$.

Let us now consider the BP log-likelihood at site i. The same arguments as above imply $h_i^{(t)} \stackrel{\mathrm{d}}{\to} h_*^{(t)}$, where the distribution of $h_*^{(t)}$ is defined by

$$h_*^{(t+1)} \stackrel{\mathrm{d}}{=} B + \sum_{b=1}^{l} u_b^{(t)}, \tag{15.12}$$

where l is a random integer distributed according to Λ_l. In particular, if we let $\mathrm{P}_{\mathrm{b}}^{(N,t)}$ be the expected (over an $\mathrm{LDPC}_N(\Lambda, P)$ ensemble) bit error rate for the decoding rule (15.10), then

$$\lim_{N \to \infty} P_b^{(N,t)} = \mathbb{P}\{h_*^{(t)} < 0\} + \frac{1}{2}\mathbb{P}\{h_*^{(t)} = 0\}. \tag{15.13}$$

The suspicious reader will notice that this statement is non-trivial, because $f(x) = \mathbb{I}(x < 0) + \frac{1}{2}\mathbb{I}(x = 0)$ is not a continuous function. We shall prove the statement below using the symmetry property of the distribution of $h_i^{(t)}$, which allows us to write the bit error rate as the expectation of a continuous function (see Exercise 15.2).

15.2.2 Basic properties: 1. Symmetry

A real random variable Z (or, equivalently, its distribution) is said to be **symmetric** if

$$\mathbb{E}\{f(-Z)\} = \mathbb{E}\{e^{-2Z} f(Z)\} \tag{15.14}$$

for any function f such that one of the expectations exists. If Z has a density $p(z)$, then the above condition is equivalent to $p(-z) = e^{-2z} p(z)$.

Symmetric variables appear naturally in the description of BMS channels.

Proposition 15.1 *Consider a BMS channel with transition probability $Q(y|x)$. Let Y be the channel output conditional on input 0 (this is a random variable with distribution $Q(y|0)$), and let $B \equiv \frac{1}{2}\log(Q(Y|0)/Q(Y|1))$. Then B is a symmetric random variable.*

Conversely, if Z is a symmetric random variable, there exists a BMS channel whose log-likelihood ratio, conditional on the input being 0, is distributed as Z.

Proof To avoid technicalities, we prove this claim when the output alphabet $\mathcal{Y}$ is a *discrete* subset of $\mathbb{R}$. In this case, using channel symmetry in the form $Q(y|0) = Q(-y|1)$, we get

$$\mathbb{E}\{f(-B)\} = \sum_y Q(y|0)\, f\left(\frac{1}{2}\log\frac{Q(y|1)}{Q(y|0)}\right) = \sum_y Q(y|1)\, f\left(\frac{1}{2}\log\frac{Q(y|0)}{Q(y|1)}\right)$$

$$= \sum_y Q(y|0)\, \frac{Q(y|1)}{Q(y|0)}\, f\left(\frac{1}{2}\log\frac{Q(y|0)}{Q(y|1)}\right) = \mathbb{E}\{e^{-2B} f(B)\}. \tag{15.15}$$

We now prove the converse. Let Z be a symmetric random variable. We build a channel with output alphabet $\mathbb{R}$ as follows. Under input 0, the output is distributed as Z, and under input 1, it is distributed as $-Z$. In terms of densities,

$$Q(z|0) = p(z), \qquad Q(z|1) = p(-z). \tag{15.16}$$

This is a BMS channel with the desired property. Of course, this construction is not unique. $\square$

Example 15.2 Consider the binary erasure channel BEC(ϵ). If the channel input is 0, then Y can take two values, either 0 (with probability $1 - \epsilon$) or $*$ (probability ϵ). The distribution of B, $\mathbb{P}_B = (1 - \epsilon)\,\delta_\infty + \epsilon\,\delta_0$, is symmetric. In particular, this is true for the two extreme cases $\epsilon = 0$ (a noiseless channel) and $\epsilon = 1$ (a completely noisy channel, where $\mathbb{P}_B = \delta_0$).

Example 15.3 Consider the binary symmetric channel BSC(p). The log-likelihood B can take two values, either $b_0 = \frac{1}{2}\log((1-p)/p)$ (input 0 and output 0) or $-b_0$ (input 0 and output 1). Its distribution, $\mathbb{P}_B = (1-p)\,\delta_{b_0} + p\,\delta_{-b_0}$, is symmetric.

Example 15.4 Finally, consider the binary white-noise additive Gaussian channel BAWGN(σ^2). If the channel input is 0, the output Y has a probability density

$$q(y) = \frac{1}{\sqrt{2\pi\sigma^2}}\exp\left\{-\frac{(y-1)^2}{2\sigma^2}\right\}, \tag{15.17}$$

i.e. it is a Gaussian of mean 1 and variance σ^2. The output density for input 1 is determined by the symmetry of the channel; it is therefore a Gaussian of mean -1 and variance σ^2. The log-likelihood for output y is easily checked to be $b = y/\sigma^2$. Therefore B also has a symmetric Gaussian density, namely

$$p(b) = \sqrt{\frac{\sigma^2}{2\pi}}\exp\left\{-\frac{\sigma^2}{2}\left(b-\frac{1}{\sigma^2}\right)^2\right\}. \tag{15.18}$$

The variables appearing in the density evolution are symmetric as well. The argument is based on the symmetry of the channel log-likelihood, and the fact that symmetry is preserved by the operations in BP evolution: if Z_1 and Z_2 are two independent symmetric random variables (not necessarily identically distributed), it is straightforward to show that $Z = Z_1 + Z_2$ and $Z' = \operatorname{atanh}[\tanh Z_1 \tanh Z_2]$ are both symmetric.

Let us now consider the communication of the all-zero codeword over a BMS channel using an LDPC code, but let us first assume that the factor graph associated with the code is a tree. We apply BP decoding with a symmetric random initial condition such as $u_{a\to i}^{(0)} = 0$. The messages passed during the decoding procedure can be regarded as random variables, because of the random received symbols y_i (which yield random log-likelihoods B_i). Furthermore, incoming messages at a given node are independent, since they are functions of B_i's (and of initial conditions) on disjoint subtrees. From the above remarks, and looking at the BP equations (15.8), it follows that the messages $u_{a\to i}^{(t)}$ and $h_{i\to a}^{(t)}$, as well as the overall log-likelihoods, $h_i^{(t)}$ are symmetric random variables at all $t \geq 0$. Therefore we can state the following proposition.

Proposition 15.5 *Consider BP decoding of an LDPC code under the above assumptions. If* $\mathsf{B}_{i\to a,t+1}(F)$ *is a tree, then* $h_{i\to a}^{(t)}$ *is a symmetric random variable. Analogously, if* $\mathsf{B}_{i,t+1}(F)$ *is a tree, then* $H_i^{(t)}$ *is a symmetric random variable.*

Proposition 15.6 *The density-evolution random variables* $\{h^{(t)}, u^{(t)}, H_*^{(t)}\}$ *are symmetric.*

Exercise 15.2 Using Proposition 15.5 and the fact that, for any finite t, $B_{i \to a, t+1}(F)$ is a tree with high probability as $N \to \infty$, show that

$$\lim_{N \to \infty} P_b^{(N,t)} = \lim_{N \to \infty} \mathbb{E} \left\{ \frac{1}{N} \sum_{i=1}^{N} f(h_i^{(t)}) \right\}, \qquad (15.19)$$

where $f(x) = 1/2$ for $x \leq 0$ and $f(x) = e^{-2x}/2$ otherwise.

Symmetry does not hold uniquely for the BP log-likelihood; it also holds for the actual (MAP) log-likelihood of a bit, as shown in the exercise below.

Exercise 15.3 Consider the actual (MAP) log-likelihood for bit i (as opposed to its BP approximation). This is defined as

$$h_i(\underline{y}) = \frac{1}{2} \log \frac{\mathbb{P}\{x_i = 0 | \underline{y}\}}{\mathbb{P}\{x_i = 1 | \underline{y}\}}. \qquad (15.20)$$

If we condition on the all-zero codeword being transmitted, so that $\mathbb{P}(\underline{y}) = \prod_i Q(y_i|0)$, then the random variable $H_i = h_i(\underline{y})$ is symmetric. This can be shown as follows.

(a) Suppose that a codeword $\underline{z} \neq \underline{0}$ has been transmitted, so that $\mathbb{P}(\underline{y}) = \prod_i Q(y_i|z_i)$, and define in this case a random variable associated with the log-likelihood of bit x_i as $H_i^{(\underline{z})} = h_i(\underline{y})$. Show that $H_i^{(\underline{z})} \overset{d}{=} H_i$ if $z_i = 0$, and $H_i^{(\underline{z})} \overset{d}{=} -H_i$ if $z_i = 1$.

(b) Consider the following process. A bit z_i is chosen uniformly at random. Then a codeword $\underline{z}$ is chosen uniformly at random conditional on the value of z_i, and transmitted through a BMS channel, yielding an output $\underline{y}$. Finally, the log-likelihood $H_i^{(\underline{z})}$ is computed. If we hide the intermediate steps in a black box, this can be seen as a communication channel $z_i \to H_i^{(\underline{z})}$. Show that this is a BMS channel.

(c) Show that H_i is a symmetric random variable.

The symmetry property is a generalization of the Nishimori condition that we encountered for spin glasses. As can be recognized from eqn (12.7), the Nishimori condition is satisfied if and only if, for each coupling constant J, βJ is a symmetric random variable. Whereas symmetry occurs only at very special values of the temperature for spin glasses, it holds generically for decoding. The common mathematical origin of these properties can be traced back to the structure discussed in Section 12.2.3.

15.2.3 Basic properties: 2. Physical degradation

It turns out that, for large block lengths, BP decoding gets better as the number of iterations t increases (although it does not necessarily converge to the correct values). This is an extremely useful result, which does not hold when BP is applied to general inference problems. A precise formulation of this statement is provided by the notion

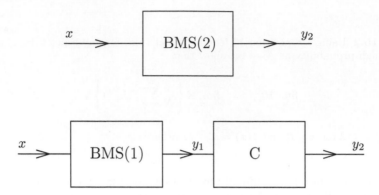

Fig. 15.2 The channel BMS(2) (*top*) is said to be physically degraded with respect to BMS(1) if it is equivalent to the concatenation of BMS(1) with a second channel C (*bottom*).

of physical degradation. We shall first define this notion in terms of BMS channels, and then extend it to all symmetric random variables. This allows us to apply it to the random variables encountered in BP decoding and density evolution.

Let us start with the case of BMS channels. Consider two such channels, denoted as BMS(1) and BMS(2); we denote their transition matrices by $\{Q_1(y|x)\}$, $\{Q_2(y|x)\}$ and the corresponding output alphabets by $\mathcal{Y}_1$, $\mathcal{Y}_2$. We say that BMS(2) is **physically degraded** with respect to BMS(1) if there exists a third channel C with input alphabet $\mathcal{Y}_1$ and output $\mathcal{Y}_2$ such that BMS(2) can be regarded as the concatenation of BMS(1) and C. By this we mean that passing a bit through BMS(1) and then feeding the output to C is statistically equivalent to passing the bit through BMS(2). If the transition matrix of C is $\{R(y_2|y_1)\}$, this can be written in formulae as

$$Q_2(y_2|x) = \sum_{y_1 \in \mathcal{Y}_1} R(y_2|y_1)\, Q_1(y_1|x)\,, \qquad (15.21)$$

where, to simplify the notation, we have assumed $\mathcal{Y}_1$ to be discrete. A pictorial representation of this relationship is provided in Fig. 15.2. A formal way of expressing the same idea is that there exists a Markov chain $X \to Y_1 \to Y_2$.

Whenever BMS(2) is physically degraded with respect to BMS(1), we shall write BMS(1) $\preceq$ BMS(2) (which is read as: 'BMS(1) is "less noisy than" BMS(2)'). Physical degradation is a partial ordering: if BMS(1) $\preceq$ BMS(2) and BMS(2) $\preceq$ BMS(3), then BMS(1) $\preceq$ BMS(3). Furthermore, if BMS(1) $\preceq$ BMS(2) and BMS(2) $\preceq$ BMS(1), then BMS(1) = BMS(2). However, given two binary memoryless symmetric channels, they are not necessarily ordered by physical degradation (i.e. it can happen that neither BMS(1) $\preceq$ BMS(2) nor BMS(2) $\preceq$ BMS(1)).

Here are a few examples of channel pairs ordered by physical degradation.

Example 15.7 Let $\epsilon_1, \epsilon_2 \in [0,1]$ with $\epsilon_1 \leq \epsilon_2$. The corresponding erasure channels are then ordered by physical degradation: $\mathrm{BEC}(\epsilon_1) \preceq \mathrm{BEC}(\epsilon_2)$.

Consider a channel C that has input and output alphabets $\mathcal{Y} = \{0, 1, *\}$ (the symbol $*$ representing an erasure). On the inputs 0, 1, it transmits the input unchanged with probability $1 - x$ and erases it with probability x. On the input $*$, it outputs an erasure. If we concatenate this channel with the output of $\mathrm{BEC}(\epsilon_1)$, we obtain a channel $\mathrm{BEC}(\epsilon)$, with $\epsilon = 1 - (1-x)(1-\epsilon)$ (the probability that a bit is *not* erased is the product of the probability that it is not erased by each of the component channels). The claim is thus proved by taking $x = (\epsilon_2 - \epsilon_1)/(1 - \epsilon_1)$.

Exercise 15.4 If $p_1, p_2 \in [0, 1/2]$ with $p_1 \leq p_2$, then $\mathrm{BSC}(p_1) \preceq \mathrm{BSC}(p_2)$. This can be proved by showing that $\mathrm{BSC}(p_2)$ is equivalent to the concatenation of $\mathrm{BSC}(p_1)$ with a second binary symmetric channel $\mathrm{BSC}(x)$. What value of the crossover probability x must one take?

Exercise 15.5 If $\sigma_1^2, \sigma_2^2 \in [0, \infty[$ with $\sigma_1^2 \leq \sigma_2^2$, show that $\mathrm{BAWGN}(\sigma_1^2) \preceq \mathrm{BAWGN}(\sigma_2^2)$.

If $\mathrm{BMS}(1) \preceq \mathrm{BMS}(2)$, most measures of the 'reliability' of the channel are ordered accordingly. Let us discuss two important such measures here: (1) the conditional entropy and (2) the bit error rate.

(1) Let Y_1 and Y_2 be the outputs of passing a uniformly random bit through channels $\mathrm{BMS}(1)$ and $\mathrm{BMS}(2)$, respectively. Then $H(X|Y_1) \leq H(X|Y_2)$ (the uncertainty in the transmitted bit is larger for the 'noisier' channel). This follows immediately from the fact that $X \rightarrow Y_1 \rightarrow Y_2$ is a Markov chain by applying the data-processing inequality (see Section 1.4).

(2) Assume that the outputs of channels $\mathrm{BMS}(1)$ and $\mathrm{BMS}(2)$ are y_1 and y_2, respectively. The MAP decision rule for x, knowing y_a, is $\widehat{x}_a(y_a) = \arg\max_x \mathbb{P}\{X = x|Y_a = y_a\}$, with $a = 1, 2$. The corresponding bit error rate is $\mathrm{P}_{\mathrm{b}}^{(a)} = \mathbb{P}\{\widehat{x}_a(y_a) \neq x\}$. Let us show that $\mathrm{P}_{\mathrm{b}}^{(1)} \leq \mathrm{P}_{\mathrm{b}}^{(2)}$. As $\mathrm{BMS}(1) \preceq \mathrm{BMS}(2)$, there is a channel C such that $\mathrm{BMS}(1)$ concatenated with C is equivalent to $\mathrm{BMS}(2)$. Then $\mathrm{P}_{\mathrm{b}}^{(2)}$ can be regarded as the bit error rate for a non-MAP decision rule, given y_1. The rule is: transmit y_1 through C, denote the output by y_2, and then compute $\widehat{x}_2(y_2)$. This non-MAP decision rule cannot be better than the MAP rule applied directly to y_1.

Since symmetric random variables can be associated with BMS channels (see Proposition 15.1), the notion of physical degradation of channels can be extended to symmetric random variables. Let Z_1, Z_2 be two symmetric random variables and let $\mathrm{BMS}(1)$, $\mathrm{BMS}(2)$ be the associated BMS channels, constructed as in the proof of Proposition 15.1. We say that Z_2 is physically degraded with respect to Z_1 (and write $Z_1 \preceq Z_2$) if $\mathrm{BMS}(2)$ is physically degraded with respect to $\mathrm{BMS}(1)$. It can be proved that this definition is in fact independent of the choice of BMS(1) and BMS(2) within the family of BMS channels associated with Z_1, Z_2.

The interesting result is that BP decoding behaves in the intuitively most natural way with respect to physical degradation. As above, we now fix a particular LDPC code and look at BP messages as random variables due to the randomness in the received vector $\underline{y}$.

Proposition 15.8 *Consider communication over a BMS channel using an LDPC code under the all-zero-codeword assumption, and BP decoding with the standard initial condition $X = 0$. If $\mathsf{B}_{i,r}(F)$ is a tree, then $h_i^{(0)} \succeq h_i^{(1)} \succeq \cdots \succeq h_i^{(t-1)} \succeq h_i^{(t)}$ for any $t \leq r-1$. Analogously, if $\mathsf{B}_{i \to a,r}(F)$ is a tree, then $h_{i \to a}^{(0)} \succeq h_{i \to a}^{(1)} \succeq \cdots \succeq h_{i \to a}^{(t-1)} \succeq h_{i \to a}^{(t)}$ for any $t \leq r-1$.*

We shall not prove this proposition in full generality here, but rather prove its most useful consequence for our purpose, namely the fact that the bit error rate decreases monotonically with t.

Proof Under the all-zero-codeword assumption, the bit error rate is $\mathbb{P}\{\widehat{x}_i^{(t)} = 1\} = \mathbb{P}\{h_i^{(t)} < 0\}$ (for the sake of simplicity, we neglect the case $h_i^{(t)} = 0$ here). Assume $\mathsf{B}_{i,r}(F)$ to be a tree, and fix $t \leq r - 1$. We want to show that $\mathbb{P}\{h_i^{(t)} < 0\} \leq \mathbb{P}\{h_i^{(t-1)} < 0\}$. The BP log-likelihood after T iterations on the original graph, $h_i^{(t)}$, is equal to the actual (MAP) log-likelihood for the reduced model defined on the tree $\mathsf{B}_{i,t+1}(F)$. More precisely, let us denote by $\mathfrak{C}_{i,t}$ the LDPC code associated with the factor graph $\mathsf{B}_{i,t+1}(F)$, and imagine the following process. A uniformly random codeword in $\mathfrak{C}_{i,t}$ is transmitted through the BMS channel, yielding an output $\underline{y}_t$. We define the log-likelihood ratio for bit x_i as

$$\widehat{h}_i^{(t)} = \frac{1}{2} \log \left\{ \frac{\mathbb{P}(x_i = 0 | \underline{y}_t)}{\mathbb{P}(x_i = 1 | \underline{y}_t)} \right\}, \tag{15.22}$$

and denote the MAP estimate for x_i by $\widehat{x}_i$. Clearly, $\mathbb{P}\{\widehat{x}_i = 1 | x_i = 0\} = \mathbb{P}\{h_i^{(t)} < 0\}$.

Instead of this MAP decoding, we can imagine that we scratch all the received symbols at a distance t from i, and then perform MAP decoding on the reduced information. We denote the resulting estimate by $\widehat{x}_i'$. The vector of non-erased symbols is $\underline{y}_{t-1}$. The corresponding log-likelihood is clearly the BP log-likelihood after $t - 1$ iterations. Therefore $\mathbb{P}\{\widehat{x}_i' = 1 | x_i = 0\} = \mathbb{P}\{h_i^{(t-1)} < 0\}$. By the optimality of the MAP decision rule, $\mathbb{P}\{\widehat{x}_i \neq x_i\} \leq \mathbb{P}\{\widehat{x}_i' \neq x_i\}$, which proves our claim. $\square$

In the case of random LDPC codes, $\mathsf{B}_{i,r}(F)$ is a tree with high probability for any fixed r, in the large-block-length limit. Therefore Proposition 15.8 has an immediate consequence in the asymptotic setting.

Proposition 15.9 *The density-evolution random variables are ordered by physical degradation. Namely, $h^{(0)} \succeq h^{(1)} \succeq \cdots \succeq h^{(t-1)} \succeq h^{(t)} \succeq \cdots$. Analogously, $h_*^{(0)} \succeq h_*^{(1)} \succeq \cdots \succeq h_*^{(t-1)} \succeq h_*^{(t)} \succeq \cdots$. As a consequence, the asymptotic bit error rate after a fixed number t of iterations, $\mathrm{P}_{\mathrm{b}}^{(t)} \equiv \lim_{N \to \infty} \mathrm{P}_{\mathrm{b}}^{(N,t)}$, decreases monotonically with t.*

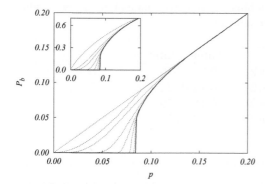 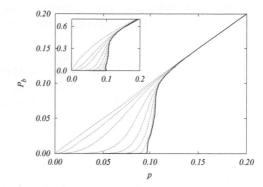

Fig. 15.3 Predicted performance of two LDPC ensembles with a binary symmetric channel. The curves were obtained through a numerical solution of density evolution, using a population dynamics algorithm with a population size of 5×10^5. *Left*: the $(3,6)$ regular ensemble. *Right*: an optimized irregular ensemble with the same design rate $R_{\rm des} = 1/2$, and degree distribution pair $\Lambda(x) = 0.4871\, x^2 + 0.3128\, x^3 + 0.0421\, x^4 + 0.1580\, x^{10}$, $P(x) = 0.6797\, x^7 + 0.3203\, x^8$. The dotted curves give the bit error rate obtained after $t = 1$, 2, 3, 6, 11, 21 and 51 iterations (from *top* to *bottom*), and the bold continuous lines refer to the limit $t \to \infty$. In the *insets* we plot the expected conditional entropy $\mathbb{E}H(X_i|\underline{Y})$.

Exercise 15.6 An alternative measure of the reliability of $h_i^{(t)}$ is provided by the conditional entropy. Assuming that a uniformly random codeword is transmitted, this is given by $H_i(t) = H(X_i|h_i^{(t)})$.

(a) Prove that if $\mathsf{B}_{i,r}(F)$ is a tree, then $H_i(t)$ decreases monotonically with t for $t \le r - 1$.

(b) Assume that, under the all-zero-codeword assumption, $h_i^{(t)}$ has a density $\mathsf{p}_t(.)$. Show that $H_i(t) = \int \log(1 + e^{-2z})\, \mathrm{d}\mathsf{p}_t(z)\,.$
[Hint: Remember that $\mathsf{p}_t(.)$ is a symmetric distribution.]

15.2.4 Numerical implementation, and threshold

Density evolution is a useful tool because it can be simulated efficiently. One can estimate numerically the distributions of the density evolution variables $\{h^{(t)}, u^{(t)}\}$, and also $\{h_*^{(t)}\}$. As we have seen, this gives access to the properties of BP decoding in the large-block-length limit, such as the bit error rate $P_{\rm b}^{(t)}$ after t iterations.

A possible approach[1] consists in representing the distributions by samples of some fixed size S. This leads to the population dynamics algorithm discussed in Section 14.6.4. This algorithm generates, at each time $t \in \{0, \ldots, T\}$, two populations $\{h_1^{(t)}, \cdots, h_{N_{\rm pop}}^{(t)}\}$

---

[1]An alternative approach is as follows. Both of the maps (15.11) can be regarded as convolutions of probability densities for an appropriate choice of the message variables. For the first map, this is immediate in terms of log-likelihoods. For the second map, one can use variables $r^{(t)} = (\mathrm{sign}\, h^{(t)}, \log|\tanh h^{(t)}|)$ and $s^{(t)} = (\mathrm{sign}\, u^{(t)}, \log|\tanh y^{(t)}|))$. By using a fast Fourier transform to implement convolutions, this can result in a significant speed-up of the calculation.

Table 15.1 Belief propagation thresholds for a binary symmetric channel, for a few regular LDPC ensembles. The third column lists the design rate $1 - l/k$.

l	k	R_{des}	p_{d}	Shannon limit
3	4	1/4	0.1669(2)	0.2145018
3	5	2/5	0.1138(2)	0.1461024
3	6	1/2	0.0840(2)	0.1100279
4	6	1/3	0.1169(2)	0.1739524

and $\{u_1^{(t)}, \cdots, u_{N_{\text{pop}}}^{(t)}\}$ which are approximately i.i.d. variables distributed as $h^{(t)}$ and $u^{(t)}$, respectively. From these populations, one can estimate the bit error rate following eqn (15.13). More precisely, the population dynamics estimate is

$$
\mathrm{P}_{\mathrm{b}}^{(t),\text{pop dyn}} = \frac{1}{R} \sum_{n=1}^{R} \varphi\left(B_n + \sum_{j=1}^{l(n)} u_{i_n(j)}^{(t-1)} \right) , \tag{15.23}
$$

where $\varphi(x) \equiv 1$ if $x > 0$, $\varphi(0) = 1/2$, and $\varphi(x) = 0$ otherwise. Here the B_n are distributed as $\frac{1}{2} \log(Q(y|0)/Q(y|1))$, $l(n)$ is distributed as Λ_l, and the indices $i_n(1), \ldots, i_n(l)$ are uniformly random in $\{1, \ldots, N_{\text{pop}}\}$. The parameter R is usually taken to be of the same order as the population size.

In Fig. 15.3, we report the results of population dynamics computations for two different LDPC ensembles used with a binary symmetric channel with crossover probability p. We consider two performance measures: the bit error rate $\mathrm{P}_{\mathrm{b}}^{(t)}$ and the conditional entropy $H^{(t)}$, which can also be easily estimated from the population.

As follows from Proposition 15.9, $\mathrm{P}_{\mathrm{b}}^{(t)}$ and $H^{(t)}$ are monotonically decreasing functions of the number of iterations. One can also show that they are monotonically increasing functions of p. Since $\mathrm{P}_{\mathrm{b}}^{(t)}$ is non-negative and decreasing in t, it has a finite limit $\mathrm{P}_{\mathrm{b}}^{\text{BP}} \equiv \lim_{t \to \infty} \mathrm{P}_{\mathrm{b}}^{(t)}$, which is itself non-decreasing in p (the limit curve $\mathrm{P}_{\mathrm{b}}^{\text{BP}}$ has been estimated in Fig. 15.3 by choosing t large enough so that $\mathrm{P}_{\mathrm{b}}^{(t)}$ is independent of t within the numerical accuracy). The **BP threshold** is defined as

$$
p_{\text{d}} \equiv \sup \left\{ p \in [0, 1/2] : \mathrm{P}_{\mathrm{b}}^{\text{BP}}(p) = 0 \right\} . \tag{15.24}
$$

Here the subscript 'd' stands for 'dynamical': its intrinsic meaning and its relation to phase transitions in other combinatorial problems will be discussed in Chapter 21. Analogous definitions can be provided for other channel families such as the erasure channel BEC(ϵ) and the Gaussian channel BAWGN(σ^2). In general, the definition (15.24) can be extended to any family of BMS channels BMS(p) indexed by a real parameter p which orders the channels with respect to physical degradation.

Numerical simulation of the density evolution allows one to determine the BP threshold p_{d} with good accuracy. We report the results of a few such simulations in Table 15.1. Let us stress that the threshold p_{d} has an important practical meaning. For any $p < p_{\text{d}}$, one can achieve an arbitrarily small bit error rate with high probability just by picking one random code from the ensemble LDPC$_N(\Lambda, P)$ with large N and decoding it using BP with a large enough (but independent of N) number of iterations. For $p > p_{\mathrm{d}}$, the bit error rate is asymptotically bounded from below by $\mathrm{P}_{\mathrm{b}}^{\mathrm{BP}}(p) > 0$ for any fixed number of iterations (in practice, it turns out that doing more iterations, say N^a, does not help). The value of p_{d} is therefore a primary measure of the performance of a code.

One possible approach to the design of good LDPC codes is to use random ensembles, and optimize the degree distribution. For instance, one can look for the degree distribution pair (Λ, P) with the largest BP threshold p_{BP}, given a certain design rate $R_{\mathrm{des}} = 1 - P'(1)/\Lambda'(1)$. In the simple case of communication over a BEC, the optimization over the degree distributions can be carried out analytically, as we shall see in Section 15.3. For general BMS channels, it can be done numerically: one computes the threshold noise level for a given degree distribution pair using density evolution, and maximizes it by a local search procedure. Figure 15.3 shows an example of an optimized irregular ensemble with rate 1/2 for a BSC, including variable nodes of degrees 2, 3, 4 and 10 and check nodes of degrees 7 and 8. Its threshold is $p_{\mathrm{d}} \approx 0.097$ (while the Shannon limit is 0.110).

Note that this ensemble has a finite fraction of variable nodes of degree 2. We can use the analysis in Chapter 11 to compute its weight enumerator function. It turns out that the parameter A in eqn (11.23) is positive. This optimized ensemble has a large number of codewords with small weight. It is surprising, and not very intuitive, that a code such that there exist codewords at a sublinear distance from the transmitted codeword nevertheless has a large BP threshold p_{d}. It turns out that this phenomenon is very general: code ensembles that approach the Shannon capacity turn out to have bad 'short-distance properties'. In particular, the weight enumerator exponent, discussed in Section 11.2, is positive for all values of the normalized weight. Low-weight codewords do not spoil the performance in terms of p_{d}. They are not harmless, though: they degrade the performance of the code at moderate block lengths N, below the threshold p_{d}. Further, they prevent the block error probability from vanishing as N goes to infinity (in each codeword, a fraction $1/N$ of the bits is decoded incorrectly). This phenomenon is referred to as the **error floor**.

Exercise 15.7 While the BP threshold (15.24) was defined in terms of the bit error rate, any other 'reasonable' measure of the error in the decoding of a single bit would give the same result. This can be shown as follows. Let Z be a symmetric random variable and $\mathrm{P}_{\mathrm{b}} \equiv \mathbb{P}\{Z < 0\} + \frac{1}{2}\mathbb{P}\{Z = 0\}$. Show that, for any $\Delta > 0$, $\mathbb{P}\{Z < \Delta\} \leq (2 + e^{2\Delta})\mathrm{P}_{\mathrm{b}}$.

Consider a sequence of symmetric random variables $\{Z^{(t)}\}$ such that the sequence of $\mathrm{P}_{\mathrm{b}}^{(t)}$ defined as before goes to 0. Show that the distribution of $Z^{(t)}$ becomes a Dirac delta at plus infinity as $t \to \infty$.

15.2.5 Local stability

In addition to numerical computation, it is useful to derive simple analytical bounds on the BP threshold. A particularly interesting bound is provided by a local stability analysis. It applies to any BMS channel, and the result depends on the specific channel only through its Bhattacharya parameter $\mathfrak{B} \equiv \sum_y \sqrt{Q(y|0)Q(y|1)} \leq 1$. This parameter, which we have already encountered in Chapter 11, is a measure of the noise level of the channel. It preserves the ordering by physical degradation (i.e. the Bhattacharya parameters of two channels BMS(1) $\preceq$ BMS(2) satisfy $\mathfrak{B}(1) \leq \mathfrak{B}(2)$), as can be checked by explicit computation.

The local stability condition depends on the LDPC code through the fraction of vertices with degree 2, $\Lambda_2 = \lambda'(0)$, and the value of $\rho'(1) = \sum_k P_k k(k-1)/\sum_k P_k k$. It is expressed as follows.

Theorem 15.10 *Consider communication of the all-zero codeword over a binary memoryless symmetric channel with Bhattacharyia parameter $\mathfrak{B}$, using random elements from the ensemble* $\mathrm{LDPC}_N(\Lambda, P)$ *and belief propagation decoding in which the initial messages* $u_{a \to i}^{(0)}$ *are i.i.d. copies of a symmetric random variable. Let* $\mathrm{P}_{\mathrm{b}}^{(t,N)}$ *be the bit error rate after t iterations, and* $\mathrm{P}_{\mathrm{b}}^{(t)} = \lim_{N \to \infty} \mathrm{P}_{\mathrm{b}}^{(t,N)}$.

1. *If* $\lambda'(0)\rho'(1)\mathfrak{B} < 1$, *then there exists* $\xi > 0$ *such that, if* $\mathrm{P}_{\mathrm{b}}^{(t)} < \xi$ *for some* ξ, *then* $\mathrm{P}_{\mathrm{b}}^{(t)} \to 0$ *as* $t \to \infty$.

2. *If* $\lambda'(0)\rho'(1)\mathfrak{B} > 1$, *then there exists* $\xi > 0$ *such that* $\mathrm{P}_{\mathrm{b}}^{(t)} > \xi$ *for any t.*

Corollary 15.11 *We define the **local stability threshold** p_{loc} as*

$$p_{\mathrm{loc}} = \inf \left\{ p \mid \lambda'(0)\rho'(1)\mathfrak{B}(p) > 1 \right\}. \tag{15.25}$$

The BP threshold p_{BP} for decoding a communication over an ordered channel family BMS(p) *using random codes from the ensemble* $\mathrm{LDPC}_N(\Lambda, P)$ *satisfies*

$$p_{\mathrm{d}} \leq p_{\mathrm{loc}}.$$

We shall not give the full proof of this theorem, but will explain the stability argument that underlies it. If the minimum variable-node degree is 2 or larger, the density evolution recursion (15.11) has as a fixed point $h, u \stackrel{\mathrm{d}}{=} Z_\infty$, where Z_∞ is a random variable that takes the value $+\infty$ with probability 1. The BP threshold p_{d} is the largest value of the channel parameter such that $\{h^{(t)}, u^{(t)}\}$ converge to this fixed point as $t \to \infty$. It is then quite natural to ask what happens if the density evolution recursion is initiated with some random initial condition that is 'close enough' to Z_∞. To this end, we consider the initial condition

$$X = \begin{cases} 0 & \text{with probability } \epsilon, \\ +\infty & \text{with probability } 1 - \epsilon. \end{cases} \tag{15.26}$$

This is nothing but the log-likelihood distribution for a bit transmitted through a binary erasure channel, with erasure probability ϵ.

Let us now apply the density evolution recursion (15.11) with the initial condition $u^{(0)} \stackrel{\mathrm{d}}{=} X$. At the first step, we have $h^{(1)} \stackrel{\mathrm{d}}{=} B + \sum_{b=1}^{l-1} X_b$, where the $\{X_b\}$ are i.i.d.

copies of X. Therefore $h^{(1)} = +\infty$ unless $X_1 = \cdots = X_{l-1} = 0$, in which case $h^{(1)} \overset{\mathrm{d}}{=} B$. We therefore have

$$\text{With probability } \lambda_l \; : h^{(1)} = \begin{cases} B & \text{with prob. } \epsilon^{l-1}, \\ +\infty & \text{with prob. } 1 - \epsilon^{l-1}, \end{cases} \tag{15.27}$$

where B is distributed as the channel log-likelihood. Since we are interested in the behaviour 'close' to the fixed point Z_∞, we linearize with respect to ϵ, thus getting

$$h^{(1)} = \begin{cases} B & \text{with prob. } \lambda_2 \epsilon + O(\epsilon^2), \\ +\infty & \text{with prob. } 1 - \lambda_2 \epsilon + O(\epsilon^2), \\ \cdots & \text{with prob. } O(\epsilon^2). \end{cases} \tag{15.28}$$

The last line has not been given in full here, but it will become necessary in subsequent iterations. It signals that $h^{(1)}$ could take some other value with a negligible probability.

Next, consider the first iteration at the check node side: $u^{(1)} = \mathrm{atanh}\{\prod_{j=1}^{k-1} \tanh h_j^{(1)}\}$. To first order in ϵ, we need to consider only two cases: Either $h_1^{(1)} = \cdots = h_{k-1}^{(1)} = +\infty$ (this happens with probability $1 - (k-1)\lambda_2\epsilon + O(\epsilon^2)$), or one of the log-likelihoods is distributed like B (with probability $(k-1)\lambda_2\epsilon + O(\epsilon^2)$). Averaging over the distribution of k, we get

$$u^{(1)} = \begin{cases} B & \text{with prob. } \lambda_2 \rho'(1)\epsilon + O(\epsilon^2), \\ +\infty & \text{with prob. } 1 - \lambda_2 \rho'(1)\epsilon + O(\epsilon^2), \\ \cdots & \text{with prob. } O(\epsilon^2). \end{cases} \tag{15.29}$$

Repeating the argument t times (and recalling that $\lambda_2 = \lambda'(0)$), we get

$$h^{(t)} = \begin{cases} B_1 + \cdots + B_t & \text{with prob. } (\lambda'(0)\rho'(1))^t \epsilon + O(\epsilon^2), \\ +\infty & \text{with prob. } 1 - (\lambda'(0)\rho'(1))^t \epsilon + O(\epsilon^2), \\ \cdots & \text{with prob. } O(\epsilon^2). \end{cases} \tag{15.30}$$

The bit error rate vanishes if and only $\mathrm{P}(t;\epsilon) = \mathbb{P}\{h^{(t)} \le 0\}$ goes to 0 as $t \to \infty$. The above calculation shows that

$$\mathrm{P}(t;\epsilon) = (\lambda'(0)\rho'(1))^t \, \epsilon \, \mathbb{P}\{B_1 + \cdots + B_t \le 0\} + O(\epsilon^2). \tag{15.31}$$

The probability of $B_1 + \cdots + B_t \le 0$ can be computed, to leading exponential order, using the large-deviations estimates of Section 4.2. In particular, we saw in Exercise 4.2 that

$$\mathbb{P}\{B_1 + \cdots + B_t \le 0\} \doteq \left\{ \inf_{z \ge 0} \mathbb{E}\,[\mathrm{e}^{-zB}] \right\}^t. \tag{15.32}$$

We leave to the reader the exercise of showing that, since B is a symmetric random variable, $\mathbb{E}\,\mathrm{e}^{-zB}$ is minimized for $z = 1$, thus yielding

$$\mathbb{P}\{B_1 + \cdots + B_t \le 0\} \doteq \mathfrak{B}^t. \tag{15.33}$$

As a consequence, the order-ϵ coefficient in eqn (15.31) behaves, to leading exponential order, as $(\lambda'(0)\rho'(1)\mathfrak{B})^t$. Depending on whether $\lambda'(0)\rho'(1)\mathfrak{B} < 1$ or $\lambda'(0)\rho'(1)\mathfrak{B} > 1$, the density evolution may or may not converge to the error-free fixed point if initiated sufficiently close to it. The full proof relies on these ideas, but it requires one to control the terms of higher order in ϵ, and other initial conditions as well.

15.3 BP decoding for an erasure channel

We now focus on an erasure channel BEC(ϵ). The analysis can be greatly simplified in this case: the BP decoding algorithm has a simple interpretation, and the density evolution equations can be studied analytically. This allows us to construct **capacity-achieving** ensembles, i.e. codes which are, in the large-N limit, error-free up to a noise level given by Shannon's threshold.

15.3.1 BP, the peeling algorithm, and stopping sets

We consider BP decoding, with initial condition $u_{a\to i}^{(0)} = 0$. As can be seen from eqn (15.7), the channel log-likelihood B_i can take three values: $+\infty$ (if a 0 has been received at position i), $-\infty$ (if a 1 has been received at position i), or 0 (if an erasure occurred at position i).

It follows from the update equations (15.8) that the messages exchanged at any subsequent time take values in $\{-\infty, 0, +\infty\}$ as well. Consider first the equation for the check nodes. If one of the incoming messages $h_{j\to a}^{(t)}$ is 0, then $u_{a\to i}^{(t)} = 0$ as well. If, on the other hand, $h_{j\to a}^{(t)} = \pm\infty$ for all incoming messages, then $u_{a\to i}^{(t)} = \pm\infty$ (the sign being the product of the incoming signs). Next, consider the update equation for the variable nodes. If $u_{b\to i}^{(t)} = 0$ for all of the incoming messages, and $B_i = 0$ as well, then of course $h_{i\to a}^{(t+1)} = 0$. If, on the other hand, some of the incoming messages or the received value B_i takes the value $\pm\infty$, then $h_{i\to a}^{(t+1)}$ takes the same value. Notice that there can never be contradictory incoming messages (i.e. both $+\infty$ and $-\infty$) at a variable node.

Exercise 15.8 Show that, if contradictory messages were sent to the same variable node, this would imply that the transmitted message was not a codeword.

The meaning of the three possible messages $\pm\infty$ and 0 and of the update equations is very clear in this case. Each time the message $h_{i\to a}^{(t)}$ or $u_{a\to i}^{(t)}$ is $+\infty$ (or $-\infty$), this means that the bit x_i is 0 (or 1, respectively) in all codewords that coincide with the channel output at the non-erased positions: the value of x_i is perfectly known. Vice versa, if $h_{i\to a}^{(t)} = 0$ (or $u_{a\to i}^{(t)} = 0$), the bit x_i is currently considered equally likely to be either 0 or 1.

The algorithm is very simple: each message changes value at most one time, either from 0 to $+\infty$, or from 0 to $-\infty$.

Exercise 15.9 To show this, consider the first time, t_1, at which a message $h_{i\to a}^{(t)}$ changes from $+\infty$ to 0. Find what has happened at time $t_1 - 1$.

Therefore a fixed point is reached after a number of updates less than or equal to the number of edges $N\Lambda'(1)$. There is also a clear stopping criterion: if, in one update round, no progress is made (i.e. if $h_{i \to a}^{(t)} = h_{i \to a}^{(t+1)}$ for all directed edges $i \to a$), then no progress will be made in subsequent rounds.

An alternative formulation of BP decoding is provided by the **peeling algorithm**. The idea is to view decoding as a linear-algebra problem. The code is defined through a linear system over $\mathbb{Z}_2$, of the form $\mathbb{H}\underline{x} = \underline{0}$. The output of an erasure channel fixes a fraction of the bits in the vector $\underline{x}$ (the non-erased bits). One is left with an inhomogeneous linear system $\mathcal{L}$ over the remaining erased bits. Decoding amounts to using this new linear system to determine the bits erased by the channel. If an equation in $\mathcal{L}$ contains a single variable x_i with a non-vanishing coefficient, it can be used to determine x_i, and replace it everywhere. One can then repeat this operation recursively until either all of the variables have been fixed (in which case the decoding is successful) or the residual linear system includes only equations over two or more variables (in which case the decoder gets stuck).

Exercise 15.10 An explicit characterization of the fixed points of the peeling algorithm can be given in terms of **stopping sets** (or **2-cores**). A stopping set is a subset of variable nodes in the factor graph such that each check node has a number of neighbours in the subset which is either zero or at least 2. Let S be the subset of undetermined bits when the peeling algorithm stops.

(a) Show that S is a stopping set.

(b) Show that the union of two stopping sets is a stopping set. Deduce that, given a subset of variable nodes U, there exists a unique 'largest' stopping set contained in U that contains any other stopping set in U.

(c) Let U be the set of erased bits. Show that S is the largest stopping set contained in U.

Exercise 15.11 We prove here that the peeling algorithm is indeed equivalent to BP decoding. As in the previous exercise, we denote by S the largest stopping set contained in the erased set U.

(a) Prove that, for any edge (i, a) with $i \in S$, $u_{a \to i}^{(t)} = h_{a \to i}^{(t)} = 0$ at all times.

(b) Vice versa, let S' be the set of bits that are undetermined by BP after a fixed point is reached. Show that S' is a stopping set.

(c) Deduce that $S' = S$ (use the maximality property of S).

15.3.2   Density evolution

Let us study BP decoding of an $\text{LDPC}_N(\Lambda, P)$ code after communication through a binary erasure channel. Under the assumption that the all-zero codeword has been transmitted, messages will take values in $\{0, +\infty\}$, and their distribution can be parameterized by a single real number. We denote by z_t the probability that $h^{(t)} = 0$, and by $\hat{z}_t$ the probability that $u^{(t)} = 0$. The density evolution recursion (15.11) translates into the following recursion on $\{z_t, \hat{z}_t\}$:

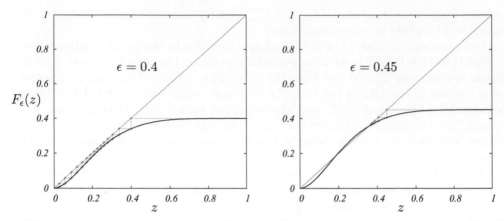

Fig. 15.4 Density evolution for the $(3,6)$ LDPC ensemble over the erasure channel $\mathrm{BEC}(\epsilon)$, for two values of ϵ, below and above the BP threshold $\epsilon_{\mathrm{d}} = 0.4294$.

$$z_{t+1} = \epsilon\lambda(\hat{z}_t), \qquad \hat{z}_t = 1 - \rho(1 - z_t). \tag{15.34}$$

We can eliminate $\hat{z}_t$ from this recursion to get $z_{t+1} = F_\epsilon(z_t)$, where we have defined $F_\epsilon(z) \equiv \epsilon\lambda(1 - \rho(1 - z))$. The bit error rate after t iterations in the large-block-length limit is $\mathrm{P}_{\mathrm{b}}^{(t)} = \epsilon\Lambda(\hat{z}_t)$.

In Fig. 15.4, we show as an illustration the recursion $z_{t+1} = F_\epsilon(z_t)$ for the $(3,6)$ regular ensemble. The edge-perspective degree distributions are $\lambda(z) = z^2$ and $\rho(z) = z^5$, so that $F_\epsilon(z) = \epsilon[1 - (1 - z)^2]^5$. Notice that $F_\epsilon(z)$ is a monotonically increasing function with $F_\epsilon(0) = 0$ (if the minimum variable-node degree is at least 2) and $F_\epsilon(1) = \epsilon < 1$. As a consequence, the sequence $\{z_t\}$ is decreasing and converges at large t to the largest fixed point of F_ϵ. In particular, $z_t \to 0$ (and consequently $\mathrm{P}_{\mathrm{b}}^{\mathrm{BP}} = 0$) if and only if $F_\epsilon(z) < z$ for all $z \in \,]0,1]$. This yields the following explicit characterization of the BP threshold:

$$\epsilon_{\mathrm{d}} = \inf\left\{ \frac{z}{\lambda(1 - \rho(1 - z))} \;:\; z \in \,]0,1] \right\}. \tag{15.35}$$

It is instructive to compare this characterization with the local stability threshold, which in this case reads $\epsilon_{\mathrm{loc}} = 1/\lambda'(0)\rho'(1)$. It is obvious that $\epsilon_{\mathrm{d}} \leq \epsilon_{\mathrm{loc}}$, since $\epsilon_{\mathrm{loc}} = \lim_{z \to 0} z/\lambda(1 - \rho(1 - z))$.

Two cases are possible, as illustrated in Fig. 15.5: either $\epsilon_{\mathrm{d}} = \epsilon_{\mathrm{loc}}$ or $\epsilon_{\mathrm{d}} < \epsilon_{\mathrm{loc}}$. Each case corresponds to a different behaviour of the bit error rate. If $\epsilon_{\mathrm{d}} = \epsilon_{\mathrm{loc}}$, then, generically,[2] $\mathrm{P}_{\mathrm{b}}^{\mathrm{BP}}(\epsilon)$ is a continuous function of ϵ at ϵ_{d}, with $\mathrm{P}_{\mathrm{b}}^{\mathrm{BP}}(\epsilon_{\mathrm{d}} + \delta) = C\delta + O(\delta^2)$ just above the threshold. If, on the other hand, $\epsilon_{\mathrm{d}} < \epsilon_{\mathrm{loc}}$, then $\mathrm{P}_{\mathrm{b}}^{\mathrm{BP}}(\epsilon)$ is discontinuous at ϵ_{d}, with $\mathrm{P}_{\mathrm{b}}^{\mathrm{BP}}(\epsilon_{\mathrm{d}} + \delta) = \mathrm{P}_{\mathrm{b}}^{\mathrm{BP},*} + C\delta^{1/2} + O(\delta)$ just above threshold.

---

[2]Other behaviours are possible, but they are not 'robust' with respect to a perturbation of the degree sequences.

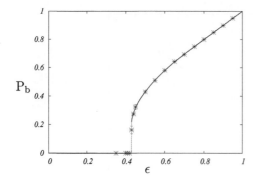
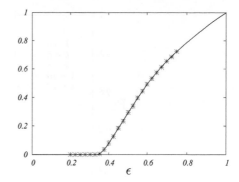

Fig. 15.5 The bit error rate under BP decoding for the $(3,6)$ (*left*) and $(2,4)$ (*right*) ensembles. The prediction of density evolution (bold lines) is compared with numerical simulations (averaged over 10 code/channel realizations with block length $N = 10^4$). For the $(3,6)$ ensemble, where $\epsilon_{\rm BP} \approx 0.4294 < \epsilon_{\rm loc} = \infty$, the transition is discontinuous. For the $(2,4)$ ensemble, where $\epsilon_{\rm BP} = \epsilon_{\rm loc} = 1/4$, the transition is continuous.

Exercise 15.12 Consider communication over a binary erasure channel using random elements from the regular (l, k) ensemble, in the limit $k, l \to \infty$, with a fixed rate $R_{\rm des} = 1 - l/k$. Prove that the BP threshold $\epsilon_{\rm d}$ tends to 0 in this limit.

15.3.3 Ensemble optimization

The explicit characterization (15.35) of the BP threshold for a binary erasure channel opens the way to the optimization of the code ensemble.

One possible set-up is the following. We fix an erasure probability $\epsilon \in \,]0,1[$: this is the estimated noise level in the channel that we are going to use. For a given degree sequence pair (λ, ρ), let $\epsilon_{\rm d}(\lambda, \rho)$ denote the corresponding BP threshold, and let $R(\lambda, \rho) = 1 - (\sum_k \rho_k/k)/(\sum_l \lambda_l/l)$ be the design rate. Our objective is to maximize the rate, while keeping $\epsilon_{\rm d}(\lambda, \rho) \leq \epsilon$. Let us assume that the check node degree distribution ρ is given. Finding the optimal variable-node degree distribution can then be recast as an (infinite-dimensional) linear programming problem:

$$\begin{cases} \text{maximize} & \sum_l \lambda_l/l\,, \\ \text{subject to} & \sum_l \lambda_l = 1 \\ & \lambda_l \geq 0 \quad \forall l\,, \\ & \epsilon\lambda(1 - \rho(1 - z)) \leq z \quad \forall z \in \,]0,1]\,. \end{cases} \tag{15.36}$$

Note that the constraint $\epsilon\lambda(1 - \rho(1 - z)) \leq z$ conflicts with the requirement of maximizing $\sum_l \lambda_l/l$, since both are increasing functions in each of the variables λ_l. As is usual with linear programming, one can show that the objective function is maximized when the constraints are satisfied with equality, i.e. $\epsilon\lambda(1 - \rho(1 - z)) = z$ for all $z \in \,]0,1]$. This 'matching condition' allows one to determine λ for a given ρ.

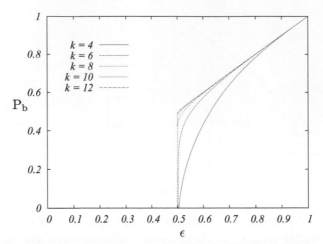

Fig. 15.6 Belief propagation bit error rate for $LDPC_N(\Lambda, P)$ ensembles from the capacity-achieving sequence $(\lambda^{(k)}, \rho^{(k)})$ defined in the main text. The sequence is constructed in such a way as to achieve capacity at a noise level $\epsilon = 0.5$ (the corresponding capacity is $C(\epsilon) = 1 - \epsilon = 0.5$). The five ensembles considered here have design rates $R_{des} = 0.42253$, 0.48097, 0.49594, 0.49894 and 0.49976 (for $k = 4, 6, 8, 10$ and 12, respectively).

We shall do this in the simple case where the check nodes have a uniform degree k, i.e. $\rho(z) = z^{k-1}$. The saturation condition implies $\lambda(z) = (1/\epsilon)[1 - (1-z)^{1/(k-1)}]$. By Taylor expanding this expression, we get, for $l \geq 2$,

$$\lambda_l = \frac{(-1)^l}{\epsilon} \frac{\Gamma(1/(k-1)+1)}{\Gamma(l)\,\Gamma(1/(k-1)-l+2)} \ . \tag{15.37}$$

In particular, $\lambda_2 = 1/((k-1)\epsilon)$, $\lambda_3 = (k-2)/(2(k-1)^2\epsilon)$, and $\lambda_l \simeq \lambda_\infty l^{-k/(k-1)}$ as $l \to \infty$. Unhappily, this degree sequence does not satisfy the normalization condition in (15.36). In fact, $\sum_l \lambda_l = \lambda(1) = 1/\epsilon$. This problem can, however, be overcome by truncating the series and letting $k \to \infty$, as shown in the exercise below. The final result is that a sequence of LDPC ensembles can be found that allows reliable communication under BP decoding, at a rate that asymptotically achieves the channel capacity $C(\epsilon) = 1 - \epsilon$. This is stated more formally below.

Theorem 15.12 *Let $\epsilon \in (0,1)$. Then there exists a sequence of degree distribution pairs $(\lambda^{(k)}, \rho^{(k)})$, with $\rho^{(k)}(x) = x^{k-1}$, such that $\epsilon_d(\lambda^{(k)}, \rho^{(k)}) > \epsilon$ and $R(\lambda^{(k)}, \rho^{(k)}) \to 1 - \epsilon$.*

The precise construction of the sequence $(\lambda^{(k)}, \rho^{(k)})$ is outlined in the next exercise. In Fig. 15.6, we show the BP error probability curves for this sequence of ensembles.

Exercise 15.13 Let $\rho^{(k)}(z) = z^{k-1}$, $\hat{\lambda}^{(k)}(z) = (1/\epsilon)[1-(1-z)^{1/(k-1)}]$, and $z_L = \sum_{l=2}^{L} \hat{\lambda}_l^{(k)}$. Define $L(k,\epsilon)$ as the smallest value of L such that $z_L \geq 1$. Finally, set $\lambda_l^{(k)} = \hat{\lambda}_l^{(k)}/z_{L(k,\epsilon)}$ if $l \leq L(k,\epsilon)$ and $\lambda_l^{(k)} = 0$ otherwise.

(a) Show that $\epsilon\lambda^{(k)}(1 - \rho^{(k)}(1 - z)) < z$ for all $z \in \,]0,1]$ and, as a consequence, $\epsilon_{\mathrm{d}}(\lambda^{(k)}, \rho^{(k)}) > \epsilon$.
   [Hint: Use the fact that the coefficients λ_l in eqn (15.37) are non-negative and hence $\lambda^{(k)}(x) \leq \hat{\lambda}^{(k)}(z)/z_{L(k,\epsilon)}$.]

(b) Show that, for any sequence $l(k)$, $\hat{\lambda}_{l(k)}^{(k)} \to 0$ as $k \to \infty$. Deduce that $L(k,\epsilon) \to \infty$ and $z_{L(k,\epsilon)} \to 1$ as $k \to \infty$.

(c) Prove that $\lim_{k\to\infty} R(\lambda^{(k)}, \rho^{(k)}) = \lim_{k\to\infty} 1 - \epsilon\, z_{L(k,\epsilon)} = 1 - \epsilon$.

15.4 The Bethe free energy and MAP decoding

So far, we have studied the performance of $\mathrm{LDPC}_N(\Lambda, P)$ ensembles under BP message-passing decoding, in the large-block-length limit. Remarkably, sharp asymptotic predictions can be obtained for optimal decoding as well, and they involve the same mathematical objects, namely distributions of messages. We shall focus here on symbol MAP decoding for a channel family $\{\mathrm{BMS}(p)\}$ ordered by physical degradation. As in Chapter 11, we can define a threshold p_{MAP} depending on the LDPC ensemble, such that MAP decoding allows one to communicate reliably at all noise levels below p_{MAP}. We shall compute p_{MAP} using the Bethe free entropy. The free entropy of our decoding problem, averaged over the received signal, is defined as $\mathbb{E}_y \log Z(y)$. Let us see how its value can be related to the properties of MAP decoding.

A crucial step to understanding MAP decoding is estimating the typical number of inputs with non-negligible probability for a given channel output. We can quantify this precisely by introducing the 'codeword entropy density' $\mathfrak{h}_N = (1/N)\,\mathbb{E}H_N(\underline{X}|\underline{Y})$, averaged over the code ensemble (throughout this section we shall use natural logarithms in the definition of the entropies, instead of logarithms to base 2). If $\mathfrak{h}_N$ is bounded away from 0 as $N \to \infty$, the typical channel output is likely to correspond to an exponential number of inputs. If, on the other hand, $\mathfrak{h}_N \to 0$, the correct input has to be searched for among a subexponential number of candidates, and one may hope to be able to decode correctly. A precise relation to the error probability is provided by Fano's inequality (1.28).

Proposition 15.13 *Denote by $\mathrm{P}_{\mathrm{b}}^N$ the bit error probability for communication using a code of block length N. Then,*

$$\mathcal{H}(\mathrm{P}_{\mathrm{b}}^N) \geq \frac{H_N(\underline{X}|\underline{Y})}{N}.$$

In particular, if the entropy density $H_N(\underline{X}|\underline{Y})/N$ is bounded away from 0, so is $\mathrm{P}_{\mathrm{b}}^N$.

Although this gives only a bound, it suggests that we can identify the MAP threshold as the largest noise level such that $\mathfrak{h}_N \to 0$ as $N \to \infty$. In other words, we define

$$p_{\mathrm{c}} \equiv \sup\left\{p : \lim_{N\to\infty} \mathfrak{h}_N = 0\right\}, \qquad (15.38)$$

and conjecture that, for LDPC ensembles, the bit error rate vanishes asymptotically if $p < p_{\mathrm{c}}$, thus implying $p_{\mathrm{MAP}} = p_{\mathrm{c}}$. Hereafter, we shall use p_{c} (or ϵ_{c} for a BEC) to denote the MAP threshold. The relation between this and similar phase transitions in other combinatorial problems will be discussed in Chapter 21.

The conditional entropy $H_N(\underline{X}|\underline{Y})$ is directly related to the free entropy of the model defined in (15.1). More precisely, we have

$$H_N(\underline{X}|\underline{Y}) = \mathbb{E}_y \log Z(\underline{y}) - N \sum_y Q(y|0) \log Q(y|0), \qquad (15.39)$$

where $\mathbb{E}_y$ denotes the expectation with respect to the output vector $\underline{y}$. In order to derive this expression, we first use the entropy chain rule to write (dropping the subscript N)

$$H(\underline{X}|\underline{Y}) = H(\underline{Y}|\underline{X}) + H(\underline{X}) - H(\underline{Y}). \qquad (15.40)$$

Since the input message is uniform over the code, $H(\underline{X}) = N \log |\mathcal{C}|$. Further, since the channel is memoryless and symmetric, $H(\underline{Y}|\underline{X}) = \sum_i H(Y_i|X_i) = NH(Y_i|X_i = 0) = -N \sum_y Q(y|0) \log Q(y|0)$. Finally, rewriting the distribution (15.1) as

$$p(\underline{x}|\underline{y}) = \frac{|\mathcal{C}|}{Z(\underline{y})} p(\underline{y}, \underline{x}), \qquad (15.41)$$

we can identify (by Bayes' theorem) $Z(\underline{y}) = |\mathcal{C}| p(\underline{y})$. Equation (15.39) follows by putting together these contributions.

The free entropy $\mathbb{E}_y \log Z(\underline{y})$ is the non-trivial term in eqn (15.39). For LDPC codes, in the large-N limit, it is natural to compute it using the Bethe approximation of Section 14.2.4. Suppose $\underline{u} = \{u_{a\to i}\}$, $\underline{h} = \{h_{i\to a}\}$ is a set of messages which solves the BP equations

$$h_{i\to a} = B_i + \sum_{b\in\partial i\setminus a} u_{b\to i}, \qquad u_{a\to i} = \operatorname{atanh}\left\{\prod_{j\in\partial a\setminus i} \tanh h_{j\to a}\right\}. \qquad (15.42)$$

Then the corresponding Bethe free entropy follows from eqn (14.28):

$$\mathbb{F}(\underline{u}, \underline{h}) = -\sum_{(ia)\in E} \log\left[\sum_{x_i} \nu_{u_{a\to i}}(x_i)\nu_{h_{i\to a}}(x_i)\right] \qquad (15.43)$$

$$+ \sum_{i=1}^{N} \log\left[\sum_{x_i} Q(y_i|x_i) \prod_{a\in\partial i} \nu_{u_{a\to i}}(x_i)\right] + \sum_{a=1}^{M} \log\left[\sum_{\underline{x}_a} \mathbb{I}_a(\underline{x}) \prod_{i\in\partial a} \nu_{h_{i\to a}}(x_i)\right].$$

where we denote the distribution of a bit x whose log-likelihood ratio is u by $\nu_u(x)$, given by $\nu_u(0) = 1/(1 + e^{-2u})$, $\nu_u(1) = e^{-2u}/(1 + e^{-2u})$.

We are interested in the expectation of this quantity with respect to the code and channel realization, in the $N \to \infty$ limit. As in Section 14.6.3, we assume that messages are asymptotically identically distributed, i.e. $u_{a \to i} \stackrel{\mathrm{d}}{=} u$, $h_{i \to a} \stackrel{\mathrm{d}}{=} h$, and that messages coming into the same node along distinct edges are asymptotically independent. Under these hypotheses, we get

$$\lim_{N \to \infty} \frac{1}{N} \mathbb{E}_y \, \mathbb{F}(\underline{u}, \underline{h}) = \mathrm{f}_{u,h}^{\mathrm{RS}} + \sum_y Q(y|0) \log Q(y|0), \qquad (15.44)$$

where the 'shifted' free-entropy density $\mathrm{f}_{u,h}^{\mathrm{RS}}$ associated with the random variables u, h is defined by

$$\mathrm{f}_{u,h}^{\mathrm{RS}} = -\Lambda'(1) \, \mathbb{E}_{u,h} \log \left[\sum_x \nu_u(x) \nu_h(x) \right] + \mathbb{E}_{l,y,\{u_i\}} \log \left[\sum_x \frac{Q(y|x)}{Q(y,0)} \prod_{i=1}^l \nu_{u_i}(x) \right]$$

$$+ \frac{\Lambda'(1)}{P'(1)} \, \mathbb{E}_k \mathbb{E}_{\{h_i\}} \log \left[\sum_{x_1 \dots x_k} \mathbb{I}\left(x_1 \oplus \cdots \oplus x_k = 0 \right) \prod_{i=1}^k \nu_{h_i}(x_i) \right]. \qquad (15.45)$$

Here k and l are distributed according to P_k and Λ_l, respectively, and $u_1, u_2, \dots$ and $h_1, h_2, \dots$ are i.i.d. and distributed as u and h, respectively.

If the Bethe free entropy is correct, the shifted Bethe free-entropy density $\mathrm{f}_{u,h}^{\mathrm{RS}}$ is equal to the codeword entropy density $\mathfrak{h}_N$. This reasonable assumption can be turned into a rigorous inequality.

Theorem 15.14 *If u, h are symmetric random variables satisfying the distributional identities $u \stackrel{\mathrm{d}}{=} \operatorname{atanh} \left\{ \prod_{i=1}^{k-1} \tanh h_i \right\}$ and $h \stackrel{\mathrm{d}}{=} B + \sum_{a=1}^{l-1} u_a$, then*

$$\lim_{N \to \infty} \mathfrak{h}_N \geq \mathrm{f}_{u,h}^{\mathrm{RS}}. \qquad (15.46)$$

It is natural to conjecture that the correct limit is obtained by optimizing the above lower bound, i.e.

$$\lim_{N \to \infty} \mathfrak{h}_N = \sup_{u,h} \, \mathrm{f}_{u,h}^{\mathrm{RS}}, \qquad (15.47)$$

where, once again, the sup is taken over the pairs of symmetric random variables u, h satisfying $u \stackrel{\mathrm{d}}{=} \operatorname{atanh} \left\{ \prod_{i=1}^{k-1} \tanh h_i \right\}$ and $h \stackrel{\mathrm{d}}{=} B + \sum_{a=1}^{l-1} u_a$.

This conjecture has indeed been proved in the case of communication over a binary erasure channel for a large class of LDPC ensembles (including, for instance, regular ones).

The above equation is interesting because it establishes a bridge between BP and MAP decoding. An example of application of this bridge is given in the next exercise.

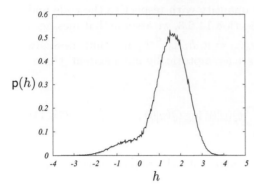
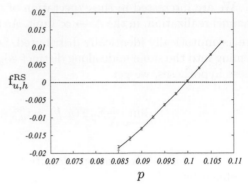

Fig. 15.7 Illustration of the RS cavity method applied to a $(3,6)$ regular code used over a binary symmetric channel. *Left*: a non-trivial distribution of the variables h found by population dynamics at a noise level $p = 0.095$. *Right*: shifted free entropy versus p, for the non-trivial solution (the normalization is such that the free entropy of the perfect-decoding phase $u = h = \infty$ is zero). When the noise level is increased, the non-trivial solution appears at p_d, and its free entropy becomes positive at p_c

Exercise 15.14 Proof that $p_\mathrm{d} \leq p_\mathrm{c}$.

(a) Recall that the pair of distributions "$u, h = +\infty$ with probability one" constitute a density-evolution fixed point for any noise level. Show that $\mathrm{f}^{\mathrm{RS}}_{h,u} = 0$ at such a fixed point.

(b) Use ordering by physical degradation to show that if any other fixed point exists, then the density evolution converges to it.

(c) Deduce that $p_\mathrm{d} \leq p_\mathrm{c}$.

Evaluating the expression in eqn (15.47) implies an a priori infinite-dimensional optimization problem. In practice, good approximations can be obtained through the following procedure:

1. Initialize h, u to a pair of symmetric random variables $h^{(0)}$, $u^{(0)}$.

2. Implement numerically the density evolution recursion (15.11) by population dynamics, and iterate it until an approximate fixed point is attained.

3. Evaluate the functional $\mathrm{f}^{\mathrm{RS}}_{u,h}$ at such a fixed point, after enforcing $u \stackrel{\mathrm{d}}{=}$
   $\mathrm{atanh} \left\{ \prod_{i=1}^{k-1} \tanh h_i \right\}$ exactly.

The above procedure can be repeated for several different initializations $u^{(0)}$, $h^{(0)}$. The largest of the corresponding values of $\mathrm{f}^{\mathrm{RS}}_{u,v}$ is then picked as an estimate for $\lim_{N \to \infty} \mathfrak{h}_N$.

While this procedure is not guaranteed to exhaust all of the possible density-evolution fixed points, it allows one to compute a sequence of lower bounds on the conditional entropy density. Further, by analogy with exactly solvable cases (such as that of a binary erasure channel), one expects a small, finite number of density-evolution fixed points. In particular, for regular ensembles and $p > p_\mathrm{d}$, a unique

Table 15.2 MAP thresholds for a binary symmetric channel compared with the BP decoding thresholds, for a few regular LDPC ensembles.

l	k	R_{des}	p_d	p_c	Shannon limit
3	4	1/4	0.1669(2)	0.2101(1)	0.2145018
3	5	2/5	0.1138(2)	0.1384(1)	0.1461024
3	6	1/2	0.0840(2)	0.1010(2)	0.1100279
4	6	1/3	0.1169(2)	0.1726(1)	0.1739524

Table 15.3 MAP thresholds for a binary erasure channel compared with the BP decoding thresholds, for a few regular LDPC ensembles.

l	k	R_{des}	ϵ_d	ϵ_c	Shannon limit
3	4	1/4	0.647426	0.746010	0.750000
3	5	2/5	0.517570	0.590989	0.600000
3	6	1/2	0.429440	0.488151	0.500000
4	6	1/3	0.506132	0.665656	0.666667

(stable) fixed point is expected to exist in addition to the no-error one $u, h = +\infty$. In Table 15.2 we present the corresponding MAP thresholds for a BSC and a few regular ensembles.

The whole approach simplifies considerably in the case of communication over a binary erasure channel, as shown in the exercise below.

Exercise 15.15 Consider the erasure channel BEC(ϵ), and look for a fixed point of the density evolution equations (15.11) such that (*i*) $h = 0$ with probability z and $h = \infty$ with probability $1 - z$, and (*ii*) $u = 0$ with probability $\hat{z}$ and $u = \infty$ with probability $1 - \hat{z}$.

(a) Show that z and $\hat{z}$ must satisfy the equations (15.34).

(b) Show that the shifted free entropy (15.45) is equal to

$$f_{u,h}^{RS} = \left[\Lambda'(1)z(1 - \hat{z}) + \frac{\Lambda'(1)}{P'(1)} \left(P(1 - z) - 1 \right) + \epsilon \Lambda(\hat{z}) \right] \log 2 \,. \tag{15.48}$$

(c) Use this expression and the conjecture (15.47) to obtain the MAP thresholds for regular ensembles listed in Table 15.3.

The two problems of computing the BP and MAP thresholds are thus unified by the use of the RS cavity method. For any noise level p, there always exists a solution to the RS cavity equations in which the distribution of u is a point mass distribution at $u = +\infty$ and the distribution of h is a point mass distribution at $h = +\infty$. This solution corresponds to a perfect decoding; its shifted free entropy density is $f_{u,h}^{RS} = 0$. When $p > p_d$, another solution to the RS cavity equations appears. Its shifted free entropy density f^{RS} can be computed from eqn (15.44): it is initially negative and increases with p. The MAP threshold is the value $p = p_\mathrm{d}$ above which f^{RS} becomes positive. Figure 15.7 illustrates this behaviour.

Still, this description leaves us with a puzzle. In the regime $p_\mathrm{d} \le p < p_\mathrm{c}$, the codeword entropy density associated with the solution $h, u < \infty$ is $\lim_{N \to \infty} \mathfrak{h}_N \ge \mathrm{f}^{\mathrm{RS}}_{u,h} < 0$. Analogously to what happens in the replica method (see Chapter 8), the solution should therefore be discarded as unphysical. It turns out that a consistent picture can be obtained only by including replica symmetry breaking, which will be the subject of Chapter 21.

Notes

Belief propagation was first applied to the decoding problem by Robert Gallager in his PhD thesis (Gallager, 1963), and called the 'sum–product' algorithm there. Several low-complexity alternative message-passing approaches were introduced in the same work, along with the basic ideas of their analysis.

The analysis of iterative decoding of irregular ensembles over an erasure channel was pioneered by Luby and co-workers (Luby *et al.*, 1997, 1998, 2001*a,b*). These papers also presented the first examples of capacity-achieving sequences.

Density evolution for general binary memoryless symmetric channels was introduced by Richardson and Urbanke (2001*b*). The whole subject is surveyed by Richardson and Urbanke (2001*a*, 2008). One important property that we have left out is 'concentration': the error probability under message-passing decoding is, for most of the codes, close to its ensemble average, which is predicted by density evolution.

The design of capacity-approaching LDPC ensembles for general BMS channels was discussed by Chung *et al.* (2001) and Richardson *et al.* (2001).

Since message passing allows efficient decoding, one may wonder whether the encoding (whose complexity is, a priori, $O(N^2)$) might become the bottleneck. Luckily this is not the case: efficient encoding schemes were discussed by Richardson and Urbanke (2001*c*).

The use of the RS replica method (equivalent to the cavity method) to characterize MAP decoding for sparse-graph codes was initiated by Kabashima and Saad (1999), who considered Sourlas's LDGM codes. MN codes (a class of sparse-graph codes defined by MacKay and Neal (1996)) and turbo codes were studied shortly after, by Kabashima *et al.* (2000*a,b*) and by Montanari and Sourlas (2000) and Montanari (2000), respectively. Plain regular LDPC ensembles were considered first by Kabashima and Saad (2000), who considered the problem on a tree, and by Nakamura *et al.* (2001). The effect of replica symmetry breaking was first investigated by Montanari (2001*b*) and standard irregular ensembles were studied by Franz *et al.* (2002).

The fact that the RS cavity method yields the exact value of the MAP threshold and that $p_{\mathrm{MAP}} = p_\mathrm{c}$ has not yet been proven rigorously in a general setting. The first proof that it gives a rigorous bound was found by Montanari (2005) and subsequently generalized by Macris (2007). An alternative proof technique uses the 'area theorem' and the related 'Maxwell construction' (Méasson *et al.*, 2005). Tightness of these bounds for a binary erasure channel was proved by Méasson *et al.* (2008). In this case the asymptotic codeword entropy density and the MAP threshold have been determined rigorously for a large family of ensembles.

The analysis that we have described in this Chapter is valid in the large-block-length limit $N \to \infty$. In practical applications, a large block length implies some communication delay. This has motivated a number of studies aimed at estimating and optimizing LDPC codes at moderate block lengths. Some pointers to this large literature can be found in Di *et al.* (2002), Amraoui *et al.* (2004, 2007), Wang *et al.* (2006), Kötter and Vontobel (2003), and Stepanov *et al.* (2005).

16

The assignment problem

Consider N 'agents' and N 'jobs', and suppose you are given an $N \times N$ matrix $\{E_{ij}\}$, where E_{ij} is the cost for having job j executed by agent i. Finding an assignment of agents to jobs that minimizes the cost is one of the most classical of combinatorial optimization problems.

The minimum-cost (also referred to as 'maximum-weight') assignment problem is important both because of its many applications and because it can be solved in polynomial time. This has motivated a number of theoretical developments, from the algorithmic as well as the probabilistic viewpoint.

Here we shall study the assignment problem as an application of message-passing techniques. It is, in fact, a success story of this approach. Given a generic instance of the assignment problem, the associated factor graph is not locally tree-like. Nevertheless, the min-sum algorithm can be proved to converge to an optimal solution in polynomial time. Belief propagation (the sum–product algorithm) can also be used for computing weighted sums over assignments, although much weaker guarantees exist in this case. A significant amount of work has been devoted to the study of random instances, mostly in the case where the costs E_{ij} are i.i.d. random variables. Typical properties (such as the cost of the optimal assignment) can be computed heuristically within the replica-symmetric cavity method. It turns out that these calculations can also be made fully rigorous.

In spite of the success of the replica-symmetric cavity method, one must be warned that apparently harmless modifications of the problem can spoil this success. One example is the generalization of minimal-cost assignment to multi-indices (say, matching agents with jobs and houses): such a 'multi-assignment' problem is not described by the replica-symmetric scenario. The more sophisticated ideas of replica symmetry breaking, described in Chapter 19 and after, are required.

After defining the problem in Section 16.1, we compute in Section 16.2 the asymptotic optimal cost for random instances using the cavity method. This approach is based on a statistical analysis of the min-sum equations. In Section 16.3, we study the algorithmic aspects of the min-sum iteration and prove its convergence to the optimal assignment. Section 16.4 contains a combinatorial study of the optimal cost that confirms the cavity result and provides sharper estimates. In Section 16.5, we discuss a generalization of the assignment problem to a multi-assignment case.

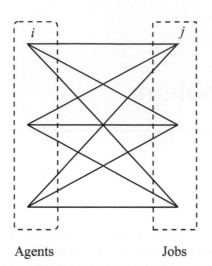 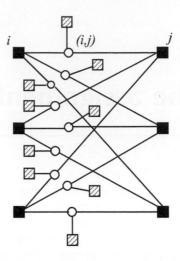

Agents   Jobs

Fig. 16.1 *Left*: graphical representation of a small assignment problem with three agents and three jobs. Each edge carries a cost (not shown); the problem is to find a perfect matching, i.e. a set of three edges which are vertex-disjoint, of minimal cost. *Right*: the factor graph corresponding to the representation (16.2) of this problem. The hatched squares indicate the function nodes associated with edge weights.

16.1 The assignment problem and random assignment ensembles

An instance of the assignment problem is determined by a cost matrix $\{E_{ij}\}$, indexed by $i \in A$ (the 'agents' set) and $j \in B$ (the 'jobs' set), with $|A| = |B| = N$. We shall often identify A and B with the set $\{1, \ldots, N\}$ and use the terms 'cost' and 'energy' interchangeably. An assignment is a one-to-one mapping of agents to jobs; that is, a permutation π of $\{1, \ldots, N\}$. The cost of an assignment π is $E(\pi) = \sum_{i=1}^{N} E_{i\pi(i)}$. The optimization problem consists in finding a permutation that minimizes $E(\pi)$.

We shall often use a graphical description of the problem as a weighted complete bipartite graph over sets of vertices A and B. Each of the N^2 edges (i, j) carries a weight E_{ij}. The problem is to find a perfect matching in this graph (a subset M of edges such that every vertex is adjacent to exactly one edge in M), of minimal weight (see Fig. 16.1).

In the following, we shall be interested in two types of questions. The first is to understand whether a minimum-cost assignment for a given instance can be found efficiently through a message-passing strategy. The second will be to analyse the typical properties of ensembles of random instances where the N^2 costs E_{ij} are i.i.d. random variables drawn from a distribution with density $\rho(E)$. One particularly convenient choice is that of exponentially distributed variables with a probability density function $\rho(E) = e^{-E}\, \mathbb{I}(E \geq 0)$. Although the cavity method allows more general distributions to be tackled, assuming exponential costs greatly simplifies rigorous combinatorial proofs.

16.2   Message passing and its probabilistic analysis

16.2.1   Statistical-physics formulation and counting

Following the general statistical-physics approach, it is of interest to relax the optimization problem by introducing a finite inverse temperature β. The corresponding computational problem associates a weight with each possible matching, as follows.

Consider the complete bipartite graph over the sets of vertices A (agents) and B (jobs). With each edge (i, j), $i \in A$, $j \in B$, we associate a variable (an 'occupation number') $n_{ij} \in \{0, 1\}$, encoding membership of edge (ij) in the matching: $n_{ij} = 1$ means that job j is done by agent i. We impose the condition that the subset of edges (i, j) with $n_{ij} = 1$ is a matching of the complete bipartite graph:

$$\sum_{j \in B} n_{ij} \leq 1 \ \ \forall i \in A\,, \qquad \sum_{i \in A} n_{ij} \leq 1 \ \ \forall j \in B\,. \tag{16.1}$$

Let us denote by $\underline{n} = \{n_{ij} : i \in A\, j \in B\}$ the matrix of occupation numbers, and define the probability distribution

$$\mu(\underline{n}) = \frac{1}{Z} \prod_{i \in A} \mathbb{I}\left(\sum_{j \in B} n_{ij} \leq 1\right) \prod_{j \in B} \mathbb{I}\left(\sum_{i \in A} n_{ij} \leq 1\right) \prod_{(ij)} e^{-\beta n_{ij}(E_{ij} - 2\gamma)}\,. \tag{16.2}$$

The support of $\mu(\underline{n})$ corresponds to matchings, thanks to the 'hard constraints' that enforce the conditions (16.1). The factor $\exp\left(2\beta\gamma \sum_{(ij)} n_{ij}\right)$ can be interpreted as a 'soft constraint': as $\gamma \to \infty$, the distribution becomes concentrated on perfect matchings (the factor 2 is introduced here for future convenience). On the other hand, in the limit $\beta \to \infty$, the distribution (16.2) becomes concentrated on the minimal-cost assignments. The optimization problem is thus recovered in the double limit of $\gamma \to \infty$ followed by $\beta \to \infty$.

There is a large degree of arbitrariness in the choice of which constraint should be 'softened' and how. The present choice makes the whole problem as similar as possible to the general class of graphical models that we are studying in this book. The factor graph obtained from eqn (16.2) has the following structure (see Fig. 16.1). It contains N^2 variable nodes, each associated with an edge (i, j) in the complete bipartite graph over the sets of vertices A and B. It also includes N^2 function nodes of degree one, one for each variable node, and $2N$ function nodes of degree N, associated with the vertices in A and B. The variable node (i, j), $i \in A$, $j \in B$, is connected to the two function nodes corresponding to i and j, as well as to the function node corresponding to the edge (i, j). The first two function nodes enforce the hard constraints (16.1); the third one corresponds to the weight $\exp\left[-\beta(E_{ij} - 2\gamma)n_{ij}\right]$.

In the case of random instances, we shall be particularly interested in the thermodynamic limit $N \to \infty$. In order for this limit to be non-trivial, the distribution (16.2) must be dominated neither by energy nor by entropy. Consider the case of i.i.d. costs $E_{ij} \geq 0$ with an exponential density $\rho(E) = e^{-E}$. One can argue in this case that low-energy assignments have, with high probability, an energy of order $O(1)$ as $N \to \infty$. The 'hand-waving' reason is that for a given agent $i \in A$ and any fixed k, the k lowest costs among those of the jobs that can be assigned to that agent (namely among $\{E_{ij} : j \in B\}$) are of order $O(1/N)$. The exercise below sketches a more formal proof. Since the entropy[1] is linear in N, we need to rescale the costs for the two contributions to be of the same order.

To summarize, throughout our cavity analysis, we shall assume the edge cost to be drawn according to the 'rescaled pdf' $\hat{\rho}(E) = (1/N)\exp(-E/N)$. This choice ensures that the occupied edges in the best assignment have a non-vanishing cost in the large-N limit.

Exercise 16.1 Assume the energies E_{ij} to be i.i.d. exponential variables of mean η. Consider the 'greedy mapping' obtained by mapping each vertex $i \in A$ to that $j = \pi_1(i) \in B$ which minimizes E_{ij}, and call the corresponding energy $E_1 = \sum_i E_{i,\pi_1(i)}$.

(a) Show that $\mathbb{E}\, E_1 = \eta$.

(b) Of course, π_1 is not necessary injective, and is therefore not a valid matching. Let C be the number of collisions (i.e. the number of vertices $j \in B$ such that there exist several i with $\pi_1(i) = j$). Show that $\mathbb{E}\, C = N(1 - 2/e) + O(1)$, and that C is tightly concentrated around its expectation.

(c) Consider the following 'fix'. Construct π_1 in the greedy fashion described above, and let $\pi_2(i) = \pi_1(i)$ whenever i is the unique vertex mapped to $\pi_1(i)$. For each collision vertex $j \in B$, and each $i \in A$ such that $\pi_1(i) = j$, let j' be the vertex in B such that $E_{ij'}$ takes the smallest value among the vertices still unmatched. What is the expectation of the resulting energy $E_2 = \sum_i E_{i,\pi_2(i)}$? What is the number of residual collisions?

(d) How can this construction be continued?

16.2.2 The belief propagation equations

The BP equations for this problem are a particular instantiation of the general ones given in eqns (14.14) and (14.15). We shall denote vertices in the sets A and B in the complete bipartite graph by i and j, respectively (see Fig. 16.1).

To be definite, let us write explicitly the equations for updating messages flowing from right to left (i.e. from vertices $j \in B$ to $i \in A$) in the graph in Fig. 16.1:

$$\nu_{ij\to i}(n_{ij}) \cong \hat{\nu}_{j\to ij}(n_{ij})\, e^{-\beta n_{ij}(E_{ij}-2\gamma)}, \qquad (16.3)$$

$$\hat{\nu}_{j\to ij}(n_{ij}) \cong \sum_{\{n_{kj}\}} \mathbb{I}\Big[n_{ij} + \sum_{k\in A\setminus i} n_{kj} \leq 1\Big] \prod_{k\in A\setminus i} \nu_{kj\to j}(n_{kj}). \qquad (16.4)$$

The equations for messages moving from A to B, i.e. $\nu_{ij\to j}$ and $\hat{\nu}_{i\to ij}$, are obtained by inverting the roles of the two sets.

Since the variables n_{ij} take values in $\{0, 1\}$, messages can be parameterized by a single real number, as usual. In the present case, it is convenient to introduce rescaled log-likelihood ratios as follows:

---

[1]The total number of assignments is $N!$, which would imply an entropy of order $N \log N$. However, if we limit the choices of $\pi(i)$ to those $j \in B$ such that the cost E_{ij} is comparable to the lowest cost, the entropy becomes $O(N)$.

$$x^{\mathrm{L}}_{j\to i} \equiv \gamma + \frac{1}{\beta} \log \left\{ \frac{\widehat{\nu}_{j\to ij}(1)}{\widehat{\nu}_{j\to ij}(0)} \right\}, \quad x^{\mathrm{R}}_{i\to j} \equiv \gamma + \frac{1}{\beta} \log \left\{ \frac{\widehat{\nu}_{i\to ij}(1)}{\widehat{\nu}_{i\to ij}(0)} \right\}. \tag{16.5}$$

Variable-to-function-node messages do not enter into this definition, but they are easily expressed in terms of the quantities $x^{\mathrm{L}}_{i\to j}$, $x^{\mathrm{R}}_{i\to j}$ using eqn (16.3). The BP equations (16.3) and (16.4) can be written as

$$
\begin{aligned}
x^{\mathrm{L}}_{j\to i} &= -\frac{1}{\beta} \log \left\{ \mathrm{e}^{-\beta\gamma} + \sum_{k\in A\backslash i} \mathrm{e}^{-\beta E_{kj} + \beta x^{\mathrm{R}}_{k\to j}} \right\}, \\
x^{\mathrm{R}}_{i\to j} &= -\frac{1}{\beta} \log \left\{ \mathrm{e}^{-\beta\gamma} + \sum_{k\in B\backslash j} \mathrm{e}^{-\beta E_{ik} + \beta x^{\mathrm{L}}_{k\to i}} \right\}.
\end{aligned}
\tag{16.6}
$$

The factor graph representation in Fig. 16.1, right frame, was necessary in order to write down the original BP equations. However, any reference to the factor graph disappears in the simplified form (16.6). This form can be regarded as a message-passing procedure operating on the original complete bipartite graph (see Fig. 16.1, left frame).

Exercise 16.2 Marginals. Consider the expectation value of n_{ij} with respect to the measure (16.2). Show that its BP estimate is $t_{ij}/(1 + t_{ij})$, where $t_{ij} \equiv \mathrm{e}^{\beta(x^{\mathrm{L}}_{j\to i} + x^{\mathrm{R}}_{i\to j} - E_{ij})}$.

The Bethe free entropy $\mathbb{F}(\underline{\nu})$ can be computed using the general formulae (14.27) and (14.28). Writing it in terms of the log-likelihood ratio messages $\{x^{\mathrm{R}}_{i\to j}, x^{\mathrm{L}}_{j\to i}\}$ is straightforward but tedious. The resulting BP estimate for the free entropy $\log Z$ is

$$\mathbb{F}(\underline{x}) = 2N\beta\gamma - \sum_{i\in A, j\in B} \log \left[1 + \mathrm{e}^{-\beta(E_{ij} - x^{\mathrm{R}}_{i\to j} - x^{\mathrm{L}}_{j\to i})} \right]$$

$$+ \sum_{i\in A} \log \left[\mathrm{e}^{-\beta\gamma} + \sum_j \mathrm{e}^{-\beta(E_{ij} - x^{\mathrm{L}}_{j\to i})} \right] + \sum_{j\in B} \log \left[\mathrm{e}^{-\beta\gamma} + \sum_i \mathrm{e}^{-\beta(E_{ij} - x^{\mathrm{R}}_{i\to j})} \right]. \tag{16.7}$$

The exercise below provides a few guidelines for this computation.

Exercise 16.3 Consider the Bethe free entropy (14.27) for the model (16.2).

(a) Show that it contains three types of function-node terms, one type of variable-node term, and three types of mixed (edge) terms.

(b) Show that the function node term associated with the weight $\mathrm{e}^{-\beta n_{ij}(E_{ij} - 2\gamma)}$ exactly cancels the mixed term involving this same factor node and the variable node (i, j).

(c) Write explicitly each of the remaining terms, express them in terms of the messages $\{x^{\mathrm{R}}_{i\to j}, x^{\mathrm{L}}_{j\to i}\}$, and derive the result (16.7).
[Hint: The calculation can be simplified by recalling that the expression (14.27) does not change value if each message is independently rescaled.]

16.2.3 Zero temperature: The min-sum algorithm

The BP equations (16.6) simplify in the double limit of $\gamma \to \infty$ followed by $\beta \to \infty$ which is relevant to the minimum-cost assignment problem. Assuming that the $\{x_{i \to j}^{\mathrm{R}}, x_{j \to i}^{\mathrm{L}}\}$ remain finite in this limit, we get

$$x_{j \to i}^{\mathrm{L}} = \min_{k \in A \backslash i} \left(E_{kj} - x_{k \to j}^{\mathrm{R}} \right) , \qquad x_{i \to j}^{\mathrm{R}} = \min_{k \in B \backslash j} \left(E_{ik} - x_{k \to i}^{\mathrm{L}} \right) . \qquad (16.8)$$

Alternatively, the same equations can be obtained directly as min-sum update rules. This derivation is outlined in the exercise below.

Exercise 16.4 Consider the min-sum equations (14.41) and (14.40), applied to the graphical model (16.2).

(a) Show that the message arriving at a variable node (ij) from the adjacent degree-1 factor node is equal to $\widehat{E}_{\to ij}(n_{ij}) = n_{ij}(E_{ij} - 2\gamma)$.

(b) Write the update equations for the other messages, and eliminate the variable-to-function-node messages $E_{ij \to i}(n_{ij})$ and $E_{ij \to j}(n_{ij})$ in favour of the function-to-variable ones. Show that the resulting equations for function-to-variable messages read as follows (see Fig. 16.1):

$$\widehat{E}_{i \to ij}(1) = \sum_{k \in B \backslash j} \widehat{E}_{k \to ik}(0) ,$$

$$\widehat{E}_{i \to ij}(0) = \sum_{k \in B \backslash j} \widehat{E}_{k \to ik}(0) + \min_{l \in B \backslash j} \left[\widehat{E}_{l \to il}(1) - \widehat{E}_{l \to il}(0) + E_{il} - 2\gamma \right]_- ,$$

where we have adopted the notation $[x]_- = x$ if $x < 0$ and $[x]_- = 0$ otherwise.

(c) Define $x_{i \to j}^{\mathrm{R}} = \widehat{E}_{i \to ij}(0) - \widehat{E}_{i \to ij}(1) + \gamma$ and, analogously $x_{i \to j}^{\mathrm{L}} = \widehat{E}_{j \to ij}(0) - \widehat{E}_{j \to ij}(1) + \gamma$. Write the above min-sum equations in terms of $\{x_{i \to j}^{\mathrm{R}}, x_{j \to i}^{\mathrm{L}}\}$.

(d) Show that, in the large-γ limit, the update equations for these x-messages coincide with eqn (16.8).

The Bethe estimate for the ground state energy (the cost of the optimal assignment) can be obtained by taking the $\gamma, \beta \to \infty$ limit of the free energy $-\mathbb{F}(x)/\beta$, where $\mathbb{F}(x)$ is the Bethe approximation for the log-partition function $\log Z$ (see eqn (16.7)). Alternatively, we can use the fact that the min-sum equations estimate the max-marginals of the graphical model (16.2). More precisely, for each pair (i, j), $i \in A$, $j \in B$, we define

$$E_{ij}(n_{ij}) \equiv n_{ij}(E_{ij} - 2\gamma) + \widehat{E}_{i \to ij}(n_{ij}) + \widehat{E}_{j \to ij}(n_{ij}) , \qquad (16.9)$$

$$n_{ij}^* \equiv \arg \min_{n \in \{0,1\}} E_{ij}(n) . \qquad (16.10)$$

The interpretation of these quantities is that $e^{-\beta E_{ij}(n)}$ is the message-passing estimate for the max-marginal of n_{ij} with respect to the distribution (16.2). Let us neglect the case of a multiple optimal assignment (in particular, the probability of such an event vanishes for the random ensemblesthat we shall consider). Under the assumption that message-passing estimates are accurate, n_{ij} necessarily takes the value n_{ij}^* in the optimal assignment; see Section 14.3. The resulting estimate of the ground state energy is $E_{\text{gs}} = \sum_{ij} n_{ij}^* E_{ij}$.

In the limit $\gamma \to \infty$, eqn (16.10) reduces to a simple **inclusion principle**: an edge ij is present in the optimal assignment (i.e. $n_{ij}^* = 1$) if and only if $E_{ij} \leq x_{i \to j}^R + x_{j \to i}^L$. We invite the reader to compare this result with that obtained in Exercise 16.2.

16.2.4   The distributional fixed point and $\zeta(2)$

Let us now consider random instances of the assignment problem. For the sake of simplicity, we assume that the edge costs E_{ij} are i.i.d. exponential random variables with mean N. We want to use the general density evolution technique of Section 14.6.2 to analyse the min-sum message-passing equations (16.8).

The sceptical reader might notice that the assignment problem does not fit into the general framework for density evolution, since the associated graph (the complete bipartite graph) is not locally tree-like. The use of density evolution can nevertheless be justified, through the following limiting procedure. Given a threshold energy E_{max}, remove from the factor graph all of the variables (ij), $i \in A$, $j \in B$, such that $E_{ij} > E_{\text{max}}$, and also remove the edges attached to them. Remembering that the typical edge costs are of order $\Theta(N)$, it is easy to check that the resulting graph is a sparse factor graph and therefore density evolution applies. On the other hand, one can prove that the error made in introducing a finite cut-off E_{max} is bounded uniformly in N by a quantity that vanishes as $E_{\text{max}} \to \infty$, which justifies the use of density evolution. In the following, we shall take the shortcut of writing density evolution equations for finite N without any cut-off and formally take the limit $N \to \infty$ for them.

Since the min-sum equations (16.8) involve minima, it is convenient to introduce the distribution function $\mathsf{A}_{N,t}(x) = \mathbb{P}\{x_{i \to j}^{(t)} \geq x\}$, where t indicates the iteration number, and $x^{(t)}$ refers to right-moving messages (from A to B) when t is even and to left moving messages when t is odd. Then, the density evolution equations read $\mathsf{A}_{N,t+1}(x) = [1 - \mathbb{E}\,\mathsf{A}_{N,t}(E - x)]^{N-1}$, where $\mathbb{E}$ denotes the expectation with respect to E (that is, an exponential random variable of mean N). In the cavity method, one seeks fixed points of this recursion. These are the distributions that solve

$$\mathsf{A}_N(x) = [1 - \mathbb{E}\,\mathsf{A}_N(E - x)]^{N-1}\,. \tag{16.11}$$

We now want to take the $N \to \infty$ limit. Assuming the fixed point $\mathsf{A}_N(x)$ has a (weak) limit $\mathsf{A}(x)$, we have

$$\mathbb{E}\,\mathsf{A}_N(E - x) = \frac{1}{N} \int_{-x}^{\infty} \mathsf{A}_N(y)\, e^{-(x+y)/N} \mathrm{d}y = \frac{1}{N} \int_{-x}^{\infty} \mathsf{A}(y)\, \mathrm{d}y + o(1/N)\,. \tag{16.12}$$

It follows from eqn (16.11) that the limit message distribution must satisfy the equation

$$\mathsf{A}(x) = \exp\left\{ -\int_{-x}^{\infty} \mathsf{A}(y)\, \mathrm{d}y \right\}\,. \tag{16.13}$$

This equation has the unique solution $\mathsf{A}(x) = 1/(1+e^x)$, corresponding to the density $\mathsf{a}(x) = \mathsf{A}'(x) = 1/[4\cosh^2(x)]$. It can be shown that density evolution does indeed converge to this fixed point.

Within the hypothesis of replica symmetry (seeSection 14.6.3), we can use the above fixed-point distribution to compute the asymptotic ground state energy (the minimum cost of the assignment). The most direct method is to use the inclusion principle: an edge (ij) is present in the optimal assignment if and only if $E_{ij} \leq x^{\mathrm{R}}_{i \to j} + x^{\mathrm{L}}_{j \to i}$. Therefore the conditional probability for (ij) to be in the optimal assignment, given its energy $E_{ij} = E$, is given by

$$q(E) = \int \mathbb{I}(x_1 + x_2 \geq E)\, \mathsf{a}(x_1)\mathsf{a}(x_2)\, \mathrm{d}x_1 \mathrm{d}x_2 = \frac{1 + (E-1)e^E}{(e^E - 1)^2} \qquad (16.14)$$

The expected cost E_* of the optimal assignment is equal to the number of edges, N^2, times the expectation of the edge cost, times the probability that the edge is in the optimal assignment. Asymptotically we have $E_* = N^2 \mathbb{E}\{Eq(E)\}$:

$$E_* = N^2 \int_0^\infty E\, e^{-E/N}\, q(E)\, \mathrm{d}E/N + o(N)$$

$$= N \int_0^\infty E \frac{1 + (E-1)e^E}{(e^E - 1)^2}\, \mathrm{d}E + o(N) = N\zeta(2) + o(N),$$

where

$$\zeta(2) \equiv \sum_{n=1}^\infty \frac{1}{n^2} = \frac{\pi^2}{6} \approx 1.64493406684823. \qquad (16.15)$$

Recall that this result holds when the edge weights are exponential random variables of mean N. If we reconsider the case of exponential random variables of mean 1, we get $E_* = \zeta(2) + o(1)$.

The reader can verify that the above derivation does not depend on the full distribution of the edge costs, but only on its behaviour near $E = 0$. More precisely, for any distribution of edge costs with a density $\rho(E)$ such that $\rho(0) = 1$, the cost of the optimal assignment converges to $\zeta(2)$.

Exercise 16.5 Suppose that the pdf of the costs $\rho(E)$ has support $\mathbb{R}_+$, and that $\rho(E) \simeq E^r$, for some $r > 0$, when $E \downarrow 0$.

(a) Show that, in order to have an optimal weight of order N, the edge costs must be rescaled by letting $E_{ij} = N^{r/r+1} \widetilde{E}_{ij}$, where the $\widetilde{E}_{ij}$ have density ρ (i.e. the typical costs must be of order $N^{r/r+1}$).

(b) Show that, within the replica-symmetric cavity method, the asymptotic $(N \to \infty)$ message distribution satisfies the following distributional equation:

$$\mathsf{A}(x) = \exp\left\{-\int_{-x}^\infty (x+y)^r\, \mathsf{A}(y)\, \mathrm{d}y\right\}. \qquad (16.16)$$

(c) Assume that the solution $\mathsf{A}(x)$ to eqn (16.16) is unique and that replica symmetry holds. Show that the expected ground state energy (in the problem with rescaled edge costs) is $E_* = N\epsilon_r + o(N)$, where $\epsilon_r \equiv -\int \mathsf{A}(x) \log(\mathsf{A}(x))\,\mathrm{d}x$. As a consequence, the optimal cost in the initial problem is $N^{r/(r+1)}\epsilon_r(1 + o(1))$.

(d) Equation (16.16) can be solved numerically by use of the population dynamics algorithm of Section 14.6.3. Write the corresponding program and show that the costs of the optimal matching for $r = 1, 2$ are $\epsilon_1 \approx 0.8086$ and $\epsilon_2 \approx 0.6382$.

16.2.5 Non-zero temperature and stability analysis

The reader may wonder whether the heuristic discussion in the previous subsections can be justified. While a rigorous justification would lead us too far, we want to discuss, still at a heuristic level, the consistency of the approach. In particular, we want to argue that BP provides good approximations to the marginals of the distribution (16.2), and that density evolution can be used to analyse its behaviour sur random instances.

Intuitively, two conditions should be satisfied for the approach to be valid: (i) the underlying factor graph should be locally tree-like; and (ii) the correlation between two variables n_{ij}, n_{kl} should decay rapidly with the distance between edges (ij) and (kl) on such a graph.

At first sight it appears that condition (i) is far from holding, since our factor graph is constructed from a complete bipartite graph. As mentioned in the previous subsection, the locally tree-like structure emerges if one notices that only edges with costs of order 1 are relevant (as above, we are assuming that the edge costs have been rescaled so that they are drawn with a probability density function $\hat{\rho}(E) = N^{-1}\exp(-E/N)$). In order to investigate this point further, we modify the model (16.2) by pruning from the original graph all edges with a cost larger than 2γ. In the large-β limit, this modification will become irrelevant since the Boltzmann weight (16.2) ensures that these 'costly' edges of the original problem are not occupied. In the modified problem, the degree of any vertex in the graph converges (as $N \to \infty$) to a Poisson random variable with mean 2γ. The costs of 'surviving' edges converge to i.i.d. uniform random variables in the interval $[0, 2\gamma]$.

For fixed β and γ, the asymptotic message distribution can be computed by the RS cavity method. The corresponding fixed-point equation reads

$$x \overset{\mathrm{d}}{=} -\frac{1}{\beta}\log\left[e^{-\beta\gamma} + \sum_{r=1}^{k}e^{-\beta(E_r - x_r)}\right], \tag{16.17}$$

where k is a Poisson random variable with mean 2γ, the E_r are i.i.d. and uniformly distributed on $[0, 2\gamma]$, and the x_r are i.i.d. with the same distribution as x. The fixed-point distribution can be estimated easily using the population dynamics algorithm of Section 14.6.3. Some results are shown in Fig. 16.2. For large β, γ, the density estimated by this algorithm converges rapidly to the analytical result for $\beta = \gamma = \infty$, namely $\mathsf{a}(x) = 1/[4\cosh^2(x/2)]$.

The distribution of messages can be used to compute the expected Bethe free entropy. Assuming that the messages entering into eqn (16.7) are independent, we get $\mathbb{E}\,\mathbb{F}(\underline{x}) = Nf^{\mathrm{RS}}(\beta, \gamma) + o(N)$, where

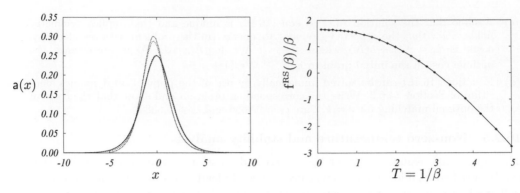

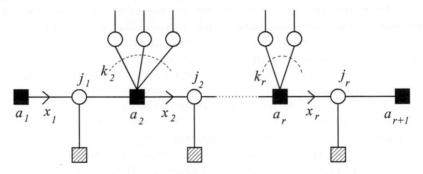

Fig. 16.2 *Left*: estimate of the probability distribution of the messages $x_{i \to j}$ obtained by population dynamics. Here we consider the modified ensemble in which costly edges (with $E_{ij} > 2\gamma$) have been removed. The three curves, from *top* to *bottom*, correspond to $(\beta = 1, \gamma = 5)$, $(\beta = 1, \gamma = 60)$, and $(\beta = 10, \gamma = 60)$. The last curve is indistinguishable from the analytical result for $(\beta = \infty, \gamma = \infty)$ namely $a(x) = 1/[4\cosh^2(x/2)]$, also shown. The curves for larger γ are indistinguishable from the curve for $\gamma = 60$ on this scale. The algorithm used a population of size 10^5, and the whole population wass updated 100 times. *Right*: free energy versus temperature $T = 1/\beta$, computed using eqn (16.18). The distribution of messages was obtained as above with $\gamma = 40$.

Fig. 16.3 Part of the factor graph used to compute the correlation between x_r and x_1.

$$f^{RS}(\beta, \gamma) = 2\beta\gamma + 2\mathbb{E} \log \left[e^{-\beta\gamma} + \sum_{j=1}^{k} e^{-\beta(E_j - x_j)} \right] - 2\gamma\mathbb{E} \log \left[1 + e^{-\beta(E_1 - x_1 - x_2)} \right].$$

(16.18)

Having a locally tree-like structure is only a necessary condition for BP to provide good approximations of the marginals. An additional condition is that the correlations of distinct variables n_{ij}, n_{kl} decay rapidly enough with the distance between the nodes (ij), (kl) in the factor graph. Let us discuss here one particular measure of these correlations, namely the spin glass susceptibility defined in Section 12.3.2. In the present case it can be written as

$$\chi_{\mathrm{SG}} \equiv \frac{1}{N} \sum_{e,f} \left(\langle n_e n_f \rangle - \langle n_e \rangle \langle n_f \rangle \right)^2 , \tag{16.19}$$

where the sum runs over all pairs of variable nodes $e = (i,j)$ and $f = (k,l)$ in the factor graph (or, equivalently, over all pairs of edges in the original bipartite graph with vertex sets A, B).

If correlations decay fast enough for the system to be stable with respect to small perturbations, χ_{SG} should remain bounded as $N \to \infty$. The intuitive explanation goes as follows. From the fluctuation–dissipation relation of Section 2.3, $\langle n_e n_f \rangle - \langle n_e \rangle \langle n_f \rangle$ is proportional to the change in $\langle n_f \rangle$ when the cost of edge e is perturbed. The sign of such a change will depend upon f, and therefore the resulting change in the expected matching size $\sum_f \langle n_f \rangle$ (namely $\sum_f (\langle n_e n_f \rangle - \langle n_e \rangle \langle n_f \rangle)$) can be either positive or negative. Assuming that this sum obeys a central limit theorem, its typical size is given by the square root of $\sum_f (\langle n_e n_f \rangle - \langle n_e \rangle \langle n_f \rangle)^2$. Averaging over the perturbed edge, we see that χ_{SG} measures the decay of correlations.

We shall thus estimate χ_{SG} using the same RS cavity assumption that we used in our computation of the expectations $\langle n_e \rangle$. If the resulting χ_{SG} is infinite, the assumption will be falsified. In the opposite case, although nothing definite can be said, the assumption will be said to be 'consistent', and the RS solution will be said to be 'locally stable' (since it is stable to small perturbations).

In order for the susceptibility to be finite, only pairs of variable nodes (e, f) whose distance r in the factor graph is bounded should give a significant contribution to the susceptibility. We can then compute

$$\chi_{\mathrm{SG}}^{(r)} \equiv \frac{1}{N} \sum_{e,f : d(e,f)=r} \left(\langle n_e n_f \rangle - \langle n_e \rangle \langle n_f \rangle \right)^2 \tag{16.20}$$

for fixed r in the $N \to \infty$ limit, and then sum the result over r. For any given r and large N, there is, with high probability, a unique path of length r joining e to f, all the others being of length $\Theta(\log N)$. On this path, we denote the variable nodes by $(j_1, j_2, \ldots, j_r)$ (with $e = j_1$, $f = j_r$), and denote the function nodes by $(a_2, \ldots, a_r)$ (see Fig. 16.3).

We consider a fixed point of the BP algorithm and denote by x_n the (log-likelihood) message passed from a_n to j_n. The BP fixed-point equations (16.6) allow us to compute x_r as a function of the message x_1 arriving at j_1, and of all the messages coming in on the path $\{a_2, \ldots, a_r\}$ from edges outside this path, which we denote by $\{y_{n,p}\}$:

$$x_2 = -\frac{1}{\beta} \log \left\{ e^{-\beta\gamma} + e^{-\beta(E_1 - x_1)} + \sum_{p=1}^{k_2} e^{-\beta(E_{2,p} - y_{2,p})} \right\},$$

$$\cdots$$

$$\cdots$$

$$x_r = -\frac{1}{\beta} \log \left\{ e^{-\beta\gamma} + e^{-\beta(E_r - x_r)} + \sum_{p=1}^{k_r} e^{-\beta(E_{r,p} - y_{r,p})} \right\}. \tag{16.21}$$

In a random instance, the k_n are i.i.d. Poisson random variables with mean 2γ, the variables E_n and $E_{n,p}$ are i.i.d. random variables uniform on $[0, 2\gamma]$, and the $y_{n,p}$ are i.i.d. random variables with the same distribution as the solution of eqn (16.17). We shall denote by $\mathbb{E}_{\text{out}}$ the expectation with respect to all of these variables outside the path. If we keep these variables fixed, a small change δx_1 in the message x_1 leads to a change $\delta x_r = (\partial x_r / \partial x_1) \delta x_1 = (\partial x_2 / \partial x_1)(\partial x_3 / \partial x_2) \dots (\partial x_r / \partial x_{r-1}) \delta x_1$ in x_r. We leave it as an exercise to the reader to show that the correlation function is given by

$$\langle n_e n_f \rangle - \langle n_e \rangle \langle n_f \rangle = C \frac{\partial x_r}{\partial x_1} = C \prod_{n=2}^{r} \frac{\partial x_n}{\partial x_{n-1}} , \qquad (16.22)$$

where the proportionality constant C is r-independent. Recalling that the expected number of variable nodes f such that $d(e, f) = r$ grows as $(2\gamma)^r$, and using eqn (16.20), we have $\mathbb{E} \chi_{\text{SG}}^{(r)} = C' \, e^{\lambda_r r}$, where

$$\lambda_r(\beta, \gamma) = \log(2\gamma) + \frac{1}{r} \log \left\{ \mathbb{E}_{\text{out}} \prod_{n=2}^{r} \left(\frac{\partial x_n}{\partial x_{n-1}} \right)^2 \right\} . \qquad (16.23)$$

Therefore, a sufficient condition for the expectation of χ_{SG} to be finite is that $\lambda_r(\beta, \gamma)$ is negative and bounded away from 0 for large enough r (when this happens, $\mathbb{E} \chi_{\text{SG}}^{(r)}$ decays exponentially with r).

The exponent $\lambda_r(\beta, \gamma)$ can be computed numerically through population dynamics: the population allows us to sample i.i.d. messages $y_{n,p}$ from the fixed-point message density, and the costs E_n and $E_{n,p}$ are sampled uniformly in $[0, 2\gamma]$. The expectation (16.23) can be estimated through a numerical average over large enough populations. Note that the quantity that we are taking the expectation of depends exponentially on r. As a consequence, its expectation becomes more difficult to compute as r grows.

In Fig. 16.4 we present some estimates of λ_r obtained by this approach. Since λ_r depends very weakly on r, we expect that λ_∞ can be safely estimated from these data. The data are compatible with the following scenario: $\lambda_\infty(\beta, \gamma)$ is negative at all finite inverse temperatures β and vanishes as $1/\beta$ as $\beta \to \infty$. This indicates that χ_{SG} is finite, so that the assumption of replica symmetry is consistent.

16.3 A polynomial message-passing algorithm

Remarkably, the min-sum message-passing algorithm introduced in Section 16.2.3 can be proved to return the minimum-cost assignment on any instance for which the minimum is unique. Let us state again the min-sum update equations of eqn (16.8), writing the iteration number explicitly:

$$x_{j \to i}^{\text{L}}(t+1) = \min_{k \in A \setminus i} \left(E_{kj} - x_{k \to j}^{\text{R}}(t) \right) , \quad x_{i \to j}^{\text{R}}(t) = \min_{k \in B \setminus j} \left(E_{ik} - x_{k \to i}^{\text{L}}(t) \right) . \quad (16.24)$$

Here, as before, A and B (with $|A| = |B| = N$) are the two sets of vertices to be matched, and we continue to denote generic vertices in A and B by i and j, respectively.

The algorithm runs as follows:

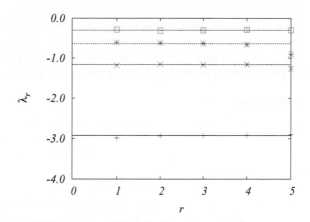

Fig. 16.4 The stability parameter λ_r defined in eqn (16.23), plotted versus r, for inverse temperatures $\beta = 10, 5, 2, 1$ (from *top* to *bottom*). The lines are guides to the eye. A negative asymptotic value of λ_r at large r shows that the spin glass susceptibility is finite. The data were obtained from a population dynamics simulation with a population of 10^6, for $\gamma = 20$.

MIN-SUM ASSIGNMENT (cost matrix E, iterations t_*)

1:	Set $x^{\mathrm{L}}_{j \to i}(0) = x^{\mathrm{R}}_{i \to j}(0) = 0$ for any $i \in A$, $j \in B$
2:	For all $t \in \{0, 1, \dots, t_*\}$:
3:	Compute the messages at time $t + 1$ using eqn (16.24)
4:	Set $\pi(i) = \arg\min_{j \in B} \left(E_{ij} - x^{\mathrm{L}}_{j \to i}(t_*) \right)$ for each $i \in A$;
5:	Output the permutation π;

This algorithm finds the optimal assignment if it is unique and if the number of iterations is large enough, as stated in the theorem below.

Theorem 16.1 *Let $W \equiv \max_{ij} |E_{ij}|$ and let ϵ be the gap between the cost E^* of the optimal assignment π^* and the next best cost: $\epsilon \equiv \min_{\pi(\neq \pi^*)}(E(\pi) - E_*)$, where $E(\pi) \equiv \sum_{i=1}^{N} E_{i\pi(i)}$. Then, for any $t_* \geq 2NW/\epsilon$, the min-sum algorithm above returns the optimal assignment π^*.*

The proof is given in Section 16.3.2, and is based on the notion of a computation tree, explained in the present context in Section 16.3.1.

For practical application of the algorithm to cases where one does not know the gap ϵ in advance, it is important to have a stopping criterion for the algorithm. This can be obtained by noticing that, after convergence, the messages become 'periodic-up-to-a-drift' functions of t. More precisely, there exists a period τ and a drift $C > 0$ such that for any $t > 2NW/\epsilon$, and any $i \in A$, $x^{\mathrm{L}}_{j \to i}(t + \tau) = x^{\mathrm{L}}_{j \to i}(t) + C$ if $j = \arg\min_{k \in B}(E_{ik} - x^{\mathrm{L}}_{k \to i}(t))$, and $x^{\mathrm{L}}_{j \to i}(t + \tau) = x^{\mathrm{L}}_{j \to i}(t) - C$ otherwise. If this happens, we write $\underline{x}^{\mathrm{L}}(t + \tau) = \underline{x}^{\mathrm{L}}(t) + \underline{C}$.

It turns out that (i) if for some time t_0, period τ, and constant $C > 0$, one has $\underline{x}^{\mathrm{L}}(t_0 + \tau) = \underline{x}^{\mathrm{L}}(t_0) + \underline{C}$, then $\underline{x}^{\mathrm{L}}(t + \tau) = \underline{x}^{\mathrm{L}}(t) + \underline{C}$ for any $t \geq t_0$; and (ii) under the same conditions, the permutation returned by the min-sum algorithm is independent of t_* for any $t_* \geq t_0$. We leave the proof of these statements as a (research-level) exercise for the reader. It can be seen immediately that they imply a clear stopping criterion: After any number of iterations t, check whether there exist $t_0 < t$ and $C > 0$, such that $\underline{x}^{\rm L}(t) = \underline{x}^{\rm L}(t_0) + \underline{C}$. If this is the case, halt the message-passing updates and return the resulting permutation as in step 4 of the above pseudocode.

16.3.1   The computation tree

As we saw in Fig. 16.1, an instance of the assignment problem corresponds to a weighted complete bipartite graph $\mathcal{G}_N$ over sets of vertices A and B, with $|A| = |B| = N$. The analysis of the min-sum algorithm described above uses the notion of a **computation tree** in a crucial way.

Given a vertex $i_0 \in A$ (the case $i_0 \in B$ is completely symmetric), the corresponding computation tree of depth t, denoted by $\mathbb{T}^t_{i_0}$, is a weighted rooted tree of depth t and degree N, which is constructed recursively as follows. First, introduce the root $\hat{i}_0$ that is in correspondence with $i_0 \in A$. For any $j \in B$, add a corresponding vertex $\hat{j}$ to $\mathbb{T}^t_{i_0}$ and connect it to $\hat{i}_0$. The weight of the resulting edge is taken to be $E_{\hat{i}_0, \hat{j}} \equiv E_{i_0, j}$. At every subsequent generation, if $\hat{i} \in \mathbb{T}^t_{i_0}$ corresponds to $i \in A$, and its direct ancestor is the $\hat{j}$ that corresponds to $j \in B$, add $N-1$ direct descendants of $\hat{i}$ to $\mathbb{T}^t_{i_0}$. Each of these descendants $\hat{k}$ corresponds to a distinct vertex $k \in B \setminus j$, and the associated weight is $E_{\hat{k}\hat{j}} = E_{kj}$. An alternative, more compact description of the computation tree $\mathbb{T}^t_{i_0}$ is to say that it is the tree of non-reversing paths on $\mathcal{G}_N$, rooted at i_0.

Imagine iterating the min-sum equations (16.24) on the computation tree $\mathbb{T}^t_{i_0}$ (starting from the initial condition $x_{\hat{i} \to \hat{j}}(0) = 0$). Since $\mathbb{T}^t_{i_0}$ has the same local structure as $\mathcal{G}_N$, for any $s \leq t$ the messages coming into the root $\hat{i}_0$ coincide with the messages along the corresponding edges in the original graph $\mathcal{G}_N$: $x_{\hat{j} \to \hat{i}_0}(s) = x_{j \to i_0}(s)$. As the min-sum algorithm correctly finds the ground state on trees (see Theorem 14.4), the following property holds.

Lemma 16.2 *We define, for any $i \in A$, $\pi^t(i) = \mathrm{argmin}_{j \in B}\left(E_{i,j} - x^L_{j \to i}(t)\right)$. Let $\hat{i}$ denote the root of the computation tree $\mathbb{T}^t_i$, and let $\hat{j}$ denote the direct descendant of $\hat{i}$ that corresponds to $\pi^t(i)$.*

*We define an **internal matching** of a tree to be a subset of the edges such that each non-leaf vertex has one adjacent edge in the subset. Then the edge $(\hat{i}, \hat{j})$ belongs to the internal matching with lowest cost in $\mathbb{T}^t_i$ (assuming that this is unique).*

Although it follows from general principles, it is instructive to rederive this result explicitly.

Exercise 16.6 Let r be an internal (non-leaf) vertex in the computation tree $\mathbb{T}^t_i$, distinct from the root. Denote the set of its direct descendants by S_r (hence $|S_r| = N - 1$), and denote the tree induced by r and all its descendants by T_r. We define a 'cavity internal matching' in T_r as a subset of the edges of T_r such that each vertex in T_r distinct from r has degree 1. Denote the cost of the optimal cavity internal matching when vertex r is not matched by A_r, and denote its cost when vertex r is matched by B_r.

Show that

$$A_r = \sum_{q \in S_r} B_q \ , \quad B_r = \min_{q \in S_r} \left[B_q + E_{rq} + \sum_{q' \in S_r \setminus \{q\}} A_{q'} \right] . \tag{16.25}$$

Show that $x_r = B_r - A_r$ satisfies the same equations as in eqn (16.24), and prove Lemma 16.2.

16.3.2 Proof of convergence of the min-sum algorithm

This subsection is a technical one, which gives a detailed proof of Theorem 16.1. It is convenient here to represent assignments as matchings, i.e. subsets of the edges such that each vertex is incident on exactly one edge in the subset. In particular, we denote the optimal matching on $\mathcal{G}$ by M^*. If π^* is the optimal assignment, then $\mathsf{M}^* \equiv \{(i, \pi_*(i)) : i \in A\}$. We denote by π the mapping returned by the min-sum algorithm. It is not necessarily injective, and therefore the subset of edges $\mathsf{M} = \{(i, \pi(i)) : i \in A\}$ is not necessarily a matching.

The proof is by contradiction. Assume that $\pi \neq \pi^*$. Then there exists at least one vertex in A, which we denote by i_0, such that $\pi(i_0) \neq \pi^*(i_0)$. Consider the depth-t computation tree of i_0, $\mathbb{T}^t_{i_0}$, denote its root by $\hat{i}_0$, and denote by $\widehat{\mathsf{M}}$ the optimal internal matching in this graph. Finally, denote by $\widehat{\mathsf{M}}^*$ the internal matching on $\mathbb{T}^t_{i_0}$ which is obtained by 'lifting' the optimal matching, M^*. Let $j = \pi(i_0) \in B$, and let $\hat{j} \in \mathbb{T}^t_{i_0}$ be the neighbour of $\hat{i}_0$ whose projection on $\mathcal{G}$ is j. By Lemma 16.2, $(\hat{i}_0, \hat{j}) \in \widehat{\mathsf{M}}$. On the other hand, since $\pi(i_0) \neq \pi_*(i_0)$, $(\hat{i}_0, \hat{j}) \notin \widehat{\mathsf{M}}^*$. The idea is to construct a new internal matching $\widehat{\mathsf{M}}'$ on $\mathbb{T}^t_{i_0}$, such that (i) $(\hat{i}_0, \hat{j}) \notin \widehat{\mathsf{M}}'$, and (ii) the cost of $\widehat{\mathsf{M}}'$ is strictly smaller than the cost of $\widehat{\mathsf{M}}$, thus leading to a contradiction.

Intuitively, the improved matching $\widehat{\mathsf{M}}'$ is constructed by modifying $\widehat{\mathsf{M}}$ in such a way as to 'get closer' to $\widehat{\mathsf{M}}^*$. In order to formalize this idea, consider the symmetric difference of $\widehat{\mathsf{M}}$ and $\widehat{\mathsf{M}}^*$, $\widehat{\mathsf{P}}' = \widehat{\mathsf{M}} \triangle \widehat{\mathsf{M}}^*$, i.e. the set of edges which are either in $\widehat{\mathsf{M}}$ or in $\widehat{\mathsf{M}}^*$ but not in both. The edge $(\hat{i}_0, \hat{j})$ belongs to $\widehat{\mathsf{P}}'$. We can therefore consider the connected component of $\widehat{\mathsf{P}}'$ that contains $(\hat{i}_0, \hat{j})$, which we shall call $\widehat{\mathsf{P}}$. A moment of thought reveals that $\widehat{\mathsf{P}}$ is a path on $\mathbb{T}^t_{i_0}$ with end-points on its leaves (see Fig. 16.5). Furthermore, its $2t$ edges alternate between edges in $\widehat{\mathsf{M}}$ and in $\widehat{\mathsf{M}}^*$. We can then define $\widehat{\mathsf{M}}' = \widehat{\mathsf{M}} \triangle \widehat{\mathsf{P}}$ (so that $\widehat{\mathsf{M}}'$ is obtained from $\widehat{\mathsf{M}}$ by deleting the edges in $\widehat{\mathsf{P}} \cap \widehat{\mathsf{M}}$ and adding those in $\widehat{\mathsf{P}} \cap \widehat{\mathsf{M}}^*$). We shall now show that, if t is large enough, the cost of $\widehat{\mathsf{M}}'$ is smaller than the cost of $\widehat{\mathsf{M}}$, in contradiction with the hypothesis.

Consider the projection of $\widehat{\mathsf{P}}$ onto the original complete bipartite graph $\mathcal{G}$; call it $\mathsf{P} \equiv \varphi(\widehat{\mathsf{P}})$ (see Fig. 16.5). This is a non-reversing path of length $2t$ on $\mathcal{G}$. As such, it can be decomposed into m simple cycles[2] $\{\mathsf{C}_1, \dots, \mathsf{C}_m\}$ (possibly with repetitions) and at most one even-length path Q, whose lengths add up to $2N$. Furthermore, the length of Q is at most $2N - 2$, and the length of each of the cycles at most $2N$. As a consequence, $m > t/N$.

---

[2]A **simple cycle** is a cycle that does not visit the same vertex twice.

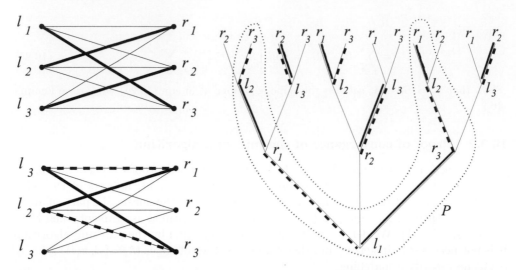

Fig. 16.5 *Top left*: an instance $\mathcal{G}$ of the assignment problem with $2N = 6$ vertices (the costs are not shown). The optimal assignment π^* is composed of the thick edges. *Right*: the computation tree $\mathbb{T}_{l_1}^2$. The matching π^* is 'lifted' to an internal matching in $\mathbb{T}_{l_1}^2$ composed of the thick edges. Notice that one edge in the original graph has many images in the computation tree. The dashed edges are those of the optimal internal matching in $\mathbb{T}_{l_1}^2$, and the alternating path P is enclosed by the dotted line. *Bottom left*: the projection of P on the original graph; here, it consists of a single cycle.

Consider now a particular cycle, say C_s. Its edges alternate between edges belonging to the optimal matching M* and edges not belonging to it. As we have assumed that the second-best matching in $\mathcal{G}$ has a cost of at least ϵ above the best one, the total cost of the edges in $\mathsf{C}_s \setminus \mathsf{M}_*$ is at least ϵ above the total cost of the edges in $\mathsf{C}_s \cap \mathsf{M}_*$.

As for the path Q, it again alternates between edges belonging to M* and edges outside of M*. We can order the edges in Q in such a way that the first edge is in M* and the last is not. By changing the last step, we can transform the path into an alternating cycle, to which the same analysis as above applies. This swapping changes the cost of edges not in Q by at most $2W$. Therefore the cost of the edges in Q $\setminus$ M* is at least the cost of the edges in Q $\cap$ M* plus $\epsilon - 2|W|$.

Let $E_{\mathbb{T}}(\widehat{\mathsf{M}})$ denote the cost of the matching $\widehat{\mathsf{M}}$ on $\mathbb{T}_{i_0}^t$. By summing the cost differences between the m cycles $\{\mathsf{C}_1, \ldots, \mathsf{C}_m\}$ and the path Q, we find that $E_{\mathbb{T}}(\widehat{\mathsf{M}}) \geq E_{\mathbb{T}}(\widehat{\mathsf{M}}') + (m+1)\epsilon - 2W$. Therefore, for $t > 2NW/\epsilon$, $E_{\mathbb{T}}(\widehat{\mathsf{M}}) > E_{\mathbb{T}}(\widehat{\mathsf{M}}')$, in contradiction with our hypothesis.$\square$

16.3.3 A few remarks

The alert reader might be puzzled by the following observation. Consider a random instance of the assignment problem with i.i.d. edge weights, for example exponentially distributed. In Section 16.2.4, we analysed the min-sum algorithm using density evolution and showed that the only fixed point is given by the x-message density

$a(x) = \frac{1}{4}\cosh^2(x/2)$. A little more work shows that, when initiated with $x = 0$ messages, the density evolution does indeed converge to such a fixed point.

On the other hand, for such a random instance the maximum weight W and the gap between the two best assignments are almost surely finite and non-vanishing, and so the hypotheses of Theorem 16.1 apply. The proof in the previous subsection implies that the min-sum messages diverge: the messages $x_{i \to \pi^*(i)}$ diverge to $+\infty$, while other ones diverge to $-\infty$ (indeed, min-sum messages are just the difference between the cost of the optimal matching on the computation tree and the cost of the optimal matching that does not include the root).

How can these two behaviours be compatible? The conundrum is that density evolution correctly predicts the distribution of messages as long as the number of iterations is kept bounded as $N \to \infty$. On the other hand, the typical number of iterations required to see the divergence of messages discussed above is NW/ϵ. If the edge weights are exponentially distributed random variables of mean N, the typical gap is $\epsilon = \Theta(1)$, while $W = \Theta(N \log N)$. Therefore the divergence sets in after $t_* = \Theta(N^2 \log N)$ iterations. The two analyses therefore describe completely distinct regimes.

16.4 Combinatorial results

It turns out that a direct combinatorial analysis allows one to prove several non-asymptotic results for ensembles of random assignment problems. Although the techniques are quite specific, the final results are so elegant that they deserve to be presented. As a by-product, they also provide rigorous proofs of some of our previous results, such as the asymptotic optimal cost $\zeta(2)$ found in eqn (16.15).

We shall consider here the case of edge weights given by i.i.d. exponential random variables with **rate** 1. Let us recall that an exponential random variable X with rate α has a density $\rho(x) = \alpha e^{-\alpha x}$ for $x \geq 0$, and therefore its expectation is $\mathbb{E}[X] = 1/\alpha$. Equivalently, the distribution of X is given by $\mathbb{P}\{X \geq x\} = e^{-\alpha x}$ for $x \geq 0$.

Exponential random variables have several special properties that make them particularly convenient in the present context. The most important one is that the minimum of two independent exponential random variables is again exponential. We shall use the following refined version of this statement.

Lemma 16.3 *Let $X_1, \ldots, X_n$ be n independent exponential random variables with respective rates $\alpha_1, \ldots, \alpha_n$. Then:*

1. *The random variable $X = \min\{X_1, \ldots, X_n\}$ is exponential with rate $\alpha \equiv \sum_{i=1}^n \alpha_i$.*
2. *The random variable $I = \arg\min_i X_i$ is independent of X, and has a distribution $\mathbb{P}\{I = i\} = \alpha_i/\alpha$.*

Proof First, note that the minimum of $\{X_1, \ldots, X_n\}$ is almost surely achieved by only one of the variables, and therefore the index I in item 2 of the lemma is well defined. An explicit computation yields, for any $x \geq 0$ and $i \in \{1, \ldots, n\}$,

$$\mathbb{P}\{I = i, X \geq x\} = \int_x^\infty \alpha_i \, e^{-\alpha_i z} \prod_{j(\neq i)} \mathbb{P}\{X_j \geq z\} \, \mathrm{d}z$$

$$= \int_x^\infty \alpha_i \, e^{-\alpha z} \, \mathrm{d}z = \frac{\alpha_i}{\alpha} \, e^{-\alpha x}. \qquad (16.26)$$

By summing over $i = 1, \ldots, n$, we get $\mathbb{P}\{X \geq x\} = e^{-\alpha x}$, which proves item 1.

By taking $x = 0$ in the above expression, we get $\mathbb{P}\{I = i\} = \alpha_i / \alpha$. Using these two results, eqn (16.26) can be rewritten as $\mathbb{P}\{I = i, X \geq x\} = \mathbb{P}\{I = i\}\,\mathbb{P}\{X \geq x\}$, which implies that X and I are independent. $\square$

16.4.1   The Coppersmith–Sorkin and Parisi formulae

The combinatorial approach is based on a recursion on the size of the problem. It is therefore natural to generalize the assignment problem by allowing for partial matching between two sets of unequal size as follows. Given a set of agents A and a set of jobs B (with $|A| = M$ and $|B| = N$), consider the complete bipartite graph $\mathcal{G}$ over the vertex sets A and B. A k-assignment between A and B is defined as a subset of k edges of G that has size k and is such that each vertex is adjacent to at most one edge. Given edge costs $\{E_{ij} : i \in A, \ j \in B\}$, the optimal k-assignment is the assignment that minimizes the sum of the costs over the edges in the matching. The assignment problem considered so far is recovered by setting $k = M = N$. Below we shall assume, without loss of generality, that $k \leq M \leq N$.

Theorem 16.4. (Coppersmith–Sorkin formula) *Assume the edge costs $\{E_{ij} : i \in A, \ j \in B\}$ to be i.i.d. exponential random variables of rate 1, with $|A| = M$ and $|B| = N$, and let $C_{k,M,N}$ denote the expected cost of the optimal k-assignment. Then,*

$$C_{k,M,N} = \sum_{i,j=0}^{k-1} \mathbb{I}(i+j < k) \, \frac{1}{(M-i)(N-j)}. \qquad (16.27)$$

This result, which we shall prove in the next two subsections, yields, as a special case, the expected cost C_N of the complete matching over a bipartite graph with $2N$ vertices.

Corollary 16.5. (Parisi formula) *Let $C_N \equiv C_{N,N,N}$ be the expected cost of the optimal complete matching between sets of vertices A, B with $|A| = |B| = N$, assuming that the edge weights are i.i.d. and exponentially distributed with a rate 1. Then*

$$C_N = \sum_{i=1}^{N} \frac{1}{i^2}. \qquad (16.28)$$

In particular, the expected cost of the optimal assignment when $N \to \infty$ is $\zeta(2)$.

Proof By Theorem 16.4, we have $C_N = \sum_{i,j=0}^{N-1} \mathbb{I}(i+j < N)\,(N-i)^{-1}(N-j)^{-1}$. By simplifying equal terms, the difference $C_{N+1} - C_N$ can be written as

$$\sum_{j=0}^{N} \frac{1}{(N+1)(N+1-j)} + \sum_{i=1}^{N} \frac{1}{(N+1-i)(N+1)} - \sum_{r=1}^{N} \frac{1}{(N+1-r)r}. \qquad (16.29)$$

By applying the identity $1/((N+1-r)r) = 1/((N+1)r) + 1/((N+1-r)(N+1))$, this implies $C_{N+1} - C_N = 1/(N+1)^2$, which establishes Parisi's formula. $\square$

The rest of this section will describe the proof of Theorem 16.4.

16.4.2 From k-assignment to $k+1$-assignment

The proof relies on two lemmas which relate the properties of the optimal k-assignment to those of the optimal $(k+1)$-assignment. Let us denote the optimal k-assignment (viewed as a subset of the edges of the complete bipartite graph) by M_k.

The first lemma applies to any realization of the edge costs, provided that no two subsets of the edges have equal cost (this happens with probability 1 within our random cost model).

Lemma 16.6. (nesting lemma) *Let $k < M \le N$, and assume that no linear combination of the edge costs $\{E_{ij} : i \in A,\, j \in B\}$ with coefficients in $\{+1, 0, -1\}$ vanishes. Then every vertex that belongs to M_k also belongs to M_{k+1}.*

The matching M_k consists of k edges which are incident on the vertices $i_1, \ldots, i_k$ in set A and on $j_1, \ldots, j_k$ in set B. We denote by $E^{(k)}$ the $k \times k$ matrix which is the restriction of E to the rows $i_1, \ldots, i_k$ and the columns $j_1, \ldots, j_k$. The nesting lemma ensures that $E^{(k+1)}$ is obtained from $E^{(k)}$ by adding one row (i_{k+1}) and one column (j_{k+1}). Therefore we have a sequence of nested matrices $E^{(1)} \subset E^{(2)} \cdots \subset E^{(M)} = E$ containing the sequence of optimal assignments $\mathsf{M}_1, \mathsf{M}_2, \ldots, \mathsf{M}_M$. **Proof** Colour all the edges in M_k red and all the edges in M_{k+1} blue, and denote by G_{k+} the bipartite graph induced by edges in $\mathsf{M}_k \cup \mathsf{M}_{k+1}$. Clearly, the maximum degree of G_{k+} is *at most* 2, and therefore its connected components are either cycles or paths.

We first show that no component of G_{k+} can be a cycle. Assume, in contradiction, that the edges $\{u_1, v_1, u_2, v_2, \ldots, u_p, v_p\} \subseteq \mathsf{G}_{k+}$ form such a cycle, with $\{u_1, \ldots, u_p\} \subseteq \mathsf{M}_k$ and $\{v_1, \ldots, v_p\} \subseteq \mathsf{M}_{k+1}$. Since M_k is the optimal k-assignment, $E_{u_1} + \cdots + E_{u_p} \le E_{v_1} + \cdots + E_{v_p}$ (in the opposite case, we could decrease its cost by replacing the edges $\{u_1, \ldots, u_p\}$ with $\{v_1, \ldots, v_p\}$, without changing its size). On the other hand, since M_{k+1} is the optimal $(k+1)$-assignment, the same argument implies $E_{u_1} + \cdots + E_{u_p} \ge E_{v_1} + \cdots + E_{v_p}$. These two inequalities imply $E_{u_1} + \cdots + E_{u_p} = E_{v_1} + \cdots + E_{v_p}$, which is impossible by the non-degeneracy hypothesis.

So far, we have proved that G_{k+} consists of a collection of disjoint simple paths, formed by alternating blue and red edges. Along such paths, all vertices have degree 2 except for the two end-points which have degree 1. Since each path alternates between red and blue edges, the difference between the numbers of red and blue edges can be either 0 or ± 1. We shall now show that there can exist only one such path, with one more blue edge than red edges.

We first notice that G_{k+} cannot contain even paths, with as many red as blue edges. This can be shown using the same argument that we used above in the case of cycles: either the cost of the blue edges along the path is lower than the cost of the red ones, which would imply that M_k is not optimal, or, vice versa, the cost of the red edges is lower, which would imply that M_{k+1} is not optimal.

We now exclude the existence of a path P of odd length with one more red edge than blue edges. Since the total number of blue edges is larger than the total number of red edges, there should exist at least one path P′ with odd length, with one more blue edge than red edges. We can then consider the double path P∪P′, which contains as many red as blue edges, and apply to it the same argument as for cycles and even paths.

We thus conclude that the symmetric difference of M_k and M_{k+1} is a path of odd length, with one end-point $i \in A$ and one end-point $j \in B$. These are the only vertices that are in M_{k+1} but not in M_k. Conversely, there is no vertex that is in M_k but not in M_{k+1}. $\square$

Lemma 16.7 *Let $\{u_i : i \in A\}$ and $\{v_j : j \in B\}$ be two collections of positive real numbers, and assume that the costs of the edges $\{E_{ij} : i \in A, j \in B\}$ are independent exponential random variables, the rate of E_{ij} being $u_i v_j$. Denote by $A_k = \{i_1, \ldots, i_k\} \subseteq A$ and $B_k = \{j_1, \ldots, j_k\} \subseteq B$ the sets of vertices appearing in the optimal k-assignment M_k. Let $I_{k+1} = A_{k+1} \setminus A_k$ and $J_{k+1} = B_{k+1} \setminus B_k$ be the extra vertices which are added in M_{k+1}. The conditional distribution of I_{k+1} and J_{k+1} is then $\mathbb{P}\{I_{k+1} = i, J_{k+1} = j | A_k, B_k\} = Q_{i,j}$, where*

$$Q_{ij} = \frac{u_i v_j}{\left(\sum_{i' \in A \setminus A_k} u_{i'}\right) \left(\sum_{j' \in B \setminus B_k} v_{j'}\right)}. \tag{16.30}$$

Proof Because of the nesting lemma, one of the following must be true: either the matching M_{k+1} contains edges (I_{k+1}, j_b) and (i_a, J_{k+1}) for some $i_a \in A_k$ and $j_b \in B_k$, or it contains the edge (I_{k+1}, J_{k+1}).

Let us fix i_a and j_b and condition on the first event

$$\mathcal{E}_1(i_a, j_b) \equiv \{A_k, B_k, (I_{k+1}, j_b), (i_a, J_{k+1}) \in M_{k+1}\}.$$

Then, necessarily, $E_{I_{k+1}, j_b} = \min\{E_{ij_b} : i \in A \setminus A_k\}$ (because, otherwise, we could decrease the cost of M_{k+1} by making a different choice for I_{k+1}). Analogously, $E_{i_a, J_{k+1}} = \min\{E_{i_a j} : j \in B \setminus B_k\}$. Since the two minima are taken over independent random variables, I_{k+1} and J_{k+1} are independent as well. Further, by Lemma 16.3,

$$\mathbb{P}\{I_{k+1} = i, J_{k+1} = j \mid \mathcal{E}_1(i_a, j_b)\} = \frac{u_i v_{j_b}}{\sum_{i' \in A \setminus A_k} u_{i'} v_{j_b}} \frac{u_{i_a} v_j}{\sum_{j' \in B \setminus B_k} u_{i_a} v_{j'}} = Q_{ij}.$$

If we instead condition on the second event

$$\mathcal{E}_2 \equiv \{A_k, B_k, (I_{k+1}, J_{k+1}) \in M_{k+1}\},$$

then $E_{I_{k+1},J_{k+1}} = \min\{E_{ij} : i \in A \setminus A_k \, j \in B \setminus B_k\}$ (because, otherwise, we could decrease the cost of M_{k+1}). By applying Lemma 16.3 again we get

$$\mathbb{P}\{I_{k+1} = i, J_{k+1} = j \mid \mathcal{E}_2\} = \frac{u_i v_j}{\sum_{i' \in A \setminus A_k, j' \in B \setminus B_k} u_{i'} v_{j'}} = Q_{ij}.$$

Since the resulting probability is Q_{ij} irrespective of the conditioning, it remains the same when we condition on the union of the events $\{\cup_{a,b}\mathcal{E}_1(i_a, j_b)\} \cup \mathcal{E}_2 = \{A_k, B_k\}$.
□

16.4.3 Proof of Theorem 16.4

In order to prove the Coppersmith–Sorkin (C–S) formula (16.27), we shall consider the difference $D_{k,M,N} \equiv C_{k,M,N} - C_{k-1,M,N-1}$, and establish in this subsection that

$$D_{k,M,N} = \frac{1}{N}\left(\frac{1}{M} + \frac{1}{M-1} + \cdots + \frac{1}{M-k+1}\right). \tag{16.31}$$

This immediately leads to the C–S formula, by recursion using as a base step the identity $C_{1,M,N-k+1} = 1/(M(N-k+1))$ (which follows from the fact that this is the minimum of $M(N-k+1)$ i.i.d. exponential random variables with rate 1).

Consider a random instance of the problem over vertex sets A and B with $|A| = M$ and $|B| = N$, whose edge costs $\{E_{ij} : i \in A, j \in B\}$ are i.i.d. exponential random variables with rate 1. Let X be the cost of its optimal k-assignment. Let Y be the cost of the optimal $(k-1)$-assignment for the new problem that is obtained by removing one fixed vertex, say the last one, from B. Our aim is to estimate the expectation value $D_{k,M,N} = \mathbb{E}(X - Y)$,

We shall use an intermediate problem with a cost matrix F of size $(M+1) \times N$, constructed as follows. The first M rows of F are identical to those of E. The elements of the matrix in its last line are N i.i.d. exponential random variables of rate λ, independent of E. We denote the cost of the edge $(M+1, N)$ by W, and call the event 'the optimal k-assignment in F uses the edge $(M+1, N)$' $\mathcal{E}$.

We claim that, as $\lambda \to 0$, $\mathbb{P}(\mathcal{E}) = \lambda\mathbb{E}[X - Y] + O(\lambda^2)$. Note first that, if $\mathcal{E}$ is true, then $W + Y < X$, and therefore

$$\mathbb{P}(\mathcal{E}) \le \mathbb{P}\{W + Y < X\} = \mathbb{E}\left[1 - e^{-\lambda(X-Y)}\right] = \lambda\mathbb{E}[X - Y] + O(\lambda^2). \tag{16.32}$$

Conversely, if $W < X - Y$, and all the edges from the vertex $M+1$ in A to $B \setminus \{N\}$ have a cost of at least X, then the optimal k-assignment in F uses the edge $(M+1, N)$. Therefore, using the independence of the edge costs,

$$\mathbb{P}(\mathcal{E}) \ge \mathbb{P}\{W < X - Y; \, E_{M+1,j} \ge X \text{ for } j \le N-1\}$$
$$= \mathbb{E}\left\{\mathbb{P}\{W < X - Y \mid X, Y\} \prod_{j=1}^{N-1} \mathbb{P}\{E_{M+1,j} \ge X \mid X\}\right\}$$
$$= \mathbb{E}\left\{\mathbb{P}\{W < X - Y \mid X, Y\} e^{-(N-1)\lambda X}\right\}$$
$$= \mathbb{E}\left\{\left(1 - e^{-\lambda(X-Y)}\right) e^{-(N-1)\lambda X}\right\} = \lambda\mathbb{E}[X - Y] + O(\lambda^2). \tag{16.33}$$

We now turn to the evaluation of $\mathbb{P}(\mathcal{E})$, and show that

$$\mathbb{P}(\mathcal{E}) = \frac{1}{N} \left[1 - \prod_{r=0}^{k-1} \frac{M-r}{M-r+\lambda} \right] . \tag{16.34}$$

Let us denote the $M+1$-th vertex in A by α. By Lemma 16.7, conditional on $\alpha \notin \mathsf{M}_{k-1}$, the probability that $\alpha \in \mathsf{M}_k$ is $\lambda/(M-(k-1)+\lambda)$. By recursion, this shows that the probability that $\alpha \notin \mathsf{M}_{k-1}$ is $\prod_{r=0}^{k-1}((M-r)/(M-r+\lambda))$. Since all of the N edges incident on α are statistically equivalent, we get eqn (16.34).

Expanding eqn (16.34) as $\lambda \to 0$, we get $\mathbb{P}(\mathcal{E}) = (\lambda/N) \sum_{r=0}^{k-1}(1/(M-r)) + O(\lambda^2)$. Since, as shown above, $\mathbb{E}[X-Y] = \lim_{\lambda\to 0}\mathbb{P}(\mathcal{E})/\lambda$, this proves eqn (16.31), which establishes the C–S formula.$\square$

16.5   An exercise: Multi-index assignment

In Section 16.2.4, we computed the asymptotic minimum cost for random instances of the assignment problem using the cavity method under the replica-symmetric (RS) assumption. The result, namely that the cost converges to $\zeta(2)$ for exponential edge weights with mean 1, was confirmed by the combinatorial analysis of Section 16.4. This suggests that the RS assumption is correct for this ensemble, an intuition that is further confirmed by the fact that the min-sum algorithm finds the optimal assignment.

Statistical physicists conjecture that there exists a broad class of random combinatorial problems which satisfy the RS assumption. On the other hand, many problems are thought not to satisfy it: the techniques developed for dealing with such problems will be presented in Chapter 19. It is important to have a feeling for the boundary separating RS from non-RS problems. This is a rather subtle point. Here we want to illustrate it by considering a generalization of random assignment: the multi-index random assignment (MIRA) problem. We shall consider studying the MIRA problem using the RS cavity method and detect the inconsistency of this approach. Since the present section is essentially an application of the methods developed above for the assignment problem, we shall skip all technical details. The reader may consider it as a long guided exercise.

An instance of the multi-index assignment problem consists of d sets $A_1, \dots, A_d$, of N vertices, and a cost E_i for every d-tuple $i = (i_1, \dots, i_d) \in A_1 \times \cdots \times A_d$. A 'hyperedge' i can be occupied ($n_i = 1$) or empty ($n_i = 0$). A matching is a set of hyperedges which are vertex disjoint; this means that $\sum_{i:\, a \in i} n_i \leq 1$ for each r and each $a \in A_r$. The cost of a matching is the sum of the costs of the hyperedges that it occupies. The problem is to find a perfect matching (i.e. a matching with N occupied hyperedges) with minimal total cost.

In order to define a random ensemble of instances of the multi-index assignment problem, we proceed as for the assignment problem, and assume that the edge costs E_a are i.i.d. exponential random variables with mean N^{d-1}. Thus the costs have a density

$$\rho(E) = N^{-d+1} e^{-E/N^{d-1}} \, \mathbb{I}(E \geq 0) . \tag{16.35}$$

The reader is invited to check that under this scaling of the edge costs, the typical optimal cost is extensive, i.e. $\Theta(N)$. The simple random assignment problem considered before corresponds to $d = 2$.

We introduce a probability distribution on matchings that naturally generalizes eqn (16.2):

$$\mu(\underline{n}) = \frac{1}{Z} \, e^{-\beta \sum_i n_i (E_i - 2\gamma)} \prod_{a \in F} \mathbb{I}\left(\sum_{i: a \in i} n_i \leq 1 \right), \qquad (16.36)$$

where $F \equiv A_1 \cup \cdots \cup A_d$. The associated factor graph has N^d variable nodes, each of degree d, corresponding to the original hyperedges, and dN factor nodes, each of degree N, corresponding to the vertices in F. As usual $i, j, \cdots \in V$ denote the variable nodes in the factor graph and $a, b, \cdots \in F$ denote the function nodes enforcing the hard constraints.

Using a parameterization analogous to that for the assignment problem, we find that the BP equations for this model take the form

$$h_{i \to a} = \sum_{b \in \partial i \setminus a} x_{b \to i},$$

$$x_{a \to i} = -\frac{1}{\beta} \log \left\{ e^{-\beta\gamma} + \sum_{j \in \partial a \setminus i} e^{-\beta(E_j - h_{j \to a})} \right\}. \qquad (16.37)$$

In the large-β, γ limit they become

$$h_{i \to a} = \sum_{b \in \partial i \setminus a} x_{b \to i} \quad , \quad x_{a \to i} = \min_{j \in \partial a \setminus i} (E_j - h_{j \to a}). \qquad (16.38)$$

Finally, the Bethe free entropy can be written in terms of x-messages, yielding

$$\mathbb{F}[\underline{x}] = Nd\beta\gamma + \sum_{a \in F} \log \left\{ e^{-\beta\gamma} + \sum_{i \in \partial a} e^{-\beta(E_i - \sum_{b \in \partial i \setminus a} x_{b \to i})} \right\}$$

$$- (d-1) \sum_{i \in V} \log \left\{ 1 + e^{-\beta(E_i - \sum_{a \in \partial i} x_{j \to a})} \right\}. \qquad (16.39)$$

Using the RS cavity method, we obtain the following equation for the distribution of x-messages in the $N \to \infty$ limit:

$$\mathsf{A}(x) = \exp \left\{ - \int \left(x + \sum_{j=1}^{d-1} t_j \right) \mathbb{I}\left(x + \sum_{j=1}^{d-1} t_j \geq 0 \right) \prod_{j=1}^{d-1} d\mathsf{A}(t_j) \right\}. \qquad (16.40)$$

This reduces to eqn (16.13) in the case of simple assignment. Under the RS assumption, the cost of the optimal assignment is $E_* = N\epsilon_* + o(N)$, where

$$\epsilon_* = \frac{1}{2} \int \left(\sum_{j=1}^{d} x_j \right)^2 \mathbb{I}\left(\sum_j x_j > 0 \right) \prod_{j=1}^{d} d\mathsf{A}(x_j). \qquad (16.41)$$

These equations can be solved numerically to high precision and allow one to derive several consequences of the RS assumption. However, the resulting predictions (in particular, the cost of the optimal assignment) are *wrong* for $d \geq 3$. There are two observations that show that the RS assumption is inconsistent:

1. Using the Bethe free-entropy expression (16.39), we can compute the asymptotic free-energy density as $\text{f}^{\text{RS}}(\beta) = \lim_{N \to \infty} \mathbb{F}/N$, for a finite $\beta = 1/T$. The resulting expression can be estimated numerically via population dynamics, for instance for $d = 3$. It turns out that the predicted entropy density $s(T) = \text{f}^{\text{RS}} - \beta(\text{df}^{\text{RS}}/\text{d}\beta)$ becomes negative for $T < T_{\text{cr}} \approx 2.43$. This implies that the prediction is incorrect: we are dealing with a statistical-mechanics model with a finite state space, and thus the entropy must be non-negative.

2. A local stability analysis can be performed analogously to what was done in Section 16.2.5. It turns out that, for $d = 3$, the stability coefficient λ_{∞} (see eqn (16.23)), becomes positive for $T \lesssim 1.6$, indicating an instability of the putative RS solution to small perturbations.

The same findings are generic for $d \geq 3$. A more satisfactory set of predictions for such problems can be developed using the RSB cavity method, which will be treated in Chapter 19.

Notes

Rigorous upper bounds on the cost of the optimal random assignment date back to Walkup (1979) and Karp (1987). The $\zeta(2)$ result for the cost was first obtained by Mézard and Parisi (1985) using the replica method. The cavity method solution was then found by Mézard and Parisi (1986) and Krauth and Mézard (1989), but the presentation in Section 16.2 is closer to that of Martin *et al.* (2005). The last of these papers deals with the multi-index assignment problem and contains answers to the exercise in Section 16.5, as well as a proper solution of this problem using the RSB cavity method.

The first rigorous proof of the $\zeta(2)$ result was derived by Aldous (2001), using a method which can be regarded as a rigorous version of the cavity method. An essential step in developing this proof was the establishment of the existence of the limit, and its description as a minimum-cost matching on an infinite tree (Aldous, 1992). An extended review of the 'objective method' on which this convergence result is based can be found in Aldous and Steele (2003). A survey of the recursive distributional equations such as eqn (16.17) that occur in the replica-symmetric cavity method can be found in Aldous and Bandyopadhyay (2005).

On the algorithmic side, the assignment problem has been a very well-studied problem for many years (Papadimitriou and Steiglitz, 1998), and there exist efficient algorithms based on ideas of network flows. The first BP algorithm was found by Bayati *et al.* (2005); it was then simplified by Bayati *et al.* (2006) into the $O(N^3)$ algorithm presented in Section 16.3. This paper also shows that the BP algorithm is basically equivalent to Bertsekas's auction algorithm (Bertsekas, 1988). The periodic-up-to-a-shift stopping criterion is due to Sportiello (2004), and the explanation for the existence of diverging time scales for the onset of the drift can be found in Grosso (2004).

Combinatorial studies of random assignments were initiated by Parisi's conjecture (Parisi, 1998). This was generalized to the Coppersmith–Sorkin conjecture by Coppersmith and Sorkin (1999). The same paper also provides a nice survey of algorithmic bounds. In 2003, these conjectures were turned into theorems by two groups independently (Nair *et al.*, 2003, 2006; Linusson and Wästlund, 2004). The simple derivation presented here is from Wästlund (2008).

17
Ising models on random graphs

In this chapter, we shall consider two statistical-physics models for magnetic systems: the Ising ferromagnet and the Ising spin glass. While for physical applications one is mainly interested in three-dimensional Euclidean lattices, we shall study models on sparse random graphs. It turns out that this is a much simpler problem, and one can hope to obtain an exact solution in the thermodynamic limit. It is expected that, for a large family of 'mean-field' graphs, the qualitative behaviour of Ising models will be similar to that for sparse random graphs. For instance, ferromagnetic Ising models on any lattice of dimension $d > 4$ share many of the features of models on random graphs. A good understanding of which graphs are mean-field is, however, still lacking.

Here we shall develop the replica-symmetric (RS) cavity approach, and study its stability. In the ferromagnetic case, this allows us to compute the asymptotic free energy per spin and the local magnetizations, at all temperatures. In the spin glass case, the 'RS solution' is correct only at high temperature, in the 'paramagnetic' phase. We shall see that the RS approach is inconsistent in the low-temperature 'spin glass' phase, and identify the critical transition temperature separating these two phases. Despite its inconsistency in the spin glass phase, the RS approach provides a very good approximation for many quantities of interest.

Our study will be based mainly on non-rigorous physical methods. The results on ferromagnets and on the high-temperature phase of spin glasses, however, can be turned into rigorous statements, and we shall briefly outline the ideas involved in the rigorous approach.

The basic notation and the BP (or cavity) formalism for Ising models are set in Section 17.1. Section 17.2 specializes to the case of random graph ensembles, and introduces the corresponding distributional equations. Finally, Sections 17.3 and 17.4 deal with the analysis of these equations and derive the phase diagram in the ferromagnetic and spin glass cases, respectively.

17.1 The BP equations for Ising spins

Given a graph $G = (V, E)$, with $|V| = N$, we consider a model for N Ising spins $\sigma_i \in \{\pm 1\}$, $i \in V$, with an energy function

$$E(\underline{\sigma}) = - \sum_{(i,j) \in E} J_{ij} \sigma_i \sigma_j. \tag{17.1}$$

The coupling constants $J_{ij} = J_{ji}$ are associated with edges of G and indicate the strength of the interaction between spins connected by an edge. The graph G and the set of coupling constants J define a 'sample'.

Given a sample, the Boltzmann distribution is

$$\mu_{G,J}(\underline{\sigma}) = \frac{1}{Z_{G,J}} e^{-\beta E(\underline{\sigma})} = \frac{1}{Z_{G,J}} \prod_{(i,j) \in E} e^{\beta J_{ij} \sigma_i \sigma_j} . \tag{17.2}$$

As in the general model (14.13), this distribution factorizes according to a factor graph that has a variable node associated with each vertex $i \in V$, and a function node associated with each edge $(i,j) \in E$. It is straightforward to apply BP to such a model.

Since the model (17.2) is pairwise (every function node has degree 2), the BP equations can be simplified following the strategy of Section 14.2.5. The message $\widehat{\nu}_{(ij) \to j}$ from function node (ij) to variable node j is related to the message $\nu_{i \to (ij)}$ through

$$\widehat{\nu}^{(t)}_{(ij) \to j}(\sigma_j) \cong \sum_{\sigma_i} e^{\beta J_{ij} \sigma_i \sigma_j} \nu^{(t)}_{i \to (ij)}(\sigma_i) . \tag{17.3}$$

We can choose to work with only one type of message on each directed edge $i \to j$ of the original graph, say the variable-to-factor message. With a slight misuse of notation, we shall write $\nu_{i \to j}(\sigma_i) \equiv \nu_{i \to (ij)}(\sigma_i)$. The BP update equations now read (see eqn(14.31))

$$\nu^{(t+1)}_{i \to j}(\sigma_i) \cong \prod_{k \in \partial i \backslash j} \sum_{\sigma_k} e^{\beta J_{ki} \sigma_k \sigma_i} \nu^{(t)}_{k \to i}(\sigma_k) . \tag{17.4}$$

Since spins are binary variables, one can parameterize messages by their log-likelihood ratio $h^{(t)}_{i \to j}$, defined through the relation

$$\nu^{(t)}_{i \to j}(\sigma_i) \cong \exp \left\{ \beta \, h^{(t)}_{i \to j} \, \sigma_i \right\} . \tag{17.5}$$

We follow here the physics convention of rescaling the log-likelihood ratio by a factor $1/(2\beta)$. The origin of this convention lies in the idea of interpreting the distribution (17.5) as an 'effective' Boltzmann distribution for the spin σ_i, with energy function $-h^{(t)}_{i \to j} \sigma_i$. In statistical-physics jargon, $h^{(t)}_{i \to j}$ is a local magnetic field.

Two messages $h^{(t)}_{i \to j}$, $h^{(t)}_{j \to i} \in \mathbb{R}$ are exchanged along each edge (i,j). In this parameterization, the BP update equations (17.4) become

$$h^{(t+1)}_{i \to j} = \sum_{k \in \partial i \backslash j} f(J_{ki}, h^{(t)}_{k \to i}) , \tag{17.6}$$

where the function f has already been encountered in Section 14.1:

$$f(J, h) = \frac{1}{\beta} \, \mathrm{atanh} \left[\tanh(\beta J) \tanh(\beta h) \right] . \tag{17.7}$$

The local marginals and the free entropy can be estimated in terms of these messages. We shall assume here that $\underline{h} \equiv \{h_{i \to j}\}$ is a set of messages which solve the fixed-point equations. The marginal of spin σ_i is then estimated as

$$\nu_i(\sigma_i) \cong e^{\beta H_i \sigma_i}\,, \qquad H_i = \sum_{k \in \partial i} f(J_{ik}, h_{k \to i})\,. \tag{17.8}$$

The free entropy $\log Z_{G,J}$ is estimated using the general formulae (14.27) and (14.28). The result can be expressed in terms of the cavity fields $\{h_{i \to j}\}$. Using the shorthand $\theta_{ij} \equiv \tanh \beta J_{ij}$, one gets

$$\mathbb{F}[\underline{h}] = \sum_{(ij) \in E} \log \cosh \beta J_{ij} - \sum_{(ij) \in E} \log \left\{ 1 + \theta_{ij} \tanh \beta h_{i \to j} \tanh \beta h_{j \to i} \right\}$$
$$+ \sum_{i \in V} \log \left\{ \prod_{j \in \partial i} (1 + \theta_{ij} \tanh \beta h_{j \to i}) + \prod_{j \in \partial i} (1 - \theta_{ij} \tanh \beta h_{j \to i}) \right\}. \tag{17.9}$$

This expression is obtained by using the parameterization (17.5) in the general expressions (14.27) and (14.28), as outlined in the exercise below. As an alternative, one can use the expression (14.32) for pairwise models.

Exercise 17.1 In order to derive eqn (17.9), it is convenient to change the normalization of the compatibility functions by letting $\psi_{ij}(\sigma_i, \sigma_j) = 1 + \theta_{ij} \sigma_i \sigma_j$. This produces an overall change in the the energy that is taken into account by the first term in eqn (17.9). To get the other terms:

(a) Show that $\widehat{\nu}_{(ij) \to j}(\sigma_j) = \frac{1}{2}(1 + \sigma_j \theta_{ij} \tanh \beta h_{i \to j})$. Note that $\mathbb{F}(\underline{\nu})$, (see eqn (14.27)), is left unchanged by a multiplicative rescaling of the messages. As a consequence, we can use $\widehat{\nu}'_{(ij) \to j}(\sigma_i) = 1 + \sigma_j \theta_{ij} \tanh \beta h_{i \to j}$.

(b) Show that, with this choice, the term $\mathbb{F}_i(\underline{\nu})$ in eqn (14.27) is equal to the last term in eqn (17.9).

(c) Note that $\sum_{\sigma_i} \sigma_i \nu_{i \to (ij)}(\sigma_i) = \tanh \beta h_{i \to j}$.

(d) Show that this implies (for $a \equiv (ij)$)

$$\mathbb{F}_a(\underline{\nu}) = \mathbb{F}_{ai}(\underline{\nu}) = \mathbb{F}_{aj}(\underline{\nu}) = \log\{1 + \theta_{ij} \tanh \beta h_{i \to j} \tanh \beta h_{j \to i}\}. \tag{17.10}$$

Exercise 17.2 Show that the fixed points of the BP update equations (17.6) are stationary points of the free-energy functional (17.9).
[Hint: Differentiate the right-hand side of eqn (17.9) with respect to $\tanh \beta h_{i \to j}$.]

Exercise 17.3 Show that the Bethe approximation to the internal energy, (see eqn (14.21)), is given by

$$U = -\sum_{(ij)} J_{ij} \frac{\theta_{ij} + \tanh \beta h_{i \to j} \tanh \beta h_{j \to i}}{1 + \theta_{ij} \tanh \beta h_{i \to j} \tanh \beta h_{j \to i}} . \tag{17.11}$$

Check that $U = -\mathrm{d}\mathbb{F}/\mathrm{d}\beta$ as it should, where $\mathbb{F}$ is given in eqn. (17.9).
[Hint: Use the result of the previous exercise]

17.2 RS cavity analysis

We shall now specialize our analysis to the case of sparse random graphs. More precisely, we assume G to be a uniformly random graph with degree profile $\{\Lambda_k\}$ (i.e., for each $k \geq 0$, the number of vertices of degree k is $N\Lambda_k$). The associated factor graph is a random factor graph from the ensemble $\mathbb{D}_N(\Lambda, P)$, where $P(x) = x^2$ and $\Lambda(x) = \sum_{k \geq 0} \Lambda_k x^k$ (see Section 9.2).

As for the couplings J_{ij}, we shall focus on two significant examples:

(i) *Ferromagnetic* models, with $J_{ij} = +1$ for any edge (i, j).

(ii) *Spin glass* models. In this case the couplings J_{ij} are i.i.d. random variables with $J_{ij} \in \{+1, -1\}$ uniformly at random.

The general case of i.i.d. random couplings J_{ij} can be treated within the same framework.

Let us emphasize that the graph G and the couplings $\{J_{ij}\}$ are *quenched* random variables. We are interested in the properties of the measure (17.2) for a typical random realization of G, J. We shall pursue this goal by analysing distributions of BP messages (cavity fields).

17.2.1 Fixed-point equations

We choose an edge (i, j) uniformly at random in the graph G. The cavity field $h_{i \to j}$ is a random variable (both because of the random choice of (i, j) and because of the randomness in the model). Within the assumptions of the RS cavity method, in the large-N limit, the distribution of $h = h_{i \to j}$ satisfies the distributional equation

$$h \stackrel{\mathrm{d}}{=} \sum_{i=1}^{K-1} f(J_i, h_i) , \tag{17.12}$$

where K is distributed according to the edge-perspective degree distribution: the probability that $K = k$ is $\lambda_k = k\Lambda_k / (\sum_{p=1}^{\infty} p\Lambda_p)$. The fields $h_1, \ldots, h_K$ are independent copies of the random variable h, and the couplings $J_1, \ldots J_K$ are i.i.d. random variables distributed as the couplings in the model.

In the physics literature, the distributional equation (17.12) is written formally in terms of the density $\mathsf{a}(\cdot)$ of the random variable h. Writing $\mathbb{E}_J$ for the expectation over the i.i.d. coupling constants $J_1, J_2, \ldots$, and enforcing the cavity equation through a Dirac delta, we have

$$a(h) = \sum_{k=1}^{\infty} \lambda_k \, \mathbb{E}_J \int \delta \left(h - \sum_{i=1}^{k} f(J_i, h_i) \right) \prod_{r=1}^{k} a(h_r) \, dh_r \, . \qquad (17.13)$$

Let us emphasize that, in writing such an equation, physicists do not assume that a genuine density $a(\cdot)$ does exist. This and similar equations should be interpreted as a proxy for expressions such as eqn (17.12).

Assuming that the distribution of h that solves eqn (17.12) has been found, the RS cavity method predicts the asymptotic free-entropy density $f^{RS} \equiv \lim_{N \to \infty} \mathbb{F}/N$ as

$$f^{RS} = -\frac{1}{2} \overline{\Lambda} \log \cosh \beta + \frac{1}{2} \overline{\Lambda} \, \mathbb{E}_{J,h} \log \left\{ 1 + \tanh \beta J \tanh \beta h_1 \tanh \beta h_2 \right\} \qquad (17.14)$$

$$+ \mathbb{E}_k \mathbb{E}_{J,h} \log \left\{ \prod_{i=1}^{k} (1 + \tanh \beta J_i \tanh \beta h_i) + \prod_{i=1}^{k} (1 - \tanh \beta J_i \tanh \beta h_i) \right\}.$$

Here k is distributed according to $\{\Lambda_k\}$, the h_i are i.i.d. random variables distributed as h, and the J_i are i.i.d. couplings (identically equal to $+1$ for ferromagnets, and uniform in $\{+1, -1\}$ for spin glasses). Finally, $\overline{\Lambda} = \sum_k k\Lambda_k = \Lambda'(1)$ denotes the average degree.

17.2.2 The paramagnetic solution

The RS distributional equation (17.12) always admits the solution '$h = 0$', meaning that the random variable h is equal to 0 with probability 1. The corresponding distribution is a Dirac delta on $h = 0$: $a = \delta_0$.

This is usually referred to as the **paramagnetic solution**. Using eqn (17.14) we obtain the corresponding prediction for the free-entropy density

$$f_{para}^{RS}(\beta) = \log 2 + \frac{1}{2} \overline{\Lambda} \log \cosh \beta \, . \qquad (17.15)$$

Exercise 17.4 In the context of this paramagnetic solution, derive expressions for the internal-energy density $(-(\overline{\Lambda}/2)\mathbb{E}_J J \tanh(\beta J))$ and the entropy density $(\log 2 + (\overline{\Lambda}/2)\mathbb{E}_J [\log \cosh(\beta J) - \beta J \tanh(\beta J)])$.

In order to interpret this solution, recall that $a(\cdot)$ is the asymptotic distribution of the cavity fields (BP messages) $h_{i \to j}$. The paramagnetic solution indicates that they vanish (apart, possibly, from a sublinear number of edges). Recalling the expression for local marginals in terms of cavity fields (see eqn (17.8)), this implies $\nu_i(\sigma_i = +1) = \nu_i(\sigma_i = -1) = 1/2$. One can similarly derive the joint distribution of two spins σ_i, σ_j connected by a single edge with a coupling J_{ij}. A straightforward calculation yields

$$\nu_{ij}(\sigma_i, \sigma_j) = \frac{1}{4 \cosh \beta} \exp \left\{ \beta J_{ij} \sigma_i \sigma_j \right\}. \qquad (17.16)$$

This is the same distribution as for two isolated spins interacting via the coupling J_{ij}. In the paramagnetic solution, the effect of the rest of the graph on local marginals is negligible.

The fact that $h = 0$ is a solution of the RS distributional equation (17.12) does not imply that the paramagnetic predictions are correct. We can distinguish three possibilities:

1. The paramagnetic predictions are correct.
2. The hypotheses of the RS cavity method do hold, but the distributional equation (17.12) admits more than one solution. In this case it is possible that behaviour of the model is correctly described by one of these solutions, although not the paramagnetic one.
3. The hypotheses of the RS method are incorrect.

It turns out that, in the high-temperature phase, the paramagnetic solution is correct for both the ferromagnetic and the spin glass model. At low temperature (the critical temperature being different in the two cases), new solutions to the RS equations appear. Wheras for the ferromagnetic model correct asymptotic predictions are obtained by selecting the appropriate RS solution, this is not the case for spin glasses.

17.3 Ferromagnetic model

For the ferromagnetic Ising model ($J_{ij} = +1$ identically), the paramagnetic solution correctly describes the asymptotic behaviour for $\beta \le \beta_{\rm c}$. The critical temperature depends on the graph ensemble only through the average degree *from the edge perspective* $\overline{\lambda}$, and is the unique solution of the equation

$$\overline{\lambda} \tanh \beta_{\rm c} = 1 \, . \qquad (17.17)$$

The edge-perspective average degree $\overline{\lambda}$ is defined in terms of the degree distribution as

$$\overline{\lambda} = \sum_k \lambda_k \, (k - 1) = \frac{\sum_k \Lambda_k k(k - 1)}{\sum_k \Lambda_k k} \, , \qquad (17.18)$$

or, more compactly, in terms of the degree generating function as $\overline{\lambda} = \lambda'(1)$. In words, we choose an edge (i, j) of G uniformly at random, and select one of its end-points, say i, also at random. Then, $\overline{\lambda}$ is the expected number of edges incident on i distinct from (i, j).

For $\beta > \beta_{\rm c}$, the RS distributional equation (17.12) admits more than one solution. While $h = 0$ is still a solution, the correct thermodynamic behaviour is obtained by selecting a different solution, the 'ferromagnetic' solution.

17.3.1 Local stability

When confronted with a distributional equation such as eqn (17.12), the simplest thing we can try to do is to take expectations. Recalling that k has a distribution λ_k and writing $\theta = \tanh \beta$, we obtain

$$\mathbb{E}\{h\} = \overline{\lambda}\,\mathbb{E}\{f(+1, h)\} = \overline{\lambda}\,\mathbb{E}\{\theta\,h + O(h^3)\} = \overline{\lambda}\theta\,\mathbb{E}\{h\} + \mathbb{E}\{O(h^3)\}\,. \quad (17.19)$$

If we neglect terms of order $\mathbb{E}\{O(h^3)\}$ (for instance, assuming that $\mathbb{E}\{h^3\} = O(\mathbb{E}\{h\}^3)$), this yields a linear iteration for the expectation of h. For $\overline{\lambda}\theta > 1$, this iteration is unstable, indicating that $h = 0$ is an unstable fixed point and new fixed distributions appear.

Indeed, a little more work shows that, for $\overline{\lambda}\theta < 1$, $h = 0$ is the unique fixed point. It is enough to take the expectation of $|h|$ and use the triangle inequality to obtain

$$\mathbb{E}|h| \le \overline{\lambda}\,\mathbb{E}|f(+1, h)|\,. \quad (17.20)$$

A little calculus shows that $f(1, x) \le \theta x$ for $x > 0$, whence $\mathbb{E}|h| \le \overline{\lambda}\theta\mathbb{E}|h|$ and therefore $h = 0$ identically.

17.3.2   Ferromagnetic susceptibility

There is a second, more physical, interpretation of the above calculation. Consider the ferromagnetic susceptibility,

$$\chi = \frac{1}{N}\sum_{i,j\in V}\left(\langle\sigma_i\sigma_j\rangle - \langle\sigma_i\rangle\langle\sigma_j\rangle\right)\,. \quad (17.21)$$

We expect this quantity to be bounded as $N \to \infty$. This corresponds to the expectation that the total magnetization $\sum_i \sigma_i$ behaves as in the central limit theorem. The idea is to assume that the paramagnetic solution is the correct one, and to check whether χ is indeed bounded.

We have already seen that, in the paramagnetic phase, $\langle\sigma_i\rangle$ vanishes. Let us now consider two randomly chosen vertices i, j at a given distance r, and let us compute the expectation $\langle\sigma_i\sigma_j\rangle$. With high probability, there is a single path with r edges joining i to j, and any finite neighbourhood of this path is a tree. Let us rename the two vertices 0 and r, and denote by σ_n, $n \in \{0, \dots, r\}$, the spins along the path joining them. Within the RS cavity approach, the marginal distribution of these spins is, (see eqn (14.18)):

$$\mu_G(\sigma_0, \dots, \sigma_r) \cong \exp\left\{\beta\sum_{p=0}^{r-1}\sigma_p\sigma_{p+1} + \beta\sum_{p=0}^{r}g_p\sigma_p\right\}\,. \quad (17.22)$$

The field g_p is the effect of the rest of the graph on the distribution of the spin σ_p. Using eqn (14.18), one gets

$$g_p = \sum_{j\in\partial p\backslash\text{path}}f(+1, h_{j\to p})\,, \quad (17.23)$$

where the sum extends over the set ∂p of neighbours of σ_p in the factor graph that are not on the path. Since, in the paramagnetic solution, the $h_{j\to p}$'s vanish, all of the g_p vanish as well. This implies $\langle\sigma_0\,\sigma_r\rangle = (\tanh\beta)^r$.

The susceptibility is then given by

$$\chi = \frac{1}{N} \sum_{i,j \in V} \langle \sigma_i \sigma_j \rangle = \sum_{r=0}^{\infty} \mathcal{N}(r) \, (\tanh \beta)^r \,, \tag{17.24}$$

where $\mathcal{N}(r)$ is the expected number of vertices j at distance r from a uniformly random vertex i. It is not hard to show that $\mathcal{N}(r) \doteq \overline{\lambda}^r$ for large r (the limit $N \to \infty$ is taken before $r \to \infty$). It follows that, for $\overline{\lambda} \tanh \beta < 1$, the susceptibility χ remains bounded. For $\overline{\lambda} \tanh \beta > 1$, the susceptibility is infinite and the paramagnetic solution cannot be correct.

17.3.3 The ferromagnetic phase

For $\beta > \beta_c$ (low temperature), one has to look for other solutions of the RS distributional equation (17.12). In general, this is done numerically, for instance using the population dynamics algorithm. In special cases, exact solutions can be found. As an example, assume that G is a random regular graph of degree $k + 1$. The corresponding factor graph is thus drawn from the ensemble $\mathbb{D}_N(\Lambda, P)$, with $\Lambda(x) = x^{k+1}$ and $P(x) = x^2$.

The local structure of the graph around a typical vertex is always the same: a regular tree of degree $(k+1)$. It is thus natural to seek a solution such that $h_{i \to j} = h_0$ for each directed edge $i \to j$. This corresponds to a solution of the distributional equation (17.12) where $h = h_0$ identically (formally $\mathsf{a} = \delta_{h_0}$). Equation (17.12) implies

$$h_0 = k f(+1, h_0) \,. \tag{17.25}$$

This equation is easily solved. For $\beta > \beta_c$, it has three solutions: the paramagnetic solution, and two opposite solutions $\pm h_0$. In order to interpret these solutions, recall that eqn (17.8) implies $\langle \sigma_i \rangle = \pm M$, with $M = \tanh((k+1)\beta f(+1, h_0))$, where h_0 is the positive solution of eqn (17.25). Figure 17.1 summarizes these results for the case $k = 2$.

The reader might be puzzled by these results since, by symmetry, one necessarily has $\langle \sigma_i \rangle = 0$. The correct interpretation of the ferromagnetic solution is similar to that of the Curie–Weiss model discussed in Section 2.5.2. In a typical configuration, the total magnetization $\sum_i \sigma_i$ is, with high probability, close to $+NM$ or $-NM$. Since each of these events occurs with probability $1/2$, the average magnetization is 0. We can decompose the Boltzmann distribution as

$$\mu(\underline{\sigma}) = \frac{1}{2}\mu_+(\underline{\sigma}) + \frac{1}{2}\mu_-(\underline{\sigma}) \,, \tag{17.26}$$

where $\mu_+(\cdot)$ and $\mu_-(\cdot)$ are supported on configurations with $\sum_i \sigma_i > 0$ and $\sum_i \sigma_i < 0$, respectively, assuming N odd for simplicity. The expectation of a typical spin with respect to $\mu_+(\cdot)$ or $\mu_-(\cdot)$ is then, asymptotically, $\langle \sigma_i \rangle_+ = +M$ or $\langle \sigma_i \rangle_- = -M$, respectively.

The two components $\mu_+(\cdot)$ and $\mu_-(\cdot)$ are often referred to as 'pure states', and are expected to have several remarkable properties. We will shall discuss decomposition into pure states further in the following chapters. It is interesting to notice that the

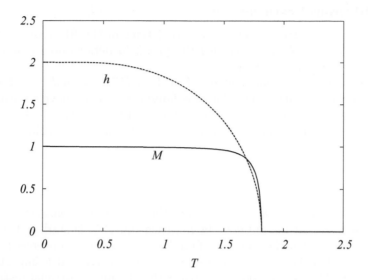

Fig. 17.1 Magnetization M and cavity field h_0 in a ferromagnetic Ising model on a random regular graph of degree $k+1=3$, as a function of the temperature. The critical temperature is $T_c = 1/\text{atanh}(1/2) \approx 1.82048$.

cavity method does not reproduce the actual averages over the Boltzmann distribution, but rather the averages with respect to pure states.

One can repeat the stability analysis for this new solution. The ferromagnetic susceptibility always involves the correlation between two spins in a one-dimensional problem, as in eqn (17.22). However, the fields g_p are now non-zero. In particular, for $p \in \{1, \dots, r-1\}$, we have $g_p = (k-1)f(+1, h_0)$. An explicit computation using the transfer matrix technique of Section 2.5.1 shows that the susceptibility of this ferromagnetic solution is finite in the whole low-temperature phase $\beta > \beta_c$.

Exercise 17.5 Consider the case of a ferromagnetic Ising model on a random regular graph in the presence of a positive external magnetic field. This is described by a term $+B \sum_i \sigma_i$ in the exponent in eqn (17.2). Show that there is no phase transition: the cavity field h_0 and the magnetization M are positive at all temperatures.

Exercise 17.6 Consider a ferromagnetic problem in which the J_{ij} are i.i.d. random variables, but always positive (i.e. their distribution is supported on $J > 0$.) Write the RS distributional equation (17.12) in this case and perform a local stability analysis of the paramagnetic solution. Show that the critical temperature is determined by the equation $\lambda \mathbb{E}_J \tanh(\beta J) = 1$. Compute the magnetization in the ferromagnetic phase (i.e. for $\beta > \beta_c$) using the population dynamics algorithm.

17.3.4 Rigorous treatment

In the case of Ising ferromagnets, many predictions of the RS cavity method can be confirmed rigorously. While describing the proofs in detail would take us to far, we shall outline the main ideas here.

To be concrete, we consider a model of the form (17.2), with $J_{ij} = +1$, on a random graph with degree profile Λ such that $\overline{\lambda}$ is finite (i.e. Λ must have a second moment). Since we know that eqn (17.12) admits more than one solution at low temperature, we need to select a particular solution. This can be done by considering the density evolution iteration

$$h^{(t+1)} \overset{\mathrm{d}}{=} \sum_{i=1}^{K-1} f(+1, h_i^{(t)}) \tag{17.27}$$

with the initial condition $h^{(0)} = +\infty$. In other words, we consider the sequence of distributions $\mathsf{a}(\cdot)$ that are obtained by iterating eqn (17.13) starting from $\delta_{+\infty}$.

It can be shown that the sequence of random variables $h^{(t)}$ converges in distribution as $t \to \infty$, and that the limit $h^{(+)}$ is a solution of eqn (17.12). It is important to stress that in general $h^{(+)}$ is a random variable with a highly non-trivial distribution. For $\overline{\lambda} \tanh \beta < 1$, the argument in Section 17.3.1 implies that $h^{(+)} = 0$ with probability one. For $\overline{\lambda} \tanh \beta > 1$, $h^{(+)}$ is supported on non-negative values.

Theorem 17.1 *Let $Z_N(\beta)$ be the partition function of a ferromagnetic Ising model on a random graph with degree profile Λ. Under the hypotheses above, the RS cavity expression for the free entropy is exact. Precisely,*

$$\lim_{N \to \infty} \frac{1}{N} \log Z_N(\beta) = \mathrm{f}^{\mathrm{RS}}(\beta), \tag{17.28}$$

the limit holding almost surely for any finite β. Here $\mathrm{f}^{\mathrm{RS}}(\beta)$ is defined as in eqn (17.14), where the h_i are i.i.d. copies of $h^{(+)}$.

Let us sketch the key ideas used in the proof, referring to the literature for a complete derivation. The first step consists in introducing a small positive magnetic field B that adds a term $-B\sum_i \sigma_i$ to the energy (17.1). One then uses the facts that the free-entropy density is continuous in B (which also applies as $N \to \infty$) and that $\log Z_N(\beta, B)$ concentrates around $\mathbb{E} \log Z_N(\beta, B)$. This allows us to focus on $\mathbb{E} \log Z_N(\beta, B)$. It is easy to check that the cavity prediction is correct at $\beta = 0$ (since, in this limit, the spins become independent). The idea is then to compute the derivative of $N^{-1}\mathbb{E} \log Z_N(\beta, B)$ with respect to β and prove that this converges to the derivative of $\mathrm{f}^{\mathrm{RS}}(\beta)$. Some calculus shows that this is equivalent to proving that, for a uniformly random edge (i, j) in the graph,

$$\lim_{N \to \infty} \mathbb{E}\langle \sigma_i \sigma_j \rangle = \mathbb{E} \left\{ \frac{\tanh \beta + \tanh \beta h_1 \tanh \beta h_2}{1 + \tanh \beta \tanh \beta h_1 \tanh \beta h_2} \right\}, \tag{17.29}$$

where h_1, h_2 are i.i.d. copies of $h^{(+)}$.

The advantage of eqn (17.29) is that $\langle \sigma_i \sigma_j \rangle$ is a local quantity. One can then try to estimate it by looking at the local structure of the graph in a large but finite neighbourhood around (i, j). The key problem is to prove that the rest of the graph 'decouples' from the expectation $\langle \sigma_i \sigma_j \rangle$.

One important ingredient in this proof is **Griffiths' inequality**. For the reader's reference, we recall it here for the case of a pairwise Ising model.

Theorem 17.2 *Consider a ferromagnetic Ising model, i.e. a Boltzmann distribution of the form*

$$\mu_{G,J}(\underline{\sigma}) = \frac{1}{Z_{G,J}} \exp \left\{ \beta \sum_{(i,j) \in G} J_{ij} \sigma_i \sigma_j + \sum_i B_i \sigma_i \right\}, \qquad (17.30)$$

with $J_{ij}, B_i \geq 0$. *Then, for any* $U \subseteq V$, $\langle \prod_{i \in U} \sigma_i \rangle$ *is non-negative and non-decreasing in all of the* J_{ij}, B_i.

The strategy is then the following:

1. Given a certain neighbourhood S of the edge (i, j), one can consider two modified measures on the spins of S given by the Boltzmann measure where the spins outside of S have been fixed. The first measure (the '+ boundary condition') has $\sigma_i = 1 \; \forall i \in \overline{S}$. The second one (the 'free boundary condition') has $\sigma_i = 0 \; \forall i \in \overline{S}$. The latter also amounts to considering the model on the subgraph induced by S. Griffiths' inequality allows one to show that the true $\langle \sigma_i \sigma_j \rangle$ is bounded by the expectations obtained with the + and with free boundary conditions.

2. It can be shown that whenever the recursion (17.27) is initialized with $h^{(0)}$ supported on non-negative values, it converges to $h^{(+)}$.

3. Finally one takes for S a ball of radius r centred on i. This is, with high probability, a tree. Using the result in item 2, one proves that the two expectations of $\langle \sigma_i \sigma_j \rangle$ with the + and free boundary conditions converge to the same value as $r \to \infty$.

17.4 Spin glass models

We now turn to the study of the spin glass problem. Again, the paramagnetic solution turns out to describe correctly the system at high temperature, and the critical temperature depends on the ensemble only through the quantity $\overline{\lambda}$. We shall see that the paramagnetic solution is unstable for $\beta > \beta_c$, where

$$\overline{\lambda} \, (\tanh \beta_c)^2 = 1. \qquad (17.31)$$

As in the previous section, we shall try to find another solution of the RS distributional equation (17.12) when $\beta > \beta_c$. Such a solution exists and can be studied numerically; it yields predictions which are better than those of the paramagnetic solution. However, we shall argue that, even in the large-system limit, this solution does not correctly describe the model: the RS assumptions do not hold for $\beta > \beta_c$, and replica symmetry breaking is needed.

17.4.1 Local stability

As with the ferromagnetic model, we begin with a local stability analysis of the paramagnetic solution, by taking moments of the RS distributional equation. Since $f(J, h)$

is antisymmetric in J, any solution of eqn (17.12) has $\mathbb{E}\{h\} = 0$ identically. The lowest non-trivial moment is therefore $\mathbb{E}\{h^2\}$.

Using the fact that $\mathbb{E}\{f(J_i, h_i)f(J_j, h_j)\} = 0$ for $i \neq j$, and the Taylor expansion of $f(J, h)$ around $h = 0$, we get

$$\mathbb{E}\{h^2\} = \overline{\lambda}\,\mathbb{E}\{f(J, h)^2\} = \overline{\lambda}\,\theta^2\,\mathbb{E}\{h^2\} + \mathbb{E}\{O(h^4)\}\,. \qquad (17.32)$$

Assuming that $\mathbb{E}\{O(h^4)\} = O(\mathbb{E}\{h^2\}^2)$ (this step can be justified through a lengthier argument), we get a linear recursion for the second moment of h which is unstable for $\overline{\lambda}\theta^2 > 1$ (i.e. $\beta > \beta_c$). On the other hand, as in Section 17.3.1, one can use the inequality $|f(J, h)| \le h(\tanh\beta)$ to show that the paramagnetic solution $h \overset{\mathrm{d}}{=} \delta_0$ is the only solution in the high-temperature phase $\overline{\lambda}\theta^2 < 1$.

17.4.2 Spin glass susceptibility

An alternative argument for the appearance of the spin glass phase at $\beta > \beta_c$ is provided by computing the spin glass susceptibility. Recalling the discussion in Section 12.3.2, and using $\langle\sigma_i\rangle = 0$ (as is the case in the paramagnetic case), we have

$$\chi_{\mathrm{SG}} = \frac{1}{N} \sum_{i,j \in V} \langle\sigma_i\sigma_j\rangle^2\,. \qquad (17.33)$$

Assuming the RS cavity approach to be correct, and using the paramagnetic solution, the computation of $\langle\sigma_i\sigma_j\rangle$ is analogous to the one we did in the ferromagnetic case. Denoting by ω the shortest path in G between i and j, the result is

$$\langle\sigma_i\sigma_j\rangle = \prod_{(kl)\in\omega} \tanh\beta J_{kl}\,. \qquad (17.34)$$

Taking the square and splitting the sum according to the distance between i and j, we get

$$\chi_{\mathrm{SG}} = \sum_{r=0}^{\infty} \mathcal{N}(r)\,(\tanh\beta)^{2r}\,, \qquad (17.35)$$

where $\mathcal{N}(r)$ is the average number of vertices at distance r from a random vertex i. We have $\mathcal{N}(r) \doteq \overline{\lambda}^r$ to the leading exponential order for large r. Therefore the above series is summable for $\overline{\lambda}\theta^2 < 1$, yielding a bounded susceptibility. Conversely, if $\overline{\lambda}\theta^2 > 1$ the series diverges, and the paramagnetic solution must be considered inconsistent.

17.4.3 Paramagnetic phase: Rigorous treatment

From a mathematical point of view, determining the free energy of the spin glass model on a sparse random graph is a challenging open problem. The only regime that is (relatively) well understood is the paramagnetic phase at zero external field.

Theorem 17.3 *Let $Z_N(\beta)$ denote the partition function of the spin glass model with couplings ± 1 on a random graph G, with a degree distribution Λ which has a finite second moment. If $\overline{\lambda}(\tanh \beta)^2 < 1$, then*

$$\lim_{N \to \infty} \frac{1}{N} \log Z_N(\beta) = f_{\text{para}}^{\text{RS}}(\beta) \equiv \log 2 + \frac{1}{2}\overline{\Lambda} \log \cosh \beta , \qquad (17.36)$$

where the limit holds almost surely.

Proof To keep things simple, we shall prove only a slightly weaker statement, namely the following. For any $\delta > 0$,

$$e^{N[f_{\text{para}}^{\text{RS}} - \delta]} \le Z_N \le e^{N[f_{\text{para}}^{\text{RS}} + \delta]} , \qquad (17.37)$$

with high probability as $N \to \infty$. In the proof, we shall denote by $\mathbb{E}_J$ the expectation with respect to the couplings $J_{ij} \in \pm 1$, and write $M = N\overline{\Lambda}/2$ for the number of edges in G, so that $e^{Nf_{\text{para}}^{\text{RS}}} = 2^N(\cosh \beta)^M$.

The probability of the event $Z_N \ge e^{N[f_{\text{para}}^{\text{RS}} + \delta]}$ is bounded from above using the Markov inequality. The annealed partition function is

$$\mathbb{E}_J\{Z_N\} = \sum_{\underline{\sigma}} \mathbb{E}_J \left\{ \prod_{(ij) \in E} e^{\beta J_{ij}\sigma_i \sigma_j} \right\} = \sum_{\underline{\sigma}}(\cosh \beta)^M = 2^N(\cosh \beta)^M , \qquad (17.38)$$

whence $Z_N \le 2^N(\cosh \beta)^M e^{N\delta}$ with high probability.

The lower bound follows by applying the second-moment method to the random realization of the couplings J_{ij}, given a typical random graph G: it is sufficient to show that $\mathbb{E}_J\{Z_N^2\} \le \mathbb{E}_J\{Z_N\}^2 e^{N\delta'}$ for any $\delta' > 0$.

Surprisingly, the computation of the second moment reduces to computing the partition function of a ferromagnetic Ising model. The key identity is

$$\mathbb{E}_J\{e^{\beta J_{ij}(\sigma_i^1 \sigma_j^1 + \sigma_i^2 \sigma_j^2)}\} = \frac{(\cosh \beta)^2}{\cosh \gamma} e^{\gamma \tau_i \tau_j} , \qquad (17.39)$$

where $\gamma \equiv \text{atanh}((\tanh \beta)^2)$; $\sigma_i^1, \sigma_i^2, \sigma_j^1, \sigma_j^2$ are Ising spins; and $\tau_i = \sigma_i^1 \sigma_i^2$ and $\tau_j = \sigma_j^1 \sigma_j^2$. Using this identity, we find

$$\mathbb{E}_J\{Z_N^2\} = \sum_{\underline{\sigma}^1, \underline{\sigma}^2} \mathbb{E}_J\left\{ \prod_{(ij) \in E} e^{\beta J_{ij}(\sigma_i^1 \sigma_j^1 + \sigma_i^2 \sigma_j^2)} \right\} \qquad (17.40)$$

$$= 2^N \left(\frac{(\cosh \beta)^2}{\cosh \gamma} \right)^M \sum_{\underline{\tau}} \prod_{(ij) \in E} e^{\gamma \tau_i \tau_j} = 2^N \left(\frac{(\cosh \beta)^2}{\cosh \gamma} \right)^M Z_N^{\text{f}}(\gamma) ,$$

where $Z_N^{\text{f}}(\gamma)$ is the partition function of a ferromagnetic Ising model on the graph G at inverse temperature γ. This can be estimated through Theorem 17.1. As $\overline{\lambda}(\tanh \gamma) = \overline{\lambda}(\tanh \beta)^2 < 1$, this ferromagnetic model is in its paramagnetic phase, and therefore $Z_N^{\text{f}}(\gamma) \le 2^N(\cosh \gamma)^M e^{N\delta}$ with high probability. This in turn implies

$$\mathbb{E}_J\{Z_N^2\} \le 2^{2N}(\cosh \beta)^{2M} e^{N\delta} , \qquad (17.41)$$

which completes the proof. $\square$

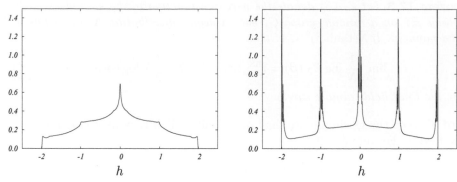

Fig. 17.2 Distribution of cavity fields approximated by the population dynamics algorithm at $T = 0.5$ (*left*) and $T = 0.1$ (*right*), for a $J_{ij} = \pm 1$ spin glass on a random regular graph with degree 3. Here, we plot a histogram with bin size $\Delta h = 0.01$. A population of 10^4 fields $\{h_i\}$ was used in the algorithm, and the resulting histogram was averaged over 10^4 iterations.

17.4.4   An attempt to describe the spin glass phase

If $\beta > \beta_c$, the RS distributional equation (17.12) admits a solution h that is not identically 0. The corresponding distribution $\mathsf{a}(\cdot)$ is symmetric under $h \to -h$ and can be approximated numerically using the population dynamics algorithm. This is usually referred to as 'the RS spin glass solution' (although it is far from obvious whether it is unique.)

In Fig. 17.2 we plot the empirical distribution of cavity fields as obtained through the population dynamics algorithm, for a random regular graph of degree $k+1 = 3$. The corresponding critical temperature can be determined through eqn (17.31), yielding $T_c = 1/\log(\sqrt{2} + 1) \approx 1.134592$. Notice that the distribution $\mathsf{a}(\cdot)$ is extremely non-trivial, and indeed is likely to be highly singular.

Once an approximation of the distribution of fields has been obtained, the free entropy can be estimated as f^{RS}, given in eqn (17.14). Figure 17.3 shows the free energy F versus temperature. It is difficult to control the solutions of the RS distributional equation (17.12) analytically, with the exception of some limiting cases. One such case is the object of the next exercise.

Exercise 17.7 Consider the spin glass model on a random regular graph with degree $k + 1$, and assume that the couplings J_{ij} take values $\{+1/\sqrt{k}, -1/\sqrt{k}\}$ independently and uniformly at random.

Argue that, as $k \to \infty$, the following limiting behaviour holds.

(a) The solution $\mathsf{a}(\cdot)$ of the RS equation (17.12) converges to a Gaussian with mean 0 and variance q, where q solves the equation

$$q = \int (\tanh \beta h)^2 \, e^{-h^2/(2q)} \, \frac{dh}{\sqrt{2\pi q}} = \mathbb{E}_h \tanh^2(\beta h) \,. \tag{17.42}$$

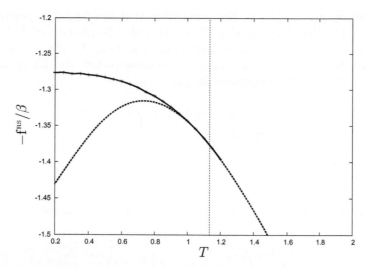

Fig. 17.3 The free-energy density $-f^{RS}/\beta$ of a spin glass model with $J_{ij} = \pm 1$ on a random regular graph with degree 3. Here, we plot results obtained for the paramagnetic solution (dashed line) and the RS spin glass solution (full black line). The critical temperature $T_c \approx 1.134592$, (see eqn (17.31)), is indicated by the vertical line.

(b) Show that the RS prediction for the free entropy per spin converges to the following value (here the limit $k \to \infty$ is taken *after* $N \to \infty$):

$$f_{SK}^{RS}(\beta) = \mathbb{E}_h \log(2\cosh(\beta h)) + \frac{\beta^2}{4}(1 - q)^2 . \qquad (17.43)$$

Notice that these results are identical to those obtained by the replica-symmetric analysis of the SK model (see eqn (8.68)).

The second limit in which a solution can be found analytically is that of zero temperature, $T \to 0$ (or, equivalently, $\beta \to \infty$). Assume that h has a finite, non-vanishing limit as $\beta \to \infty$. It is easy to show that the limiting distribution must satisfy a distributional equation that is formally identical to (17.12), but with the function $f(\cdot)$ replaced by its zero-temperature limit

$$f(J, h) = \text{sign}(J h) \, \min[|J|, |h|] . \qquad (17.44)$$

Since $J \in \{+1, -1\}$, if h takes values on the integers, then $f(J, h) \in \{+1, 0, -1\}$. As a consequence, eqn (17.12) admits a solution with support on the integers. For the sake of simplicity, we shall restrict ourselves to the case of a regular graph with degree $k + 1$. The distribution of h can be formally written as

$$a(h) = \sum_{r=-k}^{k} p_r \delta_r(h) . \qquad (17.45)$$

Let us denote by $p_+ = \sum_{r=1}^{\infty} p_r$ the probability that the field h is positive, and by p_- the probability that it is negative. Notice that the distribution of $f(J, h)$ depends only on p_+, p_0 and p_- and not on the individual weights p_r. By the symmetry of $\mathsf{a}(\cdot)$, $p_r = p_{-r}$, and therefore $p_+ = (1 - p_0)/2$. As a consequence, the RS cavity equation (17.13) implies the following equation for p_0:

$$p_0 = \sum_{q=0}^{\lfloor k/2 \rfloor} \binom{k}{2q} \binom{2q}{q} p_0^{k-2q} \left(\frac{1-p_0}{2} \right)^{2q} . \tag{17.46}$$

Exercise 17.8

(a) Show that the probability that $h = r$ (and by symmetry, the probability that $h = -r$) is given by:

$$p_r = p_{-r} = \sum_{q=0}^{\lfloor (k-r)/2 \rfloor} \binom{k}{2q+r} p_0^{k-2q-r} \left(\frac{1-p_0}{2} \right)^{2q+r} \binom{2q+r}{q} . \tag{17.47}$$

(b) Consider the distribution $\mathsf{a}_*(\cdot)$ of the local field H acting on a spin, defined by $H \stackrel{\mathrm{d}}{=} \sum_{r=1}^{k+1} f(J_r, h_r)$. Show that it takes the form $\mathsf{a}(h) = \sum_{r=-k-1}^{k+1} s_r \delta_r(h)$, where s_r is given by the same formula (17.47) as for p_r above, with k replaced by $k + 1$.

(c) Show that the ground state energy per spin is

$$E_0 = k \sum_{r=-k-1}^{k+1} s_r |r| - (k+1) \sum_{r=-k}^{k} p_r |r| - \frac{k+1}{2} p_0 (2 - p_0) . \tag{17.48}$$

As an example, consider the case of a random regular graph with degree $k+1 = 3$. In this case, solving eqn (17.46) yields $p_0 = 1/3$, $p_1 = 2/9$, and $p_2 = 1/9$. The resulting ground state energy per spin is $E_0 = -23/18$.

17.4.5 Instability of the RS spin glass solution

It turns out that the solution of the RS distributional equation discussed in the previous subsection does not correctly describe the model in the thermodynamic limit. In particular, the RS spin glass solution predicts an unacceptable negative entropy at low temperatures, and its prediction for the ground state energy per spin, $E_0 = -23/18 = -1.27777\ldots$, differs from the best numerical estimate -1.2716 ± 0.0001. Therefore the assumptions of the RS cavity method cannot be correct for the low-temperature phase $\overline{\lambda}(\tanh \beta)^2 > 1$. On the other hand, physicists think that the approach can be rescued: the asymptotic free-energy density can be computed by introducing replica symmetry breaking. This will be the subject of Chapters 19 to 22.

Here we want to show how the inconsistency of this 'RS solution' can be detected, by the computation of the spin glass susceptibility. The computation is similar to that for the paramagnetic solution, but one has to deal with the presence of non-zero values

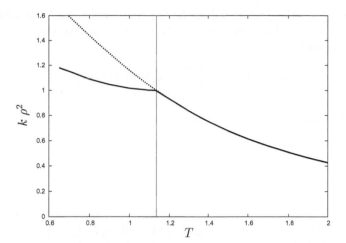

Fig. 17.4 Instability of the 'RS solution' for the Ising spin glass on a random regular graph with degree $k + 1 = 3$. The plot shows $k\rho^2$ versus the temperature, where ρ^2 is the rate of exponential decay of the spin glass correlation function $\mathbb{E}(\langle\sigma_0\sigma_r\rangle - \langle\sigma_0\rangle\langle\sigma_r\rangle)^2$ at large r. A value of $k\rho^2 < 1$ corresponds to a bounded spin glass susceptibility. This is the case for the paramagnetic solution at $T > T_c$. For $T < T_c \approx 1.134592$, the bottom curve is the result for the RS cavity solution (with a field distribution a(h) estimated from population dynamics), and the top curve is the result for the paramagnetic solution. Both 'solutions' yield $k\rho^2 > 1$, and are thus unstable.

in the support of the cavity field distribution a($\cdot$). As in Section 17.4.2, the first step is to compute the correlation between two spins σ_0 and σ_r at a distance r in G. This is found by considering the joint distribution of the spins along the shortest path between 0 and r. In the cavity method, this has the form

$$\mu_{G,J}(\sigma_0, \ldots, \sigma_r) \cong \exp\left\{\beta\sum_{p=0}^{r-1} J_p\sigma_p\sigma_{p+1} + \beta\sum_{p=0}^{r} g_p\sigma_p\right\} . \tag{17.49}$$

Here the couplings J_p are i.i.d. and uniformly random in $\{+1, -1\}$. The fields g_p are i.i.d. variables whose distribution is determined by

$$g \overset{\mathrm{d}}{=} \sum_{q=1}^{k-1} f(J_q, h_q) , \tag{17.50}$$

where the h_q are $k - 1$ independent copies of the random variable h that solves eqn (17.12), and the J_r are again independent and uniformly random in $\{+1, -1\}$. (To be precise, g_0 and g_r have a different distribution from the others, as they are sums of k i.i.d. terms instead of $k - 1$. However, this difference is irrelevant for our computation of the leading exponential order of the correlation at large r.)

The solution of the one-dimensional problem (17.22) can be obtained by the transfer matrix method discussed in Section 2.5.1 (and, in the BP context, in Section 14.1).

We denote by $z_p^+(\sigma)$ the partition function of the partial chain including the spins $\sigma_0, \sigma_1, \ldots, \sigma_p$, where we have fixed $\sigma_0 = +1$, and $\sigma_p = \sigma$. We can then immediately derive the recursion

$$z_{p+1}^+(\sigma) = \sum_{\sigma' \in \{+1, -1\}} T_p(\sigma, \sigma') z_p^+(\sigma') , \tag{17.51}$$

where the p-th transfer matrix takes the form

$$T_p = \begin{pmatrix} \mathrm{e}^{\beta J_p + \beta g_{p+1}} & \mathrm{e}^{-\beta J_p + \beta g_{p+1}} \\ \mathrm{e}^{-\beta J_p - \beta g_{p+1}} & \mathrm{e}^{\beta J_p - \beta g_{p+1}} \end{pmatrix} . \tag{17.52}$$

Similarly, we denote by $z_p^-(\sigma)$ the partition function conditional on $\sigma_0 = -1$. This satisfies the same recursion relation as z_p^+, but with a different initial condition. We have, in fact, $z_0^+(1) = 1$, $z_0^+(-1) = 0$, and $z_0^-(1) = 0$, $z_0^-(-1) = 1$.

The joint probability distribution of σ_0 and σ_r is given by

$$\mu_{G,J}(\sigma_0, \sigma_r) \cong \mathrm{e}^{\beta g_0} \mathbb{I}(\sigma_0 = +1) \, z_r^+(\sigma_r) + \mathrm{e}^{-\beta g_0} \mathbb{I}(\sigma_0 = -1) \, z_r^-(\sigma_r) . \tag{17.53}$$

From this expression, one finds, after a few lines of computation,

$$\langle \sigma_0 \sigma_r \rangle - \langle \sigma_0 \rangle \langle \sigma_r \rangle = \frac{4[z_r^+(1) z_r^-(-1) - z_r^+(-1) z_r^-(1)]}{\left[\mathrm{e}^{\beta g_0} (z_r^+(1) + z_r^+(-1)) + \mathrm{e}^{-\beta g_0} (z_r^-(1) + z_r^+(-1)) \right]^2} . \tag{17.54}$$

Approximate samples of the random variables $g_1, g_2, \ldots g_r$ can be obtained through the population dynamics algorithm. This allows one to evaluate the correlation function (17.54) and estimate its moments. In order to compute the spin glass susceptibility (see Section 17.4.2), one needs to estimate the growth rate of the second moment at large r. We define ρ through the leading exponential behaviour at large r,

$$\mathbb{E}\left(\langle \sigma_0 \sigma_r \rangle - \langle \sigma_0 \rangle \langle \sigma_r \rangle \right)^2 \doteq \rho^{2r} , \tag{17.55}$$

where $\mathbb{E}$ refers here to the expectation with respect to the fields $g_1, \ldots, g_r$ (i.e. to the graph and the couplings outside the path between 0 and r).

In a regular graph with degree $(k + 1)$, the number of neighbours at a distance r from a random vertex i grows like k^r. As a consequence, the series for the spin glass susceptibility is summable (thus suggesting that the susceptibility is bounded) if $k\rho^2 < 1$. For the paramagnetic solution, $\rho = \tanh \beta$, and this condition reduces to $\beta < \beta_c$ (see eqn (17.31)). In Figure 17.4 we plot $k\rho^2$ versus the temperature, as computed numerically for the RS spin glass field distribution $\mathsf{a}(\cdot)$. In the whole phase $T < T_c$, this is larger than one, showing that the RS 'solution' is in fact unstable. This analysis is easily generalized to irregular graph ensembles, by replacing k with the average (edge-perspective) degree $\bar{\lambda}$.

Physicists point out that the RS spin glass solution, although wrong, looks 'less wrong' than the paramagnetic solution. Indeed, $k\rho^2$ is smaller in the RS spin glass solution. Also, if one computes the entropy density $-\mathrm{d}F/\mathrm{d}T$ at zero temperature, one finds that it is negative in both solutions, but it is larger for the RS spin glass solution.

Table 17.1 Estimates of the ground state energy density of the Ising spin glass on a random regular graph of degree $k + 1$.

	Paramagnetic	RS	1RSB	2RSB	Numerics
$k = 2$	$-3/2 = -1.5$	$-23/18 \approx -1.2778$	-1.2723		$-1.2716(1)$
$k = 4$	$-5/2 = -2.5$	-1.69133	-1.6752	$-1.67316(4)$	$-1.673(1)$

In the next few chapters we shall discuss a one-step replica-symmetry-breaking formalism that goes further in the same direction. The instability parameter $k\rho^2$ becomes smaller than in the RS case, but it is still larger than one. In this respect, the Ising spin glass is a particularly complicated model. It is expected that the free-entropy density and similar asymptotic quantities wil be predicted correctly only in the limit of full replica symmetry breaking. We refer to Chapter 22 for a further discussion of this point. Table 17.1 gives the one- and two-step RSB estimates of the ground state energy in the cases $k = 2$ and 4, and compares them with the best available numerical results.

Exercise 17.9 Consider the large-degree limit as in Exercise 17.7. Show that the stability condition $k\rho^2 < 1$ becomes, in this limit, $\beta^2 \mathbb{E}_h (1 - \tanh^2(\beta h))^2 < 1$, where h is a Gaussian random variable with zero mean and a variance q satisfying eqn (17.42). This is nothing but the de Almeida–Thouless condition discussed in Chapter 8.

Let us conclude by warning the reader on one point. Although local stability is a necessary consistency check for the RS solution, it is by no means sufficient. Indeed, models with p-spin interactions and $p \geq 3$ (such as the XORSAT model treated in the next chapter) often admit a locally stable 'RS solution' that is nevertheless incorrect.

Notes

Ising ferromagnets on random graphs have appeared in several papers (e.g. Johnston and Plechác, 1998; Dorogotsev *et al.*, 2002; Leone *et al.*, 2004). The application of belief propagation to this model was considered by Looij and Kappen (2005). The rigorous cavity analysis of the Ising ferromagnet presented in this chapter can be found in Dembo and Montanari (2008c), which also proves the exponential convergence of BP.

The problem of spin glasses on random graphs was first studied using the replica method by Viana and Bray (1985), who worked out the paramagnetic solution and located the transition. The RS cavity solution that we have described here was first discussed for Erdös–Renyi graphs by Mézard and Parisi (1987); see also Kanter and Sompolinsky (1987). Expansions around the critical point were developed by Goldschmidt and De Dominicis (1990). The related (but different) problem of a spin glass on a regular tree was introduced by Thouless (1986) and further studied by Chayes *et al.* (1986) and Carlson *et al.* (1990).

One-step replica symmetry breaking for spin glasses on sparse random graphs was studied by Wong and Sherrington (1988), Goldschmidt and Lai (1990) and Monasson

(1998) using replicas, and by Goldschmidt (1991) and Mézard and Parisi (2001, 2003) using the cavity method. The 2RSB ground state energy given in Table 17.1 was obtained by Montanari (2003), and the numerical values are from Boettcher (2003).

Part V

Long-range correlations

18
Linear equations with Boolean variables

Solving a system of linear equations over a finite field $\mathbb{F}$ is arguably one of the most fundamental operations in mathematics. Several algorithms have been devised to accomplish such a task in polynomial time. The best known is Gaussian elimination, which has $O(N^3)$ complexity (here N is number of variables in the linear system, and we assume the number of equations to be $M = \Theta(N)$). As a matter of fact, one can improve over Gaussian elimination, and the best existing algorithm for general systems has complexity $O(N^{2.376\cdots})$. Faster methods also exist for special classes of instances.

The set of solutions of a linear system is an affine subspace of $\mathbb{F}^N$. Despite this apparent simplicity, the geometry of affine or linear subspaces of $\mathbb{F}^N$ can be surprisingly rich. This observation is systematically exploited in coding theory. Linear codes are just linear spaces over finite fields. Nevertheless, they are known to achieve the Shannon capacity on memoryless symmetric channels, and their structure is far from trivial, as we have already seen in Chapter 11.

From a different point of view, linear systems are a particular example of constraint satisfaction problems. We can associate with a linear system a decision problem (establishing whether it has a solution), a counting problem (counting the number of solutions), and an optimization problem (minimizing the number of violated equations). While the first two are polynomial, the latter is known to be NP-hard.

In this chapter, we consider a specific ensemble of random linear systems over $\mathbb{Z}_2$ (the field of integers modulo 2), and discuss the structure of its set of solutions. The definition of this ensemble is motivated mainly by its analogy with other random constraint satisfaction problems, which also explains the name 'XOR-satisfiability' (or 'XORSAT').

In the next section, we provide the precise definition of the XORSAT ensemble and recall a few elementary properties of linear algebra. We also introduce one of the main objects of study of this chapter: the SAT–UNSAT threshold. Section 18.2 takes a detour into the properties of belief propagation for XORSAT. These are shown to be related to the correlation structure of the uniform measure over solutions and, in Section 18.3, to the appearance of a 2-core in the associated factor graph. Sections 18.4 and 18.5 build on these results to compute the SAT–UNSAT threshold and characterize the structure of the solution space. While many results can be derived rigorously, XORSAT offers an ideal playground for understanding the non-rigorous cavity method, which will be developed further in the following chapters. This is the subject of Section 18.6.

18.1   Definitions and general remarks

18.1.1   Linear systems

Let $\mathbb{H}$ be an $M \times N$ matrix with entries $H_{ai} \in \{0,1\}$, $a \in \{1,\ldots,M\}$, $i \in \{1,\ldots,N\}$, and let $\underline{b}$ be an M-component vector with binary entries $b_a \in \{0,1\}$. An instance of the **XORSAT** problem is given by a pair $(\mathbb{H}, \underline{b})$. The decision problem requires us to find an N-component vector $\underline{x}$ with binary entries $x_i \in \{0,1\}$ which solves the linear system $\mathbb{H}\underline{x} = \underline{b}$ mod 2, or to show that the system has no solution. The name 'XORSAT' comes from the fact that the sum modulo 2 is equivalent to the 'exclusive OR' operation: the problem is whether there exists an assignment of the variables $\underline{x}$ which satisfies a set of XOR clauses. We shall thus say that an instance is SAT or UNSAT whenever the linear system has or does not have, respectively, a solution.

We shall, furthermore, be interested in the set of solutions, denoted by $\mathcal{S}$; in its size $Z = |\mathcal{S}|$; and in the properties of the uniform measure over $\mathcal{S}$. This is defined by

$$\mu(\underline{x}) = \frac{1}{Z}\,\mathbb{I}(\,\mathbb{H}\underline{x} = \underline{b} \quad \mathrm{mod}\ 2\,) = \frac{1}{Z}\prod_{a=1}^{M}\psi_a(\underline{x}_{\partial a})\,, \tag{18.1}$$

where $\partial a = (i_a(1),\ldots,i_a(K))$ is the set of non-vanishing entries in the a-th row of $\mathbb{H}$, and $\psi_a(\underline{x}_{\partial a})$ is the characteristic function for the a-th equation in the linear system (explicitly, $\psi_a(\underline{x}_{\partial a}) = \mathbb{I}(x_{i_1(a)} \oplus \cdots \oplus x_{i_K(a)} = b_a)$, where we denote the sum modulo 2 by $\oplus$ as usual). In the following, we shall omit the specification that operations are carried   mod 2 when this is clear from the context.

When $\mathbb{H}$ has a row weight p (i.e. each row has p non-vanishing entries), the problem is related to a p-spin glass model. Writing $\sigma_i = 1 - 2x_i$ and $J_a = 1 - 2b_a$, we can associate with an XORSAT instance the energy function

$$E(\underline{\sigma}) = \sum_{a=1}^{M}\left(1 - J_a\prod_{j \in \partial a}\sigma_j\right), \tag{18.2}$$

which counts (twice) the number of violated equations. This can be regarded as a p-spin glass energy function with binary couplings. The XORSAT decision problem asks whether there exists a spin configuration $\underline{\sigma}$ with zero energy or, in physical jargon, whether the above energy function is 'unfrustrated'. If there exists such a configuration, $\log Z$ is the ground state entropy of the model.

A natural generalization is the MAX-XORSAT problem. This requires us to find a configuration which maximizes the number of satisfied equations, i.e. minimizes $E(\underline{\sigma})$. In the following, we shall use the language of XORSAT but, of course, all statements have their direct counterpart in the context of p-spin glasses.

Let us recall a few well-known facts of linear algebra that will be useful in the following:

(i) The image of $\mathbb{H}$ is a vector space of dimension $\mathrm{rank}(\mathbb{H})$ ($\mathrm{rank}(\mathbb{H})$ is the number of independent rows in $\mathbb{H}$); the kernel of $\mathbb{H}$ (the set $\mathcal{S}_0$ of $\underline{x}$ which solves the homogeneous system $\mathbb{H}\underline{x} = \underline{0}$) is a vector space of dimension $N - \mathrm{rank}(\mathbb{H})$.

(ii) As a consequence, if $M \leq N$ and $\mathbb{H}$ has rank M (all of its rows are independent), then the linear system $\mathbb{H}\underline{x} = \underline{b}$ has a solution for any choice of $\underline{b}$.

(*iii*) Conversely, if rank($\mathbb{H}$) $< M$, then the linear system has a solution if and only if $\underline{b}$ is in the image of $\mathbb{H}$.

If the linear system has at least one solution $\underline{x}_*$, then the set of solutions $\mathcal{S}$ is an affine space of dimension $N - \text{rank}(\mathbb{H})$: we have $\mathcal{S} = \underline{x}_* + \mathcal{S}_0$ and $Z = 2^{N-\text{rank}(\mathbb{H})}$. We shall denote by $\mu_0(\,\cdot\,)$ the uniform measure over the set $\mathcal{S}_0$ of solutions of the homogeneous linear system:

$$\mu_0(\underline{x}) = \frac{1}{Z_0}\, \mathbb{I}(\,\mathbb{H}\underline{x} = \underline{0} \mod 2\,) = \frac{1}{Z_0} \prod_{a=1}^{M} \psi_a^0(\underline{x}_{\partial a}) \,, \tag{18.3}$$

where ψ_a^0 is given by the same expression as for ψ_a but with $b_a = 0$. Note that μ_0 is always well defined as a probability distribution, because the homogeneous system has at least the solution $\underline{x} = \underline{0}$, whereas μ is well defined only for SAT instances. The linear structure has several important consequences:

- If $\underline{y}$ is a solution of the inhomogeneous system, and if $\underline{x}$ is a uniformly random solution of the homogeneous linear system (with distribution μ_0), then $\underline{x}' = \underline{x} \oplus \underline{y}$ is a uniformly random solution of the inhomogeneous system (its probability distribution is μ).

- Under the measure μ_0, there exist only two sorts of variables x_i, those which are 'frozen to 0', (i.e. take the value 0 in all of the solutions) and those which are 'free' (they take the value 0 in one half of the solutions and 1 in the other half). Under the measure μ (when it exists), a bit can be frozen to 0, frozen to 1, or free. These facts are proved in the next exercise.

Exercise 18.1 Let $f : \{0,1\}^N \to \{0,1\}$ be a linear function (explicitly, $f(\underline{x})$ is the sum of a subset $x_{i(1)}, \ldots, x_{i(n)}$ of the bits, mod 2).

(*a*) If $\underline{x}$ is drawn from the distribution μ_0, $f(\underline{x})$ becomes a random variable taking values in $\{0,1\}$. Show that if there exists a configuration $\underline{y}$ with $\mu_0(\underline{y}) > 0$ and $f(\underline{y}) = 1$, then $\mathbb{P}\{f(\underline{x}) = 0\} = \mathbb{P}\{f(\underline{x}) = 1\} = 1/2$. In the opposite case, $\mathbb{P}\{f(\underline{x}) = 0\} = 1$.

(*b*) Suppose that there exists at least one solution to the system $\mathbb{H}\underline{x} = \underline{b}$, so that μ exists. Consider the random variable $f(\underline{x})$ obtained by drawing $\underline{x}$ from the distribution μ. Show that one of the following three cases occurs: $\mathbb{P}\{f(\underline{x}) = 0\} = 1$, $\mathbb{P}\{f(\underline{x}) = 0\} = 1/2$, or $\mathbb{P}\{f(\underline{x}) = 0\} = 0$.

These results apply in particular to the marginal of bit i, using $f(\underline{x}) = x_i$.

Exercise 18.2 Show that:

(*a*) If the number of solutions of the homogeneous system is $Z_0 = 2^{N-M}$, then the inhomogeneous system is satisfiable (SAT), and has 2^{N-M} solutions, for any $\underline{b}$.

(*b*) Conversely, if the number of solutions of the homogeneous system is $Z_0 > 2^{N-M}$, then the inhomogeneous system is SAT only for a fraction $2^{N-M}/Z_0$ of the $\underline{b}$'s.

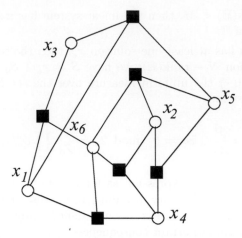

Fig. 18.1 Factor graph for a 3-XORSAT instance with $N = 6$, $M = 6$ (see Exercise 18.3).

The distribution μ admits a natural factor graph representation: variable nodes are associated with variables and factor nodes with linear equations; see Fig. 18.1. Given an XORSAT formula F (i.e. a pair $\mathbb{H}, \underline{b}$), we denote the associated factor graph by $G(F)$. It is remarkable that one can identify subgraphs of $G(F)$ that serve as witnesses of the satisfiability or unsatisfiability of F. By this we mean that the existence of such subgraphs implies the satisfiability or unsatisfiability of F. The existence of a simple witness for unsatisfiability is intimately related to the polynomial nature of XORSAT. Such a witness is obtained as follows. Given a subset L of the clauses, draw the factor graph including all of the clauses in L, all of the adjacent variable nodes, and the edges between them. If this subgraph has *even degree at each of the variable nodes*, and if $\oplus_{a \in L} b_a = 1$, then L is a witness for unsatisfiability. Such a subgraph is sometimes called a frustrated hyperloop (in analogy with the frustrated loops that appear in spin glasses, where function nodes have degree 2).

Exercise 18.3 Consider a 3-XORSAT instance defined through the 6×6 matrix

$$
\mathbb{H} = \begin{bmatrix} 0 & 1 & 0 & 1 & 1 & 0 \\ 1 & 0 & 0 & 1 & 0 & 1 \\ 0 & 1 & 0 & 0 & 1 & 1 \\ 1 & 0 & 1 & 0 & 0 & 1 \\ 0 & 1 & 0 & 1 & 0 & 1 \\ 1 & 0 & 1 & 0 & 1 & 0 \end{bmatrix}. \tag{18.4}
$$

(a) Compute rank($\mathbb{H}$) and find the two solutions of the homogeneous linear system.

(b) Show that the linear system $\mathbb{H}x = \underline{b}$ has a solution if and only if $b_1 \oplus b_4 \oplus b_5 \oplus b_6 = 0$. How many solutions does it have in this case?

(c) Consider the factor graph associated with this linear system, (see Fig. 18.1). Show that each solution of the homogeneous system must correspond to a subset U of variable nodes with the following property: the subgraph induced by U and including all of the adjacent function nodes has even degree at the function nodes. Find one subgraph with this property.

18.1.2 Random XORSAT

The **random K-XORSAT** ensemble is defined by taking $\underline{b}$ uniformly at random in $\{0,1\}^M$, and $\mathbb{H}$ uniformly at random from the $N \times M$ matrices with entries in $\{0,1\}$ which have exactly K non-vanishing elements per row. Each equation thus involves K distinct variables chosen uniformly from the $\binom{N}{K}$ K-tuples, and the resulting factor graph is distributed according to the ensemble $\mathbb{G}_N(K, M)$.

A slightly different ensemble is defined by including each of the $\binom{N}{K}$ possible rows with K non-zero entries independently with probability $p = N\alpha/\binom{N}{K}$. The corresponding factor graph is then distributed according to the ensemble $\mathbb{G}_N(K, \alpha)$.

Given the relation between homogeneous and inhomogeneous systems described above, it is quite natural to introduce an ensemble of homogeneous linear systems. This is defined by taking $\mathbb{H}$ to be distributed as above, but with $\underline{b} = \underline{0}$. Since a homogeneous linear system always has at least one solution, this ensemble is sometimes referred to as **SAT K-XORSAT** or, in its spin interpretation, as the **ferromagnetic K-spin model**. Given a K-XORSAT formula F, we shall denote by F_0 the formula corresponding to the homogeneous system.

We are interested in the limit of large systems $N, M \to \infty$ with $\alpha = M/N$ fixed. By applying Friedgut's theorem (see Section 10.5), it is possible to show that, for $K \geq 3$, the probability for a random formula F to be SAT has a *sharp threshold*. More precisely, there exists an $\alpha_s^{(N)}(K)$ such that for $\alpha > (1+\delta)\alpha_s^{(N)}(K)$ or $\alpha < (1-\delta)\alpha_s^{(N)}(K)$, $\mathbb{P}\{F \text{ is SAT}\} \to 0$ or $\mathbb{P}\{F \text{ is SAT}\} \to 1$, respectively, as $N \to \infty$.

A moment of thought reveals that $\alpha_s^{(N)}(K) = \Theta(1)$. Let us give two simple bounds to convince the reader of this statement.

Upper bound. The relation between the homogeneous system and the original linear system derived in Exercise 18.2 implies that $\mathbb{P}\{F \text{ is SAT}\} = 2^{N-M}\mathbb{E}\{1/Z_0\}$. As $Z_0 \geq 1$, we get $\mathbb{P}\{F \text{ is SAT}\} \leq 2^{-N(\alpha-1)}$ and therefore $\alpha_s^{(N)}(K) \leq 1$.

Lower bound. For $\alpha < 1/K(K-1)$, the factor graph associated with F is formed, with high probability, by finite trees and unicyclic components. This corresponds to the matrix $\mathbb{H}$ being decomposable into blocks, each one corresponding to a connected component. The reader can show that, for $K \geq 3$, both a tree formula and a unicyclic component correspond to a linear system of full rank. Since each block has full rank, $\mathbb{H}$ has full rank as well. Therefore $\alpha_s^{(N)}(K) \geq 1/K(K-1)$.

Exercise 18.4 There is no sharp threshold for $K = 2$.

(a) Let $c(G)$ be the cyclic number of the factor graph G (the number of edges, minus the number of vertices, plus the number of connected components) of a random 2-XORSAT formula. Show that $\mathbb{P}\{F \text{ is SAT}\} = \mathbb{E}\, 2^{-c(G)}$.

(b) Argue that this implies that $\mathbb{P}\{F \text{ is SAT}\}$ is bounded away from 1 for any $\alpha > 0$.

(c) Show that $\mathbb{P}\{F \text{ is SAT}\}$ is bounded away from 0 for any $\alpha < 1/2$.

[Hint: Remember the geometrical properties of G discussed in Sections 9.3.2 and 9.4.]

In the next sections we shall show that $\alpha_{\mathrm{s}}^{(N)}(K)$ has a limit $\alpha_{\mathrm{c}}(K)$ and compute it explicitly. Before delving into this, it is instructive to derive two improved bounds.

Exercise 18.5 In order to obtain a better upper bound on $\alpha_{\mathrm{s}}^{(N)}(K)$, proceed as follows:

(a) Assume that, for any α, $Z_0 \geq 2^{N f_K(\alpha)}$ with probability larger than some $\varepsilon > 0$ at large N. Show that $\alpha_{\mathrm{s}}^{(N)}(K) \leq \alpha^*(K)$, where $\alpha^*(K)$ is the smallest value of α such that $1 - \alpha - f_K(\alpha) \leq 0$.

(b) Show that the above assumption holds with $f_K(\alpha) = \mathrm{e}^{-K\alpha}$, and that this yields $\alpha^*(3) \approx 0.941$. What is the asymptotic behaviour of $\alpha^*(K)$ for large K? How can you improve the exponent $f_K(\alpha)$?

Exercise 18.6 A better lower bound on $\alpha_{\mathrm{s}}^{(N)}(K)$ can be obtained through a first-moment calculation. In order to simplify the calculations, we consider here a modified ensemble in which the K variables entering into each equation are chosen independently and uniformly at random (they do not need to be distinct). The scrupulous reader can check at the end that returning to the original ensemble needs only small changes.

(a) Show that for a positive random variable Z, $(\mathbb{E}Z)(\mathbb{E}[1/Z]) \geq 1$. Deduce that $\mathbb{P}\{F \text{ is SAT}\} \geq 2^{N-M}/\mathbb{E}\, Z_{F_0}$.

(b) Prove that

$$\mathbb{E}\, Z_{F_0} = \sum_{w=0}^{N} \binom{N}{w} \left[\frac{1}{2}\left(1 + \left(1 - \frac{2w}{N}\right)^K\right) \right]^M. \tag{18.5}$$

(c) Let $g_K(x) = \mathcal{H}(x) + \alpha \log\left[\frac{1}{2}\left(1 + (1-2x)^K\right)\right]$, and define $\alpha_*(K)$ to be the largest value of α such that the maximum of $g_K(x)$ is achieved at $x = 1/2$. Show that $\alpha_{\mathrm{s}}^{(N)}(K) \geq \alpha_*(K)$. For $K = 3$, we get $\alpha_*(3) \approx 0.889$.

18.2 Belief propagation

18.2.1 BP messages and density evolution

Equation (18.1) provides a representation of the uniform measure over solutions of an XORSAT instance as a graphical model. This suggests that we should apply message-passing techniques. We shall describe belief propagation here and analyse its behaviour. While this may seem at first sight a detour from the objective of computing $\alpha_s^{(N)}(K)$, it will instead provide some important insight.

Let us assume that the linear system $\mathbb{H}\underline{x} = \underline{b}$ admits at least one solution, so that the model (18.1) is well defined. We shall first study the homogeneous version $\mathbb{H}\underline{x} = 0$, i.e. the measure μ_0, and then pass to μ. Applying the general definitions of Chapter 14, the BP update equations (14.14) and (14.15) for the homogeneous problem read

$$\nu_{i\to a}^{(t+1)}(x_i) \cong \prod_{b\in\partial i\backslash a} \widehat{\nu}_{b\to i}^{(t)}(x_i), \qquad \widehat{\nu}_{a\to i}^{(t)}(x_i) \cong \sum_{\underline{x}_{\partial a\backslash i}} \psi_a^0(\underline{x}_{\partial a}) \prod_{j\in\partial a\backslash i} \nu_{j\to a}^{(t)}(x_j).$$

$$(18.6)$$

These equations can be considerably simplified using the linear structure. We have seen that under μ_0, there are two types of variables, those 'frozen to 0' (i.e. equal to 0 in all solutions) and those which are 'free' (equally likely to be 0 or 1). BP aims at determining whether any single bit belongs to one class or the other. Now consider BP messages, which are also distributions over $\{0,1\}$. Suppose that at time $t = 0$ they also take one of two possible values that we denote by $*$ (corresponding to the uniform distribution) and 0 (distribution entirely supported on 0). It is not hard to show that this remains true at all subsequent times. The BP update equations (18.6) simplify under this initialization (they reduce to the erasure decoder of Sect. 15.3):

- At a variable node, the outgoing message is 0 unless all the incoming messages are $*$.
- At a function node, the outgoing message is $*$ unless all the incoming messages are 0.

(The message coming out of a degree-1 variable node is always $*$.)

These rules preserve a natural partial ordering. Given two sets of messages $\nu = \{\nu_{i\to a}\}$, $\widetilde{\nu} = \{\widetilde{\nu}_{i\to a}\}$, let us say that $\nu^{(t)} \succeq \widetilde{\nu}^{(t)}$ if, for each directed edge $i \to a$ where the message $\widetilde{\nu}_{i\to a}^{(t)} = 0$, then $\nu_{i\to a}^{(t)} = 0$ as well. It follows immediately from the update rules that if at some time t the messages are ordered as $\nu^{(t)} \succeq \widetilde{\nu}^{(t)}$, then this order is preserved at all later times: $\nu^{(s)} \succeq \widetilde{\nu}^{(s)}$ for all $s > t$.

This partial ordering suggests that we should pay special attention to the two 'extremal' initial conditions, namely $\nu_{i\to a}^{(0)} = *$ for all directed edges $i \to a$ or $\nu_{i\to a}^{(0)} = 0$ for all $i \to a$. The fraction of edges Q_t that carry a message 0 at time t is a deterministic quantity in the $N \to \infty$ limit. It satisfies the recursion

$$Q_{t+1} = 1 - \exp\{-K\alpha Q_t^{K-1}\},$$

$$(18.7)$$

with $Q_0 = 1$ for the '0' initial condition and $Q_0 = 0$ for the '$*$' initial condition. The density evolution recursion (18.7) is represented pictorially in Fig. 18.2.

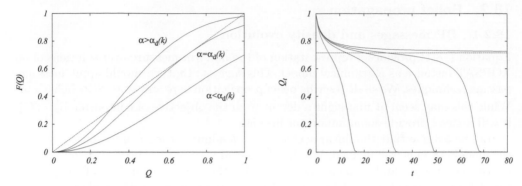

Fig. 18.2 Density evolution for the fraction of messages 0 for 3-XORSAT. *Left*: the mapping $F(Q) = 1 - \exp(-K\alpha Q^{K-1})$ below, at, and above the critical point $\alpha_{\mathrm{d}}(K = 3) \approx 0.818468$. *Right*: evolution of Q_t for (from *bottom to top*) $\alpha = 0.75$, 0.8, 0.81, 0.814, 0.818468.

Under the '*' initial condition, we have $Q_t = 0$ at all times t. In fact, the all-$*$ message configuration is always a fixed point of BP. On the other hand, when $Q_0 = 1$, we find two possible asymptotic behaviours: $Q_t \to 0$ for $\alpha < \alpha_{\mathrm{d}}(K)$, and $Q_t \to Q > 0$ for $\alpha > \alpha_{\mathrm{d}}(K)$. Here $Q > 0$ is the largest positive solution of $Q = 1 - \exp\{-K\alpha Q^{K-1}\}$. The critical value $\alpha_{\mathrm{d}}(K)$ of the density of equations $\alpha = M/N$ separating these two regimes is given by

$$\alpha_{\mathrm{d}}(K) = \sup \left\{ \alpha \ \text{such that} \ \ \forall x \in]0,1] : \ x < 1 - \mathrm{e}^{-K\alpha x^{K-1}} \right\}. \tag{18.8}$$

We get, for instance, $\alpha_{\mathrm{d}}(K) \approx 0.818469$, 0.772280, and 0.701780 for $K = 3$, 4, and 5, respectively, and $\alpha_{\mathrm{d}}(K) = \log K / K[1 + o(1)]$ as $K \to \infty$.

We have therefore found two regimes for the homogeneous random XORSAT problem in the large-N limit. For $\alpha < \alpha_{\mathrm{d}}(K)$, there is a unique BP fixed point with all messages[1] equal to $*$. The BP prediction for single-bit marginals that corresponds to this fixed point is $\nu_i(x_i = 0) = \nu_i(x_i = 1) = 1/2$.

For $\alpha > \alpha_{\mathrm{d}}(K)$, there exists more than one BP fixed point. We have found two of them: the all-$*$ one, and one with a density of $*$'s equal to Q. Other fixed points of the inhomogeneous problem can be constructed as follows for $\alpha \in]\alpha_{\mathrm{d}}(K), \alpha_{\mathrm{s}}(K)[$. Let $\underline{x}^{(*)}$ be a solution of the inhomogeneous problem, and let $\nu, \widehat{\nu}$ be a BP fixed point in the homogeneous case. Then the messages $\nu^{(*)}, \widehat{\nu}^{(*)}$, defined by

$$
\begin{aligned}
\nu_{j\to a}^{(*)}(x_j = 0) = \nu_{j\to a}^{(*)}(x_j = 1) = 1/2 && \text{if } \nu_{j\to a} = *, \\
\nu_{j\to a}^{(*)}(x_j) = \mathbb{I}(x_j = x_j^{(*)}) && \text{if } \nu_{j\to a} = 0, \tag{18.9}
\end{aligned}
$$

(and similarly for $\widehat{\nu}^{(*)}$) are a BP fixed point for the inhomogeneous problem.

For $\alpha < \alpha_{\mathrm{d}}(K)$, the inhomogeneous problem admits, with high probability, a unique BP fixed point. This is a consequence of the result obtained in the following exercise:

---

[1] While a vanishing fraction of messages $\nu_{i\to a} = 0$ is not excluded by our argument, it can be ruled out by a slightly lengthier calculation.

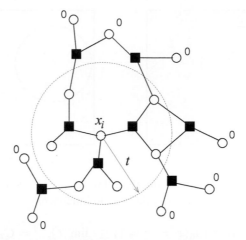

Fig. 18.3 Factor graph for a 3-XORSAT instance with the depth $t = 1$ neighbourhood of vertex i, $B_{i,t}(G)$, indicated. Fixing all of the variables outside $B_{i,t}(G)$ to 0 does not imply that x_i must be 0 in order to satisfy the homogeneous linear system.

Exercise 18.7 Consider a BP fixed point $\nu^{(*)}, \widehat{\nu}^{(*)}$ for the inhomogeneous problem, and assume all the messages to be of one of three types: $\nu_{j\to a}^{(*)}(x_j = 0) = 1$, $\nu_{j\to a}^{(*)}(x_j = 0) = 1/2$, and $\nu_{j\to a}^{(*)}(x_j = 0) = 0$. Assume furthermore that the messages are not 'contradictory,' i.e. that there exists no variable node i such that $\widehat{\nu}_{a\to i}^{(*)}(x_i = 0) = 1$ and $\widehat{\nu}_{b\to i}^{(*)}(x_i = 0) = 0$.
    Construct a non-trivial BP fixed point for the homogeneous problem.

18.2.2   Correlation decay

The BP prediction is that for $\alpha < \alpha_d(K)$, the marginal distribution of any bit x_i is uniform under either of the measures μ_0 or μ. The fact that the BP estimates do not depend on the initialization is an indication that the prediction is correct. Let us prove that this is indeed the case. To be definite, we consider the homogeneous problem (i.e. μ_0). The inhomogeneous case follows, using the general remarks in Section 18.1.1.

We start from an alternative interpretation of Q_t. Let $i \in \{1, \dots, N\}$ be a uniformly random variable index, and consider the ball of radius t around i in the factor graph G, $B_{i,t}(G)$ (see Fig. 18.3). Set all of the variables x_j outside this ball to $x_j = 0$, and let $Q_t^{(N)}$ be the probability that, under this condition, all of the solutions of the linear system $\mathbb{H}\underline{x} = \underline{0}$ have $x_i = 0$. The convergence of $B_{i,t}(G)$ to the tree model $\mathbb{T}(K, \alpha)$ discussed in Section 9.5 then implies that, for any given t, $\lim_{N\to\infty} Q_t^{(N)} = Q_t$. It also determines the initial condition to be $Q_0 = 1$.

Consider now the marginal distribution $\mu_0(x_i)$. If $x_i = 0$ in all of the solutions of $\mathbb{H}\underline{x} = \underline{0}$, then, a fortiori, $x_i = 0$ in all of the solutions that fulfil the additional condition $x_j = 0$ for $j \notin B_{i,t}(G)$. Therefore we have $\mathbb{P}\{\mu_0(x_i = 0) = 1\} \leq Q_t^{(N)}$. By taking the $N \to \infty$ limit, we get

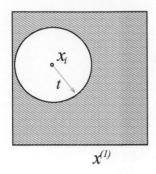
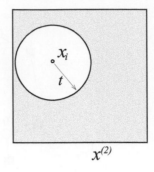

$x^{(1)}$ $x^{(2)}$

Fig. 18.4 A thought experiment: fix variables 'far' from i to two different assignments and check their influence on x_i. For $\alpha < \alpha_d$, there is no influence.

$$\lim_{N \to \infty} \mathbb{P}\{\mu_0(x_i = 0) = 1\} \leq \lim_{N \to \infty} Q_t^{(N)} = Q_t. \tag{18.10}$$

Letting $t \to \infty$ and noticing that the left-hand side does not depend on t, we get $\mathbb{P}\{\mu_0(x_i = 0) = 1\} \to 0$ as $N \to \infty$. In other words, all but a vanishing fraction of the bits are free for $\alpha < \alpha_d(K)$.

The number Q_t also has another interpretation, which generalizes to the inhomogeneous problem. We choose a solution $\underline{x}^{(*)}$ of the homogeneous linear system and, instead of fixing the variables outside the ball of radius t to 0, we fix them to $x_j = x_j^{(*)}$, $j \notin \mathsf{B}_{i,t}(G)$. Then $Q_t^{(N)}$ is the probability that $x_i = x_i^{(*)}$, under this condition. The same argument holds in the inhomogeneous problem, with the measure μ: if $\underline{x}^{(*)}$ is a solution of $\mathbb{H}\underline{x} = \underline{b}$ and we fix the variables outside $\mathsf{B}_{i,t}(G)$ to $x_j = x_j^{(*)}$, the probability that $x_i = x_i^{(*)}$ under this condition is again $Q_t^{(N)}$. The fact that $\lim_{t \to \infty} Q_t = 0$ when $\alpha < \alpha_d(K)$ thus means that a spin decorrelates from the whole set of variables at distance larger than t, when t is large. This formulation of the correlation decay is rather specific to XORSAT, because it relies on the dichotomous nature of this problem: either the 'far away' variables completely determine x_i, or they have no influence on it and it is uniformly random. A more generic formulation of the correlation decay, which generalizes to other problems which do not have this dichotomy property, consists in comparing two different choices $\underline{x}^{(1)}$, $\underline{x}^{(2)}$ of the reference solution (see Fig. 18.4). For $\alpha < \alpha_d(K)$, the correlations decay even in the worst case:

$$\lim_{N \to \infty} \mathbb{E}\left\{ \sup_{\underline{x}^{(1)}, \underline{x}^{(2)}} |\mu(x_i|\underline{x}_{\sim i,t}^{(1)}) - \mu(x_i|\underline{x}_{\sim i,t}^{(2)})| \right\} = Q_t \to 0 \tag{18.11}$$

as $t \to \infty$. In Chapter 22, we shall discuss weaker (non-worst-case) definitions of correlation decay, and their relation to phase transitions.

18.3   Core percolation and BP

18.3.1   The 2-core and peeling

What happens for $\alpha > \alpha_d(K)$? A first hint is provided by the instance in Fig. 18.1. In this case, the configuration of messages $\nu_{i \to a}^{(t)} = 0$ on all directed edges $i \to a$ is a fixed point of the BP update for the homogeneous system. A moment of thought shows that this happens because G has the property that each variable node has degree at least 2. We shall now see that, for $\alpha > \alpha_d(K)$, G has with high probability a subgraph (called the 2-core) with the same property.

We have already encountered similar structures in Section 15.3, where we identified them as being responsible for errors in the iterative decoding of LDPC codes over an erasure channel. Let us recall the relevant points[2] from that discussion. Given a factor graph G, a stopping set is a subset of the function nodes such that all of the variable nodes have a degree greater than or equal to 2 in the induced subgraph. The 2-core is the largest stopping set. It is unique and can be found by the peeling algorithm, which amounts to iterating the following procedure: find a variable node of degree 0 or 1 (a 'leaf'), and erase it together with the factor node adjacent to it, if there is one. The resulting subgraph, the 2-core, will be denoted as $K_2(G)$.

The peeling algorithm is of direct use for solving the linear system: if a variable has degree 1, the unique equation where it appears allows one to express it in terms of other variables. It can thus be eliminated from the problem. The 2-core of G is the factor graph associated with the linear system obtained by iterating this procedure, which we shall refer to as the 'core system'. The original system has a solution if and only if the core does. We shall refer to solutions of the core system as **core solutions**.

18.3.2 Clusters

Core solutions play an important role, as the set of solutions can be partitioned according to their core values. Given an assignment $\underline{x}$, we denote by $\pi_*(\underline{x})$ its projection onto the core, i.e. the vector of those entries in $\underline{x}$ that corresponds to vertices in the core. Suppose that the factor graph has a non-trivial 2-core, and let $\underline{x}^{(*)}$ be a core solution. We define the **cluster** associated with $\underline{x}^{(*)}$ as the set of solutions to the linear system such that $\pi_*(\underline{x}) = \underline{x}^{(*)}$ (the reason for the name 'cluster' will become clear in Section 18.5). If the core of G is empty, we shall adopt the convention that the entire set of solutions forms a unique cluster.

Given a solution $\underline{x}^{(*)}$ of the core linear system, we shall denote the corresponding cluster by $\mathcal{S}(\underline{x}^{(*)})$. One can obtain the solutions in $\mathcal{S}(\underline{x}^{(*)})$ by running the peeling algorithm in the reverse direction, starting from $\underline{x}^{(*)}$. In this process, one finds variables which are uniquely determined by $\underline{x}^{(*)}$; they form what is called the 'backbone' of the graph. More precisely, we define the **backbone** $B(G)$ as the subgraph of G that is obtained by augmenting $K_2(G)$ as follows. Set $B_0(G) = K_2(G)$. For any $t \geq 0$, pick a function node a which is not in $B_t(G)$ and which has at least $K-1$ of its neighbouring variable nodes in $B_t(G)$, and build $B_{t+1}(G)$ by adding a (and its neighbouring variables) to $B_t(G)$. If no such function node exists, set $B(G) = B_t(G)$ and halt the procedure. This definition of $B(G)$ does not depend on the order in which function nodes are added. The backbone contains the 2-core, and is such that any two solutions of the linear system which belong to the same cluster coincide on the backbone.

We have thus found that the variables in a linear system naturally divide into three possible types: the variables in the 2-core $K_2(G)$, those in $B(G) \setminus K_2(G)$ which are not

---

[2]Note that the structure causing decoding errors was the 2-core of the *dual* factor graph that is obtained by exchanging variable and function nodes.

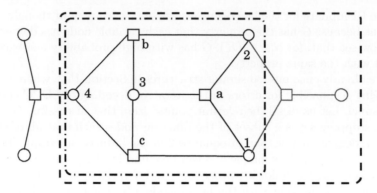

Fig. 18.5 The factor graph of an XORSAT problem, its core (enclosed by the dash–dotted line), and its backbone (enclosed by the dashed line).

in the core but are fixed by the core solution, and the variables which are not uniquely determined by $\underline{x}^{(*)}$. This distinction is based on the geometry of the factor graph, i.e. it depends only the matrix $\mathbb{H}$, and not on the value of the right-hand side $\underline{b}$ in the linear system. We shall now see how BP finds these structures.

18.3.3 Core, backbone, and belief propagation

Consider the homogeneous linear system $\mathbb{H}\underline{x} = 0$, and suppose that the BP algorithm is run with the initial condition $\nu_{i\to a}^{(0)} = 0$. Denote by $\nu_{i\to a}$, $\widehat{\nu}_{a\to i}$ the fixed point reached by BP (with the measure μ_0) under this initialization (the reader is invited to show that such a fixed point is indeed reached after a number of iterations at most equal to the number of messages).

The fixed-point messages $\nu_{i\to a}$, $\widehat{\nu}_{a\to i}$ can be exploited to find the 2-core $K_2(G)$, using the following properties (which can be proved by induction over t): (*i*) $\nu_{i\to a} = \widehat{\nu}_{a\to i} = 0$ for each edge (i,a) in $K_2(G)$; (*ii*) a variable i belongs to the core $K_2(G)$ if and only if it receives messages $\widehat{\nu}_{a\to i} = 0$ from at least two of the neighbouring function nodes $a \in \partial i$; and (*iii*) if a function node $a \in \{1, \ldots, M\}$ has $\nu_{i\to a} = 0$ for all of the neighbouring variable nodes $i \in \partial a$, then $a \in K_2(G)$.

The fixed-point BP messages also contain information about the backbone: a variable i belongs to the backbone $B(G)$ if and only if it receives at least one message $\widehat{\nu}_{a\to i} = 0$ from its neighbouring function nodes $a \in \partial i$.

Exercise 18.8 Consider an XORSAT problem described by the factor graph in Fig. 18.5.

(*a*) Using the peeling and backbone construction algorithms, check that the core and backbone are those described in the caption.

(*b*) Compute the BP messages found for the homogeneous problem as a fixed point of the BP iteration starting from the all-0 configuration. Check the core and backbone that you obtain from these messages.

(c) Consider the general inhomogeneous linear system with the same factor graph. Show that there exist two solutions to the core system: $x_1 = 0, x_2 = b_b \oplus b_c, x_3 = b_a \oplus b_b \oplus b_c, x_4 = b_a \oplus b_b$ and $x_1 = 0, x_2 = b_b \oplus b_c \oplus 1, x_3 = b_a \oplus b_b \oplus b_c, x_4 = b_a \oplus b_b \oplus 1$. Identify the two clusters of solutions.

18.4 The SAT–UNSAT threshold in random XORSAT

We shall now see how a sharp characterization of the core size in random linear systems provides the clue to the determination of the satisfiability threshold. Remarkably, this characterization can again be achieved through an analysis of BP.

18.4.1 The size of the core

Consider an homogeneous linear system over N variables drawn from the random K-XORSAT ensemble, and let $\{\nu_{i \to a}^{(t)}\}$ denote the BP messages obtained from the initialization $\nu_{i \to a}^{(0)} = 0$. The density evolution analysis of Section 18.2.1 implies that the fraction of edges carrying a message 0 at time t (which we called Q_t) satisfies the recursion equation (18.7). This recursion holds for any given t asymptotically as $N \to \infty$.

It follows from the same analysis that, in the large-N limit, the messages $\widehat{\nu}_{a \to i}^{(t)}$ entering a variable node i are i.i.d. with $\mathbb{P}\{\widehat{\nu}_{a \to i}^{(t)} = 0\} = \widehat{Q}_t \equiv Q_t^{K-1}$. Let us assume for a moment that the limits $t \to \infty$ and $N \to \infty$ can be exchanged without much harm. This means that the fixed-point messages $\widehat{\nu}_{a \to i}$ entering a variable node i are asymptotically i.i.d. with $\mathbb{P}\{\widehat{\nu}_{a \to i} = 0\} = \widehat{Q} \equiv Q^{K-1}$, where Q is the largest solution of the fixed-point equation

$$Q = 1 - e^{-K\alpha\widehat{Q}}, \qquad \widehat{Q} = Q^{K-1}. \tag{18.12}$$

The number of incoming messages with $\widehat{\nu}_{a \to i} = 0$ therefore converges to a Poisson random variable with mean $K\alpha\widehat{Q}$. The expected number of variable nodes in the core will be $\mathbb{E}|K_2(G)| = NV(\alpha, K) + o(N)$, where $V(\alpha, K)$ is the probability that such a Poisson random variable is greater than or equal to 2; that is,

$$V(\alpha, K) = 1 - e^{-K\alpha\widehat{Q}} - K\alpha\widehat{Q}\, e^{-K\alpha\widehat{Q}}. \tag{18.13}$$

In Fig. 18.6, we plot $V(\alpha)$ as a function of α. For $\alpha < \alpha_d(K)$, the peeling algorithm erases the whole graph, and there is no core. The size of the core jumps to some finite value at $\alpha_d(K)$, and when $\alpha \to \infty$ the core is the full graph.

Is $K_2(G)$ a random factor graph or does it have any particular structure? By construction, it cannot contain variable nodes of degree zero or one. Its expected degree profile (the expected fraction of nodes of any given degree) is asymptotically $\widehat{\Lambda} \equiv \{\widehat{\Lambda}_l\}$, where $\widehat{\Lambda}_l$ is the probability that a Poisson random variable of parameter $K\alpha\widehat{Q}$, conditioned to be at least 2, is equal to l. Explicitly, $\widehat{\Lambda}_0 = \widehat{\Lambda}_1 = 0$, and

$$\widehat{\Lambda}_l = \frac{1}{e^{K\alpha\widehat{Q}} - 1 - K\alpha\widehat{Q}} \frac{1}{l!} (K\alpha\widehat{Q})^l \qquad \text{for } l \geq 2. \tag{18.14}$$

Somewhat surprisingly, $K_2(G)$ does not have any more structure than that determined by its degree profile. This fact is stated more formally in the following theorem.

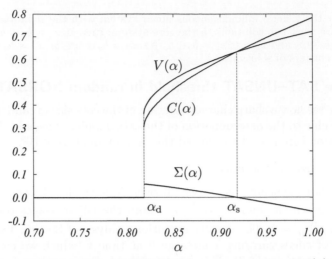

Fig. 18.6 The core of the random 3-XORSAT formulae contains $NV(\alpha)$ variables and $NC(\alpha)$ equations. These numbers are plotted versus the number of equations per variable of the original formula, α. The number of solutions to the XORSAT linear system is $\Sigma(\alpha) = V(\alpha) - C(\alpha)$. The core appears for $\alpha \geq \alpha_{\mathrm{d}}$, and the system becomes UNSAT for $\alpha > \alpha_{\mathrm{s}}$, where α_{s} is determined by $\Sigma(\alpha_{\mathrm{s}}) = 0$.

Theorem 18.1 *Consider a factor graph G from the ensemble $\mathbb{G}_N(K, N\alpha)$ with $K \geq 3$. Then*

(i) $K_2(G) = \emptyset$ *with high probability for* $\alpha < \alpha_{\mathrm{d}}(K)$.

(ii) *For* $\alpha > \alpha_{\mathrm{d}}(K)$, $|K_2(G)| = NV(\alpha, K) + o(N)$ *with high probability.*

(iii) *The fraction of vertices of degree l in $K_2(G)$ is between $\widehat{\Lambda}_l - \varepsilon$ and $\widehat{\Lambda}_l + \varepsilon$ with probability greater than $1 - \mathrm{e}^{-\Theta(N)}$.*

(iv) *Conditional on the number of variable nodes $n = |K_2(G)|$, the degree profile being $\widehat{\Lambda}$, $K_2(G)$ is distributed according to the ensemble $\mathbb{D}_n(\widehat{\Lambda}, x^K)$.*

We shall not provide a proof of this theorem. The main ideas have already been presented in the previous pages, except for one important mathematical point: how to exchange the limits $N \to \infty$ and $t \to \infty$. The basic idea is to run BP for a large but fixed number of steps t. At this point the resulting graph is 'almost' a 2-core, and one can show that a sequential peeling procedure stops in fewer than $N\varepsilon$ steps.

In Fig. 18.7, we compare the statement in this theorem with numerical simulations. The probability that G contains a 2-core $\mathrm{P}_{\mathrm{core}}(\alpha)$ increases from 0 to 1 as α ranges from 0 to ∞, with a threshold that becomes sharper and sharper as the size N increases. The threshold behaviour can be accurately described using finite-size scaling. Setting $\alpha = \alpha_{\mathrm{d}}(K) + \beta(K)\, z\, N^{-1/2} + \delta(K)\, N^{-2/3}$ (with properly chosen $\beta(K)$ and $\delta(K)$), one can show that $\mathrm{P}_{\mathrm{core}}(\alpha)$ approaches a K-independent non-trivial limit that depends smoothly on z.

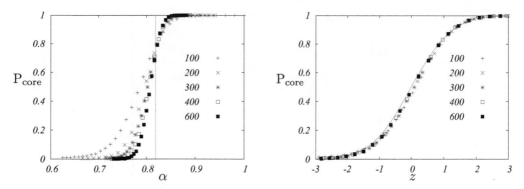

Fig. 18.7 The probability that a random graph from the ensemble $\mathbb{G}_N(K, \alpha)$ with $K = 3$ (or equivalently, the factor graph of a random 3-XORSAT formula) contains a 2 core. *Left*: the outcome of numerical simulations is compared with the asymptotic threshold $\alpha_{\mathrm{d}}(K)$. *Right*: scaling plot (see text).

18.4.2   The threshold

Knowing that the core is a random graph with degree distribution $\widehat{\Lambda}_l$, we can compute the expected number of equations in the core. This is given by the number of vertices times their average degree, divided by K, which yields $NC(\alpha, K) + o(N)$, where

$$C(\alpha, K) = \alpha\widehat{Q}(1 - \mathrm{e}^{-K\alpha\widehat{Q}}). \qquad (18.15)$$

In Fig. 18.6, we plot $C(\alpha, K)$ versus α. If $\alpha < \alpha_{\mathrm{d}}(K)$, there is no core. For $\alpha \in \,]\alpha_{\mathrm{d}}, \alpha_{\mathrm{s}}[$, the number of equations in the core is smaller than the number of variables $V(\alpha, K)$. Above α_c, there are more equations than variables.

A linear system has a solution if and only if the associated core problem has a solution. In a large random XORSAT instance, the core system involves approximately $NC(\alpha, K)$ equations between $NV(\alpha, K)$ variables. We shall show that these equations are, with high probability, linearly independent as long as $C(\alpha, K) < V(\alpha, K)$, which implies the following result.

Theorem 18.2. (XORSAT satisfiability threshold) *For $K \geq 3$, let*

$$\Sigma(K, \alpha) = V(K, \alpha) - C(K, \alpha) = Q - \alpha\widehat{Q}(1 + (K - 1)(1 - Q)), \qquad (18.16)$$

where $Q, \widehat{Q}$ are the largest solution of eqn (18.12). Let $\alpha_{\mathrm{s}}(K) = \inf\{\alpha : \Sigma(K, \alpha) < 0\}$. Consider a random K-XORSAT linear system with N variables and $N\alpha$ equations. The following results hold with a probability going to 1 in the large-N limit:

(i)   The system has a solution when $\alpha < \alpha_{\mathrm{s}}(K)$.

(ii) The system has no solution when $\alpha > \alpha_{\mathrm{s}}(K)$.

(iii) For $\alpha < \alpha_{\mathrm{s}}(K)$, the number of solutions is $2^{N(1-\alpha)+o(N)}$, and the number of clusters is $2^{N\Sigma(K,\alpha)+o(N)}$.

Note that the the last expression in eqn (18.16) can be obtained from eqns (18.13) and (18.15) using the fixed-point condition (18.12).

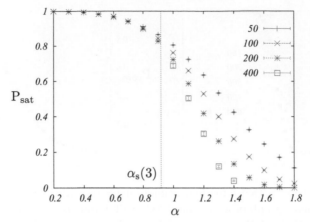

Fig. 18.8 The probability that a random 3-XORSAT formula with N variables and $N\alpha$ equations is SAT, estimated numerically by generating 10^3–10^4 random instances.

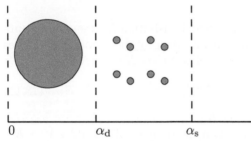

K	3	4	5
α_{d}	0.81847	0.77228	0.70178
α_{s}	0.91794	0.97677	0.99244

Fig. 18.9 *Left*: a pictorial view of the phase transitions in random XORSAT systems. The satisfiability threshold is α_{s}. In the 'Easy-SAT' phase, for $\alpha < \alpha_{\mathrm{d}}$, there is a single cluster of solutions. In the 'Hard-SAT' phase, for $\alpha_{\mathrm{d}} < \alpha < \alpha_{\mathrm{s}}$, the solutions of the linear system are grouped into well-separated clusters. *Right*: the thresholds α_{d}, α_{s} for various values of K. At large K, one has $\alpha_{\mathrm{d}}(K) \simeq \log K / K$ and $\alpha_{\mathrm{s}}(K) = 1 - \mathrm{e}^{-K} + O(\mathrm{e}^{-2K})$.

The predictions of this theorem are compared with numerical simulations in Fig. 18.8, and Fig. 18.9 summarizes the results for the thresholds for XORSAT.

Proof We shall convey only the basic ideas of the proof here and refer to the literature for technical details.

Let us start by proving (*ii*), namely that for $\alpha > \alpha_{\mathrm{s}}(K)$, random XORSAT instances are with high probability UNSAT. This follows from a linear-algebra argument. Let $\mathbb{H}_*$ denote the 0–1 matrix associated with the core, i.e. a matrix including those rows/columns such that the associated function/variable nodes belong to $K_2(G)$. Notice that if a given row is included in $\mathbb{H}_*$ then all the columns corresponding to non-zero entries of that row are also in $\mathbb{H}_*$. As a consequence, a necessary condition for the rows of $\mathbb{H}$ to be independent is that the rows of $\mathbb{H}_*$ are independent. This is, in turn, impossible if the number of columns in $\mathbb{H}_*$ is smaller than its number of rows.

Quantitatively, one can show that $M - \mathrm{rank}(\mathbb{H}) \geq \mathrm{rows}(\mathbb{H}_*) - \mathrm{cols}(\mathbb{H}_*)$ (with the obvious meanings of $\mathrm{rows}(\,\cdot\,)$ and $\mathrm{cols}(\,\cdot\,)$). For large random XORSAT systems, Theo-

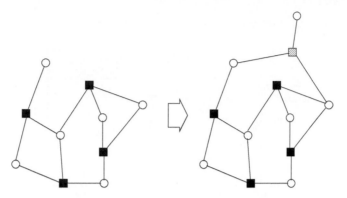

Fig. 18.10 Adding a function node involving a variable node of degree one. The corresponding linear equation is independent of the other equations.

rem 18.1 implies that $\mathrm{rows}(\mathbb{H}_*) - \mathrm{cols}(\mathbb{H}_*) = -N\Sigma(K,\alpha) + o(N)$ with high probability. According to our discussion in Section 18.1.1, among the 2^M possible choices of the right-hand side vector $\underline{b}$, only $2^{\mathrm{rank}(\mathbb{H})}$ are in the image of $\mathbb{H}$ and thus lead to a solvable system. In other words, conditional on $\mathbb{H}$, the probability that random XORSAT is solvable is $2^{\mathrm{rank}(\mathbb{H})-M}$. By the above argument this is, with high probability, smaller than $2^{N\Sigma(K,\alpha)+o(N)}$. Since $\Sigma(K,\alpha) < 0$ for $\alpha > \alpha_{\mathrm{s}}(K)$, it follows that the system is UNSAT with high probability.

In order to show that a random system is satisfiable with high probability when $\alpha < \alpha_{\mathrm{s}}(K)$, one has to prove the following facts: (i) if the core matrix $\mathbb{H}_*$ has maximum rank, then $\mathbb{H}$ has maximum rank as well; (ii) if $\alpha < \alpha_{\mathrm{s}}(K)$, then $\mathbb{H}_*$ has maximum rank with high probability. As a by-product, the number of solutions is $2^{N-\mathrm{rank}(\mathbb{H})} = 2^{N-M}$.

(i) The first step follows from the observation that G can be constructed from $K_2(G)$ through an inverse peeling procedure. At each step, one adds a function node which involves at least a degree-one variable (see Fig. 18.10). Obviously this newly added equation is linearly independent of the previous ones, and therefore $\mathrm{rank}(\mathbb{H}) = \mathrm{rank}(\mathbb{H}_*) + M - \mathrm{rows}(\mathbb{H}_*)$.

(ii) Let $n = \mathrm{cols}(\mathbb{H}_*)$ be the number of variable nodes and $m = \mathrm{rows}(\mathbb{H}_*)$ the number of function nodes in the core $K_2(G)$. Let us consider the homogeneous system on the core, $\mathbb{H}_* \underline{x} = \underline{0}$, and denote by Z_* the number of solutions to this system. We shall show that, with high probability, this number is equal to 2^{n-m}. This means that the dimension of the kernel of $\mathbb{H}_*$ is $n - m$ and therefore $\mathbb{H}_*$ has full rank.

We know from linear algebra that $Z_* \geq 2^{n-m}$. To prove the reverse inequality, we use a first-moment method. According to Theorem 18.1, the core is a uniformly random factor graph with $n = NV(K,\alpha) + o(N)$ variables and degree profile $\Lambda = \widehat{\Lambda} + o(1)$. We denote the expectation value with respect to this ensemble by $\mathbb{E}$. We shall use a first-moment analysis below to show that, when $\alpha < \alpha_{\mathrm{c}}(K)$,

$$\mathbb{E}\{Z_*\} = 2^{n-m}[1 + o_N(1)]. \qquad (18.17)$$

The Markov inequality $\mathbb{P}\{Z_* > 2^{n-m}\} \leq 2^{-n+m}\mathbb{E}\{Z_*\}$ then implies the bound.

The surprise is that eqn (18.17) holds, and thus a simple first-moment estimate allows us to establish that $\mathbb{H}_*$ has full rank. We saw in Exercise 18.6 that the same

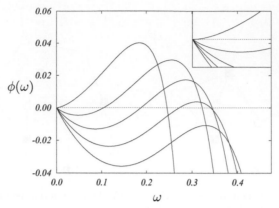

Fig. 18.11 The exponential rate $\phi(\omega)$ of the weight enumerator of the core of a random 3-XORSAT formula. From *top* to *bottom*, $\alpha = \alpha_{\mathrm{d}}(3) \approx 0.818469$, 0.85, 0.88, 0.91, and 0.94 (recall that $\alpha_{\mathrm{s}}(3) \approx 0.917935$). *Inset*: blow-up of the small-ω region.

approach, when applied directly to the original linear system, fails above some $\alpha_*(K)$ which is strictly smaller than $\alpha_{\mathrm{s}}(K)$. Reducing the original graph to its 2-core has drastically reduced the fluctuations of the number of solutions, thus allowing a successful application of the first-moment method.

We now turn to the proof of eqn (18.17), and we shall limit ourselves to the computation of $\mathbb{E}\{Z_*\}$ to the leading exponential order, when the core size and degree profiles take their typical values $n = NV(K,\alpha)$, $\Lambda = \widehat{\Lambda}$, and $P(x) = x^K$. This problem is equivalent to computing the expected number of codewords in the LDPC code defined by the core system, which we have already done in Section 11.2. The result takes the typical form

$$\mathbb{E}\{Z_*\} \doteq \exp\left\{N \sup_{\omega \in [0, V(K,\alpha)]} \phi(\omega)\right\}. \tag{18.18}$$

Here, $\phi(\omega)$ is the exponential rate for the number of solutions with weight $N\omega$. Adapting eqn (11.18) to the present case, we obtain the following parametric expression:

$$\phi(\omega) = -\omega \log x - \eta(1 - e^{-\eta}) \log(1 + yz) \tag{18.19}$$

$$+ \sum_{l \geq 2} e^{-\eta} \frac{\eta^l}{l!} \log(1 + xy^l) + \frac{\eta}{K}(1 - e^{-\eta}) \log q_K(z),$$

$$\omega = \sum_{l \geq 2} e^{-\eta} \frac{\eta^l}{l!} \frac{xy^l}{1 + xy^l}, \tag{18.20}$$

where $\eta = K\alpha\widehat{Q}_*$ and $q_K(z) = [(1+z)^K + (1-z)^K]/2$, and $y = y(x)$, $z = z(x)$ are the solution of

$$z = \frac{\sum_{l \geq 1}[\eta^l/l!]\,[xy^{l-1}/(1+xy^l)]}{\sum_{l \geq 1}[\eta^l/l!]\,[1/(1+xy^l)]}, \qquad y = \frac{(1+z)^{K-1} - (1-z)^{K-1}}{(1+z)^{K-1} + (1-z)^{K-1}}. \tag{18.21}$$

With a little work one can see that $\omega_* = V(K,\alpha)/2$ is a local maximum of $\phi(\omega)$, with $\phi(\omega_*) = \Sigma(K,\alpha)\log 2$. As long as ω_* is a global maximum, $\mathbb{E}\{Z_*|n,\Lambda\} \doteq \exp\{N\phi(\omega_*)\} \doteq 2^{n-m}$. It turns out (seeFig. 18.11) that the only other local maximum is at $\omega = 0$, corresponding to $\phi(0) = 0$. Therefore $\mathbb{E}\{Z_*|n,\Lambda\} \doteq 2^{n-m}$ as long as $\phi(\omega_*) = \Sigma(K,\alpha) > 0$, i.e. for any $\alpha < \alpha_{\rm s}(K)$

Note that the actual proof of eqn (18.17) is more complicated, because it requires estimating the subexponential factors. Nevertheless, it can be carried out successfully. $\square$

18.5   The Hard-SAT phase: Clusters of solutions

In random XORSAT, the whole regime $\alpha < \alpha_{\rm s}(K)$ is SAT. This means that, with high probability, there exist solutions to the random linear system, and the number of solutions is in fact $Z \doteq e^{N(1-\alpha)}$. Note that the number of solutions does not show any precursor of the SAT–UNSAT transition at $\alpha_{\rm s}(K)$ (recall that $\alpha_{\rm s}(K) < 1$), nor does it carry any trace of the sudden appearance of a non-empty 2-core at $\alpha_{\rm d}(K)$.

On the other hand, the threshold $\alpha_{\rm d}(K)$ separates two phases, which we shall call the **Easy-SAT** phase (for $\alpha < \alpha_{\rm d}(K)$) and the **Hard-SAT** phase (for $\alpha \in]\alpha_{\rm d}(K), \alpha_{\rm s}(K)[$). These two phases differ in the structure of the solution space, as well as in the behaviour of some simple algorithms.

In the Easy-SAT phase there is no core, solutions can be found in (expected) linear time using the peeling algorithm, and they form a unique cluster. In the Hard-SAT phase, the factor graph has a large 2-core, and no algorithm is known that finds a solution in linear time. The solutions are partitioned into $2^{N\Sigma(K,\alpha)+o(N)}$ clusters. Until now, we have used the name 'cluster' fairly arbitrarily, and it has only denoted a subset of solutions that coincide in the core. The next result shows that distinct clusters are 'far apart' in Hamming space.

Proposition 18.3 *In the Hard-SAT phase, there exists a $\delta(K,\alpha) > 0$ such that, with high probability, any two solutions in distinct clusters have a Hamming distance larger than $N\delta(K,\alpha)$.*

Proof The proof follows from the computation of the weight enumerator exponent $\phi(\omega)$ (see eqn (18.20) and Fig. 18.11). One can see that for any $\alpha > \alpha_{\rm d}(K)$, $\phi'(0) < 0$, and, as a consequence, there exists a $\delta(K,\alpha) > 0$ such that $\phi(\omega) < 0$ for $0 < \omega < \delta(K,\alpha)$. This implies that if $\underline{x}_*$, $\underline{x}'_*$ are two distinct solutions of the core linear system, then either $d(\underline{x}_*, \underline{x}'_*) = o(N)$ or $d(\underline{x}, \underline{x}') > N\delta(K,\alpha)$. It turns out that the first case can be excluded along the lines of the minimal-distance calculation of Section 11.2. Therefore, if $\underline{x}$, $\underline{x}'$ are two solutions belonging to distinct clusters, then $d(\underline{x}, \underline{x}') \geq d(\pi_*(\underline{x}), \pi_*(\underline{x}')) \geq N\delta(K,\alpha)$. $\square$

This result suggests that we can regard clusters as 'lumps' of solutions well separated from each other. One aspect which has been conjectured, but not proved, concerns the assertion that clusters form 'well-connected components'. By this we mean that any two solutions in a cluster can be joined by a sequence of other solutions, whereby two successive solutions in the sequence differ in at most s_N variables, with $s_N = o(N)$ (a reasonable expectation is $s_N = \Theta(\log N)$).

18.6   An alternative approach: The cavity method

The analysis of random XORSAT in the previous sections relied heavily on the linear structure of the problem, as well as on the very simple instance distribution. This section describes an alternative approach that is potentially generalizable to more complex situations. The price to be paid is that this second derivation relies on some assumptions about the structure of the solution space. The observation that our final results coincide with those obtained in the previous section gives some credibility to these assumptions.

The starting point is the observation that BP correctly computes the marginals of $\mu(\cdot)$ (the uniform measure over the solution space) for $\alpha < \alpha_{\rm d}(K)$, i.e. as long as the set of solutions forms a single cluster. We want to extend its domain of validity to $\alpha > \alpha_{\rm d}(K)$. If we index the clusters by $n \in \{1, \ldots, \mathcal{N}\}$, the uniform measure $\mu(\cdot)$ can be decomposed into a convex combination of uniform measures over each single cluster:

$$\mu(\cdot) = \sum_{n=1}^{\mathcal{N}} w_n \, \mu^n(\cdot). \tag{18.22}$$

Note that in the present case, $w_n = 1/\mathcal{N}$ is independent of n and the measures $\mu^n(\cdot)$ are obtained from each other via a translation, but this will not be true in more general situations.

Consider an inhomogeneous XORSAT linear system, and denote by $\underline{x}^{(*)}$ one of its solutions in cluster n. The distribution μ^n has single-variable marginals $\mu^n(x_i) = \mathbb{I}(x_i = x_i^{(*)})$ if node i belongs to the backbone, and $\mu^n(x_i = 0) = \mu^n(x_i = 1) = 1/2$ on the other nodes.

In fact, we can associate with each solution $\underline{x}^{(*)}$ a fixed point of the BP equations. We have already described this in Section 18.2.1 (see eqn (18.9)). At this fixed point, messages take one of the following three values: $\nu_{i \to a}^{(*)}(x_i) = \mathbb{I}(x_i = 0)$ (which we shall denote by $\nu_{i \to a}^{(*)} = 0$), $\nu_{i \to a}^{(*)}(x_i) = \mathbb{I}(x_i = 1)$ (denoted by $\nu_{i \to a}^{(*)} = 1$), and $\nu_{i \to a}^{(*)}(x_i = 0) = \nu_{i \to a}^{(*)}(x_i = 1) = 1/2$ (denoted by $\nu_{i \to a}^{(*)} = *$). An analogous notation will be used for function-to-variable-node messages. The solution can be written most easily in terms of the latter,

$$\widehat{\nu}_{a \to i}^{(*)} = \begin{cases} 1 \text{ if } x_i^{(*)} = 1 \text{ and } i, a \in B(G), \\ 0 \text{ if } x_i^{(*)} = 0 \text{ and } i, a \in B(G), \\ * \text{ otherwise.} \end{cases} \tag{18.23}$$

Note that these messages depend only on the value of $x_i^{(*)}$ on the backbone of G, and hence they depend on $\underline{x}^{(*)}$ only through the cluster that it belongs to. Conversely, for any two distinct clusters, the above definition gives two distinct fixed points. Because of this observation, we shall denote these fixed points by $\{\nu_{i \to a}^{(n)}, \widehat{\nu}_{a \to i}^{(n)}\}$, where n is a cluster index.

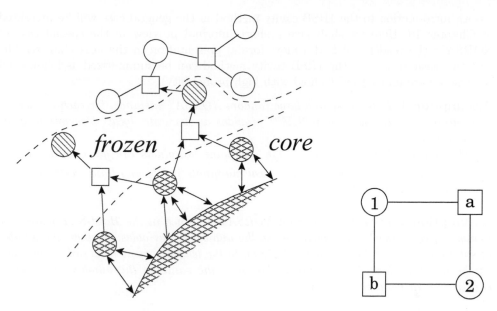

Fig. 18.12 *Left*: a set of BP messages associated with one cluster (cluster number n) of solutions. An arrow along an edge means that the corresponding message (either $\nu_{i \to a}^{(n)}$ or $\widehat{\nu}_{a \to i}^{(n)}$) takes a value in $\{0, 1\}$. The other messages are equal to $*$. *Right*: a small XORSAT instance. The core is the whole graph. In the homogeneous problem, there are two solutions, which form two clusters: $x_1 = x_2 = 0$ and $x_1 = x_2 = 1$. Besides the two corresponding BP fixed points described in Proposition 18.4, and the all-$*$ fixed point, there exist other fixed points such as $\widehat{\nu}_{a \to 1} = \nu_{1 \to b} = \widehat{\nu}_{b \to 2} = \nu_{2 \to a} = 0$, $\widehat{\nu}_{a \to 2} = \nu_{2 \to b} = \widehat{\nu}_{b \to 1} = \nu_{1 \to a} = *$.

Let us recall the BP fixed-point condition:

$$\nu_{i \to a} = \begin{cases} * & \text{if } \widehat{\nu}_{b \to i} = * \text{ for all } b \in \partial i \backslash a, \\ \text{any 'non-}*\text{' } \widehat{\nu}_{b \to i} & \text{otherwise,} \end{cases} \tag{18.24}$$

$$\widehat{\nu}_{a \to i} = \begin{cases} * & \text{if } \exists j \in \partial a \backslash i \text{ such that } \widehat{\nu}_{j \to a} = *, \\ b_a \oplus \nu_{j_1 \to a} \oplus \cdots \oplus \nu_{j_l \to a} & \text{otherwise.} \end{cases} \tag{18.25}$$

Below, we shall denote these equations symbolically as

$$\nu_{i \to a} = \mathsf{f}\{\widehat{\nu}_{b \to i}\}, \qquad \widehat{\nu}_{a \to i} = \widehat{\mathsf{f}}\{\nu_{j \to a}\}. \tag{18.26}$$

Let us summarize our findings.

Proposition 18.4 *With each cluster n, we can associate a distinct fixed point of the BP equations (18.25) $\{\nu_{i \to a}^{(n)}, \widehat{\nu}_{a \to i}^{(n)}\}$, such that $\widehat{\nu}_{a \to i}^{(n)} \in \{0, 1\}$ if i, a are in the backbone and $\widehat{\nu}_{a \to i}^{(n)} = *$ otherwise.*

Note that the converse of this proposition is false: there may exist solutions to the BP equations which are not of the above type. One of them is the all-$*$ solution. Non-trivial solutions exist as well, as shown in Fig. 18.12.

An introduction to the 1RSB cavity method in the general case will be presented in Chapter 19. Here we shall give a short informal preview in the special case of XORSAT: the reader will find a more formal presentation in the next chapter. The first two assumptions of the 1RSB cavity method can be summarized as follows (all statements are understood to hold with high probability).

Assumption 1 *In the case of a large random XORSAT instance, for each cluster 'n' of solutions, the BP solution $\nu^{(n)}, \widehat{\nu}^{(n)}$ provides an accurate 'local' description of the measure $\mu^n(\cdot)$.*

This means that, for instance, the one-point marginals are given by $\mu^n(x_j) \cong \prod_{a\in\partial j} \widehat{\nu}_{a\to j}^{(n)}(x_j) + o(1)$, but also that local marginals inside any finite cavity are well approximated by eqn (14.18).

Assumption 2 *For a large random XORSAT instance in the Hard-SAT phase, the number of clusters $e^{N\Sigma}$ is exponential in the number of variables. Further, the number of solutions of the BP equations (18.25) is, to the leading exponential order, the same as the number of clusters. In particular, it is the same as the number of solutions constructed in Proposition 18.4.*

A priori, one might have hoped to identify the set of messages $\{\nu_{i\to a}^{(n)}\}$ for each cluster. The cavity method gives up this ambitious objective and aims instead to compute the distribution of $\nu_{i\to a}^{(n)}$ for any fixed edge $i \to a$, when n is a cluster index drawn with distribution $\{w_n\}$. We thus want to compute the quantities

$$Q_{i\to a}(\nu) = \mathbb{P}\left\{\nu_{i\to a}^{(n)} = \nu\right\}, \qquad \widehat{Q}_{a\to i}(\widehat{\nu}) = \mathbb{P}\left\{\widehat{\nu}_{a\to i}^{(n)} = \widehat{\nu}\right\}, \qquad (18.27)$$

for $\nu, \widehat{\nu} \in \{0, 1, *\}$. Computing these probabilities rigorously is still a challenging task. In order to proceed, we make some assumptions about the joint distribution of the messages $\nu_{i\to a}^{(n)}$ when n is a random cluster index (chosen with probability w_n).

The simplest idea would be to assume that messages on 'distant' edges are independent. For instance, let us consider the set of messages entering a given variable node i. Their only correlations are induced through BP equations along the loops to which i belongs. Since in random K-XORSAT formulae such loops have, with high probability, a length of order $\log N$, one might think that messages coming into a given node are asymptotically independent. Unfortunately, this assumption is false. The reason is easily understood if we assume that $\widehat{Q}_{a\to i}(0), \widehat{Q}_{a\to i}(1) > 0$ for at least two of the function nodes a adjacent to a given variable node i. This would imply that, with positive probability, a randomly sampled cluster has $\nu_{a\to i}^{(n)} = 0$ and $\nu_{b\to i}^{(n)} = 1$. But there does not exist any such cluster, because in such a situation there is no consistent prescription for the marginal distribution of x_i under $\mu^n(\cdot)$.

Our assumption will be that the next simplest thing happens: messages are independent conditional on the fact that they do not contradict each other.

Assumption 3 *Consider the Hard-SAT phase of a random XORSAT problem. Denote by $i \in G$ a uniformly random node, denote by n a random cluster index with distribution $\{w_n\}$, and let ℓ be an integer ≥ 1. Then the messages $\{\nu_{j\to b}^{(n)}\}$, where*

(j, b) *are all of the edges at distance ℓ from i directed towards i, are asymptotically independent under the condition of being* **compatible**.

Here 'compatible' means the following. Consider the linear system $\mathbb{H}_{i,\ell}\underline{x}_{i,\ell} = \underline{0}$ for the neighbourhood of radius ℓ around node i. If this admits a solution under the boundary condition $x_j = \nu_{j\to b}$ for all of the boundary edges (j, b) on which $\{\nu_{j\to b}\} \in \{0, 1\}$, then the messages $\{\nu_{j\to b}\}$ are said to be compatible.

Given the messages $\nu_{j\to b}$ at the boundary of a radius-ℓ neighbourhood, the BP equations (18.24) and (18.25) allow one to determine the messages inside this neighbourhood. Consider, in particular, two nested neighbourhoods at distances ℓ and $\ell+1$ from i. The inward messages on the boundary of the larger neighbourhood completely determines those on the boundary of the smaller neighbourhood. A little thought shows that if the messages on the outer boundary are distributed according to Assumption 3, then the distribution of the resulting messages on the inner boundary also satisfies the same assumption. Further, the distributions are consistent if and only if the following 'survey propagation' equations are satisfied by the one-message marginals:

$$Q_{i\to a}(\nu) \cong \sum_{\{\widehat{\nu}_b\}} \prod_{b\in\partial i\backslash a} \widehat{Q}_{b\to i}(\widehat{\nu}_b)\ \mathbb{I}(\nu = \mathsf{f}\{\widehat{\nu}_b\})\ \mathbb{I}(\{\widehat{\nu}_b\}_{b\in\partial i\backslash a} \in \mathsf{COMP})\,, \quad (18.28)$$

$$\widehat{Q}_{a\to i}(\widehat{\nu}) = \sum_{\{\nu_j\}} \prod_{j\in\partial a\backslash i} Q_{j\to a}(\nu_j)\ \mathbb{I}(\widehat{\nu} = \hat{\mathsf{f}}\{\nu_j\})\,. \quad (18.29)$$

Here, $\{\widehat{\nu}_b\} \in \mathsf{COMP}$ if and only if the messages are compatible (i.e. they do not contain both a 0 and a 1). Since Assumptions 1, 2, and 3 above hold only with high probability, and asymptotically in the system size, the equalities in eqns (18.28) and (18.29) must also be interpreted as approximate. The equations should be satisfied within any given accuracy ε, with high probability as $N \to \infty$.

Exercise 18.9 Show that eqns (18.28) and (18.29) can be written explicitly as

$$Q_{i\to a}(0) \cong \prod_{b\in\partial i\backslash a} (\widehat{Q}_{b\to i}(0) + \widehat{Q}_{b\to i}(*)) - \prod_{b\in\partial i\backslash a} \widehat{Q}_{b\to i}(*)\,, \quad (18.30)$$

$$Q_{i\to a}(1) \cong \prod_{b\in\partial i\backslash a} (\widehat{Q}_{b\to i}(1) + \widehat{Q}_{b\to i}(*)) - \prod_{b\in\partial i\backslash a} \widehat{Q}_{b\to i}(*)\,, \quad (18.31)$$

$$Q_{i\to a}(*) \cong \prod_{b\in\partial i\backslash a} \widehat{Q}_{b\to i}(*)\,, \quad (18.32)$$

where the symbol $\cong$ hides a global normalization constant, and

$$\widehat{Q}_{a\to i}(0) = \frac{1}{2}\left\{\prod_{j\in\partial a\backslash i}(Q_{j\to a}(0) + Q_{j\to a}(1)) + \prod_{j\in\partial a\backslash i}(Q_{j\to a}(0) - Q_{j\to a}(1))\right\},$$

(18.33)

$$\widehat{Q}_{a\to i}(1) = \frac{1}{2}\left\{\prod_{j\in\partial a\backslash i}(Q_{j\to a}(0) + Q_{j\to a}(1)) - \prod_{j\in\partial a\backslash i}(Q_{j\to a}(0) - Q_{j\to a}(1))\right\},$$

(18.34)

$$\widehat{Q}_{a\to i}(*) = 1 - \prod_{j\in\partial a\backslash i}(Q_{j\to a}(0) + Q_{j\to a}(1)).$$

(18.35)

The final step of the 1RSB cavity method consists in looking for a solution of eqns (18.28) and (18.29). There are no rigorous results on the existence or number of such solutions. Further, since these equations are only approximate, approximate solutions should be considered as well. In the present case, a very simple (and somewhat degenerate) solution can be found that yields correct predictions for all of the quantities of interest. In this solution, the message distributions take one of two possible forms: on some edges, one has $Q_{i\to a}(0) = Q_{i\to a}(1) = 1/2$ (with some misuse of notation, we shall write $Q_{i\to a} = 0$ in this case), and on some other edges, $Q_{i\to a}(*) = 1$ (we shall then write $Q_{i\to a} = *$). Analogous forms hold for $\widehat{Q}_{a\to i}$. A little algebra shows that this is a solution if and only if the η's satisfy

$$Q_{i\to a} = \begin{cases} * & \text{if } \widehat{Q}_{b\to i} = * \text{ for all } b \in \partial i\backslash a, \\ 0 & \text{otherwise,} \end{cases}$$

(18.36)

$$\widehat{Q}_{a\to i} = \begin{cases} * & \text{if } \exists j \in \partial a\backslash i \text{ such that } \widehat{Q}_{j\to a} = *, \\ 0 & \text{otherwise.} \end{cases}$$

(18.37)

These equations are identical to the original BP equations for the homogeneous problem (this feature is very specific to XORSAT and will not generalize to more advanced applications of the method). However, the interpretation is now completely different. On the edges where $Q_{i\to a} = 0$, the corresponding message $\nu_{i\to a}^{(n)}$ depends on the cluster n; $\nu_{i\to a}^{(n)} = 0$ in half of the clusters, and $\nu_{i\to a}^{(n)} == 1$ in the other half. These edges are those inside the core, or in the backbone but directed 'outward' with respect to the core, as shown in Fig. 18.12. On the other edges, the message does not depend upon the cluster, and $\nu_{i\to a}^{(n)} = *$ for all n's.

A concrete interpretation of these results is obtained if we consider the one-bit marginals $\mu^n(x_i)$ under the single-cluster measure. According to Assumption 1 above, we have $\mu^n(x_i = 0) = \mu^n(x_i = 1) = 1/2$ if $\widehat{\nu}_{a\to i}^{(n)} = *$ for all $a \in \partial i$. If, on the other hand, $\widehat{\nu}_{a\to i}^{(n)} = 0$ or 1 for at least one $a \in \partial i$, then $\mu^n(x_i = 0) = 1$ or $\mu^n(x_i = 0) = 0$, respectively. We thus recover the full solution discussed in the previous sections: inside a given cluster n, the variables in the backbone are completely frozen, either to 0 or to 1. The other variables have equal probability to be 0 or 1 under the measure μ^n.

The cavity approach allows one to compute the complexity $\Sigma(K, \alpha)$, as well as many other properties of the measure $\mu(\,\cdot\,)$. We shall see this in the next chapter.

Notes

Random XORSAT formulae were first studied as a simple example of random satisfiability by Creignou and Daudé (1999). This work considered the case of 'dense formulae' where each clause includes $O(N)$ variables. In this case the SAT–UNSAT threshold is at $\alpha = 1$. In the field of coding theory, this model had been characterized since the work by Elias (1955) (see Chapter 6).

The case of sparse formulae was addressed using moment bounds by Creignou *et al.* (2003). The replica method was used by Ricci-Tersenghi *et al.* (2001), Franz *et al.* (2001*a*), and Franz *et al.* (2001*b*) to derive the clustering picture, determine the SAT–UNSAT threshold, and study the glassy properties of the clustered phase.

The fact that, after reducing the linear system to its core, the first-moment method provides a sharp characterization of the SAT–UNSAT threshold was discovered independently by two groups: Cocco *et al.* (2003) and Mézard *et al.* (2003). The latter also discussed the application of the cavity method to the problem. The full second-moment calculation that completes the proof can be found for the case $K = 3$ in Dubois and Mandler (2002).

Papers by Montanari and Semerjian (2005), Montanari and Semerjian (2006*a*), and Mora and Mézard (2006) were devoted to finer geometrical properties of the set of solutions of random K-XORSAT formulae. Despite these efforts, it remains to be proved that clusters of solutions are indeed 'well connected'.

Since the locations of various transitions are known rigorously, a natural question is to study the critical window. Finite-size scaling of the SAT–UNSAT transition was investigated numerically by Leone *et al.* (2001). A sharp characterization of finite-size scaling for the appearence of a 2-core, corresponding to the clustering transition, was achieved by Dembo and Montanari (2008*a*).

19

The 1RSB cavity method

The effectiveness of belief propagation depends on one basic assumption: when a function node is pruned from the factor graph, the adjacent variables become weakly correlated with respect to the resulting distribution. This hypothesis may break down either because of the existence of small loops in the factor graph or because variables are correlated at large distances. In factor graphs with a locally tree-like structure, the second scenario is responsible for the failure of BP. The emergence of such long-range correlations is a signature of a phase transition separating a 'weakly correlated' and a 'highly correlated' phase. The latter is often characterized by the decomposition of the (Boltzmann) probability distribution into well-separated 'lumps' (pure Gibbs states).

We considered a simple example of this phenomenon in our study of random XOR-SAT. A similar scenario holds in a variety of problems, from random graph colouring to random satisfiability and spin glasses. The reader should be warned that the structure and organization of the pure states in such systems is far from being fully understood. Furthermore, the connection between long-range correlations and pure-state decomposition is more subtle than is suggested by the above remarks.

Despite these complications, physicists have developed a non-rigorous approach to deal with this phenomenon: the 'one-step replica symmetry breaking' (1RSB) cavity method. This method postulates a few properties of the pure-state decomposition, and, on this basis, allows one to derive a number of quantitative predictions ('conjectures' from a mathematical point of view). Examples include the satisfiability threshold for random K-SAT and for other random constraint satisfaction problems.

The method is rich enough to allow some self-consistency checks of such assumptions. In several cases in which the 1RSB cavity method has passed this test, its predictions have been confirmed by rigorous arguments (and there is no case in which they have been falsified so far). These successes encourage the quest for a mathematical theory of Gibbs states on sparse random graphs.

This chapter explains the 1RSB cavity method. It alternates between a general presentation and a concrete illustration of the XORSAT problem. We strongly encourage readers to read the previous chapter on XORSAT before the present one. This should help them to gain some intuition about the whole scenario.

We start with a general description of the 1RSB glass phase, and the decomposition into pure states, in Section 19.1. Section 19.2 introduces an auxiliary constraint satisfaction problem in order to count the number of solutions of BP equations. The 1RSB analysis amounts to applying belief propagation to this auxiliary problem. One can then apply the methods of Chapter 14 (for instance, density evolution) to the auxiliary problem. Section 19.3 illustrates this approach with the XORSAT problem and shows how the 1RSB cavity method recovers the rigorous results of the previous chapter.

In Section 19.4, we show how the 1RSB formalism, which in general is rather complicated, simplifies considerably when the temperature of the auxiliary constraint satisfaction problem takes the value x = 1. Section 19.5 explains how to apply this formalism to optimization problems (leveraging the min-sum algorithm), leading to the survey propagation algorithm. Section 19.6 describes the physical intuition which underlies the whole method. The appendix in Section 19.7 contains some technical aspects of the survey propagation equations applied to XORSAT, and their statistical analysis.

19.1   Beyond BP: Many states

19.1.1   Bethe measures

The main lesson of the previous chapters is that in many cases, the probability distribution specified by graphical models with a locally tree-like structure takes a relatively simple form, which we shall call a Bethe measure (or Bethe state). Let us first define precisely what we mean by this, before we proceed to discuss what kinds of other scenarios can be encountered.

As in Chapter 14, we consider a factor graph $G = (V, F, E)$, with variable nodes $V = \{1, \ldots, N\}$, factor nodes $F = \{1, \ldots, M\}$, and edges E. The joint probability distribution over the variables $\underline{x} = (x_1, \ldots, x_N) \in \mathcal{X}^N$ takes the form

$$\mu(\underline{x}) = \frac{1}{Z} \prod_{a=1}^{M} \psi_a(\underline{x}_{\partial a}) \,. \tag{19.1}$$

Given a subset of variable nodes $U \subseteq V$ (which we shall call a 'cavity'), the **induced subgraph** $G_U = (U, F_U, E_U)$ is defined as the factor graph that includes all of the factor nodes a such that $\partial a \subseteq U$, and the adjacent edges. We also write $(i, a) \in \partial U$ if $i \in U$ and $a \in F \setminus F_U$. Finally, a **set of messages** $\{\widehat{\nu}_{a \to i}\}$ is a set of probability distributions over $\mathcal{X}$, indexed by directed edges $a \to i$ in E with $a \in F$, $i \in V$.

Definition 19.1. (Informal) *The probability distribution μ is a **Bethe measure** (or **Bethe state**) if there exists a set of messages $\{\widehat{\nu}_{a \to i}\}$ such that, for 'almost all' of the 'finite-size' cavities U, the distribution $\mu_U(\cdot)$ of the variables in U can be approximated as*

$$\mu_U(\underline{x}_U) \cong \prod_{a \in F_U} \psi_a(\underline{x}_{\partial a}) \prod_{(ia) \in \partial U} \widehat{\nu}_{a \to i}(x_i) \; + \; \mathrm{err}(\underline{x}_U) \,, \tag{19.2}$$

where $\mathrm{err}(\underline{x}_U)$ is a 'small' error term, and $\cong$ denotes, as usual, equality up to a normalization.

A formal definition should specify what is meant by 'almost all', 'finite-size' and 'small'. This can be done by introducing a tolerance ϵ_N (with $\epsilon_N \downarrow 0$ as $N \to \infty$) and a size L_N (where L_N is bounded as $N \to \infty$). One then requires that some norm of $\mathrm{err}(\cdot)$ (e.g. an L_p norm) is smaller than ϵ_N for a fraction larger than $1 - \epsilon_N$ of all possible

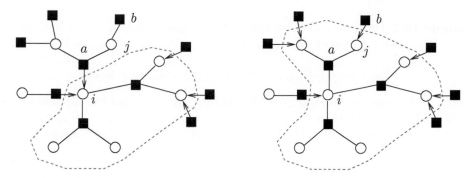

Fig. 19.1 Two examples of cavities. The *right*-hand one is obtained by adding the extra function node a. The consistency of the Bethe measure in these two cavities implies the BP equation for $\widehat{\nu}_{a\to i}$; see Exercise 19.1.

cavities U of size $|U| < L_N$. The underlying intuition is that the measure $\mu(\cdot)$ is well approximated locally by the given set of messages. In the following, we shall follow the physicists' habit of leaving implicit the various approximation errors.

Notice that the above definition does not make use of the fact that μ factorizes as in eqn (19.1). It thus applies to any distribution over $\underline{x} = \{x_i : i \in V\}$.

If $\mu(\cdot)$ is a Bethe measure with respect to the message set $\{\widehat{\nu}_{a\to i}\}$, then the consistency of eqn (19.2) for different choices of U implies some non-trivial constraints on the messages. In particular, if the loops in the factor graph G are not too small (and under some technical conditions on the functions $\psi_a(\cdot)$), then the messages must be close to satisfying the BP equations. More precisely, we define a **quasi-solution** of the BP equations as a set of messages which satisfy almost all of the equations within some accuracy. The reader is invited to prove this statement in the exercise below.

Exercise 19.1 Assume that $G = (V, F, E)$ has girth larger than 2, and that $\mu(\cdot)$ is a Bethe measure with respect to the message set $\{\widehat{\nu}_{a\to i}\}$, where $\widehat{\nu}_{a\to i}(x_i) > 0$ for any $(i, a) \in E$, and $\psi_a(\underline{x}_{\partial a}) > 0$ for any $a \in F$. For $U \subseteq V$ and $(i, a) \in \partial U$, define a new subset of variable nodes as $W = U \cup \partial a$ (see Fig. 19.1).

Applying eqn (19.2) to the subsets of variables U and W, show that a message must satisfy (up to an error term of the same order as $\mathrm{err}(\cdot)$)

$$\widehat{\nu}_{a\to i}(x_i) \cong \sum_{\underline{x}_{\partial a \setminus i}} \psi_a(\underline{x}_{\partial a}) \prod_{j \in \partial a \setminus i} \left\{ \prod_{b \in \partial j \setminus a} \widehat{\nu}_{b \to j}(x_j) \right\}. \tag{19.3}$$

Show that these equations are equivalent to the BP equations (14.14) and (14.15). [Hint: Define, for $k \in V$, $c \in F$ and $(k, c) \in E$, $\nu_{k \to c}(x_k) \cong \prod_{d \in \partial k \setminus c} \widehat{\nu}_{d \to k}(x_k)$.]

It would be pleasant if the converse was true, i.e. if each quasi-solution of the BP equations corresponded to a distinct Bethe measure. In fact, such a relation will be at the heart of the assumptions of the 1RSB method. However, one should keep in mind that this is not always true, as the following example shows.

Example 19.2 Let G be a factor graph with the same degree $K \geq 3$ at both factor and variable nodes. Consider binary variables ($\mathcal{X} = \{0, 1\}$) and, for each $a \in F$, let

$$\psi_a(x_{i_1(a)}, \ldots, x_{i_K(a)}) = \mathbb{I}(x_{i_1(a)} \oplus \cdots \oplus x_{i_K(a)} = 0). \qquad (19.4)$$

Given a perfect matching $\mathsf{M} \subseteq E$, a solution of the BP equations can be constructed as follows. If $(i, a) \in \mathsf{M}$, then let $\widehat{\nu}_{a \to i}(x_i) = \mathbb{I}(x_i = 0)$ and $\nu_{i \to a}(0) = \nu_{i \to a}(1) = 1/2$. If, on the other hand, $(i, a) \notin \mathsf{M}$, then let $\widehat{\nu}_{a \to i}(0) = \widehat{\nu}_{a \to i}(1) = 1/2$ and $\nu_{i \to a}(0) = \mathbb{I}(x_i = 0)$.

Check that this is a solution of the BP equations and that all the resulting local marginals coincide with the ones of the measure $\mu(\underline{x}) \cong \mathbb{I}(\underline{x} = \underline{0})$, independently of M. If we take, for instance, G to be a random regular graph with degree $K \geq 3$, both at factor nodes and at variable nodes, then the number of perfect matchings of G is, with high probability, exponential in the number of nodes. Therefore we have constructed an exponential number of solutions of the BP equations that describe the same Bethe measure.

19.1.2 A few generic scenarios

Bethe measures are a conceptual tool for describing distributions of the form (19.1). Inspired by the study of glassy phases (see Section 12.3), statistical mechanics studies have singled out a few generic scenarios in this respect, that we informally describe below.

RS (replica-symmetric). This is the simplest possible scenario: the distribution $\mu(\,\cdot\,)$ is a Bethe measure. A slightly more complicated situation (which we still place in the 'replica-symmetric' family) arises when $\mu(\,\cdot\,)$ decomposes into a finite set of Bethe measures related by 'global symmetries', as in the Ising ferromagnet discussed in Section 17.3.

d1RSB (dynamic one-step replica symmetry breaking). There exist an exponentially large (in the system size N) number of Bethe measures. The measure μ decomposes into a convex combination of these Bethe measures, i.e.

$$\mu(\underline{x}) = \sum_n w_n \, \mu^n(\underline{x}), \qquad (19.5)$$

with weights w_n exponentially small in N. Furthermore $\mu(\,\cdot\,)$ is itself a Bethe measure.

s1RSB (static one-step replica symmetry breaking). As in the d1RSB case, there exist an exponential number of Bethe measures, and μ decomposes into a convex combination of such states. However, a finite number of the weights w_n are of order 1 as $N \to \infty$, and (unlike in the previous case) μ is not itself a Bethe measure.

In the following, we shall focus on the d1RSB and s1RSB scenarios, which are particularly interesting and can be treated in a unified framework (we shall sometimes refer to both of them as '1RSB'). More complicated scenarios, such as 'full RSB', are also possible. We shall not discuss such scenarios here because, so far, we have a relatively poor control of them in sparse graphical models.

In order to proceed further, we shall make a series of assumptions about the structure of Bethe states in the 1RSB case. While further research work is required to formalize these assumptions completely, they are precise enough for deriving several interesting quantitative predictions.

To avoid technical complications, we assume that the compatibility functions $\psi_a(\,\cdot\,)$ are strictly positive. (The cases with $\psi_a(\,\cdot\,) = 0$ should be treated as limiting cases of such models.) Let us index the various quasi-solutions $\{\nu_{i\to a}^n, \widehat{\nu}_{a\to i}^n\}$ of the BP equations by n. With each of them we can associate a Bethe measure, and we can compute the corresponding Bethe free entropy $\mathbb{F}_n = \mathbb{F}(\underline{\nu}^n)$. The three postulates of the 1RSB scenario are listed below.

Assumption 1 *There exist exponentially many quasi-solutions of the BP equations. The number of such solutions with free entropy $\mathbb{F}(\underline{\nu}^n) \approx N\phi$ is (to leading exponential order) $\exp\{N\Sigma(\phi)\}$, where $\Sigma(\,\cdot\,)$ is the **complexity** function.*[1]

This can be expressed more formally as follows. There exists a function $\Sigma : \mathbb{R} \to \mathbb{R}_+$ (the complexity) such that, for any interval $[\phi_1, \phi_2]$, the number of quasi-solutions of the BP equations with $\mathbb{F}(\underline{\nu}^n) \in [N\phi_1, N\phi_2]$ is $\exp\{N\Sigma_* + o(N)\}$, where $\Sigma_* = \sup\{\Sigma(\phi) : \phi_1 \le \phi \le \phi_2\}$. We shall also assume in the following that $\Sigma(\phi)$ is 'regular enough', without going into the details.

Among Bethe measures, a special role is played by those which have short range correlations (are *extremal*). We have already mentioned this point in Chapter 12, and shall discuss the relevant notion of correlation decay in Chapter 22. We denote the set of extremal measures by E.

Assumption 2 *The 'canonical' measure μ, defined as in eqn (19.1), can be written as a convex combination of extremal Bethe measures*

$$\mu(\underline{x}) = \sum_{n \in \mathsf{E}} w_n\, \mu^n(\underline{x}) \,, \tag{19.6}$$

with weights related to the Bethe free entropies $w_n = \mathrm{e}^{\mathbb{F}_n}/\Xi$, $\Xi \equiv \sum_{n \in \mathsf{E}} \mathrm{e}^{\mathbb{F}_n}$.

Note that Assumption 1 characterizes the number of (approximate) BP fixed points, and Assumption 2 expresses the measure $\mu(\,\cdot\,)$ in terms of extremal Bethe measures. While each such measure gives rise to a BP fixed point by the arguments in the previous subsection, it is not clear that the converse holds. The next assumption implies that this is the case, to the leading exponential order.

Assumption 3 *To leading exponential order, the number of extremal Bethe measures equals the number of quasi-solutions of the BP equations: the number of extremal Bethe measures with free entropy $\approx N\phi$ is also given by $\exp\{N\Sigma(\phi)\}$.*

---

[1] As we are interested only in the leading exponential behaviour, the details of the definitions of quasi-solutions become irrelevant, as long as (for instance) the fraction of violated BP equations vanishes in the large-N limit.

19.2   The 1RSB cavity equations

Within the three assumptions described above, the complexity function $\Sigma(\phi)$ provides basic information on how the measure μ decomposes into Bethe measures. Since the number of extremal Bethe measures with a given free-entropy density is exponential in the system size, it is natural to treat them within a statistical-physics formalism. The BP messages of the original problem will be the new variables, and Bethe measures will be the new configurations. This is what 1RSB is about.

We introduce the auxiliary statistical-physics problem through the definition of a canonical distribution over extremal Bethe measures: we assign to a measure $n \in \mathsf{E}$ the probability $w_n(\mathbf{x}) = e^{\mathbf{x} \mathbb{F}_n}/\Xi(\mathbf{x})$. Here $\mathbf{x}$ plays the role of an inverse temperature (and is often called the **Parisi 1RSB parameter**). [2] The partition function of this generalized problem is

$$\Xi(\mathbf{x}) = \sum_{n \in \mathsf{E}} e^{\mathbf{x} \mathbb{F}_n} \doteq \int e^{N[\mathbf{x}\phi + \Sigma(\phi)]} \, d\phi \,. \tag{19.7}$$

According to Assumption 2 above, extremal Bethe measures contribute to μ with a weight $w_n = e^{\mathbb{F}_n}/\Xi$. Therefore the original problem is described by the choice $\mathbf{x} = 1$. But varying $\mathbf{x}$ will allow us to recover the full complexity function $\Sigma(\phi)$.

If $\Xi(\mathbf{x}) \doteq e^{N \mathfrak{F}(\mathbf{x})}$, a saddle point evaluation of the integral in eqn (19.7) gives Σ as the Legendre transform of $\mathfrak{F}$:

$$\mathfrak{F}(\mathbf{x}) = \mathbf{x}\phi + \Sigma(\phi) \,, \qquad \frac{\partial \Sigma}{\partial \phi} = -\mathbf{x} \,. \tag{19.8}$$

19.2.1   Counting BP fixed points

In order to actually estimate $\Xi(\mathbf{x})$, we need to consider the distribution induced by $w_n(\mathbf{x})$ in the messages $\underline{\nu} = \{\nu_{i \to a}, \widehat{\nu}_{a \to i}\}$, which we shall denote by $\mathsf{P}_x(\underline{\nu})$. The fundamental observation is that this distribution can be written as a graphical model, whose variables are BP messages. A first family of function nodes enforces the BP equations, and a second one implements the weight $e^{\mathbf{x} \mathbb{F}(\underline{\nu})}$. Furthermore, it turns out that the topology of the factor graph in this **auxiliary graphical model** is very close to that of the original factor graph. This suggests that we should use the BP approximation in this auxiliary model in order to estimate $\Sigma(\phi)$.

The 1RSB approach can be therefore summarized in one sentence:

Introduce a Boltzmann distribution over Bethe measures, write it in the form of a graphical model, and use BP to study this model.

This programme is straightforward, but one must be careful not to confuse the two models (the original one and the auxiliary one), and their messages. Let us first simplify the notations for the original messages. The two types of messages entering the BP equations of the original problem will be denoted by $\widehat{\nu}_{a \to i} = \widehat{m}_{ai}$ and $\nu_{i \to a} = m_{ia}$; we shall denote the set of all of the m_{ia} by $\underline{m}$ and the set of all of the $\widehat{m}_{ai}$ by $\underline{\widehat{m}}$. Each

---

[2]It turns out that the present approach is equivalent to the cloning method discussed in Chapter 12, where $\mathbf{x}$ is the number of clones.

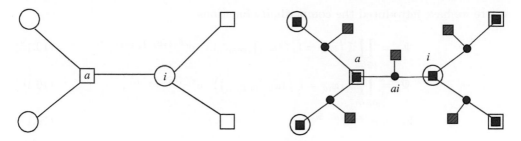

Fig. 19.2 A part of the original factor graph (*left*), and the corresponding auxiliary factor graph (*right*).

of these $2|\mathcal{E}|$ messages is a normalized probability distribution over the alphabet $\mathcal{X}$. With this notation, the original BP equations read

$$\mathsf{m}_{ia}(x_i) \cong \prod_{b \in \partial i \backslash a} \widehat{\mathsf{m}}_{bi}(x_i), \qquad \widehat{\mathsf{m}}_{ai}(x_i) \cong \sum_{\{x_j\}_{j \in \partial a \backslash i}} \psi_a(\underline{x}_{\partial a}) \prod_{j \in \partial a \backslash i} \mathsf{m}_{ja}(x_j). \qquad (19.9)$$

Hereafter, we shall write them in the compact form

$$\mathsf{m}_{ia} = \mathsf{f}_i\left(\{\widehat{\mathsf{m}}_{bi}\}_{b \in \partial i \backslash a}\right), \qquad \widehat{\mathsf{m}}_{ai} = \widehat{\mathsf{f}}_a\left(\{\mathsf{m}_{ja}\}_{j \in \partial a \backslash i}\right). \qquad (19.10)$$

Each message set $(\underline{\mathsf{m}}, \underline{\widehat{\mathsf{m}}})$ is given a weight proportional to $e^{x\mathbb{F}(\underline{\mathsf{m}},\underline{\widehat{\mathsf{m}}})}$, where the free entropy $\mathbb{F}(\underline{\mathsf{m}}, \underline{\widehat{\mathsf{m}}})$ is written in terms of BP messages as

$$\mathbb{F}(\underline{\mathsf{m}}, \underline{\widehat{\mathsf{m}}}) = \sum_{a \in F} \mathbb{F}_a\left(\{\mathsf{m}_{ja}\}_{j \in \partial a}\right) + \sum_{i \in V} \mathbb{F}_i\left(\{\widehat{\mathsf{m}}_{bi}\}_{b \in \partial i}\right) - \sum_{(ia) \in E} \mathbb{F}_{ia}\left(\mathsf{m}_{ia}, \widehat{\mathsf{m}}_{ai}\right). \qquad (19.11)$$

The functions $\mathbb{F}_a, \mathbb{F}_i, \mathbb{F}_{ia}$ were obtained in (14.28). Let us copy them here for convenience:

$$\mathbb{F}_a(\{\mathsf{m}_{ja}\}_{j \in \partial a}) = \log \left[\sum_{\underline{x}_{\partial a}} \psi_a(\underline{x}_{\partial a}) \prod_{j \in \partial a} \mathsf{m}_{ja}(x_j) \right],$$

$$\mathbb{F}_i(\{\widehat{\mathsf{m}}_{bi}\}_{b \in \partial i}) = \log \left[\sum_{x_i} \prod_{b \in \partial i} \widehat{\mathsf{m}}_{bi}(x_i) \right], \qquad (19.12)$$

$$\mathbb{F}_{ia}(\mathsf{m}_{ia}, \widehat{\mathsf{m}}_{ai}) = \log \left[\sum_{x_i} \mathsf{m}_{ia}(x_i)\widehat{\mathsf{m}}_{ai}(x_i) \right]. \qquad (19.13)$$

We now consider the $2|\mathcal{E}|$ messages $\underline{\mathsf{m}}$ and $\underline{\widehat{\mathsf{m}}}$ as variables in our auxiliary graphical model. The distribution induced by $w_n(\mathbf{x})$ in such messages takes the form

$$\mathsf{P}_\mathsf{x}(\underline{\mathsf{m}}, \underline{\widehat{\mathsf{m}}}) = \frac{1}{\Xi(x)} \prod_{a \in F} \Psi_a(\{\mathsf{m}_{ja}, \widehat{\mathsf{m}}_{ja}\}_{j \in \partial a}) \prod_{i \in V} \Psi_i(\{\mathsf{m}_{ib}, \widehat{\mathsf{m}}_{ib}\}_{b \in \partial i}) \prod_{(ia) \in E} \Psi_{ia}(\mathsf{m}_{ia}, \widehat{\mathsf{m}}_{ia}),$$

$$(19.14)$$

where we have introduced the compatibility functions

$$\Psi_a = \prod_{i \in \partial a} \mathbb{I} \left(\widehat{\mathsf{m}}_{ai} = \hat{\mathsf{f}}_a \left(\{\mathsf{m}_{ja}\}_{j \in \partial a \setminus i} \right) \right) \; e^{\mathsf{x} \mathbb{F}_a \left(\{\mathsf{m}_{ja}\}_{j \in \partial a} \right)}, \tag{19.15}$$

$$\Psi_i = \prod_{a \in \partial i} \mathbb{I} \left(\mathsf{m}_{ia} = \mathsf{f}_i \left(\{\widehat{\mathsf{m}}_{bi}\}_{b \in \partial i \setminus a} \right) \right) \; e^{\mathsf{x} \mathbb{F}_i \left(\{\widehat{\mathsf{m}}_{bi}\}_{b \in \partial i} \right)}, \tag{19.16}$$

$$\Psi_{ia} = e^{-\mathsf{x} \mathbb{F}_{ia} (\mathsf{m}_{ia}, \widehat{\mathsf{m}}_{ai})}. \tag{19.17}$$

The corresponding factor graph is depicted in Fig. 19.2 and can be described as follows:

- *For each edge (i, a) of the original factor graph*, we introduce a variable node in the auxiliary factor graph. The associated variable is the pair $(\mathsf{m}_{ia}, \widehat{\mathsf{m}}_{ai})$. Furthermore, we introduce a function node connected to this variable, contributing to the weight through a factor $\Psi_{ia} = e^{-\mathsf{x} \mathbb{F}_{ai}}$.

- *For each function node a of the original graph*, we introduce a function node in the auxiliary graph and connect it to all of the variable nodes corresponding to edges (i, a), $i \in \partial a$. The compatibility function Ψ_a associated with this function node has two roles: (i) it enforces the $|\partial a|$ BP equations expressing the variables $\{\widehat{\mathsf{m}}_{ai}\}_{i \in \partial a}$ in terms of the $\{\mathsf{m}_{ia}\}_{i \in \partial a}$ (see eqn (19.9)); (ii) it contributes to the weight through a factor $e^{\mathsf{x} \mathbb{F}_a}$.

- *For each variable node i of the original graph*, we introduce a function node in the auxiliary graph, and connect it to all of the variable nodes corresponding to edges (i, a), $a \in \partial i$. The compatibility function Ψ_i has two roles: (i) it enforces the $|\partial i|$ BP equations expressing the variables $\{\mathsf{m}_{ib}\}_{b \in \partial i}$ in terms of $\{\widehat{\mathsf{m}}_{bi}\}_{b \in \partial i}$ (see eqn (19.9)); (ii) it contributes to the weight through a factor $e^{\mathsf{x} \mathbb{F}_i}$.

Note that we were a bit sloppy in eqns (19.15)–(19.17). The messages m_{ia}, $\widehat{\mathsf{m}}_{ai}$ are in general continuous, and indicator functions should therefore be replaced by delta functions. This might, in turn, pose some definition problems (what is the reference measure on the messages? can we hope for *exact* solutions of BP equations?). One should consider the above as a shorthand for the following procedure. First, we discretize the messages (and BP equations) in such a way that they can take a finite number q of values. We compute the complexity by letting $N \to \infty$ at fixed q, and take the limit $q \to \infty$ at the end. It is easy to define several alternative, and equally reasonable, limiting procedures. We expect all of them to yield the same result. In practice, the ambiguities in eqns (19.15)–(19.17) are resolved on a case-by-case basis.

19.2.2   Message passing in the auxiliary model

The problem of counting the number of Bethe measures (more precisely, computing the complexity function $\Sigma(\phi)$) has been reduced to one of estimating the partition function $\Xi(\mathsf{x})$ of the auxiliary graphical model (19.14). Since we are interested in the case of locally tree-like factor graphs G, the auxiliary factor graph is locally tree-like as well. We can therefore apply BP to estimate its free-entropy density $\mathfrak{F}(\mathsf{x}) = \lim_N N^{-1} \log \Xi(\mathsf{x})$. This will give us the complexity through the Legendre transform of eqn (19.8).

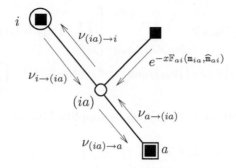

Fig. 19.3 Messages in the auxiliary graphical model.

In the following, we denote by $i \in V$ and $a \in F$ a generic variable node and function node in the graph G, and by $(ia) \in E$ an edge in G. By extension, we denote the corresponding nodes in the auxiliary graph in the same way. The messages appearing in the BP analysis of the auxiliary model can be classified as follows (see Fig. 19.3):

- From the variable node (ia), two messages are issued: $\nu_{(ia)\to a}(\mathbf{m}_{ia}, \widehat{\mathbf{m}}_{ai})$ and $\nu_{(ia)\to i}(\mathbf{m}_{ia}, \widehat{\mathbf{m}}_{ai})$.
- From the function node a, $|\partial a|$ messages to nodes $i \in \partial a$ are issued: $\widehat{\nu}_{a\to(ai)}(\mathbf{m}_{ia}, \widehat{\mathbf{m}}_{ai})$.
- From the function node i, $|\partial i|$ messages to nodes $a \in \partial i$ are issued: $\widehat{\nu}_{i\to(ui)}(\mathbf{m}_{ia}, \widehat{\mathbf{m}}_{ai})$.
- From the degree-one function node connected to the variable node (ia), a message is issued towards this variable node. This message is simply $e^{-\mathsf{x}\mathbb{F}_{ia}(\mathbf{m}_{ia}, \widehat{\mathbf{m}}_{ai})}$.

The BP equations for the variable node (ia) take a simple form:

$$\nu_{(ia)\to a}(\mathbf{m}_{ia}, \widehat{\mathbf{m}}_{ai}) \cong \widehat{\nu}_{i\to(ia)}(\mathbf{m}_{ia}, \widehat{\mathbf{m}}_{ai})\, e^{-\mathsf{x}\mathbb{F}_{ia}(\mathbf{m}_{ia}, \widehat{\mathbf{m}}_{ai})},$$
$$\nu_{(ia)\to i}(\mathbf{m}_{ia}, \widehat{\mathbf{m}}_{ai}) \cong \widehat{\nu}_{a\to(ia)}(\mathbf{m}_{ia}, \widehat{\mathbf{m}}_{ai})\, e^{-\mathsf{x}\mathbb{F}_{ia}(\mathbf{m}_{ia}, \widehat{\mathbf{m}}_{ai})}. \tag{19.18}$$

We can use these equations to eliminate the messages $\widehat{\nu}_{i\to(ia)}$, $\widehat{\nu}_{a\to(ia)}$ in favour of $\nu_{(ia)\to a}$, $\nu_{(ia)\to i}$. In order to emphasize this choice (and to simplify the notation) we define

$$Q_{ia}(\mathbf{m}_{ia}, \widehat{\mathbf{m}}_{ai}) \equiv \nu_{(ia)\to a}(\mathbf{m}_{ia}, \widehat{\mathbf{m}}_{ai}), \quad \widehat{Q}_{ai}(\mathbf{m}_{ia}, \widehat{\mathbf{m}}_{ai}) \equiv \nu_{(ia)\to i}(\mathbf{m}_{ia}, \widehat{\mathbf{m}}_{ai}). \tag{19.19}$$

We can now write the remaining BP equations of the auxiliary graphical model in terms of $Q_{ia}(\cdot,\cdot)$ and $\widehat{Q}_{ai}(\cdot,\cdot)$. The BP equation associated with the function node corresponding to $i \in V$ reads

$$Q_{ia}(\mathbf{m}_{ia}, \widehat{\mathbf{m}}_{ai}) \cong \sum_{\{\mathbf{m}_{ib}, \widehat{\mathbf{m}}_{bi}\}_{b\in\partial i\setminus a}} \left[\prod_{c\in\partial i} \mathbb{I}\left(\mathbf{m}_{ic} = \mathsf{f}_i(\{\widehat{\mathbf{m}}_{di}\}_{d\in\partial i\setminus c})\right)\right]$$
$$\times \exp\left\{\mathsf{x}\left[\mathbb{F}_i\left(\{\widehat{\mathbf{m}}_{bi}\}_{b\in\partial i}\right) - \mathbb{F}_{ai}(\mathbf{m}_{ia}, \widehat{\mathbf{m}}_{ai})\right]\right\} \prod_{b\in\partial i\setminus a} \widehat{Q}_{bi}(\mathbf{m}_{ib}, \widehat{\mathbf{m}}_{bi}), \tag{19.20}$$

and the equation associated with the function node corresponding to $a \in F$ is

$$\widehat{Q}_{ai}(\mathtt{m}_{ia},\widehat{\mathtt{m}}_{ai}) \cong \sum_{\{\mathtt{m}_{ja},\widehat{\mathtt{m}}_{aj}\}_{j\in\partial a\backslash i}} \left[\prod_{j\in\partial a} \mathbb{I}\left(\widehat{\mathtt{m}}_{aj} = \widehat{\mathsf{f}}_a(\{\mathtt{m}_{ka}\}_{k\in\partial a\backslash j}) \right) \right]$$

$$\times \exp\left\{ \mathtt{x}\left[\mathbb{F}_a\left(\{\mathtt{m}_{ja}\}_{j\in\partial a}\right) - \mathbb{F}_{ai}\left(\mathtt{m}_{ia},\widehat{\mathtt{m}}_{ai}\right) \right] \right\} \prod_{j\in\partial a\backslash i} Q_{ja}(\mathtt{m}_{ja},\widehat{\mathtt{m}}_{aj}). \quad (19.21)$$

Equations (19.20) and (19.21) can be simplified further, using the following lemma.

Lemma 19.3 *Assume that* $\sum_{x_i} \mathtt{m}_{ia}(x_i)\widehat{\mathtt{m}}_{ai}(x_i) > 0$. *Under the condition* $\mathtt{m}_{ia} = \mathsf{f}_i(\{\widehat{\mathtt{m}}_{di}\}_{d\in\partial i\backslash a})$ *(which holds when the indicator functions in eqn (19.20) evaluate to 1), the difference* $\mathbb{F}_i\left(\{\widehat{\mathtt{m}}_{bi}\}_{b\in\partial i}\right) - \mathbb{F}_{ai}\left(\mathtt{m}_{ia},\widehat{\mathtt{m}}_{ai}\right)$ *can be expressed in terms of* $\{\widehat{\mathtt{m}}_{bi}\}_{b\in\partial i\backslash a}$. *Explicitly, we have*

$$\mathrm{e}^{\mathbb{F}_i - \mathbb{F}_{ia}} = z_{ia}(\{\widehat{\mathtt{m}}_{bi}\}_{b\in\partial i\backslash a}) \equiv \sum_{x_i} \prod_{b\in\partial i\backslash a} \widehat{\mathtt{m}}_{bi}(x_i). \quad (19.22)$$

Analogously, under the condition $\widehat{\mathtt{m}}_{ai} = \widehat{\mathsf{f}}_a(\{\mathtt{m}_{ka}\}_{k\in\partial a\backslash i})$ *(which holds when the indicator functions in eqn (19.21) evaluate to 1), the difference* $\mathbb{F}_a\left(\{\mathtt{m}_{ja}\}_{j\in\partial a}\right) - \mathbb{F}_{ai}\left(\mathtt{m}_{ia},\widehat{\mathtt{m}}_{ai}\right)$ *depends only on* $\{\mathtt{m}_{ja}\}_{j\in\partial a\backslash i}$. *Explicitly,*

$$\mathrm{e}^{\mathbb{F}_a - \mathbb{F}_{ia}} = \widehat{z}_{ai}(\{\mathtt{m}_{ja}\}_{j\in\partial a\backslash i}) \equiv \sum_{\underline{x}_{\partial a}} \psi_a(\underline{x}_{\partial a}) \prod_{j\in\partial a\backslash i} \mathtt{m}_{ja}(x_j). \quad (19.23)$$

Proof Let us consider eqn (19.22) first. From the definition (14.28), it follows that

$$\mathrm{e}^{\mathbb{F}_i - \mathbb{F}_{ia}} = \frac{\sum_{x_i} \prod_{b\in\partial i} \widehat{\mathtt{m}}_{bi}(x_i)}{\sum_{x_i} \mathtt{m}_{ia}(x_i)\widehat{\mathtt{m}}_{ai}(x_i)}. \quad (19.24)$$

Substituting $\mathtt{m}_{ia} = \mathsf{f}_i(\{\widehat{\mathtt{m}}_{ci}\}_{c\in\partial i\backslash a})$ in the denominator, and keeping track of the normalization constant, we get

$$\sum_{x_i} \mathtt{m}_{ia}(x_i)\widehat{\mathtt{m}}_{ai}(x_i) = \frac{\sum_{x_i} \prod_{b\in\partial i} \widehat{\mathtt{m}}_{bi}(x_i)}{\sum_{x_i} \prod_{b\in\partial i\backslash a} \widehat{\mathtt{m}}_{ai}(x_i)}, \quad (19.25)$$

which implies eqn (19.22).

The derivation of eqn (19.23) is completely analogous and is left as an exercise for the reader. □

Note that the functions $z_{ia}(\cdot)$ and $\widehat{z}_{ai}(\cdot)$ appearing in eqns (19.22) and (19.23) are in fact the normalization constants hidden by the '$\cong$' notation in eqns (19.9).

Because of this lemma, we can seek a solution of eqns (19.20) and (19.21) where Q_{ia} depends only on $\mathtt{m}_{ia}$, and $\widehat{Q}_{ai}$ depends only on $\widehat{\mathtt{m}}_{ai}$. Hereafter, we shall focus on this case, and, with some misuse of notation, we shall write

$$Q_{ia}(\mathtt{m}_{ia},\widehat{\mathtt{m}}_{ai}) = Q_{ia}(\mathtt{m}_{ia}), \quad \widehat{Q}_{ia}(\mathtt{m}_{ia},\widehat{\mathtt{m}}_{ai}) = \widehat{Q}_{ai}(\widehat{\mathtt{m}}_{ai}). \quad (19.26)$$

The BP equations for the auxiliary graphical model (19.20), (19.21) then become

$$Q_{ia}(\mathtt{m}_{ia}) \cong \sum_{\{\widehat{\mathtt{m}}_{bi}\}} \mathbb{I}\left(\mathtt{m}_{ia} = g_i(\{\widehat{\mathtt{m}}_{bi}\})\right) \left[z_{ia}(\{\widehat{\mathtt{m}}_{bi}\})\right]^{\mathtt{x}} \prod_{b \in \partial i \backslash a} \widehat{Q}_{bi}(\widehat{\mathtt{m}}_{bi}), \qquad (19.27)$$

$$\widehat{Q}_{ai}(\widehat{\mathtt{m}}_{ai}) \cong \sum_{\{\mathtt{m}_{ja}\}} \mathbb{I}\left(\widehat{\mathtt{m}}_{ai} = f_a(\{\mathtt{m}_{ja}\})\right) \left[\hat{z}_{ai}(\{\mathtt{m}_{ja}\})\right]^{\mathtt{x}} \prod_{j \in \partial a \backslash i} Q_{ja}(\mathtt{m}_{ja}), \qquad (19.28)$$

where $\{\widehat{\mathtt{m}}_{bi}\}$ is a shorthand for $\{\widehat{\mathtt{m}}_{bi}\}_{b \in \partial i \backslash a}$ and $\{\mathtt{m}_{ja}\}$ is a shorthand for $\{\mathtt{m}_{ja}\}_{j \in \partial a \backslash i}$. The expressions for $z_{ia}(\{\widehat{\mathtt{m}}_{bi}\})$ and $\hat{z}_{ai}(\{\mathtt{m}_{ja}\})$ are given in eqns (19.22) and (19.23).

Equations (19.27) and (19.28) are the **1RSB cavity equations**. As we did in the case of the ordinary BP equations, we can consider them as an update rule for a message-passing algorithm. This will be discussed further in the following sections. Sometimes the notation $Q_{i \to a}(\,\cdot\,)$, $\widehat{Q}_{a \to i}(\,\cdot\,)$ is used, to emphasize the fact that 1RSB messages are associated with *directed* edges.

Notice that our derivation was based on the assumption that $\sum_{x_i} \mathtt{m}_{ia}(x_i)\widehat{\mathtt{m}}_{ai}(x_i) > 0$. This condition holds if, for instance, the compatibility functions of the original model are bounded away from 0. Under this condition, we have shown the following.

Proposition 19.4 *If the 1RSB cavity equations (19.27) and (19.28) have a solution* $\widehat{Q}, Q$, *this corresponds to a solution to the BP equations of the auxiliary graphical model. Conversely, if the BP equations of the auxiliary graphical model admit a solution satisfying the condition (19.26), then the resulting messages must be a solution of the 1RSB cavity equations.*

The assumption (19.26)—which is suggestive of a form of 'causality'— cannot be further justified within the present approach, but alternative derivations of the 1RSB equations confirm its validity.

19.2.3 Computing the complexity

We now compute the free entropy of the auxiliary graphical model within the BP approximation. We expect the result of this procedure to be asymptotically exact for a wide class of locally tree-like graphs, thus yielding the correct free-entropy density $\mathfrak{F}(\mathtt{x}) = \lim_N N^{-1} \log \Xi(\mathtt{x})$.

Assume $\{Q_{ia}, \widehat{Q}_{ai}\}$ to be a solution (or a quasi-solution) of the 1RSB cavity equations (19.27) and (19.28). We use the general form (14.27) of the Bethe free entropy, but take into account the degree-one factor nodes using the simplified expression derived in Exercise 14.6. The various contributions to the free entropy are:

- The contribution from the function node a (here $\{\mathtt{m}_{ia}\}$ is a shorthand for $\{\mathtt{m}_{ia}\}_{i \in \partial a}$):

$$\mathbb{F}_a^{\mathrm{RSB}} = \log \left\{ \sum_{\{\mathtt{m}_{ia}\}} e^{\mathtt{x}\mathbb{F}_a(\{\mathtt{m}_{ia}\})} \prod_{i \in \partial a} Q_{ia}(\mathtt{m}_{ia}) \right\}. \qquad (19.29)$$

- The contribution from the function node i (here $\{\widehat{\mathtt{m}}_{ai}\}$ is a shorthand for $\{\widehat{\mathtt{m}}_{ai}\}_{a \in \partial i}$):

$$\mathbb{F}_i^{\mathrm{RSB}} = \log \left\{ \sum_{\{\widehat{\mathtt{m}}_{ai}\}} e^{\mathtt{x}\mathbb{F}_i(\{\widehat{\mathtt{m}}_{ai}\})} \prod_{a \in \partial i} \widehat{Q}_{ai}(\widehat{\mathtt{m}}_{ai}) \right\}. \qquad (19.30)$$

- The contribution from the variable node (ia):

$$\mathbb{F}_{ia}^{\mathrm{RSB}} = \log \left\{ \sum_{\mathtt{m}_{ia}, \widehat{\mathtt{m}}_{ai}} \mathrm{e}^{\mathtt{x}\mathbb{F}_{ia}(\mathtt{m}_{ia}, \widehat{\mathtt{m}}_{ai})} Q_{ia}(\mathtt{m}_{ia}) \widehat{Q}_{ai}(\widehat{\mathtt{m}}_{ai}) \right\}. \tag{19.31}$$

- The contributions from the two edges a–(ai) and i–(ai) are both equal to $-\mathbb{F}_{ia}^{\mathrm{RSB}}$.
  The Bethe free entropy of the auxiliary graphical model is equal to

$$\mathbb{F}^{\mathrm{RSB}}(\{Q, \widehat{Q}\}) = \sum_{a \in F} \mathbb{F}_a^{\mathrm{RSB}} + \sum_{i \in V} \mathbb{F}_i^{\mathrm{RSB}} - \sum_{(ia) \in E} \mathbb{F}_{ia}^{\mathrm{RSB}}. \tag{19.32}$$

19.2.4   Summary

The 1RSB cavity equations (19.27) and (19.28) are the BP equations for the auxiliary graphical model defined in eqn (19.14). They relate $2|\mathcal{E}|$ messages $\{Q_{ia}(\mathtt{m}_{ia}), \widehat{Q}_{ai}(\widehat{\mathtt{m}}_{ai})\}$. Each such message is a probability distribution of ordinary BP messages $\mathtt{m}_{ia}(x_i)$ and $\widehat{\mathtt{m}}_{ai}(x_i)$, respectively. These elementary messages are, in turn, probability distributions on variables $x_i \in \mathcal{X}$.

Given a solution (or an approximate solution) $\{Q_{ia}, \widehat{Q}_{ai}\}$, one can estimate the free-entropy density of the auxiliary model as

$$\log \Xi(\mathtt{x}) = \mathbb{F}^{\mathrm{RSB}}(\{Q, \widehat{Q}\}) + \mathrm{err}_N, \tag{19.33}$$

where $\mathbb{F}^{\mathrm{RSB}}(\{Q, \widehat{Q}\})$ is given by eqn (19.32). For a large class of locally tree-like models, we expect the BP approximation to be asymptotically exact for the auxiliary model. This means that the error term err_N is $o(N)$.

For such models, the free-entropy density is given by its 1RSB cavity expression $\mathfrak{F}(\mathtt{x}) = \mathtt{f}^{\mathrm{RSB}}(\mathtt{x}) \equiv \lim_{N \to \infty} \mathbb{F}^{\mathrm{RSB}}(\{Q, \widehat{Q}\})/N$. The complexity $\Sigma(\phi)$ is then computed through the Legendre transform (19.8).

19.2.5   Random graphical models and density evolution

Let us consider the case where G is a random graphical model as defined in Section 14.6.1. The factor graph is distributed according to one of the ensembles $\mathbb{G}_N(K, \alpha)$ or $\mathbb{D}_N(\Lambda, P)$. The function nodes are taken from a finite list $\{\psi^{(k)}(x_1, \ldots, x_k; \widehat{J})\}$ indexed by a label $\widehat{J}$ with distribution $P_{\widehat{J}}^{(k)}$. Each factor $\psi_a(\cdot)$ is taken to be equal to $\psi^{(k)}(\cdots; \widehat{J}_a)$ independently with the same distribution. We also introduce explicitly a degree-one factor $\psi_i(x_i)$ connected to each variable node $i \in V$. This is also drawn independently from a list of possible factors $\{\psi(x; J)\}$, indexed by a label J with distribution P_J.

For a random graphical model, the measure $\mu(\cdot)$ becomes random, and so does its decomposition into extremal Bethe states in particular the probabilities $\{w_n\}$ and the message sets $\{\nu_{i \to a}^n, \widehat{\nu}_{a \to i}^n\}$. In particular, the 1RSB messages $\{Q_{ia}, \widehat{Q}_{ai}\}$ become random. It is important to keep in mind the 'two levels' of randomness. Given an edge (ia), the message $\nu_{i \to a}^n$ is random if the Bethe state n is drawn from the distribution w_n. The resulting distribution $Q_{ia}(\mathtt{m})$ becomes a random variable when the graphical model is itself random.

The distributions of $Q_{ia}(\mathtt{m})$ and $\widehat{Q}_{ai}(\widehat{\mathtt{m}})$ can then be studied through the density evolution method of Section 14.6.2. Let us assume an i.i.d. initialization $Q_{ia}^{(0)}(\,\cdot\,) \stackrel{\mathrm{d}}{=} Q^{(0)}(\,\cdot\,)$ (and $\widehat{Q}_{ai}^{(0)}(\,\cdot\,) \stackrel{\mathrm{d}}{=} \widehat{Q}^{(0)}(\,\cdot\,)$), and denote by $Q_{ia}^{(t)}(\,\cdot\,)$ and $\widehat{Q}_{ai}^{(t)}(\,\cdot\,)$ the 1RSB messages along edge (ia) after t parallel updates using the 1RSB equations (19.27) and (19.28). If (ia) is a uniformly random edge then, as $N \to \infty$, $Q_{ia}^{(t)}(\,\cdot\,)$ converges in distribution[3] to $Q^{(t)}(\,\cdot\,)$ (and $\widehat{Q}_{ia}^{(t)}(\,\cdot\,)$ converges in distribution to $\widehat{Q}^{(t)}(\,\cdot\,)$). The distributions $Q^{(t)}(\,\cdot\,)$ and $\widehat{Q}^{(t)}(\,\cdot\,)$ are themselves random variables that satisfy the following equations:

$$Q^{(t+1)}(\mathtt{m}) \stackrel{\mathrm{d}}{\cong} \sum_{\{\widehat{\mathtt{m}}_b\}} \mathbb{I}\left(\mathtt{m} = \mathsf{f}(\{\widehat{\mathtt{m}}_b\}; J)\right) z(\{\widehat{\mathtt{m}}_b\}; J)^{\mathtt{x}} \prod_{b=1}^{l-1} \widehat{Q}_b^{(t)}(\widehat{\mathtt{m}}_b), \qquad (19.34)$$

$$\widehat{Q}^{(t)}(\widehat{\mathtt{m}}) \stackrel{\mathrm{d}}{\cong} \sum_{\{\mathtt{m}_j\}} \mathbb{I}\left(\widehat{\mathtt{m}} = \widehat{\mathsf{f}}(\{\mathtt{m}_j\}; \widehat{J})\right) \widehat{z}(\{\mathtt{m}_j\}; \widehat{J})^{\mathtt{x}} \prod_{j=1}^{k-1} Q_j^{(t)}(\mathtt{m}_j), \qquad (19.35)$$

where k and l are distributed according to the edge-perspective degree profiles ρ_k and λ_l, the $\left\{\widehat{Q}_b^{(t)}\right\}$ are $k-1$ independent copies of $\widehat{Q}^{(t)}(\,\cdot\,)$, and the $\left\{Q_j^{(t)}\right\}$ are $l-1$ independent copies of $Q^{(t)}(\,\cdot\,)$. The functions z and $\widehat{z}$ are given by

$$z(\{\widehat{\mathtt{m}}_b\}; J) = \sum_x \psi(x, J) \prod_{b=1}^{l-1} \widehat{\mathtt{m}}_b(x),$$

$$\widehat{z}(\{\mathtt{m}_j\}; \widehat{J}) = \sum_{x_1, \ldots, x_k} \psi^{(k)}(x_1, \ldots, x_k; \widehat{J}) \prod_{j=1}^{k-1} \mathtt{m}_j(x_j). \qquad (19.36)$$

In the 1RSB cavity method, the actual distribution of $Q_{i \to a}$ is assumed to coincide with one of the fixed points of the above density evolution equations. As in the RS case, one hopes that, on large enough instances, the message-passing algorithm will converge to messages distributed according to this fixed-point equation (meaning that there is no problem in exchanging the limits $t \to \infty$ and $N \to \infty$). This can be checked numerically.

For random graphical models, the 1RSB free-entropy density converges to a finite limit $\mathsf{f}^{\mathrm{RSB}}(\mathtt{x})$. This can be expressed in terms of the distributions of Q and $\widehat{Q}$. by taking the expectation of eqns (19.29)–(19.31) and assuming that the 1RSB messages coming into the same node are i.i.d. As in eqn (14.77), the result takes the form

$$\mathsf{f}^{\mathrm{RSB}} = \mathsf{f}_{\mathrm{v}}^{\mathrm{RSB}} + n_{\mathrm{f}} \mathsf{f}_{\mathrm{f}}^{\mathrm{RSB}} - n_{\mathrm{e}} \mathsf{f}_{\mathrm{e}}^{\mathrm{RSB}}. \qquad (19.37)$$

Here n_{f} is the average number of function nodes per variable (equal to $\Lambda'(1)/P'(1)$ for a graphical model in the ensemble $\mathbb{D}_N(\Lambda, P)$, and to α for a graphical model in the ensemble $\mathbb{G}_N(K, \alpha)$), and n_{e} is the number of edges per variable (equal to $\Lambda'(1)$ and

---

[3] We shall not discuss the measure-theoretic subtleties related to this statement. Let us just mention that weak topology is understood on the space of messages $Q^{(t)}$.

to $K\alpha$ for these two ensembles). The contributions from variable nodes $\mathfrak{f}_{\mathrm{v}}^{\mathrm{RSB}}$, function nodes $\mathfrak{f}_{\mathrm{f}}^{\mathrm{RSB}}$, and edges $\mathfrak{f}_{\mathrm{e}}^{\mathrm{RSB}}$ are

$$
\mathfrak{f}_{\mathrm{v}}^{\mathrm{RSB}} = \mathbb{E}_{l,J,\{\widehat{Q}\}} \log \left\{ \sum_{\{\widehat{\mathfrak{m}}_1,\dots,\widehat{\mathfrak{m}}_l\}} \widehat{Q}_1(\widehat{\mathfrak{m}}_1) \dots \widehat{Q}_l(\widehat{\mathfrak{m}}_l) \left[\sum_{x\in\mathcal{X}} \widehat{\mathfrak{m}}_1(x)\dots\widehat{\mathfrak{m}}_l(x)\psi(x;J) \right]^{\mathrm{x}} \right\},
$$

$$
\mathfrak{f}_{\mathrm{f}}^{\mathrm{RSB}} = \mathbb{E}_{k,\widehat{J},\{Q\}} \log \left\{ \sum_{\{\mathfrak{m}_1,\dots,\mathfrak{m}_k\}} Q_1(\mathfrak{m}_1)\dots Q_k(\mathfrak{m}_k) \right.
$$
$$
\left. \times \left[\sum_{x_1,\dots,x_k\in\mathcal{X}} \mathfrak{m}_1(x_1)\dots\mathfrak{m}_k(x_k)\psi^{(k)}(x_1,\dots,x_k;\widehat{J}) \right]^{\mathrm{x}} \right\},
$$

$$
\mathfrak{f}_{\mathrm{e}}^{\mathrm{RSB}} = \mathbb{E}_{\widehat{Q},Q} \log \left\{ \sum_{\widehat{\mathfrak{m}},\mathfrak{m}} \widehat{Q}(\widehat{\mathfrak{m}})Q(\mathfrak{m}) \left[\sum_{x\in\mathcal{X}} \widehat{\mathfrak{m}}(x)\mathfrak{m}(x) \right]^{\mathrm{x}} \right\}. \tag{19.38}
$$

19.2.6 Numerical implementation

Needless to say, it is extremely challenging to find a fixed point of the density evolution equations (19.34) and (19.35), and thus determine the distributions of Q and $\widehat{Q}$. A simple numerical approach consists in generalizing the population dynamics algorithm described in the context of the RS cavity method (see Section 14.6.3).

There are two important issues related to such a generalization:

(i) We seek the distribution of $Q(\,\cdot\,)$ (and $\widehat{Q}(\,\cdot\,)$), which is itself a distribution of messages. If we approximate $Q(\,\cdot\,)$ by a sample (a 'population'), we thus need two levels of populations. In other words, we seek a population $\{\mathfrak{m}_r^s\}$ with NM items. For each $r \in \{1,\dots,N\}$, the set of messages $\{\mathfrak{m}_r^s\}$, $s \in \{1,\dots,M\}$, represents a distribution $Q_r(\,\cdot\,)$ (ideally, it would be an i.i.d. sample from this distribution). At the next level, the population $\{Q_r(\,\cdot\,)\}$, $r \in \{1,\cdots,N\}$, represents the distribution of $Q(\,\cdot\,)$ (ideally, an i.i.d. sample). Analogously, for function-to-variable messages, we use a population $\{\widehat{\mathfrak{m}}_r^s\}$, with $r \in \{1,\dots,N\}$ and $s \in \{1,\dots,M\}$.

(ii) The reweighting factors $z(\{\widehat{\mathfrak{m}}_b\};J)^{\mathrm{x}}$ and $\hat{z}(\{\mathfrak{m}_j\};\widehat{J})^{\mathrm{x}}$ appearing in eqns (19.34) and (19.35) do not have any analogue in the RS context. How can one take such factors into account when $Q(\,\cdot\,)$ and $\widehat{Q}(\,\cdot\,)$ are represented as populations? One possibility is to generate an intermediate weighted population, and then sample from it with a probability proportional to the weight.

This procedure is summarized in the following pseudocode.

1RSB POPULATION DYNAMICS (model ensemble, sizes N, M, iterations T)

1:     Initialize $\{\mathsf{m}_r^s\}$;
2:     **for** $t = 1, \ldots, T$:
3:        **for** $r = 1, \ldots, N$:
4:           Draw an integer k with distribution ρ;
5:           Draw $i(1), \ldots, i(k-1)$ uniformly in $\{1, \ldots, N\}$;
6:           Draw $\widehat{J}$ with distribution $P_{\widehat{J}}^{(k)}$;
7:           **for** $s = 1, \ldots, M$:
8:              Draw $s(1), \ldots, s(k-1)$ uniformly in $\{1, \ldots, M\}$;
9:              Compute $\widehat{\mathsf{m}}_{\text{temp}}^s = \widehat{\mathsf{f}}(\mathsf{m}_{i(1)}^{s(1)}, \ldots, \mathsf{m}_{i(k-1)}^{s(k-1)}; \widehat{J})$
10:            Compute $W^s = \widehat{z}(\mathsf{m}_{i(1)}^{s(1)}, \ldots, \mathsf{m}_{i(k-1)}^{s(k-1)}; \widehat{J})^{\mathsf{x}}$
11:           **end;**
12:           Generate the new population
                 $\{\widehat{\mathsf{m}}_r^s\}_{s \in [M]} = \text{REWEIGHT}(\{\widehat{\mathsf{m}}_{\text{temp}}^s, W^s\}_{s \in [M]})$
13:        **end;**
14:        **for** $r = 1, \ldots, N$:
15:           Draw an integer l with distribution λ;
16:           Draw $i(1), \ldots, i(l-1)$ uniformly in $\{1, \ldots, N\}$;
17:           Draw J with distribution P;
18:           **for** $s = 1, \ldots, M$:
19:              Draw $s(1), \ldots, s(l-1)$ uniformly in $\{1, \ldots, M\}$;
20:              Compute $\mathsf{m}_{\text{temp}}^s = \mathsf{f}(\widehat{\mathsf{m}}_{i(1)}^{s(1)}, \ldots, \widehat{\mathsf{m}}_{i(l-1)}^{s(k-1)}; J)$
21:              Compute $W^s = z(\widehat{\mathsf{m}}_{i(1)}^{s(1)}, \ldots, \widehat{\mathsf{m}}_{i(l-1)}^{s(l-1)}; J)^{\mathsf{x}}$
22:           **end;**
23:           Generate the new population
                 $\{\mathsf{m}_r^s\}_{s \in [M]} = \text{REWEIGHT}(\{\mathsf{m}_{\text{temp}}^s, W^s\}_{s \in [M]})$
24:     **end;**
25:     **return** $\{\widehat{\mathsf{m}}_r^s\}$ and $\{\mathsf{m}_r^s\}$.

The reweighting procedure is given by the following pseudocode:

REWEIGHT (population of messages and weights $\{(\mathsf{m}_{\text{temp}}^s, W^s)\}_{s \in [M]}$)

1:     **for** $s = 1, \ldots, M$, set $p^s \equiv W^s / \sum_{s'} W^{s'}$;
2:     **for** $s = 1, \ldots, M$:
3:        Draw $i \in \{1, \ldots, M\}$ with distribution p^s;
4:        Set $\mathsf{m}_{\text{new}}^s = \mathsf{m}_{\text{temp}}^i$;
5:     **end;**
6:     **return** $\{\mathsf{m}_{\text{new}}^s\}_{s \in [M]}$.

In the large-N, M limit, the populations generated by this algorithm should converge to i.i.d. samples distributed as $Q^{(T)}(\cdot)$ and $\widehat{Q}^{(T)}(\cdot)$ (see eqns (19.34) and (19.35)). If we let T grow, they should represent accurately the fixed points of the density evolution, although the caveats expressed in the RS case should be repeated here.

Among the other quantities, the populations generated by this algorithm allow one to estimate the 1RSB free-entropy density (19.37). Suppose we have generated a population of messages $\{\widehat{\mathtt{m}}_r^s(\cdot)\}$, where each message is a probability distribution on $\mathcal{X}$. The corresponding estimate of $\mathfrak{f}_v^{\mathrm{RSB}}$ is

$$
\widehat{\mathfrak{f}}_v^{\mathrm{RSB}} = \mathbb{E}_{l,J} \frac{1}{N^l} \sum_{r(1)\ldots r(l)=1}^{N} \log \left\{ \frac{1}{M^l} \sum_{s(1),\ldots,s(l)=1}^{M} \left[\sum_{x \in \mathcal{X}} \widehat{\mathtt{m}}_{r(1)}^{s(1)}(x) \cdots \widehat{\mathtt{m}}_{r(l)}^{s(l)}(x)\, \psi(x; J) \right]^x \right\} .
$$

Similar expressions are easily written for $\mathfrak{f}_f^{\mathrm{RSB}}$ and $\mathfrak{f}_e^{\mathrm{RSB}}$. Their (approximate) evaluation can be accelerated considerably by summing over a random subset of the l-tuples $r(1),\ldots,r(l)$ and $s(1),\ldots,s(l)$. Further, as in the RS case, it is beneficial to average over iterations (or, equivalently, over T) in order to reduce statistical errors at small computational cost.

19.3   A first application: XORSAT

Let us apply the 1RSB cavity method to XORSAT. This approach was introduced in Section 18.6, but we now want to show how it follows as a special case of the formalism developed in the previous sections of this chapter. Our objective is to exemplify the general ideas with a well understood problem, and to build some basic intuition that will be useful in more complicated applications.

As in Chapter 18, we consider the distribution over $\underline{x} = (x_1,\ldots,x_N) \in \{0,1\}^N$ specified by

$$
\mu(\underline{x}) = \frac{1}{Z} \prod_{a=1}^{M} \mathbb{I}\left(x_{i_1(a)} \oplus \cdots \oplus x_{i_k(a)} = b_a\right) . \tag{19.39}
$$

As usual, $\oplus$ denotes the sum modulo 2 and, for each $a \in \{1,\ldots,M\}$, $\partial a = \{i_1(a),\ldots,i_K(a)\}$ is a subset of $\{1,\cdot,N\}$, and $b_a \in \{0,1\}$. Random K-XORSAT formulae are generated by choosing both the index set $\{i_1(a),\ldots,i_K(a)\}$ and the right-hand side b_a uniformly at random.

19.3.1   BP equations

The BP equations read

$$
\mathtt{m}_{ia}(x_i) = \frac{1}{z_{ia}} \prod_{b \in \partial i \setminus a} \widehat{\mathtt{m}}_{bi}(x_i) , \tag{19.40}
$$

$$
\widehat{\mathtt{m}}_{ai}(x_i) = \frac{1}{\hat{z}_{ai}} \sum_{\underline{x}_{\partial a \setminus i}} \mathbb{I}\left(x_{i_1(a)} \oplus \cdots \oplus x_{i_K(a)} = b_a\right) \prod_{j \in \partial a \setminus i} \mathtt{m}_{ja}(x_j). \tag{19.41}
$$

As in Section 18.6, we shall assume that messages can take only three values, which we denote by the shorthands $\mathtt{m}_{ia} = 0$ if $(\mathtt{m}_{ia}(0) = 1,\ \mathtt{m}_{ia}(1) = 0)$, $\mathtt{m}_{ia} = 1$ if $(\mathtt{m}_{ia}(0) = 0,\ \mathtt{m}_{ia}(1) = 1)$, and $\mathtt{m}_{ia} = *$ if $(\mathtt{m}_{ia}(0) = \mathtt{m}_{ia}(1) = 1/2)$.

Consider the first BP equation (19.40), and denote by n_0, n_1, and n_* the numbers of messages of type 0, 1, and *, respectively, in the set of incoming messages $\{\widehat{\mathtt{m}}_{bi}\}$, $b \in \partial i \backslash a$. Equation (19.40) can then be rewritten as

$$
\mathtt{m}_{ia} = \begin{cases} 0 & \text{if } n_0 > 0,\ n_1 = 0, \\ 1 & \text{if } n_0 = 0,\ n_1 > 0, \\ * & \text{if } n_0 = 0,\ n_1 = 0, \\ ? & \text{if } n_0 > 0,\ n_1 > 0, \end{cases} \qquad z_{ia} = \begin{cases} 2^{-n_*} & \text{if } n_0 > 0,\ n_1 = 0, \\ 2^{-n_*} & \text{if } n_0 = 0,\ n_1 > 0, \\ 2^{1-n_*} & \text{if } n_0 = 0,\ n_1 = 0, \\ 0 & \text{if } n_0 > 0,\ n_1 > 0. \end{cases} \qquad (19.42)
$$

The computation of the normalization constant z_{ia} will be useful in the 1RSB analysis. Notice that if $n_0 > 0$ and $n_1 > 0$, a contradiction arises at node i and therefore $\mathtt{m}_{ia}$ is not defined. However, we shall see that, because $z_{ia} = 0$ in this case, this situation does not create any problem within 1RSB.

In the second BP equation (19.41), we denote by $\widehat{n}_0$, $\widehat{n}_1$, and $\widehat{n}_*$ the numbers of messages of type 0, 1, and *, respectively, among $\{\mathtt{m}_{ja}\}$, $j \in \partial a \backslash i$. We then obtain

$$
\widehat{\mathtt{m}}_{ai} = \begin{cases} 0 & \text{if } n_* = 0 \text{ and } n_1 \text{ has the same parity as } b_a, \\ 1 & \text{if } n_* = 0 \text{ and } n_1 \text{ does not have the same parity as } b_a, \\ * & \text{if } n_* > 0. \end{cases} \qquad (19.43)
$$

In all three cases, $\widehat{z}_{ai} = 1$.

In Section 18.6 we studied the equations (19.40) and (19.41) above and deduced that, for typical random instances with $\alpha = M/N < \alpha_{\mathrm{d}}(K)$, they have a unique solution, with $\mathtt{m}_{ia} = \widehat{\mathtt{m}}_{ai} = *$ on each edge.

Exercise 19.2 Evaluate the Bethe free entropy for this solution, and show that it yields the free-entropy density $f^{\mathrm{RS}} = (1 - \alpha) \log 2$.

19.3.2 The 1RSB cavity equations

We now assume that the BP equations (19.42) and (19.43) have many solutions, and apply the 1RSB cavity method to study their statistics.

The 1RSB messages Q_{ia} and $\widehat{Q}_{ai}$ are distributions over $\{0, 1, *\}$. A little effort shows that eqn (19.27) yields

$$
Q_{ia}(0) = \frac{1}{Z_{ia}} \left\{ \prod_{b \in \partial i \backslash a} \left(\widehat{Q}_{bi}(0) + 2^{-x} \widehat{Q}_{bi}(*) \right) - \prod_{b \in \partial i \backslash a} \left(2^{-x} \widehat{Q}_{bi}(*) \right) \right\}, \qquad (19.44)
$$

$$
Q_{ia}(1) = \frac{1}{Z_{ia}} \left\{ \prod_{b \in \partial i \backslash a} \left(\widehat{Q}_{bi}(1) + 2^{-x} \widehat{Q}_{bi}(*) \right) - \prod_{b \in \partial i \backslash a} \left(2^{-x} \widehat{Q}_{bi}(*) \right) \right\}, \qquad (19.45)
$$

$$
Q_{ia}(*) = \frac{1}{Z_{ia}} 2^{x} \prod_{b \in \partial i \backslash a} 2^{-x} \widehat{Q}_{bi}(*). \qquad (19.46)
$$

For instance, eqn (19.44) follows from the first line of eqn (19.42): $\mathtt{m}_{ia} = 0$ if all of the incoming messages are $\widehat{\mathtt{m}}_{bi} \in \{*, 0\}$ (first term), unless they are all equal to * (subtracted term). The reweighting $z_{ia}^{\mathsf{x}} = 2^{-\mathsf{x} n_*}$ decomposes into factors associated with the incoming $*$ messages.

The second group of 1RSB equations, eqn (19.28), takes the form

$$
\widehat{Q}_{ai}(0) = \frac{1}{2} \left\{ \prod_{j \in \partial a \backslash i} (Q_{ja}(0) + Q_{ja}(1)) + s(b_a) \prod_{j \in \partial a \backslash i} (Q_{ja}(0) - Q_{ja}(1)) \right\} ,
$$

(19.47)

$$
\widehat{Q}_{ai}(1) = \frac{1}{2} \left\{ \prod_{j \in \partial a \backslash i} (Q_{ja}(0) + Q_{ja}(1)) - s(b_a) \prod_{j \in \partial a \backslash i} (Q_{ja}(0) - Q_{ja}(1)) \right\} ,
$$

(19.48)

$$
\widehat{Q}_{ai}(*) = 1 - \prod_{j \in \partial a \backslash i} (Q_{ja}(0) + Q_{ja}(1)) ,
$$

(19.49)

where $s(b_a) = +1$ if $b_a = 0$, and $s(b_a) = -1$ otherwise.

Notice that, if one takes $\mathsf{x} = 0$, the two sets of equations coincide with those obtained in Section 18.6 (see eqn (18.35)) (the homogeneous linear system, $b_a = 0$, was considered there). As in that section, we look for solutions such that the messages $Q_{ia}(\cdot)$ (and similarly $\widehat{Q}_{ai}(\cdot)$) take two possible values: either $Q_{ia}(0) = Q_{ia}(1) = 1/2$, or $Q_{ia}(*) = 1$. This assumption is consistent with the 1RSB cavity equations (19.44) and (19.49). Under this assumption, the x dependency drops from these equations and we recover the analysis in Section 18.6. In particular, we can repeat the density evolution analysis discussed there. If we denote by Q_* the probability that a randomly chosen edge carries the 1RSB message $Q_{ia}(0) = Q_{ia}(1) = 1/2$, then the fixed-point equation of density evolution reads

$$
Q_* = 1 - \exp\{-k\alpha Q_*^{k-1}\} .
$$

(19.50)

For $\alpha < \alpha_{\mathrm{d}}(K)$, this equation admits only the solution $Q_* = 0$, implying $Q_{ia}(*) = 1$ with high probability. This indicates (once more) that the only solution of the BP equations in this regime is $\mathbf{m}_{ia} = *$ for all $(i, a) \in E$.

For $\alpha > \alpha_{\mathrm{d}}$, a pair of non-trivial solutions (with $Q_* > 0$) appear, indicating the existence of a large number of BP fixed points (and hence Bethe measures). Considerations of stability under density evolution suggest that we should select the largest one. It will also be useful in the following to introduce the probability

$$
\widehat{Q}_* = Q_*^{k-1}
$$

(19.51)

that a uniformly random edge carries a message $\widehat{Q}_{ai}(0) = \widehat{Q}_{ai}(1) = 1/2$.

19.3.3   Complexity

We can now compute the Bethe free entropy (19.32) of the auxiliary graphical model. The complexity will be computed through the Legendre transform of the 1RSB free entropy (see eqn (19.8)).

Let us start by computing the contribution $\mathbb{F}_a^{\mathrm{RSB}}$ defined in eqn (19.29). Consider the weight

$$e^{\mathbb{F}_a(\{\mathtt{m}_{ia}\})} = \sum_{\underline{x}_{\partial a}} \mathbb{I}(x_{i_1(a)} \oplus \cdots \oplus x_{i_K(a)} = b_a) \prod_{i \in \partial a} \mathtt{m}_{ia}(x_i). \tag{19.52}$$

Let $\widehat{n}_0$, $\widehat{n}_1$, and $\widehat{n}_*$ denote the number of variable nodes $i \in \partial a$ such that $\mathtt{m}_{ia} = 0, 1$, and $*$, respectively, for $i \in \partial a$. We then obtain

$$e^{\mathbb{F}_a(\{\mathtt{m}_{ia}\})} = \begin{cases} 1/2 & \text{if } \widehat{n}_* > 0, \\ 1 & \text{if } \widehat{n}_* = 0 \text{ and } \widehat{n}_1 \text{ has the same parity as } b_a, \\ 0 & \text{if } \widehat{n}_* = 0 \text{ and } \widehat{n}_1 \text{ does not have the same parity as } b_a. \end{cases} \tag{19.53}$$

Taking the expectation of $e^{\mathtt{x}\mathbb{F}_a(\{\mathtt{m}_{ia}\})}$ with respect to $\{\mathtt{m}_{ia}\}$, distributed independently according to $Q_{ia}(\cdot)$, and assuming that $Q_{ia}(0) = Q_{ia}(1)$ (which is the case in our solution), we get

$$\mathbb{F}_a^{\mathrm{RSB}} = \log \left\{ \frac{1}{2} \prod_{i \in \partial a} (1 - Q_{ia}(*)) + \frac{1}{2^{\mathtt{x}}} \left[1 - \prod_{i \in \partial a} (1 - Q_{ia}(*)) \right] \right\}. \tag{19.54}$$

The first term corresponds to the case $\widehat{n}_* = 0$ (the factor $1/2$ being the probability that the parity of $\widehat{n}_1$ is b_a), and the second to $\widehat{n}_* > 0$. Within our solution, either $Q_{ia}(*) = 0$ or $Q_{ia}(*) = 1$. Therefore only one of the above terms survives: either $Q_{ia}(*) = 0$ for all $i \in \partial a$, yielding $\mathbb{F}_a^{\mathrm{RSB}} = -\log 2$, or $Q_{ia}(*) = 1$ for some $i \in \partial a$, implying $\mathbb{F}_a^{\mathrm{RSB}} = -\mathtt{x} \log 2$.

Until now, we have considered a generic K-XORSAT instance. For random instances, we can take the expectation with respect to $Q_{ia}(*)$, independently distributed as in the density-evolution fixed point. The first case, namely $Q_{ia}(*) = 0$ for all $i \in \partial a$ (and thus $\mathbb{F}_a^{\mathrm{RSB}} = -\log 2$), occurs with probability Q_*^k. The second, i.e. $Q_{ia}(*) = 1$ for some $i \in \partial a$ (and $\mathbb{F}_a^{\mathrm{RSB}} = -\mathtt{x} \log 2$), occurs with probability $1 - Q_*^k$. Altogether, we obtain

$$\mathbb{E}\{\mathbb{F}_a^{\mathrm{RSB}}\} = -\left[Q_*^k + \mathtt{x}(1 - Q_*^k)\right] \log 2 + o_N(1). \tag{19.55}$$

Assuming the messages $Q_{ia}(\cdot)$ to be short-range correlated, $\sum_{a \in F} \mathbb{F}_a^{\mathrm{RSB}}$ will concentrate around its expectation. We then have, with high probability,

$$\frac{1}{N} \sum_{a \in F} \mathbb{F}_a^{\mathrm{RSB}} = -\alpha \left[Q_*^k + x(1 - Q_*^k)\right] \log 2 + o_N(1). \tag{19.56}$$

The contributions from variable-node and edge terms can be computed along similar lines. We shall just sketch these computations, and invite the reader to work out the details.

Consider the contribution $\mathbb{F}_i^{\mathrm{RSB}}$, $i \in V$, defined in eqn (19.30). Assume that $\widehat{Q}_{ai}(*) = 1$ and $\widehat{Q}_{ai}(0) = \widehat{Q}_{ai}(1) = 1/2$ for n_* and n_0, respectively, of the neighbouring function nodes $a \in \partial i$. Then $\mathbb{F}_i^{\mathrm{RSB}} = -(n_* \mathtt{x} + n_0 - 1) \log 2$ if $n_0 \geq 1$, and

$\mathbb{F}_i^{\text{RSB}} = -(n_* - 1)\text{x} \log 2$ otherwise. Averaging these expressions over n_0 (a Poisson-distributed random variable with mean $k\alpha \widehat{Q}_*$) and n_* (Poisson with mean $k\alpha(1 - \widehat{Q}_*)$), we obtain

$$\frac{1}{N} \sum_{i \in V} \mathbb{F}_i^{\text{RSB}} = -\left\{ \left[k\alpha \widehat{Q}_* - 1 + \text{e}^{-k\alpha \widehat{Q}_*} \right] + \left[k\alpha(1 - \widehat{Q}_*) - \text{e}^{-k\alpha \widehat{Q}_*} \right] \text{x} \right\} \log 2 + o_N(1).$$
(19.57)

Let us consider, finally, the edge contribution $\mathbb{F}_{(ia)}^{\text{RSB}}$ defined in eqn (19.31). If $Q_{ia}(0) = Q_{ia}(1) = 1/2$ and $\widehat{Q}_{ai}(0) = \widehat{Q}_{ai}(1) = 1/2$, then either $\text{e}^{\mathbb{F}_{ai}} = 1$ or $\text{e}^{\mathbb{F}_{ia}} = 0$, each with probability $1/2$. As a consequence, $\mathbb{F}_{(ia)}^{\text{RSB}} = -\log 2$. If either $Q_{ia}(*) = 1$ or $\widehat{Q}_{ai}(*) = 1$ (or both), $\text{e}^{\mathbb{F}_{ia}^{\text{RSB}}} = 1/2$ with probability 1, and therefore $\mathbb{F}_{(ia)}^{\text{RSB}} = -\text{x} \log 2$. Altogether, we obtain, with high probability,

$$\frac{1}{N} \sum_{(ia) \in E} \mathbb{F}_{(ia)}^{\text{RSB}} = -k\alpha \left\{ Q_* \widehat{Q}_* + (1 - Q_* \widehat{Q}_*)\text{x} \right\} \log 2 + o_N(1).$$
(19.58)

The free entropy (19.32) of the auxiliary graphical model is obtained by collecting the various terms. We obtain $\mathbb{F}^{\text{RSB}}(\text{x}) = N\text{f}^{\text{RSB}}(\text{x}) + o(N)$, where $\text{f}^{\text{RSB}}(\text{x}) = [\Sigma_{\text{tot}} + \text{x}\,\phi_{\text{typ}}] \log 2$ and

$$\Sigma_{\text{tot}} = k\alpha Q_* \widehat{Q}_* - k\alpha \widehat{Q}_* - \alpha Q_*^k + 1 - \text{e}^{-k\alpha \widehat{Q}_*},$$
(19.59)

$$\phi_{\text{typ}} = -k\alpha Q_* \widehat{Q}_* + k\alpha \widehat{Q}_* + \alpha Q_*^k - \alpha + \text{e}^{-k\alpha \widehat{Q}_*}.$$
(19.60)

Here Q_* is the largest solution of eqn (19.50), and $\widehat{Q}_* = Q_*^{k-1}$, a condition that has a pleasing interpretation, as shown in the exercise below.

Exercise 19.3 Consider the function $\Sigma_{\text{tot}}(Q_*, \widehat{Q}_*)$ defined in eqn (19.59). Show that the stationary points of this function coincide with the solutions of eqn (19.50) and that $\widehat{Q}_* = Q_*^{k-1}$.

Because of the linear dependence on x, the Legendre transform (19.8) is straightforward:

$$\Sigma(\phi) = \begin{cases} \Sigma_{\text{tot}} & \text{if } \phi = \phi_{\text{typ}}, \\ -\infty & \text{otherwise.} \end{cases}$$
(19.61)

This means that there are $2^{N\Sigma_{\text{tot}}}$ Bethe measures, and these all have entropy $N\phi_{\text{typ}} \log 2$. Furthermore, $\Sigma_{\text{tot}} + \phi_{\text{typ}} = 1 - \alpha$, confirming that the total entropy is $(1 - \alpha) \log 2$. This identity can be also written in the form

$$\frac{1}{2^{N(1-\alpha)}} = \frac{1}{2^{N\Sigma_{\text{tot}}}} \times \frac{1}{2^{N\phi_{\text{typ}}}},$$
(19.62)

which is nothing but the decomposition (19.6) into extremal Bethe measures. Indeed, if $\underline{x}$ is a solution of the linear system, then $\mu(\underline{x}) = 1/2^{N(1-\alpha)}$, $w_n \approx 1/2^{N\Sigma_{\text{tot}}}$, and (assuming the μ^n to have disjoint supports) $\mu^n(\underline{x}) \approx 1/2^{N\phi_{typ}}$ for the state n which contains $\underline{x}$.

Note that the value of Σ that we find here coincides with the result that we obtained in Section 18.5 for the logarithm of the number of clusters in random XORSAT formulae. This provides an independent check of our assumptions and, in particular, it shows that the number of clusters is, to leading order, the same as the number of Bethe measures. In particular, the SAT–UNSAT transition occurs at the value of α where the complexity Σ_{tot} vanishes. At this value, each cluster still contains a large number, $2^{N(1-\alpha_s)}$, of configurations.

Exercise 19.4 Repeat this 1RSB cavity analysis for a linear Boolean system described by a factor graph from the ensemble $\mathbb{D}_N(\Lambda, P)$. (This means a random system of linear equations, where the fraction of equations involving k variables is P_k, and the fraction of variables which appear in exactly ℓ equations is Λ_ℓ.)

(a) Show that Q_* and $\widehat{Q}_*$ satisfy

$$\widehat{Q}_* = \rho(Q_*) \ , \quad Q_* = 1 - \lambda(1 - \widehat{Q}_*) \ , \tag{19.63}$$

where λ and ρ are the edge-perspective degree profiles.

(b) Show that the complexity is given by

$$\Sigma_{tot} = 1 - \frac{\Lambda'(1)}{P'(1)} P(Q_*) - \Lambda(1 - \widehat{Q}_*) - \Lambda'(1)(1 - Q_*)\widehat{Q}_* \ , \tag{19.64}$$

and the internal entropy of the clusters is $\phi_{typ} = 1 - \Lambda'(1)/P'(1) - \Sigma_{tot}$.

(c) In the case where all variables have a degree strictly larger than 1 (so that $\lambda(0) = 0$), argue that the relevant solution is $Q_* = \widehat{Q}_* = 1$, $\Sigma_{tot} = 1 - \Lambda'(1)/P'(1)$, $\phi_{typ} = 0$. What is the interpretation of this result in terms of the core structure discussed in Section 18.3?

19.4 The special value x $= 1$

Let us now return to the general formalism. The case x $= 1$ plays a special role, in that the weights $\{w_n(\mathrm{x})\}$ of various Bethe measures in the auxiliary model coincide with those appearing in the decomposition (19.6). This fact manifests itself in some remarkable properties of the 1RSB formalism.

19.4.1 Back to BP

Consider the general 1RSB cavity equations (19.27) and (19.28). Using the explicit form of the reweighting factors $e^{\mathbb{F}_i - \mathbb{F}_{ia}}$ and $e^{\mathbb{F}_a - \mathbb{F}_{ia}}$ provided in eqns (19.22) and (19.23), they can be written, for x $= 1$, as

$$Q_{ia}(\mathtt{m}_{ia}) \cong \sum_{x_i} \sum_{\{\widehat{\mathtt{m}}_{bi}\}} \mathbb{I}\left(\mathtt{m}_{ia} = g_i(\{\widehat{\mathtt{m}}_{bi}\})\right) \prod_{b \in \partial i \setminus a} \widehat{Q}_{bi}(\widehat{\mathtt{m}}_{bi}) \, \widehat{\mathtt{m}}_{bi}(x_i) \,, \tag{19.65}$$

$$\widehat{Q}_{ai}(\widehat{\mathtt{m}}_{ai}) \cong \sum_{\underline{x}_{\partial a}} \psi_a(\underline{x}_{\partial a}) \sum_{\{\mathtt{m}_{ja}\}} \mathbb{I}\left(\widehat{\mathtt{m}}_{ai} = f_a(\{\mathtt{m}_{ja}\})\right) \prod_{j \in \partial a \setminus i} Q_{ja}(\mathtt{m}_{ja}) \, \mathtt{m}_{ja}(x_j) \,. \tag{19.66}$$

We introduce the messages obtained by taking the averages of the 1RSB messages $\{Q_{ia}, \widehat{Q}_{ai}\}$:

$$\nu^{\mathrm{av}}_{i \to a}(x_i) \equiv \sum_{\mathtt{m}_{ia}} Q_{ia}(\mathtt{m}_{ia}) \, \mathtt{m}_{ia}(x_i) \,, \quad \widehat{\nu}^{\mathrm{av}}_{a \to i}(x_i) \equiv \sum_{\widehat{\mathtt{m}}_{ai}} \widehat{Q}_{ai}(\widehat{\mathtt{m}}_{ai}) \, \widehat{\mathtt{m}}_{ai}(x_i) \,.$$

The interpretation of these quantities is straightforward. Given an extremal Bethe measure sampled according to the distribution w_n, let $\nu^n_{i \to a}(\,\cdot\,)$ (or $\widehat{\nu}^n_{a \to i}(\,\cdot\,)$) be the corresponding message along the directed edge $i \to a$ (or $a \to i$, respectively). Its expectation, with respect to the random choice of the measure, is $\nu^{\mathrm{av}}_{i \to a}(\,\cdot\,)$ (or $\widehat{\nu}^{\mathrm{av}}_{a \to i}(\,\cdot\,)$, respectively).

Using the eqns (19.9), one finds that eqns (19.65) and (19.66) imply

$$\nu^{\mathrm{av}}_{i \to a}(x_i) \cong \prod_{b \in \partial i \setminus a} \widehat{\nu}^{\mathrm{av}}_{b \to i}(x_i) \,, \tag{19.67}$$

$$\widehat{\nu}^{\mathrm{av}}_{a \to i}(x_i) \cong \sum_{\{x_j\}_{j \in \partial a \setminus i}} \psi_a(\underline{x}_{\partial a}) \prod_{j \in \partial a \setminus i} \nu^{\mathrm{av}}_{j \to a}(x_j) \,, \tag{19.68}$$

which are nothing but the ordinary BP equations. This suggests that even if $\mu(\,\cdot\,)$ decomposes into an exponential number of extremal Bethe measures $\mu^n(\,\cdot\,)$, it is itself a (non-extremal) Bethe measure. In particular, there exists a quasi-solution of the BP equations associated with it that allows one to compute its marginals.

The reader might be disappointed by these remarks. Why should we insist on the 1RSB cavity approach if, when the 'correct' weights are used, one recovers the much simpler BP equations? There are at least two answers:

1. The 1RSB approach provides a much more refined picture: decomposition into extremal Bethe states, long-range correlations, and complexity. This is useful and interesting per se.
2. In the case of a static (s1RSB) phase, it turns out that the region $\mathtt{x} = 1$ corresponds to an 'unphysical' solution of the 1RSB cavity equations, and that (asymptotically) correct marginals are instead obtained by letting $\mathtt{x} = \mathtt{x}_*$, for some $\mathtt{x}_* \in [0, 1)$. In such cases it is essential to resort to the full 1RSB formalism (see Section 19.6 below).

19.4.2   A simpler recursion

As we stressed above, controlling (either numerically or analytically) the 1RSB distributional recursions (19.34) and (19.35) is a difficult task. In the case $\mathtt{x} = 1$, they simplify considerably and lend themselves to a much more accurate numerical study. This observation can be very useful in practice.

As in Section 19.2.5, we consider a random graphical model. We shall also assume a 'local uniformity condition'. More precisely, the original model $\mu(\,\cdot\,)$ is a Bethe measure for the message set $\nu^{\mathrm{av}}_{i\to a}(x_i) = 1/q$ and $\widehat{\nu}^{\mathrm{av}}_{a\to i}(x_i) = 1/q$, where $q = |\mathcal{X}|$ is the size of the alphabet. While such a local uniformity condition is not necessary, it considerably simplifies the derivation below. The reader can find a more general treatment in the literature.

Consider eqns (19.34) and (19.35) at x $= 1$. The normalization constants can be easily computed using the uniformity condition. We can then average over the structure of the graph and the function node distribution: let us denote the averaged distributions by Q_{av} and $\widehat{Q}_{\mathrm{av}}$. They satisfy the following equations:

$$Q^{(t+1)}_{\mathrm{av}}(\mathtt{m}) = \mathbb{E}\left\{ q^{l-2} \sum_{\{\widehat{\mathtt{m}}_b\}} \mathbb{I}\left(\mathtt{m} = \mathsf{f}(\{\widehat{\mathtt{m}}_b\}; J)\right) z(\{\widehat{\mathtt{m}}_b\}) \prod_{b=1}^{l-1} \widehat{Q}^{(t)}_{\mathrm{av}}(\widehat{\mathtt{m}}_b) \right\}, \qquad (19.69)$$

$$\widehat{Q}^{(t)}_{\mathrm{av}}(\widehat{\mathtt{m}}) = \mathbb{E}\left\{ \frac{q^{k-2}}{\overline{\psi}_k} \sum_{\{\mathtt{m}_j\}} \mathbb{I}\left(\widehat{\mathtt{m}} = \widehat{\mathsf{f}}(\{\mathtt{m}_j\}; \widehat{J})\right) \widehat{z}(\{\mathtt{m}_j\}; \widehat{J}) \prod_{j=1}^{k-1} Q^{(t)}_{\mathrm{av}}(\mathtt{m}_j) \right\}, \qquad (19.70)$$

where the expectations are taken over $l, k, J, \widehat{J}$, distributed according to the random graphical model. Here, $\overline{\psi}_k = \sum_{x_1,\dots,x_{k-1}} \psi(x_1,\dots,x_{k-1},x;\widehat{J})$ can be shown to be independent of x (this is necessary for the uniformity condition to hold).

Equations (19.69) and (19.70) are considerably simpler than the original distributional equations (19.34) and (19.35) in that $Q^{(t)}_{\mathrm{av}}(\,\cdot\,)$ and $\widehat{Q}^{(t)}_{\mathrm{av}}(\,\cdot\,)$ are non-random. On the other hand, they still involve a reweighting factor that is difficult to handle. It turns out that this reweighting can be eliminated by introducing a new pair of distributions for each $x \in \mathcal{X}$:

$$\widehat{R}^{(t)}_x(m) \equiv q\, m(x)\, \widehat{Q}^{(t)}_{\mathrm{av}}(m)\,, \qquad R^{(t)}_x(m) = q\, m(x)\, Q^{(t)}_{\mathrm{av}}(m)\,. \qquad (19.71)$$

One can show that eqns (19.69) and (19.70) imply the following recursions for $R^{(t)}_x$ and $\widehat{R}^{(t)}_x$:

$$R^{(t+1)}_x(\mathtt{m}) = \mathbb{E}\left\{ \sum_{\{\widehat{\mathtt{m}}_b\}} \mathbb{I}\left(\mathtt{m} = g(\{\widehat{\mathtt{m}}_b\}; J)\right) \prod_{b=1}^{l-1} \widehat{R}^{(t)}_x(\widehat{\mathtt{m}}_b) \right\}, \qquad (19.72)$$

$$\widehat{R}^{(t)}_x(\widehat{\mathtt{m}}) = \mathbb{E}\left\{ \sum_{\{x_j\}} \pi(\{x_j\}|x; \widehat{J}) \sum_{\{\mathtt{m}_j\}} \mathbb{I}\left(\widehat{\mathtt{m}} = f(\{\mathtt{m}_j\}; \widehat{J})\right) \prod_{j=1}^{k-1} R^{(t)}_{x_j}(\mathtt{m}_j) \right\}. \qquad (19.73)$$

Here $\mathbb{E}$ denotes the expectation with respect to $l, \widehat{J}, k, J$ and, for any $x, \widehat{J}$, the distribution $\pi(\{x_j\}|x; \widehat{J})$ is defined by

$$\pi(x_1,\dots,x_{k-1}|x; \widehat{J}) = \frac{\psi(x_1,\dots,x_{k-1},x;\widehat{J})}{\sum_{y_1,\dots,y_{k-1}} \psi(y_1,\dots,y_{k-1},x;\widehat{J})}\,. \qquad (19.74)$$

Exercise 19.5 Prove eqns (19.72) and (19.73). It might be useful to recall the following explicit expressions for the reweighting factors z and $\hat{z}$:

$$z(\{\widehat{\mathtt{m}}_b\})\,\mathtt{m}(x) = \prod_{b=1}^{l-1} \widehat{\mathtt{m}}_b(x)\,, \tag{19.75}$$

$$\hat{z}(\{\mathtt{m}_j\};\widehat{J})\,\widehat{\mathtt{m}}(x) = \sum_{\{x_i\},x} \psi(x_1,\dots,x_{k-1},x;\widehat{J}) \prod_{j=1}^{k-1} \mathtt{m}_j(x_j)\,. \tag{19.76}$$

Equations (19.72) and (19.73) have a simple operational description. Let $\widehat{J}$ and k be drawn according to their distribution, and, given x, let us generate $x_1,\dots,x_{k-1}$ according to the kernel $\pi(x_1,\dots,x_k|x;\widehat{J})$. Then we draw independent messages $\mathtt{m}_1,\dots,\mathtt{m}_{k-1}$ with distributions $R^{(t)}_{x_1},\dots,R^{(t)}_{x_{k-1}}$, respectively. According to eqn (19.73), $\widehat{\mathtt{m}} = f(\{\mathtt{m}_j\};\widehat{J})$ then has the distribution $\widehat{R}^{(t)}_x$. For eqn (19.72), we draw J and l according to their distribution. Given x, we draw $l-1$ i.i.d. messages $\widehat{\mathtt{m}}_1,\dots,\widehat{\mathtt{m}}_{l-1}$ with distribution $\widehat{R}^{(t)}_x$. Then $\mathtt{m} = g(\{\widehat{\mathtt{m}}_b\};J)$ has the distribution $R^{(t+1)}_x$.

We shall see in Chapter 22 that this procedure does indeed coincide with that for computing 'point-to-set correlations' with respect to the measure $\mu(\cdot)$.

To summarize, for $\mathtt{x} = 1$ we have succeeded in simplifying the 1RSB density evolution equations in two directions: (i) the resulting equations do not involve 'distributions of distributions'; and (ii) we have got rid of the reweighting factor. A third crucial simplification is the following.

Theorem 19.5 *The 1RSB equations have a non-trivial solution (meaning a solution different from the RS one) if and only if eqns (19.72) and (19.73), when initialized so that $R^{(0)}_x$ is a singleton distribution on $\mathtt{m}(y) = \mathbb{I}(y = x)$, converge as $t \to \infty$ to a non-trivial distribution.*

This theorem resolves (in the case $\mathtt{x} = 1$) the ambiguity in the initial condition of the 1RSB iteration. In other words, if the 1RSB equations admit a non-trivial solution, this solution can be reached if we iterate the equations starting from the initial condition mentioned in the theorem. We refer the reader to the literature for a proof.

Exercise 19.6 Show that the free entropy of the auxiliary model $\mathbb{F}^{\mathrm{RSB}}(\mathtt{x})$, evaluated at $\mathtt{x} = 1$, coincides with the RS Bethe free entropy.

Further, its derivative with respect to $\mathtt{x}$ at $\mathtt{x} = 1$ can be expressed in terms of the fixed-point distributions $R^{(\infty)}_x$ and $\widehat{R}^{(\infty)}_x$. In particular, the complexity and internal free entropy can be computed from the fixed points of the simplified equations (19.72) and (19.73).

The conclusion of this section is that 1RSB calculations at $\mathsf{x} = 1$ are not technically harder than RS ones. In view of the special role played by the value $\mathsf{x} = 1$, this observation can be exploited in a number of contexts.

19.5 Survey propagation

The 1RSB cavity method can be applied to other message-passing algorithms whenever these have many fixed points. A particularly important case is the min-sum algorithm of Section 14.3. This approach (in both its RS and its 1RSB version) is sometimes referred to as the **energetic cavity method** because, in physics terms, the min-sum algorithm aims at computing the ground state configuration and its energy. We shall call the corresponding 1RSB message-passing algorithm $\mathsf{SP(y)}$ ('survey propagation at finite y').

19.5.1 The $\mathsf{SP(y)}$ equations

The formalism follows closely that used for counting the solutions of BP equations. To emphasize the similarities, let us adopt the same notation for the min-sum messages as for the BP ones. We define

$$\mathsf{m}_{ja}(x_j) \equiv E_{i \to a}(x_i), \quad \widehat{\mathsf{m}}_{ai}(x_i) \equiv \widehat{E}_{a \to i}(x_i), \tag{19.77}$$

and write the min-sum equations (14.41) and (14.40) as

$$\mathsf{m}_{ia} = \mathsf{f}_i^{\mathsf{e}}\left(\{\widehat{\mathsf{m}}_{bi}\}_{b \in \partial i \setminus a}\right), \qquad \widehat{\mathsf{m}}_{ai} = \widehat{\mathsf{f}}_a^{\mathsf{e}}\left(\{\mathsf{m}_{ja}\}_{j \in \partial a \setminus i}\right). \tag{19.78}$$

The functions $\mathsf{f}_i^{\mathsf{e}}$ and $\widehat{\mathsf{f}}_a^{\mathsf{e}}$ are defined by eqns (14.41) and (14.40), which we reproduce here:

$$\mathsf{m}_{ia}(x_i) = \sum_{b \in \partial i \setminus a} \widehat{\mathsf{m}}_{bi}(x_i) - u_{ia}, \tag{19.79}$$

$$\widehat{\mathsf{m}}_{ai}(x_i) = \min_{\underline{x}_{\partial a \setminus i}} \left[E_a(\underline{x}_{\partial a}) + \sum_{j \in \partial a \setminus i} \mathsf{m}_{ja}(x_j) \right] - \hat{u}_{ai}, \tag{19.80}$$

where $u_{ia}, \hat{u}_{ai}$ are normalization constants (independent of x_i) which ensure that $\min_{x_i} \widehat{\mathsf{m}}_{ai}(x_i) = 0$ and $\min_{x_i} \mathsf{m}_{ia}(x_i) = 0$.

With any set of messages $\{\mathsf{m}_{ia}, \widehat{\mathsf{m}}_{ai}\}$, we associate the Bethe energy

$$\mathbb{F}^{\mathsf{e}}(\underline{\mathsf{m}}, \underline{\widehat{\mathsf{m}}}) = \sum_{a \in F} \mathbb{F}_a^{\mathsf{e}}(\{\mathsf{m}_{ia}\}_{i \in \partial a}) + \sum_{i \in V} \mathbb{F}_i^{\mathsf{e}}(\{\widehat{\mathsf{m}}_{ai}\}_{a \in \partial i}) - \sum_{(ia) \in E} \mathbb{F}_{ia}^{\mathsf{e}}(\mathsf{m}_{ia}, \widehat{\mathsf{m}}_{ai}), \tag{19.81}$$

where the various terms are (see eqn (14.45))

$$\mathbb{F}_a^{\mathsf{e}} = \min_{\underline{x}_{\partial a}} \left[E_a(\underline{x}_{\partial a}) + \sum_{j \in \partial a} \mathsf{m}_{ia}(x_i) \right], \qquad \mathbb{F}_i^{\mathsf{e}} = \min_{x_i} \left[\sum_{a \in \partial i} \widehat{\mathsf{m}}_{ai}(x_i) \right],$$

$$\mathbb{F}_{ia}^{\mathsf{e}} = \min_{x_i} \left[\mathsf{m}_{ia}(x_i) + \widehat{\mathsf{m}}_{ai}(x_i) \right]. \tag{19.82}$$

Having set up the message-passing algorithm and the associated energy functional, we can repeat the programme developed in the previous sections. In particular, in analogy with Assumption 1, we have the following.

Assumption 4 *There exist exponentially many quasi-solutions* $\{\underline{m}^n\}$ *of the min-sum equations. The number of such solutions with Bethe energy* $\mathbb{F}^e(\underline{m}^n) \approx N\epsilon$ *is (to leading exponential order)* $\exp\{N\Sigma^e(\epsilon)\}$, *where* $\Sigma^e(\epsilon)$ *is the* **energetic complexity function**.

In order to estimate $\Sigma^e(\epsilon)$, we introduce an auxiliary graphical model, whose variables are the min-sum messages $\{m_{ia}, \widehat{m}_{ai}\}$. These are forced to satisfy (within some accuracy) the min-sum equations (19.79) and (19.80). Each solution is given a weight $e^{-y\mathbb{F}^e(\underline{m},\widehat{\underline{m}})}$, with $y \in \mathbb{R}$. The corresponding distribution is

$$P_y(\underline{m}, \widehat{\underline{m}}) = \frac{1}{\Xi(y)} \prod_{a \in F} \Psi_a(\{m_{ja}, \widehat{m}_{ja}\}_{j \in \partial a}) \prod_{i \in V} \Psi_i(\{m_{ib}, \widehat{m}_{ib}\}_{b \in \partial i}) \prod_{(ia) \in E} \Psi_{ia}(m_{ia}, \widehat{m}_{ia}),$$

(19.83)

where

$$\Psi_a = \prod_{i \in \partial a} \mathbb{I}\left(\widehat{m}_{ai} = \widehat{f}^e_a\left(\{m_{ja}\}_{j \in \partial a \setminus i}\right)\right) \; e^{-y\mathbb{F}^e_a(\{m_{ja}\}_{j \in \partial a})},$$

(19.84)

$$\Psi_i = \prod_{a \in \partial i} \mathbb{I}\left(m_{ia} = f^e_i\left(\{\widehat{m}_{bi}\}_{b \in \partial i \setminus a}\right)\right) \; e^{-y\mathbb{F}^e_i(\{\widehat{m}_{bi}\}_{b \in \partial i})},$$

(19.85)

$$\Psi_{ia} = e^{y\mathbb{F}^e_{ia}(m_{ia}, \widehat{m}_{ai})}.$$

(19.86)

Since the auxiliary graphical model is again locally tree-like, we can hope to derive asymptotically exact results through belief propagation. The messages of the auxiliary problem, which will be denoted as $Q_{ia}(\cdot)$ and $\widehat{Q}_{ai}(\cdot)$, are distributions over the min-sum messages. The $\mathsf{SP}(y)$ equations are obtained by making, further, the independence assumption (19.26).

The reader will certainly have noticed that the whole procedure is extremely close to our study in Section 19.2.2. We can apply our previous analysis verbatim to derive the $\mathsf{SP}(y)$ update equations. The only step that requires some care is the formulation of the proper analogue of Lemma 19.3. This becomes the following.

Lemma 19.6 *Assume that* $m_{ia}(x_i) + \widehat{m}_{ai}(x_i) < \infty$ *for at least one value of* $x_i \in \mathcal{X}$. *If* $m_{ia} = f^e_i(\{\widehat{m}_{bi}\}_{b \in \partial i \setminus a})$, *then*

$$\mathbb{F}^e_i - \mathbb{F}^e_{ia} = u_{ia}(\{\widehat{m}_{bi}\}_{b \in \partial i \setminus a}) \equiv \min_{x_i}\left\{ \sum_{b \in \partial i \setminus a} \widehat{m}_{bi}(x_i)\right\}.$$

(19.87)

Analogously, if $\widehat{m}_{ai} = f^e_a(\{m_{ja}\}_{j \in \partial a \setminus i})$, *then*

$$\mathbb{F}^e_a - \mathbb{F}^e_{ia} = \widehat{u}_{ai}(\{m_{ja}\}_{j \in \partial a \setminus i}) \equiv \min_{\underline{x}_{\partial a}}\left\{ E_a(\underline{x}_{\partial a}) + \sum_{k \in \partial a \setminus i} m_{ka}(x_k)\right\}.$$

(19.88)

Using this lemma, the same derivation as in Section 19.2.2 leads to the following result.

Proposition 19.7 *The* SP(y) *equations are (with the shorthands* $\{\widehat{m}_{bi}\}$ *for* $\{\widehat{m}_{bi}\}_{b\in\partial i\setminus a}$ *and* $\{m_{ja}\}$ *for* $\{m_{ja}\}_{j\in\partial a\setminus i}$)

$$Q_{ia}(m_{ia}) \cong \sum_{\{\widehat{m}_{bi}\}} \mathbb{I}\left(m_{ia} = g_i^e(\{\widehat{m}_{bi}\})\right) e^{-yu_{ia}(\{\widehat{m}_{bi}\})} \prod_{b\in\partial i\setminus a} \widehat{Q}_{bi}(\widehat{m}_{bi}), \qquad (19.89)$$

$$\widehat{Q}_{ai}(\widehat{m}_{ai}) \cong \sum_{\{m_{ja}\}} \mathbb{I}\left(\widehat{m}_{ai} = f_a^e(\{m_{ja}\})\right) e^{-y\hat{u}_{ai}(\{m_{ja}\})} \prod_{j\in\partial a\setminus i} Q_{ja}(m_{ja}). \qquad (19.90)$$

In the following we shall reserve the name **survey propagation** (SP) for the $y = \infty$ case of these equations.

19.5.2 Energetic complexity

The Bethe free entropy for the auxiliary graphical model is given by

$$\mathbb{F}^{\text{RSB},e}(\{Q,\widehat{Q}\}) = \sum_{a\in F} \mathbb{F}_a^{\text{RSB},e} + \sum_{i\in V} \mathbb{F}_i^{\text{RSB},e} - \sum_{(ia)\in E} \mathbb{F}_{ia}^{\text{RSB},e}, \qquad (19.91)$$

and allows us to count the number of min-sum fixed points. The various terms are formally identical to those in eqns (19.29), (19.30), and (19.31), provided $\mathbb{F}.(\cdot)$ is replaced everywhere by $-\mathbb{F}^e.(\cdot)$, and x by y. We reproduce these equations here for convenience:

$$\mathbb{F}_a^{\text{RSB},e} = \log\left\{ \sum_{\{m_{ia}\}} e^{-y\mathbb{F}_a^e(\{m_{ia}\})} \prod_{i\in\partial a} Q_{ia}(m_{ia}) \right\}, \qquad (19.92)$$

$$\mathbb{F}_i^{\text{RSB},e} = \log\left\{ \sum_{\{\widehat{m}_{ai}\}} e^{-y\mathbb{F}_i^e(\{\widehat{m}_{ai}\})} \prod_{a\in\partial i} \widehat{Q}_{ai}(\widehat{m}_{ai}) \right\}, \qquad (19.93)$$

$$\mathbb{F}_{ia}^{\text{RSB},e} = \log\left\{ \sum_{m_{ia},\widehat{m}_{ai}} e^{-y\mathbb{F}_{ia}^e(m_{ia},\widehat{m}_{ai})} Q_{ia}(m_{ia})\widehat{Q}_{ai}(\widehat{m}_{ai}) \right\}. \qquad (19.94)$$

Assuming that the Bethe free entropy gives the correct free entropy of the auxiliary model, the energetic complexity function $\Sigma^e(\epsilon)$ can be computed from $\mathbb{F}^{\text{RSB},e}(y)$ through a Legendre transform: in the large-N limit, we expect $\mathbb{F}^{\text{RSB},e}(\{Q,\widehat{Q}\}) = N\mathfrak{F}^e(y) + o(N)$, where

$$\mathfrak{F}^e(\{Q,\widehat{Q}\}) = \Sigma^e(\epsilon) - y\epsilon, \qquad \frac{\partial\Sigma^e}{\partial\epsilon} = y. \qquad (19.95)$$

Finally, the 1RSB population dynamics algorithm can be used to sample (approximately) the SP(y) messages in random graphical models.

19.5.3   Constraint satisfaction and binary variables

In Section 14.3.3, we noticed that the min-sum messages simplify significantly when we are dealing with constraint satisfaction problems. In such problems, the energy function takes the form $E(\underline{x}) = \sum_a E_a(\underline{x}_{\partial a})$, where $E_a(\underline{x}_{\partial a}) = 0$ if constraint a is satisfied by the assignment $\underline{x}$, and $E_a(\underline{x}_{\partial a}) = 1$ otherwise. As discussed in Section 14.3.3, the min-sum equations then admit solutions with $\widehat{\mathfrak{m}}_{ai}(x_i) \in \{0,1\}$. Furthermore, we do not need to keep track of the variable-to-function-node messages $\mathfrak{m}_{ia}(x_i)$, but only of their 'projection' on $\{0,1\}$.

In other words, in constraint satisfaction problems the min-sum messages take $2^{|\mathcal{X}|} - 1$ possible values (the all-1 message cannot appear). As a consequence, the SP(y) messages $\widehat{Q}_{ai}(\cdot)$ and $Q_{ia}(\cdot)$ simplify considerably: they are points in a $(2^{|\mathcal{X}|}-1)$-dimensional simplex.

If the min-sum messages are interpreted in terms of warnings, as we did in Section 14.3.3, then SP(y) messages keep track of the warnings' statistics (over pure states). One can use this interpretation to derive the SP(y) update equations directly without going through the whole 1RSB formalism. Let us illustrate this approach with the important case of binary variables $|\mathcal{X}| = 2$.

The min-sum messages $\widehat{\mathfrak{m}}$ and $\mathfrak{m}$ (once projected) can take three values: $(\widehat{\mathfrak{m}}(0), \widehat{\mathfrak{m}}(1)) \in \{(0,1), (1,0), (0,0)\}$. We shall denote them by 0 (interpreted as a warning 'take value 0'), 1 (interpreted as a warning 'take value 1') and $*$ (interpreted as a warning 'you can take any value'), respectively. Warning propagation (WP) can be described in words as follows.

Consider the message from variable node i to function node a. This depends on all of the messages to i from function nodes $b \in \partial i \setminus a$. Suppose that $\widehat{n}_0$, $\widehat{n}_1$, and $\widehat{n}_*$ of these messages are of type 0, 1, and $*$, respectively, for $i \in \partial a$. If $\widehat{n}_0 > \widehat{n}_1$, i sends a 0 message to a. If $\widehat{n}_1 > \widehat{n}_0$, it sends a 1 message to a. If $\widehat{n}_1 = \widehat{n}_0$, it send a $*$ message to a. The 'number of contradictions' among the messages that it receives is $\mathbb{F}_i^e - \mathbb{F}_{ia}^e = u_{ia} = \min(\widehat{n}_1, \widehat{n}_0)$.

Now consider the message from function node a to variable node i. This depends on the messages coming from the neighbouring variables $j \in \partial a \setminus i$. We partition the neighbours into subsets $\mathcal{P}_*, \mathcal{P}_0, \mathcal{P}_1$, where $\mathcal{P}_{\mathfrak{m}}$ is the set of indices j such that $\mathfrak{m}_{ja} = \mathfrak{m}$. For each value of $x_i \in \{0,1\}$, the algorithm computes the minimal value of $E_a(\underline{x}_{\partial a})$ such that the variables in $\mathcal{P}_0$ and $\mathcal{P}_1$ are fixed to 0 and to 1, respectively. More explicitly, let us define a function $\Delta_{\mathcal{P}}(x_i)$ as follows:

$$\Delta_{\mathcal{P}}(x_i) = \min_{\{x_j\}_{j \in \mathcal{P}_*}} E_a(x_i, \{x_j\}_{j \in \mathcal{P}_*}, \{x_k = 0\}_{k \in \mathcal{P}_0}, \{x_l = 1\}_{l \in \mathcal{P}_1}). \qquad (19.96)$$

The following table then gives the outgoing message $\widehat{\mathfrak{m}}_{ai}$ and the number of contradictions at a, $\mathbb{F}_a^e - \mathbb{F}_{ai}^e = \widehat{u}_{ai}$, as a function of the values of $\Delta_{\mathcal{P}}(0)$ and $\Delta_{\mathcal{P}}(1)$:

$\Delta_{\mathcal{P}}(0)$	$\Delta_{\mathcal{P}}(1)$	$\widehat{\mathfrak{m}}_{ai}$	$\widehat{u}_{ai}$
0	0	$*$	0
0	1	0	0
1	0	1	0
1	1	$*$	1

Having established the WP update rules, we can immediately write the $\mathsf{SP(y)}$ equations. Consider a node, and one of its neighbours to which it sends messages. For each possible configuration of incoming warnings at this node, denoted by input, we have found the rules to compute the outgoing warning $\texttt{output} = \widehat{\texttt{OUT}}(\texttt{input})$ and the number of contradictions $\delta\mathbb{F}^{\mathrm{e}}(\texttt{input})$. $\mathsf{SP(y)}$ messages are distributions over $(0, 1, *)$, namely $(Q_{ia}(0), Q_{ia}(1), Q_{ia}(*))$ and $(\widehat{Q}_{ai}(0), \widehat{Q}_{ai}(1), \widehat{Q}_{ai}(*))$. Notice that these messages are only marginally more complicated than ordinary BP messages. Let $\mathbb{P}(\texttt{input})$ denote the probability of a given input assuming independent warnings with distribution $Q_{ia}(\cdot)$ or $\widehat{Q}_{ai}(\cdot)$. The probability of an outgoing message $\texttt{output} \in \{0, 1, *\}$ is then

$$\mathbb{P}(\texttt{output}) \cong \sum_{\texttt{input}} \mathbb{P}(\texttt{input})\mathbb{I}(\widehat{\texttt{OUT}}(\texttt{input}) = \texttt{output})e^{-\mathrm{y}\delta\mathbb{F}^{\mathrm{e}}(\texttt{input})} . \qquad (19.97)$$

Depending on whether the node we are considering is a variable or function node, this probability distribution corresponds to the outgoing message $Q_{ia}(\cdot)$ or $\widehat{Q}_{ai}(\cdot)$.

It can be shown that the Bethe energy (19.82) associated with a given fixed point of the WP equations coincides with the total number of contradictions. This can be expressed as the number of contradictions at function nodes, plus those at variable nodes, minus the number of edges (i, a) such that the warning in direction $a \to i$ contradicts the warning in direction $i \to a$ (the last term avoids double counting). It follows that the Bethe free entropy of the auxiliary graphical model $\mathbb{F}^{\mathrm{RSB,e}}(\mathrm{y})$ weights each WP fixed point depending on its number of contradictions, as it should.

19.5.4 XORSAT again

Let us now apply the $\mathsf{SP(y)}$ formalism to random K-XORSAT instances. We let the energy function $E(\underline{x})$ count the number of unsatisfied linear equations:

$$E_a(\underline{x}_{\partial a}) = \begin{cases} 0 & \text{if } x_{i_1(a)} \oplus \cdots \oplus x_{i_K(a)} = b_a, \\ 1 & \text{otherwise.} \end{cases} \qquad (19.98)$$

The simplifications discussed in the previous subsection apply to this case. The 1RSB population dynamics algorithm can be used to compute the free-entropy density $\mathfrak{F}^{\mathrm{e}}(\mathrm{y})$. Here we limit ourselves to describing the results of this calculation for the case $K = 3$.

Let us stress that the problem we are considering here is different from the one investigated in Section 19.3. Whereas there we were interested in the uniform measure over solutions (thus focusing on the satisfiable regime $\alpha < \alpha_{\mathrm{s}}(K)$), here we are estimating the minimum number of unsatisfied constraints (which is most interesting in the unsatisfiable regime $\alpha > \alpha_{\mathrm{s}}(K)$).

It is easy to show that the $\mathsf{SP(y)}$ equations always admit a solution in which $Q_{ia}(*) = 1$ for all (i, a), indicating that the min-sum equations have a unique solution. This corresponds to a density-evolution fixed point where $Q(*) = 1$ with probability 1, yielding an $\mathfrak{F}^{\mathrm{e}}(\mathrm{y})$ independent of y. For y smaller than an α-dependent threshold $\mathrm{y}^*(\alpha)$, this is the only solution that we find. For larger values of y, the $\mathsf{SP(y)}$ equations have a non-trivial solution. Figure 19.4 shows the result for the free-entropy density $\mathfrak{F}^{\mathrm{e}}(\mathrm{y})$ for three values of α.

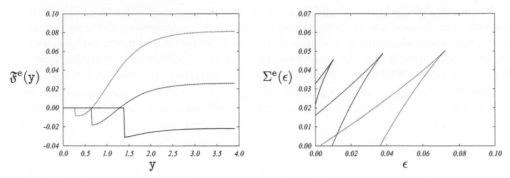

Fig. 19.4 Random 3-XORSAT at $\alpha = 0.87$, 0.97 and 1.07. Recall that, for $K = 3$, $\alpha_{\rm d}(K) \approx 0.818$ and $\alpha_{\rm s}(K) \approx 0.918$. *Left*: free-entropy density $\mathfrak{F}^{\rm e}(y)$ as a function of y, obtained using the population dynamics algorithm, with $N = 2 \times 10^4$ and $t = 5 \times 10^3$ (α increases from *bottom* to *top*). *Right*: complexity $\Sigma^{\rm e}(\epsilon)$ as a function of energy density (equal to the number of violated constraints per variable); α increases from left to right. Only the concave part of $\Sigma^{\rm e}(\epsilon)$ is physically meaningful.

Above this threshold, the density evolution converges to a 'non-trivial' 1RSB fixed point. The complexity functions $\Sigma^{\rm e}(\epsilon)$ can be deduced by taking the Legendre transform (see eqn (19.95)-, which requires differentiating $\mathfrak{F}^{\rm e}(y)$ and plotting $(\epsilon, \Sigma^{\rm e})$ in parametric form. The derivative can be computed numerically in a number of ways:

1. Compute analytically the derivative of $\mathbb{F}^{\rm RSB,e}(y)$ with respect to y. This turns out to be a functional of the fixed-point distributions of Q and $\widehat{Q}$, and can therefore be easily evaluated.

2. Fit the numerical results for the function $\mathfrak{F}^{\rm e}(y)$ and differentiate the fitting function

3. Approximate the derivative as a difference at nearby values of y.

In the present case, we have followed the second approach using the parametric form $\mathfrak{F}^{\rm fit}(y) = a + b\,{\rm e}^{-y} + c\,{\rm e}^{-2y} + d\,{\rm e}^{-3y}$. As shown in Fig. 19.4, the resulting parametric curve $(\epsilon, \Sigma^{\rm e})$ is multiple-valued (this is a consequence of the fact that $\mathfrak{F}^{\rm e}(y)$ is not concave). Only the concave part of $\Sigma^{\rm e}(\epsilon)$ is retained as physically meaningful. Indeed, the convex branch is 'unstable' (in the sense that further RSB would be needed), and it is not yet understood whether it has any meaning.

For $\alpha \in [\alpha_{\rm d}(K), \alpha_{\rm s}(K)[$, $\Sigma^{\rm e}(\epsilon)$ remains positive down to $\epsilon = 0$. The intercept $\Sigma^{\rm e}(\epsilon = 0)$ coincides with the complexity of clusters of SAT configurations, as computed in Chapter 18 (see Theorem 18.2). For $\alpha > \alpha_{\rm s}(K)$ (in the UNSAT phase), $\Sigma^{\rm e}(\epsilon)$ vanishes at $\epsilon_{\rm gs}(K, \alpha) > 0$. The energy density $\epsilon_{\rm gs}(K, \alpha)$ is the minimal fraction of violated equations in a random XORSAT linear system. Note that $\Sigma^{\rm e}(\epsilon)$ is not defined above a second energy density $\epsilon_{\rm d}(K, \alpha)$. This indicates that we should take $\Sigma^{\rm e}(\epsilon) = -\infty$ there: above $\epsilon_{\rm d}(K, \alpha)$, one recovers a simple problem with a unique Bethe measure.

Figure 19.5 shows the values of $\epsilon_{\rm gs}(K, \alpha)$ and $\epsilon_{\rm d}(K, \alpha)$ as functions of α for $K = 3$ (random 3-XORSAT).

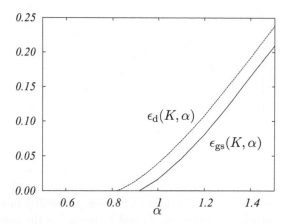

Fig. 19.5 Asymptotic ground state energy (= minimal number of violated constraints) per variable, $\epsilon_{gs}(K,\alpha)$, for random $K = 3$-XORSAT formulae; $\epsilon_{gs}(K,\alpha)$ vanishes for $\alpha < \alpha_s(K)$. The dashed line shows $\epsilon_d(K,\alpha)$, which is the highest energy density e such that configurations with $E(\underline{x}) < Ne$ are clustered. It vanishes for $\alpha < \alpha_d(K)$.

19.6 The nature of 1RSB phases

In the previous sections, we discussed how to compute the complexity function $\Sigma(\phi)$ (or its 'zero-temperature' version, the energetic complexity $\Sigma^e(\epsilon)$). Here we want to come back to the problem of determining some qualitative properties of the measure $\mu(\cdot)$ for random graphical models, on the basis of its decomposition into extremal Bethe measures,

$$\mu(\underline{x}) = \sum_{n \in \mathsf{E}} w_n \mu^n(\underline{x}) \,. \tag{19.99}$$

Assumptions 2 and 3 imply that, in this decomposition, we introduce a negligible error if we drop all of the states n except the ones with free entropy $\phi_n \approx \phi_*$, where

$$\phi_* = \mathrm{argmax} \left\{ \phi + \Sigma(\phi) : \Sigma(\phi) \ge 0 \right\} \,. \tag{19.100}$$

In general, $\Sigma(\phi)$ is strictly positive and continuous in an interval $[\phi_{\min}, \phi_{\max}]$ with $\Sigma(\phi_{\max}) = 0$, and

$$\Sigma(\phi) = \mathsf{x}_*(\phi_{\max} - \phi) + O((\phi_{\max} - \phi)^2) \tag{19.101}$$

for ϕ close to $\phi_{\max}$.

It turns out that the decomposition (19.99) has different properties depending on the result of the optimization (19.100). One can distinguish two phases (see Fig. 19.6): d1RSB (dynamic one-step replica symmetry breaking), when the maximum is achieved in the interior of $[\phi_{\min}, \phi_{\max}]$ and, as a consequence $\Sigma(\phi_*) > 0$; and s1RSB (static one-step replica symmetry breaking), when the maximum is achieved at $\phi_* = \phi_{\max}$ and therefore $\Sigma(\phi_*) = 0$ (this case occurs iff $\mathsf{x}_* \le 1$).

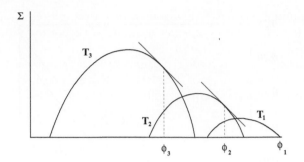

Fig. 19.6 A sketch of the complexity Σ versus the free-entropy density ϕ in a finite-temperature problem with a 1RSB phase transition, at three temperatures $T_1 < T_2 < T_3$. A random configuration $\underline{x}$ with distribution $\mu(\underline{x})$ is found with high probability in a cluster with free-entropy density ϕ_1, ϕ_2, or phi_3, respectively. T_2 and T_3 are above the condensation transition: ϕ_2 and ϕ_3 are the points where $\partial\Sigma/\partial\phi = -1$. T_1 is below the condensation transition: ϕ_1 is the largest value of ϕ where Σ is positive.

19.6.1   Dynamical 1RSB

We assume here that $\Sigma_* = \Sigma(\phi_*) > 0$. We can then restrict the sum (19.99) to those states n such that $\phi_n \in [\phi_* - \varepsilon, \phi_* + \varepsilon]$, if we allow an exponentially small error. To the leading exponential order, there are $e^{N\Sigma_*}$ such states whose weights are $w_n \in [e^{-N(\Sigma_* + \varepsilon')}, e^{-N(\Sigma_* - \varepsilon')}]$.

Different states are expected to have essentially disjoint support. By this we mean that there exist subsets $\{\Omega_n\}_{n\in\mathsf{E}}$ of the configuration space $\mathcal{X}^N$ such that, for any $m \in \mathsf{E}$,

$$\mu^m(\Omega_m) \approx 1 \,, \qquad \sum_{n\in\mathsf{E}\backslash m} w_n \mu^n(\Omega_m) \approx 0 \,. \qquad (19.102)$$

Further, different states are separated by 'large free-energy barriers'. This means that one can choose the above partition in such a way that only an exponentially small (in N) fraction of the probability measure is on its boundaries.

This structure has two important consequences:

Glassy dynamics. Let us consider a local Markov chain dynamics that satisfies detailed balance with respect to the measure $\mu(\,\cdot\,)$. As an example, we can consider the Glauber dynamics introduced in Chapter 4 (in order to avoid trivial reducibility effects, we can assume in this discussion that the compatibility functions $\psi_a(\underline{x}_{\partial a})$ are bounded away from 0).

Imagine initiating the dynamics at time 0 with an equilibrated configuration $\underline{x}(0)$ distributed according to $\mu(\,\cdot\,)$. This is essentially equivalent to picking a state n uniformly at random from among the typical states, and then sampling $\underline{x}(0)$ from $\mu^n(\,\cdot\,)$. Because of the exponentially large barriers, the dynamics will stay confined in Ω_n for an exponentially large time, and equilibrate among states only on larger time scales.

This can be formalized as follows. We denote by $D(\underline{x}, \underline{x}')$ the Hamming distance in $\mathcal{X}^N$. Take two i.i.d. configurations with distribution μ, and let $N d_0$ be the expectation value of their Hamming distance. Analogously, take two i.i.d. configurations with distribution μ^n, and let Nd_1 be the expectation value of their Hamming distance. When the state n is chosen randomly with distribution w_n, we expect d_1 not to depend on the state n, asymptotically, for large sizes. Furthermore, we expect $d_1 < d_0$. We can then consider the (normalized) expected Hamming distance between configurations at time t in Glauber dynamics, $d(t) = \langle D(\underline{x}(0), \underline{x}(t)) \rangle / N$. For any $\varepsilon < d_0 - d_1$, the correlation time $\tau(\varepsilon) \equiv \inf\{t : d(t) \geq d_0 - \varepsilon\}$ is expected to be exponentially large in N.

Short-range correlations in finite-dimensional projections. We described the motivation for the 1RSB cavity method by considering the emergence of long-range correlations due to decomposition of $\mu(\cdot)$ into many extremal Bethe measures. Surprisingly, such correlations cannot be detected by probing a bounded (when $N \to \infty$) number of variables. More precisely, if $i(1), \ldots, i(k) \in \{1, \cdots, N\}$ are uniformly random variable indices, then, in the d1RSB phase,

$$\mathbb{E}|\langle f_1(x_{i(1)})f_2(x_{i(2)}) \cdots f_k(x_{i(k)})\rangle - \langle f_1(x_{i(1)})\rangle\langle f_2(x_{i(2)})\rangle \cdots \langle f_k(x_{i(k)})\rangle| \overset{N\to\infty}{\to} 0\,.$$

(Here $\langle \cdot \rangle$ denotes the expectation with respect to the measure μ, and $\mathbb{E}$ denotes the expectation with respect to the graphical model for a random ensemble.) This finding can be understood intuitively as follows. If there are long-range correlations among subsets of k variables, then it must be true that conditioning on the values of $k - 1$ of those variables changes the marginal distribution of the k-th one. On the other hand, we think that long-range correlations arise because far-apart variables 'know' that the whole system is in the same state n. But conditioning on a bounded number $(k - 1)$ of variables cannot select in any significant way from among the $e^{N\Sigma_*}$ relevant states, and thus cannot change the marginal of the k-th variable.

An alternative argument makes use of the observation that if $\underline{x}^{(1)}$ and $\underline{x}^{(2)}$ are two i.i.d. configurations with distribution $\mu(\cdot)$, then their distance $D(\underline{x}^{(1)}, \underline{x}^{(2)})$ concentrates in probability. This is due to the fact that the two configurations will be, with high probability, in different states $n_1 \neq n_2$ (the probability of $n_1 = n_2$ being $e^{-N\Sigma_*}$), whose distance depends weakly on the pair of states.

Let us note, finally, that the absence of long range correlations among bounded subsets of variables is related to the observation that $\mu(\cdot)$ is itself a Bethe measure (although a non-extremal one) in a d1RSB phase (see Section 19.4.1). Indeed, each BP equation involves a bounded subset of the variables and can be violated only because of correlations among them.

As we shall discuss in Section 22.1.2, long-range correlations in a d1RSB phase can be probed through more sophisticated 'point-to-set' correlation functions.

19.6.2 Static 1RSB

In this case the decomposition (19.99) is dominated by a few states of near-to-maximal free entropy $\phi_n \approx \phi_{\max}$. If we 'zoom in' near the edge by letting $\phi_n = \phi_{\max} + s_n/N$, then the 'free-entropy shifts' s_n form a point process with density $\exp(-\mathrm{x}_*s)$.

The situation is analogous to the one we found in the random energy model for $T < T_{\mathrm{c}}$. Indeed, it is expected that the weights $\{w_n\}$ will converge to the same universal

Poisson-Dirichlet process found there, and to depend on the details of the model only through the parameter x_* (we have already discussed this universality using replicas in Chapter 8). In particular, if $\underline{x}^{(1)}$ and $\underline{x}^{(2)}$ are two i.i.d. replicas with distribution μ, and n_1 and n_2 are the states that they belong to, then the probability for them to belong to the same state is

$$\mathbb{E}\left\{\mathbb{P}_\mu(n_1 = n_2)\right\} = \mathbb{E}\left\{\sum_{n\in\mathsf{E}} w_n^2\right\} = 1 - x_* . \tag{19.103}$$

Here $\mathbb{E}$ denotes the expectation with respect to the graphical-model distribution.

As a consequence, the distance $D(\underline{x}^{(1)}, \underline{x}^{(2)})$ between two i.i.d. replicas does not concentrate (the overlap distribution is non-trivial). This, in turn, can only be true if the two-point correlation function does not vanish at large distances. Long-range correlations of this type make BP break down. The original graphical model $\mu(\cdot)$ is no longer a Bethe measure: its local marginals cannot be described in terms of a set of messages. The 1RSB description, according to which $\mu(\cdot)$ is a convex combination of Bethe measures, is unavoidable.

At this point, we are left with a puzzle. How do we circumvent the argument given in Section 19.4.1 that, if the 'correct' weight $x = 1$ is used, then the marginals as computed within 1RSB still satisfy the BP equations? The conundrum is that, within an s1RSB phase, the parameter $x = 1$ is *not* the correct one to use in the 1RSB cavity equations (although it is the correct one for weighting states). In order to explain this, let us first note that, if the complexity is convex and behaves as in eqn (19.101) near its edge, with a slope $-x_* > -1$, then the optimization problem (19.100) has the same result as

$$\phi_* = \operatorname{argmax}\left\{x\phi + \Sigma(\phi) : \Sigma(\phi) \geq 0\right\} , \tag{19.104}$$

for any $x \geq x_*$. Therefore, in the 1RSB cavity equations, we could in principle use any value of x greater than or equal to x_* (this would select the same states). However, the constraint $\Sigma(\phi) \geq 0$ cannot be enforced locally, and does not show up in the cavity equations. If one performs the computation of Σ within the cavity method using a value $x > x_*$, then one finds a negative value of Σ, which must be rejected (it is believed to be related to the contribution of some exponentially rare instances). Therefore, in order to ensure that one studies the interval of ϕ such that $\Sigma(\phi) \geq 0$, one must *impose* $x \leq x_*$ in the cavity method. In order to select the states with free-entropy density $\phi_{\max}$, we must thus choose the Parisi parameter that corresponds to $\phi_{\max}$, namely $x = x_*$.

19.6.3   When does 1RSB fail?

The 1RSB cavity method is a powerful tool, but does not always provide correct answers, even for locally tree-like models, in the large-system limit. The main assumption of the 1RSB approach is that once we pass to the auxiliary graphical model (which 'enumerates' BP fixed points), a simple BP procedure is asymptotically exact. In other words, the auxiliary problem has a simple 'replica-symmetric' structure and no glassy phase. This is correct in some cases, such as random XORSAT and SAT close to their SAT–UNSAT threshold, but it may fail in others.

One mechanism leading to a failure of the 1RSB approach is that the auxiliary graphical model is incorrectly described by BP. This may happen because the measure of the auxiliary model decomposes into many Bethe states. In such a case, one must introduce a second auxiliary model, dealing with the multiplicity of BP fixed points in the first auxiliary model. This is usually referred to as 'two-step replica symmetry breaking' (2RSB). Obviously one can find situations in which it is necessary to iterate this construction, leading to an R-th level auxiliary graphical model (R-RSB). Continuous (or full) RSB corresponds to the large-R limit.

While such developments are conceptually clear (at least from a heuristic point of view), they are technically challenging. So far, only limited results have been obtained beyond 1RSB. A brief survey is given in Chapter 22.

19.7 Appendix: The SP(y) equations for XORSAT

This appendix provides some technical details of the 1RSB treatment of random K-XORSAT within the 'energetic' formalism. The results of this approach were discussed in Section 19.5.4. In particular, we shall derive the behaviour of the auxiliary free entropy $\mathfrak{F}^e(\mathsf{y})$ at large y, and deduce the behaviour of the complexity $\Sigma^e(\epsilon)$ at small ϵ. This section can be regarded as an exercise in applying the SP(y) formalism. We shall skip many details and just give the main intermediate results of the computation.

XORSAT is a constraint satisfaction problem with binary variables. We can thus apply the simplified method of Section 19.5.3. The projected min-sum messages can take three values: 0, 1, $*$. By exploiting the symmetry of XORSAT between 0 and 1, SP(y) messages can be parameterized by a single number, for instance the sum of their weights on 0 and 1. We shall therefore write $Q_{ia}(0) = Q_{ia}(1) = \zeta_{ia}/2$ (thus implying $Q_{ia}(*) = 1 - \zeta_{ia}$) and $\widehat{Q}_{ai}(0) = \widehat{Q}_{ai}(1) = \eta_{ai}/2$ (whence $\widehat{Q}_{ai}(*) = 1 - \eta_{ai}$).

In terms of these variables, the SP(y) equation at function node a reads

$$\eta_{ai} = \prod_{j \in \partial a \setminus i} \zeta_{ja} \,. \tag{19.105}$$

The SP(y) equation at variable node i is a little more complicated. Let us consider all of the $|\partial i| - 1$ incoming messages $\widehat{Q}_{bi}$, $b \in \partial i \setminus a$. Each of them is parameterized by a number η_{bi}. We let $\underline{\eta} = \{\eta_{bi}, \ b \in \partial i \setminus a\}$ and define the function $B_q(\underline{\eta})$ as follows:

$$B_q(\underline{\eta}) = \sum_{S \subset \{\partial i \setminus a\}} \mathbb{I}(|S| = q) \prod_{b \in \partial i \setminus \{S \cup \{a\}\}} (1 - \eta_{bi}) \prod_{c \in S} \eta_{cj} \,. \tag{19.106}$$

Let $A_{q,r}(\underline{\eta}) = B_{q+r}(\underline{\eta}) \binom{q+r}{q} 2^{-(q+r)}$. After some thought, one can obtain the following update equation:

$$\zeta_{ia} = \frac{2 \sum_{q=0}^{|\partial i|-2} \sum_{r=q+1}^{|\partial i|-1} A_{q,r}(\underline{\eta}) \mathrm{e}^{-\mathsf{y}q}}{\sum_{q=0}^{\lfloor (|\partial i|-1)/2 \rfloor} A_{q,q}(\underline{\eta}) \mathrm{e}^{-\mathsf{y}q} + 2 \sum_{q=0}^{|\partial i|-2} \sum_{r=q+1}^{|\partial i|-1} A_{q,r}(\underline{\eta}) \mathrm{e}^{-\mathsf{y}q}} \,. \tag{19.107}$$

The auxiliary free entropy $\mathbb{F}^{\mathrm{RSB},e}(y)$ has the general form (19.91), with the various contributions expressed as follows in terms of the parameters $\{\zeta_{ia}, \eta_{ai}\}$:

$$e^{\mathbb{F}_a^{\mathrm{RSB},e}} = 1 - \frac{1}{2}(1 - e^{-y}) \prod_{i \in \partial a} \zeta_{ia} \,, \qquad e^{\mathbb{F}_{ai}^{\mathrm{RSB},e}} = 1 - \frac{1}{2}\eta_{ai}\zeta_{ia}(1 - e^{-y}) \,,$$

$$e^{\mathbb{F}_i^{\mathrm{RSB},e}} = \sum_{q=0}^{d_i} \sum_{r=0}^{d_i-q} A_{q,r}\left(\{\eta_{ai}\}_{a \in \partial i}\right) e^{-y \min(q,r)} \,. \tag{19.108}$$

Let us consider random K-XORSAT instances with a constraint density α. Equations (19.105) and (19.107) are promoted to distributional relations that determine the asymptotic distribution of η and ζ on a randomly chosen edge (i,a). The 1RSB population dynamics algorithm can be used to approximate these distributions. We encourage the reader to implement it, and obtain a numerical estimate of the auxiliary free-entropy density $\mathfrak{F}^e(y)$.

It turns out that, at large y, one can control the distributions of η and ζ analytically, provided their qualitative behaviour satisfies the following assumptions (which can be checked numerically):

- With probability t one has $\eta = 0$, and with probability $1-t$, $\eta = 1 - e^{-y}\hat{\eta}$, where t has a limit in $]0,1[$, and $\hat{\eta}$ converges to a random variable with support on $[0,\infty[$, as $y \to \infty$.
- With probability s, one has $\zeta = 0$, and with probability $1 - s$, $\zeta = 1 - e^{-y}\hat{\zeta}$, where s has a limit in $]0,1[$, and $\hat{\zeta}$ converges to a random variable with support on $[0,\infty[$, as $y \to \infty$.

Under these assumptions, we shall expand the distributional version of eqns (19.105) and (19.107) keeping terms up to first order in e^{-y}. We shall use $t, s, \hat{\eta}, \hat{\zeta}$ to denote the limit quantities mentioned above.

It is easy to see that t and s must satisfy the equations $(1 - t) = (1 - s)^{k-1}$ and $s = e^{-K\alpha(1-t)}$. These are identical to eqns (19.50) and (19.51), whence $t = 1 - \widehat{Q}_*$ and $s = 1 - Q_*$.

Equation (19.105) leads to the distributional equation

$$\hat{\eta} \stackrel{\mathrm{d}}{=} \hat{\zeta}_1 + \cdots + \hat{\zeta}_{K-1} \,, \tag{19.109}$$

where $\hat{\zeta}_1, \ldots, \hat{\zeta}_{K-1}$ are $K - 1$ i.i.d. copies of the random variable $\hat{\zeta}$.

The update equation (19.107) is more complicated. There are, in general, l inputs to a variable node, where l is Poisson-distributed with mean $K\alpha$. Let us denote by m the number of incoming messages with $\eta = 0$. The case $m = 0$ yields $\zeta = 0$ and is taken care of in the relation between t and s. If we condition on $m \geq 1$, the distribution of m is

$$\mathbb{P}(m) = \frac{\lambda^m}{m!}e^{-\lambda}\frac{1}{1 - e^{-\lambda}} \mathbb{I}(m \geq 1) \,, \tag{19.110}$$

where $\lambda = K\alpha(1 - t)$. Conditional on m, eqn (19.107) simplifies as follows:

- If $m = 1$, $\hat{\zeta} \stackrel{\mathrm{d}}{=} \hat{\eta}$.

- If $m = 2$, $\hat{\zeta} = 1$ identically.
- If $m \geq 3$, $\hat{\zeta} = 0$ identically.

The various contributions to the free entropy (19.37) are given by

$$f_f^{\mathrm{RSB,e}} = (1-s)^k \left[-\log 2 + e^{-y}(1 + K\langle\hat{\zeta}\rangle) \right] + o(e^{-y}), \tag{19.111}$$

$$f_v^{\mathrm{RSB,e}} = \frac{\lambda^2}{2} e^{-\lambda} \left[-\log 2 + e^{-y}(1 + 2\langle\hat{\eta}\rangle) \right]$$
$$+ \sum_{m=3}^{\infty} \frac{\lambda^m}{m!} e^{-\lambda} \left[(1-m)\log 2 + e^{-y}m(1 + \langle\hat{\eta}\rangle) \right] + o(e^{-y}), \tag{19.112}$$

$$f_e^{\mathrm{RSB,e}} = (1-t)(1-s) \left[-\log 2 + e^{-y}(1 + \langle\hat{\eta}\rangle + \langle\hat{\zeta}\rangle) \right] + o(e^{-y}), \tag{19.113}$$

where $\langle\hat{\eta}\rangle$ and $\langle\hat{\zeta}\rangle$ are the expectation values of $\hat{\eta}$ and $\hat{\zeta}$. This gives the free-entropy density $\mathfrak{F}^{\mathrm{e}}(y) = f_f^{\mathrm{RSB,e}} + \alpha f_v^{\mathrm{RSB,e}} - K\alpha f_e^{\mathrm{RSB,e}} = \Sigma_0 + e^{-y}\epsilon_0 + o(e^{-y})$, where

$$\Sigma_0 = \left[1 - \frac{\lambda}{k} - e^{-\lambda}\left(1 + \frac{k-1}{k}\lambda\right) \right] \log 2, \tag{19.114}$$

$$\epsilon_0 = \frac{\lambda}{k} \left[1 - e^{-\lambda}\left(1 + \frac{k}{2}\lambda\right) \right]. \tag{19.115}$$

Taking the Legendre transform (see eqn (19.95)), we obtain the following behaviour of the energetic complexity as $\epsilon \to 0$:

$$\Sigma^{\mathrm{e}}(\epsilon) = \Sigma_0 + \epsilon \log \frac{\epsilon_0 e}{\epsilon} + o(\epsilon), \tag{19.116}$$

This shows, in particular, that the ground state energy density is proportional to $(\alpha - \alpha_s)/|\log(\alpha - \alpha_s)|$ close to the SAT–UNSAT transition (when $0 < \alpha - \alpha_s \ll 1$).

Exercise 19.7 In the other extreme case, show that at large α one gets $\epsilon_{\mathrm{gs}}(K, \alpha) = \alpha/2 + \sqrt{2\alpha}\epsilon_*(K) + o(\sqrt{\alpha})$, where the positive constant $\epsilon_*(K)$ is the absolute value of the ground state energy of the fully connected K-spin model studied in Section 8.2. This indicates that there is no interesting intermediate asymptotic regime between $M = \Theta(N)$ (discussed in the present chapter) and $M = \Theta(N^{K-1})$ (discussed with the use of the replica method in Chapter 8).

Notes

The cavity method originated as an alternative to the replica approach in the study of the Sherrington–Kirkpatrick model (Mézard *et al.*, 1985*b*). The 1RSB cavity method for locally tree-like factor graphs was developed in the context of spin glasses by Mézard and Parisi (2001). Its application to zero-temperature problems (counting solutions of the min-sum equations) was also first described in the context of spin glasses, by Mézard and Parisi (2003). The presentation in this chapter differs in its scope from those studies, which were more focused on computing averages over random instances. For a rigorous treatment of the notion of a Bethe measure, we refer to Dembo and Montanari (2008*b*).

The idea that the 1RSB cavity method is in fact equivalent to applying BP on an auxiliary model appeared in several papers treating the cases of colouring and satisfiability with $y = 0$ (Parisi, 2002; Braunstein and Zecchina, 2004; Maneva *et al.*, 2005). The treatment presented here generalizes these studies, with the important difference that the variables of our auxiliary model are messages on the edges rather than quantities defined on vertices.

The analysis of the $x = 1$ case is strictly related to the problem of reconstruction on a tree. This has been studied by Mézard and Montanari (2006), where the reader will find the proof of Theorem 19.5 and the expression for the free entropy obtained in Exercise 19.6.

The SP(y) equations for one single instance were written down first in the context of the K-satisfiability problem by Mézard and Zecchina (2002); see also Mézard *et al.* (2003*a*). The direct derivation of the SP(y) equations for binary-variable problems described in Section 19.5.3 was done originally for the satisfiability problem by Braunstein *et al.* (2005); see also Braunstein and Zecchina (2004) and Maneva *et al.* (2005). The application of the 1RSB cavity method to the random XORSAT problem, and its comparison with the exact results, was done by Mézard *et al.* (2003*b*).

An alternative to the cavity approach followed throughout this book is provided by the replica method of Chapter 8. As we saw, the replica method was first invented in order to treat fully connected models (i.e. models on complete graphs) (see Mézard *et al.*, 1987), and was subsequently developed in the context of sparse random graphs (Mézard and Parisi, 1985; De Dominicis and Mottishaw, 1987; Mottishaw and De Dominicis, 1987; Wong and Sherrington, 1988; Goldschmidt and Lai, 1990). The technique was further improved in Monasson (1998), which offers a very lucid presentation of the method.

20
Random K-satisfiability

This chapter applies the cavity method to the random K-satisfiability problem. We shall study both the phase diagram (in particular, we shall determine the SAT–UNSAT threshold $\alpha_s(K)$) and the algorithmic applications of message passing. The whole chapter is based on heuristic derivations: the rigorization of the whole approach is still in its infancy. Neither the conjectured phase diagram nor the efficiency of message-passing algorithms has yet been confirmed rigorously. But the computed value of $\alpha_s(K)$ is conjectured to be exact, and the low-complexity message-passing algorithms that we shall describe turn out to be particularly efficient in finding solutions.

We start in Section 20.1 by writing the BP equations, following the approach set out in Chapter 14. The statistical analysis of such equations provides a first (replica-symmetric) estimate of $\alpha_s(K)$. This, however, turns out to be incorrect. The reason for this failure is traced back to the incorrectness of the replica-symmetric assumption close to the SAT–UNSAT transition. The system undergoes a 'structural' phase transition at a clause density smaller than $\alpha_s(K)$. Nevertheless, BP empirically converges for a wide range of clause densities, and it can be used to find SAT assignments for large instances provided the clause density α is not too close to $\alpha_s(K)$. The key idea is to use BP as a heuristic guide in a sequential decimation procedure.

In Section 20.2, we apply the 1RSB cavity method developed in Chapter 19. A statistical analysis of the 1RSB equations gives the values for $\alpha_s(K)$ summarized in Table 20.1. From the algorithmic point of view, one can use SP instead of BP as a guide in the decimation procedure. We shall explain and study numerically the corresponding 'survey-guided decimation' algorithm, which is presently the most efficient algorithm for finding SAT assignments in large random satisfiable instances with a clause density close to the threshold $\alpha_s(K)$.

This chapter focuses on K-SAT with $K \geq 3$. The $K = 2$ problem is quite different: satisfiability can be proved in polynomial time, the SAT–UNSAT phase transition is driven by a very different mechanism, and the threshold is known to be $\alpha_s(2) = 1$. It turns out that a (more subtle) qualitative difference also distinguishes $K = 3$ from $K \geq 4$. In order to illustrate this point, we shall use both 3-SAT and 4-SAT as running examples.

The problem of colouring random graphs turns out to be very similar to that of random K-satisfiability. Section 20.4 presents a few highlights in the study of random graph colourings. In particular, we emphasize how the techniques used for K-satisfiability are successful in this case as well.

20.1   Belief propagation and the replica-symmetric analysis

We have already studied some aspects of random K-SAT in Chapter 10, where we derived, in particular, some rigorous bounds on the SAT–UNSAT threshold $\alpha_s(K)$. Here we shall study the problem using message-passing approaches. Let us start by summarizing our notation.

An instance of the K-satisfiability problem is defined by M clauses (indexed by $a, b, \dots \in \{1, \dots, M\}$) over N Boolean variables $x_1, \dots, x_N$ taking values in $\{0,1\}$. We denote by ∂a the set of variables in clause a, and by ∂i the set of clauses in which variable x_i appears. Further, for each $i \in \partial a$, we introduce a number J_{ai}, which takes the value 1 if x_i occurs negated in clause a, and takes the value 0 if the variable occurs unnegated.

It will be convenient to distinguish elements of ∂a according to the value of J_{ai}. We let $\partial_0 a \equiv \{i \in \partial a$ such that $J_{ai} = 0\}$ and $\partial_1 a = \{i \in \partial a$ such that $J_{ai} = 1\}$. Similarly we denote by $\partial_0 i$ and $\partial_1 i$ the neighbourhoods of i: $\partial_0 i = \{a \in \partial i$ such that $J_{ai} = 0\}$ and $\partial_1 i = \{a \in \partial i$ such that $J_{ai} = 1\}$.

As usual, the indicator function over clause a being satisfied is denoted by $\psi_a(\cdot)$: $\psi_a(\underline{x}_{\partial a}) = 1$ if clause a is satisfied by the assignment $\underline{x}$, and $\psi_a(\underline{x}_{\partial a}) = 0$ if it is not. Given a SAT instance, we begin by studying the uniform measure over SAT assignments,

$$\mu(\underline{x}) = \frac{1}{Z} \prod_{a=1}^{M} \psi_a(\underline{x}_{\partial a}) \,. \tag{20.1}$$

We shall represent this distribution with a factor graph, as in Fig. 10.1, and in this graph we draw dashed edges when $J_{ai} = 1$ and full edges when $J_{ai} = 0$.

20.1.1   The BP equations

The BP equations for a general model of the form (20.1) have already been written down in Chapter 14. Here we want to rewrite them in a more compact form that is convenient for both analysis and implementation. They are best expressed using the following notation. Consider a variable node i connected to a factor node a and partition its neighbourhood as $\partial i = \{a\} \cup \mathcal{S}_{ia} \cup \mathcal{U}_{ia}$, where (see Fig. 20.1)

$$
\begin{aligned}
&\text{if } J_{ai} = 0 \text{ then} \quad \mathcal{S}_{ia} = \partial_0 i \setminus \{a\},\ \mathcal{U}_{ia} = \partial_1 i\,, \\
&\text{if } J_{ai} = 1 \text{ then} \quad \mathcal{S}_{ia} = \partial_1 i \setminus \{a\},\ \mathcal{U}_{ai} = \partial_0 i\,.
\end{aligned} \tag{20.2}
$$

Since the variables x_i are binary, the BP messages $\nu_{i \to a}(\cdot)$ and $\widehat{\nu}_{a \to i}(\cdot)$ at any time can be parameterized by a single real number. We fix the parameterization by letting $\zeta_{ia} \equiv \nu_{i \to a}(x_i = J_{ai})$ (which obviously implies $\nu_{i \to a}(x_i = 1 - J_{ai}) = 1 - \zeta_{ia}$) and $\widehat{\zeta}_{ai} \equiv \widehat{\nu}_{a \to i}(x_i = J_{ai})$ (yielding $\widehat{\nu}_{a \to i}(x_i = 1 - J_{ai}) = 1 - \widehat{\zeta}_{ai}$).

A straightforward calculation allows us to express the BP equations (here in fixed-point form) in terms of these variables:

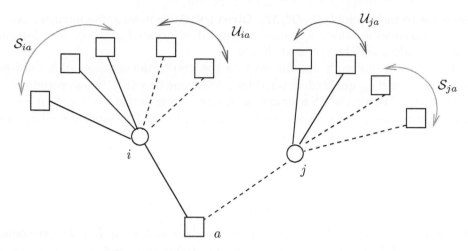

Fig. 20.1 The set $\mathcal{S}_{ia}$ contains all checks b in $\partial i \setminus a$ such that $J_{bi} = J_{ai}$; the set $\mathcal{U}_{ia}$ contains all checks b in $\partial i \setminus a$ such that $J_{bi} = 1 - J_{ai}$.

$$\zeta_{ia} = \frac{\left[\prod_{b \in \mathcal{S}_{ia}} \hat{\zeta}_{bi}\right]\left[\prod_{b \in \mathcal{U}_{ia}}(1 - \hat{\zeta}_{bi})\right]}{\left[\prod_{b \in \mathcal{S}_{ia}} \hat{\zeta}_{bi}\right]\left[\prod_{b \in \mathcal{U}_{ia}}(1 - \hat{\zeta}_{bi})\right] + \left[\prod_{b \in \mathcal{U}_{ia}} \hat{\zeta}_{bi}\right]\left[\prod_{b \in \mathcal{S}_{ia}}(1 - \hat{\zeta}_{bi})\right]},$$

$$\hat{\zeta}_{ai} = \frac{1 - \prod_{j \in \partial a \setminus i} \zeta_{ja}}{2 - \prod_{j \in \partial a \setminus i} \zeta_{ja}}, \tag{20.3}$$

with the convention that a product over zero terms is equal to 1. Note that evaluating the right-hand sides takes $O(|\partial i|)$ and $O(|\partial a|)$ operations, respectively. This should be contrasted with the general implementation of the BP equations (see Chapter 14), which requires $O(|\partial i|)$ operations at variable nodes but $O(2^{|\partial a|})$ at function nodes.

The Bethe free entropy takes the usual form, $\mathbb{F} = \sum_{a \in F} \mathbb{F}_a + \sum_{i \in V} \mathbb{F}_i - \sum_{(ia) \in E} \mathbb{F}_{ia}$ (see eqn (14.27)). The various contributions can be expressed in terms of the parameters ζ_{ia}, $\hat{\zeta}_{ai}$ as follows:

$$\mathbb{F}_a = \log\left[1 - \prod_{i \in \partial a} \zeta_{ia}\right], \quad \mathbb{F}_i = \log\left[\prod_{a \in \partial_0 i} \hat{\zeta}_{ai} \prod_{b \in \partial_1 i}(1 - \hat{\zeta}_{bi}) + \prod_{a \in \partial_0 i}(1 - \hat{\zeta}_{ai}) \prod_{b \in \partial_1 i} \hat{\zeta}_{bi}\right],$$

$$\mathbb{F}_{ai} = \log\left[(1 - \zeta_{ia})(1 - \hat{\zeta}_{ai}) + \zeta_{ia}\hat{\zeta}_{ai}\right]. \tag{20.4}$$

Given the messages, the BP estimate for the marginal on site i is

$$\nu_i(x_i) \cong \prod_{a \in \partial i} \hat{\nu}_{a \to i}(x_i). \tag{20.5}$$

20.1.2 Statistical analysis

Let us now consider a random K-SAT formula, i.e. a uniformly random formula with N variables and $M = N\alpha$ clauses. The resulting factor graph will be distributed according to the ensemble $\mathbb{G}_N(K, M)$. Given a variable index i, the numbers $|\partial_0 i|$ and $|\partial_1 i|$ of variables in which x_i occurs unnegated or negated converge to independent Poisson random variables of mean $K\alpha/2$.

If (i, a) is a uniformly random edge in the graph, the corresponding fixed-point messages ζ_{ia} and $\hat{\zeta}_{ai}$ are random variables (we assume here that an 'approximate' fixed point exists). Within the RS assumption, they converge in distribution, as $N \to \infty$, to random variables ζ and $\hat{\zeta}$ whose distributions satisfy the RS distributional equations

$$\hat{\zeta} \overset{\mathrm{d}}{=} \frac{1 - \zeta_1 \dots \zeta_{K-1}}{2 - \zeta_1 \dots \zeta_{K-1}} , \tag{20.6}$$

$$\zeta \overset{\mathrm{d}}{=} \frac{\hat{\zeta}_1 \dots \hat{\zeta}_p (1 - \hat{\zeta}_{p+1}) \dots (1 - \hat{\zeta}_{p+q})}{\hat{\zeta}_1 \dots \hat{\zeta}_p (1 - \hat{\zeta}_{p+1}) \dots (1 - \hat{\zeta}_{p+q}) + (1 - \hat{\zeta}_1) \dots (1 - \hat{\zeta}_p) \hat{\zeta}_{p+1} \dots \hat{\zeta}_{p+q}} . \tag{20.7}$$

Here p and q are two i.i.d. Poisson random variables with mean $K\alpha/2$ (corresponding to the sizes of $\mathcal{S}$ and $\mathcal{U}$), $\zeta_1, \dots, \zeta_{K-1}$ are i.i.d. copies of ζ, and $\hat{\zeta}_1, \dots, \hat{\zeta}_{p+q}$ are i.i.d. copies of $\hat{\zeta}$.

The distributions of ζ and $\hat{\zeta}$ can be approximated using the population dynamics algorithm. The resulting samples can then be used to estimate the free-entropy density, as outlined in the exercise below.

Exercise 20.1 Argue that, within the RS assumptions, the large-N limit of the Bethe free-entropy density is given by $\lim_{N \to \infty} \mathbb{F}/N = \mathrm{f}^{\mathrm{RS}} = \mathrm{f}_{\mathrm{v}}^{\mathrm{RS}} + \alpha \mathrm{f}_{\mathrm{c}}^{\mathrm{RS}} - K\alpha \mathrm{f}_{\mathrm{e}}^{\mathrm{RS}}$, where

$$\mathrm{f}_{\mathrm{v}}^{\mathrm{RS}} = \mathbb{E} \log \left[\prod_{a=1}^{p} \hat{\zeta}_a \prod_{a=p+1}^{p+q} (1 - \hat{\zeta}_a) + \prod_{a=1}^{p} (1 - \hat{\zeta}_a) \prod_{a=p+1}^{p+q} \hat{\zeta}_a \right] ,$$

$$\mathrm{f}_{\mathrm{c}}^{\mathrm{RS}} = \mathbb{E} \log \left[1 - \zeta_1 \cdots \zeta_{K-1} \right] ,$$

$$\mathrm{f}_{\mathrm{e}}^{\mathrm{RS}} = \mathbb{E} \log \left[(1 - \zeta_1)(1 - \hat{\zeta}_1) + \zeta_1 \hat{\zeta}_1 \right] . \tag{20.8}$$

Here $\mathbb{E}$ denotes the expectation with respect to $\zeta_1, \dots, \zeta_K$, which are i.i.d. copies of ζ; $\hat{\zeta}_1, \dots, \hat{\zeta}_{p+q}$, which are i.i.d. copies of $\hat{\zeta}$; and p and q, which are i.i.d. Poisson random variables with mean $K\alpha/2$.

Figure 20.2 shows an example of the entropy density found within this approach for 3-SAT. For each value of α in a mesh, we used a population of size 10^4, and ran the algorithm for 3×10^3 iterations. Messages were initialized uniformly in $]0, 1[$, and the first 10^3 iterations were not used for computing the free entropy.

The predicted entropy density is strictly positive and decreasing in α for $\alpha \leq \alpha_*(K)$, with $\alpha_*(3) \approx 4.6773$. Above $\alpha_*(K)$, the RS distributional equations do not seem to admit any solution with $\zeta, \hat{\zeta} \in [0, 1]$. This is revealed numerically by the fact that the denominator of eqn (20.7) vanishes during the population updates. Since we find an RS entropy density which is positive for all $\alpha < \alpha_*(K)$, the value $\alpha_*(K)$ is the RS prediction for the SAT–UNSAT threshold. It turns out that $\alpha_*(K)$ can be computed without population dynamics, as outlined by the exercise below.

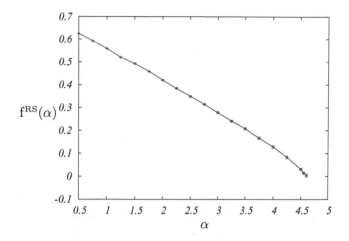

Fig. 20.2 RS prediction for the asymptotic entropy density of random 3-SAT formulae, plotted versus the clause density α for 3-SAT. The result is expected to be correct for $\alpha \leq \alpha_c(3) = \alpha_d(3) \approx 3.86$.

Exercise 20.2 How to compute $\alpha_*(K)$? The idea is that, above this value of the clause density, any solution of the RS distributional equations has $\hat{\zeta} = 0$ with positive probability. In this case the denominator of eqn (20.7) vanishes with positive probability, leading to a contradiction.

We start by regularizing eqn (20.7) with a small parameter ϵ. Each $\hat{\zeta}_i$ is replaced by $\max(\hat{\zeta}_i, \epsilon)$. Let us denote by x the probability that $\hat{\zeta}$ is of order ϵ, and by y the probability that ζ is of order $1 - \epsilon$. Consider the limit $\epsilon \to 0$.

(a) Show that $x = y^{K-1}$

(b) Show that $1 - 2y = e^{-K\alpha x} I_0(K\alpha x)$, where $I_0(z)$ is the Bessel function with Taylor expansion $I_0(t) = \sum_{p=0}^{\infty} (1/p!^2)(t/2)^{2p}$.

[Hint: Suppose that there are p' variables among $\hat{\zeta}_1 \ldots \hat{\zeta}_p$ and q' among $\hat{\zeta}_{p+1} \ldots \hat{\zeta}_{p+q}$ that are of order ϵ. Show that this update equation gives $\zeta = O(\epsilon)$ if $p' > q'$, $\zeta = 1 - O(\epsilon)$ if $p' < q'$, and $\zeta = O(1)$ if $p' = q'$.]

(c) Let $\alpha_*(K)$ be the largest clause density such that the two equations derived in (a) and (b) admit the unique solution $x = y = 0$. Show that, for $\alpha \geq \alpha_*(K)$, a new solution appears with $x, y > 0$.

(d) By solving the above equations numerically, show that $\alpha_*(3) \approx 4.6673$ and $\alpha_*(4) \approx 11.83$.

Unhappily, this RS computation is incorrect when α is large enough, and, as a consequence, the prediction for the SAT–UNSAT phase transition is wrong as well. In particular, it contradicts the upper bound $\alpha_{UB,2}(K)$ found in Chapter 10 (for instance, in the two cases $K = 3$ and 4, one has $\alpha_{UB,2}(3) \approx 4.66603 < \alpha_*(3)$ and $\alpha_{UB,2}(4) \approx 10.2246 < \alpha_*(4)$). The largest α such that the RS entropy density is correct is nothing but the value for the condensation transition $\alpha_c(K)$. We shall discuss

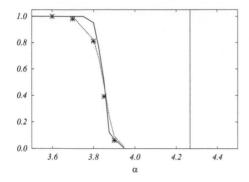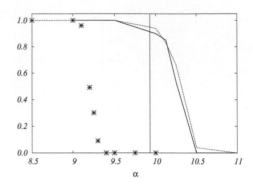

Fig. 20.3 Empirical probability that BP converges to a fixed point, plotted versus the clause density α, for 3-SAT (*left plot*) and 4-SAT (*right plot*). The statistics are over 100 instances, with $N = 5 \times 10^3$ variables (dashed curve) and $N = 10^4$ variables (full curve). There is an indication of a phase transition occurring for $\alpha_{\text{BP}} \approx 3.85$ ($K = 3$) and $\alpha_{\text{BP}} \approx 10.3$ ($K = 4$). The data points show the empirical probability that BP-guided decimation finds a SAT assignment, computed over 100 instances with $N = 5 \times 10^3$. The vertical lines correspond to the SAT–UNSAT threshold.

this phase transition further below and in Chapter 22.

There is another way to realize that something is wrong with the RS assumption close to the SAT–UNSAT phase transition. The idea is to look at the BP iteration.

20.1.3   BP-guided decimation

The simplest experiment consists in iterating the BP equations (20.3) on a randomly generated K-SAT instance. We start from uniformly random messages, and choose the following convergence criterion, defined in terms of a small number δ: the iteration is halted at the first time $t_*(\delta)$ such that no message has changed by more than δ over the last iteration.

If a large time t_{max} is fixed, one can estimate the probability of convergence within t_{max} iterations by repeating the same experiment many times. Figure 20.3 shows this probability for $\delta = 10^{-2}$ and $t_{\text{max}} = 10^3$, plotted versus α. The probability curves show a sharp decrease around a critical value of α, α_{BP}, which is robust to variations of t_{max} and δ. This numerical result is indicative of a threshold behaviour: the typical convergence time $t_*(\delta)$ stays finite (or grows moderately) with N when $\alpha < \alpha_{\text{BP}}$. Above α_{BP}, BP fails to converge in a time t_{max} on a typical random instance.

When it converges, BP can be used to find a SAT assignment, using it as a heuristic guide for a sequential decimation procedure. Each time the value of a new variable has to be fixed, BP is iterated until the convergence criterion, with parameter δ, is met (alternatively, one may be more bold and use the BP messages after a time t_{max} even when they have not converged). Then one uses the BP messages in order to decide (i) which variable to fix, and (ii) what value the variable should take.

In the present implementation, these decisions were taken on the basis of a simple statistic: the bias of the variables. Given the BP estimate $\nu_i(\,\cdot\,)$ of the marginal of x_i, we define the bias as $\pi_i \equiv \nu_i(0) - \nu_i(1)$. The algorithm is given below.

BP-GUIDED DECIMATION (SAT formula $\mathcal{F}$, accuracy ϵ, iterations $t_{\max}$)

1:     For all $n \in \{1, \ldots, N\}$:
2:         Call BP($\mathcal{F}$, ϵ, $t_{\max}$);
3:         If BP does not converge, return 'NOT found' and exit;
4:         For each variable node j, compute the bias π_j;
5:         Find a variable $i(n)$ with the largest absolute bias $|\pi_{i(n)}|$;
6:         If $\pi_{i(n)} \geq 0$, fix $x_{i(n)}$ to $x^*_{i(n)} = 0$;
7:         Otherwise, fix $x_{i(n)}$ to $x^*_{i(n)} = 1$;
8:         Replace $\mathcal{F}$ by the formula obtained after this reduction
9:     End-For;
10:    Return the assignment $\underline{x}^*$

A pseudocode for BP was given in Section 14.2. Let us emphasize that the same decimation procedure could be used not only with BP but also with other types of guidance, if we had some way to estimate the marginals.

The empirical success probability of the BP-guided decimation algorithm on random formulae is shown in Fig. 20.3 (estimated from 100 instances of size $N = 5 \times 10^4$) for several values of α. The qualitative difference between 3-SAT and 4-SAT emerges clearly from this data. For 3-SAT, the decimation procedure returns a SAT assignment almost every time it converges, i.e. with probability close to one for $\alpha \lesssim 3.85$. For 4-SAT, BP converges most of the time if $\alpha \lesssim 10.3$. This value is larger than the conjectured SAT–UNSAT threshold $\alpha_s(4) \approx 9.931$ (and also larger than the best rigorous upper bound $\alpha_{\mathrm{UB},2}(4) \approx 10.2246$.) On the other hand, the BP-guided decimation finds SAT assignments only when $\alpha \lesssim 9.25$. It is believed that the cases $K \geq 5$ behave like $K = 4$.

20.1.4   On the validity of the RS analysis

These experiments suggest that something is not correct in the RS assumptions when α is large enough. The precise mechanism by which they are incorrect depends, however, on the value of K. For $K = 3$, the BP fixed point become unstable, and this leads to errors in decimations. In fact, the local stability of the BP fixed point can be computed along the lines of Section 17.4.2. The result is that it become unstable at $\alpha_{\mathrm{st}}(3) \approx 3.86$. In contrast, for $K \geq 4$ the fixed point remains stable but does not correspond to the correct marginals. Local stability is not a good enough test in this case.

Correspondingly, one can define two types of threshold:

(*i*) A stability threshold $\alpha_{\mathrm{st}}(K)$, beyond which BP does not have a locally stable fixed point.

(*ii*) A 1RSB condensation threshold $\alpha_{\mathrm{c}}(K)$, beyond which there is no BP fixed point that gives a correct estimate of the local marginals and free entropy.

Clearly, one should have $\alpha_{\mathrm{c}}(K) \leq \alpha_{\mathrm{st}}(K)$. Our study suggests that $\alpha_{\mathrm{c}}(3) = \alpha_{\mathrm{st}}(3) \simeq 3.86$, whereas, for $K \geq 4$, there is a strict inequality $\alpha_{\mathrm{c}}(K) < \alpha_{\mathrm{st}}(K)$.

The reason for the failure of BP is the decomposition of the measure (20.1) into many pure states. This happens at a third critical value $\alpha_{\mathrm{d}}(K) \leq \alpha_{\mathrm{c}}(K)$, referred to as the dynamical transition, in accordance with our discussion of spin glasses in

Section 12.3: $\alpha_{\mathrm{d}}(K)$ is the critical clause density above which Glauber dynamics will become inefficient. If $\alpha_{\mathrm{d}}(K) < \alpha < \alpha_{\mathrm{c}}(K)$, one expects, as we discussed in Section 19.4.1, that there will exist many pure states, and many quasi-solutions to the BP equations, among which one will give the correct marginals.

At this point, the reader might well be discouraged. This is understandable: we started by seeking one threshold (the SAT–UNSAT transition $\alpha_{\mathrm{s}}(K)$) and rapidly ended up defining a number of other thresholds, $\alpha_{\mathrm{d}}(K) \le \alpha_{\mathrm{c}}(K) \le \alpha_{\mathrm{st}}(K) \le \alpha_{\mathrm{s}}(K)$, to describe a zoology of exotic phenomena. It turns out that, while an understanding of the proliferation of pure states is necessary to get the correct value of $\alpha_{\mathrm{s}}(K)$, one does not need a detailed description of the clusters, which is a challenging task. Luckily, there exists a shortcut, through the use of the energetic cavity method. It turns out that the sketchy description of clusters that we get from this method, as if looking at them *from far away*, is enough to determine α_{s}. Even more than that, the sketch is a very useful and interesting one. In Section 20.3, we shall discuss a more detailed picture obtained through the fully fledged 1RSB cavity method applied to the model (20.1).

20.2 Survey propagation and the 1RSB phase

The use of the energetic 1RSB cavity method can be justified in two ways. From the first point of view, we are changing the problem. Instead of computing marginals of the distribution (20.1), we consider the problem of minimizing the energy function

$$E(\underline{x}) = \sum_{a=1}^{M} E_a(\underline{x}_{\partial a}) \,. \tag{20.9}$$

Here $E_a(\underline{x}_{\partial a}) = 0$ if clause a is satisfied by the assignment $\underline{x}$, and $E_a(\underline{x}_{\partial a}) = 1$ otherwise. The SAT–UNSAT threshold $\alpha_{\mathrm{s}}(K)$ is thus identified as the critical value above which the ground state energy $\min E(\underline{x})$ vanishes.

With the cavity method, we shall estimate the ground state energy density, and find that it vanishes below some threshold. This is then identified as $\alpha_{\mathrm{s}}(K)$. This identification amounts to assuming that, for generic large random K-SAT problems, there is no interval of α where the ground state energy is positive but sublinear in N. This assumption is reasonable, but of course it does not hold in more general situations. If, for instance, we added to a random K-SAT formula a small unsatisfiable subformula (including $o(N)$ variables), our approach would not detect the change, whereas the formula would always be unsatisfiable.

For $\alpha < \alpha_{\mathrm{s}}(K)$, the cavity method provides a rough picture of zero-energy pure states. This brings us to the second way of justifying this 'sketch'. We saw that describing a pure (Bethe) state in a locally tree-like graph amounts to assigning a set of cavity messages, i.e. of marginal distributions for the variables. The simplified description of the energetic 1RSB method only distinguishes between marginals that are concentrated on a single value and marginals that are not. The concentrated marginals are described exactly, while the other ones are just summarized by a single statement, 'not concentrated'.

20.2.1 The SP(y) equations

The satisfiability problem involves only hard constraints and binary variables. We can thus use the simplified $\mathsf{SP}(\mathrm{y})$ equations of Section 19.5.3. The messages are triples: $(Q_{ia}(0), Q_{ia}(1), Q_{ia}(*))$ for variable-to-function messages, and $(\widehat{Q}_{ai}(0), \widehat{Q}_{ai}(1), \widehat{Q}_{ai}(*))$ for function-to-variable messages.

In the case of K-satisfiability, these can be further simplified. The basic observation is that if $J_{ai} = 0$, then $\widehat{Q}_{ai}(1) = 0$, and if $J_{ai} = 1$, then $\widehat{Q}_{ai}(0) = 0$. This can be shown either by starting from the general formalism in Section 19.5.3 or by reconsidering the interpretation of warning propagation messages. Recall that a '0' message means that the constraint a 'forces' variable x_i to take the value 0 in order to minimize the system's energy. In K-SAT this can happen only if $J_{ai} = 0$, because $x_i = 0$ is then the value that satisfies the clause a. With this remark in mind, the function-to-variable-node message can be parameterized by a single real number. We shall choose this number to be $\widehat{Q}_{ai}(0)$ if $J_{ai} = 0$ and $\widehat{Q}_{ai}(1)$ if $J_{ai} = 1$, and denote it by $\widehat{Q}_{ai}$. This number $\widehat{Q}_{ai}$ is the probability that there is a warning sent from a to i which forces the value of variable x_i.

Analogously, it is convenient to adopt a parameterization of the variable-to-function message $Q_{ia}(\mathrm{m})$ which takes into account the value of J_{ai}. More precisely, recall that Q_{ia} is supported on three types of messages: $\mathrm{m}(0) = 0, \mathrm{m}(1) > 0$; $\mathrm{m}(0) = \mathrm{m}(1) = 0$; or $\mathrm{m}(0) > 0, \mathrm{m}(1) = 0$. Let us denote the corresponding weights by $Q_{ia}(0)$, $Q_{ia}(*)$ and $Q_{ia}(1)$. If $J_{ai} = 0$, we then define $Q_{ia}^{\mathrm{S}} \equiv Q_{ia}(0)$, $Q_{ia}^{*} \equiv Q_{ia}(*)$, and $Q_{ia}^{\mathrm{U}} \equiv Q_{ia}(1)$. Conversely, if $J_{ai} = 1$, we let $Q_{ia}^{\mathrm{S}} \equiv Q_{ia}(1)$, $Q_{ia}^{*} \equiv Q_{ia}(*)$, and $Q_{ia}^{\mathrm{U}} \equiv Q_{ia}(0)$.

We summarize this notation below with the corresponding interpretations. We emphasize that 'probability' refers here to the random choice of a pure state (see Section 19.1).

- Q_{ia}^{S}: probability that x_i is forced by the clauses $b \in \partial i \setminus a$ to satisfy a;
- Q_{ia}^{U}: probability that x_i is forced by the clauses $b \in \partial i \setminus a$ to violate a;
- Q_{ia}^{*}: probability that x_i is not forced by the clauses $b \in \partial i \setminus a$;
- $\widehat{Q}_{ai}$: probability that x_i is forced by clause a to satisfy it.

The 1RSB cavity equations have been written in Section 19.5.3.

Exercise 20.3 Write the 1RSB equations explicitly in terms of the messages $Q^{\mathrm{S}}, Q^{\mathrm{U}}, Q^{*}, \widehat{Q}$ by applying the procedure of Section 19.5.3.

Alternatively, the 1RSB cavity equations can be guessed, bearing the above interpretation in mind. Clause a forces variable x_i to satisfy it if and only if all of the other variables entering into clause a are forced (by some other clause) not to satisfy a. This means that

$$\widehat{Q}_{ai} = \prod_{j \in \partial a \setminus i} Q_{ja}^{\mathrm{U}} . \tag{20.10}$$

Consider, on the other hand, variable node i, and assume for definiteness that $J_{ia} = 0$ (the opposite case gives rise to identical equations). Remember that, in this case, $\mathcal{S}_{ia}$ denotes the subset of clauses $b \neq a$ in which $J_{ib} = 0$, and $\mathcal{U}_{ia}$ the subset in which $J_{ib} = 1$. Assume that the clauses in $\Omega^{\mathrm{S}} \subseteq \mathcal{S}_{ia}$ and $\Omega^{\mathrm{U}} \subseteq \mathcal{U}_{ia}$ force x_i to satisfy them. Then x_i is forced to either satisfy or violate a, depending whether $|\Omega^{\mathrm{S}}| > |\Omega^{\mathrm{U}}|$ or $|\Omega^{\mathrm{S}}| < |\Omega^{\mathrm{U}}|$. Finally, x_i is not forced if $|\Omega^{\mathrm{S}}| = |\Omega^{\mathrm{U}}|$. The energy shift is equal to the number of 'forcing' clauses in $\partial i \setminus a$ that are violated when x_i is chosen to satisfy the largest number of them, namely $\min(|\Omega^{\mathrm{U}}|, |\Omega^{\mathrm{S}}|)$. We thus get the equations

$$Q_{ia}^{\mathrm{U}} \cong \sum_{|\Omega^{\mathrm{U}}|>|\Omega^{\mathrm{S}}|} e^{-y|\Omega^{\mathrm{S}}|} \prod_{b\in\Omega^{\mathrm{U}}\cup\Omega^{\mathrm{S}}} \widehat{Q}_{bi} \prod_{b\notin\Omega^{\mathrm{U}}\cup\Omega^{\mathrm{S}}} (1-\widehat{Q}_{bi}), \tag{20.11}$$

$$Q_{ia}^{\mathrm{S}} \cong \sum_{|\Omega^{\mathrm{S}}|>|\Omega^{\mathrm{U}}|} e^{-y|\Omega^{\mathrm{U}}|} \prod_{b\in\Omega^{\mathrm{U}}\cup\Omega^{\mathrm{S}}} \widehat{Q}_{bi} \prod_{b\notin\Omega^{\mathrm{U}}\cup\Omega^{\mathrm{S}}} (1-\widehat{Q}_{bi}), \tag{20.12}$$

$$Q_{ia}^{*} \cong \sum_{|\Omega^{\mathrm{U}}|=|\Omega^{\mathrm{S}}|} e^{-y|\Omega^{\mathrm{U}}|} \prod_{b\in\Omega^{\mathrm{U}}\cup\Omega^{\mathrm{S}}} \widehat{Q}_{bi} \prod_{b\notin\Omega^{\mathrm{U}}\cup\Omega^{\mathrm{S}}} (1-\widehat{Q}_{bi}). \tag{20.13}$$

The overall normalization is fixed by the condition $Q_{ia}^{\mathrm{U}} + Q_{ia}^{*} + Q_{ia}^{\mathrm{S}} = 1$.

As usual, eqns (20.10–20.13) can be understood either as defining a mapping from the space of messages $\{\widehat{Q}_{ai}, Q_{ia}\}$ onto itself or as a set of fixed-point conditions. In both cases, they are referred to as the SP(y) equations for the satisfiability problem. From the computational point of view, these equations involve a sum over $2^{|\partial i|-1}$ terms. This is often too much if we want to iterate the SP(y) equations on large K-SAT formulae: the average degree of a variable node in a random K-SAT formula with clause density α is $K\alpha$. Further, in the most interesting regime—close to the SAT–UNSAT threshold—$\alpha = \Theta(2^K)$, and the sum is over $2^{\Theta(K2^K)}$ terms, which rapidly becomes impractical. It is thus important to notice that the sums can be computed efficiently by interpreting them as convolutions.

Exercise 20.4 Consider a sequence of independent Bernoulli random variables $X_1, \ldots, X_n, \ldots$, with means $\eta_1, \ldots, \eta_n, \ldots$, respectively. Let $W_n(m)$ be the probability that the sum $\sum_{b=1}^{n} X_b$ is equal to m.

(a) Show that these probabilities satisfy the recursion

$$W_n(m) = \eta_n W_{n-1}(m-1) + (1-\eta_n)W_{n-1}(m),$$

for $m \in \{0, \ldots, n\}$. Argue that these identities can be used, together with the initial condition $W_0(m) = \mathbb{I}(m=0)$, to compute $W_n(m)$ in $O(n^2)$ operations.

(b) How can one compute the right-hand sides of eqns (20.11)–(20.13) in $O(|\partial i|^2)$ operations?

20.2.2   The free entropy $\mathbb{F}^{\mathrm{RSB,e}}$

Within the 1RSB energetic cavity method, the free entropy $\mathbb{F}^{\mathrm{RSB,e}}(\{Q, \widehat{Q}\})$ provides detailed information on the minimal energy of (Bethe) pure states. These pure states are nothing but metastable minima of the energy function (i.e. minima whose energy cannot be decreased with a bounded number of spin flips).

The 1RSB free entropy is expressed in terms of a set of messages $\{Q_{ia}, \widehat{Q}_{ai}\}$ that provide a (quasi-)solution of the SP(y) equations (20.10)–(20.13). Following the general theory in Section 19.5.2, it can be written in the form

$$\mathbb{F}^{\mathrm{RSB,e}}(\{Q, \widehat{Q}\}) = \sum_{a \in C} \mathbb{F}_a^{\mathrm{RSB,e}} + \sum_{i \in V} \mathbb{F}_i^{\mathrm{RSB,e}} - \sum_{(i,a) \in E} \mathbb{F}_{ia}^{\mathrm{RSB,e}} . \qquad (20.14)$$

Equation (19.94) yields

$$\mathrm{e}^{\mathbb{F}_{ia}^{\mathrm{RSB,e}}} = 1 - (1 - \mathrm{e}^{-\mathsf{y}})\widehat{Q}_{ai}Q_{ia}^{\mathrm{U}} . \qquad (20.15)$$

The contribution $\mathbb{F}_a^{\mathrm{RSB,e}}$ defined in eqn (19.92) can be computed as follows. The reweighting $\mathbb{F}_a^{\mathrm{e}}(\{\mathsf{m}_{ia}\})$ is always equal to 0, except in the case where all of the variables in clause a receive a warning requesting that they point in the 'wrong direction', namely the direction which does not satisfy the clause. Therefore,

$$\mathrm{e}^{\mathbb{F}_a^{\mathrm{RSB,e}}} = 1 - (1 - \mathrm{e}^{-\mathsf{y}}) \prod_{i \in \partial a} Q_{ia}^{\mathrm{U}} .$$

Finally, the contribution $\mathbb{F}_i^{\mathrm{RSB,e}}$ defined in eqn (19.93) depends on the messages sent from the check nodes $b \in \partial i$. Let us denote by $\Omega^{\mathrm{S}} \subseteq \partial_0 i$ the subset of check nodes $b \in \partial_0 i$ such that clause b forces x_i to satisfy it. Similarly, we denote by $\Omega^{\mathrm{U}} \subseteq \partial_1 i$ the subset of $\partial_1 i$ such that clause b forces x_i to satisfy it. We then have

$$\mathrm{e}^{\mathbb{F}_i^{\mathrm{RSB,e}}} = \sum_{\Omega^{\mathrm{U}}, \Omega^{\mathrm{S}}} \mathrm{e}^{-\mathsf{y}\min(\Omega^{\mathrm{S}}, \Omega^{\mathrm{U}})} \left[\prod_{b \in \Omega^{\mathrm{U}} \cup \Omega^{\mathrm{S}}} \widehat{Q}_{bi} \right] \left[\prod_{b \notin \Omega^{\mathrm{U}} \cup \Omega^{\mathrm{S}}} (1 - \widehat{Q}_{bi}) \right] . \qquad (20.16)$$

Exercise 20.5 Show that, for any $i \in \partial a$, $\mathbb{F}_{ia}^{\mathrm{RSB,e}} = \mathbb{F}_a^{\mathrm{RSB,e}}$.

20.2.3 The large-y limit: The SP equations

Consider now the case of satisfiable instances. A crucial problem is to characterize satisfying assignments and to find them efficiently. This amounts to focusing on zero-energy assignments, which are selected by taking the $\mathsf{y} \to \infty$ limit within the energetic cavity method.

We can take the limit $\mathsf{y} \to \infty$ in the SP(y) equations (20.11)–(20.13). This yields

$$\widehat{Q}_{ai} = \prod_{j \in \partial a \backslash i} Q_{ja}^{\mathrm{U}}, \tag{20.17}$$

$$Q_{ja}^{\mathrm{U}} \cong \prod_{b \in \mathcal{S}_{ja}} (1 - \widehat{Q}_{bj}) \left[1 - \prod_{b \in \mathcal{U}_{ja}} (1 - \widehat{Q}_{bj}) \right], \tag{20.18}$$

$$Q_{ja}^{\mathrm{S}} \cong \prod_{b \in \mathcal{U}_{ja}} (1 - \widehat{Q}_{bj}) \left[1 - \prod_{b \in \mathcal{S}_{ja}} (1 - \widehat{Q}_{bj}) \right], \tag{20.19}$$

$$Q_{ja}^{*} \cong \prod_{b \in \partial j \backslash a} (1 - \widehat{Q}_{bj}), \tag{20.20}$$

where the normalization is always fixed by the condition $Q_{ja}^{\mathrm{U}} + Q_{ja}^{\mathrm{S}} + Q_{ja}^{*} = 1$.

The $\mathrm{y} = \infty$ equations have a simple interpretation. Consider a variable x_j appearing in clause a, and assume that it receives a warning from clause $b \neq a$ independently with probability $\widehat{Q}_{bj}$. Then $\prod_{b \in \mathcal{S}_{ja}} (1 - \widehat{Q}_{bj})$ is the probability that variable j receives no warning forcing it in the direction which satisfies clause a. The product $\prod_{b \in \mathcal{U}_{ja}} (1 - \widehat{Q}_{bj})$ is the probability that variable j receives no warning forcing it in the direction which violates clause a. Therefore Q_{ja}^{U} is the probability that variable j receives at least one warning forcing it in the direction which violates clause a, *conditional on the fact that there are no contradictions in the warnings received by j from clauses $b \neq a$*. Analogous interpretations hold for Q_{ja}^{S} and Q_{ja}^{*}. Finally, $\widehat{Q}_{ai}$ is the probability that all variables in $\partial a \backslash i$ are forced in the direction which violates clause a, under the same *condition of no contradiction*.

Note that the $\mathrm{y} = \infty$ equations are a relatively simple modification of the BP equations in eqn (20.3). However, the interpretation of the messages is very different in the two cases.

Finally, the free entropy in the $\mathrm{y} = \infty$ limit is obtained as

$$\mathbb{F}^{\mathrm{RSB,e}} = \sum_{a \in C} \mathbb{F}_a^{\mathrm{RSB,e}} + \sum_{i \in V} \mathbb{F}_i^{\mathrm{RSB,e}} - \sum_{(i,a) \in E} \mathbb{F}_{ia}^{\mathrm{RSB,e}}, \tag{20.21}$$

where

$$\mathbb{F}_{ia}^{\mathrm{RSB,e}} = \log \left\{ 1 - Q_{ia}^{\mathrm{U}} \widehat{Q}_{ai} \right\}, \tag{20.22}$$

$$\mathbb{F}_i^{\mathrm{RSB,e}} = \log \left\{ \prod_{b \in \partial_0 i} (1 - \widehat{Q}_{bi}) + \prod_{b \in \partial_1 i} (1 - \widehat{Q}_{bi}) - \prod_{b \in \partial i} (1 - \widehat{Q}_{bi}) \right\}, \tag{20.23}$$

$$\mathbb{F}_a^{\mathrm{RSB,e}} = \log \left\{ 1 - \prod_{j \in \partial a} Q_{ja}^{\mathrm{U}} \right\}. \tag{20.24}$$

Exercise 20.6 Show that, if the SP messages satisfy the fixed-point equations (20.17)–(20.20), the free entropy can be rewritten as $\mathbb{F}^{\mathrm{RSB,e}} = \sum_i \mathbb{F}_i^{\mathrm{RSB,e}} + \sum_a (1 - |\partial a|) \mathbb{F}_a^{\mathrm{RSB,e}}$.

20.2.4   The SAT–UNSAT threshold

The SP(y) equations (20.10)–(20.13) always admit a 'no warning' fixed point corresponding to $\widehat{Q}_{ai} = 0$, $Q_{ia}^{\mathrm{S}} = Q_{ia}^{\mathrm{U}} = 0$, and $Q_{ia}^{*} = 1$ for each $(i, a) \in E$. Other fixed points can be explored numerically by iterating the equations on large random formulae.

Within the cavity approach, the distribution of the message associated with a uniformly random edge (i, a) satisfies a distributional equation. As explained in Section 19.2.5, this distributional equation is obtained by promoting $\widehat{Q}_{ai}$ and $(Q_{ia}^{\mathrm{U}}, Q_{ia}^{\mathrm{S}}, Q_{ia}^{*})$ to random variables and reading eqns (20.10)–(20.13) as equalities in distribution. The distribution can then be studied by the population dynamics of Section 19.2.6. It obviously admits a no-warning (or 'replica-symmetric') fixed point, with $\widehat{Q} = 0$ and $(Q^{\mathrm{U}}, Q^{\mathrm{S}}, Q^{*}) = (0, 0, 1)$ identically, but (as we shall see) in some cases one also finds a different, 'non-trivial' fixed-point distribution.

Given a fixed point, the 1RSB free-entropy density $\mathfrak{F}^{\mathrm{e}}(\mathrm{y})$ can be estimated by taking the expectation of eqn (20.14) (with respect to both degrees and fields) and dividing by N. When evaluated at the no-warning fixed point, the free-entropy density $\mathfrak{F}^{\mathrm{e}}(\mathrm{y})$ vanishes. This means that the number of clusters of SAT assignments is subexponential, so that the corresponding complexity density vanishes. To a first approximation, this solution corresponds to low-energy assignments forming a single cluster. Note that the energetic cavity method counts the number of clusters of SAT assignments, and not the number of SAT assignments itself (which is actually exponentially large).

Figure 20.4 shows the outcome of a population dynamics computation. We plot the free-entropy density $\mathfrak{F}^{\mathrm{e}}(\mathrm{y})$ as a function of y for random 3-SAT, for a few values of the clause density α. These plots were obtained by initializing the population dynamics recursion with i.i.d. messages $\{\widehat{Q}_{i}\}$ uniformly random in $[0, 1]$. For $\alpha < \alpha_{\mathrm{d,SP}} \simeq 3.93$, the iteration converges to the 'no-warning' fixed point where all of the messages $\widehat{Q}$ are equal to 0.

For $\alpha > \alpha_{\mathrm{d,SP}}$, and when y is larger than a critical value $\mathrm{y}_{\mathrm{d}}(\alpha)$, the iteration converges to a non-trivial fixed point. This second solution has a non-vanishing value of the free-entropy density $\mathfrak{F}^{\mathrm{e}}(\mathrm{y})$. The energetic complexity $\Sigma^{\mathrm{e}}(\epsilon)$ is obtained from $\mathfrak{F}^{\mathrm{e}}(\mathrm{y})$ via the Legendre transform (19.95).

In practice, the Legendre transform is computed by fitting the population dynamics data, and then transforming the fitting curve. Good results were obtained with a fit of the form $\mathfrak{F}_{\mathrm{fit}}^{\mathrm{e}}(\mathrm{y}) = \sum_{r=0}^{r_{*}} \psi_{r} \, \mathrm{e}^{-r\mathrm{y}}$ with r_{*} between 2 and 4. The resulting curves $\Sigma^{\mathrm{e}}(\epsilon)$ (or, more precisely, their concave branches) are shown in Fig. 20.5.

Exercise 20.7 Show that $\Sigma^{\mathrm{e}}(\epsilon = 0) = \lim_{\mathrm{y} \to \infty} \mathfrak{F}^{\mathrm{e}}(\mathrm{y})$

The energetic complexity $\Sigma^{\mathrm{e}}(\epsilon)$ is the exponential growth rate of the number of (quasi-) solutions of the min-sum equations with energy density u. As can be seen in Fig. 20.5, for $\alpha = 4.1$ or 4.2 (and in general, in an interval above $\alpha_{\mathrm{d}}(3)$), one finds $\Sigma^{\mathrm{e}}(\epsilon = 0) > 0$. The interpretation is that there exist exponentially many solutions of the min-sum equations with zero energy density.

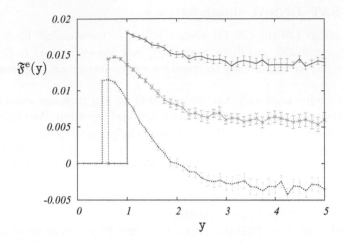

Fig. 20.4 1RSB free-entropy density for 3-SAT, computed from a population dynamics analysis of the SP equation, at $\alpha = 4.1$, 4.2, and 4.3 (from *top* to *bottom*). For each α, y, a population of size 12 000 was iterated 12×10^6 times. The resulting $\mathfrak{F}^e$ was computed by averaging over the last 8×10^6 iterations.

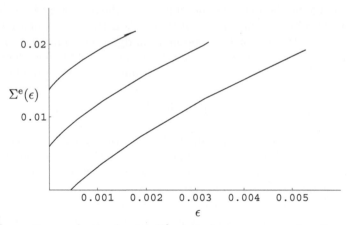

Fig. 20.5 Energetic complexity density Σ^e plotted versus energy density ϵ, for the 3-SAT problem at $\alpha = 4.1$, 4.2, and 4.3 (from *top* to *bottom*). These curves were obtained from Fig. 20.4, using a Legendre transform of free-entropy fits. .

In contrast, when $\alpha = 4.3$ the curve starts at a positive ϵ or, equivalently, the 1RSB complexity curve has $\Sigma^e(\epsilon = 0) < 0$. Of course, the typical number of min-sum solutions cannot decrease exponentially. The result $\Sigma^e(\epsilon = 0) < 0$ is interpreted as a consequence of the fact that a typical random formula does not admit any (approximate) solutions of the min-sum equations with energy density $\epsilon = 0$. Given the correspondence between min-sum fixed points and clusters of low-energy assignments, this in turn implies that a typical random formula does not have any SAT assignment.

Table 20.1 Predictions of the 1RSB cavity method for the SAT–UNSAT threshold of random K-satisfiability.

K	3	4	5	6	7	8	9	10
$\alpha_\mathrm{s}(K)$	4.2667	9.931	21.117	43.37	87.79	176.5	354.0	708.9

From Fig. 20.5, one would expect that the SAT–UNSAT transition would lie between $\alpha = 4.2$ and $\alpha = 4.3$. A more precise estimate can be obtained by plotting $\mathfrak{F}^\mathrm{e}(\mathsf{y} \to \infty)$ versus α, and locating the value of α where it vanishes. For 3-SAT, we obtain the estimate for the SAT–UNSAT threshold $\alpha_\mathrm{s}(3) = 4.26675 \pm 0.00015$. The predictions of this method for $\alpha_\mathrm{s}(K)$ are shown in Table 20.1. In practice, reliable estimates can be obtained with population dynamics only for $K \leq 7$. The reason is that $\alpha_\mathrm{s}(K)$ increases exponentially with K, and the size of the population needed in order to achieve a given precision will increase accordingly (the average number of independent messages entering into the distributional equations is $K\alpha$).

For large K, one can formally expand the distributional equations, which yields a series for $\alpha_\mathrm{s}(K)$ in powers of 2^{-K}. The first two terms (seven terms have been computed) of this expansion are

$$\alpha_\mathrm{s}(K) = 2^K \log 2 - \frac{1}{2}(1 + \log 2) + O(2^{-K}K^2) \,. \tag{20.25}$$

20.2.5 SP-guided decimation

The analysis in the last few pages provides a refined description of the set of solutions of random formulae. This knowledge can be exploited to efficiently find some solutions, much in the same way as we used belief propagation in Section 20.1.3. The basic strategy is again to use the information provided by the SP messages as a clever heuristic in a decimation procedure.

The first step consists in finding an approximate solution of the SP(y) equations (20.10)–(20.13), or of their simplified $\mathsf{y} = \infty$ version (eqns (20.17)–(20.20)), for a given instance of the problem. To be definite, we shall focus on the latter case, since $\mathsf{y} = \infty$ selects zero-energy states. We can seek solutions of the SP equations by iteration, exactly as we would do with BP. We initialize the SP messages, generally as i.i.d. random variables with some common distribution, and then update them according to eqns (20.17)–(20.20). Updates can, for instance, be implemented in parallel until a convergence criterion has been met.

Figure 20.6 shows the empirical probability that the iteration converges before $t_\mathrm{max} = 1000$ iterations on random formulae as a function of the clause density α. As a convergence criterion, we required that the maximal difference between any two subsequent values of a message was smaller than $\delta = 10^{-2}$. Messages were initialized by drawing, for each edge, $\widehat{Q}_{ai} \in [0, 1]$ independently and uniformly at random. It is clear that SP has better convergence properties than BP for $K = 3$ and, indeed, it converges even for α larger than the SAT–UNSAT threshold.

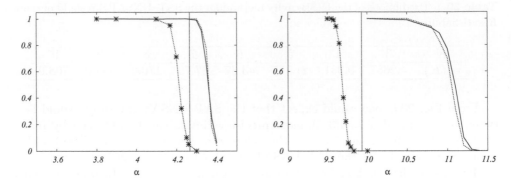

Fig. 20.6 Empirical convergence probability of SP (initialized with uniformly random messages) plotted versus the clause density α for 3-SAT (*left*), and 4-SAT (*right*). The average is over 100 instances, with $N = 5 \times 10^3$ (solid line) and $N = 10^4$ (dashed line) variables. The data points show the empirical probability that SP-guided decimation finds a SAT assignment, computed over 100 instances with $N = 5 \times 10^3$. The vertical lines are the predicted SAT–UNSAT thresholds.

The numerics suggest the existence of two thresholds $\alpha_{d,SP}(K)$ and $\alpha_{u,SP}(K)$ that characterize the convergence behaviour as follows (all of the statements below should be interpreted as holding with high probability in the large-N limit):

- For $\alpha < \alpha_{d,SP}$, the iteration converges to the trivial fixed point defined by $\widehat{Q}_{ai} = 0$ for all edges $(i, a) \in G$.
- For $\alpha_{d,SP} < \alpha < \alpha_{u,SP}$, the iteration converges to a 'non-trivial' fixed point.
- For $\alpha_{u,SP} < \alpha$, the iteration does not converge.

In the interval $\alpha_{d,SP}(K) < \alpha < \alpha_{U,SP}(K)$, it is expected that an exponential number of fixed points will exist, but most of them will be degenerate and correspond to 'disguised' WP fixed points. In particular, $\widehat{Q}_{ai} = 0$ or 1 for all of the edges (i, a). On the other hand, the fixed point actually reached by iteration is stable with respect to changes in the initialization. This suggests the existence of a unique non-degenerate fixed point. The threshold $\alpha_{d,SP}(K)$ is conjectured to be the same as that defined for the distributional equation in the previous section; this is why we have used a similar symbol. In particular, $\alpha_{d,SP}(K = 3) \approx 3.93$ and $\alpha_{d,SP}(K = 4) \approx 8.30$. We have also obtained $\alpha_{u,SP}(K = 3) \approx 4.36$ and $\alpha_{u,SP}(K = 4) \approx 9.7$.

SP can be used in a decimation procedure. After iterating the SP equations until convergence, one computes the **SP marginal** for each variable $i \in \{1, \dots, N\}$:

$$w_i(1) \cong \prod_{a \in \partial_0 i} (1 - \widehat{Q}_{ai}) \left[1 - \prod_{a \in \partial_1 i} (1 - \widehat{Q}_{ai}) \right],$$

$$w_i(0) \cong \prod_{a \in \partial_1 i} (1 - \widehat{Q}_{ai}) \left[1 - \prod_{a \in \partial_0 i} (1 - \widehat{Q}_{ai}) \right],$$

$$w_i(*) \cong \prod_{a \in \partial i} (1 - \widehat{Q}_{ai}), \tag{20.26}$$

with the normalization condition $w_i(1) + w_i(0) + w_i(*) = 1$. The interpretation of these SP marginals is the following: $w_i(1)$ and $w_i(0)$ are the probabilities that the variable i receives a warning forcing it to take the value $x_i = 1$ or $x_i = 0$, respectively, *conditional* on the fact that it does not receive contradictory warnings. The bias of the variable is then defined as $\pi_i \equiv w_i(0) - w_i(1)$. The variable with the largest absolute bias is selected, and fixed according to the sign of the bias. This procedure is then iterated as with BP-guided decimation.

It typically happens that, after some fraction of the variables have been fixed with this method, the SP iteration on the reduced instance converges to the trivial fixed point $\widehat{Q}_{ai} = 0$. According to our interpretation, this means that the resulting problem is described by a unique Bethe measure, and the SAT assignments are no longer clustered. In fact, in agreement with this interpretation, it is found that, typically, simple algorithms are able to solve the reduced problem. A possible approach is to run BP-guided decimation. An even simpler alternative is to apply a simple local search algorithm, such as Walksat or simulated annealing.

The pseudocode for this algorithm is as follows.

SP-GUIDED DECIMATION (formula $\mathcal{F}$, SP parameter ϵ, $t_{\max}$,
                              WalkSAT parameters f, p)

1:   Set $U = \emptyset$;
2:   Repeat until FAIL or $U = V$:
3:       Call SP$(\mathcal{F}, \epsilon, t_{\max})$. If it does not converge, FAIL;
4:       For each $i \in V \setminus U$, compute the bias π_i;
5:       Let $j \in V \setminus U$ have the largest value of $|\pi_i|$;
6:       If $|\pi_j| \leq 2K\epsilon$ call WalkSAT$(\mathcal{F}, f, p)$;
7:         Else fix x_j according to the sign of π_j,
                 and define $\mathcal{F}$ as the new formula obtained after fixing x_j;
8:   End-Repeat;
9:   Return the current assignment;

SP (formula $\mathcal{F}$, accuracy ϵ, iterations $t_{\max}$)

1:    Initialize SP messages to i.i.d. random variables;
2:    For $t \in \{0, \ldots, t_{\max}\}$
3:        For each $(i, a) \in E$
4:          Compute the new value of $\widehat{Q}_{ai}$ using eqn (20.10)
5:        For each $(i, a) \in E$
6:          Compute the new value of Q_{ai} using eqns (20.11)–(20.13)
7:        Let Δ be the maximum difference from the previous iteration;
8:          If $\Delta < \epsilon$ return current messages;
9:    End-For;
10:   Return 'Not Converged';

The pseudocode of WalkSAT was given in Section 10.2.3.

In Fig. 20.6, we plot the empirical success probability of the SP-guided-decimation

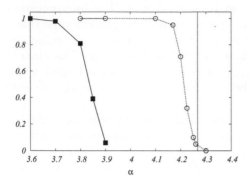 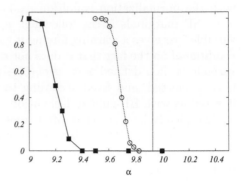

Fig. 20.7 Performance of BP-guided decimation and SP-guided decimation on 3-SAT (*left plot*) and 4-SAT (*right plot*) problems. The plots show the probability of finding a SAT assignment versus the clause density, averaged over 100 instances with $N = 5 \times 10^3$ variables. The SP-based algorithm (dotted line) performs better than the BP-based one (full line). The vertical lines are the SAT–UNSAT thresholds.

algorithm for random 3-SAT and 4-SAT formulae as a function of the clause density α. A careful study suggests that the algorithm finds a satisfying assignment with high probability when $\alpha \lesssim 4.252$ (for $K = 3$) and $\alpha \lesssim 9.6$ (for $K = 4$). These values are slightly smaller than the conjectured locations of the SAT–UNSAT threshold $\alpha_s(3) \approx 4.2667$ and $\alpha_s(4) \approx 9.931$.

Apart from the SP routine (which builds upon the statistical-mechanics insight), the above algorithm is quite naive and could be improved in a number of directions. One possibility is to allow the algorithm to backtrack, i.e. to release some variables that were fixed at a previous stage of the decimation. Further, we have not used, at any step, the information provided by the free entropy $\mathfrak{F}^e(y = \infty)$, which can be computed at little extra cost. Since this gives an estimate of the logarithm of the number of clusters of solutions, it can also be reasonable to make choices that maximize the value of $\mathfrak{F}^e$ in the resulting formula.

As can be deduced from Fig. 20.7, SP-guided decimation outperforms BP-guided Decimation. Empirically, this algorithm, or small variations of it, provides the most efficient procedure for solving large random K-SAT formulae close to the SAT–UNSAT threshold. Furthermore, it has extremely low complexity. Each SP iteration requires $O(N)$ operations, which yields $O(Nt_{\max})$ operations per SP call. In the implementation outlined above, this implies an $O(N^2 t_{\max})$ complexity. This can, however, be reduced to $O(Nt_{\max})$ by noticing that fixing a single variable does not affect the SP messages significantly. As a consequence, SP can be called every $N\delta$ decimation steps for some small δ. Finally, the number of iterations required for convergence seems to grow very slowly with N, if it does at all. One should probably think of $t_{\max}$ as a big constant, or $t_{\max} = O(\log N)$

In order to get a better understanding of how SP-guided decimation works, it is useful to monitor the evolution of the energetic-complexity curve $\Sigma^e(\epsilon)$ while decimating. When SP iteration has converged for a given instance, one can use eqn (20.21) to compute the free entropy, and then compute the curve $\Sigma^e(\epsilon)$ by means of a Legendre

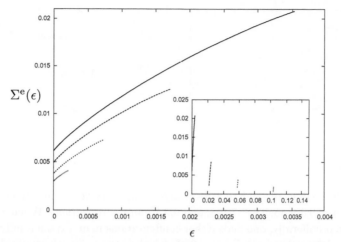

Fig. 20.8 Decimation process: the complexity versus energy density ($1/N$ times the number of violated clauses) measured on a single instance of random 3-SAT with $N = 10\,000$ and $\alpha = 4.2$ (*top curve*), and on the decimated instances obtained after fixing $1000, 2000$, and 3000 variables with the SP-guided decimation procedure (from *top* to *bottom*). For comparison, the *inset* shows the same complexity versus the total energy after fixing $1000, 2000$, and 3000 randomly chosen variables to arbitrary values

transform.

In Fig. 20.8, we consider a run of the SP-guided decimation algorithm on one random 3-SAT formula with $N = 10^4$ at $\alpha = 4.2$. Here, the complexity curve of the residual formula ($N\Sigma^e(\epsilon)$ versus the number of violated clauses $N\epsilon$) is plotted every 1000 decimation steps. One can notice two main effects: (i) the zero-energy complexity $N\Sigma^e(0)$ decreases, showing that some clusters of solutions are lost during the decimation; and (ii) the number of violated clauses in the most numerous metastable clusters, the 'threshold energy', decreases as well,[1] implying that the problem becomes simpler: the true solutions are less and less hidden among metastable minima.

The important point is that the effect (ii) is much more pronounced than (i). After about half of the variables have been fixed, the threshold energy vanishes. SP converges to the trivial fixed point, the resulting instance becomes 'simple' and is solved easily by Walksat.

20.3 Some ideas about the full phase diagram

20.3.1 Entropy of clusters

The energetic 1RSB cavity method has given two important results: on the one hand, a method to locate the SAT–UNSAT transition threshold α_s, which is conjectured to be exact, and on the other, a powerful message-passing algorithm: SP. These results

---

[1]Because of the instability of the 1RSB solution at large energies (see Chapter 22), the threshold energies obtained within the 1RSB approach are not exact. However, one expects the actual behaviour to be quantitatively close to the 1RSB description.

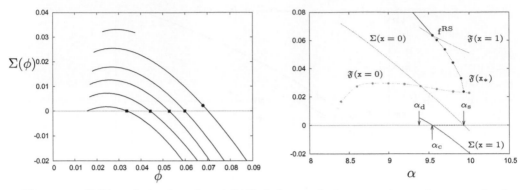

Fig. 20.9 1RSB analysis of random 4-SAT. *Left*: complexity versus internal entropy density of clusters, for $\alpha = 9.3, 9.45, 9.6, 9.7, 9.8$, and 9.9 (from *top* to *bottom*). When SAT configurations are sampled uniformly, one finds either configurations in an exponentially large number of clusters (the dot on the curve for $\alpha = 9.45$, which is the point where $d\Sigma/d\phi = -1$), or a condensed phase where the measure is dominated by a few clusters (the squares on the curves for $\alpha \geq 9.6$). *Right*: complexity $\Sigma(\mathbf{x})$ and free-entropy density $\mathfrak{F}(\mathbf{x})$ at a few key values of $\mathbf{x}$: $\mathbf{x} = 0$ corresponds to the maximum of $\Sigma(\phi)$, $\mathbf{x} = 1$ to the point where $d\Sigma/d\phi = -1$, and $\mathbf{x} = \mathbf{x}_*$ to $\Sigma(\phi) = 0$. The dynamical transition is at $\alpha_d \approx 9.38$, the condensation transition at $\alpha_c \approx 9.547$, and the SAT–UNSAT transition at $\alpha_s \approx 9.931$.

were obtained at a cost: we completely forgot about the size of the clusters of SAT assignments, their 'internal entropy'.

In order to get a finer understanding of the geometry of the set of solutions in the SAT phase, we need to get back to the uniform measure over SAT assignments of eqn (20.1), and use the 1RSB method of Section 19.2. Our task is, in principle, straightforward: we need to estimate the 1RSB free entropy $\mathfrak{F}(\mathbf{x})$, and evaluate the Legendre transform (19.8) in order to get the complexity function $\Sigma(\phi)$. Recall that $\Sigma(\phi)$ is the exponential growth rate of the number of clusters with free entropy $N\phi$ (in the present case, since we are restricting ourselves to SAT configurations, the free entropy of a cluster is equal to its entropy).

This is a rather demanding task from the numerical point of view. Let us understand why: each BP message is parameterized by one real number in $[0, 1]$, as we saw in eqn (20.3). A 1RSB message characterizes the distribution of this number, so it is a pdf on $[0, 1]$. One such distribution is associated with each directed edge of the factor graph. For a study of the phase diagram, one needs to perform a statistical analysis of the 1RSB messages. Within the population dynamics approach, this means that we must use a (large) population of distribution functions. For each value of $\mathbf{x}$, the algorithm must be run for a large enough number of iterations to estimate $\mathfrak{F}(\mathbf{x})$. This is at the limit of what can be done numerically. Fortunately, it can be complemented by two simpler computations: the SP approach, which gives the results corresponding to $\mathbf{x} = 0$, and a study of the $\mathbf{x} = 1$ case using the simplification described in Section 19.4.

20.3.2   The condensation transition for $K \geq 4$

We shall not provide any technical details of these computations, but focus instead on the main results using $K = 4$-SAT as a running example. As shown by Fig. 20.9, this system displays the full scenario of phase transitions described in Section 19.6. Upon increasing the clause density α, one finds first an RS phase for $\alpha < \alpha_d$, then a d1RSB phase with exponentially many relevant states for $\alpha_d < \alpha < \alpha_c$, and then an s1RSB phase with condensation of the measure on a few states for $\alpha_c < \alpha < \alpha_s$. The system becomes UNSAT for $\alpha > \alpha_s$.

Figure 20.9 shows the evolution of the complexity versus the internal entropy density of the clusters as α increases (note that increasing α plays the same role as decreasing the temperature in the general scenario sketched in Fig. 19.6). For a given α, almost all clusters have an internal entropy density ϕ_0 corresponding to the maximum of $\Sigma(\phi)$. The complexity at the maximum, $\Sigma(\phi_0) = \mathfrak{F}(\mathrm{x} = 0)$, is equal to the complexity at zero energy density that we found by the energetic 1RSB cavity method. When SAT configurations are sampled uniformly, almost all of them are found in clusters of internal entropy density ϕ_1 such that $\Sigma(\phi) + \phi$ is maximum, conditional on the fact that $\Sigma(\phi) \geq 0$. In the d1RSB phase, one has $\Sigma(\phi_1) > 0$, and in the s1RSB phase one has $\Sigma(\phi_1) = 0$. The condensation point α_c can therefore be found through a direct (and more precise) study at $\mathrm{x} = 1$. Indeed it is identified as the value of the clause density such that the two equations $\Sigma(\phi) = 0$ and $\mathrm{d}\Sigma/\mathrm{d}\phi = -1$ admit a solution.

Exercise 20.8 Using the Legendre transform in eqn (19.8), show that this condensation point α_c is the point where the 1RSB free-entropy function $\mathfrak{F}(\mathrm{x})$ satisfies $\mathfrak{F}(1) - \mathfrak{F}'(1) = 0$ (where the prime means a derivative with respect to x). As we saw in Section 19.4, the value of $\mathfrak{F}(1)$ is equal to the RS free entropy. The value of the internal entropy $\mathfrak{F}'(1)$ can be obtained explicitly from the $\mathrm{x} = 1$ formalism. Writing down the full $\mathrm{x} = 1$ formalism for random satisfiability, including this computation of $\mathfrak{F}'(1)$, is an interesting (non-trivial) exercise.

The dynamical transition point α_d is defined as the smallest value of α such that there exists a non-trivial solution to the 1RSB equations at $\mathrm{x} = 1$ (in practice, it is best studied using the point-to-set correlation, which will be described in Chapter 22). Notice from Fig. 20.9 that there can exist clusters of SAT assignments even at $\alpha < \alpha_d$: for $\alpha = 4.3$, there exists a branch of $\Sigma(\phi)$ around the point ϕ_0 where it is maximum, but this branch disappears, if one increases ϕ, before one can find a point where $\mathrm{d}\Sigma/\mathrm{d}\phi = -1$. The interpretation of this regime is that an exponentially small fraction of the solutions are grouped into well-separated clusters. The vast majority of the solutions belong instead to a single, well-connected 'replica-symmetric' cluster. As we saw in the case of the energetic cavity method, the first occurrence of the clusters around ϕ_0 occurs at the value $\alpha_{d,SP}$, which is around 8.3 for 4-SAT.

The same scenario has been found in studies of random K-SAT with $K = 5$ and 6, and it is expected to hold for all $K \geq 4$. The situation is somewhat different for $K = 3$, as the condensation point α_c coincides with α_d: the 1RSB phase is always condensed. Table 20.2 summarizes the values of the thresholds.

Table 20.2 Predictions of the 1RSB cavity method for the non-trivial SP, dynamical, condensation, and SAT–UNSAT thresholds of random K-satisfiability.

K	α_d	α_c	α_s
3	3.86	3.86	4.2667
4	9.38	9.547	9.931
5	19.16	20.80	21.117
6	36.53	43.08	43.37

20.4   An exercise: Colouring random graphs

Recall that a proper q-colouring of a graph $\mathcal{G} = (\mathcal{V}, \mathcal{E})$ is an assignment of colours $\{1, \ldots, q\}$ to the vertices of q in such a way that no edge has its two adjacent vertices of the same colour. Hereafter, we shall refer to a proper q-colouring as simply a 'colouring' of $\mathcal{G}$. Colourings of a random graph can be studied following the approach just described for satisfiability, and reveal a strikingly similar behaviour. Here we shall just present some key steps of this analysis: this section can be seen as a long exercise in applying the cavity method. We shall focus on the case of random regular graphs, which is technically simpler. In particular, many results can be derived without resorting to a numerical solution of the cavity equations. The reader is encouraged to work out the many details which have been left out.

We shall adopt the following description of the problem: with each vertex $i \in \mathcal{V}$ of a graph $G = (\mathcal{V}, \mathcal{E})$, we associate a variable $x_i \in \{1, \ldots, q\}$. The energy of a colour assignment $\underline{x} = \{x_1, \ldots, x_N\}$ is given by the number of edges whose vertices have the same colour:

$$E(\underline{x}) = \sum_{(ij) \in \mathcal{E}} \mathbb{I}(x_i = x_j) \, . \tag{20.27}$$

If the graph is colourable, we are also interested in the uniform measure over proper colourings:

$$\mu(\underline{x}) = \frac{1}{Z} \, \mathbb{I}(E(\underline{x}) = 0) = \frac{1}{Z} \prod_{(ij) \in \mathcal{E}} \mathbb{I}(x_i \neq x_j) \, , \tag{20.28}$$

where Z is the number of proper colourings of $\mathcal{G}$. The factor graph associated with $\mu(\cdot)$ is easily constructed. We associate one variable node with each vertex of $i \in \mathcal{G}$ and one function node with each edge $(ij) \in \mathcal{C}$, and connect this function node to the two variable nodes corresponding to i and j. The probability distribution $\mu(\underline{x})$ is therefore a pairwise graphical model.

We shall assume that $\mathcal{G}$ is a random regular graph of degree c. Equivalently, the corresponding factor graph is distributed according to the ensemble $\mathbb{D}_N(\Lambda, P)$, with $\Lambda(x) = x^c$ and $P(x) = x^2$. The important technical simplification is that, for any fixed r, the radius-r neighbourhood around a random vertex i is, with high probability, a tree of degree c, i.e. it is non-random. In other words, the neighbourhood of most of the nodes is the same.

Let us start with the RS analysis of the graphical model (20.28). As we saw in Section 14.2.5, we can get rid of function-to-variable-node messages, and work with variable-to-function-node messages $\nu_{i \to j}(x_i)$. The BP equations read

$$\nu_{i\to j}(x) \cong \prod_{k\in\partial i\setminus j}(1-\nu_{k\to i}(x)) \ . \tag{20.29}$$

Because of the regularity of the graph, there exist solutions of these equations such that messages take the same value on all edges. In particular, eqn (20.29) admits the solution $\nu_{i\to j}(\,\cdot\,) = \nu_{\text{unif}}(\,\cdot\,)$, where $\nu_{\text{unif}}(\,\cdot\,)$ is the uniform message: $\nu_{\text{unif}}(x) = 1/q$ for $x \in \{1,\dots,q\}$. The corresponding free-entropy density (equal here to the entropy density) is

$$f^{\text{RS}} = \log q + \frac{c}{2}\log\left(1-\frac{1}{q}\right) \ . \tag{20.30}$$

It can be shown that this coincides with the 'annealed' estimate $N^{-1}\log\mathbb{E}Z$. It decreases with the degree c of the graph and becomes negative for c larger than $c_{\text{UB}}(q) \equiv 2\log q/\log(q/(q-1))$, similarly to what we saw in Fig. 20.2. The Markov inequality implies that, with high probability, a random c-regular graph does not admit a proper q-colouring for $c > c_{\text{UB}}(q)$. Further, the RS solution is certainly incorrect for $c > c_{\text{UB}}(q)$.

A stability analysis of this solution shows that the spin glass susceptibility diverges as $c \uparrow c_{\text{st}}(q)$, with $c_{\text{st}}(q) = q^2 - 2q + 2$. For $q \geq 4$, $c_{\text{st}}(q) > c_{\text{UB}}(q)$.

In order to correct the above inconsistencies, we have to resort to the energetic 1RSB approach. Let us focus on the $y \to \infty$ limit (or, equivalently, the zero-energy limit). In this limit, we obtain the SP equations. These can be written in terms of messages $Q_{i\to j}(\,\cdot\,)$ that have the following interpretation:

- $Q_{i\to j}(x) = $ probability that, in the absence of (i,j), x_i is forced to the value x,
- $Q_{i\to j}(*) = $ probability that, in the absence of (i,j), x_i is not forced.

Recall that 'probability' is interpreted here with respect to a random Bethe state.

An SP equation expresses the message $Q_{i\to j}(\,\cdot\,)$ in terms of the $c-1$ incoming messages $Q_{k\to i}(\,\cdot\,)$, with $k \in \partial i \setminus j$. To keep the notation simple, we fix an edge $i \to j$ and denote it by 0, and use $1,\dots,c-1$ to label the edges $k \to i$ with $k \in \partial i \setminus j$. Then, for any x in $\{1,\dots,q\}$, we have

$$Q_0(x) = \frac{\sum_{(x_1\dots x_{c-1})\in\mathcal{N}(x)} Q_1(r_1)Q_2(x_2)\cdots Q_{c-1}(x_{c-1})}{\sum_{(x_1\dots x_{c-1})\in\mathcal{D}} Q_1(r_1)Q_2(x_2)\cdots Q_{c-1}(x_{c-1})} \ , \tag{20.31}$$

where:

- $\mathcal{D}$ is the set of tuples $(x_1,\dots,x_{c-1}) \in \{*,1,\dots,q\}^n$ such that there exists $z \in \{1,\dots,q\}$ with $z \neq x_1,\dots,x_{c-1}$. According to the interpretation above, this means that there is no contradiction among the warnings to i.
- $\mathcal{N}(x)$ is the set of tuples $(x_1,\dots,x_{c-1}) \in \mathcal{D}$ such that, for any $z \neq x$, there exists $k \in \{1,\dots,c-1\}$ such that $x_k = z$. In other words, x is the only colour for vertex i that is compatible with the warnings.

$Q_0(*)$ is determined by the normalization condition $Q_0(*) + \sum_x Q_0(x) = 1$.

On a random regular graph of degree c, these equations admit a solution with $Q_{i\to j}(\,\cdot\,) = Q(\,\cdot\,)$ independent of the edge (i,j). Furthermore, if we assume this solution

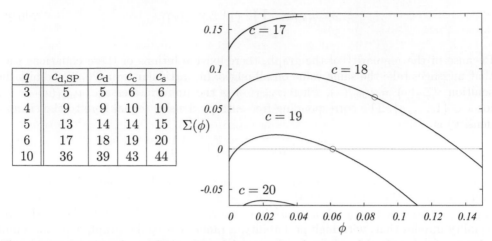

q	$c_{\mathrm{d,SP}}$	c_{d}	c_{c}	c_{s}
3	5	5	6	6
4	9	9	10	10
5	13	14	14	15
6	17	18	19	20
10	36	39	43	44

Fig. 20.10 Results of a 1RSB analysis of proper q-colourings of random regular graphs. The table gives the thresholds: the threshold for the appearance of non-trivial SP solutions $c_{\mathrm{d,SP}}$, the dynamical threshold c_{d}, the condensation threshold c_{c}, and the threshold for the colourable/uncolourable transition c_{s}. The figure on the *right* shows the complexity of the clusters as a function of their internal entropy density. Here, $q = 6$ and the degrees of the graphs are $c = 17$ (RS), $c = 18$ (d1RSB), $c = 19$ (s1RSB) and $c = 20$ (uncolourable). The circles denote the points of slope -1 on the complexity curves.

to be symmetric under permutation of colours, the corresponding message can be parameterized by a single number $a \in [0, 1/q]$:

$$Q(x) = a \text{ for } x \in \{1, \ldots, q\} ,$$
$$Q(*) = 1 - qa . \tag{20.32}$$

Plugging this Ansatz into eqn (20.31), we get

$$a = \frac{\sum_{r=0}^{q-1} (-1)^r \binom{q-1}{r} (1 - (r+1)a)^{c-1}}{\sum_{r=0}^{q-1} (-1)^r \binom{q}{r+1} (1 - (r+1)a)^{c-1}} . \tag{20.33}$$

The complexity $\Sigma^{\mathrm{e}}(\epsilon = 0)$, yielding the exponential growth rate of the number of clusters of proper colourings, is given by $\Sigma^{\mathrm{e}}(e = 0) = \lim_{y \to \infty} \mathfrak{F}^{\mathrm{e}}(y)$. We find that

$$\Sigma^{\mathrm{e}}(\epsilon = 0; c, q) = \log \left(\sum_{r=0}^{q-1} (-1)^r \binom{q}{r+1} (1 - (r+1)a)^c \right) - \frac{c}{2} \log(1 - qa^2) . \tag{20.34}$$

Given the number of colours q, one can study what happens when the degree c grows (which amounts to increasing the density of constraints). The situation is very similar to that found for the satisfiability problem. For $c \geq c_{\mathrm{d,SP}}(q)$, there exists a pair of non-trivial solution to eqn (20.33) with $a > 0$. The complexity $\Sigma^{\mathrm{e}}(e = 0)$ can be computed from eqn (20.34) (evaluated for the largest solution a of eqn (20.33)), and is decreasing in c. It becomes negative for $c \geq c_s(q)$. The degree $c_s(q)$ is thus the 1RSB prediction for the SAT–UNSAT threshold.

When $c < c_s(q)$, the uniform measure over valid colourings can be studied, and, in particular, one can characterize the distribution of the entropy of the clusters. Figure 20.10 shows the complexity as a function of the internal entropy density of the clusters. The similarity to Fig. 20.9 is obvious. One can define two particularly relevant thresholds: c_d is the smallest degree such that the 1RSB equations at $x = 1$ have a non-trivial solution, and c_c is the smallest degree such that the uniform measure over proper colourings is 'condensed'. The table in Fig. 20.10 gives some examples of these thresholds. An asymptotic analysis for large q shows that

$$c_{d,SP} = q(\log q + \log \log q + 1 - \log 2 + o(1)) , \tag{20.35}$$

$$c_d = q(\log q + \log \log q + O(1)) , \tag{20.36}$$

$$c_c = 2q \log q - \log q - 2 \log 2 + o(1) , \tag{20.37}$$

$$c_s = 2q \log q - \log q - 1 + o(1) . \tag{20.38}$$

These predictions can be rephrased into a statement on the **chromatic number**, i.e. the minimal number of colours needed to colour a graph. Because of the heuristic nature of the approach, we formulate it as a conjecture.

Conjecture 20.1 *With high probability, the chromatic number of a random regular graph with N vertices and degree $c \geq 4$ is equal to $\chi_{\mathrm{chrom}}(c)$, where*

$$\chi_{\mathrm{chrom}}(c) = \max\{q : \ \Sigma^e(\epsilon = 0; c, q) > 0\} . \tag{20.39}$$

Here $\Sigma^e(\epsilon = 0; c, q)$ is given by eqn (20.34), where a is the largest solution of eqn (20.33) in the interval $[0, 1/q]$.

Using the numbers in Fig. 20.10, this conjecture predicts, for instance, that $\chi_{\mathrm{chrom}}(c) = 3$ for $c = 4, 5$, $\chi_{\mathrm{chrom}}(c) = 4$ for $c = 6, 7, 8, 9$, and $\chi_{\mathrm{chrom}}(c) = 5$ for $10 \leq c \leq 14$.

On the side of rigorous results, a clever use of the first- and second- moment methods allows one to prove the following result.

Theorem 20.2 *With high probability, the chromatic number of a random regular graph with N vertices and degree c is either k, $k + 1$ or $k + 2$, where k is the smallest integer such that $c < 2k \log k$. Furthermore, if $c > (2k - 1) \log k$, then with high probability the chromatic number is either k or $k + 1$.*

One can check explicitly that the results of the 1RSB cavity conjecture agree with this theorem, which proves the correct leading behaviour at large c.

Although this presentation was focused on random regular graphs, a large class of random graph ensembles can be analysed along the same lines.

Notes

Random K-satisfiability was first analysed using the replica-symmetric cavity method by Monasson and Zecchina (1996, 1997). The resulting equations are equivalent to a density evolution analysis of belief propagation. BP was used as an algorithm for finding SAT assignments by Pumphrey (2001). That study concluded that BP was ineffective in solving satisfiability problems, mainly because it assigned variables in a one-shot fashion, unlike the case when it is used in decimation.

The 1RSB cavity method was applied to random satisfiability in Mézard et al. (2002) and Mézard and Zecchina (2002), where the value of α_c was computed for 3-SAT. This approach was applied to larger K by Mertens et al. (2006), who also derived the large-K asymptotics. The SP(y) and SP equations for satisfiability were first written down in Mézard and Zecchina (2002), where SP-guided decimation was introduced. A more algorithmic presentation of SP was then developed by Braunstein et al. (2005), together with an optimized source code for SP and decimation (Braunstein et al., 2004). The idea of backtracking was suggested by Parisi (2003), but its performance has not been systematically studied yet.

The condensation phenomenon was discussed by Krzakala et al. (2007), in relation to studies of the entropic complexity in the colouring problem (Mézard et al., 2005b; Krzakala and Zdeborová, 2007) and in the satisfiability problem (Montanari et al., 2008).

The analysis in this chapter was heuristic, and is waiting for a rigorous proof. Let us point out that one important aspect of the whole scenario has been established rigorously for $K \geq 8$: it has been shown that in some range of clause density below $\alpha_s(K)$, the SAT assignments are grouped into exponentially many clusters, well separated from each other (Mézard et al., 2005a; Achlioptas and Ricci-Tersenghi, 2006; Daudé et al., 2008). This result can be obtained by a study of the 'x-satisfiability' problem, which requires one to determine whether a formula has *two* SAT assignments differing in xN variables. Bounds on the x-satisfiability threshold can be obtained through the first- and second- moment methods.

The colouring problem was first studied with the energetic 1RSB cavity method by Mulet et al. (2002) and Braunstein et al. (2003): these papers contain the derivation of the SAT/UNSAT threshold and the SP equations. A detailed study of the entropy of clusters, and the computation of the other thresholds, was carried out by Krzakala and Zdeborová (2007). These papers also study the case of Erdös–Rényi graphs. Theorem 20.2 was proven by Achlioptas and Moore (2004), and its analogue for Erdös–Rényi graphs by Achlioptas and Naor (2005). The validity of the RS solution for colouring graphs of low enough degree was proven by Bandyopadhyay and Gamarnik (2006). Linear relaxations of satisfiability have been studied by Gamarnik (2004).

21

Glassy states in coding theory

In Chapter 15, we studied the problem of decoding random LDPC codes, and found two phase transitions that characterize the performance of the code in the large-block-length limit. Consider, for instance, communication over a binary symmetric channel with crossover probability p. Under belief propagation decoding, the bit error rate vanishes in the large-block-length limit below a first threshold p_d and remains strictly positive for $p > p_d$. On the other hand, the minimal bit error rate achievable with the same ensemble (i.e. the bit error rate under symbol MAP decoding) vanishes up to a larger noise level p_c and is bounded away from 0 for $p > p_c$.

In principle, one should expect every decoding algorithm to have a different threshold. This suggests that we should not attach too much importance to the BP threshold p_d. On the contrary, we shall see in this chapter that p_d is, in some sense, a 'universal' characteristic of the code ensemble: above p_d, the decoding problem is plagued by an exponential number of metastable states (Bethe measures). In other words, the phase transition which takes place at p_d is not only algorithmic, it is a *structural* phase transition. This transition turns out to be a dynamical 1RSB glass transition, and this suggests that p_d is the largest possible threshold for a large class of local decoding algorithms.

We have already seen, in the last section of Chapter 15, that the two thresholds p_d and p_c are closely related and can both be computed formally within the RS cavity method, i.e. in terms of the density-evolution fixed point. The analysis below will provide a detailed explanation of this connection in terms of the glass transition studied in Chapter 19.

In Section 21.1, we start by considering a numerical investigation of the role of metastable states in decoding. Section 21.2 considers the particularly instructive case of a binary erasure channel, where the glassy states can be analysed relatively easily using the energetic 1RSB cavity method. The analysis of general memoryless channels is described in Section 21.3. Finally, Section 21.4 describes the connection between metastable states, which are one of the main subjects of study in this chapter, and trapping sets (subgraphs of the original factor graph that are often regarded as responsible for coding failures).

21.1 Local search algorithms and metastable states

The codewords of an LDPC code are solutions of a constraint satisfaction problem. The variables are the bits of a word $\underline{x} = (x_1, x_2, \ldots, x_N)$, with $x_i \in \{0, 1\}$, and the constraints are the parity check equations, i.e. a set of linear equations mod 2. This is analogous to the XORSAT problem considered in Chapter 18, although the ensembles of linear systems used in coding are different.

An important difference from XORSAT is that we are looking for a *specific* solution of the linear system, namely the transmitted codeword. The received message $\underline{y}$ gives us a hint of where to look for this solution. For notational simplicity, we shall assume that the output alphabet $\mathcal{Y}$ is discrete, and the channel is a binary-input, memory-less, output-symmetric (BMS) channel (see Chapter 15)) with transition probability[1] $\mathcal{Q}(y|x)$. The probability that $\underline{x}$ is the transmitted codeword, given the received message $\underline{y}$, is given by the usual formula (15.1), $\mathbb{P}(\underline{x}|\underline{y}) = \mu_{\underline{y}}(\underline{x})$, where

$$\mu_{\underline{y}}(\underline{x}) \cong \prod_{i=1}^{N} \mathcal{Q}(y_i|x_i) \prod_{a=1}^{M} \mathbb{I}(x_{i_1^a} \oplus \cdots \oplus x_{i_{k(a)}^a} = 0). \qquad (21.1)$$

It is natural to associate an optimization problem with the code. We define the energy $E(\underline{x})$ of a word $\underline{x}$ (also called a 'configuration') as *twice* the number of parity check equations violated by $\underline{x}$ (the factor 2 is introduced for future simplification). Codewords coincide with the global minima of this energy function, with zero energy.

We already know that decoding consists in computing marginals of the distribution $\mu_{\underline{y}}(\underline{x})$ (symbol MAP decoding) or finding its argmax (word MAP decoding). In the following, we shall discuss two closely related problems: (*i*) optimizing the energy function $E(\underline{x})$ within a subset of the configuration space defined by the received word and the channel properties; and (*ii*) sampling from a 'tilted' Boltzmann distribution associated with $E(\underline{x})$.

21.1.1 Decoding by constrained optimization

Let us start by considering the word MAP decoding problem. We shall exploit our knowledge of BMS channels. Conditional on the received word $\underline{y} = (y_1, y_2, \ldots, y_N)$, the log-likelihood for $\underline{x}$ to be the channel input is

$$L_{\underline{y}}(\underline{x}) = \sum_{i=1}^{N} \log \mathcal{Q}(y_i|x_i). \qquad (21.2)$$

We shall later use the knowledge that the input word was a codeword, but $L_{\underline{y}}(\underline{x})$ is well defined for any $\underline{x} \in \{0, 1\}^N$, regardless of whether it is a codeword or not, so let us first characterize its properties.

Assume, without loss of generality, that the codeword $\underline{0}$ was transmitted. By the law of large numbers, for large N the log-likelihood of this codeword is close to $-Nh$, where h is the channel entropy: $h = -\sum_y \mathcal{Q}(y|0) \log \mathcal{Q}(y|0)$. The probability of an order-N deviation away from this value is exponentially small in N. This suggests that we should look for the transmitted codeword among those $\underline{x}$ such that $L_{\underline{y}}(\underline{x})$ is close to h.

The corresponding **typical-pairs** decoding strategy goes as follows. Given the channel output $\underline{y}$, look for a codeword $\underline{x} \in \mathfrak{C}$ such that $L_{\underline{y}}(\underline{x}) \geq -N(h + \delta)$. We shall

---

[1]Throughout this chapter, we adopt a different notation for the channel transition probability from that in the rest of the book, in order to avoid confusion with 1RSB messages.

refer to this condition as the 'distance constraint'. For instance, in the case of a binary symmetric channel, it amounts to constraining the Hamming distance between the codeword $\underline{x}$ and the received codeword $\underline{y}$ to be small enough. If exactly one codeword satisfies the distance constraint, return it. If there is no such codeword, or if there are several of them, declare an error. Here $\delta > 0$ is a parameter of the algorithm, which should be thought of as going to 0 *after* $N \to \infty$.

Exercise 21.1 Show that the block error probability of typical-pairs decoding is independent of the transmitted codeword.
[Hint: Use the linear structure of LDPC codes, and the symmetry property of a BMS channel.]

Exercise 21.2 This exercise is aimed at convincing the reader that typical-pairs decoding is 'essentially' equivalent to maximum-likelihood (or MAP) decoding.

(a) Show that the probability that no codeword exists with $L_{\underline{y}}(\underline{x}) \in [-N(h+\delta), -N(h-\delta)]$ is exponentially small in N.
[Hint: Apply Sanov's Theorem, (see Section 4.2), to the type of the received codeword.]

(b) Obtain an upper bound for the probability that maximum-likelihood decoding succeeds and typical-pairs decoding fails in terms of the probability that there exists an incorrect codeword $\underline{x}$ with $L_{\underline{y}}(\underline{x}) \geq -N(h+\delta)$, but no incorrect codeword $L_{\underline{y}}(\underline{x}) \geq -N(h-\delta)$.

(c) Estimate the last probability for Shannon's random code ensemble. Show, in particular, that it is exponentially small for all noise levels strictly smaller than the MAP threshold and δ small enough.

Since codewords are global minima of the energy function $E(\underline{x})$, we can rephrase typical-pairs decoding as an optimization problem:

$$\text{Minimize} \quad E(\underline{x}) \quad \text{subject to} \quad L_{\underline{y}}(\underline{x}) \geq -N(h + \delta) \,. \tag{21.3}$$

Neglecting exponentially rare events, we know that there always exists at least one solution with cost $E(\underline{x}) = 0$, corresponding to the transmitted codeword. Therefore, typical-pairs decoding is successful if and only if the minimum is non-degenerate. This happens with high probability for $p < p_{\mathrm{c}}$. In contrast, for $p > p_{\mathrm{c}}$, the optimization admits other minima with zero cost (incorrect codewords). We have already explored this phenomenon in Chapters 11 and 15, and we shall discuss it further below. For $p > p_{\mathrm{c}}$, there exists an exponential number of codewords whose likelihood is greater than or equal to the likelihood of the transmitted one.

Similarly to what we have seen in other optimization problems (such as MAX-XORSAT and MAX-SAT), there exists generically an intermediate regime $p_{\mathrm{d}} < p < p_{\mathrm{c}}$, which is characterized by an exponentially large number of metastable states. For these values of p, the global minimum of $E(\underline{x})$ is still the transmitted codeword, but is 'hidden' by the proliferation of deep local minima. Remarkably, the threshold for the appearance of an exponential number of metastable states coincides with the BP threshold p_d. Thus, for $p \in]p_\mathrm{d}, p_\mathrm{c}[$, MAP decoding would be successful, but message-passing decoding fails. In fact, no practical algorithm which succeeds in this regime is known. A 'cartoon' of this geometrical picture is presented in Fig. 21.1.

At this point, the reader might be puzzled by the observation that finding configurations with $E(\underline{x}) = 0$ is, per se, a polynomial task. Indeed, it amounts to solving a linear system modulo 2, and can be done by Gaussian elimination. However, the problem (21.3) involves the condition $L_y(\underline{x}) \geq -N(h + \delta)$, which is *not* a linear constraint modulo 2. If one resorts to local-search-based decoding algorithms, the proliferation of metastable states for $p > p_\mathrm{d}$ can block the algorithm. We shall discuss this phenomenon for two local search strategies: Δ-local search and simulated annealing.

21.1.2 Δ-local-search decoding

A simple local search algorithm consists in starting from a word $\underline{x}(0)$ such that $L_y(\underline{x}(0)) \geq -N(h + \delta)$ and then recursively constructing $\underline{x}(t + 1)$ by optimizing the energy function within a neighbourhood of radius Δ around $\underline{x}(t)$, as in the following pseudocode.

Δ LOCAL SEARCH (channel output y, search size Δ, likelihood resolution δ)
1:   Find $\underline{x}(0)$ such that $L_y(\underline{x}(0)) \geq -N(h + \delta)$;
2:   **for** $t = 0, \dots, t_\mathrm{max} - 1$:
3:       Choose a uniformly random connected set $U \subset \{1, \dots, N\}$
         of variable nodes in the factor graph with $|U| = \Delta$;
4:       Find the configuration $\underline{x}'$ that minimizes the energy subject
         to $x'_j = x_j$ for all $j \notin U$;
5:       If $L_y(\underline{x}') \geq -N(h + \delta)$, set $\underline{x}(t + 1) = \underline{x}'$;
         otherwise, set $\underline{x}(t + 1) = \underline{x}(t)$;
6:   **end;**
7:   return $\underline{x}(t_\mathrm{max})$.

(Recall that a set of variable nodes U is 'connected' if, for any $i, j \in U$, there exists a path in the factor graph connecting i to j such that all variable nodes along the path are in U as well.)

Exercise 21.3 A possible implementation of step 1 consists in setting $x_i(0) = \arg\max_x \mathcal{Q}(y_i | x)$. Show that this choice meets the likelihood constraint.

If the factor graph has bounded degree (which is the case with LDPC ensembles), and Δ is bounded as well, each execution of the cycle above implies a bounded number of operations. As a consequence, if we let $t_\mathrm{max} = O(N)$, the algorithm has linear complexity. A computationally heavier variant consists in choosing U at step 3 greedily. This means going over all such subsets and then taking the one that maximizes the decrease in energy $|E(\underline{x}(t + 1)) - E(\underline{x}(t))|$.

Fig. 21.1 Three possible 'cartoon' landscapes for the energy function $E(\underline{x})$ (the number of violated checks), plotted in the space of all configurations $\underline{x}$ with $L_y(\underline{x}) \geq N(h - \delta)$. *Left*: the energy has a unique global minimum with $E(\underline{x}) = 0$ (the transmitted codeword) and no (deep) local minima. *Centre*: many deep local minima appear, although the global minimum remains non-degenerate. *Right*: more than one codeword is compatible with the likelihood constraint, and the global minimum $E(\underline{x}) = 0$ becomes degenerate.

Obviously, the energy $E(\underline{x}(t))$ of the configuration produced after t iterations is a non-increasing function of t. If it vanishes at some time $t \leq t_{\max}$, then the algorithm implements a typical-pairs decoder. Ideally, one would like a characterization of the noise levels and code ensembles such that $E(\underline{x}(t_{\max})) = 0$ with high probability.

The case $\Delta = 1$ was analysed in Chapter 11, under the name of the 'bit-flipping' algorithm, for communicating over the channel BSC(p). We saw that there exists a threshold noise level p_1 such that if $p < p_1$, the algorithm returns the transmitted codeword with high probability. It is reasonable to expect that the algorithm will be unsuccessful with high probability for $p > p_1$.

Analogously, one can define thresholds p_Δ for each value of Δ. Determining these thresholds analytically is an extremely challenging problem.

One line of approach could consist in first studying Δ**-stable configurations**. We say that a configuration $\underline{x}$ is Δ-stable if, for any configuration $\underline{x}'$ such that $L_y(\underline{x}') \geq -N(h + \delta)$ and $d(\underline{x}, \underline{x}') \leq \Delta$, $E(\underline{x}') \geq E(\underline{x})$.

Exercise 21.4 Show that, if no Δ-stable configuration exists, then the greedy version of the algorithm above will find a codeword after at most M steps (M being the number orf parity checks).

While this exercise hints at a connection between the energy landscape and the difficulty of decoding, one should be aware that the problem of determining p_Δ cannot be reduced to determining whether Δ-stable states exist or to estimating their number. The algorithm in fact fails if, after a number t of iterations, the distribution of $\underline{x}(t)$ is (mostly) supported in the basin of attraction of Δ-stable states. The key difficulty is, of course, to characterize the distribution of $\underline{x}(t)$.

21.1.3 Decoding by simulated annealing

A more detailed understanding of the role of metastable configurations in the decoding problem can be obtained through an analysis of the MCMC decoding procedure that we discussed in Section 13.2.1. We thus soften the parity check constraints through the introduction of an inverse temperature $\beta = 1/T$ (this should not be confused with the temperature introduced in Chapter 6, which instead multiplied the log-likelihood of the codewords). Given the received word y, we define the following distribution over the transmitted message x (see eqn (13.10)):

$$\mu_{y,\beta}(\underline{x}) \equiv \frac{1}{Z(\beta)} \exp\{-\beta E(\underline{x})\} \prod_{i=1}^{N} \mathcal{Q}(y_i|x_i) . \tag{21.4}$$

This is the 'tilted Boltzmann form' that we alluded to before. In the low-temperature limit, it reduces to the familiar a posteriori distribution which we would like to sample: $\mu_{y,\beta=\infty}(\underline{x})$ is supported on the codewords, and gives each of them a weight proportional to its likelihood. At infinite temperature, $\beta = 0$, the distribution factorizes over the bits x_i. More precisely, under $\mu_{y,\beta=0}(\underline{x})$, the bits x_i are independent random variables with marginals $\mathcal{Q}(y_i|x_i)/(\mathcal{Q}(y_i|0) + \mathcal{Q}(y_i|1))$. Sampling from this measure is very easy.

For $\beta \in\ ]0, \infty[$, $\mu_{y,\beta}(\cdot)$ can be regarded as a distribution of possible channel inputs for a code with 'soft' parity check constraints. Note that, unlike the $\beta = \infty$ case, this distribution depends in general on the actual parity check matrix and not just on the codebook $\mathcal{C}$. This is actually a good feature of the tilted measure: the performance of practical algorithms does indeed depend upon the representation of the parity check matrix of $\mathcal{C}$. It is therefore necessary to take it into account.

We sample from $\mu_{y,\beta}(\cdot)$ using Glauber dynamics, (see Section 13.2.1). We have already seen in that section that decoding through sampling at a fixed β fails above a certain noise level. Let us now try to improve on that method using a simulated-annealing procedure in which β is increased gradually according to an annealing schedule $\beta(t)$, with $\beta(0) = 0$. The following decoder uses as input the received word y, the annealing schedule, and some maximal numbers of iterations $t_{\max}$, n.

SIMULATED ANNEALING DECODER $(y, \{\beta(t)\}, t_{\max}, n)$

1:   Generate $\underline{x}_*(0)$ from $\mu_{y,0}(\cdot)$;
2:   **for** $t = 0, \ldots t_{\max} - 1$:
3:       Set $\underline{x}(0; t) = \underline{x}_*(t - 1)$;
4:       Let $\underline{x}(j; t)$, $j \in \{1, \ldots, n\}$ be the configurations produced by
           n successive Glauber updates at $\beta = \beta(t)$;
5:       Set $\underline{x}_*(t) = \underline{x}(n; t)$;
6:   **end**
7:   return $\underline{x}(t_{\max})$.

The algorithmic complexity of the decoder is proportional to the total number of Glauber updates $nt_{\max}$. If we want the algorithm to be efficient, this should grow linearly or slightly superlinearly with N. The intuition is that the first (small-β) steps allow the Markov chain to equilibrate across the configuration space, whereas, as β gets larger, the sample becomes concentrated onto (or near to) codewords. Hopefully, at each stage $\underline{x}_*(t)$ will be approximately distributed according to $\mu_{y,\beta(t)}(\cdot)$.

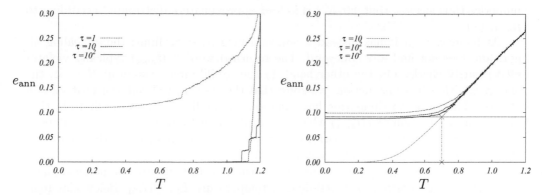

Fig. 21.2 Decoding random codes from the $(5, 6)$ LDPC ensemble using simulated annealing. Here we consider a block length $N = 12\,000$ and transmission over BSC(p) with $p = 0.12$ (*left*) and 0.25 (*right*). The system was annealed through $t_{\max} = 1200$ temperature values equally spaced between $T = 1.2$ and $T = 0$. At each temperature, $n = N\tau$ updates were executed. The statistical errors are comparable to the size of the jumps along the curves.

Figure 21.2 shows the result obtained by use of this simulated-annealing decoder, using random LDPC codes from the $(5, 6)$ regular ensemble, over a binary symmetric channel with crossover probabilities $p = 0.12$ and 0.25 (for this ensemble, $p_d \approx 0.139$ and $p_c \approx 0.264$). The annealing schedule was linear in the temperature, namely $\beta(t) = 1/T(t)$, where

$$T(t) = T(0) - \left\{ T(0) - T(t_{\max}) \right\} \left(\frac{t}{t_{\max}} \right), \tag{21.5}$$

with $T(0) = 1.2$ and $T(t_{\max}) = 0$. The performance of the decoding can be evaluated through the number of violated checks in the final configuration, which is half of $E(\underline{x}(t_{\max}))$. The figure shows the energy density averaged over 10 repetitions of the decoding experiment (each time with a new code randomly chosen from the ensemble), $e(t) = (1/N)\langle E(\underline{x}(t)) \rangle$, versus the temperature $T(t)$. As the number of updates performed at each temperature increases, the number of violated checks per variable seems to converge to a well-defined limiting value that depends on t only through the corresponding temperature:

$$\frac{1}{N} \langle E(\underline{x}(t)) \rangle \to e_{\mathrm{ann}}(\beta(t)). \tag{21.6}$$

Further, $E(\underline{x}(t))/N$ seems to concentrate around its mean as $N \to \infty$.

At small p, $e_{\mathrm{ann}}(\beta)$ quickly converges to 0 as $\beta \to \infty$: a codeword (the transmitted one) is found efficiently. In fact, even at $\beta = 1$, the numerical result for $e_{\mathrm{ann}}(\beta)$ is indistinguishable from 0. We expect that $e_{\mathrm{ann}}(\beta)$ will coincide within the numerical accuracy with the theoretical prediction for the equilibrium average

$$e_{\mathrm{eq}}(\beta) \equiv \frac{1}{N} \lim_{N \to \infty} \langle E(\underline{x}) \rangle_\beta. \tag{21.7}$$

This agrees with the above observations, since $e_{\mathrm{eq}}(\beta) = O(\mathrm{e}^{-10\beta})$ (the lowest excitation over the ground state amounts to flipping a single bit; its energy is equal to 10). The numerics thus suggest that $\underline{x}(t_{\max})$ is indeed approximately distributed according to $\mu_{y,\beta(t)}(\,\cdot\,)$.

At large p, $e_{\mathrm{ann}}(\beta)$ has instead a non-vanishing $\beta \to \infty$ limit: the annealing algorithm does not find any codeword. The returned word $\underline{x}_*(t_{\max})$ typically violates $\Theta(N)$ parity checks. On the other hand, in the equilibrated system at $\beta = \infty$, the energy vanishes by construction (we know that the transmitted codeword satisfies all checks). Therefore the simulation has fallen out of equilibrium at some finite β, thus yielding a distribution of $\underline{x}(t_{\max})$ which is very different from $\mu_{y,\beta=\infty}(\,\cdot\,)$. The data in Fig. 21.2 show that the energy varies very slowly at low temperatures, which confirms the fact that the system is out of equilibrium.

We shall argue below that this slowing down is in fact due to a dynamical glass-phase transition occuring at a well-defined temperature $T_{\mathrm{d}} = 1/\beta_{\mathrm{d}}$. Below this temperature, $\underline{x}(t_{\max})$ gets trapped with high probability into a pure state corresponding to a deep local minimum of $E(\underline{x})$ with positive energy, and never reaches a global minimum of the energy (i.e. a codeword).

This is related to the 'energy landscape' picture discussed in the previous subsection. Indeed, the success of the simulated-annealing decoder for $p \leq p_{\mathrm{d}}$ can be understood as follows. At small noise, the 'tilting' factor $\prod_i \mathcal{Q}(y_i|x_i)$ effectively selects a portion of the configuration space around the transmitted codeword (more or less like the likelihood constraint above), and this portion is small enough that there is no metastable state inside it. An interesting aspect of simulated-annealing decoding is that it can be analysed on the basis of a purely static calculation. Indeed, for any $\beta \leq \beta_{\mathrm{d}}$, the system is still in equilibrium and its distribution is simply given by eqn (21.4). Its study, and the determination of β_{d}, will be the subject of the next sections.

Before moving to this analysis, let us make a last remark about simulated annealing: for any finite β, the MCMC algorithm is able to equilibrate if it is iterated a large number of times (a direct consequence of the fact that Glauber dynamics is irreducible and aperiodic). This raises a paradox, as it seems to imply that the annealing energy always coincides with the equilibrium energy, and the system never falls out of equilibrium during the annealing process. The solution to the conundrum is that in the previous discussion, we tacitly assumed that the number of Monte Carlo steps cannot grow exponentially with the system size. To be more precise, one can, for instance, define the annealing energy as

$$e_{\mathrm{ann}}(\beta) \equiv \lim_{t_{\max} \to \infty} \lim_{N \to \infty} \frac{1}{N} \langle E_N(\underline{x}(t_\beta = \lfloor (1 - \beta(0)/\beta) t_{\max} \rfloor))) \rangle, \qquad (21.8)$$

where we have assumed $\beta(t_{\max}) = \infty$ The important point is that the limit $N \to \infty$ is taken before $t_{\max} \to \infty$: in such a case, simulated annealing can be trapped in metastable states.

21.2 The binary erasure channel

If communication takes place over the binary erasure channel $\mathrm{BEC}(\epsilon)$, the analysis of metastable states can be carried out in detail by adopting the point of view of constrained optimization introduced in Section 21.1.1.

Suppose that the all-zero codeword $\underline{x}_* = (0, \ldots, 0)$ has been sent, and let $\underline{y} \in \{0, *\}^N$ be the channel output. We shall denote by $U = U(\underline{y})$ the set of erased bits. The log-likelihood for the word $\underline{x}$ to be the input can take two possible values: $L_{\underline{y}}(\underline{x}) = |U| \log \epsilon$ if $x_i = 0$ for all $i \notin U$, and $L_{\underline{y}}(\underline{x}) = -\infty$ otherwise. Of course, the input codeword belongs to the first set: $L_{\underline{y}}(\underline{x}_*) = |U| \log \epsilon$. The strategy of Section 21.1.1 reduces therefore to minimizing $E(\underline{x})$ (i.e. minimizing the number of violated parity checks) among all configurations $\underline{x}$ such that $x_i = 0$ on all of the non-erased positions.

When the noise ϵ is smaller than the MAP threshold, there is a unique minimum with energy 0, namely the transmitted codeword $\underline{x}_*$. Our aim is to study the possible existence of metastable states, using the energetic cavity method of Section 19.5. This problem is closely related to XORSAT, whose analysis was presented in Chapters 18 and Chapter 19. Once all the non-erased bits have been fixed to $x_i = 0$, decoding amounts to solving a homogeneous system of linear equations for the remaining bits. If one uses a code from the ensemble $\mathrm{LDPC}_N(\Lambda, P)$, the degree profiles of the remaining nodes are $\Lambda(x)$ and $R(x)$, where the probability of a check node to have degree k, R_k, is given in terms of the original P_k by

$$R_k = \sum_{k'=k}^{k_{\max}} P_{k'} \binom{k'}{k} \epsilon^k (1 - \epsilon)^{k'-k} , \qquad (21.9)$$

and the corresponding edge-perspective degree profile is given as usual by $r_k = kR_k / \sum_p pR_p$.

Exercise 21.5 Show that $r(u) = \sum_k r_k u^{k-1} = \rho(1 - \epsilon(1 - u))$.

Assuming as usual that the number of metastable states—solutions of the min-sum equations—of energy Ne grows like $\exp(N\Sigma^{\mathrm{e}}(e))$, we shall use the 1RSB energetic cavity method to compute the energetic complexity $\Sigma^{\mathrm{e}}(e)$. This can be done using the $\mathsf{SP}(y)$ equations on the original factor graph. As our problem involves only hard constraints and binary variables, we can use the simplified formalism of Section 19.5.3. Each min-sum message can take three possible values, 0 (the meaning of which is 'take value 0'), 1 ('take value 1') and * ('you can take any value'). The $\mathsf{SP}(y)$ messages are distributions on these three values or, equivalently, normalized triplets.

21.2.1 The energetic 1RSB equations

Let us now turn to the statistical analysis of these messages. We denote by $Q = (Q_0, Q_1, Q_*)$ the messages from variable to check nodes, and by $\widehat{Q}$ the messages from check to variable nodes. We first note that, if a bit is not erased, then it sends a sure 0 message $Q = (1, 0, 0)$ to all its neighbouring checks. This means that the distribution of Q has a mass of at least $1 - \epsilon$ on sure 0 messages. We can write

$$Q = \begin{cases} (1, 0, 0) & \text{with probability } (1 - \epsilon) , \\ \widehat{Q} & \text{with probability } \epsilon . \end{cases} \qquad (21.10)$$

The distributional equations for $\widetilde{Q}$ and $\widehat{Q}$ can then be obtained exactly as in Sections 19.5 and 19.7.

Exercise 21.6 Show that the distributions of $\widetilde{Q}$ and $\widehat{Q}$ satisfy the equations

$$\widetilde{Q}_\sigma \stackrel{\mathrm{d}}{=} \mathsf{F}_{l,\sigma}(\widehat{Q}^1, \cdots, \widehat{Q}^{l-1}), \tag{21.11}$$

$$\begin{pmatrix} \widehat{Q}_0 \\ \widehat{Q}_1 \\ \widehat{Q}_* \end{pmatrix} \stackrel{\mathrm{d}}{=} \begin{pmatrix} \frac{1}{2}\prod_{i=1}^{k-1}(\widetilde{Q}_0^i + \widetilde{Q}_1^i) + \frac{1}{2}\prod_{i=1}^{k-1}(\widetilde{Q}_0^i - \widetilde{Q}_1^i) \\ \frac{1}{2}\prod_{i=1}^{k-1}(\widetilde{Q}_0^i + \widetilde{Q}_1^i) - \frac{1}{2}\prod_{i=1}^{k-1}(\widetilde{Q}_0^i - \widetilde{Q}_1^i) \\ 1 - \prod_{i=1}^{k-1}(1 - \widetilde{Q}_{*,i}) \end{pmatrix}, \tag{21.12}$$

where we have defined, for $\sigma \in \{0, 1, *\}$,

$$\mathsf{F}_{l,\sigma}(\widehat{Q}^1, \ldots, \widehat{Q}^{l-1}) \equiv \frac{\mathsf{Z}_{l,\sigma}(\{\widehat{Q}^a\})}{\mathsf{Z}_{l,0}(\{\widehat{Q}^a\}) + \mathsf{Z}_{l,1}(\{\widehat{Q}^a\}) + \mathsf{Z}_{l,*}(\{\widehat{Q}^a\})}, \tag{21.13}$$

$$\mathsf{Z}_{l,\sigma}(\{\widehat{Q}^a\}) \equiv \sum_{\Omega_0,\Omega_1,\Omega_*}^{(\sigma)} e^{-y\min(|\Omega_0|,|\Omega_1|)} \prod_{a\in\Omega_0} \widehat{Q}_0^a \prod_{a\in\Omega_1} \widehat{Q}_1^a \prod_{a\in\Omega_*} \widehat{Q}_*^a. \tag{21.14}$$

Here we have denoted by $\sum_{\Omega_0,\Omega_1,\Omega_*}^{(\sigma)}$ the sum over partitions of $\{1,\ldots,l-1\} = \Omega_0 \cup \Omega_1 \cup \Omega_*$ such that $|\Omega_0| > |\Omega_1|$ (for the case $\sigma = 0$), $|\Omega_0| = |\Omega_1|$ (for $\sigma = *$), or $|\Omega_0| < |\Omega_1|$ (for $\sigma = 1$). Furthermore, k and l are random integers, with distributions r_k and λ_l respectively, the $\{\widetilde{Q}^i\}$ are $l-1$ i.i.d. copies of $\widetilde{Q}$, and the $\{\widehat{Q}^a\}$ are $k-1$ i.i.d. copies of $\widehat{Q}$.

Given a solution of the 1RSB equations, we can compute the Bethe free-entropy density $\mathbb{F}^{\mathrm{RSB},\mathrm{e}}(Q, \widehat{Q})$ of the auxiliary problem. Within the 1RSB cavity method, we can estimate the free-entropy density of the auxiliary model using the Bethe approximation as $\mathfrak{F}^{\mathrm{e}}(y) = (1/N)\mathbb{F}^{\mathrm{RSB},\mathrm{e}}(Q, \widehat{Q})$. This gives access to the energetic complexity function $\Sigma^{\mathrm{e}}(e)$ through the Legendre transform $\mathfrak{F}^{\mathrm{e}}(y) = \Sigma^{\mathrm{e}}(e) - ye$. Within the 1RSB cavity method, we can estimate the latter using the Bethe approximation: $\mathfrak{F}^{\mathrm{e}}(y) = f^{\mathrm{RSB},\mathrm{e}}(y)$.

Exercise 21.7 *Computation of the free entropy.* Using eqn (19.91), show that the Bethe free entropy of the auxiliary graphical model is $Nf^{\mathrm{RSB},\mathrm{e}} + o(N)$, where

$$f^{\mathrm{RSB},\mathrm{e}} = -\Lambda'(1)\epsilon\, \mathbb{E}\log z_{\mathrm{e}}(\widetilde{Q}, \widehat{Q}) + \epsilon\, \mathbb{E}\log z_{\mathrm{v}}(\{\widehat{Q}^a\}; l)$$
$$+ \frac{\Lambda'(1)}{P'(1)}\mathbb{E}\log z_{\mathrm{f}}(\{\widetilde{Q}^i\}; k). \tag{21.15}$$

Here the expectations are taken over l (with distribution Λ_l), k (with distribution R_k, defined in eqn (21.9)), $\widetilde{Q}, \widehat{Q}$, and their i.i.d. copies $\widetilde{Q}^i, \widehat{Q}^a$. The contributions of edges (z_{e}), variable nodes (z_{v}) and function nodes (z_{f}) take the form

$$z_{\mathrm{e}}(\widetilde{Q}, \widehat{Q}) = 1 + (\mathrm{e}^{-y} - 1)\left(\widetilde{Q}_0\widehat{Q}_1 + \widetilde{Q}_1\widehat{Q}_0\right),\tag{21.16}$$

$$z_{\mathrm{v}}(\{\widehat{Q}^i\}; l) = \sum_{\Omega_0,\Omega_1,\Omega_*} \prod_{b\in\Omega_0} \widehat{Q}_0^b \prod_{b\in\Omega_1} \widehat{Q}_1^b \prod_{b\in\Omega_*} \widehat{Q}_*^b \, \mathrm{e}^{-y\,\min(|\Omega_0|,|\Omega_1|)},\tag{21.17}$$

$$z_{\mathrm{f}}(\{\widetilde{Q}^i\}; k) = 1 + \frac{1}{2}(\mathrm{e}^{-y} - 1)\left\{\prod_{i=1}^{k}(\widetilde{Q}_0^i + \widetilde{Q}_1^i) - \prod_{i=1}^{k}(\widetilde{Q}_0^i - \widetilde{Q}_1^i)\right\},\tag{21.18}$$

where the sum in the second equation runs over the partitions $\Omega_0 \cup \Omega_1 \cup \Omega_* = [l]$.

21.2.2 The BP threshold and the onset of metastability

A complete study of the distributional equations (21.11) and (21.12) is a rather challenging task. On the other hand, they can be solved approximately through population dynamics. It turns out that the distribution obtained numerically shows different symmetry properties depending on the value of ϵ. We define a distribution $\widetilde{Q}$ (or $\widehat{Q}$) to be 'symmetric' if $\widetilde{Q}_0 = \widetilde{Q}_1$, and 'positive' if $\widetilde{Q}_0 > \widetilde{Q}_1$. We know from the analysis of BP decoding that directed edges in the graph can be divided into two classes: those that eventually carry a message 0 under BP decoding, and those that instead carry a message $*$ even after a BP fixed point has been reached. It is natural to think that edges of the first class correspond to a positive 1RSB message $\widetilde{Q}$ (i.e., even among metastable states, the corresponding bits are biased towards 0), while edges of the second class correspond instead to a symmetric message $\widetilde{Q}$.

This suggests the following hypothesis concerning the distributions of $\widetilde{Q}$ and $\widehat{Q}$. We assume that there exist weights $\xi, \hat{\xi} \in [0,1]$ and random distributions b, $\hat{\mathsf{b}}$, c, $\hat{\mathsf{c}}$, such that b, $\hat{\mathsf{b}}$ are symmetric, c, $\hat{\mathsf{c}}$ are positive, and

$$\widetilde{Q} \stackrel{\mathrm{d}}{=} \begin{cases} \mathsf{b} & \text{with probability } \xi, \\ \mathsf{c} & \text{with probability } 1-\xi, \end{cases}\tag{21.19}$$

$$\widehat{Q} \stackrel{\mathrm{d}}{=} \begin{cases} \hat{\mathsf{b}} & \text{with probability } \hat{\xi}, \\ \hat{\mathsf{c}} & \text{with probability } 1-\hat{\xi}. \end{cases}\tag{21.20}$$

In other words ξ and $\hat{\xi}$ denote the probabilities that Q and $\widehat{Q}$, respectively, are symmetric.

Equation (21.11) shows that, in order for $\widetilde{Q}$ to be symmetric, all of the inputs $\widehat{Q}^i$ must be symmetric. On the other hand, eqn (21.12) implies that $\widehat{Q}$ is symmetric if at least one of the inputs $\widetilde{Q}^a$ is symmetric. Using the result of Exercise 21.5, we thus find that our Ansatz is consistent only if the weights $\xi, \hat{\xi}$ satisfy the equations

$$\xi = \lambda(\hat{\xi}), \quad \hat{\xi} = 1 - \rho(1 - \epsilon\xi).\tag{21.21}$$

If we define $z \equiv \epsilon\xi$ and $\hat{z} \equiv \hat{\xi}$, these coincide with the density-evolution fixed-point conditions for BP (see eqns (15.34)). This is not surprising, in view of the physical discussion which lead us to introduce the Ansatz in eqns (21.19) and (21.20): ξ corresponds to the fraction of edges that remain erased at the BP fixed point. On the

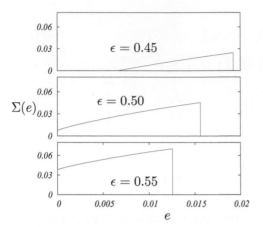
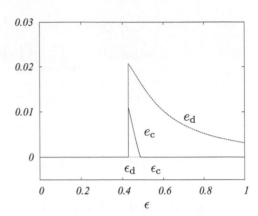

Fig. 21.3 Metastable states for random elements of the $(3,6)$ regular ensemble used over $\mathrm{BEC}(\epsilon)$ (for this ensemble, $\epsilon_{\mathrm{d}} \approx 0.4294$ and $\epsilon_{\mathrm{c}} \approx 0.4882$). *Left*: complexity as a function of the energy density for three values of the channel parameter above ϵ_{d}. *Right*: the maximum and minimum energy densities e_{d} and e_{c} of the metastable states as a function of the erasure probability.

other hand, we shall see that this observation implies that BP ceases to converge to the correct fixed point at the same threshold noise ϵ_{d} where metastable states start to appear.

For $\epsilon \leq \epsilon_{\mathrm{d}}$, eqns (21.21) admit the unique solution $\xi = \hat{\xi} = 0$, corresponding to the fact that BP decoding recovers the full transmitted message. As a consequence, we can take $Q(\cdot) \overset{\mathrm{d}}{=} c(\cdot)$, $\widehat{Q}(\cdot) \overset{\mathrm{d}}{=} \hat{c}(\cdot)$ to have an almost surely positive mean. In fact, it is not hard to check that a consistent solution of eqns (21.11) and (21.12) is obtained by taking

$$\widehat{Q} = \widetilde{Q} = (1, 0, 0) \qquad \text{almost surely.} \tag{21.22}$$

Since the cavity fields do not fluctuate from state to state (their distribution is almost surely a point mass), the structure of this solution indicates that no metastable state is present for $\epsilon \leq \epsilon_{\mathrm{d}}$. This is confirmed by the fact that the free-entropy density of this solution, $\mathfrak{F}^{\mathrm{e}}(y)$, vanishes for all y.

Above a certain noise threshold, i.e. for $\epsilon > \epsilon_{\mathrm{d}}$, eqns (21.21) still possess the solution $\xi = \hat{\xi} = 0$, but a new solution with $\xi, \hat{\xi} > 0$ appears as well. We have discussed this new solution in the density evolution analysis of BP decoding: it is associated with the fact that the BP iterations have a fixed point at which a finite fraction of the bits remain undetermined. Numerical calculations show that that, for $\epsilon > \epsilon_{\mathrm{d}}$, the iteration of eqns (21.11) and (21.12) converges to a non-trivial distribution. In particular, $\widetilde{Q}$ and $\widehat{Q}$ are found to be symmetric with probability $\xi > 0$ and $\hat{\xi} > 0$, respectively, where the values of $\xi, \hat{\xi}$ are the non-trivial solution of eqns (21.21). The free entropy of the auxiliary model, $\mathfrak{F}^{\mathrm{e}}(y)$, can be computed using eqn (21.15). Its Legendre transform is the energetic-complexity curve $\Sigma^{\mathrm{e}}(e)$.

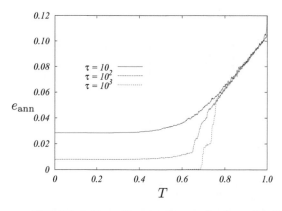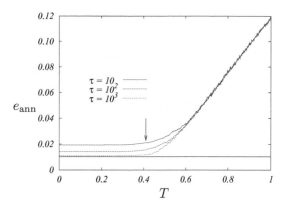

Fig. 21.4 Decoding random codes from the $(3,6)$ regular ensemble used over BEC(ϵ). In both cases, $N = 10^4$, and the annealing schedule consisted of $t_{\max} = 10^3$ equidistant temperatures in $T = 1/\beta \in [0, 1]$. At each value of the temperature, $n = N\tau$ Monte Carlo updates were performed. *Left*, $\epsilon = 0.4 < \epsilon_d$. *Right*, $\epsilon = 0.6 > \epsilon_d$; the horizontal line corresponds to the energy density of the highest metastable states $e_d(\epsilon = 0.6)$.

Figure 21.3 shows a typical outcome of such a calculation for LDPC ensembles when $\epsilon > \epsilon_d$. In this whole regime, there exists a zero-energy word, the transmitted (all-0) codeword. This is described by the solution $\xi = \hat{\xi} = 0$. On top of this, the non-trivial solution gives a complexity curve $\Sigma^e(e)$ which is positive in an interval of energy densities (e_c, e_d). A positive complexity means that an exponential number of metastable states is present. But, when their energy density e is strictly positive, these metastable states violate a finite fraction of the parity checks.

As ϵ increases, both e_d and e_c decrease. At ϵ_c, e_c vanishes continuously, and $e_c = 0$ and $e_d > 0$ for all $\epsilon \geq \epsilon_c$. In other words, at noise levels larger than ϵ_c, there appears an exponential number of zero-energy states. These are codewords that are separated by energy barriers with a height $\Theta(N)$. Consistently with this interpretation, $\Sigma(e = 0) = f_{h,u}^{\mathrm{RS}}$, where $f_{h,u}^{\mathrm{RS}}$ is the RS free-entropy density (15.48) estimated for the non-trivial fixed point of density evolution.

The notion of metastable states thus allows us to compute the BP and MAP thresholds within a unified framework. The BP threshold is the noise level where an exponential number of metastable states appears. This shows that this threshold not only is associated with a specific decoding algorithm, but also has a structural, geometric meaning. On the other hand, the MAP threshold coincides with the noise level where the energy of the lowest-lying metastable states vanishes.

Figure 21.4 shows the results of some numerical experiments with the simulated-annealing algorithm of Section 21.1.3. Below the BP threshold, and for a slow enough annealing schedule, the algorithm succeeds in finding a codeword (a zero-energy state) in linear time. Above the threshold, even at the slowest annealing rate, we could not find a codeword. Furthermore, the residual energy density at zero temperature is close to e_d, suggesting that the optimization procedure is indeed trapped among the highest metastable states. This suggestion is confirmed by Fig. 21.5, which compares the ϵ dependence of e_d with the residual energy under simulated annealing. Once again,

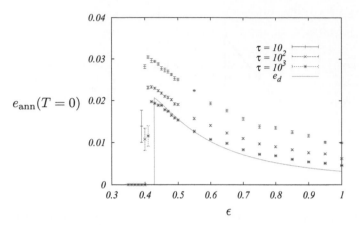

Fig. 21.5 Decoding random codes from the $(3,6)$ regular ensemble used over BEC(ϵ). Here, we plot the minimum energy density achieved through simulated annealing versus the channel parameter. The continuous line shows the energy of the highest-lying metastable states. The size and annealing schedule were as in Fig. 21.4.

there is rough agreement between the two (let us stress that one should not expect perfect agreement between the residual energy in Fig. 21.5 and e_{d}: the former does indeed depend on the whole dynamical annealing process).

21.3   General binary memoryless symmetric channels

One would like to generalize to other channel models the above analysis of metastable states in the constrained-optimization formulation of decoding. In general, the computation is technically more intricate than for a BEC. The reason is that, for general channels, the distance condition $L_{\underline{y}}(\underline{x}) \geq -N(h+\delta)$ cannot be written in terms of 'local' binary constraints. As a consequence, one cannot use the simplified approach of Section 19.5.3, and the general 1RSB formalism is required.

We shall follow this line of approach, but rather than push it to the point of determining the full complexity function, we shall only determine whether the model (21.4) undergoes a dynamical phase transition as β increases from 0 to ∞, and locate the critical point $\beta_{\mathrm{d}}(p)$ (here p denotes the channel parameter). This is, indeed, the most important piece of information for our purposes. If a dynamical phase transition occurs at some $\beta_{\mathrm{d}} < \infty$, then for $\beta > \beta_{\mathrm{d}}$ the measure (21.4) decomposes into an exponential number of metastable pure states. As β crosses β_{d}, the system is trapped in one of these and falls out of equilibrium. Upon further cooling (increase of β), the energy density of the annealed system remains higher than the equilibrium energy density and does not vanish as $\beta \to \infty$. This analysis allows us to determine the noise threshold of the simulated-annealing decoder as the largest noise level p such that there is no finite β_{d}.

In the following, we first write the general 1RSB equations at finite β, and present some results obtained by solving them numerically. Finally, we give a heuristic argument showing that $\beta_{\mathrm{d}}(p)$ goes to infinity for exactly $p \downarrow p_{\mathrm{d}}$.

21.3.1 The 1RSB cavity approach

We shall apply the 1RSB cavity approach of Chapter 19 to the decoding problem. Given a code and a received message $\underline{y}$, we want to study the probability distribution $\mu_{y,\beta}(\underline{x})$ defined in eqn (21.4), and understand whether it decomposes into exponentially many extremal Bethe measures. The BP equations are simple generalizations of those written in Chapter 15 for the case $\beta = \infty$. In terms of the log-likelihoods

$$h_{i \to a} = \frac{1}{2} \log \frac{\nu_{i \to a}(0)}{\nu_{i \to a}(1)}, \qquad u_{a \to i} = \frac{1}{2} \log \frac{\widehat{\nu}_{a \to i}(0)}{\widehat{\nu}_{a \to i}(1)},$$

$$B_i = \frac{1}{2} \log \frac{\mathcal{Q}(y_i|0)}{\mathcal{Q}(y_i|1)} \equiv B(y_i), \tag{21.23}$$

they read

$$h_{i \to a} = B_i + \sum_{b \in \partial i \setminus a} u_{b \to i} \equiv \mathsf{f}_i(\{u_{b \to i}\}), \tag{21.24}$$

$$u_{a \to i} = \operatorname{atanh}\left\{ \tanh \beta \prod_{j \in \partial a \setminus i} \tanh h_{j \to a} \right\} \equiv \widehat{\mathsf{f}}_a(\{h_{j \to a}\}). \tag{21.25}$$

The corresponding Bethe free entropy is given by the following equation (unlike in Chapter 15, we use natural logarithms here):

$$\mathbb{F}(\underline{u}, \underline{h}) = - \sum_{(ia) \in E} \log \left[\sum_{x_i} \widehat{\nu}_{u_{a \to i}}(x_i) \nu_{h_{i \to a}}(x_i) \right] + \sum_{i=1}^{N} \log \left[\sum_{x_i} \mathcal{Q}(y_i|x_i) \prod_{a \in \partial i} \widehat{\nu}_{u_{a \to i}}(x_i) \right]$$

$$+ \sum_{a=1}^{M} \log \left[\sum_{\underline{x}_{\partial a}} \exp(-\beta E_a(\underline{x}_{\partial a})) \prod_{i \in \partial a} \nu_{h_{i \to a}}(x_i) \right]. \tag{21.26}$$

As in eqn (15.44), we introduce a 'shifted' free-entropy density ϕ, defined as

$$\phi = \frac{1}{N} \mathbb{F}(\underline{u}, \underline{h}) - \sum_{y} \mathcal{Q}(y|0) \log \mathcal{Q}(y|0). \tag{21.27}$$

Recall that the 1RSB cavity approach assumes that, to leading exponential order, the number $\mathcal{N}(\phi)$ of Bethe measures with a shifted free-entropy density equal to ϕ is equal to the number of quasi-solutions of eqns (21.24) and (21.25). We shall write, as usual, $\mathcal{N}(\phi) \doteq \exp(N\Sigma(\phi))$, and our aim is to compute the complexity $\Sigma(\phi)$, using as in Chapter 19 an auxiliary graphical model which counts the number of solutions of the BP equations, weighted by a factor $\exp(N\mathsf{x}\phi)$. If the free entropy of the auxiliary model is $\mathfrak{F}(\mathsf{x}) = \lim_{N \to \infty} \mathbb{F}^{\mathrm{RSB}}(\mathsf{x})/N$, then $\Sigma(\phi)$ is given by the Legendre transform $\mathfrak{F}(\mathsf{x}) = \mathsf{x}\phi + \Sigma(\phi)$, $\partial \Sigma/\partial \phi = -\mathsf{x}$.

For a given code and received $\underline{y}$, the basic objects involved in the 1RSB approach are the distributions of the fields $h_{i \to a}$ and $u_{b \to j}$, denoted respectively by Q_{ia} and $\widehat{Q}_{bj}$. They satisfy the following 1RSB equations:

$$Q_{ia}(h_{i \to a}) \cong \int \delta\left(h_{i \to a} = \mathsf{f}_i(\{u_{b \to i}\})\right) (z_{ia})^{\mathsf{x}} \prod_{b \in \partial i \setminus a} \mathrm{d}\widehat{Q}_{bi}(u_{b \to i}), \tag{21.28}$$

$$\widehat{Q}_{ai}(u_{a \to i}) \cong \int \delta\left(u_{a \to i} = \hat{\mathsf{f}}_a(\{h_{j \to a}\})\right) (\hat{z}_{ai})^{\mathsf{x}} \prod_{j \in \partial a \setminus i} \mathrm{d}Q_{ja}(h_{j \to a}). \tag{21.29}$$

Exercise 21.8 Show that the factors z_{ia} and $\hat{z}_{ai}$ in these equations, defined in eqns (19.22) and (19.23), are given by

$$z_{ia}(\{u_{b \to i}\}, B_i) = \frac{2\cosh(B_i + \sum_{b \in \partial i \setminus a} u_{b \to i})}{\prod_{b \in \partial i \setminus a}(2\cosh(u_{b \to i}))}, \tag{21.30}$$

$$\hat{z}_{ai}(\{h_{j \to a}\}) = 1 + e^{-2\beta}. \tag{21.31}$$

Although in this case $\hat{z}_{ai}$ is a constant and can be absorbed in the normalization, we shall keep it explicitly in the following.

We now turn to the statistical analysis of the above equations. If we pick out a uniformly random edge in the factor graph of a code from the ensemble $\mathrm{LDPC}_N(\Lambda, P)$, the densities $\widehat{Q}$ and Q themselves become random objects, which satisfy the distributional equations

$$Q(h) \stackrel{\mathrm{d}}{=} \frac{1}{Z} \int z(\{u_a\}; B(y))^{\mathsf{x}} \, \delta\left(h - \mathsf{f}_{l-1}(\{u_a\}; B(y))\right) \prod_{a=1}^{l-1} \mathrm{d}\widehat{Q}_a(u_a), \tag{21.32}$$

$$\widehat{Q}(u) \stackrel{\mathrm{d}}{=} \frac{1}{\widehat{Z}} \int \hat{z}(\{h_i\})^{\mathsf{x}} \, \delta\left(u - \hat{\mathsf{f}}_{k-1}(\{h_i\})\right) \prod_{i=1}^{k-1} \mathrm{d}Q_i(h_i) \tag{21.33}$$

where k, l, y are random variables, the $\{\widehat{Q}_a\}$ are $l-1$ i.i.d. copies of $\widehat{Q}$, and the $\{Q_i\}$ are $k-1$ i.i.d. copies of Q. Further, l is drawn from the edge-perspective variable-degree profile λ, k is drawn from the edge-perspective check-degree profile ρ, and y is drawn from $\mathcal{Q}(\cdot \,|0)$, the distribution of the channel output upon input 0. The functions $\hat{\mathsf{f}}_{k-1}(\{h_i\}) = \mathrm{atanh}(\tanh\beta \prod_{i=1}^{k-1} \tanh(h_i))$ and $\mathsf{f}_{l-1}(\{u_a\}; B) = B - \sum_{a=1}^{l-1} u_a$ are defined analogously to eqns (21.24) and (21.25). The functions $z(\cdot)$ and $\hat{z}(\cdot)$ are given similarly by the expressions in eqns (21.30) and (21.31).

The 1RSB free-entropy density (i.e. the entropy density of the auxiliary model) is estimated as $\mathfrak{F}(\mathsf{x}) = \mathsf{f}^{\mathrm{RSB}}(Q, \widehat{Q})$, where $\mathsf{f}^{\mathrm{RSB}}(Q, \widehat{Q})$ is the expected free-entropy density, and Q and $\widehat{Q}$ are distributed according to the 'correct' solution of the distributional equations (21.32) and (21.33):

$$\mathsf{f}^{\mathrm{RSB}}(Q, \widehat{Q}) = -\Lambda'(1)\, \mathbb{E}\, \log z_{\mathrm{e}}(Q, \widehat{Q}) + \mathbb{E}\, \log z_{\mathrm{v}}(\{\widehat{Q}_a\}; l, y) + \frac{\Lambda'(1)}{P'(1)}\, \mathbb{E}\, \log z_{\mathrm{f}}(\{Q_i\}; k).$$

Here the expectation is taken with respect to k i.i.d. copies of $\widehat{Q}$ and l i.i.d. copies of Q, and with respect to $k \stackrel{\mathrm{d}}{=} P$., $l \stackrel{\mathrm{d}}{=} \Lambda$., and $y \stackrel{\mathrm{d}}{=} \mathcal{Q}(\cdot \,|0)$. Finally, $z_{\mathrm{e}}, z_{\mathrm{v}}, z_{\mathrm{f}}$ read

$$z_e(Q, \widehat{Q}) = \int dQ(h) \, d\widehat{Q}(u) \left[\sum_{x=0}^{1} \nu_h(x)\nu_u(x) \right]^{\mathbf{x}}, \tag{21.34}$$

$$z_v(\{\widehat{Q}_a\}; l, y) = \int \prod_{a=1}^{l} d\widehat{Q}_a(u_a) \left[\sum_{x=0}^{1} \frac{Q(y|x)}{Q(y|0)} \prod_{a=1}^{l} \nu_{u_a}(x) \right]^{\mathbf{x}}, \tag{21.35}$$

$$z_f(\{Q_i\}; k) = \int \prod_{i=1}^{l} dQ_i(h_i) \left[\sum_{\{x_1, \cdots, x_k\}} \prod_{i=1}^{k} \nu_{h_i}(x_i) \right.$$
$$\left. \times \left(\mathbb{I}\left(\sum_i x_i = \text{even}\right) + e^{-2\beta} \, \mathbb{I}\left(\sum_i x_i = \text{odd}\right) \right) \right]^{\mathbf{x}}. \tag{21.36}$$

A considerable amount of information is contained in the 1RSB free-energy density $\mathfrak{F}(\mathbf{x})$. For instance, one could deduce the energetic complexity from it by taking the appropriate $\beta \to \infty$ limit. Here, we shall not attempt to develop a full solution of the 1RSB distributional equations, but shall use them to detect the occurrence of a dynamical phase transition.

21.3.2 The dynamical phase transition

The location of the dynamical phase transition $\beta_d(p)$ is determined by the smallest value of β such that the distributional equations (21.32) and (21.33) have a non-trivial solution at $\mathbf{x} = 1$. For $\beta > \beta_d(p)$, the distribution (21.4) decomposes into an exponential number of pure states. As a consequence, we expect simulated annealing to fall out of equilibrium when $\beta_d(p)$ is crossed.

In the top part of Fig. 21.6, we show the result of applying such a technique to the $(5, 6)$ regular ensemble used for communication over BSC(p). At small p, no dynamic phase transition is revealed through this procedure at any positive temperature. Above a critical value of the noise level p, the behaviour changes dramatically, and a phase transition is encountered at a critical point $\beta_d(p)$ that decreases monotonically for larger p. By changing both β and p, one can identify a phase transition line that separates the ergodic and non-ergodic phases. Remarkably, the noise level at which a finite β_d appears is numerically indistinguishable from $p_d \approx 0.145$.

Does the occurrence of a dynamical phase transition for $p \gtrsim p_d$ indeed influence the behaviour of the simulated-annealing decoder? Some numerical confirmation was presented earlier in Fig. 21.2. Further support in favour of this thesis is provided by the bottom part of Fig. 21.6, which plots the residual energy density of the configuration produced by the decoder as $\beta \to \infty$. Above p_d, this becomes strictly positive and only slowly dependent on the cooling rate. It is compared with the equilibrium value of the internal energy at $\beta_d(p)$. This would be the correct prediction if the energy of the system did not decrease any more after the system fell out of equilibrium at $\beta_d(p)$. Although we do not expect this to be strictly true, the resulting curve provides a good first estimate.

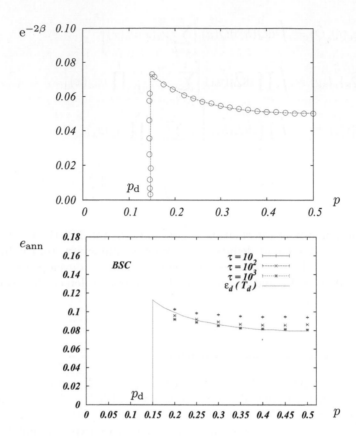

Fig. 21.6 *Top*: location of the dynamic phase transition for random codes from the $(5, 6)$ ensemble used over BSC(p) (the circles were obtained through sampled density evolution; the dashed line is a guide for the eye). *Bottom*: residual energy density after simulated annealing, as measured in numerical simulations. The dashed line gives the equilibrium energy at the dynamical transition temperature T_d.

21.3.3   Metastable states and the BP threshold

One crucial element of this picture can be confirmed analytically, for a generic BMS channel family ordered by physical degradation with respect to p: at zero temperature, the dynamical transition, which signals the proliferation of metastable Bethe states, occurs exactly at the decoding threshold p_d. More precisely, the argument below proves that at $\beta = \infty$ there cannot exist any non-trivial $\mathbf{x} = 1$ solution of eqns (21.32) and (21.33) for $p < p_d$, while there exists one for $p > p_d$. We expect that, for most channel families, the same situation should hold for β large enough (dependent on p), but this has not been proven yet.

Let us consider the 1RSB equations (21.32) and (21.33) in the case $\beta = \infty$. Assuming that the degree profiles are such that $l \geq 2$ and $k \geq 2$ (a reasonable requirement for useful code ensembles), it is clear that they have a special 'no-error' solution associated with the sent codeword in which $Q(h) = \delta_{\infty}(h)$ and $\widehat{Q}(u) = \delta_{\infty}(h)$ almost surely. It is a simple exercise to check that the (shifted) free-entropy density of this solution is equal to 0.

The important question is whether there exist other solutions beyond the 'no-error' one. We can make use of the simplification occuring at $x = 1$. As we saw in Section 19.4.1, the expectation values of the messages, $\nu^{av}_{i \to a}(x_i) \equiv \sum_{\nu_{ia}} Q_{ia}(\nu_{ia})\nu_{ia}(x_i)$ and $\widehat{\nu}^{av}_{a \to i}(x_i) \equiv \sum_{\widehat{m}_{ai}} \widehat{Q}_{ai}(\widehat{\nu}_{ai})\widehat{\nu}_{ai}(x_i)$, satisfy the BP equations.

Let us first study the case $p < p_d$. We have seen in Chapter 15 that there is a unique solution of the BP equations: the no-error solution. This shows that in this low-noise regime, there cannot exist any non-trivial 1RSB solution. We conclude that there is no glass phase in the regime $p < p_d$

We now turn to the case $p > p_d$ (always with $\beta = \infty$), and use the analysis of BP presented in Chapter 15. That analysis revealed that, when $p > p_d$, the density evolution of the BP messages admits at least one 'replica-symmetric' fixed point distinct from the no-error one.

We shall now use this replica-symmetric fixed point in order to construct a non-trivial 1RSB solution. The basic intuition behind this construction is that each Bethe measure consists of a single configuration, well separated from other ones. Indeed, we expect that each Bethe measure can be identified with a zero-energy configuration, i.e. with a codeword. If this is true, then, with respect to each of these Bethe measures, the local distribution of a variable is deterministic, either a unit mass on 0 or a unit mass on 1. Therefore we seek a solution where the distribution of Q and $\widehat{Q}$ is supported on functions of the form

$$Q(h) = \frac{1}{2}(1 + \tanh \tilde{h}) \, \delta_{+\infty}(h) + \frac{1}{2}(1 - \tanh \tilde{h}) \, \delta_{-\infty}(h), \qquad (21.37)$$

$$\widehat{Q}(u) = \frac{1}{2}(1 + \tanh \tilde{u}) \, \delta_{+\infty}(u) + \frac{1}{2}(1 + \tanh \tilde{u}) \, \delta_{-\infty}(u), \qquad (21.38)$$

where $\tilde{h}$ and $\tilde{u}$ are random variables.

Exercise 21.9 Show that this Ansatz solves eqns (21.32) and (21.33) at $\beta = \infty$ if and only if the distributions of $\tilde{h}$ and $\tilde{u}$ satisfy

$$\tilde{h} \stackrel{d}{=} B(y) + \sum_{a=1}^{l-1} \tilde{u}, \qquad \tilde{u} \stackrel{d}{=} \operatorname{atanh}\left[\prod_{i=1}^{k-1} \tanh \tilde{h}_i\right]. \qquad (21.39)$$

It is easy to check that the random variables $\tilde{h}$ and $\tilde{u}$ satisfy the same equations as the fixed point of the density evolution for BP (see eqn (15.11)). We conclude that, for $p > p_d$ and $x = 1$, a solution to the 1RSB equations is given by the Ansatz in eqns (21.37) and (21.38) if $\tilde{h}$ and $\tilde{u}$ are drawn from the fixed-point distributions of eqn (15.11).

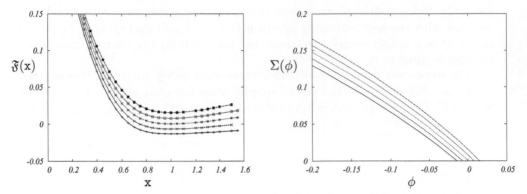

Fig. 21.7 *Left*: the free entropy of the auxiliary model, $\mathfrak{F}(\mathtt{x})$, as a function of the weight parameter $\mathtt{x}$, for a $(3, 6)$ code on a BSC (recall that $p_{\mathrm{d}} \approx 0.084$ and $p_{\mathrm{c}} \approx 0.101$ in this case). From *bottom* to *top*, $p = 0.090, 0.095, 0.100, 0.105, 0.110$. *Right*: the complexity $\Sigma(\phi)$ plotted versus the shifted free-entropy density ϕ. From *left* to *right*, $p = 0.090, 0.095, 0.100, 0.105, 0.110$.

It turns out that a similar solution is easily found for any value of $\mathtt{x} > 0$, provided $\beta = \infty$. The only place where $\mathtt{x}$ plays a role is in the reweighting factor of eqn (21.35): when $\mathtt{x} \neq 1$, the only modification in the distributional equations (21.39) is that $B(y)$ must be multiplied by $\mathtt{x}$. Therefore one can obtain the 1RSB solution for any $\mathtt{x} > 0$ if one knows the solution to the RS cavity equations (i.e. the fixed point of the density evolution for BP) for a slightly modified problem in which $B(y)$ is changed to $\mathtt{x}B(y)$. Technically, this is equivalent to studying the modified measure

$$\mu_y(\underline{x}) \cong \prod_{a=1}^{M} \mathbb{I}(x_{i_1^a} \oplus \cdots \oplus x_{i_{k(a)}^a} = 0) \prod_{i=1}^{N} \mathcal{Q}(y_i|x_i)^{\mathtt{x}}, \qquad (21.40)$$

within the RS approach of Chapter 15 (such a modified measure has already been introduced in Chapter 6).

Let us assume that we have found a non-trivial fixed point for this auxiliary problem, characterized by the distributions $\mathsf{a}_{\mathrm{RS}}^{(\mathtt{x})}(h)$ and $\hat{\mathsf{a}}_{\mathrm{RS}}^{(\mathtt{x})}(u)$, and denote by $\mathsf{f}^{\mathrm{RS}}(\mathtt{x})$ the corresponding value of the free-entropy density defined in eqn (15.45). The 1RSB equations with reweighting parameter $\mathtt{x}$ have a solution of the type (21.37) and (21.38), provided $\tilde{h}$ is distributed according to $\mathsf{a}_{\mathrm{RS}}^{(\mathtt{x})}(\cdot)$, and $\tilde{u}$ is distributed according to $\hat{\mathsf{a}}_{\mathrm{RS}}^{(\mathtt{x})}(\cdot)$. The 1RSB free-entropy density $\mathfrak{F}(\mathtt{x}) = \mathbb{E}\, \mathbb{F}^{\mathrm{RSB}}(\mathtt{x})/N$ is given simply by

$$\mathfrak{F}(\mathtt{x}) = \mathsf{f}^{\mathrm{RS}}(\mathtt{x}). \qquad (21.41)$$

Therefore the problem of computing $\mathfrak{F}(\mathtt{x})$, and its Legendre transform, the complexity $\Sigma(\phi)$, reduce to a replica-symmetric computation. This is a simple generalization of the problem considered in Chapter 15, where the decoding measure is modified by raising it to the power $\mathtt{x}$, as in eqn (21.40). Notice, however, that the interpretation is now different. In particular, $\mathtt{x}$ has to be properly chosen in order to focus on dominant pure states.

The problem can easily be studied numerically using the population dynamics algorithm. Figure 21.7 shows an example of the complexity $\Sigma(\phi)$ for a binary symmetric

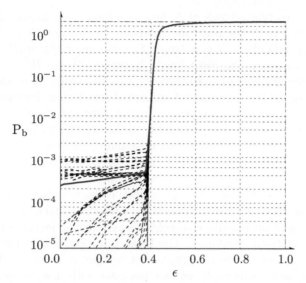

Fig. 21.8 Bit error probability for 40 random elements of the $(3,6)$ regular ensemble with $N = 2500$ used over the $\mathrm{BEC}(\epsilon)$. The continuous curve corresponds to the average error probability.

channel. The regime $p_\mathrm{d} < p < p_\mathrm{c}$ is characterized by the existence of a band of metastable states with a negative shifted free entropy $\phi \leq \phi_0 < 0$. They are, in principle, irrelevant when compared with the 'no-error' solution, which has $\phi = 0$, confirming that MAP decoding will return the transmitted codeword. In fact, they are even unphysical: ϕ is nothing but the conditional entropy density of the transmitted codeword given the received message. As a consequence, it must be non-negative. However, the solution extends to $\beta < \infty$, where it makes perfect sense (it describes non-codeword metastable configurations), thus solving the puzzle.

The appearance of metastable states coincides with the noise threshold above which BP decoding fails. When $p > p_\mathrm{c}$, the top end of the band ϕ_0 becomes positive: the 'glassy' states dominate the measure and MAP decoding fails.

21.4   Metastable states and near-codewords

In a nutshell, the failure of BP decoding for $p > p_\mathrm{d}$ can be traced back to configurations (words) $\underline{x}$ that (*i*) are deep local minima of the energy function $E(\underline{x})$ (which counts the number of violated parity checks), and (*ii*) have a significant weight under the measure $\prod_i Q(y_i | x_i)$.

Typically, such configurations are not codewords, although they can be very close to codewords from the energy point of view. An interesting qualitative analogy can be drawn between this analysis, and various notions that have been introduced to characterize the **error floor**.

Let us start by describing the error floor problem. We saw that, for $p < p_\mathrm{d}$, the bit error rate under BP decoding vanishes when the block length $N \to \infty$. Unhappily, the block length cannot be taken arbitrarily large, because of two types of practical considerations. First, coding a block of N bits simultaneously implies a communication delay proportional to N. Second, any hardware implementation of BP decoding becomes increasingly difficult as N gets larger. Depending on the application, one can be forced to consider a maximum block length between 10^3 and 10^5.

This brings up the problem of characterizing the bit error rate at moderate block length. Figure 21.8 shows the outcomes of numerical simulations for random elements of the $(3, 6)$ ensemble used over an erasure channel. One can clearly distinguish two regimes: a rapid decrease of the error probability in the 'waterfall region' $\epsilon \lesssim \epsilon_d \approx 0.429$ (in physics terms, the 'critical regime'); and a flattening at lower noise values, in the 'error floor'. It is interesting to note that the error floor level is small but highly dependent (in relative terms) on the graph realization.

We know that the error floor should vanish when codes with larger and larger block length are taken, but we would like a prediction of its value *given* the graph G. With the notable exception of an erasure channel, this problem is largely open. However, several heuristics have been developed. The basic intuition is that the error floor is due to small subgraphs of the factor graph that are prone to error. If U is the set of variable nodes in such a subgraph, we can associate with it a configuration $\underline{x}$ that takes the value 1 on U and 0 otherwise (throughout our analysis, we are assuming that the codeword $\underline{0}$ has been transmitted). This $\underline{x}$ need not be a codeword but should in some sense be 'close' to one.

Once a class $\mathcal{F}$ of such subgraphs has been identified, the error probability is estimated by assuming that any type of error is unlikely, and errors in different subsets are roughly independent:

$$P_B(G) \approx \sum_{U \in \mathcal{F}} \mathbb{P}\left\{\text{BP decoder fails on } U\right\}. \tag{21.42}$$

If the subsets U are small, each of the terms on the right-hand side can be evaluated efficiently via importance sampling.

It is interesting to take a look at some definitions of the class of subgraphs $\mathcal{F}$ that have been introduced in the literature. In each case, the subgraph is characterized by two integers (w, e) that describe how dangerous/close to codewords it is (small w or e corresponding to dangerous subgraphs). In practice, the sum in eqn (21.42) is restricted to small w, e.

• **Trapping sets** (or **near-codewords**). A trapping set is a subgraph including the variable nodes in U, all of the adjacent check nodes, and the edges that connect them. It is a (w, e) near-codeword if the number of variable nodes is $|U| = w$ and the number of check nodes of odd degree is e. In our framework, a trapping set is simply a configuration $\underline{x}$ with a weight (number of non-zero entries) equal to w and an energy $E(\underline{x}) = 2e$. Notice that hardly any restriction is imposed on trapping sets. Special constraints are sometimes added, depending on the channel model and on the decoding algorithm (if not BP).

• **Absorbing sets.** A (w, e) absorbing set is a (w, e) trapping set that satisfies two further requirements: (i) each variable node is adjacent to more check nodes of even degree (with respect to the subgraph) than of odd degree, and (ii) it does not contain a (w', e) absorbing set with $w' < w$. The first condition implies that the corresponding configuration $\underline{x}$ is a local minimum of $E(\underline{x})$, stable with respect to one flip.

The connection between small weak subgraphs and the error probability is still somewhat vague. The 'energy landscape' $E(\underline{x})$ might provide some hints towards bridging this gap.

Notes

This chapter is largely based on the analysis of metastable states given by Montanari (2001*a*, *b*) and Franz *et al.* (2002). One-step replica symmetry breaking was also investigated by Migliorini and Saad (2006). The approach was extended to asymmetric channels by Neri *et al.* (2008).

The typical-pairs decoding presented here is slightly different from the original procedure of Aji *et al.* (2001).

Stopping sets were introduced by Di *et al.* (2002), and inspired much of the subsequent research on error floors. The idea that small subgraphs of the factor graph are responsible for error floors was first convincingly demonstrated for general channel models by MacKay and Postol (2003) and Richardson (2003). Absorbing sets were defined by Dolecek *et al.* (2007).

After its invention, simulated annealing was the object of a significant amount of work in the fields of operations research and probability. A review can be found in Aarts *et al.* (2003). A detailed comparison between 1RSB analysis and simulated-annealing experiments for models on sparse graphs was presented by Montanari and Ricci-Tersenghi (2004).

22

An ongoing story

This book describes a unified approach to a number of important problems in information theory, physics, and computer science. We have presented a consistent set of methods to address these problems, but the field is far from being fully understood, and there remain many open challenges. This chapter provides a concise description of some of these challenges, as well as a survey of recent progress. Our ambition is to set an agenda for the newly developed field that we have been describing. We shall distinguish roughly three types of directions.

The first one, to be discussed in Section 22.1, is the main challenge. It aims at a better qualitative understanding of models on sparse random graphs. At the core of the cavity method lies the postulate that such systems can have only a limited number of 'behaviours' (phases). Each phase corresponds to a different pattern of replica symmetry breaking (replica-symmetric (RS), one-step replica symmetry breaking (1RSB), etc.). In turn, these phases also have a description in terms of pure states decomposition, as well as in terms of long-range correlations. Understanding the fundamental reasons and conditions for the universality of these phases, as well as the equivalence between their characterizations, would be extremely important.

The second direction, described in Section 22.2, concerns the development of the cavity formalism itself. We have focused mainly on systems in which either the RS or the 1RSB cavity method is expected to be asymptotically exact in the large-size limit. This expectation is, in part, based on some internal consistency checks of the 1RSB approach. An important such check is to verify that the 1RSB 'solution' is stable with respect to small perturbations. Whenever this test is passed, physicists feel confident enough that the cavity method provides exact conjectures (thresholds, minimum cost per variable, etc.). If the test is not passed, higher-order RSB is thought to be needed. The situation is much less satisfactory in this case, and the cavity method poses some technical problems even at the heuristic level.

Section 22.3 lists a number of fascinating questions that arise in the connection between the existence of glassy-phase transitions and algorithmic slowdown. These are particularly important in view of applications in computer science and information theory: sparse graphical models can be useful for a number of practically relevant tasks, as the example of the use of LDPC codes in channel coding has shown. There is some empirical evidence that phase transitions have an impact on the behaviour and efficiency of algorithms. Physicists hope that this impact can be understood (to some extent) in a unified way, and that it is ultimately related to the geometric structure of the set of solutions and to correlation properties of the measure. While some general arguments in favour of this statement have been put forward, the actual understanding is still very poor.

22.1   Gibbs measures and long-range correlations

At an abstract level, the cavity method explored in the last few chapters relies on a (yet unproven) *structural theorem*. Consider a generic graphical model, for a probability distribution on N variables, $\underline{x}$, taking values in a discrete space $\mathcal{X}^N$:

$$\mu(\underline{x}) = \frac{1}{Z} \prod_{a \in F} \psi_a(\underline{x}_{\partial a}) \,. \qquad (22.1)$$

The cavity method postulates that, for large classes of models taken from appropriate ensembles, the model is qualitatively described in the large-N limit by one out of a small number of generic scenarios, or phases. The postulated qualitative features of such phases are then cleverly used to derive quantitative predictions (e.g. the locations of phase transitions).

Needless to say, we are not able to state precisely, let alone prove, such a structural theorem in this generality. The complete set of necessary hypotheses is unknown. However, we have discussed several examples, from XORSAT to diluted spin glasses and error-correcting codes. In principle, it is not necessary that the factor graph be locally tree-like, but, in practice, locally tree-like models are the ones that we can control most effectively. Such a structure implies that when one digs a cavity in the graph, the variables on the boundary of the cavity are far apart. This leads to a simple structure of their correlation in the large-system limit, and hence to the possibility of writing asymptotically exact recursion equations.

Here, we do not want to discuss the necessary hypotheses in any more detail. It would certainly be a significant achievement to prove such a structural theorem even in a restricted setting (say, for the uniform measure over solutions of random K-SAT formulae). We want, instead, to convey some important features of the phases postulated within the cavity approach. In particular, there is a key aspect that we want to stress. Each of the various phases mentioned can be characterized from two complementary points of view:

1. In terms of decomposition of the distribution $\mu(\,\cdot\,)$ into 'lumps' or 'clusters'. Below, we shall propose a precise definition of these lumps, and they will be called **pure states**.
2. In terms of correlations among far-apart variables on the factor graph. We shall introduce two notions of *correlation decay* that differ in a rather subtle way but correspond to different phases.

These two characterizations are, in turn, related to the various aspects of the cavity method.

22.1.1   On the definition of pure states

The notion of a pure state is a crucial one in rigorous statistical mechanics. Unfortunately, standard definitions are tailored to translation-invariant models on infinite graphs. The graphical models that we have in mind are sparse random graphs (in this class we include labelled random graphs, where the labels specify the nature of func-tion nodes), and standard approaches do not apply to them. In particular, we need a concrete definition that is meaningful for finite graphs.

Consider a sequence of *finite* graphical models $\{\mu_N(\cdot)\}$, indexed by the number of variable nodes N. A **pure-state decomposition** is defined by assigning, for each N, a partition of the configuration space $\mathcal{X}^N$ into $\mathcal{N}_N$ subsets $\Omega_{1,N}, \dots, \Omega_{\mathcal{N}_N,N}$:

$$\mathcal{X}^N = \Omega_{1,N} \cup \cdots \cup \Omega_{\mathcal{N}_N,N}. \tag{22.2}$$

The pure-state decomposition must meet the following conditions:

1. The measure of each subset in the partition is bounded away from 1:

$$\max\{\mu_N(\Omega_{1,N}), \dots, \mu_N(\Omega_{\mathcal{N},N})\} \leq 1 - \delta. \tag{22.3}$$

2. The subsets are separated by 'bottlenecks.' More precisely, for $\Omega \subseteq \mathcal{X}^N$, we define its ϵ-boundary as

$$\partial_\epsilon \Omega \equiv \{x \in \mathcal{X}^N : 1 \leq d(x, \Omega) \leq N\epsilon\}. \tag{22.4}$$

where $d(x, \Omega)$ is the minimum Hamming distance between x and any configuration $x' \in \Omega$. Then we require

$$\lim_{N \to \infty} \max_r \frac{\mu_N(\partial_\epsilon \Omega_{r,N})}{\mu_N(\Omega_{r,N})} = 0, \tag{22.5}$$

for some $\epsilon > 0$. Note that the measure of $\partial_\epsilon \Omega_{r,N}$ can be small for two reasons, either because $\Omega_{r,N}$ is small itself (and therefore has a small boundary) or because the boundary of $\Omega_{r,N}$ is much smaller than its interior. Only the last situation corresponds to a true bottleneck, as is enforced by the denominator $\mu_N(\Omega_{r,N})$ in eqn (22.5).

3. The conditional measure on the subset $\Omega_{r,N}$, defined by

$$\mu_N^r(\underline{x}) \equiv \frac{1}{\mu_N(\Omega_{r,N})} \, \mu_N(\underline{x}) \mathbb{I}(\underline{x} \in \Omega_{r,N}), \tag{22.6}$$

cannot be decomposed further according to the two conditions above.

Given such a partition, the distribution $\mu_N(\cdot)$ can be written as a convex combi-nation of distributions with disjoint support

$$\mu_N(\cdot) = \sum_{r=1}^{\mathcal{N}_N} w_r \, \mu_N^r(\cdot), \quad w_r \equiv \mu_N(\Omega_{r,N}). \tag{22.7}$$

Note that this decomposition is not necessarily unique, as shown by the example below. The non-uniqueness is due to the fact that sets of configurations of $\mathcal{X}^N$ with negligeable weight can be attributed to one state or another. On the other hand, the conditional measures $\mu_N^r(\cdot)$ should depend only weakly on the precise choice of decomposition.

Example 22.1 Consider a ferromagnetic Ising model on a random regular graph of degree $(k+1)$. The Boltzmann distribution reads

$$\mu_N(\underline{x}) = \frac{1}{Z_N(\beta)} \exp\left\{ \beta \sum_{(i,j)\in E} x_i x_j \right\}, \qquad (22.8)$$

with $x_i \in \mathcal{X} = \{+1, -1\}$. To avoid irrelevant complications, let us assume that N is odd. Following the discussion in Section 17.3, we expect this distribution to admit a non-trivial pure-state decomposition for $k \tanh \beta > 1$, with a partition $\Omega_+ \cup \Omega_- = \mathcal{X}^N$. Here Ω_+ and Ω_-) are the sets of configurations for which $\sum_i x_i$ is positive or negative, respectively. With respect to this decomposition, $w_+ = w_- = 1/2$.

Of course, an (asymptotically) equivalent decomposition is obtained by letting Ω_+ be the set of configurations with $\sum_i x_i \geq C$ for some fixed C.

It is useful to recall that the condition (22.5) implies that any 'local' Markov dynamics that satisfies detailed balance with respect to $\mu_N(\cdot)$ is slow. More precisely, assume that

$$\frac{\mu_N(\partial_\epsilon \Omega_{r,N})}{\mu_N(\Omega_{r,N})} \leq \exp\{-\Delta(N)\}. \qquad (22.9)$$

Then any Markov dynamics that satisfies detailed balance with respect to μ_N and flips at most $N\epsilon$ variables at each step has a relaxation time larger than $C \exp\{\Delta(N)\}$ (where C is an N-independent constant that depends on the details of the model). Moreover, if the dynamics is initialized in $\underline{x} \in \Omega_{r,N}$, it will take a time of order $C \exp\{\Delta(N)\}$ to get to a distance $N\epsilon$ from $\Omega_{r,N}$.

In many cases based on random factor graph ensembles, we expect eqn (22.9) to hold with a $\Delta(N)$ which is linear in N. In fact, in the definition of pure-state decomposition, we might require a bound of the form (22.9) to hold, for some function $\Delta(N)$ (e.g. $\Delta(N) = N^\psi$, with some appropriately chosen ψ). This implies that pure states are stable on time scales shorter than $\exp\{\Delta(N)\}$.

22.1.2   Notions of correlation decay

The above discussion of relaxation times brings up a second key concept: **correlation decay**. According to an important piece of wisdom in statistical mechanics, physical systems that have only short-range correlations should relax rapidly to their equilibrium distribution. The hand-waving reason is that, if different degrees of freedom (particles, spins, etc.) are independent, then the system relaxes on microscopic time scales (namely the relaxation time of a single particle, spin, etc.). If the degrees of freedom are not independent, but their correlations are short-ranged, they can be coarse-grained in such a way that they become nearly independent. Roughly speaking, this means that one can construct 'collective' variables from blocks of original variables. Such conditional variables take $|\mathcal{X}|^B$ values, where B is the block size, and are nearly independent under the original (Boltzmann) distribution.

As we are interested in models on non-Euclidean graphs, the definition of correlation decay must be specified. We shall introduce two distinct types of criteria. Although they may look similar at first sight, it turns out that they are not, and each of them will characterize a distinct generic phase.

The simplest approach, widely used in physics, consists in considering two-point correlation functions. Averaging them over the two positions defines a susceptibility. For instance, in the case of Ising spins $x_i \in \mathcal{X} = \{1, -1\}$, we have already discussed the spin glass susceptibility

$$\chi^{\mathrm{SG}} = \frac{1}{N} \sum_{i,j \in V} \left(\langle x_i x_j \rangle - \langle x_i \rangle \langle x_j \rangle \right)^2, \tag{22.10}$$

where $\langle \cdot \rangle$ denotes the expectation value with respect to μ. When χ^{SG} is bounded as $N \to \infty$, this is an indication of short-range correlations. Through the fluctuation–dissipation theorem (see Section 2.3), this is equivalent to stability with respect to local perturbations. Let us recall the mechanism of this equivalence. Imagine a perturbation of the model (22.16) that acts on a single variable x_i. Stability requires that the effect of such a perturbation on the expectation of a global observable $\sum_j f(x_j)$ should be bounded. The change in the marginal at node j due to a perturbation at i is proportional to the covariance $\langle x_i x_j \rangle - \langle x_i \rangle \langle x_j \rangle$. As in Section 12.3.2, the average effect of the perturbation at i on the variables x_j, $j \neq i$, often vanishes (more precisely, $\lim_{N \to \infty} (1/N) \sum_{j \in V} \left(\langle x_i x_j \rangle - \langle x_i \rangle \langle x_j \rangle \right) = 0$) because terms related to different vertices j cancel. The *typical* effect of the perturbation is captured by the spin glass susceptibility.

Generalizing this definition to arbitrary alphabets is easy. We need to use a measure of how much the joint distribution $\mu_{ij}(\cdot, \cdot)$ of x_i and x_j is different from the product of the marginals $\mu_i(\cdot)$ times $\mu_j(\cdot)$. One such measure is provided by the variation distance

$$||\mu_{ij}(\cdot, \cdot) - \mu_i(\cdot)\mu_j(\cdot)|| \equiv \frac{1}{2} \sum_{x_i, x_j} |\mu_{ij}(x_i, x_j) - \mu_i(x_i)\mu_j(x_j)|. \tag{22.11}$$

We then define the two-point correlation by averaging this distance over the vertices i, j:

$$\chi^{(2)} \equiv \frac{1}{N} \sum_{i,j \in V} ||\mu_{ij}(\cdot, \cdot) - \mu_i(\cdot)\mu_j(\cdot)||. \tag{22.12}$$

Exercise 22.1 Consider again the case of Ising variables, where $\mathcal{X} = \{+1, -1\}$. Show that $\chi^{\mathrm{SG}} = o(N)$ if and only if $\chi^{(2)} = o(N)$.

[Hint: Let $C_{ij} \equiv \langle x_i x_j \rangle - \langle x_i \rangle \langle x_j \rangle$. Show that $C_{ij} = 2||\mu_{ij}(\cdot, \cdot) - \mu_i(\cdot)\mu_j(\cdot)||$. Then use $\chi^{\mathrm{SG}} = N\mathbb{E}\{C_{ij}^2\}$ and $\chi^{(2)} = N\mathbb{E}\{|C_{ij}|\}/2$, the expectation $\mathbb{E}$ being over uniformly random $i, j \in V$.]

Of course, one can define l-point correlations in an analogous manner:

$$\chi^{(l)} \equiv \frac{1}{N^{l-1}} \sum_{i(1),\dots,i(l)\in V} ||\mu_{i(1)\dots i(l)}(\cdots) - \mu_{i(1)}(\cdot)\cdots\mu_{i(l)}(\cdot)||. \qquad (22.13)$$

The l-point correlation $\chi^{(l)}$ has a useful interpretation in terms of a thought experiment. Suppose you are given an N-dimensional distribution $\mu(\underline{x})$ and have access to the marginal $\mu_{i(1)}(\cdot)$ at a uniformly random variable node $i(1)$. You want to test how stable this marginal is with respect to small perturbations. Perturbations affect $l-1$ randomly chosen variable nodes $i(2)$, ..., $i(l)$, changing $\mu(\underline{x})$ into $\mu'(\underline{x}) \cong \mu(\underline{x})(1 + \delta_2(x_{i(2)}))\cdots(1 + \delta_l(x_{i(l)}))$. The effect of the resulting perturbation on $\mu_{i(1)}$, to first order in the product $\delta_2\cdots\delta_l$, is bounded in terms of the expectation by $\chi^{(l)}$ (this is again a version of the fluctuation–dissipation theorem).

Definition 22.2. (First type of correlation decay) *The graphical model given by $\mu(\cdot)$ is said to be* **stable to small perturbations** *if, for all finite l, $\chi^{(l)}/N \to 0$ as $N \to \infty$.*

In practice, in sufficiently homogeneous (mean-field) models, this type of stability is equivalent to that found using only $l=2$.

Let us now introduce another type of criterion for correlation decay. Again we look at a variable node i, but now we want to check how strongly x_i is correlated with all of the 'far apart' variables. Of course, we must define what 'far apart' means. We fix an integer ℓ, define $\mathsf{B}(i,\ell)$ as the ball of radius ℓ centred on i, and define $\overline{\mathsf{B}}(i,\ell)$ as its complement, i.e. the subset of variable nodes j such that $d(i,j) \geq \ell$. We then want to estimate the correlation between x_i and $\underline{x}_{\overline{\mathsf{B}}(i,\ell)} = \{x_j : j \in \overline{\mathsf{B}}(i,\ell)\}$. This amounts to measuring the distance between the joint distribution $\mu_{i,\overline{\mathsf{B}}(,\ell)}(\cdot,\cdot)$ and the product of the marginals $\mu_i(\cdot)\mu_{\overline{\mathsf{B}}(,\ell)}(\cdot)$. If we use the total variation distance defined in eqn (22.11), we obtain the following **point-to-set correlation function**:

$$G_i(\ell) \equiv ||\mu_{i,\overline{\mathsf{B}}(i,\ell)}(\cdot,\cdot) - \mu_i(\cdot)\mu_{\overline{\mathsf{B}}(i,\ell)}(\cdot)||. \qquad (22.14)$$

The function $G_i(\ell)$ can be interpreted according to two distinct but equally suggestive thought experiments. The first one comes from the theory of structural glasses (it is meant to elucidate the kind of long-range correlations arising in a fragile glass). Imagine that we draw a reference configuration $\underline{x}^*$ from the distribution $\mu(\cdot)$. Now, we generate a second configuration $\underline{x}$ as follows. Variables outside the ball, with $i \in \overline{\mathsf{B}}(i,\ell)$, are forced to the reference configuration, $x_i = x_i^*$. Variables at a distance smaller than ℓ (denoted by $\underline{x}_{\mathsf{B}(i,\ell)}$) are instead drawn from the conditional distribution $\mu(\underline{x}_{\mathsf{B}(i,\ell)}|\underline{x}^*_{\overline{\mathsf{B}}(i,\ell)})$. If the model $\mu(\cdot)$ has some form of *rigidity* (long-range correlations), then x_i should be close to x_i^*. The correlation $G_i(\ell)$ measures how much the distributions of x_i and x_i^* differ.

The second experiment is closely related to the first one, but has the flavour of a statistics (or computer science) question. Someone draws the configuration $\underline{x}^*$ as above from the distribution $\mu(\cdot)$. They then reveal to you the values of far-apart variables in the reference configuration, i.e. the values of x_j^* for all $j \in \overline{\mathsf{B}}(i,\ell)$. They ask you to *reconstruct* the value of x_i^*, or to guess it as well as you can. The correlation function $G_i(\ell)$ measures how likely you are to guess correctly (assuming unbounded computational power), compared with the case in which no variable has been revealed to you.

This discussion suggests the following definition.

Definition 22.3. (Second type of correlation decay) *The graphical model $\mu(\,\cdot\,)$ is said to satisfy the **non-reconstructibility** (or **extremality**) condition if, for all i's, $G_i(\ell) \to 0$ as $\ell \to \infty$. (More precisely, we require that there exists a function $\delta(\ell)$, with $\lim_{\ell\to\infty} \delta(\ell) = 0$, such that $G_i(\ell) \le \delta(\ell)$ for all i and N.) In the opposite case, i.e. if $G_i(\ell)$ remains bounded away from zero at large distance, the model is said to be **reconstructible**.*

22.1.3 Generic scenarios

We shall now describe the correlation decay properties and the pure-state decomposition for the three main phases that we have encountered in the previous chapters: the RS, dynamical 1RSB, and static 1RSB phases. When we are dealing with models on locally tree-like random graphs, each of these phases can also be studied using the appropriate cavity approach, as we shall recall.

Here we focus on phases that appear 'generically'. This means that we exclude (i) critical points, which are obtained by fine-tuning some parameters of the model; and (ii) multiplicities due to global symmetries, for instance in the zero-field ferromagnetic Ising model. Of course, there also exist other types of generic phases, such as the higher-order RSB phases that will be discussed in the next section, and maybe more that have not been explored yet.

- *Replica-symmetric.* In this phase, there exists no non-trivial decomposition into pure states of the form (22.7). In other words, $\mathcal{N}_N = 1$ with high probability. Correlations decay according to both criteria: the model is stable to small perturbations and it satisfies the non-reconstructibility condition. Therefore it is short-range correlated in the strongest sense. Finally, the replica-symmetric cavity method of Chapter 14 yields asymptotically exact predictions.

- *Dynamical 1RSB.* In this phase, the measure $\mu(\,\cdot\,)$ admits a non-trivial decomposition of the form (22.7) into an exponential number of pure states: $\mathcal{N}_N = e^{N\Sigma + o(N)}$ with high probability for some $\Sigma > 0$. Furthermore, most of the measure is carried by states of equal size. More precisely, for any $\delta > 0$, all but an exponentially small fraction of the measure is contained in states $\Omega_{r,N}$ such that

$$-\Sigma - \delta \le \frac{1}{N} \log \mu(\Omega_{r,N}) \le -\Sigma + \delta\,. \tag{22.15}$$

From the correlation point of view, this phase is stable to small perturbations, but it is reconstructible. In other words, a finite number of probes would fail to reveal long-range correlations. But long-range correlations of the point-to-set type are in fact present, and they are revealed, for instance, by a slowdown of reversible Markov dynamics. The glass order parameter, namely the overlap distribution $P(q)$, is trivial in this phase (as implied by eqn (12.31)), but its glassy nature

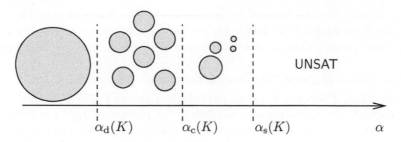

Fig. 22.1 A pictorial view of the different phases in K-SAT with $K \geq 4$, depending on the number of clauses per variable α. From *left* to *right*: replica-symmetric, dynamical 1RSB, static 1RSB, and UNSAT.

can be found through the ϵ-coupling method of Section 12.3.4. The model can be solved exactly (in the sense of determining its asymptotic free-energy density) within the 1RSB cavity method. The thermodynamically dominant states, i.e. those satisfying eqn (22.15), correspond to the 1RSB parameter $x = 1$.

- *Static 1RSB*. This is the 'genuine' 1RSB phase analogous to the low-temperature phase of the random energy model. The model admits a non-trivial pure-state decomposition with wildly varying weights. For any $\delta > 1$, a fraction $1 - \delta$ of the measure is contained in the $k(N, \delta)$ pure states with the largest weights. The number $k(N, \delta)$ converges, when $N \to \infty$, to a *finite* random variable (taking integer values). If we order the weights according to their magnitudes $w^{(1)} \geq w^{(2)} \geq w^{(3)} \geq \cdots$, they converge to a Poisson–Dirichlet process (see Chapter 8). This phase is not stable to small perturbations, and it is reconstructible: it has long-range correlations according to both criteria. The asymptotic overlap distribution function $P(q)$ has two delta-function peaks, as in Fig. 12.3. Again, it can be solved exactly within the 1RSB cavity method.

These three phases are present in a variety of models, and are often separated by phase transitions. The 'clustering' or 'dynamical' phase transition separates the RS and dynamical 1RSB phases, and a condensation phase transition separates the dynamical 1RSB from the static 1RSB phase. Figure 22.1 describes the organization of the various phases in random K-SAT with $K \geq 4$, as discussed in Section 20.3. For $\alpha < \alpha_d(K)$, the model is RS; for $\alpha_d(K) < \alpha < \alpha_c(K)$, it is dynamically 1RSB; for $\alpha_c(K) < \alpha < \alpha_s(K)$, it is statically 1RSB; and for $\alpha_s(K) < \alpha$, it is UNSAT. Figure 22.2 shows the point-to-set correlation function in random 4-SAT. It clearly develops long-range correlations at $\alpha \geq \alpha_d \approx 9.38$. Notice the peculiar development of correlations through a plateau whose width increases with α, and diverges at α_d. This is typical of the dynamical 1RSB transition.

22.2 Higher levels of replica symmetry breaking

For some of the models studied in this book, the RS or the 1RSB cavity method is thought to yield asymptotically exact predictions. However, in general, higher orders of RSB are necessary. We shall sketch how to construct these higher-order solutions

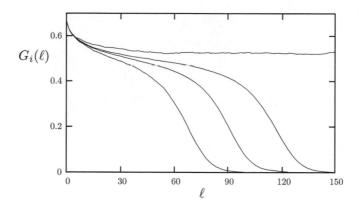

Fig. 22.2 The point-to-set correlation function defined in eqn (22.14), plotted versus distance for random 4-satisfiability, at clause densities $\alpha = 9.30$, 9.33, 9.35, and 9.40 (from *bottom* to *top*).

hierarchically in locally tree-like graphical models. In particular, understanding the structure of the two-step replica-symmetry-breaking (2RSB) solution allows us to derive a 'stability criterion' for the 1RSB approach. It is on the basis of this criterion that, for instance, our derivation of the SAT–UNSAT threshold in Chapter 20 is conjectured to give an exact result.

22.2.1   The high-level picture

Let us first briefly summarize the RS/1RSB approach. Consider an ensemble of graphical models defined through the distribution (22.1) with a locally tree-like factor graph structure. Within the RS cavity method, the local marginals of $\mu(\cdot)$ are accurately described in terms of the message sets $\{\nu_{i \to a}\}$ and $\{\widehat{\nu}_{a \to i}\}$. Given a small (tree-like) subgraph induced by the vertex set $U \subset V$, the effect of the rest of the graph $G \setminus G_U$ on U is described by a factorized measure on the boundary of U.

One-step replica symmetry breaking relaxes this assumption, by allowing long-range correlations, with a peculiar structure. Namely, the probability distribution $\mu(\cdot)$ is assumed to decompose into a convex combination of Bethe measures $\mu_r(\cdot)$. Within each 'state' r, the local marginals of the measure restricted to this state are well described in terms of a set of messages $\{\nu^r_{i \to a}\}$ (by 'well described', we mean that the description becomes asymptotically exact at large N). Sampling a state r at random defines a probability distribution $\mathsf{P}(\{\nu\}, \{\widehat{\nu}\})$ over messages. This distribution is then found to be described by an 'auxiliary' graphical model, which is easily deduced from the original one. In particular, the auxiliary factor graph inherits the structure of the original one, and therefore it is again locally tree-like. 1RSB amounts to using the RS cavity method to study this auxiliary graphical model over messages.

In some cases, 1RSB is expected to be asymptotically exact in the thermodynamic limit. However, this is not always the case: it may fail because the measure $\mathsf{P}(\{\nu\}, \{\widehat{\nu}\})$ decomposes into multiple pure states. Higher-order RSB is used to study this type of situation by iterating the above construction.

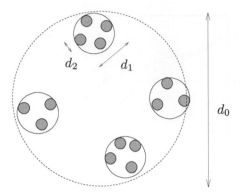

Fig. 22.3 Cartoon of the distribution $\mu(\underline{x})$ for a model described by two-step replica symmetry breaking. The probability mass is concentrated in the grey 'lumps' of radius d_2, which are organized into 'clouds' of radius $d_1 > d_2$. The dashed circle corresponds to the typical distance d_0 between clouds.

More precisely, the 2RSB method starts from the 'auxiliary' distribution $\mathsf{P}(\{\nu\}, \{\widehat{\nu}\})$. Instead of studying this with the RS method as we havedone so far, we use instead the 1RSB method to study $\mathsf{P}(\{\nu\}, \{\widehat{\nu}\})$ (therefore introducing an auxiliary auxiliary model, which is studied by the RS method).

The 2RSB Ansatz admits a hand-waving interpretation in terms of the qualitative features of the original model $\mu(\,\cdot\,)$. Consider 1RSB again. The interpretation was that $\mu(\,\cdot\,)$ is a convex combination of 'pure states' $\mu^r(\,\cdot\,)$, each forming a well-separated lump in configuration space. Within 2RSB, the lumps have a hierarchical organization, i.e. they are grouped into 'clouds'. Each lump is addressed by giving a 'cloud index' r_1, and, within the cloud, a 'lump index' r_2. The measure thus decomposes as

$$\mu(\underline{x}) = \sum_{r_1 \in S_1,\, r_2 \in S_2(r_1)} w_{r_1, r_2}\, \mu^{r_1, r_2}(\underline{x})\,. \tag{22.16}$$

Here $S_2(r_1)$ is the set of indices of the lumps inside cloud r_1. A pictorial sketch of this interpretation is shown in Fig. 22.3.

Even the most forgiving reader should be puzzled by all this. For instance, what is the difference between $\mathcal{N}_1$ clouds, each involving $\mathcal{N}_1$ lumps, and just $\mathcal{N}_1 \mathcal{N}_2$ lumps? In order to distinguish between these two cases, one can look at a properly defined distance, say the Hamming distance divided by N, between two i.i.d. configurations drawn with distribution $\mu(\,\cdot\,)$ (in physics jargon, two replicas). If the two configurations are conditioned to belong to the same lump, to different lumps within the same cloud, or to different clouds, the normalized distances concentrate around three values d_2, d_1, d_0, respectively, with $d_2 < d_1 < d_0$. As in the case of 1RSB, one could in principle distinguish dynamic and static 2RSB phases depending on the numbers of relevant clouds and lumps within clouds. For instance, in the most studied case of static 2RSB, these numbers are subexponential. As a consequence, the asymptotic distribution of the distance between two replicas has non-zero weight on each of the three values d_0,

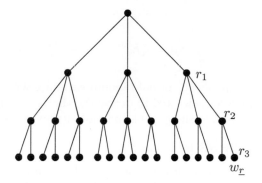

Fig. 22.4 Hierarchical structure of the distribution $\mu(\underline{x})$ within k-step replica symmetry breaking. Here $k = 3$.

d_1, d_2 (in other words, the overlap distribution $P(q)$ is a combination of three delta functions).

Of course, this whole construction can be bootstrapped further, by having clouds grouped into larger structures, etc. Within k-RSB, the probability distribution $\mu(\,\cdot\,)$ is a convex combination of 'states' $\mu^{\underline{r}}(\,\cdot\,)$, where $\underline{r} = (r_1, r_2, \ldots, r_k)$ indexes the leaves of a k-generation tree. The indices $r_1, r_2, \ldots, r_k$ correspond to the nodes encountered along the path between the root and the leaf. This translates into a hierarchy of auxiliary graphical models. If k is allowed to be arbitrarily large, this hierarchy is expected to determine the asymptotic properties of a large class of models. In particular, one can use it to compute the free entropy per variable $\phi \equiv \lim_{N \to \infty} N^{-1} \log Z_N$.

The resulting description of $\mu(\underline{x})$ has a natural ultrametric structure, as discussed in Chapter 8 and recalled in Fig. 22.4. This structure is captured by the generalized random energy model (GREM), a simple model that generalizes the REM discussed in Chapter 5. While presenting the solution of the GREM would take us too far, it is instructive to give its definition.

Example 22.4 The GREM is a simple model for the probability distribution $\mu(\,\cdot\,)$ within k-step RSB. Its definition involves one parameter $N \in \mathbb{N}$ that corresponds to the system size, and several others (denoted by $\{a_0, a_1, \ldots, a_{k-1}\}$, $\{d_0, d_2, \ldots, d_{k-1}\}$, and $\{\Sigma_0, \Sigma_1, \ldots, \Sigma_{k-1}\}$) that are thought of as being fixed as $N \to \infty$. States are associated with the leaves of a k-generation tree. Each leaf is indexed by the path $\underline{r} = (r_0, \ldots, r_{k-1})$ that connects it to the root (see Fig. 22.4).

The GREM does not describe the structure of each state $\mu_{\underline{r}}(\,\cdot\,)$ (which can be thought of as supported on a single configuration). It describes only the distribution of distances between the states, and the distribution of the weights $w_{\underline{r}}$ appearing in the decomposition (22.16).

A node at level i has $\exp\{N\Sigma_i\}$ offspring. The total number of states is therefore $\exp\{N(\Sigma_0 + \cdots + \Sigma_{k-1})\}$. Two random configurations drawn from states $\underline{r}$ and $\underline{s}$ have a distance $d_{i(\underline{r},\underline{s})}$, where $i(\underline{r}, \underline{s})$ is the largest integer i such that $r_i = s_i$. Finally, the weight of state $\underline{r}$ has the form

$$w_{\underline{r}} = \frac{1}{Z} \exp\{-\beta(E_{r_0}^{(0)} + \cdots + E_{r_{k-1}}^{(k-1)})\}, \qquad (22.17)$$

where the $E_r^{(i)}$ are independent normal random variables with mean 0 and variance Na_i. The interested reader is invited to derive the thermodynamic properties of the GREM, for instance the free energy as a function of the temperature.

22.2.2 What does 2RSB look like?

Higher-order RSB has been studied in some detail in many 'fully connected' models such as the p-spin Ising model considered in Chapter 8. In contrast, if one considers models on sparse graphs as we do here, any cavity calculation beyond 1RSB is technically very challenging. In order to understand why, it is interesting to have a superficial look at how a 2RSB cavity calculation would be formally set up, without any attempt at justifying it.

For the sake of simplicity, we shall consider a model of the form (22.1) with pairwise interactions. Therefore all of the factor nodes have degree 2, and BP algorithms can be simplified by using only one type of message passed along the edges of an ordinary graph (see Section 14.2.5). Consider a variable node $0 \in V$ of degree $(l + 1)$, and denote l of its neighbours by $\{1, \ldots, l\}$. We let $\nu_1, \ldots, \nu_l$ be the messages from $1, \ldots, l$, respectively, and let ν_0 be the message from 0 to its $(l + 1)$-th neighbour.

As we saw in Section 14.2.5, the RS cavity equation (i.e. the BP fixed-point equation) at node 0 reads

$$\nu_0(x_0) = \frac{1}{z\{\nu_i\}} \prod_{i=1}^{k} \sum_{x_i} \psi_{0i}(x_0, x_i)\nu_i(x_i), \qquad (22.18)$$

where $z\{\nu_i\}$ is determined by the normalization condition of $\nu_0(\cdot)$. In order to lighten the notation, it is convenient to introduce a function f_0 that, when evaluated on l messages $\nu_1, \ldots, \nu_l$, returns the message ν_0 as above. We shall therefore write eqn (22.18) in shorthand form as $\nu_0 = f_0\{\nu_i\}$. Each ν_i is a point in a $(|\mathcal{X}|-1)$-dimensional simplex.

The 1RSB cavity equations are obtained from eqn (22.18) by promoting the messages ν_i to random variables with distribution $Q_i(\cdot)$ (see Chapter 19). The equations depend on the 1RSB parameter (a real number), which we denote here by x_1. Adopting a continuous notation for the distributions of messages, we get

$$Q_0(\nu_0) = \frac{1}{Z\{Q_i\}} \int z\{\nu_i\}^{x_1} \delta(\nu_0 - f_0\{\nu_i\}) \prod_{i=1}^{l} dQ_i(\nu_i), \qquad (22.19)$$

Analogously to the replica-symmetric case (eqn (22.18)), we shall write $Q_0 = F_0\{Q_i\}$ as a shorthand for this equation. The function F_0 takes as its argument l distributions $Q_1, \ldots, Q_l$ and evaluates a new distribution Q_0 (each of the Q_i's is a distribution over a $(|\mathcal{X}| - 1)$-dimensional simplex).

At this point, the formal similarity of eqns (22.18) and (22.19) should be clear. The 2RSB cavity equations are obtained by promoting the distributions Q_i to random variables (taking values in the set of distributions over a $|\mathcal{X}|$-dimensional simplex).[1] Their probability distributions are denoted by $\mathcal{Q}_i$, and the resulting equations depend on one further real parameter x_2. Formally, the 2RSB equation can be written as

$$\mathcal{Q}_0(Q_0) = \frac{1}{\mathcal{Z}\{\mathcal{Q}_i\}} \int Z\{Q_i\}^{\mathrm{x}_2/\mathrm{x}_1} \, \delta(Q_0 - \mathsf{F}_0\{Q_i\}) \prod_{i=1}^{l} \mathrm{d}\mathcal{Q}_i(Q_i) \,. \tag{22.20}$$

This equation might look scary, as the $\mathcal{Q}_i(\cdot)$ are distributions over distributions over a compact subset of the reals. It is useful to rewrite it in a mathematically more correct form. This is done by requiring, for any measurable set of distributions $\mathcal{A}$ (see the footnote), the following equality to hold:

$$\mathcal{Q}_0(\mathcal{A}) = \frac{1}{\mathcal{Z}\{\mathcal{Q}_i\}} \int Z\{Q_i\}^{\mathrm{x}_2/\mathrm{x}_1} \, \mathbb{I}(\mathsf{F}_0\{Q_i\} \in \mathcal{A}) \prod_{i=1}^{l} \mathrm{d}\mathcal{Q}_i(Q_i) \,. \tag{22.21}$$

The interpretation of the 2RSB messages $\mathcal{Q}_i$ is obtained by analogy with the 1RSB case. Let α_1 be the index of a particular cloud of states, and let $Q_i^{\alpha_1}(\cdot)$ be the distribution of the message ν_i over the lumps in cloud α_1. Then $\mathcal{Q}_i$ is the distribution of $Q_i^{\alpha_1}$ when one picks out a cloud index α_1 randomly (each cloud being sampled with a weight that depends on x_1.)

In principle, eqn (22.20) can be studied numerically by generalizing the population dynamics approach of Chapter 19. In the present case, one can think of two implementations: for one given instance, one can generalize the SP algorithm, but this generalization involves, on each directed edge of the factor graph, a population of populations. If, instead, one wants to perform a statistical analysis of these messages, seeking a fixed point of the corresponding density evolution, one must use a population of populations of populations! This is obviously challenging from the point of view of computer resources (both memory and time). To the best of our knowledge, it has been tried only once, in order to compute the ground state energy of the spin glass on random 5-regular graphs. Because the graph is regular, it looks identical at any finite distance from any given point. One can therefore seek a solution such that the $\mathcal{Q}_i$ on all edges are the same, and one is back to the study of populations of populations. The results have been summarized in Table 17.1: if one looks at the ground state energy, the 2RSB method provides a small correction of order 10^{-4} to the 1RSB value, and this correction seems to be in agreement with the numerical estimates of the ground state.

22.2.3 Local stability of the 1RSB phase

The above discussion of 2RSB will help us to check the stability of the 1RSB phase. The starting point consists in understanding the various ways in which the 2RSB formalism can reduce to the 1RSB one.

---

[1]The mathematically inclined reader might be curious about the precise definition of a probability distribution over the space of distributions. It turns out that, given a measure space Ω (in our case a $(|\mathcal{X}| - 1)$-dimensional simplex), the set of distributions over Ω can be given a measurable structure that makes 2RSB equations well defined. This is done by using the smallest σ-field under which the mapping $Q \mapsto Q(A)$ is measurable for any $A \subseteq \Omega$ that is measurable.

The first obvious reduction consists in taking the 2RSB distribution $\mathcal{Q}_i$ to be a Dirac delta at Q_i^*. In other words, for any continuous functional $\mathcal{F}$ on the space of distributions,

$$\int \mathcal{F}(Q_i) \, d\mathcal{Q}_i(Q_i) = \mathcal{F}(Q_i^*) \,. \tag{22.22}$$

It is not hard to check that, if $\{Q_i^*\}$ solves the 1RSB equation (22.19), this choice of $\{\mathcal{Q}_i\}$ solves eqn (22.20) independently of x_2.

There exists, however, a second reduction, which corresponds to taking $\mathcal{Q}_i(\cdot)$ to be a non-trivial distribution, but supported on Dirac deltas. Let us denote by δ_{ν^*} a 1RSB distribution which is a Dirac delta on the message $\nu = \nu^*$. Given a set of messages $\{Q_i^*\}$ that solves the 1RSB equation (22.19), we construct $\mathcal{Q}_i(\cdot)$ as a superposition of Dirac deltas over all values of ν^*, each one appearing with a weight $Q_i^*(\nu^*)$. Again, this distribution is defined more precisely by its action on a continuous functional $\mathcal{F}(Q)$:

$$\int \mathcal{F}(Q_i) \, d\mathcal{Q}_i(Q_i) = \int \mathcal{F}(\delta_{\nu^*}) \, dQ_i^*(\nu^*) \,. \tag{22.23}$$

Exercise 22.2 Suppose that $\{Q_i^*\}$ solves the analogue of the 1RSB equation (22.19) in which the parameter x_1 has been changed to x_2. Show that the $\mathcal{Q}_i$ defined by eqn (22.23) solves eqn (22.20) independently of x_1.

[Hint: Show that, when evaluated on Dirac deltas, the normalization Z appearing in eqn (22.19) is related to the normalization z in eqn (22.18) by $Z\{\delta_{\nu_i}\} = (z\{\nu_i\})^{x_1}$.]

In view of the interpretation of the 2RSB messages $\mathcal{Q}_i$ outlined in the previous subsection, and illustrated schematically in Fig. 22.3, these two reductions correspond to qualitatively different limiting situations. In the first case, described by eqn (22.22), the distribution over clouds becomes degenerate: there is essentially one cloud (by this we mean that the number of clouds is not exponentially large in N: the corresponding complexity vanishes). In the second case, described by eqn (22.23), it is the distribution within each cloud that trivializes: there is only one cluster (in the same sense as above) in each cloud.

What are the implications of these remarks? Within the 1RSB approach, one needs to solve eqn (22.19) in the space of distributions over BP messages: let us call this the '1RSB space'. When passing to 2RSB, one seeks a solution of eqn (22.20) within a larger '2RSB space', namely the space of distributions over distributions over BP messages. Equations (22.22) and (22.23) provide two ways of embedding the 1RSB space in the 2RSB space.

When one finds a 1RSB solution, one should naturally ask whether there exists a proper 2RSB solution as well (i.e. a solution outside the 1RSB subspace). If this is not the case, physicists usually conjecture that the 1RSB solution is asymptotically correct (for instance, it yields the correct free energy per spin). This check has been carried out for models on complete graphs (e.g. the fully connected p-spin glasses). So far, the difficulties of studying the 2RSB equations have prevented its implementation for sparse factor graphs.

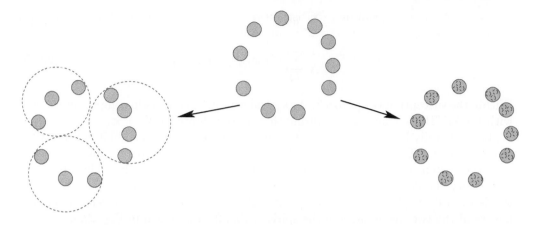

Fig. 22.5 Cartoon of the two types of local instabilities from a 1RSB solution towards 2RSB.

Luckily there is a convenient (albeit less ambitious) alternative: check the **local stability of 1RSB** solutions with respect to higher-order RSB. Given a 1RSB solution, we look at it as a point in the 2RSB space according to the two possible embeddings, and we study the effect of a small perturbation. More precisely, consider the iteration of the 2RSB equations (22.20):

$$\mathcal{Q}_{i\to j}^{(t+1)}(Q_0) = \frac{1}{\mathcal{Z}\{\mathcal{Q}_{l\to i}\}} \int Z\{Q_{l\to i}\}^r\, \delta(Q_{i\to j} - \mathsf{F}_i\{Q_{l\to i}\}) \prod_{l\in\partial i\setminus j} \mathrm{d}\mathcal{Q}_{l\to i}^{(t)}(Q_{l\to i})\,.$$

Given the factor graph G, we initiate this iteration from a point close to the 1RSB solution described by either of the embeddings (22.22) or (22.23) and see if the iteration converges back to the 1RSB fixed point. This is studied by linearizing the iteration in terms of an appropriate 'perturbation' parameter. If the iteration does not converge to the 1RSB fixed point, the 1RSB solution is said to be unstable. The instability is said to be of 'type I' if it occurs when the embedding (22.22) is used, and of 'type II' if it occurs for the embedding (22.23).

An alternative approach to checking the local stability of a 1RSB solution consists in computing the spin glass susceptibility, which describes the reaction of the model (22.16) to a perturbation that acts on a single variable x_i. As we discussed above, the effect of this perturbation (studied to linear order) remains finite when the spin glass susceptibility $\chi^{(2)}$ is finite. One should therefore compute $\chi^{(2)}$ assuming that the 1RSB solution is correct, and check that it is finite. However, the 1RSB picture implies a second condition: each single lump r should also be stable to small perturbations. More precisely, we define $\chi^{\mathrm{SG},r}$ as the spin glass susceptibility with respect to the measure $\mu^r(\cdot)$ restricted to the state r. If we denote by $\langle\cdot\rangle_r$ the expectation value with respect to μ^r, the 'intra-state' susceptibility $\chi^{\mathrm{SG},\mathrm{intra}}$ is a weighted average of $\chi^{\mathrm{SG},r}$ over the states:

$$\chi^{\mathrm{SG,intra}} = \sum_r w_r \, \chi^{\mathrm{SG},r}, \qquad\qquad (22.24)$$

$$\chi^{\mathrm{SG},r} = \frac{1}{N} \sum_{i,j} \left(\langle x_i x_j \rangle_r - \langle x_i \rangle_r \langle x_j \rangle_r \right)^2. \qquad\qquad (22.25)$$

Within the susceptibility approach, the second condition consists in computing $\chi^{\mathrm{SG,intra}}$ with the 1RSB approach and requiring that it stays finite as $N \to \infty$.

It is generally believed that these two approaches to the local stability of the 1RSB phase coincide. Type I stability should be equivalent to $\chi^{(2)}$ being finite; it means that the system is stable with respect to the grouping of states into clusters. Type II stability should be equivalent to $\chi^{\mathrm{SG,intra}}$ being finite; it means that the system is stable towards a splitting of the states into substates. A pictorial representation of the nature of the two instabilities in the spirit of Fig. 22.3 is shown in Fig. 22.5.

The two approaches to stability computations have been developed in several special cases, and are conjectured to coincide in general. Remarkably, 1RSB is unstable in several interesting cases, and higher-order RSB would be needed to obtain exact predictions.

Stability computations are somewhat involved, and a detailed description is beyond our scope. Nevertheless, we would like to give an example of the results that can be obtained through a local stability analysis. Consider random K-SAT formulae, with N variables and $M = N\alpha$ clauses. Let $e_{\mathrm{s}}(\alpha)$ denote the minimum number of unsatisfied clauses per variable, in the large-system limit. The limit $e_{\mathrm{s}}(\alpha)$ can be computed along the lines of Chapter 20 using the 1RSB cavity method: for a given α, one computes the energetic complexity density $\Sigma^{\mathrm{e}}(e)$ versus the density of violated clauses e. Then $e_{\mathrm{s}}(\alpha)$ is found as the minimal value of u such that $\Sigma^{\mathrm{e}}(e) > 0$. It vanishes for $\alpha < \alpha_{\mathrm{s}}(K)$ (the SAT–UNSAT threshold) and, for $\alpha > \alpha_{\mathrm{s}}(K)$, departs continuously from 0, increasing monotonically.

The stability computation shows that, for a given α, there is in general an instability of type II which appears above some value $e = e_{\mathrm{G}}(\alpha)$: only the part of $\Sigma^{\mathrm{e}}(e)$ with $e \leq e_{\mathrm{G}}(\alpha)$ is in a locally stable 1RSB phase. When $\alpha < \alpha_{\mathrm{m}}(K)$, $e_{\mathrm{G}}(\alpha) = 0$ and the whole 1RSB computation is unstable. For $\alpha > \alpha_{\mathrm{G}}(K)$, $e_{\mathrm{G}}(\alpha) < e_{\mathrm{s}}(\alpha)$ (the ground state energy density), and again 1RSB is unstable (this implies that the 1RSB prediction for $e_{\mathrm{s}}(\alpha)$ is not correct). The conclusion is that the 1RSB calculation is stable only in an interval $]\alpha_{\mathrm{m}}(K), \alpha_{\mathrm{G}}(K)[$. Figure 22.6 summarizes this discussion for 3-SAT. For all values of K, the stable interval $]\alpha_{\mathrm{m}}(K), \alpha_{\mathrm{G}}(K)[$ contains the SAT–UNSAT threshold $\alpha_{\mathrm{s}}(K)$.

The stability check leads to the conjecture that the 1RSB prediction for $\alpha_{\mathrm{s}}(K)$ is exact. Let us stress, however, that stability has been checked only with respect to small perturbations. A much stronger argument would be obtained if one could do the 2RSB computation and show that it has no solution apart from the two 'embedded 1RSB solutions' that we discussed above.

22.2.4   Open problems within the cavity method

The main open problem is, of course, to prove that the 1RSB cavity approach yields correct predictions for some models. This has been achieved so far only for a class of

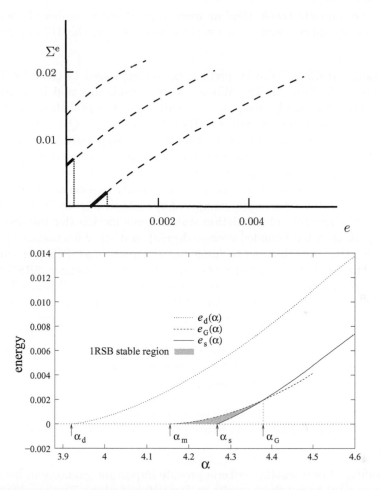

Fig. 22.6 *Top*: the energetic complexity Σ^e in a random 3-SAT problem, computed within the 1RSB cavity method, plotted versus the density e of violated clauses, for $\alpha = 4.1$, 4.2, and 4.3 (from *top* to *bottom*). The curves reproduce Fig. 20.5, but now the stable and unstable regions are shown. The thick lines, below $e_G(\alpha)$, give the part of the complexity curve for which the 1RSB computation is locally stable (absent for $\alpha = 4.1 < \alpha_m(3)$, where the whole curve is unstable). This is the only part that is computed reliably by 1RSB; the parts shown by dashed lines are unstable. *Bottom*: for the same random 3-SAT problem, plotted versus the clause density α, the continuous line gives the minimum density of unsatisfied clauses as predicted within 1RSB (this is the value of e where $\Sigma^e(e)$ starts to become positive). The dotted line gives the threshold energy density as predicted within 1RSB (the maximal value of e where $\Sigma^e(e)$ exists). The grey area indicates the region of local stability of the 1RSB solution. The ground state energy density predicted by 1RSB is wrong for $\alpha > \alpha_G$ (although probably very close to the actual value), because in this region there is an instability towards higher-order RSB. It is conjectured that the stable region, $\alpha_m < \alpha < \alpha_s$, is in a 1RSB phase: if this conjecture holds, the 1RSB prediction of α_s for the SAT–UNSAT threshold is correct. For $K = 3$, one has $\alpha_m(3) = 4.153(1)$, $\alpha_s(3) = 4.2667(1)$, and $\alpha_G(3) = 4.390(5)$.

models on the complete graph. Here we want to point out a number of open questions that wait for an answer, even at a heuristic level, within the 1RSB cavity method itself.

Distributional equations.   Cavity predictions are expressed in terms of fixed points of equations of the form (22.19). When we are considering models on ensembles of random graphs, this can be read as an equation for the probability distribution of $Q_0(\cdot)$ (which is taken to be identical to those of $Q_1(\cdot), \ldots, Q_k(\cdot)$.)

Currently such equations are mostly studied using the population dynamics method of Section 14.6.4. The main alternative explored so far has been to formally expand the equations for large degrees. Population dynamics is powerful and versatile. However, in many cases, this approach is too coarse, particularly as soon as one wants to study k-RSB with $k \geq 2$. It is intrinsically hampered by statistical errors that are of the order of the inverse square root of population size. In some models (for instance, for graph ensembles with large but bounded average degree), statistical fluctuations are too large for the population sizes that can be implemented on ordinary PCs (typically 10^7–10^8 elements). This limits the possibility of distinguishing, for instance, 2RSB from 1RSB effects, because high precision is generally required to see the difference. Furthermore, metastability is the crux (and the limit) of the whole population dynamics approach. Therefore it would be interesting to make progress in two directions:

- Analytical tools and generic results for the cavity equations; this could provide important guiding principles for any numerical study.
- New efficient and stable numerical methods.

A step forward has been made with the reconstruction algorithm discussed in Theorem 19.5, but unfortunately it is limited to one value of the rescaling parameter, $x = 1$.

Local stability.   Local stability criteria provide important guidance in heuristic studies. It is important to put these results on firmer foundations. Two specific tasks could be, for instance:

- Prove that if all 1RSB solutions of the cavity equations are locally unstable, then there must exist a 2RSB solution outside the 1RSB subspace.
- Prove that if a solution of the cavity equations is locally unstable, it does not describe the model correctly.

Occurrence of k-RSB.   A number of random graphical models have been studied within the cavity (or replica) method. In most cases, it has been found that the system is either RS, 1RSB, or FRSB. The cases in which a 2RSB phase is found are rare, and they always involve some kind of special construction of the compatibility function (for instance, a fully connected model which is a superposition of two p-spin glass interactions, with $p_1 = 3$ and $p_2 = 16$, displays 2RSB). Therefore one should:

- Find a 'natural' model for which 2RSB is asymptotically exact, or understand why this is impossible.

Full replica symmetry breaking. We saw that k-RSB provides, as k increases, a sequence of 'nested' schemes aimed at computing various quantities such as local marginals and the free-entropy density in the large-system limit. A k-th order scheme includes all of the lower l-RSB schemes with $l < k$ as nested subspaces of the set of feasible solutions to the cavity equations. On the other hand, as the number of steps increases, the description of the set of feasible solutions becomes more and more complicated (distributions of distributions of . . .).

Surprisingly, in the case of fully connected models, there exists a compact description of the space of feasible solutions in the FRSB limit $k \to \infty$. An outstanding problem is to find an analogous description in the case of models on sparse graphs. This would allow us to look for the best solution in the k-RSB space for all k. Therefore we need to:

- Find a description of the space of full replica-symmetry-breaking messages for models on sparse graphs.

Variational aspects. It is widely believed that if one finds a consistent solution of the cavity k-RSB equations, the free-energy density computed with this solution is always a lower bound on the correct free-energy density of the model (in particular, the k-RSB ground state energy density prediction is a lower bound on the true value). This should hold for a large class of models.While this has been proven in some specific cases, one would like to:

- Find a general proof that the free energy computed with the cavity method is a lower bound on the correct free energy of the model.

22.3 Phase structure and the behaviour of algorithms

A good part of this book has been devoted to the connection between the various phases in random graphical models and the behaviour of algorithms. There exists by now substantial evidence (empirical, heuristic, and in some cases rigorous) that such a connection exists. For instance, we have seen with the example of codes in Chapter 21 how the appearance of a 1RSB phase, and the corresponding proliferation of metastable states, determines the noise threshold where BP decoding fails. Developing a broader understanding of this connection, and determining the class of algorithms to which it applies, is a very important problem.

We propose here a list of broad research problems, whose advancement will probably help to clarify this issue. We always have in mind a graphical model of the form (22.1), with a locally tree-like factor graph.

Impact of the dynamical transition on Monte Carlo dynamics. Consider the problem of sampling from the distribution (22.1) using a Monte Carlo Markov chain (MCMC) algorithm. The Markov chain is assumed to flip a sublinear ($o(N)$) number of variables at each step, and to satisfy detailed balance with respect to the probability distribution $\mu(\cdot)$.

One expects that, if the system is in a 1RSB phase, the relaxation time of this algorithm will increase rapidly (probably exponentially) with the system size. Intuitive arguments in favour of this statement can be obtained from each of the two characterizations of the 1RSB phases introduced in Section 22.1. The argument is different depending on whether we start from the pure-state decomposition or from the characterization in terms of correlations. In the first case, the relaxation time is estimated through the time to cross a bottleneck (see also Chapter 13). In the second case, one can define a correlation length ℓ_i^* through the point-to-set correlation function $G_i(\ell)$ (see eqn (22.14)). In order for the system to relax, information has to travel a distance ℓ_i^*. But if ℓ_i^* diverges with the size, so must the relaxation time.

This picture is intuitively satisfying, but it is far from being proved, and needs to be formulated more precisely. For instance, it often happens that in RS phases there exist small isolated metastable states that make the relaxation time (the inverse spectral gap of the MCMC) formally large. But even in such cases, numerical simulations indicate that Glauber dynamics equilibrates rapidly within the RS phase. This observation is probably related to the fact that the initial condition is chosen uniformly at random, and that equilibration is only checked for on local observables. A number of questions arise:

- Why is metastability irrelevant 'in practice' in an RS phase? Is it because of local measurements? Or because of the uniform initial condition? If the latter is true, what is so special about the uniform initial condition?

- Within an RS phase, can one approximate partition functions efficiently?

Message passing and the estimation of marginals.   For a number of models on sparse random graphs within the RS and (sometimes) dynamical 1RSB phases, message-passing methods such as belief propagation and survey propagation show good performance empirically. More precisely, they return good approximations of local expectation values if initialized from uniform messages.

Current rigorous techniques for analysing BP often aim at proving that it is accurate regardless of the initialization. As a consequence, results are dominated by the behaviour under *worst-case* initializations that are not used in practice. As an illustration, consider applying BP to the uniform measure over solutions of a random K-SAT formula. An analysis under worst-case initialization allows one to prove that BP is accurate only for $\alpha \leq (2 \log K)/K[1 + o(1)]$. This threshold is embarrassingly small when compared with the dynamical transition point that terminates the RS phase at $\alpha_d(K) = 2^K \log K/K[1 + o(1)]$.

In general, we have no good mathematical control of when BP or SP converges and/or gives good approximations of marginals. Empirically, it seems that SP is able to converge in some regions of 1RSB phases where BP does not. We have no real understanding of this fact beyond the hand-waving argument that 1RSB correctly captures the structure of correlations in these phases.

Here are a number of open questions about these issues:

- Why is the performance of BP and SP on random instances, with uniformly random initialization, much better than in the worst case? What is special about the uniform initialization? What are the features of random instances that make them easier? Can these features be characterized and checked efficiently?

- Under what conditions do BP (and SP) algorithms converge and give good approximations to local marginals? When their naive iteration does not converge, can one systematically either force convergence or use time averages of the messages?

- It seems that, on sparse random graphical models, BP and SP outperform local MCMC algorithms. In particular, these message-passing algorithms can have (at least in principle), good performance within the dynamical 1RSB phase. Can one demonstrate this possibility convincingly for some model?

Message-passing algorithms and optimization. If one is seeking a solution to a random constraint satisfaction problem using message passing, the main approach so far has been the use of decimation: one first computes all local marginals, then decides, based on this knowledge, how to fix a variable, and then iterates the procedure. In general, this procedure converges when the number of constraints per variable is not too large, but it fails above a critical value of this number, which is strictly smaller than the SAT–UNSAT threshold. No one knows how to determine this threshold analytically.

An alternative to decimation is the reinforcement method: instead of fixing a variable based on a knowledge of local marginals, it modifies some local factors applying to each of the individual variables, based on this same information. So far, optimizing this modification is an art, and its critical threshold cannot be estimated either. Some of the questions that arise are the following:

- How can we *predict* the performance of BP+decimation or SP+decimation? For instance, empirically these methods find solutions to random K-SAT formulae with high probability for $\alpha < \alpha_{\mathrm{BP}}(K)$ (or $\alpha < \alpha_{\mathrm{SP}}(K)$), but we have no prediction for these algorithmic thresholds. In what class of problems is SP better than BP?

- Similar questions for BP+reinforcement and SP+reinforcement.

- We need to find new ways to use the information about local marginals found by message passing in order to exhibit solutions.

- In an UNSAT phase, the message-passing procedure is able to give an estimate of the minimal number of violated constraints. Is it possible to use this information, and that contained in the messages, in order to prove unsatisfiability for one given instance?

The above questions focus on sparse random instances. Message-passing techniques have been (partially) understood and sharpened for this type of instance. They naturally arise in a large class of applications where the graphical model is random or pseudorandom *by design*. The theory of sparse-graph codes is a clear example in this direction. In the limit of large block lengths, random constructions have proved to be generally superior to deterministic ones. More recently, sparse-graph constructions have been proposed for data compression (both lossless and lossy), online network measurements, multiterminal communications, distributed storage, group testing, etc.

On the other hand, being able to deal with structured graphs would open up an even broader class of applications. When applied to structured problems, message-passing algorithms often fail to converge. This is typically the reason why the decimation method may fail, even when the marginals of the original problem are well estimated by message passing: the instance found after fixing many variables is no longer random.

Finding appropriate modifications of message passing for structured graphs would therefore be very interesting. The question that arises, therefore, is:

- How can we use message passing in order to improve the solution of some general classes of (non-random) constraint satisfaction problems? Can it be coupled efficiently to other general methods (such as MCMC)?

Notes

The present chapter has inevitably been elliptic. We shall provide a few pointers to recent research here, without any ambition to be comprehensive.

The connection between correlation lengths and phase transitions is a classical topic in statistical mechanics which has recently been revived by the interest in the glass transition. A good starting point for learning about this subject in the context of glasses is the paper by Bouchaud and Biroli (2004), which describes the point-to-set correlation function using a thought-experiment where one freezes the variables outside a ball (see Section 22.1.2).

The description of point-to-set correlations in terms of 'reconstruction' problems is taken from Evans *et al.* (2000). This paper studied the reconstruction phase transition for Ising models on trees. Results for a wide class of models on trees were surveyed by Mossel and Peres (2003) and Mossel (2004). We also refer to Gerschenfeld and Montanari (2007) for the generalization to non-tree graphs. The connection between 'reconstruction' and the 'dynamical' 1RSB phase transition was first pointed out by Mézard and Montanari (2006). The implications of this phase transition for dynamics were explored by Berger *et al.* (2005), Martinelli *et al.* (2004), and Montanari and Semerjian (2006*b*). The definition of pure states presented in this chapter and the location of the dynamical and condensation phase transitions for random K-SAT and the colouring of random graphs are from Krzakala *et al.* (2007).

The GREM was introduced by Derrida (1985) and studied in detail by Derrida and Gardner (1986). A 2RSB phase in fully connected models has been found by Crisanti and Leuzzi (2007). There are very few results about higher-order RSB in models on sparse random graphs. For spin glasses, one can use perturbative expansions close to the critical point (Viana and Bray, 1985) or for large degrees (Goldschmidt and De Dominicis, 1990). The 2RSB computation of the ground-state energy for spin glasses mentioned in Section 22.2 is from Montanari (2003). The method for verifying the local stability of the 1RSB solution for sparse systems was first devised by Montanari and Ricci-Tersenghi (2003), and applied to random satisfiability problems by Montanari *et al.* (2004). A complete list of stability thresholds, including their asymptotic behaviour, for random K-SAT can be found in Mertens *et al.* (2006). The interpretation of 1RSB instability in terms of susceptibilities was discussed by Rivoire *et al.* (2003).

The fact that the free energy computed with the cavity (or replica) method is a lower bound on the true value can be proven in some fully connected models using the inequalities of Guerra (2003). The same strategy also yields rigorous bounds in some diluted systems (Franz and Leone, 2003; Franz *et al.*, 2003; Panchenko and Talagrand, 2004) but it still relies on some details of the structure of the models, and a general proof applicable to all cases is lacking.

The reinforcement algorithm was introduced and discussed for SAT by Chavas *et al.* (2005).

There exist only scarce results on the algorithmic consequences of the structure of the solution space. Some recent analyses can be found in Altarelli *et al.* (2007), Montanari *et al.* (2007), Ardelius and Aurell (2006) and Alava *et al.* (2007). The convergence and correctness of BP for random K-satisfiability at small enough α was proven by Montanari and Shah (2007).

This book has covered only a small subset of the problems that lie at the intersection between information theory, computer science and statistical physics. It would be difficult to provide an exhaustive list of references on the topics that we did not touch: we shall instead limit ourselves to a few 'access points'.

The vertex-covering problem has been studied in detail by the replica and the cavity method (Weigt and Hartmann, 2000; Zhou, 2003; Hartmann and Weigt, 2005).

As we mentioned, channel coding is only one of the fundamental problems addressed by information theory. Data compression, in particular in its 'lossy' version, is a key component in many modern technologies, and presents a number of open problems (Ciliberti *et al.*, 2005; Wainwright and Maneva, 2005). Some other statistical problems such as group testing are similar in spirit to data compression (Mézard *et al.*, 2007).

Modern wireless and wireline communication systems are intrisically multi-user systems. Finding optimal coding schemes in a multi-user context is a very open subject of great practical interest. Even the information-theoretic capacity of such systems is unknown. Two fields that have benefited from tools or from analogies with statistical mechanics are multi-user detection (Tanaka, 2002; Guo and Verdú, 2002) and networking (Kelly, 1991). Also within a communications context, a large effort has been devoted to characterizing large communication networks such as the Internet. A useful review was provided by Kleinberg *et al.* (1999).

Statistical mechanics concepts have been applied to the analysis of fluctuations in financial markets (Bouchaud and Potters, 2003) and for modelling interactions among economic agents (Challet *et al.*, 2005). Finally, biology presents a number of problems in which randomness, interaction between different components, and robustness play important roles. Stochastic models on networks, and inference algorithms have been studied in a number of contexts, from neural networks (Baldassi *et al.*, 2007; Coolen *et al.*, 2000) to phylogeny (Mossel, 2003) and gene expression (Friedman *et al.*, 2000).

A few of these topics, and others, are reviewed in recent summer school proceedings (Bouchaud *et al.*, 2007).

Appendix A
Symbols and notation

In this appendix, we summarize the conventions that we have adopted throughout the book for symbols and notation. Secs. A.1 and A.2 deal with equivalence relations and orders of growth. Sec. A.3 presents the notation used in combinatorics and probability. The table in Section A.4 gives the main mathematical notation, and that in Section A.5 gives the notation for information theory. The table in Section A.6 summarizes the notation used for factor graphs and graph ensembles. The table in Section A.7 focuses on the notation used in message passing, belief propagation and survey propagation, and the cavity method.

A.1 Equivalence relations

As usual, the symbol '=' denotes equality. We also use '≡' for definitions and '≈' for 'numerically close to'. For instance, we may say that the Euler–Mascheroni constant is given by

$$\gamma_{\mathrm{E}} \equiv \lim_{n \to \infty} \left(\sum_{k=1}^{n} \frac{1}{k} - \log n \right) \approx 0.5772156649 \,. \tag{A.1}$$

When dealing with two random variables X and Y, we write $X \overset{\mathrm{d}}{=} Y$ if X and Y have the same distribution. For instance, given $n + 1$ i.i.d. Gaussian variables $X_0, \ldots, X_n$, with zero mean and unit variance, we can write

$$X_0 \overset{\mathrm{d}}{=} \frac{1}{\sqrt{n}} \left(X_1 + \cdots + X_n \right) . \tag{A.2}$$

We have adopted several equivalence symbols to denote the asymptotic behavior of functions as their argument tends to some limit. For the sake of simplicity, we assume here that the argument is an integer $n \to \infty$. The limit to be considered in each particular case should be clear from the context. We write $f(n) \doteq g(n)$ if f and g are equal 'to the leading exponential order' as $n \to \infty$, i.e. if

$$\lim_{n \to \infty} \frac{1}{n} \log \frac{f(n)}{g(n)} = 0 \,. \tag{A.3}$$

For instance, we may write

$$\binom{n}{\lfloor n/2 \rfloor} \doteq 2^n \,. \tag{A.4}$$

We write instead $f(n) \sim g(n)$ if f and g are asymptotically equal 'up to a constant', i.e. if

$$\lim_{n \to \infty} \frac{f(n)}{g(n)} = C \,, \tag{A.5}$$

for some constant $C \neq 0$. For instance, we have

$$\frac{1}{2^n} \binom{n}{\lfloor n/2 \rfloor} \sim n^{-1/2} \,. \tag{A.6}$$

Finally, the symbol '$\simeq$' is reserved for asymptoric equality, i.e. if

$$\lim_{n \to \infty} \frac{f(n)}{g(n)} = 1 \,. \tag{A.7}$$

For instance, we have

$$\frac{1}{2^n} \binom{n}{\lfloor n/2 \rfloor} \simeq \sqrt{\frac{2}{\pi n}} \,. \tag{A.8}$$

The symbol '$\cong$' denotes equality up to a constant. If $p(\,\cdot\,)$ and $q(\,\cdot\,)$ are two measures on the same finite space $\mathcal{X}$ (not necessarily normalized), we write $p(x) \cong q(x)$ if there exists $C > 0$ such that

$$p(x) = C \, q(x) \,, \tag{A.9}$$

for any $x \in \mathcal{X}$. The definition generalizes straightforwardly to infinite sets $\mathcal{X}$: the Radon–Nikodyn derivative between p and q is a positive constant.

A.2 Orders of growth

We have used a few symbols to denote the order of growth of functions when their arguments tend to some definite limit. For the sake of definiteness, we refer here to functions of an integer $n \to \infty$. As above, the adaptation to any particular context should be straightforward.

We write $f(n) = \Theta(g(n))$, and say that $f(n)$ is of order $g(n)$, if there exist two positive constants C_1 and C_2 such that

$$C_1 \, g(n) \leq |f(n)| \leq C_2 g(n) \,, \tag{A.10}$$

for any n large enough. For instance, we have

$$\sum_{k=1}^{n} k = \Theta(n^2) \,. \tag{A.11}$$

We write instead $f(n) = o(g(n))$ if

$$\lim_{n \to \infty} \frac{f(n)}{g(n)} = 0 \,. \tag{A.12}$$

For instance,

$$\sum_{k=1}^{n} k - \frac{1}{2} n^2 = o(n^2) \,. \tag{A.13}$$

Finally, $f(n) = O(g(n))$ if there exist a constant C such that

$$|f(n)| \le C \, g(n) \tag{A.14}$$

for any n large enough. For instance,

$$n^3 \sin(n/10) = O(n^3) \,. \tag{A.15}$$

Note that both $f(n) = \Theta(g(n))$ and $f(n) = o(g(n))$ imply $f(n) = O(g(n))$. As the last example shows, the converse is not necessarily true.

A.3 Combinatorics and probability

The standard notation is used for multinomial coefficients. For any $n \ge 0$, $l \ge 2$, and $n_1, \ldots, n_l \ge 0$ such that $n_1 + \cdots + n_l = n$, we have

$$\binom{n}{n_1, n_2, \ldots, n_l} \equiv \frac{n!}{n_1! n_2! \ldots n_l!} \,. \tag{A.16}$$

For binomial coefficients (i.e. for $l = 2$), the usual shorthand is

$$\binom{n}{k} \equiv \binom{n}{k, l - k} = \frac{n!}{k!(n - k)!} \,. \tag{A.17}$$

In combinatorics, certain quantities are most easily described in terms of their generating functions. Given a formal power series $f(x)$, $\mathsf{coeff}\{f(x), x^n\}$ denotes the coefficient of the monomial x^n in the series. More formally,

$$f(x) = \sum_n f_n x^n \quad \Rightarrow \quad f_n = \mathsf{coeff}\{f(x), x^n\} \,. \tag{A.18}$$

For instance,

$$\mathsf{coeff}\{(1 + x)^m, x^n\} = \binom{m}{n} \,. \tag{A.19}$$

Some standard random variables are as follows:

- A Bernoulli p variable is a random variable X taking values in $\{0, 1\}$ such that $\mathbb{P}(X = 1) = p$.
- $B(n, p)$ denotes a binomial random variable with parameters n and p. This is defined as a random variable taking values in $\{0, \ldots, n\}$ and having a probability distribution

$$\mathbb{P}\{B(n, p) = k\} = \binom{n}{k} p^k (1 - p)^{n-k} \,. \tag{A.20}$$

- A Poisson random variable X with parameter λ takes integer values and has the following probability distribution:

$$\mathbb{P}\{X = k\} = \frac{\lambda^k}{k!} e^{-\lambda}. \tag{A.21}$$

The parameter λ is the mean of X.

Finally, we have used the symbol δ_a for the Dirac 'delta function'. This is in fact a measure that attributes unit mass to the point a. In formulae, for any set A,

$$\delta_a(A) = \mathbb{I}(a \in A). \tag{A.22}$$

A.4   Summary of mathematical notation

$=$	Equal.
$\equiv$	Defined as.
$\approx$	Numerically close to.
$\overset{\mathrm{d}}{=}$	Equal in distribution.
$\doteq$	Equal to the leading exponential order.
$\sim$	Asymptotically equal up to a constant.
$\cong$	Equal up to a normalization constant (for probabilities; see Eq. (14.3)).
$\Theta(f)$	Of the same order as f (see Sec. A.2).
$o(f)$	Grows more slowly than f (see Sec. A.2).
$\mathrm{argmax} f(x)$	The set of values of x where the real-valued function f reaches its maximum.
$\lfloor \cdot \rfloor$	Integer part. $\lfloor x \rfloor$ is the largest integer n such that $n \leq x$.
$\lceil \cdot \rceil$	$\lceil x \rceil$ is the smallest integer n such that $n \geq x$.
$\mathbb{N}$	The set of integer numbers.
$\mathbb{R}$	The set of real numbers.
$\beta \downarrow \beta_{\mathrm{c}}$	β goes to β_{c} through values $> \beta_{\mathrm{c}}$.
$\beta \uparrow \beta_{\mathrm{c}}$	β goes to β_{c} through values $< \beta_{\mathrm{c}}$.
$]a, b[$	Open interval of real numbers x such that $a < x < b$.
$]a, b]$	Interval of real numbers x such that $a < x \leq b$.
$\mathbb{Z}_2$	The field of integers modulo 2.
$a \oplus b$	The sum modulo 2 of the two integers a and b.
$\mathbb{I}(\cdot)$	Indicator function: $\mathbb{I}(A) = 1$ if the logical statement A is true, and $\mathbb{I}(A) = 0$ if the statement A is false.
$A \succeq 0$	The matrix A is positive semidefinite.

A.5 Information theory

H_X	Entropy of the random variable X (See Eq. (1.7)).
I_{XY}	Mutual information of the random variables X and Y (see Eq. (1.25)).
$\mathcal{H}(p)$	Entropy of a Bernoulli variable with parameter p.
$\mathfrak{M}(\mathcal{X})$	Space of probability distributions over a finite set $\mathcal{X}$.
$\mathfrak{C}$	Codebook.
$\preceq$	BMS(1) $\preceq$ BMS(2): channel BMS(2) is physically degraded with respect to BMS(1).
$\mathfrak{B}$	Bhattacharya parameter of a channel.

A.6 Factor graphs

$\mathbb{G}_N(k, M)$	Random k-factor graph with M function nodes and N variable nodes.
$\mathbb{G}_N(k, \alpha)$	Random k-factor graph with N variable nodes. Each function node is present independently with probability $N\alpha / \binom{N}{k}$.
$\mathbb{D}_N(\Lambda, P)$	Degre- constrained random factor graph ensemble.
$\mathbb{T}_r(\Lambda, P)$	Degree-constrained random tree factor graph ensemble.
$\mathbb{T}_r(k, \alpha)$	Shorthand for the random tree factor graph $\mathbb{T}_r(\Lambda(x) = e^{k\alpha(x-1)}, P(x) = x^k)$.
$\Lambda(x)$	Degree profile of variable nodes.
$P(x)$	Degree profile of function nodes.
$\lambda(x)$	Edge-perspective degree profile of variable nodes.
$\rho(x)$	Edge-perspective degree profile of function nodes.
$\mathsf{B}_{i,r}(F)$	Neighbourhood of radius r of variable node i.
$\mathsf{B}_{i \to a, t}(F)$	Directed neigbourhood of an edge.

A.7 Cavity and message-passing methods

$\nu_{i \to a}(x_i)$	BP messages (variable to function node).
$\widehat{\nu}_{a \to i}(x_i)$	BP messages (function to variable node).
Φ	Free entropy.
$\mathbb{F}(\underline{\nu})$	Bethe free entropy (as a function of messages).
$\mathbb{F}^{\mathrm{e}}(\underline{\nu})$	Bethe energy (as a function of min-sum messages).
f^{RS}	Bethe (RS) free-entropy density.
$Q_{i \to a}(\nu)$	1RSB cavity message/SP message (variable to function node).
$\widehat{Q}_{a \to i}(\widehat{\nu})$	1RSB cavity message/SP message (function to variable node).
x	Parisi 1RSB parameter.
$\mathfrak{F}(\mathsf{x})$	Free-entropy density of the auxiliary model counting BP fixed points.

$\Sigma(\phi)$	Complexity.
$\mathbb{F}^{\mathrm{RSB}}(Q)$	1RSB cavity free entropy (Bethe free entropy of the auxiliary model, a function of the messages).
$\mathfrak{f}^{\mathrm{RSB}}$	1RSB cavity free-entropy density.
y	Zero-temperature Parisi 1RSB parameter ($\mathrm{y} = \lim_{\beta \to \infty} \beta \mathrm{x}$).
$\mathfrak{F}^{\mathrm{e}}(\mathrm{y})$	Free-entropy density of the auxiliary model counting min-sum fixed points.
$\Sigma^{\mathrm{e}}(e)$	Energetic complexity.
$\mathbb{F}^{\mathrm{RSB,e}}(Q)$	Energetic 1RSB cavity free entropy (Bethe free entropy of the auxiliary model, a function of the messages).
$\mathfrak{f}^{\mathrm{RSB,e}}$	Energetic 1RSB cavity free-entropy density.

References

Aarts, E. H. L., Korst, J. H. M., and van Laarhoven, P. J. M. (2003). Simulated annealing. In *Local Search in Combinatorial Optimization*, pp. 91–120. Princeton University Press.

Abou-Chacra, R., Anderson, P. W., and Thouless, D. J. (1973). A self-consistent theory of localization. *J. Phys. C*, **6**, 1734–1752.

Achlioptas, D. (2001). Lower bounds for random 3-SAT via differential equations. *Theor. Comput. Sci.*, **265**, 159–185.

Achlioptas, D. (2007). Personal communication.

Achlioptas, D. and Moore, C. (2004). The chromatic number of random regular graphs. In *Proc. RANDOM'04*, Lecture Notes in Computer Science, Vol. 3122, pp. 219–228. Springer, Berlin.

Achlioptas, D. and Moore, C. (2007). Random k-SAT: Two moments suffice to cross a sharp threshold. *SIAM J. Comput.*, **36**, 740–762.

Achlioptas, D. and Naor, A. (2005). The two possible values of the chromatic number of a random graph. *Ann. Math.*, **162**, 1333–1349.

Achlioptas, D. and Peres, Y. (2004). The threshold for random k-SAT is $2k \log 2 - O(k)$. *J. Am. Math. Soc.*, **17**, 947–973.

Achlioptas, D. and Ricci-Tersenghi, F. (2006). On the solution-space geometry of random constraint satisfaction problems. In *Proc. 38th ACM Symposium on Theory of Computing, STOC*, Seattle, WA.

Achlioptas, D., Naor, A., and Peres, Y. (2005). Rigorous location of phase transitions in hard optimization problems. *Nature*, **435**, 759–764.

Aizenman, M., Sims, R., and Starr, S. L. (2005). Mean-field spin glass models from the Cavity-ROSt perspective. In *Mathematical Physics of Spin Glasses*, Cortona. Eprint `arXiv:math-ph/0607060`.

Aji, S.M. and McEliece, R.J. (2000). The generalized distributive law. *IEEE Trans. Inf. Theory*, **46**, 325–343.

Aji, A., Jin, H., Khandekar, A., MacKay, D., and McEliece, R. (2001). BSC thresholds for code ensembles based on 'Typical Pairs' decoding. In *Codes, Systems, and Graphical Models*, IMA Volumes in Mathematics and Its Applications, Vol.123, pp. 1–37. Springer, Berlin.

Alava, M., Ardelius, J., Aurell, E., Kaski, P., Krishnamurthy, S., Orponen, P., and Seitz, S. (2007). Circumspect descent prevails in solving random constraint satisfaction problems. Eprint `arXiv:0711.4902`.

Aldous, D. (1992). Asymptotics in the random assignment problem. *Probab. Theory Relat. Fields*, **93**, 507–534.

Aldous, D. (2001). The $\zeta(2)$ limit in the random assignment problem. *Random Struct. Algorithms*, **18**, 381–418.

Aldous, D. and Bandyopadhyay, A. (2005). A survey of max-type recursive distributional equations. *Ann. Appl. Probab.*, **15**, 1047–1110.

Aldous, D. and Fill, J. (2008). Reversible Markov chains and random walks on graphs. Book in preparation. Available online at `http://www.stat.berkeley.edu/users/aldous/RWG/book.html`.

Aldous, D. and Steele, J. M. (2003). The objective method: Probabilistic combinatorial optimization and local weak convergence. In *Probability on Discrete Structures* (ed. H. Kesten), pp. 1–72. Springer, Berlin.

Altarelli, F., Monasson, R., and Zamponi, F. (2007). Relationship between clustering and algorithmic phase transitions in the random k-XORSAT model and its NP-complete extensions. *J. Phys. A*, **40**, 867–886. Proc. International Workshop on Statistical-Mechanical Informatics, Kyoto.

Amraoui, A., Montanari, A., Richardson, T. J., and Urbanke, R. (2004). Finite-length scaling for iteratively decoded LDPC ensembles. *IEEE Trans. Inf. Theory*, in press. Eprint `http://arxiv.org/abs/cs/0406050`.

Amraoui, A., Montanari, A., and Urbanke, R. (2007). How to find good finite-length codes: From art towards science. *Eur. Trans. Telecommun.*, **18**, 491–508.

Applegate, D., Bixby, R., Chvátal, V., and Cook, W. The traveling salesman problem. Available online at `http://www.tsp.gatech.edu/`.

Ardelius, J. and Aurell, E. (2006). Behavior of heuristics on large and hard satisfiability problems. *Phys. Rev. E*, **74**, 037702.

Baldassi, C., Braunstein, A., Brunel, N., and Zecchina, R. (2007). Efficient supervised learning in networks with binary synapses. *Proc. Natl. Acad. Sci.*, **1049**, 11079–11084.

Balian, R. (1992). *From Microphysics to Macrophysics: Methods and Applications of Statistical Physics*. Springer, New York.

Bandyopadhyay, A. and Gamarnik, D. (2006). Counting without sampling: New algorithms for enumeration problems using statistical physics. In *Proc. 17th annual ACM-SIAM Symposium on Discrete Algorithms*. ACM, New York.

Barg, A. (1998). Complexity issues in coding theory. In *Handbook of Coding Theory* (ed. V. S. Pless and W. C. Huffman), Chapter 7. Elsevier Science, Amsterdam.

Barg, A. and Forney, G. D. (2002). Random codes: minimum distances and error exponents. *IEEE Trans. Inf. Theory*, **48**, 2568–2573.

Bauke, H. (2002). Statistische Mechanik des Zahlenaufteilungsproblems. Available at: `http://tina.nat.uni-magdeburg.de/heiko/Publications/index.php`.

Baxter, R. J. (1982). *Exactly Solved Models in Statistical Mechanics*. Academic Press, London.

Bayati, M., Shah, D., and Sharma, M. (2005). Maximum weight matching via max-product belief propagation. In *Proc. IEEE International Symposium on Information Theory*, Adelaide, pp. 1763–1767.

Bayati, M., Shah, D., and Sharma, M. (2006). A simpler max-product maximum weight matching algorithm and the auction algorithm. In *Proc. IEEE International Symposium on Information Theory*, Seattle, pp. 557–561.

Bender, E. A. and Canfield, E. R. (1978). The asymptotic number of labeled graphs with given degree sequence. *J. Comb. Theory (A)*, **24**, 296–307.

Berger, N., Kenyon, C., Mossel, E., and Peres, Y. (2005). Glauber dynamics on trees and hyperbolic graphs. *Probab. Theory Relat. Fields*, **131**, 311–340.

Berlekamp, E., McEliecee, R. J., and van Tilborg, H. C. A. (1978). On the inherent intractability of certain coding problems. *IEEE Trans. Inf. Theory*, **29**, 384–386.

Berrou, C. and Glavieux, A. (1996). Near optimum error correcting coding and decoding: Turbo codes. *IEEE Trans. Commun.*, **44**, 1261–1271.

Bertsekas, D. P. (1988). The auction algorithm: A distributed relaxation method for the assignment problem. *Ann. Oper. Res.*, **14**, 105–123.

Bethe, H. A. (1935). Statistical theory of superlattices. *Proc. R. Soc. London A*, **150**, 552–558.

Binder, K. and Young, A. P. (1986). Spin glasses: Experimental facts, theoretical concepts, and open questions. *Rev. Mod. Phys.*, **58**, 801–976.

Biroli, G. and Mézard, M. (2002). Lattice glass models. *Phys. Rev. Lett.*, **88**, 025501.

Boettcher, S. (2003). Numerical results for ground states of mean-field spin glasses at low connectivities. *Phys. Rev. B*, **67**, 060403.

Bollobás, B. (1980). A probabilistic proof of an asymptotic formula for the number of labelled regular graphs. *Eur. J. Combinatorics*, **1**, 296–307.

Bollobás, B. (2001). *Random Graphs*. Cambridge University Press, Cambridge.

Bollobas, B., Borgs, C., Chayes, J. T., Kim, J. H., and Wilson, D. B. (2001). The scaling window of the 2-SAT transition. *Random Struct. Algorithms*, **18**, 201–256.

Borgs, C., Chayes, J., and Pittel, B. (2001). Phase transition and finite-size scaling for the integer partitioning problem. *Random Struct. Algorithms*, **19**, 247.

Borgs, C., Chayes, J., Mertens, S., and Pittel, B. (2003). Phase diagram for the constrained integer partitioning problem. *Random Struct. Algorithms*, **24**, 315–380.

Bouchaud, J.-P. and Biroli, G. (2004). On the Adam–Gibbs–Kirkpatrick–Thirumalai–Wolynes scenario for the viscosity increase in glasses. *J. Chem. Phys*, **121**, 7347–7354.

Bouchaud, J.-P. and Mézard, M. (1997). Universality classes for extreme value statistics. *J. Phys. A: Math. Gen.*, **30**, 7997–8015.

Bouchaud, J.-P. and Potters, M. (2003). *Theory of Financial Risk and Derivative Pricing*. Cambridge University Press, Cambridge.

Bouchaud, J.-P., Cugliandolo, L., Kurchan, J., and Mézard, M. (1997). Out of equilibrium dynamics in spin glasses and other glassy systems. In *Spin Glasses and Random Fields* (ed. A.P. Young), pp.161–224. World Scientific, Singapore.

Bouchaud, J.-P., Mézard, M., and Dalibard, J. (eds.) (2007). *Complex Systems: Lecture Notes of the Les Houches Summer School 2006*. Elsevier, Amsterdam.

Boyd, S. P. and Vandenberghe, L. (2004). *Convex Optimization*. Cambridge University Press, Cambridge.

Braunstein, A. and Zecchina, R. (2004). Survey propagation as local equilibrium equations. *J. Stat. Mech.*, 06007.

Braunstein, A., Mulet, R., Pagnani, A., Weigt, M., and Zecchina, R. (2003). Polynomial iterative algorithms for coloring and analyzing random graphs. *Phys. Rev. E*, **68**, 036702.

Braunstein, A., Mézard, M., and Zecchina, R. (2004). C code for the SP algorithm. Available online at `http://www.ictp.trieste.it/~zecchina/SP/`.

Braunstein, A., Mézard, M., and Zecchina, R. (2005). Survey propagation: an algorithm for satisfiability. *Random Struct. Algorithms*, **27**, 201–226.

Burshtein, D. and Miller, G. (2004). Asymptotic enumeration methods for analyzing LDPC codes. *IEEE Trans. Inf. Theory*, **50**, 1115–1131.

Burshtein, D., Krivelevich, M., Litsyn, S., and Miller, G. (2002). Upper bounds on the rate of LDPC codes. *IEEE Trans. Inf. Theory*, **48**, 2437–2449.

Caracciolo, S., Parisi, G., Patarnello, S., and Sourlas, N. (1990). 3d Ising spin glass in a magnetic field and mean-field theory. *Europhys. Lett.*, **11**, 783.

Carlson, J. M., Chayes, J. T., Chayes, L., Sethna, J. P., and Thouless, D. J. (1990). Bethe lattice spin glass: The effects of a ferromagnetic bias and external fields. I. Bifurcation. *J. Stat. Phys.*, **61**, 987–1067.

Challet, D., Marsili, M., and Zhang, Y.-C. (2005). *Minority Games*. Oxford University Press, Oxford.

Chao, M.-T. and Franco, J. (1986). Probabilistic analysis of two heuristics for the 3-satisfiability problem. *SIAM J. Comput*, **15**, 1106–1118.

Chao, M. T. and Franco, J. (1990). Probabilistic analysis of a generalization of the unit-clause literal selection heuristics for the k-satisfiability problem. *Inf. Sci*, **51**, 289–314.

Chavas, J., Furtlehner, C., Mézard, M., and Zecchina, R. (2005). Survey-propagation decimation through distributed local computations. *J. Stat. Mech.*, 11016.

Chayes, J. T., Chayes, L., Sethna, J. P., and Thouless, D. J. (1986). A mean field spin glass with short-range interactions. *Commun. Math. Phys.*, **106**, 41–89.

Chung, S.-Y., Forney, G. D., Richardson, T. J., and Urbanke, R. (2001). On the design of low-density parity-check codes within 0.0045 dB of the Shannon limit. *IEEE Commun. Lett.*, **5**, 58–60.

Chvátal, V. and Reed, B. (1992). Mick gets some (and the odds are on his side). In *Proc. 33rd IEEE Symposium on Foundations of Computer Science, FOCS*, Pittsburgh, pp. 620–627.

Ciliberti, S., Mézard, M., and Zecchina, R. (2005). Lossy data compression with random gates. *Phys. Rev. Lett.*, **95**, 038701.

Clifford, P. (1990). Markov random fields in statistics. In *Disorder in Physical Systems: A Volume in Honour of John M. Hammersley* (eds. G. Grimmett and D. Welsh), pp. 19–32. Oxford University Press, Oxford.

Cocco, S. and Monasson, R. (2001*a*). Statistical physics analysis of the computational complexity of solving random satisfiability problems using backtrack algorithms. *Eur. Phys. J. B*, **22**, 505–531.

Cocco, S. and Monasson, R. (2001*b*). Trajectories in phase diagrams, growth processes, and computational complexity: How search algorithms solve the 3-satisfiability problem. *Phys. Rev. Lett.*, **86**, 1654–1657.

Cocco, S., Dubois, O., Mandler, J., and Monasson, R. (2003). Rigorous decimation-based construction of ground pure states for spin-glass models on random lattices. *Phys. Rev. Lett.*, **90**, 047205.

Cocco, S., Monasson, R., Montanari, A., and Semerjian, G. (2006). Approximate analysis of search algorithms with physical methods. In *Computational Complexity and Statistical Physics* (eds. A. Percus, G. Istrate, and C. Moore), Santa Fe Studies in the Science of Complexity, pp. 1–37. Oxford University Press, Oxford.

Conway, J. H. and Sloane, N. J. A. (1998). *Sphere Packings, Lattices and Groups.* Springer, New York.

Cook, S. A. (1971). The complexity of theorem-proving procedures. In *Proc. 3rd ACM Symposium on the Theory of Computing, STOC*, Shaker Heights, OH, pp. 151–158.

Coolen, A.C.C., Kuehn, R., and Sollich, P. (2005). *Theory of Neural Information Processing Systems.* Oxford University Press, Oxford.

Cooper, G. F. (1990). The computational complexity of probabilistic inference using Bayesian belief networks. *Artif. Intell./*, **42**, 393–405.

Coppersmith, D. and Sorkin, G. B. (1999). Constructive bounds and exact expectations for the random assignment problem. *Random Struct. Algorithms*, **15**, 133–144.

Cover, T. M. and Thomas, J. A. (1991). *Elements of Information Theory.* Wiley, New York.

Creignou, N. and Daudé, H. (1999). Satisfiability threshold for random XOR-CNF formulas. *Discrete Appl. Math.*, **96–97**, 41–53.

Creignou, N., Daudé, H., and Dubois, O. (2003). Approximating the satisfiability threshold for random k-XOR-formulas. *Combinatorics, Probab. Comput.*, **12**, 113–126.

Crisanti, A. and Leuzzi, L. (2007). Amorphous–amorphous transition and the two-step replica symmetry breaking phase. *Phys. Rev. B*, **76**, 184417.

Csiszár, I. and Körner, J. (1981). *Information Theory: Coding Theorems for Discrete Memoryless Systems.* Academic Press, New York.

Dagum, P. and Luby, M. (1993). Approximating probabilistic inference in Bayesian belief networks is NP-hard. *Artif. Intell.*, **60**, 141–153.

Darling, R. W. R. and Norris, J. R. (2005). Structure of large random hypergraphs. *Ann. Appl. Probab.*, **15**, 125–152.

Daudé, H., Mézard, M., Mora, T., and Zecchina, R. (2008). Pairs of SAT assignments in random boolean formulae. *Theor. Comput. Sci.*, **393**, 260–279.

Davis, M. and Putnam, H. (1960). A computing procedure for quantification theory. *J. Assoc. Comput. Mach.*, **7**, 201–215.

Davis, M., Logemann, G., and Loveland, D. (1962). A machine program for theorem-proving. *Commun. ACM*, **5**, 394–397.

de Almeida, J. R. L. and Thouless, D. (1978). Stability of the Sherrington–Kirkpatrick solution of a spin glass model. *J. Phys. A*, **11**, 983–990.

De Dominicis, C. and Mottishaw, P. (1987). Replica symmetry breaking in weak connectivity systems. *J. Phys. A*, **20**, L1267–L1273.

de la Vega, W. F. (1992). On random 2-SAT. Unpublished manuscript.

de la Vega, W. F. (2001). Random 2-SAT: Results and problems. *Theor. Comput. Sci.*, **265**, 131–146.

Dembo, A. and Montanari, A. (2008*a*). Finite size scaling for the core of large random hypergraphs. *Ann. Appl. Probab..*

Dembo, A. and Montanari, A. (2008*b*). Graphical models, Bethe states and all that. In preparation.

Dembo, A. and Montanari, A. (2008*c*). Ising models on locally tree-like graphs.

Eprint `arxiv:0804.4726`.

Dembo, A. and Zeitouni, O. (1998). *Large Deviations Techniques and Applications.* Springer, New York.

Derrida, B. (1980). Random-energy model: Limit of a family of disordered models. *Phys. Rev. Lett.*, **45**, 79.

Derrida, B. (1981). Random-energy model: An exactly solvable model of disordered systems. *Phys. Rev. B*, **24**, 2613–2626.

Derrida, B. (1985). A generalization of the random energy model which includes correlations between energies. *J. Physique Lett.*, **46**, L401–L407.

Derrida, B. and Gardner, E. (1986). Solution of the generalised random energy model. *J. Phys. C*, **19**, 2253–2274.

Derrida, B. and Toulouse, G. (1985). Sample to sample fluctuations in the random energy model. *J. Physique Lett.*, **46**, L223–L228.

Di, C., Proietti, D., Richardson, T. J., Telatar, E., and Urbanke, R. (2002). Finite length analysis of low-density parity-check codes on the binary erasure channel. *IEEE Trans. Inf. Theory*, **48**, 1570–1579.

Di, C., Montanari, A., and Urbanke, R. (2004). Weight distribution of LDPC code ensembles: Combinatorics meets statistical physics. In *Proc. IEEE International Symposium on Information Theory*, Chicago, p. 102.

Di, C., Richardson, T. J., and Urbanke, R. (2006). Weight distribution of low-density parity-check codes. *IEEE Trans. Inf. Theory*, **52**, 4839–4855.

Diaconis, P. and Saloff-Coste, L. (1993). Comparison theorems for reversible Markov chains. *Ann. Appl. Probab.*, **3**, 696–730.

Diaconis, P. and Stroock, D. (1991). Geometric bounds for eigenvalues of Markov chains. *Ann. Appl. Probab.*, **1**, 36–61.

Dolecek, L., Zhang, Z., Anantharam, V., and Nikolić, B. (2007). Analysis of absorbing sets for array-based LDPC codes. In *Proc. IEEE International Conference on Communications, ICC*, Glasgow.

Dorogotsev, S. N., Goltsev, A. V., and Mendes, J. F. F. (2002). Ising models on networks wth arbitrary distribution of connections. *Phys. Rev. E*, **66**, 016104.

Dubois, O. and Boufkhad, Y. (1997). A general upper bound for the satisfiability threshold of random r-sat formulae. *J. Algorithms*, **24**, 395–420.

Dubois, O. and Mandler, J. (2002). The 3-XORSAT threshold. In *Proc. 43rd IEEE Symposium on Foundations of Computer Science, FOCS*, pp. 769–778.

Duchet, P. (1995). Hypergraphs. In *Handbook of Combinatorics* (eds. R. Graham, M. Grotschel, and L. Lovasz), pp. 381–432. MIT Press, Cambridge, MA.

Durrett, R. (1995). *Probability: Theory and Examples.* Duxbury Press, New York.

Edwards, S. F. and Anderson, P. W. (1975). Theory of spin glasses. *J. Phys. F*, **5**, 965–974.

Elias, P. (1955). Coding for two noisy channels. In *Third London Symposium on Information Theory*, pp. 61–76.

Ellis, R. S. (1985). *Entropy, Large Deviations and Statistical Mechanics.* Springer, New York.

Erdös, P. and Rényi, A. (1960). On the evolution of random graphs. *Publ. Math. Sci. Hung. Acad. Sci*, **5**, 17–61.

Euler, L. (1736). Solutio problematis ad geometriam situs pertinentis. *Comment. Acad. Sci. U. Petrop.*, **8**, 128–140. Reprinted in *Opera Omnia Ser. I-7*, pp. 1–10, 1766.

Evans, W., Kenyon, C., Peres, Y., and Schulman, L. J. (2000). Broadcasting on trees and the Ising model. *Ann. Appl. Probab.*, **10**, 410–433.

Feller, W. (1968). *An Introduction to Probability Theory and Its Applications*. Wiley, New York.

Ferreira, F. F. and Fontanari, J. F. (1998). Probabilistic analysis of the number partitioning problem. *J. Phys. A*, **31**, 3417–3428.

Fischer, K. H. and Hetz, J. A. (1993). *Spin Glasses*. Cambridge University Press, Cambridge.

Flajolet, P. and Sedgewick, R. (2008). *Analytic Combinatorics*. Cambridge University Press, Cambridge.

Forney, G. D. (2001). Codes on graphs: Normal realizations. *IEEE Trans. Inf. Theory*, **47**, 520–548.

Forney, G. D. and Montanari, A. (2001). On exponential error bounds for random codes on the DMC. Available online at `http://www.stanford.edu/~montanar/PAPERS/`.

Franco, J. (2000). Some interesting research directions in satisfiability. *Ann. Math. Artif. Intell.*, **28**, 7–15.

Franz, S. and Leone, M. (2003). Replica bounds for optimization problems and diluted spin systems. *J. Stat. Phys*, **111**, 535–564.

Franz, S. and Parisi, G. (1995). Recipes for metastable states in spin glasses. *J. Physique I*, **5**, 1401.

Franz, S., Leone, M., Ricci-Tersenghi, F., and Zecchina, R. (2001*a*). Exact solutions for diluted spin glasses and optimization problems. *Phys. Rev. Lett.*, **87**, 127209.

Franz, S., Mézard, M., Ricci-Tersenghi, F., Weigt, M., and Zecchina, R. (2001*b*). A ferromagnet with a glass transition. *Europhys. Lett.*, **55**, 465.

Franz, S., Leone, M., Montanari, A., and Ricci-Tersenghi, F. (2002). Dynamic phase transition for decoding algorithms. *Phys. Rev. E*, **22**, 046120.

Franz, S., Leone, M., and Toninelli, F. (2003). Replica bounds for diluted non-Poissonian spin systems. *J. Phys. A*, **36**, 10967–10985.

Friedgut, E. (1999). Sharp thresholds of graph proprties, and the k-sat problem. *J. Am. Math. Soc.*, **12**, 1017–1054.

Friedman, N., Linial, M., Nachman, I., and Peér, D. (2000). Using bayesian networks to analyze expression data. *J. Comput. Biol.*, **7**, 601–620.

Galavotti, G. (1999). *Statistical Mechanics: A Short Treatise*. Springer, New York.

Gallager, R. G. (1962). Low-density parity-check codes. *IEEE Trans. Inf. Theory*, **8**, 21–28.

Gallager, R. G. (1963). *Low-Density Parity-Check Codes*. MIT Press, Cambridge, MA. Available online at `http://web./gallager/www/pages/ldpc.pdf`.

Gallager, R. G. (1965). A simple derivation of the coding theorem and some applications. *IEEE Trans. Inf. Theory*, **IT-11**, 3–18.

Gallager, R. G. (1968). *Information Theory and Reliable Communication*. Wiley, New York.

Gamarnik, D. (2004). Linear phase transition in random linear constraint satisfaction problems. *Probab. Theory Relat. Fields*, **129**, 410–440.

Gardner, E. (1985). Spin glasses with p-spin interactions. *Nucl. Phys. B*, **257**, 747–765.

Garey, M. R. and Johnson, D. S. (1979). *Computers and Intractability: A Guide to the Theory of NP-Completeness*. W. H. Freeman, New York.

Garey, M. R., Johnson, D. S., and Stockmeyer, L. (1976). Some simplified NP-complete graph problems. *Theor. Comput. Sci.*, **1**, 237–267.

Gent, I. P. and Walsh, T. (1998). Analysis of heuristics for number partitioning. *Comput. Intell.*, **14(3)**, 430–451.

Georgii, H.-O. (1988). *Gibbs Measures and Phase Transitions*. Walter de Gruyter, Berlin.

Gerschenfeld, A. and Montanari, A. (2007). Reconstruction for models on random graph. In *Proc. 48th IEEE Symposium on Foundations of Computer Science, FOCS*, Providence, RI, pp. 194–205.

Goerdt, A. (1996). A threshold for unsatisfiability. *J. Comput. Syst. Sci.*, **53**, 469–486.

Goldschmidt, Y. Y. (1991). Spin glass on the finite-connectivity lattice: The replica solution without replicas. *Phys. Rev. B*, **43**, 8148–8152.

Goldschmidt, Y. Y. and De Dominicis, C. (1990). Replica symmetry breaking in the spin-glass model on lattices with finite connectivity: Application to graph partitioning. *Phys. Rev. B*, **41**, 2184–2197.

Goldschmidt, Y. Y. and Lai, P. Y. (1990). The finite connectivity spin glass: Investigation of replica symmetry breaking of the ground state. *J. Phys. A*, **23**, L775–L782.

Gomes, C. P. and Selman, B. (2005). Can get satisfaction. *Nature*, **435**, 751–752.

Gross, D. J. and Mézard, M. (1984). The simplest spin glass. *Nucl. Phys.*, **B240** [**FS12**], 431–452.

Grosso, C. (2004). Cavity method analysis for random assignment problems. Tesi di Laurea, Università degli Studi di Milano.

Gu, J., Purdom, P. W., Franco, J., and Wah, B. W. (1996). Algorithms for the satisfiability (SAT) problem: A survey. In *Satisfiability Problem: Theory and Applications* (eds. D. Du, J. Gu, and P. M. Pardalos), pp. 19–151. American Mathematical Society, Providence, RI.

Guerra, F. (2003). Broken replica symmetry bounds in the mean field spin glass model. *Commun. Math. Phys.*, **233**, 1–12.

Guerra, F. (2005). Spin glasses. Eprint `arXiv:cond-mat/0507581`.

Guo, D. and Verdú, S. (2002). Multiuser detection and statistical mechanics. In *Communications Information and Network Security* (eds. V. Bhargava, H. Poor, V. Tarokh, and S. Yoon), Chapter 13, pp. 229–277. Kluwer Academic, Dordrecht.

Hartmann, A. K. and Rieger, H. (eds.) (2002). *Optimization Algorithms in Physics*. Wiley-VCH, Berlin.

Hartmann, A. K. and Rieger, H. (2004). *New Optimization Algorithms in Physics*. Wiley-VCH, Berlin.

Hartmann, A. K. and Weigt, M. (2005). *Phase Transitions in Combinatorial Optimization Problems*. Wiley-VCH, Weinheim.

Hayes, B. (2002). The easiest hard problem. *Am. Sci.*, **90**(2), 113–117.

Henley, C. L. (1986). Ising domain growth barriers on a Cayley tree at percolation. *Phys. Rev. B*, **33**, 7675–7682.

Huang, K. (1987). *Statistical Mechanics*. Wiley, New York.

Janson, S., Luczak, T., and Ruciński, A. (2000). *Random Graphs*. Wiley, New York.

Jaynes, E. T. (1957). Information theory and statistical mechanics. *Phys. Rev.*, **106**, 620–630.

Jensen, F. V. (ed.) (1996). *An Introduction to Bayesian Networks*. UCL Press, London.

Jerrum, M. and Sinclair, A. (1996). The Markov chain Monte Carlo method: An approach to approximate counting and integration. In *Approximation Algorithms for NP-Hard Problems*, pp. 482–520. PWS Publishing, Boston, MA.

Johnston, D. A. and Plecháč, P. (1998). Equivalence of ferromagnetic spin models on trees and random graphs. *J. Phys. A*, **31**, 475–482.

Jordan, M. (ed.) (1998). *Learning in Graphical Models*. MIT Press, Boston, MA.

Kabashima, Y. and Saad, D. (1998). Belief propagation vs. TAP for decoding corrupted messages. *Europhys. Lett.*, **44**, 668–674.

Kabashima, Y. and Saad, D. (1999). Statistical mechanics of error correcting codes. *Europhys. Lett.*, **45**, 97–103.

Kabashima, Y. and Saad, D. (2000). Error-correcting code on a cactus: A solvable model. *Europhys. Lett.*, **51**, 698–704.

Kabashima, Y., Murayama, T., and Saad, D. (2000*a*). Typical performance of Gallager-type error-correcting codes. *Phys. Rev. Lett.*, **84**, 1355–1358.

Kabashima, Y., Murayama, T., Saad, D., and Vicente, R. (2000*b*). Regular and irregular Gallager-type error correcting codes. In *Advances in Neural Information Processing Systems 12* (eds. S. A. Solla, T. K. Leen, and K.R. Müller). MIT Press, Cambridge, MA.

Kanter, I. and Sompolinsky, H. (1987). Mean field theory of spin-glasses with finite coordination number. *Phys. Rev. Lett.*, **58**, 164–167.

Karmarkar, N. and Karp, R. M. (1982). The differencing method of set partitioning. Technical Report CSD 82/113, University of California, Berkeley.

Karmarkar, N., Karp, R. M., Lueker, G. S., and Odlyzko, A. M. (1986). Probabilistic analysis of optimum partitioning. *J. Appl. Probab.*, **23**, 626–645.

Karoński, M. and Luczak, T. (2002). The phase transition in a random hypergraph. *J. Comput. Appl. Math.*, **142**, 125–135.

Karp, R. M. (1987). An upper bound on the expected cost of an optimal assignment. In *Proc. Japan–U.S. Joint Seminar*, New York (eds D. S. Johnson, T. Nishizeki, A. Nozaki, and H. S. Wilf), Perspectives in Computing, Vol. 15, pp 1–4, Academic Press, Orlando.

Katsura, S., Inawashiro, S., and Fujiki, S. (1979). Spin glasses for the infinitely long ranged bond Ising model without the use of the replica method. *Physica*, **99A**, 193–216.

Kelly, F. P. (1991). Network routing. *Phil. Trans. R. Soc. London A*, **337**, 343–367.

Kikuchi, R. (1951). A theory of cooperative phenomena. *Phys. Rev.*, **81**, 988–1003.

Kirkpatrick, S. and Selman, B. (1994). Critical behavior in the satisfiability of random boolean expressions. *Science*, **264**, 1297–1301.

Kirkpatrick, S. and Sherrington, D. (1978). Infinite ranged models of spin glasses. *Phys. Rev. B*, **17**, 4384–4403.

Kirkpatrick, T. R. and Thirumalai, D. (1987). *p*-spin interaction spin glass models: Connections with the structural glass problem. *Phys. Rev. B*, **36**, 5388–5397.

Kirkpatrick, T. R. and Wolynes, P. G. (1987). Connections between some kinetic and equilibrium theories of the glass transition. *Phys. Rev. A*, **35**, 3072–3080.

Kirkpatrick, S., Gelatt, C. D., and Vecchi, M. P. (1983). Optimization by simulated annealing. *Science*, **220**, 671–680.

Kirousis, L. M., Kranakis, E., Krizanc, D., and Stamatiou, Y. (1998). Approximating the unsatisfiability threshold of random formulas. *Random Struct. Algorithms*, **12**, 253–269.

Klein, M. W., Schowalter, L. J., and Shukla, P. (1979). Spin glasses in the Bethe–Peierls–Weiss and other mean field approximations. *Phys. Rev. B*, **19**, 1492–1502.

Kleinberg, J. M., Kumar, R., Raghavan, P., Rajagopalan, S., and Tomkins, A. S. (1999). The Web as a graph: Measurements, models, and methods. In *5th Conference on Computing and Combinatorics*, Tokyo, pp. 1–17.

Korf, R. E. (1998). A complete anytime algorithm for number partitioning. *Artif. Intell.*, **106**, 181:203.

Kötter, R. and Vontobel, P. O. (2003). Graph covers and iterative decoding of finite-length codes. In *Proc. 3rd Int. Conf. on Turbo Codes and Related Topics*, Brest, France, pp. 75–82.

Krauth, W. (2006). *Statistical Mechanics: Algorithms and Computations*. Oxford University Press, Oxford.

Krauth, W. and Mézard, M. (1989). The cavity method and the Traveling Salesman Problem. *Europhys. Lett.*, **8**, 213–218.

Krzakala, F. and Zdeborova, L. (2007). Phase transitions in the coloring of random graphs. *Phys. Rev. E*, **76**, 031131.

Krzakala, F., Montanari, A., Ricci-Tersenghi, F., Semerjian, G., and Zdeborova, L. (2007). Gibbs states and the set of solutions of random constraint satisfaction problems. *Proc. Natl. Acad. Sci.*, **104**, 10318–10323.

Kschischang, F. R., Frey, B. J., and Loeliger, H.-A. (2001). Factor graphs and the sum–product algorithm. *IEEE Trans. Inf. Theory*, **47**, 498–519.

Lauritzen, S. L. (1996). *Graphical Models*. Oxford University Press, Oxford.

Leone, M., Ricci-Tersenghi, F., and Zecchina, R. (2001). Phase coexistence and finite-size scaling in random combinatorial problems. *J. Phys. A*, **34**, 4615–4626.

Leone, M., Vázquez, A., Vespignani, A., and Zecchina, R. (2004). Ferromagnetic ordering in graphs with arbitrary degree distribution. *Eur. Phys. J. B*, **28**, 191–197.

Linusson, S. and Wästlund, J. (2004). A proof of Parisi's conjecture on the random assignment problem. *Probab. Theory Relat. Fields*, **128**, 419–440.

Litsyn, S. and Shevelev, V. (2003). Distance distributions in ensembles of irregular low-density parity-check codes. *IEEE Trans. Inf. Theory*, **49**, 3140–3159.

Luby, M., Mitzenmacher, M., Shokrollahi, A., Spielman, D. A., and Stemann, V. (1997). Practical loss-resilient codes. In *Proc. 29th ACM Symposium on Theory of*

Computing, STOC, pp. 150–159.

Luby, M., Mitzenmacher, M., Shokrollahi, A., and Spielman, D. A. (1998). Analysis of low density codes and improved designs using irregular graphs. In *Proc. 30th ACM Symposium on Theory of Computing, STOC*, pp. 249–258.

Luby, M., Mitzenmacher, M., Shokrollahi, A., and Spielman, D. A. (2001*a*). Efficient erasure correcting codes. *IEEE Trans. Inf. Theory*, **47**(2), 569–584.

Luby, M., Mitzenmacher, M., Shokrollahi, A., and Spielman, D. A. (2001*b*). Improved low-density parity-check codes using irregular graphs. *IEEE Trans. Inf. Theory*, **47**, 585–598.

Ma, S.-K. (1985). *Statistical Mechanics*. World Scientific, Singapore.

MacKay, D. J. C. (1999). Good error correcting codes based on very sparse matrices. *IEEE Trans. Inf. Theory*, **45**, 399–431.

MacKay, D. J. C. (2002). *Information Theory, Inference & Learning Algorithms*. Cambridge University Press, Cambridge.

MacKay, D. J. C. and Neal, R. M. (1996). Near Shannon limit performance of low density parity check codes. *Electron. Lett.*, **32**, 1645–1646.

MacKay, D. J. C. and Postol, M. S. (2003). Weaknesses of Margulis and Ramanujan–Margulis low-density parity check codes. *Elect. Notes in Theor. Computer Sci.*, **74**, 97–104.

Macris, N. (2007). Sharp bounds on generalised EXIT function. *IEEE Trans. Inf. Theory*, **53**, 2365–2375.

Maneva, E., Mossel, E., and Wainwright, M. J. (2005). A new look at survey propagation and its generalizations. In *Proc. 16th ACM–SIAM Symposium on Discrete Algorithms, SODA*, Vancouver, pp. 1089–1098.

Marinari, E., Parisi, G., and Ruiz-Lorenzo, J. (1997). Numerical simulations of spin glass systems. In *Spin Glasses and Random Fields* (ed. A. Young). World Scientific, Singapore.

Martin, O. C., Mézard, M., and Rivoire, O. (2005). Random multi-index matching problems. *J. Stat. Mech.*, 09006.

Martinelli, F. (1999). Lectures on Glauber dynamics for discrete spin models. In *Lectures on Probability Theory and Statistics, Saint-Flour 1997*, Lecture Notes in Mathematics, pp. 93–191. Springer, Berlin.

Martinelli, F., Sinclair, A., and Weitz, D. (2004). Glauber dynamics on trees: Boundary conditions and mixing time. *Commun. Math. Phys*, **250**, 301–334.

McEliece, R. J., MacKay, D. J. C., and Cheng, J.-F. (1998). Turbo decoding as an instance of Pearl's 'belief propagation' algorithm. *IEEE J. Sel. Areas Commun.*, **16**, 140–152.

Méasson, C., Montanari, A., and Urbanke, R. (2008). Maxwell construction: The hidden bridge between iterative and maximum a posteriori decoding. *IEEE Trans. Inf. Theory*,, **54**, 5277–5307.

Méasson, C., Montanari, A., Richardson, T., and Urbanke, R. (2005). The generalized area theorem and some of its consequences. Submitted. Eprint http://arxiv.org/abs/cs/0511039.

Mertens, S. (1998). Phase transition in the number partitioning problem. *Phys. Rev. Lett.*, **81**(20), 4281–4284.

Mertens, S. (2000). Random costs in combinatorial optimization. *Phys. Rev. Lett.*, **84**(7), 1347–1350.

Mertens, S. (2001). A physicist's approach to number partitioning. *Theor. Comput. Sci.*, **265**, 79–108.

Mertens, S., Mézard, M., and Zecchina, R. (2006). Threshold values of random K-SAT from the cavity method. *Random Struct. Algorithms*, **28**, 340–373.

Mézard, M. and Montanari, A. (2006). Reconstruction on trees and spin glass transition. *J. Stat. Phys.*, **124**, 1317–1350.

Mézard, M. and Parisi, G. (1985). Replicas and optimization. *J. Physique Lett.*, **46**, L771–L778.

Mézard, M. and Parisi, G. (1986). Mean-field equations for the matching and the traveling salesman problems. *Europhys. Lett.*, **2**, 913–918.

Mézard, M. and Parisi, G. (1987). Mean-field theory of randomly frustrated systems with finite connectivity. *Europhys. Lett.*, **3**, 1067–1074.

Mézard, M. and Parisi, G. (1999). Thermodynamics of glasses: A first principles computation. *Phys. Rev. Lett.*, **82**, 747–751.

Mézard, M. and Parisi, G. (2001). The Bethe lattice spin glass revisited. *Eur. Phys. J. B*, **20**, 217–233.

Mézard, M. and Parisi, G. (2003). The cavity method at zero temperature. *J. Stat. Phys.*, **111**, 1–34.

Mézard, M. and Zecchina, R. (2002). The random K-satisfiability problem: From an analytic solution to an efficient algorithm. *Phys. Rev. E*, **66**, 056126.

Mézard, M., Parisi, G., and Virasoro, M. A. (1985a). Random free energies in spin glasses. *J. Physique Lett.*, **46**, L217–L222.

Mézard, M., Parisi, G., and Virasoro, M. A. (1985b). SK model: The replica solution without replicas. *Europhys. Lett.*, **1**, 77–82.

Mézard, M., Parisi, G., Sourlas, N., Toulouse, G., and Virasoro, M. A. (1985c). Replica symmetry breaking and the nature of the spin glass phase. *J. Physique*, **45**, 843–854.

Mézard, M., Parisi, G., and Virasoro, M. A. (1987). *Spin Glass Theory and Beyond.* World Scientific, Singapore.

Mézard, M., Parisi, G., and Zecchina, R. (2003a). Analytic and algorithmic solution of random satisfiability problems. *Science*, **297**, 812–815.

Mézard, M., Ricci-Tersenghi, F., and Zecchina, R. (2003b). Two solutions to diluted p-spin models and XORSAT problems. *J. Stat. Phys.*, **111**, 505–533.

Mézard, M., Mora, T., and Zecchina, R. (2005a). Clustering of solutions in the random satisfiability problem. *Phys. Rev. Lett.*, **94**, 197205.

Mézard, M., Palassini, M., and Rivoire, O. (2005b). Landscape of solutions in constraint satisfaction problems. *Phys. Rev. Lett.*, **95**, 200202.

Mézard, M., Tarzia, M., and Toninelli, C. (2007). Statistical physics of group testing. *J. Phys. A*, **40**, 12019–12029. Proc. International Workshop on Statistical-Mechanical Informatics, Kyoto.

Migliorini, G. and Saad, D. (2006). Finite-connectivity spin-glass phase diagrams and low-density parity check codes. *Phys. Rev. E*, **73**, 026122.

Molloy, M. and Reed, B. (1995). A critical point for random graphs with a given degree sequence. *Random Struct. Algorithms*, **6**, 161–180.

Monasson, R. (1995). Structural glass transition and the entropy of metastable states. *Phys. Rev. Lett.*, **75**, 2847–2850.

Monasson, R. (1998). Optimization problems and replica symmetry breaking in finite connectivity spin glasses. *J. Phys. A*, **31**, 513–529.

Monasson, R. and Zecchina, R. (1996). Entropy of the K-satisfiability problem. *Phys. Rev. Lett.*, **76**, 3881–3885.

Monasson, R. and Zecchina, R. (1997). Statistical mechanics of the random K-satisfiability problem. *Phys. Rev. E*, **56**, 1357–1370.

Monasson, R. and Zecchina, R. (1998). Tricritical points in random combinatorics: The $(2+p)$-sat case. *J. Phys. A*, **31**, 9209–9217.

Monasson, R., Zecchina, R., Kirkpatrick, S., Selman, B., and Troyansky, L. (1999). Determining computational complexity from characteristic phase transitions. *Nature*, **400**, 133–137.

Monod, P. and Bouchiat, H. (1982). Equilibrium magnetization of a spin glass: is mean-field theory valid? *J. Physique Lett.*, **43**, 45–54.

Montanari, A. (2000). Turbo codes: The phase transition. *Eur. Phys. J. B*, **18**, 121–136.

Montanari, A. (2001*a*). Finite size scaling and metastable states of good codes. In *Proc. 39th Allerton Conference on Communications, Control and Computing*, Monticello, IL.

Montanari, A. (2001*b*). The glassy phase of Gallager codes. *Eur. Phys. J. B*, **23**, 121–136.

Montanari, A. (2003). 2RSB population dynamics for spin glasses. Unpublished.

Montanari, A. (2005). Tight bounds for LDPC and LDGM codes under MAP decoding. *IEEE Trans. Inf. Theory*, **51**, 3221–3246.

Montanari, A. and Ricci-Tersenghi, F. (2003). On the nature of the low-temperature phase in discontinuous mean-field spin glasses. *Eur. Phys. J. B*, **33**, 339–346.

Montanari, A. and Ricci-Tersenghi, F. (2004). Cooling-schedule dependence of the dynamics of mean-field glasses. *Phys. Rev. B*, **70**, 134406.

Montanari, A. and Semerjian, G. (2005). From large scale rearrangements to mode coupling phenomenology in model glasses. *Phys. Rev. Lett.*, **94**, 247201.

Montanari, A. and Semerjian, G. (2006*a*). On the dynamics of the glass transition on Bethe lattices. *J. Stat. Phys.*, **124**, 103–189.

Montanari, A. and Semerjian, G. (2006*b*). Rigorous inequalities between length and time scales in glassy systems. *J. Stat. Phys.*, **125**, 23–54.

Montanari, A. and Shah, D. (2007). Counting good truth assignments for random satisfiability formulae. In *Proc. 18th Symposium on Discrete Algorithms, SODA*, New Orleans, pp. 1255–1265.

Montanari, A. and Sourlas, N. (2000). The statistical mechanics of turbo codes. *Eur. Phys. J. B*, **18**, 107–119.

Montanari, A., Parisi, G., and Ricci-Tersenghi, F. (2004). Instability of one-step replica-symmetry breaking in satisfiability problems. *J. Phys. A*, **37**, 2073–2091.

Montanari, A., Ricci-Tersenghi, F., and Semerjian, G. (2007). Solving constraint satisfaction problems through belief-propagation-guided decimation. In *Proc. 45th*

Allerton Conference on Communications, Control and Computing, Monticello, IL.

Montanari, A., Ricci-Tersenghi, F., and Semerjian, G. (2008). Cluster of solutions and replica symmetry breaking in random k-satisfiability. *J. Stat. Mech.*, 04004.

Mooij, J. M. and Kappen, H. K. (2005). On the properties of Bethe aproximation and loopy belief propagation on binary networks. *J. Stat. Mech.*, 11012.

Mora, T. and Mézard, M. (2006). Geometrical organization of solutions to random linear Boolean equations. *J. Stat. Mech.*, 10007.

Mora, T. and Zdeborová, L. (2007). Random subcubes as a toy model for constraint satisfaction problems. Eprint `arXiv:0710.3804`.

Morita, T. (1979). Variational principle for the distribution function of the effective field for the random Ising model in the Bethe approximation. *Physica*, **98A**, 566–572.

Mossel, E. (2003). Phase transitions in phylogeny. *Trans. Am. Math. Soc.*, **356**, 2379–2404.

Mossel, E. (2004). Survey: Information flow on trees. In *Graphs, Morphisms, and Statistical Physics*, pp. 155–170. American Mathematical Society, Providence, RI.

Mossel, E. and Peres, Y. (2003). Information flow on trees. *Ann. Appl. Probab.*, **13**, 817–844.

Mottishaw, P. and De Dominicis, C. (1987). On the stability of randomly frustrated systems with finite connectivity. *J. Phys. A*, **20**, L375–L379.

Mulet, R., Pagnani, A., Weigt, M., and Zecchina, R. (2002). Coloring random graphs. *Phys. Rev. Lett.*, **89**, 268701.

Nair, C., Prabhakar, B., and Sharma, M. (2003). Proofs of the Parisi and Coppersmith–Sorkin conjectures for the finite random assignment problem. In *Proc. 44th IEEE Symposium on Foundations of Computer Science, FOCS*, pp. 168–178.

Nair, C., Prabhakar, B., and Sharma, M. (2006). Proofs of the Parisi and Coppersmith-Sorkin random assignment conjectures. *Random Struct. Algorithms*, **27**, 413–444.

Nakamura, K., Kabashima, Y., and Saad, D. (2001). Statistical mechanics of low-density parity check error-correcting codes over Galois fields. *Europhys. Lett.*, **56**, 610–616.

Nakanishi, K. (1981). Two- and three-spin cluster theory of spin glasses. *Phys. Rev. B*, **23**, 3514–3522.

Neri, I., Skantzos, N. S., and Bollé, D. (2008). Gallager error correcting codes for binary asymmetric channels. Eprint `arXiv:0803.2580`.

Nishimori, H. (2001). *Statistical Physics of Spin Glasses and Information Processing*. Oxford University Press, Oxford.

Norris, J. R. (1997). *Markov Chains*. Cambridge University Press, Cambridge.

Panchenko, D. and Talagrand, M. (2004). Bounds for diluted mean-field spin glass models. *Probab. Theor. Relat. Fields*, **130**, 319–336.

Papadimitriou, C. H. (1991). On selecting a satisfying truth assignment. In *Proc. 32nd IEEE Symposium on Foundations of Computer Science, FOCS*, pp. 163–169.

Papadimitriou, C. H. (1994). *Computational Complexity*. Addison-Wesley, Reading, MA.

Papadimitriou, C. H. and Steiglitz, K. (1998). *Combinatorial Optimization*. Dover

Publications, Mineola, NY.

Parisi, G. (1979). Toward a mean field theory for spin glasses. *Phys. Lett.*, **73A**, 203–205.

Parisi, G. (1980*a*). A sequence of approximated solutions to the SK model for spin glasses. *J. Phys. A*, **13**, L115–L121.

Parisi, G. (1980*b*). The order parameter for spin glasses: A function on the interval [0, 1]. *J. Phys. A*, **13**, 1101–1112.

Parisi, G. (1983). Order parameter for spin glasses. *Phys. Rev. Lett.*, **50**, 1946–1948.

Parisi, G. (1988). *Statistical Field Theory*. Addison-Wesley, Reading, MA.

Parisi, G. (1998). A conjecture on random bipartite matching. Eprint arXiv:cond-mat/9801176.

Parisi, G. (2002). On local equilibrium equations for clustering states. Eprint arXiv:cs/0212047v1.

Parisi, G. (2003). A backtracking survey propagation algorithm for *K*-satisfiability. Eprint arxiv:cond-mat/0308510.

Pearl, J. (1988). *Probabilistic Reasoning in Intelligent Systems: Networks of Plausible Inference*. Morgan Kaufmann, San Francisco.

Pitman, J. and Yor, M. (1997). The two-parameter Poisson–Dirichlet distribution derived from a stable subordinator. *Ann. Probab.*, **25**, 855–900.

Prim, R. C. (1957). Shortest connection networks and some generalizations. *Bell Syst. Tech. J.*, **36**, 1389–1401.

Pumphrey, S. J. (2001). Solving the satisfiability problem using message passing techniques. Part III Physics Project Report, Department of Physics, University of Cambridge.

Reif, F. (1965). *Fundamentals of Statistical and Thermal Physics*. McGraw-Hill, New York.

Ricci-Tersenghi, F., Weigt, M., and Zecchina, R. (2001). Exact solutions for diluted spin glasses and optimization problems. *Phys. Rev. E*, **63**, 026702.

Richardson, T. J. (2003). Error floors of LDPC codes. In *Proc. 41st Allerton Conference on Communications, Control and Computing*, Monticello, IL.

Richardson, T. and Urbanke, R. (2001*a*). An introduction to the analysis of iterative coding systems. In *Codes, Systems, and Graphical Models*, IMA Volumes in Mathematics and Its Applications, Vol. 123, pp. 1–37. Springer, Berlin.

Richardson, T. J. and Urbanke, R. (2001*b*). The capacity of low-density parity check codes under message-passing decoding. *IEEE Trans. Inf. Theory*, **47**, 599–618.

Richardson, T. J. and Urbanke, R. (2001*c*). Efficient encoding of low-density parity-check codes. *IEEE Trans. Inf. Theory*, **47**, 638–656.

Richardson, T. J. and Urbanke, R. (2008). *Modern Coding Theory*. Cambridge University Press, Cambridge. Available online at http://lthcwww.epfl.ch/mct/index.php.

Richardson, T. J., Shokrollahi, A., and Urbanke, R. (2001). Design of capacity-approaching irregular low-density parity-check codes. *IEEE Trans. Inf. Theory*, **47**, 610–637.

Rivoire, O., Biroli, G., Martin, O. C., and Mézard, M. (2003). Glass models on Bethe lattices. *Eur. Phys. J. B*, **37**, 55–78.

Ruelle, D. (1987). A mathematical reformulation of Derrida's REM and GREM. *Commun. Math. Phys.*, **108**, 225–239.

Ruelle, D. (1999). *Statistical Mechanics: Rigorous Results*. World Scientific, Singapore.

Rujan, P. (1993). Finite temperature error-correcting codes. *Phys. Rev. Lett.*, **70**, 2968–2971.

Schmidt-Pruzan, J. and Shamir, E. (1985). Component structure in the evolution of random hypergraphs. *Combinatorica*, **5**, 81–94.

Schöning, U. (1999). A probabilistic algorithm for *k*-SAT and constraint satisfaction problems. In *Proc. 40th IEEE Symposium on Foundations of Computer Science, FOCS*, New York, pp. 410–414.

Schöning, U. (2002). A probabilistic algorithm for *k*-SAT based on limited local search and restart. *Algorithmica*, **32**, 615–623.

Selman, B. and Kautz, H. A. (1993). Domain independent extensions to gsat: Solving large structural satisfiability problems. In *Proc. IJCAI-93*, Chambery, France.

Selman, B. and Kirkpatrick, S. (1996). Critical behavior in the computational cost of satisfiability testing. *Artif. Intell.*, **81**, 273–295.

Selman, B., Kautz, H. A., and Cohen, B. (1994). Noise strategies for improving local search. In *Proc. AAAI-94*, Seattle, WA.

Selman, B., Mitchell, D., and Levesque, H. (1996). Generating hard satisfiability problems. *Artif. Intell.*, **81**, 17–29.

Shannon, C. E. (1948). A mathematical theory of communication. *Bell Syst. Tech. J.*, **27**, 379–423, 623–655. Available online at `http://cm.bell-labs.com /cm/ms/what/shannonday/paper.html`.

Sherrington, D. and Kirkpatrick, S. (1975). Solvable model of a spin glass. *Phys. Rev. Lett.*, **35**, 1792–1796.

Shohat, J. A. and Tamarkin, J. D. (1943). *The Problem of Moments*, Mathematical Surveys No. 1. American Mathematical Society, New York.

Sinclair, A. (1997). Convergence rates for Monte Carlo experiments. In *Numerical Methods for Polymeric Systems*, IMA Volumes in Mathematics and Its Applications, Vol. 102,pp. 1–18. Springer, Berlin.

Sipser, M. and Spielman, D. A. (1996). Expander codes. *IEEE Trans. Inf. Theory*, **42**, 1710–1722.

Sokal, A. D. (1996). Monte Carlo methods in statistical mechanics: foundations and new algorithms. In *Lectures at the Cargèse Summer School 'Functional Integration: Basics and Applications'*. Available online at `http://www.math.nyu.edu/faculty/goodman/teaching/Monte_Carlo/ monte_carlo.html`.

Sourlas, N. (1989). Spin-glass models as error-correcting codes. *Nature*, **339**, 693.

Spielman, D. A. (1997). The complexity of error-correcting codes. In *Fundamentals of Computation Theory* (eds Chlebus, B. S. and Craja, L.), Lecture Notes in Computer Science, Vol. 1279, pp. 67–84. Springer, Berlin.

Sportiello, A. (2004). Personal communication.

Stepanov, M. G., Chernyak, V. Y., Chertkov, M., and Vasic, B. (2005). Diagnosis of weaknesses in modern error correction codes: A physics approach. *Phys.*

Rev. Lett., **95**, 228701–228704.

Svenson, P. and Nordahl, M. G. (1999). Relaxation in graph coloring and satisfiability problems. *Phys. Rev. E*, **59**, 3983–3999.

Talagrand, M. (2000). Rigorous low temperature results for the mean field p-spin interaction model. *Probab. Theor. Relat. Fields*, **117**, 303–360.

Talagrand, M. (2003). *Spin Glasses: A Challenge for Mathematicians*. Springer, Berlin.

Tanaka, T. (2002). A statistical-mechanics approach to large-system analysis of CDMA multiuser detectors. *IEEE Trans. Inf. Theory*, **48**, 2888–2910.

Tanner, R.M. (1981). A recursive approach to low complexity codes. *IEEE Trans. Inf. Theory*, **27**, 533–547.

Thouless, D. J. (1986). Spin-glass on a Bethe lattice. *Phys. Rev. Lett.*, **6**, 1082–1085.

Thouless, D. J., Anderson, P. W., and Palmer, R. G. (1977). Solution of 'Solvable model of a spin glass'. *Phil. Mag.*, **35**, 593–601.

Toulouse, G. (1977). Theory of the frustration effect in spin glasses: I. *Comm. Phys.*, **2**, 115–119.

Viana, L. and Bray, A. J. (1985). Phase diagrams for dilute spin glasses. *J. Phys. C*, **18**, 3037–3051.

Wainwright, M. J. and Maneva, E. (2005). Lossy source encoding via message-passing and decimation over generalized codewords of LDGM codes. In *IEEE International Symposium on Information Theory*, Adelaide, pp. 1493–1497.

Wainwright, M. J., Jaakkola, T. S., and Willsky, A. S. (2005a). A new class of upper bounds on the log partition function. *IEEE Trans. Inf. Theory*, **51**(7), 2313–2335.

Wainwright, M. J., Jaakkola, T. S., and Willsky, A. S. (2005b). MAP estimation via agreement on trees: Message-passing and linear programming. *IEEE Trans. Inf. Theory*, **51**(11), 3697–3717.

Walkup, D. W. (1979). On the expected value of a random assignment problem. *SIAM J. Comput.*, **8**, 440–442.

Wang, C. C., Kulkarni, S. R., and Poor, H. V. (2006). Exhausting error-prone patterns in ldpc codes. *IEEE Trans. Inf. Theory*, submitted. Eprint http://arxiv.org/abs/cs/0609046

Wästlund, J. (2008). An easy proof of the $\zeta(2)$ limit in the random assignment problem. *Electron. Commmun. Probab.*, **13**, 258–265.

Weigt, M. and Hartmann, A. K. (2000). Number of guards needed by a museum: A phase transition in vertex covering of random graphs. *Phys. Rev. Lett*, **84**, 6118–6121.

Wong, K. Y. M. and Sherrington, D. (1988). Intensively connected spin glasses: Towards a replica symmetry breaking solution of the ground state. *J. Phys. A*, **21**, L459–L466.

Wormald, N. C. (1999). Models of random regular graphs. In *Surveys in Combinatorics, 1999* (eds. J. D. Lamb and D. A. Preece), London Mathematical Society Lecture Note Series, pp. 239–298. Cambridge University Press, Cambridge.

Yakir, B. (1996). The differencing algorithm LDM for partitioning: A proof of a conjecture of Karmarkar and Karp. *Math. Oper. Res.*, **21**, 85.

Yedidia, J. S., Freeman, W. T., and Weiss, Y. (2001). Generalized belief propagation.

In *Advances in Neural Information Processing Systems, NIPS*, pp. 689–695. MIT Press, Cambridge, MA.

Yedidia, J. S., Freeman, W. T., and Weiss, Y. (2005). Constructing free energy approximations and generalized belief propagation algorithms. *IEEE Trans. Inf. Theory*, **51**, 2282–2313.

Yuille, A. L. (2002). CCCP algorithms to minimize the Bethe and Kikuchi free energies: Convergent alternatives to belief propagation. *Neural Comput.*, **14**, 691–1722.

Zhou, H. (2003). Vertex cover problem studied by cavity method: Analytics and population dynamics. *Eur. Phys. J. B*, **32**, 265–270.

Index